T0205351

Lecture Notes in Computer Science   11411

Commenced Publication in 1973
Founding and Former Series Editors:
Gerhard Goos, Juris Hartmanis, and Jan van Leeuwen

Editorial Board

David Hutchison
  Lancaster University, Lancaster, UK
Takeo Kanade
  Carnegie Mellon University, Pittsburgh, PA, USA
Josef Kittler
  University of Surrey, Guildford, UK
Jon M. Kleinberg
  Cornell University, Ithaca, NY, USA
Friedemann Mattern
  ETH Zurich, Zurich, Switzerland
John C. Mitchell
  Stanford University, Stanford, CA, USA
Moni Naor
  Weizmann Institute of Science, Rehovot, Israel
C. Pandu Rangan
  Indian Institute of Technology Madras, Chennai, India
Bernhard Steffen
  TU Dortmund University, Dortmund, Germany
Demetri Terzopoulos
  University of California, Los Angeles, CA, USA
Doug Tygar
  University of California, Berkeley, CA, USA

More information about this series at http://www.springer.com/series/7407

Kalyanmoy Deb · Erik Goodman
Carlos A. Coello Coello · Kathrin Klamroth
Kaisa Miettinen · Sanaz Mostaghim
Patrick Reed (Eds.)

Evolutionary Multi-Criterion Optimization

10th International Conference, EMO 2019
East Lansing, MI, USA, March 10–13, 2019
Proceedings

 Springer

Editors
Kalyanmoy Deb (iD)
Michigan State University
East Lansing, MI, USA

Erik Goodman
Michigan State University
East Lansing, MI, USA

Carlos A. Coello Coello (iD)
CINVESTAV-IPN
Mexico, Mexico

Kathrin Klamroth
University of Wuppertal
Wuppertal, Germany

Kaisa Miettinen (iD)
University of Jyvaskyla
Jyväskylä, Finland

Sanaz Mostaghim (iD)
Otto von Guericke University Magdeburg
Magdeburg, Germany

Patrick Reed (iD)
Cornell University
Ithaca, NY, USA

ISSN 0302-9743   ISSN 1611-3349 (electronic)
Lecture Notes in Computer Science
ISBN 978-3-030-12597-4   ISBN 978-3-030-12598-1 (eBook)
https://doi.org/10.1007/978-3-030-12598-1

Library of Congress Control Number: 2019930730

LNCS Sublibrary: SL1 – Theoretical Computer Science and General Issues

© Springer Nature Switzerland AG 2019
This work is subject to copyright. All rights are reserved by the Publisher, whether the whole or part of the material is concerned, specifically the rights of translation, reprinting, reuse of illustrations, recitation, broadcasting, reproduction on microfilms or in any other physical way, and transmission or information storage and retrieval, electronic adaptation, computer software, or by similar or dissimilar methodology now known or hereafter developed.
The use of general descriptive names, registered names, trademarks, service marks, etc. in this publication does not imply, even in the absence of a specific statement, that such names are exempt from the relevant protective laws and regulations and therefore free for general use.
The publisher, the authors and the editors are safe to assume that the advice and information in this book are believed to be true and accurate at the date of publication. Neither the publisher nor the authors or the editors give a warranty, express or implied, with respect to the material contained herein or for any errors or omissions that may have been made. The publisher remains neutral with regard to jurisdictional claims in published maps and institutional affiliations.

This Springer imprint is published by the registered company Springer Nature Switzerland AG
The registered company address is: Gewerbestrasse 11, 6330 Cham, Switzerland

Preface

Over the past 25 years, there has been a phenomenal growth in research and application of multi-criterion optimization. Among the many approaches, evolutionary multi-criterion optimization (EMO) has received significant attention for many reasons. EMO methods use a population of evolving solutions in their iterative steps by establishing an implicit parallel search. The population approach also allows a number of diverse non-dominated solutions to be processed simultaneously, thereby enabling populations to converge near the Pareto-optimal set with a good distribution.

A search with certain keywords related to EMO topics found 76,358 articles in the SCOPUS database from 1990 to the present. Figure 1 shows the numbers of articles per year, indicating an exponential growth until about 2010. During this initial phase, a number of efficient two- and three-objective EMO algorithms, their extensions, and their applications were reported. While the field seemed to have steadily produced about 4,500 articles per year during 2010–2012, another steady increase in the number of published articles is observed from 2012, with a renewed interest in handling

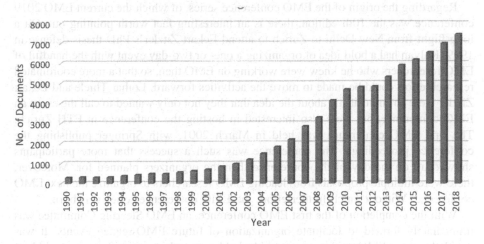

Fig. 1. Number of articles of EMO papers since 1990.

more-than-three-objective optimization problems. The renewed effort for handling these problems has been significant, and EMO researchers have labeled this effort "evolutionary many-objective optimization" or EMaO. Besides algorithmic developments, EMO and EMaO methods have also been applied to various practical problems. A number of software companies have also helped to make the field popular. Figure 2 shows the distribution of 76,358 articles according to the disciplines in which they are applied. While engineering and computer science fields cover about half of the total

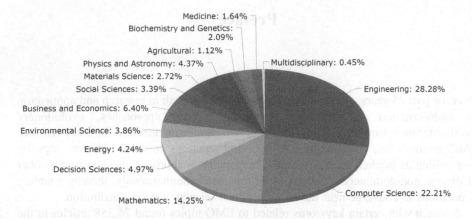

Fig. 2. Distribution of EMO-related articles across various disciplines.

articles, mostly in the area of algorithmic developments, it is interesting to observe how EMO has infiltrated many other science and technology areas. Among them, the use of EMO in mathematics, social sciences, agriculture, and medicine is worth mentioning. It is clear from the chart that EMO has taken the evolutionary computation field beyond its usual habitat and proliferated into various application areas.

Regarding the origin of the EMO conference series, of which the current EMO 2019 conference was the 10th edition, there is an interesting fact worth pointing out. On a long flight from New Delhi to Zurich to attend Eckart Zitzler's PhD thesis defense in 1999, Kalyan had a bold idea of organizing a one- or two-day event with the handful of EMO researchers who he knew were working on EMO then, so that a more coordinated research effort could be made to move the activities forward. Lothar Thiele and Eckart Zitzler were so enthusiastic about the idea that they not only wanted to call the event an EMO conference, but were also interested in hosting the conference in ETH Zurich. The first EMO conference was held in March 2001, with Springer publishing the conference proceedings. The conference was such a success that more participants showed up on the day of the conference than the organizers planned for. Moreover, there were multiple proposals from leading EMO researchers to organize the next EMO conference. Without much further effort, the EMO conference series was born.

With the completion of the first EMO conference, an EMO Steering Committee was immediately formed to facilitate organization of future EMO-related events. It was decided that the EMO conference would be held every other year and that it would be a single-track event in order to have a more focused emphasis on the topic.

EMO conferences have been held on four continents so far: EMO 2001 in Switzerland (LNCS 1993), EMO 2003 in Portugal (LNCS 2632), EMO 2005 in Mexico (LNCS 3410), EMO 2007 in Japan (LNCS 4403), EMO 2009 in France (LNCS 5467), EMO 2011 in Brazil (LNCS 6576), EMO 2013 in the UK (LNCS 7811), EMO 2015 in Portugal (LNCS 9019), and EMO 2017 in Germany (LNCS 10173). EMO 2019 (LNCS 11411) was the first EMO conference to be held in the USA.

A multi-objective optimization task is not complete without the use of a multi-criterion decision-making (MCDM) effort. EMO organizers recognized this fact early on, and invited key MCDM researchers to present their latest methodologies. However, since the EMO 2009 conference, a separate MCDM track was arranged to promote further participation of MCDM researchers in EMO conferences. It is also worth mentioning that a separate EMO track has also been held during the bi-annual MCDM conferences since 2009.

This volume contains the papers presented at EMO 2019: the Evolutionary Multi-Criterion Optimization Conference held during March 10–13, 2019, in East Lansing, Michigan, USA. There were 76 full paper submissions, of which 69 papers were in the EMO track and seven papers in the MCDM track. Each submission was reviewed by at least three, and on average 3.6, Program Committee members. After the rigorous review process, the committee decided to accept 59 papers for presentations at the conference. The accepted papers are divided into eight categories, each representing a key area of current interest in the EMO field today. They include theoretical developments, algorithmic developments, issues in many-objective optimization, performance metrics, knowledge extraction and surrogate-based EMO, multi-objective combinatorial problem solving, MCDM and interactive EMO methods, and applications.

Following the trends of the past few EMO conferences, this conference included an Industry Session in which industrial participants were encouraged to attend and present their work in EMO and MCDM areas without the need to submit a full paper. The abstracts of the accepted industry session presentations were printed in the conference program booklet.

The conference arranged two plenary talks, one by Prof. Prabhat Hajela, Provost at Rensselaer Polytechnic Institute in Troy, New York, USA and one of the proposers of an early EMO algorithm, called the Weight-Based Genetic Algorithms in 1993, and the other by Dr. Matthew Ferringer, Principal Director of National GEOINT Programs, National Intelligence Division, The Aerospace Corporation, USA, and an EMO-MCDM specialist in space-related system design, build, and launch activities.

For the first time, this EMO conference organized two tutorials, one on EMO delivered by Prof. Hisao Ishibuchi, Chair Professor, Southern University of Science Technology (SUSTech), Shenzhen, China, and Editor-in-Chief of the *IEEE Computational Intelligence Magazine*, and another on MCDM delivered by Prof. Jyrki Wallenius, former Dean of Aalto University School of Business, Helsinki, Finland, and former Editor-in-Chief of the *European Journal of Operations Research*, so that participants of each group could benefit from the recent key research developments of the other group.

We express our sincere gratitude and appreciation to the plenary and tutorial speakers for accepting our invitations. We thank all the authors for submitting their excellent work and responding to reviewers' comments in a short time-frame. We cannot thank enough each and every Program Committee member who reviewed three to six papers in the short time allotted to them. We appreciate all the hard work put by the various organization-related committee members, a list of whom is provided herein.

Finally, it is time to thank our sponsors. This conference was organized in partnership with HEEDS Design Space Exploration and sponsored partially by ArcelorMittal and FinnOPT. We appreciate their contributions. Awards from the first two organizations are highly appreciated. We also thank Springer for supporting an award and also for publishing the proceedings. The help from the staff of the Kellogg Hotel and Conference Center in setting up the conference venue and amenities is highly appreciated. The support provided by students and associated members of the Computational Optimization and Innovation (COIN) Laboratory at Michigan State University, the staff of the NSF BEACON Center for the Study of Evolution in Action at Michigan State University, and the facilities of the College of Engineering and Michigan State University are appreciated. It would be unfair not to recognize here the dedication, time, and efforts of three persons: Connie James, Yashesh Dhebar, and Julian Blank, who helped us at every step of the way since the very beginning.

The paper submission and review process were executed using the EasyChair system, which helped to make our tasks much simpler.

December 2018

Kalyanmoy Deb
Erik Goodman
Carlos A. Coello Coello
Kathrin Klamroth
Kaisa Miettinen
Sanaz Mostaghim
Patrick Reed

Organization

General Chairs

Kalyanmoy Deb   Michigan State University, USA
Erik Goodman   Michigan State University, USA

Program Chairs

Carlos A. Coello Coello   CINVESTAV-IPN, Mexico
Kaisa Miettinen   University of Jyväskylä, Finland
Sanaz Mostaghim   University of Magdeburg, Germany
Patrick Reed   Cornell University, USA

MCDM Chairs

Kathrin Klamroth   University of Wuppertal, Germany
Kaisa Miettinen   University of Jyväskylä, Finland

Industry Session Chairs

Amos Ng   University of Skövde, Sweden
Ranny Sidhu   Siemens PLM Software, USA

Publicity Chairs

Salvatore Greco   University of Catania, Italy
Robin Purshouse   University of Sheffield, UK

Competitions Chairs

Xiaodong Li   RMIT University, Australia
Qingfu Zhang   City University of Hong Kong, SAR China

Archival Advisory Chairs

Hisao Ishibuchi   SUSTech, China
Kay Chen Tan   City University of Hong Kong, SAR China

Local Organizing Committee Chairs

Connie James   Michigan State University, USA
Sriram Narayanan   Michigan State University, USA
Pouyan Nejadhashemi   Michigan State University, USA

Local Organizing Committee

Chunteng Bao	Tongji University, China
Julian Blank	Michigan State University, USA
Yashesh Dhebar	Michigan State University, USA
Abhinav Gaur	Michigan State University, USA
Abhiroop Ghosh	Michigan State University, USA
Rayan Hussein	Michigan State University, USA
Ian Kropp	Michigan State University, USA
Jose Llera	Michigan State University, USA
Zhichao Lu	Michigan State University, USA
M. Melissa Rojas-Downing	Michigan State University, USA
Proteek Roy	Michigan State University, USA
Haitham Seada	Michigan State University, USA
Khaled Talukder	Michigan State University, USA
Yash Vesikar	Michigan State University, USA
Rahul Yalamanchili	Michigan State University, USA

International Advisory Committee Members

Juergen Branke	Warwick Business School, UK
Dimo Brockhoff	Inria, France
Andries Engelbrecht	University of Pretoria, South Africa
Zhun Fan	Shantou University, China
António Gaspar-Cunha	University of Minho, Portugal
Jeff Horn	Northern Michigan University, USA
Yaochu Jin	University of Surrey, UK
Bin Li	USTC, China
Henri Luchian	Alexandru Ioan Cuza University of Iasi, Romania
Masaharu Munetomo	Hokkaido University, Japan
Gisele Lobo Pappa	Universidade Federal de Minas Gerais, Brazil
Shahryar Rahmanayan	UOIT, Canada
Tapabrata Ray	University of New South Wales, Australia
Oliver Schütze	CINVESTAV-IPN, Mexico
Ankur Sinha	Indian Institute of Management Ahmedabad, India
Theodor Stewart	University of Cape Town, South Africa
P. N. Suganthan	Nanyang Technological University, Singapore
Ricardo Takahashi	Universidade Federal de Minas Gerais, Brazil
Santosh Tiwari	General Motors, USA
Jyrki Wallenius	Aalto University, Finland
Lyndon While	University of Western Australia, Australia
Xin Yao	SUSTech, China
Mengjie Zhang	Victoria University of Wellington, New Zealand

EMO Steering Committee Members

Carlos A. Coello Coello	CINVESTAV-IPN, Mexico
David Corne	Heriot-Watt University, UK
Kalyanmoy Deb	Michigan State University, USA
Michael Emmerich	LIACS, Leiden University, The Netherlands
Carlos Fonseca	University of Coimbra, Portugal
Hisao Ishibuchi	SUSTech, China
Joshua Knowles	University of Birmingham, UK
Kaisa Miettinen	University of Jyväskylä, Finland
Robin Purshouse	University of Sheffield, UK
David Schaffer	Binghamton University, USA
Lothar Thiele	ETH Zurich, Switzerland

Program Committee

Janos Abonyi	University of Pannonia, Hungary
Richard Allmendinger	University of Manchester, UK
Maria João Alves	University of Coimbra, Portugal
Sunith Bandaru	University of Skövde, Sweden
Helio Barbosa	LNCC, Brazil
Julian Blank	Michigan State University, USA
Juergen Branke	University of Warwick, UK
Dimo Brockhoff	Inria, France
Marco Chiarandini	University of Southern Denmark, Denmark
Sung-Bae Cho	Yonsei University, South Korea
João Clímaco	University of Coimbra, Portugal
Leandro Coelho	Pontifícia Universidade Católica do Parana, Brazil
Carlos A. Coello Coello	CINVESTAV-IPN, Mexico
Fernanda Costa	University of Minho, Portugal
Lino Costa	University of Minho, Portugal
Adiel Teixeira de Almeida	Federal University of Pernambuco, Brazil
Yves De Smet	Université Libre de Bruxelles, Belgium
Kalyanmoy Deb	Michigan State University, USA
Alexandre Delbem	University of Sao Paulo, Brazil
Clarisse Dhaenens	Université de Lille, France
Yashesh Dhebar	Michigan State University, USA
Michael Doumpos	Technical University of Crete, Greece
Rolf Drechsler	University of Bremen, Germany
Matthias Ehrgott	Lancaster University, UK
Michael Emmerich	Leiden University, The Netherlands
Alexander Engau	University of Colorado Denver, USA
Andries Engelbrecht	University of Stellenbosch, South Africa
Isabel Espírito	University of Minho, Portugal
Jonathan Fieldsend	University of Exeter, UK

José Rui Figueira	Instituto Superior Técnico, Lisbon, Portugal
Peter Fleming	University of Sheffield, UK
Joerg Fliege	University of Southampton, UK
Xavier Gandibleux	Université de Nantes, France
Antonio Gaspar-Cunha	University of Minho, Portugal
Abhiroop Ghosh	Michigan State University, USA
Erik Goodman	Michigan State University, USA
Bernard Grabot	ENIT, France
Salvatore Greco	University of Catania, Italy
Jussi Hakanen	University of Jyväskylä, Finland
Jin-Kao Hao	University of Angers, France
Carlos Henggeler Antunes	University of Coimbra, Portugal
Rayan Hussein	Michigan State University, USA
Masahiro Inuiguchi	Osaka University, Japan
Hisao Ishibuchi	Osaka Prefecture University, Japan
Johannes Jahn	University of Erlangen-Nuremberg, Germany
Andrzej Jaszkiewicz	Poznan University of Technology, Poland
Laetitia Jourdan	Inria, France
Kathrin Klamroth	University of Wuppertal, Germany
Arnaud Liefooghe	Université de Lille, France
Mariano Luque	University of Malaga, Spain
Robert Lygoe	Ford Motor Company Limited, UK
Manuel López-Ibáñez	University of Manchester, UK
Basseur Matthieu	LERIA Angers, France
Efrén Mezura-Montes	University of Veracruz, Mexico
Martin Middendorf	University of Leipzig, Germany
Kaisa Miettinen	University of Jyväskylä, Finland
Julian Molina	University of Malaga, Spain
Sanaz Mostaghim	University of Magdeburg, Germany
Vincent Mousseau	LAMSADE, France
Boris Naujoks	Cologne University of Applied Sciences, Germany
Antonio J. Nebro	University of Malaga, Spain
Frank Neumann	University of Adelaide, Australia
Amos Ng	University of Skovde, Sweden
Luis Paquete	University of Coimbra, Portugal
Dmitry Podkopaev	Polish Academy of Sciences, Poland
Aurora Pozo	Federal University of Paraná, Brazil
Robin Purshouse	University of Sheffield, UK
Tapabrata Ray	University of New South Wales, Australia
Patrick Reed	Cornell University, USA
Peter Rockett	University of Sheffield, UK
Proteek Roy	Michigan State University, USA
Guenter Rudolph	TU Dortmund, Germany
Francisco Ruiz	University of Malaga, Spain
Dhish Saxena	Indian Institute of Technology Roorkee, India

Serpil Sayin	Koc University, Turkey
Hartmut Schmeck	Karlsruhe Institute of Technology, Germany
Marc Schoenauer	Inria, France
Oliver Schuetze	CINVESTAV-IPN, Mexico
Marc Sevaux	Université de Bretagne-Sud, France
Deepak Sharma	Indian Institute of Technology Guwahati, India
Pradyumn Kumar Shukla	Karlsruhe Institute of Technology, Germany
Patrick Siarry	Université de Paris, France
Ranny Sidhu	Siemens PLM Software Inc., USA
Johannes Siebert	University of Bayreuth, Germany
Karthik Sindhya	FINNOPT Oy, Finland
Hemant Singh	University of New South Wales, Australia
Ankur Sinha	Indian Institute of Management Ahmedabad, India
Michael Stiglmayr	University of Wuppertal, Germany
Ricardo H. C. Takahashi	Universidade Federal de Minas Gerais, Brazil
Khaled Talukdar	Michigan State University, USA
Jürgen Teich	University of Erlangen-Nuremberg, Germany
Lothar Thiele	ETH Zurich, Switzerland
Gregorio Toscano-Pulido	CINVESTAV-IPN, Mexico
Heike Trautmann	University of Münster, Germany
Vipin Tripathi	College of Engineering Pune, India
Alexis Tsoukias	LAMSADE, France
Eiji Uchino	Yamaguchi University, Japan
Daniel Vanderpooten	LAMSADE, France
Elizabeth Wanner	Universidade Federal de Minas Gerais, Brazil
Farouk Yalaoui	University of Technology of Troyes, France
Saúl Zapotecas Martínez	UAM-Cuajimalpa, Mexico
Heiner Zille	University of Magdeburg, Germany

Sponsors

The EMO 2019 conference thanks the following organizations for sponsoring the conference.

Contents

Many-Objective EMO

Performance Metrics and Indicators

Innovization and Surrogates

Combinatorial EMO

MCDM and Interactive EMO

Applications

Theory

On Bi-objective Convex-Quadratic Problems

Cheikh Toure[✉], Anne Auger, Dimo Brockhoff, and Nikolaus Hansen

Inria and CMAP, Ecole Polytechnique, Palaiseau, France
cheikh.toure@polytechnique.edu

Abstract. In this paper, we analyze theoretical properties of bi-objective convex-quadratic problems. We give a complete description of their Pareto set and prove the convexity of their Pareto front. We show that the Pareto set is a line segment when both Hessian matrices are proportional. We then propose a novel set of convex-quadratic test problems, describe their theoretical properties and the algorithm abilities required by those test problems. This includes in particular testing the sensitivity with respect to separability, ill-conditioned problems, rotational invariance, and whether the Pareto set is aligned with the coordinate axis.

Keywords: Bi-objective optimization · Pareto set · Convex front · Convex-quadratic problems

1 Introduction

Convex-quadratic functions are among the simplest yet very useful test functions in optimization. Given a positive definite matrix Q of $\mathbb{R}^{n \times n}$, a convex quadratic function is defined as

$$f(x) = \frac{1}{2}(x - x^*)^\top Q(x - x^*)$$

where x^* is the unique optimum of the function. The Hessian of f coincides with the matrix Q. The level-sets of f defined as $\{x \in \mathbb{R}^n : (x - x^*)^\top Q(x - x^*) = c, c \geq 0\}$ are hyper-ellipsoids whose main axes are the eigenvectors of the matrix Q with length proportional to the inverse of the eigenvalues of Q.

By changing the eigenvalues and eigenvectors of Q, one can model different essential difficulties in numerical optimization: if the eigenvectors are not aligned with the coordinate axes (if the matrix Q is not diagonal), then the associated function is non-separable: it cannot be efficiently optimized by coordinate-wise search. In practice, difficult optimization problems are non-separable. Having a large condition number for Q, that is a large ratio between the largest and smallest eigenvalue of Q models ill-conditioned problems where the characteristic scale along different directions is very different. Ill-conditioning is very frequent in real-world problems. They arise naturally as one often optimizes quantities that have different natures and different intrinsic scales (some variables can be

© Springer Nature Switzerland AG 2019
K. Deb et al. (Eds.): EMO 2019, LNCS 11411, pp. 3–14, 2019.
https://doi.org/10.1007/978-3-030-12598-1_1

akin to time, others to weights, ...) such that a unit change along each variable can have a completely different impact on the function optimized. More generally, the eigenspectrum of Q entirely characterizes the scale among the different axes of the hyper-ellipsoidal level sets and parametrizes the difficulty of the function: from the arguably easiest function, the sphere function $f(x) = \sum_{i=1}^{n} x_i^2$, to very difficult ill-conditioned functions where condition numbers of Q of up to 10^{10} have been observed in real-world problems, for example in [3].

Convex-quadratic functions have been central to the design of several important classes of optimization algorithms for single-objective optimization. Newton or quasi-Newton methods use or learn a second order approximation of the objective function optimized [11]. This second order approximation is done by convex-quadratic functions (assuming that the function is twice continuously differentiable and convex). Introduced more recently, the class of derivative-free-optimization (DFO) trust-region based algorithms builds a second-order approximation of the objective function by interpolation [12]. In the evolutionary computation (EC) context, convex-quadratic functions have also played a central role for the design of algorithms like CMA-ES: they have been intensively used for designing the algorithm and the performance of the method has been carefully quantified on different eigenspectra of the matrix Q for different condition numbers [6].

Given that a multiobjective problem is "simply" the simultaneous optimization of single-objective problems, the typical difficulties of each objective function are the same as the typical difficulties of single-objective problems. In particular non-separability and ill-conditioning are important difficulties that the single functions have. Therefore, combining convex-quadratic problems seems natural for testing and designing multiobjective algorithms. This has already been done in the past for instance for the design of multiobjective versions of CMA-ES [7] or as a subset of the biobjective BBOB test function suite [2,13].

Yet, while the difficulties encoded and parametrized within a convex-quadratic problem are well-understood for single-objective optimization, the situation is different for multiobjective optimization, starting from bi-objective optimization. Simple properties like convexity of the Pareto front associated to bi-objective convex-quadratic problems as well as properties of the Pareto set have not been systematically investigated. Additionally, convex-quadratic bi-objective test problems used in the literature do not capture all important properties one could be testing with convex-quadratic problems. There is more degree of freedom than for single objective optimization that is not exploited: we can combine two functions having the same Hessian matrix, place the optima on the functions both on one axis of the search space, ... and this will affect how the Pareto set and Pareto front look like.

This paper aims at filling the gaps from the literature on multiobjective optimization with respect to convex-quadratic problems. More precisely the objectives are twofold: clarify theoretical Pareto properties of bi-objective problems where each function is convex-quadratic and define sets of bi-objective convex-quadratic problems that allow to test different (well-understood) difficulties of bi-objective problems. The paper is organized as follows: in Sect. 2 we present theoretical properties of convex-quadratic problems and discuss new test functions in Sect. 3.

2 Theoretical Properties of Bi-objective Convex-Quadratic Problems

2.1 Preliminaries

We consider bi-objective problems (f_1, f_2) defined on the search space $\mathbb{R}^n$.

The Pareto set of (f_1, f_2) is defined as the set of all non-dominated (or efficient) solutions $\{x \in \mathbb{R}^n \mid \nexists y \in \mathbb{R}^n$ such that $f_1(y) \leq f_1(x)$ and $f_2(y) \leq f_2(x)$ and at least one inequality is strict$\}$. The image of the Pareto set (in the objective space $\mathbb{R}^2$) is called the Pareto front of (f_1, f_2). We first remark that the Pareto set remains unchanged if we compose the objective functions with a strictly increasing function. More precisely the following lemma holds.

Lemma 1 (Invariance of the Pareto set to strictly increasing transformations of the objectives). *Given a bi-objective problem* $x \mapsto (f_1(x), f_2(x))$ *and* $g_1 : \mathrm{Im}(f_1) \longmapsto \mathbb{R}$, $g_2 : \mathrm{Im}(f_2) \longmapsto \mathbb{R}$ *two strictly increasing functions, then* (f_1, f_2) *and* $(g_1 \circ f_1, g_2 \circ f_2)$ *have the same Pareto set.*

Proof. If x is not in the Pareto set of $(g_1 \circ f_1, g_2 \circ f_2)$, then their exists y such that $g_1 \circ f_1(y) \leq g_1 \circ f_1(x)$ and $g_2 \circ f_2(y) \leq g_2 \circ f_2(x)$ with one inequality being strict, which is equivalent to the fact that $f_1(y) \leq f_1(x)$ and $f_2(y) \leq f_2(x)$, with one inequality being strict. And vice versa. Hence x is not in the Pareto set of $(g_1 \circ f_1, g_2 \circ f_2)$ if and only if it is not in the Pareto set of (f_1, f_2), which shows that both problems have the same Pareto set. $\square$

From now on (f_1, f_2) denote a bi-objective convex-quadratic problem. More precisely, let x_1, x_2 be two *different* vectors in $\mathbb{R}^n$, and $\alpha, \beta > 0$. Let Q_1 and Q_2 (in $\mathbb{R}^{n^2}$) be two positive definite matrices and consider the bi-objective *minimization* problem (f_1, f_2) defined for $x \in \mathbb{R}^n$ as

$$f_1(x) = \frac{1}{\alpha} (x - x_1)^\top Q_1 (x - x_1), f_2(x) = \frac{1}{\beta} (x - x_2)^\top Q_2 (x - x_2). \quad (1)$$

We denote this general bi-objective convex-quadratic problem by $\mathcal{P}$, and assume that the optimization goal is to find (an approximation of) the Pareto set of $\mathcal{P}$.

2.2 Pareto Set

We characterize in this section the Pareto set of $\mathcal{P}$. We use the linear scalarization method to obtain the whole Pareto set. This is doable, whenever f_1 and f_2 are strict convex functions (see [8]). Then the Pareto set of $\mathcal{P}$ is described by the solutions of

$$\min_{x \in \mathbb{R}^n} (1 - t) f_1(x) + t f_2(x), \text{ for } t \in [0, 1].$$

We prove in the next proposition that the Pareto set of $\mathcal{P}$ is a continuous and differentiable parametric curve of $\mathbb{R}^n$ whose extremes are x_1 and x_2.

Proposition 1. *The Pareto set of* $\mathcal{P}$ *is the image of the function* φ *defined as*

$$\varphi : t \in [0,1] \mapsto [(1-t)Q_1 + tQ_2]^{-1}[(1-t)Q_1x_1 + tQ_2x_2] . \tag{2}$$

The function φ *is differentiable and verifies for any* t *in* $[0,1]$

$$(1-t)Q_1\left(\varphi(t) - x_1\right) = tQ_2\left(x_2 - \varphi(t)\right), \tag{3}$$
$$t\left[(1-t)Q_1 + tQ_2\right]\varphi'(t) = Q_1\left(\varphi(t) - x_1\right). \tag{4}$$

Hence, the Pareto set is a continuous (differentiable) curve of $\mathbb{R}^n$ *whose extremes are* $x_1 = \varphi(0)$ *and* $x_2 = \varphi(1)$.

Proof. For any s in $[0,1]$, define $g_s \stackrel{\text{def}}{=} (1-s)f_1 + sf_2$. We observe that g_s, like f_1 and f_2, is strictly convex, differentiable, and diverges to ∞ when $\|x\|$ goes to ∞ (where $\|x\|$ denotes the Euclidean norm). Then its critical point minimizes g_s. Let us now compute the gradient of g_s times $\alpha\beta$ for x in $\mathbb{R}^n$:

$$\alpha\beta\nabla g_s(x) = (1-s)\alpha\beta\,\nabla f_1(x) + s\alpha\beta\,\nabla f_2(x) = 2(1-s)\beta\,Q_1(x - x_1) + 2s\alpha\,Q_2(x - x_2)$$

Thus, $\alpha\beta\nabla g_s(x) = 2\left[(1-s)\beta\,Q_1 + s\alpha\,Q_2\right]x - 2(1-s)\beta\,Q_1x_1 - 2s\alpha\,Q_2x_2.$

Then it follows that for any s in $[0,1]$, the point that minimizes g_s (its critical point), denoted by $\tilde{x}_s$ verifies $\frac{(1-s)\beta\,Q_1 + s\alpha\,Q_2}{(1-s)\beta + s\alpha}\,\tilde{x}_s = \frac{(1-s)\beta\,Q_1x_1 + s\alpha\,Q_2x_2}{(1-s)\beta + s\alpha}$. Since $[0,1] \ni s \longmapsto \frac{s\alpha}{(1-s)\beta + s\alpha} \in [0,1]$ is bijective (its derivative is $s \longmapsto \frac{\alpha\beta}{((1-s)\beta + s\alpha)^2}$), then it is equivalent to parametrize the Pareto set with $t \stackrel{\text{def}}{=} \frac{s\alpha}{(1-s)\beta + s\alpha}$. Hence, the Pareto set is fully described by $(\varphi(t))_{t \in [0,1]}$ such that:

$$[(1-t)Q_1 + tQ_2]\,\varphi(t) = (1-t)Q_1x_1 + tQ_2x_2, \tag{5}$$
$$(1-t)Q_1\left(\varphi(t) - x_1\right) = tQ_2\left(x_2 - \varphi(t)\right). \tag{6}$$

The function $t \to [(1-t)Q_1 + tQ_2]^{-1}$ is differentiable as inverse of a differentiable and invertible matrix function. Then φ is differentiable.

We differentiate (5) and multiply by t to obtain $t\left[(1-t)Q_1 + tQ_2\right]\varphi'(t) = tQ_2x_2 - tQ_1x_1 + tQ_1\varphi(t) - tQ_2\varphi(t)$. Injecting in (6) gives $t\left[(1-t)Q_1 + tQ_2\right]\varphi'(t) = Q_1\left(\varphi(t) - x_1\right)$, for any $t \in [0,1]$. $\square$

We obtain as corollary that when f_1 and f_2 have proportional Hessian matrices, then the Pareto set is the line segment between the optima of the functions f_1 and f_2.

Corollary 1. *In the case where* f_1 *and* f_2 *have proportional Hessian matrices, the Pareto set of* $\mathcal{P}$ *is the line segment between* x_1 *and* x_2.

Proof. In that case, their exists a real γ such that $\frac{Q_1}{\alpha} = \gamma\frac{Q_2}{\beta}$. Then, Proposition 1 implies that for any $t \in [0,1]$,

$$\varphi(t) = \left[(1-t)\gamma\frac{\alpha}{\beta}Q_2 + tQ_2\right]^{-1}\left[(1-t)\gamma\frac{\alpha}{\beta}Q_2x_1 + tQ_2x_2\right] = \frac{\gamma\alpha(1-t)x_1 + t\beta x_2}{(1-t)\alpha\gamma + t\beta},$$

which is $[x_1, x_2]$, since $[0,1] \ni t \longmapsto \frac{t\beta}{(1-t)\alpha\gamma + t\beta} \in [0,1]$ is a bijection. $\square$

Using Lemma 1, we directly deduce the following corollary.

Corollary 2. *If f_1 and f_2 have proportional Hessian matrices, $g_1 : \mathrm{Im}(f_1) \longmapsto \mathbb{R}$, $g_2 : \mathrm{Im}(f_2) \longmapsto \mathbb{R}$ are two strictly increasing functions, then the Pareto set of the problem $(g_1 \circ f_1, g_2 \circ f_2)$ is the line segment between x_1 and x_2.*

As an example, the double-norm problem defined as: $(x \to \|x - x_1\|_2, x \to \|x - x_2\|_2)$ can be seen as: $(g \circ f_1, g \circ f_2)$ where $g(x) = \sqrt{x}$, $f_1(x) = \|x - x_1\|_2^2$ and $f_2(x) = \|x - x_2\|_2^2$.

Then $(g \circ f_1, g \circ f_2)$ has the same Pareto set than the double-sphere problem (f_1, f_2), which is the line segment between x_1 and x_2. Therefore the Pareto front of the double-norm problem is described by $(t\|x_2 - x_1\|_2, (1 - t)\|x_2 - x_1\|_2)_{t \in [0,1]}$. Thereby, the front is described by the function $u \longmapsto \|x_2 - x_1\|_2 - u$. We recover the well-known result that the double-norm problem has a linear front.

Corollary 2 allows also to recover the Pareto set description for the one-peak scenario in the Mixed-Peak Bi-Objective Problem (see [9] and [10]).[1]

In general, the Pareto set of a bi-objective convex-quadratic problem is not necessarily a line segment. Consider for instance for $n = 2$ the case where $x_1 = (0, 0)^\top$, $x_2 = (1, 1)^\top$ and where we generate two different matrices Q_1 and Q_2 by randomly rotating a diagonal matrix with eigenvalues 1 and 10. Two resulting Pareto fronts associated to different random rotations are depicted in Fig. 1.

For $n = 10$, we also define $\mathcal{P}_{10}$ setting $x_1 = (0, \ldots, 0)^\top$, $x_2 = (1, \ldots, 1)^\top$ and Q_1 and Q_2 as diagonal matrices such that for $i = 1, \ldots, 10$

$$Q_1(i, i) = 100^{\frac{i-1}{9}}, \text{ and } Q_2(i, i) = 10^{\frac{i-1}{9}}. \tag{7}$$

The different coordinates of the Pareto set given in (3) are depicted in Fig. 1.

2.3 Convexity of the Pareto Front

Corollary 1 proves that in the case where we have proportional Hessian matrices in problem $\mathcal{P}$, the Pareto set is a line segment. Then it is reasonable to expect a simple analytic expression for the corresponding Pareto front. In what follows, we will express the Pareto front of a bi-objective problem as a one-dimensional function $u \in \mathbb{R} \mapsto g(u)$. Formally, if $t \in \mathbb{R} \mapsto \varphi(t) \in \mathbb{R}^n$ is a parametrization of the Pareto set, then the function g satisfies $f_2(\varphi(t)) = g(f_1(\varphi(t)))$. It is well-known that when (f_1, f_2) is the double-sphere, that is $f_1(x) = \frac{1}{n} \sum_{i=1}^n x_i^2$ and $f_2(x) = \frac{1}{n} \sum_{i=1}^n (x_i - 1)^2$, then the Pareto front expression is given by $g(u) = (1 - \sqrt{u})^2$ [4]. In the next proposition, we show that this expression of the Pareto front holds (up to a normalization) for all bi-objective convex-quadratic problems, provided the Hessians of f_1 and f_2 are proportional.

---

[1] In that scenario, we set $f_1(x) = (x - c)^\top \Sigma (x - c)$, $f_2(x) = (x - c')^\top \Sigma' (x - c')$ (f_1 and f_2 are seen as squares of the Mahalanobis distance to the optima, with respect to the Hessian matrices), $g_1(u) = 1 - \frac{h_1}{1 + \frac{\sqrt{u}}{r_1}}$, $g_2(u) = 1 - \frac{h_2}{1 + \frac{\sqrt{u}}{r_2}}$.

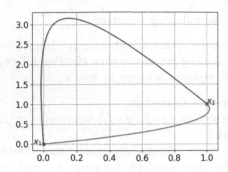
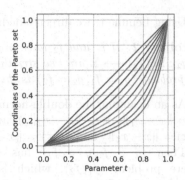

Fig. 1. Left: Two Pareto sets for $n = 2$ represented in $\mathbb{R}^2$ with Q_1 and Q_2 randomly sampled and different. Right: Pareto set for $n = 10$ with matrices given in (7) represented as the function of the parameter t given in (3). The coordinates are ordered, the first one is on top and last one below.

Proposition 2. *When we have proportional Hessian matrices in the problem $\mathcal{P}$, the Pareto front is described by the following continuous and convex function:*

$$u \in [0, \kappa_\alpha] \mapsto \kappa_\beta \left(1 - \sqrt{\frac{u}{\kappa_\alpha}}\right)^2, \text{ where } \begin{cases} \kappa_\alpha = \dfrac{(x_2-x_1)^\top Q_1 (x_2-x_1)}{\alpha} \\ \kappa_\beta = \dfrac{(x_2-x_1)^\top Q_2 (x_2-x_1)}{\beta} \end{cases} \tag{8}$$

Proof. Denote $u \overset{\text{def}}{=} f_1 \circ \varphi$ and $v \overset{\text{def}}{=} f_2 \circ \varphi$, where $\varphi : [0,1] \ni t \longmapsto (1-t)x_1 + tx_2 \in [x_1, x_2]$ is the line segment between x_1 and x_2.

For any $t \in [0,1]$, $u(t) = f_1(\varphi(t)) = \frac{1}{\alpha}(x_2-x_1)^\top Q_1 (x_2-x_1) t^2, v(t) = f_2(\varphi(t)) = \frac{1}{\beta}(x_2-x_1)^\top Q_2 (x_2-x_1)(1-t)^2$. It follows that for any $t \in [0,1]$:

$$v(t) = \frac{(x_2-x_1)^\top Q_2 (x_2-x_1)}{\beta}\left(1 - \sqrt{\frac{\alpha u(t)}{(x_2-x_1)^\top Q_1 (x_2-x_1)}}\right)^2. \qquad \square$$

From Proposition 2, we deduce that if we set $\kappa_\alpha = \kappa_\beta = 1$, then the Pareto front will be independent from the Hessian matrix and will be described by the front of the double-sphere problem: $u \mapsto (1 - \sqrt{u})^2$.

We investigate now the general case where the Hessians of the functions f_1 and f_2 are not necessarily proportional. Yet, before digging into the general convex-quadratic problems, we show a result on the shape of the Pareto front of a larger class of bi-objective problems.

Theorem 1. *Let $f_1 : \mathbb{R}^n \longmapsto \mathbb{R}$ and $f_2 : \mathbb{R}^n \longmapsto \mathbb{R}$ be strict convex differentiable functions such that the problem (f_1, f_2) has, as Pareto set, the image of a differentiable function $\varphi : [0,1] \longmapsto \mathbb{R}^n$.*
*Assume that: (i) $f_1 \circ \varphi$ is strictly monotone, (ii) $\lim\limits_{t \to 0} \frac{(f_1 \circ \varphi)'(t)}{t} \neq 0$ and (iii) $\lim\limits_{t \to 1} \frac{(f_2 \circ \varphi)'(t)}{1-t} \neq 0$. Then, the Pareto front is a **convex** curve, with **vertical** tangent at $t = 0$ and horizontal tangent at $t = 1$.*

Proof. Denote by $u \overset{\text{def}}{=} f_1 \circ \varphi$ and $v \overset{\text{def}}{=} f_2 \circ \varphi$. Then the Pareto front is described by the parametric equation $(u(t), v(t))$, for $t \in [0, 1]$. We will show that $u'v'' - u''v' > 0$ which implies the convexity of the curve.

By linear scalarization (see [8], or weighted sum method in [5]), as in the proof of Proposition 1, we have $(1-t)\nabla f_1(\varphi(t)) + t\nabla f_2(\varphi(t)) = 0$. If we take the scalar product of the former equation with $\varphi'(t)$, we obtain that

$$(1-t)\langle \nabla f_1(\varphi(t)), \varphi'(t)\rangle + t\langle \nabla f_2(\varphi(t)), \varphi'(t)\rangle = 0. \tag{9}$$

Moreover, for any differentiable function f with suitable domains,

$$(f \circ \varphi)'(t) = \mathrm{d}(f \circ \varphi)_t(1) = \mathrm{d}f_{\varphi(t)}(\mathrm{d}\varphi_t(1)) = \langle \nabla f(\varphi(t)), \varphi'(t)\rangle. \tag{10}$$

Inserting this in (9) shows $(1-t)(f_1 \circ \varphi)'(t) + t(f_2 \circ \varphi)'(t) = 0$, which is the same as:

$$(1-t)u'(t) + tv'(t) = 0, \text{ for any } t \in [0, 1]. \tag{11}$$

Since $\lim\limits_{t \to 0} \frac{(f_1 \circ \varphi)'(t)}{t}$ exists, (11) implies that:

$$v'(t) = \left(1 - \frac{1}{t}\right)u'(t), \text{ for any } t \in [0, 1]. \tag{12}$$

By deriving (12) and multiplying by $u'(t)$ in a suitable way, we obtain

$$u'(t)v''(t) = \frac{1}{t^2}u'(t)^2 + \left(1 - \frac{1}{t}\right)u'(t)u''(t), \text{ for any } t \in [0, 1]. \tag{13}$$

Using (12) in (13) gives $u'(t)v''(t) = \frac{1}{t^2}u'(t)^2 + v'(t)u''(t)$. Thanks to the assertions on $f_1 \circ \varphi$, we have that $u'(t)v''(t) - u''(t)v'(t) > \frac{1}{t^2}u'(t)^2 > 0$, for any $t \in [0, 1]$. Thus, the Pareto front is a **convex** curve.

Evaluating (11) at $t = 0$ and at $t = 1$ implies that $u'(0) = 0, v'(1) = 0$. And if we divide (11) by t (resp. $1-t$) and take the limit to 0 (resp. 1), it follows that $v'(0) \neq 0$ (resp. $u'(1) \neq 0$). Thereby we also obtain the derivative assumptions on the extremal points. □

Remark 1. Note that the above result about the tangents in the extremal points have additional consequences: according to [1], the assumptions of Theorem 1 imply that the extremal points are never included in any optimal μ-distributions of the Hypervolume indicator.

We now deduce the convexity of the Pareto front for convex-quadratic bi-objective problems and characterize the derivatives at the extremes of the front.

Corollary 3. *For the problem $\mathcal{P}$, the Pareto front is a **convex** curve, with **vertical** tangent at $(0, f_2(x_1))$ and **horizontal** tangent at $(f_1(x_2), 0)$.*

Proof. We will show that $f_1 \circ \varphi$ verifies the assumptions of Theorem 1. From (10) we know that

$$(f_1 \circ \varphi)'(t) = \langle \nabla f_1(\varphi(t)), \varphi'(t) \rangle. \tag{14}$$

In addition, $\nabla f_1(\varphi(t)) = \frac{2}{\alpha} Q_1 (\varphi(t) - x_1)$ and Eq. (4) of Proposition 1 gives $t[(1-t)Q_1 + tQ_2]\varphi'(t) = Q_1 (\varphi(t) - x_1)$. Multiplying (14) by $t \in [0,1]$ shows

$$t(f_1 \circ \varphi)'(t) = \frac{2}{\alpha} \left\langle [(1-t)Q_1 + tQ_2]^{-1} Q_1 (\varphi(t) - x_1), Q_1 (\varphi(t) - x_1) \right\rangle. \tag{15}$$

Since $[(1-t)Q_1 + tQ_2]^{-1}$ is a positive definite matrix, then $t(f_1 \circ \varphi)'(t) \geq 0$. Let us prove that $\varphi(t) \neq x_1$, for $t \in (0,1]$. By contradiction, assume that there exists $t \in (0,1]$ such that $\varphi(t) = x_1$. Then Eq. (3) in Proposition 1 shows that: $tQ_2(x_2 - \varphi(t)) = (1-t)Q_1(\varphi(t) - x_1) = 0$, which implies that $x_2 = \varphi(t) = x_1$: that is impossible since $x_1 \neq x_2$. Hence, by *reductio ad absurdum*, $\varphi(t) \neq x_1$, for $t \in (0,1]$. From (15), it follows that

$$(f_1 \circ \varphi)'(t) > 0, \text{ for any } t \in (0,1]. \tag{16}$$

If we use again the relation from Proposition 1, we obtain $\lim\limits_{t \to 0} \frac{Q_1(\varphi(t) - x_1)}{t} = Q_2(x_2 - \varphi(0)) = Q_2(x_2 - x_1)$. Injecting this result in (15), it follows that:

$$\lim_{t \to 0} \frac{(f_1 \circ \varphi)'(t)}{t} = \frac{2}{\alpha} \langle Q_1^{-1} Q_2 (x_2 - x_1), Q_2 (x_2 - x_1) \rangle$$

$$> 0, \text{ since } (Q_1^{-1} \text{ is a positive definite matrix}) \tag{17}$$

In the same way as above, we obtain that

$$\lim_{t \to 1} \frac{(f_2 \circ \varphi)'(t)}{1-t} = -\frac{2}{\beta} \langle Q_2^{-1} Q_1 (x_1 - x_2), Q_1 (x_1 - x_2) \rangle < 0. \tag{18}$$

Equations (16), (17), and (18) allow us to apply Theorem 1.   □

We illustrate the previous corollary by taking three random instances of our general problem $\mathcal{P}$, with the scalings always chosen as $\alpha = \beta = \max(f_1(x_2), f_2(x_1))$. The Pareto fronts are presented in Fig. 2. We observe that the Pareto fronts are convex and their derivatives are infinite on the left and zero on the right.

3   New Classes of Bi-objective Test Functions

Bi-objective problems using convex-quadratic functions have been used to test MO algorithms (see for example [7]). Problems where both Hessian matrices have the same eigenvalues have been used in particular. Yet, test problems considered so far do not explore the full possibilities of properties that can be tested. We therefore extend the test problems from the literature to be able to capture more

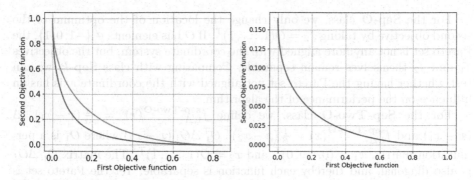

Fig. 2. Left: Two Pareto fronts for $n = 2$ represented in $\mathbb{R}^2$ with Q_1, Q_2 randomly sampled and different. Right: Pareto front for $n = 10$ with matrices given in (7).

properties. To do so we present seven classes of bi-objective convex-quadratic problems where the eigenspectra of both Hessian matrices are equal. A natural extension of these classes is to use in each objective different eigenspectra, Δ, which leads in general to a nonlinear Pareto set.

The proposed construction parametrizes, apart from search space translations, *all* bi-objective convex-quadratic functions with identical Hessian eigenspectrum in seven classes with increasing difficulty. The particular focus is on problems with a linear Pareto set in five of the seven classes. Some classes represent essentially different problems, hence we do not expect uniform performance over all problems within each class. Independently of the given construction, invariance to search space rotation can be tested by applying an orthogonal transformation to the input argument.

We start from a diagonal matrix Δ with positive entries that define a separable convex-quadratic function $f(x) = \frac{1}{\alpha} x^\top \Delta x$. For instance, Δ can be equal to the identity and we recover the sphere function. If $\Delta(1,1) = 1$, $\Delta(n, n) = 10^8$ and $\Delta(i, i) = 10^4$, we recover the separable cig-tab function and if $\Delta(i, i) = 10^{6\frac{i-1}{n-1}}$, we recover the separable ellipsoid function.

In the sequel, O and O_2 denote orthogonal matrices. O_1 is either a permutation matrix, or an orthogonal matrix, depending on the context. The classes of problems proposed are summarized in Tables 1 and 2.

The Sep Problem Classes. We define the **Sep-**k class by considering two separable functions and place the optimum of f_1 in 0 and of f_2 in the kth unit vector: $f_{1,\Delta}^{\mathsf{sep}\text{-}k}(x) = \frac{1}{\alpha} (x - x_1)^\top \Delta (x - x_1)$ and $f_{2,\Delta}^{\mathsf{sep}\text{-}k}(x) = \frac{1}{\beta} (x - x_2)^\top \Delta (x - x_2)$, where $x_1 = (0, \ldots, 0)^\top$ and $x_2 = (0, \ldots, 0, \sqrt{n}, 0, \ldots, 0)^\top$ where $\sqrt{n}$ is at coordinate k. According to Corollary 1, the Pareto set of this class of problems is the line segment between the optima of the single-objective problems. These problems allow to test the performance on separable problems with a Pareto set aligned with the coordinate axis and check the sensibility with respect to different axes (by varying k).

For the **Sep-O** class, we only change the location of the optimum of the second objective by taking $x_2 = O(1, \ldots, 1)^\top$. If O has elements $\notin \{-1, 0, 1\}$, the Pareto set is not anymore aligned with the coordinate system, but the objectives f_1 and f_2 themselves remain separable. Comparing with class **Sep-k**, we can test whether having the Pareto set not aligned with the coordinate axis has an influence on the performance of the algorithm.

For the **Sep-Two-O** class, we define $f_{1,\Delta}^{\textbf{sep-Two-O}}(x) = \frac{1}{\alpha}(x - x_1)^\top \Delta$ $(x - x_1)$ and $f_{2,\Delta}^{\textbf{sep-Two-O}}(x) = \frac{1}{\beta}(x - x_2)^\top O_1^\top \Delta O_1 (x - x_2)$ where O_1 is a permutation matrix, $x_1 = (0, \ldots, 0)^\top$ and $x_2 = O(1, \ldots, 1)^\top$. The matrix $O_1^\top \Delta O_1$ is also diagonal, and thereby each function is separable. Yet the Pareto set is generally not a line segment anymore since we have different Hessian matrices. We can test here the difficulty of having a nonlinear Pareto set on separable functions.

The One and the One-O Problem Classes. We now consider non-separable problems with a line segment as Pareto set. We define $f_{1,\Delta}^{\textbf{one}}(x) = \frac{1}{\alpha}(x - x_1)^\top O_1^\top$ $\Delta O_1 (x - x_1)$ and $f_{2,\Delta}^{\textbf{one}}(x) = \frac{1}{\alpha}(x - x_2)^\top O_1^\top \Delta O_1 (x - x_2)$, where O_1 is an orthogonal matrix, $x_1 = (0, \ldots, 0)^\top$ and $x_2 = (1, \ldots, 1)^\top$. We replace x_2 by Ox_2 to obtain the **One-O** problems.

These two problem classes allow to test the performance on non-separable problems that have a line segment as Pareto set comparing in particular to class **Sep-O**. Up to a reformulation, the problems ELLI1 and CIGTAB1 from [7] are from the **One-O** problem class. Generally, we do not expect different performance over all problems of the **One** vs the **One-O** class.

The Two and the Two-O Problem Classes. For these classes, we rotate each function independently; then the Pareto set is generally not a line segment

Table 1. Unconstrained quadratic bi-objective test problems: Δ is a positive diagonal matrix, O is an orthogonal matrix, O_1 is a permutation matrix.

	Sep-k	**Sep-O**	**Sep-Two-O**
x_1	$(0, \ldots, 0)^\top$	$(0, \ldots, 0)^\top$	$(0, \ldots, 0)^\top$
x_2	$\underbrace{(0, .., \sqrt{n}, .., 0)}_{\sqrt{n} \text{ is at row } k}^\top$	$O(1, \ldots, 1)^\top$	$O(1, \ldots, 1)^\top$
Q_1, Q_2	Δ, Δ	Δ, Δ	$\Delta, O_1^\top \Delta O_1$
Level sets			

Table 2. Unconstrained quadratic bi-objective test problems: Δ is a positive diagonal matrix, O, O_1 and O_2 are three independent orthogonal matrices.

	One	One-O	Two	Two-O
x_1	$(0,\dots,0)^\top$	$(0,\dots,0)^\top$	$(0,\dots,0)^\top$	$(0,\dots,0)^\top$
x_2	$(1,\dots,1)^\top$	$O(1,\dots,1)^\top$	$(1,\dots,1)^\top$	$O(1,\dots,1)^\top$
Q_1,Q_2	$O_1^\top\Delta O_1,\, O_1^\top\Delta O_1$	$O_1^\top\Delta O_1,\, O_1^\top\Delta O_1$	$O_1^\top\Delta O_1,\, O_2^\top\Delta O_2$	$O_1^\top\Delta O_1,\, O_2^\top\Delta O_2$
Level sets				

anymore. We define $f_{1,\Delta}^{\mathbf{two}}(x) = \frac{1}{\alpha}(x - x_1)^\top O_1^\top \Delta O_1 (x - x_1)$ and $f_{2,\Delta}^{\mathbf{two}}(x) = \frac{1}{\alpha}(x - x_2)^\top O_2^\top \Delta O_2 (x - x_2)$, with O_1 orthogoanal, $x_1 = (0,\dots,0)^\top$ and $x_2 = (1,\dots,1)^\top$. The corresponding **O** problems are obtained with Ox_2 replacing x_2. All presented classes are subsets of the **Two-O** class. ELLI2 and CIGTAB2 from [7] fall within the **Two-O** class. Compared to the respective **One** classes, we can test the impact of having a nonlinear Pareto set.

4   Summary

We have presented an analytic description of the Pareto set for quadratic bi-objective problems. We have shown that the Pareto set is a line segment when both objectives have proportional Hessian matrices and deduced a complete description of the Pareto front in that case. We have also proven that some properties of the double-sphere are conserved in a wider framework that includes the general quadratic bi-objective problem: the Pareto front remains convex and its vertical and horizontal tangents remain at the extremal points of the front. Such assumptions on the derivatives imply that when looking at the optimal μ-distributions of the Hypervolume indicator, the extremal points are always excluded [1]. We have also presented several classes of problems, where each one tests a specific capability of the multiobjective algorithm.

Acknowledgments. The Ph.D. of Cheikh Touré is funded by Inria and Storengy. We particularly thank F. Huguet and A. Lange from Storengy for their strong support, practical ideas and expertise.

References

1. Auger, A., Bader, J., Brockhoff, D., Zitzler, E.: Theory of the hypervolume indicator: optimal μ-distributions and the choice of the reference point. In: Foundations of Genetic Algorithms (FOGA 2009), pp. 87–102. ACM (2009)

2. Brockhoff, D., Tran, T.D., Hansen, N.: Benchmarking numerical multiobjective optimizers revisited. In: Genetic and Evolutionary Computation Conference (GECCO 2015), pp. 639–646. ACM (2015). https://doi.org/10.1145/2739480.2754777

3. Collange, G., Delattre, N., Hansen, N., Quinquis, I., Schoenauer, M.: Multidisciplinary optimisation in the design of future space launchers. In: Multidisciplinary Design Optimization in Computational Mechanics, pp. 487–496. Wiley (2010)

4. Emmerich, M.T.M., Deutz, A.H.: Test problems based on Lamé superspheres. In: Obayashi, S., Deb, K., Poloni, C., Hiroyasu, T., Murata, T. (eds.) EMO 2007. LNCS, vol. 4403, pp. 922–936. Springer, Heidelberg (2007). https://doi.org/10.1007/978-3-540-70928-2_68

5. Grodzevich, O., Romanko, O.: Normalization and other topics in multi-objective optimization. In: Proceedings of the Fields-MITACS Industrial Problems Workshop 2006, pp. 89–101. Fields-MITACS (2006)

6. Hansen, N., Ostermeier, A.: Completely derandomized self-adaptation in evolution strategies. Evol. Comput. 9(2), 159–195 (2001)

7. Igel, C., Hansen, N., Roth, S.: Covariance matrix adaptation for multi-objective optimization. Evol. Comput. 15(1), 1–28 (2007)

8. Jahn, J.: Vector Optimization: Theory, Applications and Extensions. Springer, Heidelberg (2004). https://doi.org/10.1007/978-3-642-17005-8

9. Kerschke, P., et al.: Search dynamics on multimodal multi-objective problems. Evol. Comput. 1–30 (2018)

10. Kerschke, P., et al.: Towards analyzing multimodality of continuous multiobjective landscapes. In: Handl, J., Hart, E., Lewis, P.R., López-Ibáñez, M., Ochoa, G., Paechter, B. (eds.) PPSN 2016. LNCS, vol. 9921, pp. 962–972. Springer, Cham (2016). https://doi.org/10.1007/978-3-319-45823-6_90

11. Nocedal, J., Wright, S.J.: Numerical Optimization. Springer, New York (2006). https://doi.org/10.1007/b98874

12. Powell, M.J.: The NEWUOA software for unconstrained optimization without derivatives. Technical report. DAMTP 2004/NA05, CMS, University of Cambridge, Cambridge CB3 0WA, UK, November 2004

13. Tušar, T., Brockhoff, D., Hansen, N., Auger, A.: COCO: the bi-objective black box optimization benchmarking (bbob-biobj) test suite. CoRR abs/1604.00359 (2016). http://arxiv.org/abs/1604.00359

An Empirical Investigation of the Optimality and Monotonicity Properties of Multiobjective Archiving Methods

Miqing Li[1](✉) and Xin Yao[1,2]

[1] CERCIA, School of Computer Science, University of Birmingham,
Birmingham B15 2TT, UK
limitsing@gmail.com, x.yao@cs.bham.ac.uk
[2] Shenzhen Key Laboratory of Computational Intelligence (SKyLoCI),
Department of Computer Science and Engineering,
Southern University of Science and Technology, Shenzhen, People's Republic of China

Abstract. Most evolutionary multiobjective optimisation (EMO) algorithms explicitly or implicitly maintain an archive for an approximation of the Pareto front. A question arising is whether existing archiving methods are reliable with respect to their convergence and approximation ability. Despite theoretical results available, it remains unknown how these archivers actually perform in practice. In particular, what percentage of solutions in their final archive are Pareto optimal? How frequently do they experience deterioration during the archiving process? Deterioration means archiving a new solution which is dominated by some solution discarded previously. This paper answers the above questions through a systematic investigation of eight representative archivers on 37 test instances with two to five objectives. We have found that (1) deterioration happens to all the archivers; (2) the deterioration degree can vary dramatically on different problems; (3) some archivers clearly perform better than others; and (4) several popular archivers sometime return a population with most solutions being the non-optimal. All of these suggest the need of improvement of current archiving methods.

Keywords: Multi-objective optimisation · Archive · Optimality · Monotonicity · Empirical investigation · Evolutionary computation

1 Introduction

Most evolutionary multiobjective optimisation (EMO) algorithms, and other multiobjective search techniques, keep an archive[1] to capture the output of the search process. Such an archiver is typically used to approximate the Pareto

---

[1] For EMO algorithms without considering an external archive (e.g., NSGA-II [8]), their population can also be seen as an implicit archive where the selection operation is performed to preserve the best solutions ever produced [40].

© Springer Nature Switzerland AG 2019
K. Deb et al. (Eds.): EMO 2019, LNCS 11411, pp. 15–26, 2019.
https://doi.org/10.1007/978-3-030-12598-1_2

front and/or as a collection of the current most promising solutions to guide next step search. Archiving can be seen as a process of taking new points from a point sequence, comparing them with the old points in the archive and deciding how to update the archive.

An archive of bounded size is of importance due to not only the consideration of computational resource but also search performance and later decision-making process. As such, numerous archiving methods (or archivers) emerge, known as elite preservation or environmental selection in evolutionary algorithms. They all serve the purpose of maintaining a set of well-converged and well-diversified solutions to represent the Pareto front.

However, an important issue of archiving has received relatively little attention—the optimality/monotonicity properties of archivers. In particular, one may be curious about whether an archiving method is able to return a subset of the Pareto optimal solutions discovered so far. This matters as the decision maker certainly does not want to face a situation that s/he has to select an inferior solution in the archive but misses a Pareto optimal solution once produced. In fact, many papers have observed that EMO algorithms whose archiving has no theoretical quality guarantee can suffer from dramatic performance oscillation during the search process on various instances, such as synthetic input sequences [21,31], benchmark test problems [2,11,25], and real life scenarios [10,32].

Unfortunately, most modern archivers do not have such optimality/monotonicity properties. They fail to ensure a subset of the Pareto optimal points with respect to an input sequence. Points can be preserved even when they are dominated by the points eliminated previously in the archiving process. A subsequent archive can be worse than an earlier archive. These drawbacks have been well illustrated in the literature, on different types of archiving methods, such as Pareto-based archiver [11,14,25], indicator-based archiver [22,31], and decomposition-based archiver [10]. López-Ibáñez et al. [31] have made a comprehensive summary of the approximation properties for popular archiving methods.

On the other hand, some work focused on development of monotonic archiving methods, including theoretical analysis [6,14,20,33] and algorithm design [18,25,26,35]. However, without problem-specific knowledge available a priori, monotonic archiving methods often fail to maintain a diverse solution set and may end up with very few solutions in the archive (see [21,31]). As such, non-monotonic archiving methods are still dominantly used everywhere.

Given the above, one interesting question raised is how current state-of-the-art archiving methods, despite their theoretical drawbacks, perform in practice. In particular, do archivers, in most cases, actually return a subset of the Pareto optimal solutions discovered so far; in other words, what percentage of solutions in the final archive are Pareto optimal? How frequently do archivers experience deterioration during the search process, in the sense that a point will still be preserved even if dominated by points which were discarded in the previous archiving? In this paper, we aim to answer the above questions. These correspond to two properties defined in [31]: (1) $\subseteq Y^*$ (i.e., the returning archive is a subset of the Pareto optimal solutions found so far) and (2) monotone (i.e., the deterioration never happens in the archiving process) We systematically investigate archiving methods associated with eight representative EMO algorithms on 37 test problems with from two to five objectives.

2 Experimental Design

2.1 Assessment Indexes

We consider two indexes, optimal ratio (OR) and deterioration ratio (DR). OR is, for one run of an EMO algorithm, the percentage of the nondominated solutions in the final archive/population are Pareto optimal with respect to all the solutions produced in this run. DR is, for one run of an EMO algorithm, the ratio of the times of deterioration occurring in the archiving process to the number of solutions considered to enter the archive, where the deterioration means that archiving a solution which is dominated by some solution discarded in the previous archiving process.

2.2 Archivers Investigated

We consider four classes of eight archiving methods: (1) Pareto-based archivers used in NSGA-II [8] and SPEA2 [41]; (2) indicator-based archivers in IBEA [39] and SMS-EMOA[2] [3]; (3) decomposition-based archivers in MOEA/D [37] and NSGA-III [7]; (4) enhanced Pareto-based archivers for many-objective optimisation (i.e., modifying Pareto dominance or density estimation) used in NSGA-II+ϵ [23] and SPEA2+SDE [28].

Pareto-based archivers first compare the Pareto dominance relation between solutions, and when the solutions have the same Pareto-based fitness (e.g., the non-dominated front in NSGA-II and the Pareto strength in SPEA2) their estimated density values are used to further distinguish between them. Indicator-based archivers adopt a performance indicator to optimise a certain preference of the solution set. In IBEA, the ϵ or dominated hypervolume indicator, based on solutions' pairwise comparison, is used, while in SMS-EMOA the set-based dominated hypervolume is used. Decomposition-based archivers decompose the space into a set of subspaces, ideally each solution representing one subspace. One difference between MOEA/D and NSGA-III is that the latter first sorts all solutions on the basis of Pareto dominance, and then decomposes the solutions on the same layer. Enhanced Pareto-based archivers increase the selection pressure of the Pareto-based archiving by either modifying the Pareto dominance criterion or modifying the crowding degree of solutions. NSGA-II+ϵ belongs to the former where the ϵ dominance [25] is used to replace crowding distance in NSGA-II, and SPEA2+SDE belongs to the latter where a position shift strategy is used to estimate solutions' density in order to make it cover both convergence and diversity.

---

[2] The method of computing the dominated hypervolume in SMS-EMOA was from [13], available at http://iridia.ulb.ac.be/~manuel/hypervolume.

2.3   Test Problems

A set of 37 problem instances were tested, including popular benchmark suites, early-developed problems and recently-developed ones. Specifically, we considered three popular suites, ZDT [38], WFG [15] and DTLZ [9]; seven early-developed problems, SCH1–SCH2 [34], FON [12], KUR [24] and VNT1–VNT3 [36]; seven recently-developed problems, convex DTLZ2 (denoted by CDTLZ2), inverted DTLZ1 (IDTLZ1), inverted DTLZ2 (IDTLZ2), scaled DTLZ1 (IDTLZ1), scaled DTLZ2 (IDTLZ2) [7,17], multiple point distance minimisation problem (MPDMP) [16,23], and multiple line distance minimisation problem (MLDMP) [27]. As to objective dimensionality settings of the scalable problems, the 2-objective WFG, the 3-objective DTLZ, and the 4-objective MPDMP and MLDMP (aka the rectangle problem [29]) were used; we also considered the 5-objective DTLZ1 and DTLZ2.

2.4   General Experimental Settings

All the results presented were obtained by executing 30 independent runs of each algorithm on each problem with the termination criterion of 30,000 evaluations. The population/archive size was set to 100 for all the algorithms except MOEA/D and NSGA-III where a closest number to 100 amongst the possible values was selected. To perform variation, simulated binary crossover with probability $p_c = 1.0$ and polynomial mutation with probability $p_m = 1/d$ (d denotes the number of decision variables) were considered in all the algorithms. The indicator ϵ was used in IBEA, and the PBI scalarising function was used in MOEA/D. All the parameters of the algorithms were configured as the same as in their original papers.

3   Results

3.1   Optimal Ratio

Table 1 shows the average optimal ratio (OR) of 30 runs of the eight algorithms on all the 37 problems. As can be seen, SMS-EMOA performs best, followed by SPEA2+SDE and NSGA-III; MOEA/D, NSGA-II and SPEA2 are among the worst algorithms, with only over 70% solutions being Pareto optimal[3] in their final archive on average. Taking a particular look at SMS-EMOA, unlike other algorithms whose OR varies on different problems, SMS-EMOA always achieves over 99.9% OR values on all the problems. This excellent ability may be attributed to the fact that the hypervolume value of the SMS-EMOA's archive never (or very rarely [19]) decreases and the archiving is of ◁-monotonicity (see [31]) when the reference point is stable, leading to the dominated solutions hard to stay in the archive.

---

[3] Here, "Pareto optimal" means being nondominated to all the solutions found during the run, rather than the problem's Pareto optimal solutions.

Table 1. The average optimal ratio (OR) of 30 runs of the eight algorithms. The higher the better; 100% (in boldface) means that all the solutions in the final archive or population are Pareto optimal with respect to the produced solutions (i.e., their input sequence).

Problem	NSGA-II	NSGA-II+ε	SPEA2	SPEA2+SDE	IBEA	SMS-EMOA	MOEA/D	NSGA-III
SCH1	**100.0%**	99.8%	**100.0%**	99.8%	99.8%	**100.0%**	**100.0%**	**100.0%**
SCH2	**100.0%**	99.4%	99.9%	99.4%	91.0%	**100.0%**	98.4%	**100.0%**
FON	33.6%	36.5%	43.8%	60.8%	92.5%	99.4%	43.1%	86.8%
KUR	51.9%	48.1%	67.0%	69.5%	85.2%	99.9%	47.1%	75.3%
ZDT1	81.7%	86.7%	91.8%	96.4%	97.3%	**100.0%**	54.4%	99.6%
ZDT2	87.6%	91.1%	93.3%	98.4%	85.8%	**100.0%**	62.8%	99.3%
ZDT3	80.5%	86.9%	89.9%	92.1%	97.8%	99.9%	72.6%	93.5%
ZDT4	96.8%	97.2%	97.1%	99.3%	80.7%	99.9%	64.1%	96.9%
ZDT6	97.7%	98.2%	96.6%	99.6%	94.4%	**100.0%**	36.8%	98.7%
WFG1	95.7%	97.8%	98.3%	99.7%	71.8%	**100.0%**	43.2%	96.8%
WFG2	83.6%	86.8%	90.4%	92.8%	88.5%	**100.0%**	72.9%	83.9%
WFG3	61.2%	75.3%	70.3%	86.4%	93.9%	99.9%	45.0%	92.4%
WFG4	55.9%	69.1%	65.7%	83.9%	78.6%	99.7%	52.2%	91.1%
WFG5	57.8%	67.9%	72.8%	84.8%	80.1%	**100.0%**	49.0%	95.1%
WFG6	69.4%	80.6%	79.9%	92.3%	79.2%	99.9%	55.2%	93.8%
WFG7	51.3%	61.4%	60.8%	84.6%	79.0%	**100.0%**	48.9%	88.5%
WFG8	68.0%	72.0%	81.9%	94.5%	59.1%	99.8%	54.2%	85.8%
WFG9	50.8%	61.3%	55.6%	77.2%	81.0%	**100.0%**	40.4%	83.6%
VNT1	71.4%	94.7%	64.8%	96.8%	97.9%	**100.0%**	81.3%	82.3%
VNT2	56.5%	85.8%	59.0%	90.2%	91.7%	**100.0%**	67.0%	78.8%
VNT3	53.5%	37.6%	72.1%	66.0%	90.3%	99.8%	92.6%	57.5%
DTLZ1	96.2%	99.1%	90.2%	99.9%	40.7%	**100.0%**	90.1%	98.3%
DTLZ2	63.1%	69.8%	67.0%	88.9%	94.6%	**100.0%**	90.3%	77.8%
DTLZ3	94.4%	97.9%	97.6%	99.8%	27.0%	99.8%	87.3%	96.4%
DTLZ4	60.8%	71.4%	70.1%	89.8%	90.4%	**100.0%**	94.0%	78.7%
DTLZ5	58.1%	66.6%	72.0%	87.1%	80.9%	**100.0%**	97.8%	52.0%
DTLZ6	**100.0%**	99.1%	94.1%	**100.0%**	99.9%	**100.0%**	90.7%	93.9%
DTLZ7	56.2%	82.5%	67.6%	94.2%	97.1%	**100.0%**	49.8%	68.2%
CDTLZ2	59.2%	78.0%	63.8%	94.5%	97.9%	**100.0%**	84.4%	79.7%
IDTLZ1	93.3%	99.5%	97.1%	99.9%	15.9%	**100.0%**	97.7%	98.0%
IDTLZ2	64.9%	80.3%	71.6%	93.3%	97.4%	**100.0%**	96.6%	67.3%
SDTLZ1	96.2%	99.0%	93.8%	99.3%	42.9%	**100.0%**	75.1%	98.3%
SDTLZ2	60.3%	74.6%	62.5%	82.2%	94.6%	99.3%	71.6%	77.6%
MPDMP	90.0%	94.8%	85.0%	99.4%	72.8%	**100.0%**	98.9%	93.5%
MLDMP	99.5%	99.1%	98.0%	**100.0%**	74.4%	**100.0%**	82.0%	98.6%
DTLZ1-5	57.6%	99.7%	15.9%	**100.0%**	83.9%	**100.0%**	96.8%	98.3%
DTLZ2-5	66.3%	90.1%	20.2%	94.2%	98.3%	**100.0%**	91.1%	86.6%
Average	73.54%	82.05%	76.15%	91.54%	81.62%	99.93%	72.31%	87.65%

The other seven archivers do not have these desirable properties. They can reach/approach 100% OR values on some problems (e.g., SCH1, SCH2 and DTLZ6), but perform rather poorly on some other problems (e.g., FON, KUR, WFG8 and WFG9). In addition, some archivers appear to behave quite distinctly from others on a couple of problems. For example, SPEA2 and NSGA-II perform considerably worse than the other archivers on the 5-objective DTLZ1 and DTLZ2; IBEA performs on DTLZ1, DTLZ3, IDTLZ1, and SDTLZ1; MOEA/D performs on ZDT6 and WFG1. Figure 1 shows the final population obtained by MOEA/D in a typical run on WFG1 and also all the solutions produced in this run and the Pareto optimal ones. As can be seen in the figure, many solutions of the final population of MOEA/D are not Pareto optimal of the whole

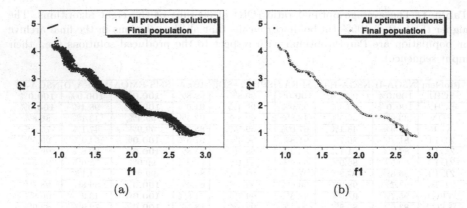

Fig. 1. Final population obtained by MOEA/D in a typical run on WFG1, coupled with (a) all the solutions produced in this run and (b) all the Pareto optimal solutions in this run.

set of solutions produced, particularly in the bottom right and top left of the figure, where they are dominated by some solutions which are eliminated in the archiving process of MOEA/D.

Since all the archivers (except SMS-EMOA) are the same in terms of theoretical properties, the observations of the different behaviours are from specific problems and archiving methodologies. In general, there are several situations that lead to an archive to only/mostly contain the Pareto optimal solutions. The first is that the newly produced solutions are typically dominated by some solutions in the archive. This happens on the test instance SCH1. The second situation is that a certain amount of newly produced solutions dominate some solutions in the archive even at the end of the evolution process. This happens often when the archive does not approach the Pareto front yet, such as Pareto-based algorithms on DTLZ1 and DTLZ3. The last situation is that the newly produced solutions are usually nondominated to the solutions in the archive, and also nondominated to any previously produced one. This happens either when the produced solutions are already Pareto optimal to the given problem (such as SCH2 and DTLZ6), or when they are stuck in the local optimum (such as Pareto-based algorithms on MLDMP).

Now, comparing different classes of the archiving methods, enhanced Pareto-based archivers generally outperform Pareto-based ones, with NSGA-II+ϵ and SPEA2+SDE improving the original NSGA-II and SPEA2 on average by around 10% and 15% respectively. This means that the density-based secondary archiving criterion (without incorporating convergence information) leads to OR degenerating. As to the three mainstream archiving classes, the Pareto-based, the indicator-based, and the decomposition-based, there is no clear pattern between their OR values. But we can infer the importance of Pareto dominance as the first archiving criterion in decomposition-based archiver, as NSGA-III, equipped with the Pareto nondominated sorting, performs significantly better than MOEA/D.

3.2 Deterioration Ratio

Next, let us move to the deterioration ratio (DR) results. DR denotes the ratio of the times of deterioration occurring in the archiving process to the total times of the archiving operations, and DR = 0% implies OR = 100%. It is then expected that a similar pattern to OR will be observed. Table 2 gives the average DR of 30 runs of the eight algorithms. Surprisingly, IBEA, which takes the fifth place on the average OR result, performs best here, slightly better than SMS-EMOA. One possible explanation is that the deterioration occurs mainly during the late phase of IBEA's evolutionary process, thereby some dominated solutions (in a global sense) remaining in the final archive. In contrast, in SMS-EMOA the deterioration occurs mainly during the evolutionary phase when the archive does not approach the Pareto front. This is also supported by the poor DR values of SMS-EMOA on WFG1 and DTLZ3 where the final archive is still far from the Pareto front.

It is noticed that MOEA/D reaches nearly 10% DR on average, significantly higher than the other algorithms, indicating that its archiving process preserves many dominated solutions with respect to the input sequence. This, interestingly, is contrary to the observations in [2], where the authors have seen that MOEA/D perform well (against Pareto-based, indicator-based and enhance Pareto-based EMO algorithms) in archiving the Pareto optimal solutions found on the MNK-landscape problem [1]. One possible reason for this could be different behaviours of MOEA/D between on continuous problems and on combinatorial problems. Another more likely explanation is the different performances of MOEA/D in exploration and archiving. The matting selection which considers neighbouring solutions in MOEA/D could be promising in generating Pareto optimal solutions, but it is difficult for the archive (here the population) to always keep them; i.e., good solutions can be easily generated and easily discarded as well.

3.3 Summary

Now we make a summary of the above observations.

- Consistent with the theoretical results, deterioration can happen to all the archivers in practice. However, the deterioration degree may vary dramatically on different test problems.
- SMS-EMOA performs best, especially in preserving the Pareto optimal solutions in the final archive. This is probably due to the desirable property of its hypervolume-based archiving—the ⊲-monotonicity [31]. Such a hypervolume-based bounded-size archiving, originally proposed in [22], can significantly reduce the occurrences of deterioration.
- IBEA does well in preventing the dominated solutions (with respect to the input sequence) from entering the archive, but it works mainly at the early phase of the evolution. This leads to the archive often ending up not being a subset of the Pareto optimal solutions.

Table 2. The average deterioration ratio (DR) of 30 runs of the eight algorithms. The lower the better; 0.00% (in boldface) means that there is no archived solution which is dominated by the solutions eliminated in the previous archiving process.

Problem	NSGA-II	NSGA-II+ϵ	SPEA2	SPEA2+SDE	IBEA	SMS-EMOA	MOEA/D	NSGA-III
SCH1	**0.00%**	0.10%	**0.00%**	0.08%	0.20%	**0.00%**	0.25%	**0.00%**
SCH2	**0.00%**	0.22%	**0.00%**	0.14%	0.36%	**0.00%**	0.50%	**0.00%**
FON	10.21%	10.66%	4.05%	5.13%	1.02%	0.02%	2.99%	0.66%
KUR	5.06%	6.12%	1.83%	2.05%	0.49%	0.05%	5.86%	3.74%
ZDT1	3.79%	2.78%	1.90%	0.94%	0.05%	0.67%	21.65%	0.68%
ZDT2	2.42%	1.74%	1.23%	0.47%	0.11%	0.42%	29.26%	0.77%
ZDT3	3.46%	2.42%	1.92%	1.48%	0.06%	0.85%	22.98%	2.64%
ZDT4	0.58%	0.41%	0.40%	0.21%	0.26%	0.73%	12.25%	0.53%
ZDT6	0.97%	0.67%	0.62%	0.20%	**0.00%**	0.72%	19.00%	0.60%
WFG1	1.11%	0.60%	1.13%	0.30%	**0.00%**	3.96%	16.16%	2.11%
WFG2	1.81%	1.37%	1.20%	0.75%	0.03%	1.02%	17.45%	2.64%
WFG3	7.48%	4.75%	4.22%	1.71%	0.05%	0.45%	16.89%	1.72%
WFG4	8.34%	5.25%	4.70%	2.50%	0.08%	0.45%	15.81%	3.81%
WFG5	10.46%	8.00%	5.06%	4.12%	0.09%	0.31%	18.16%	2.13%
WFG6	4.88%	3.32%	2.57%	1.49%	0.04%	0.51%	17.97%	2.26%
WFG7	10.22%	7.97%	5.86%	3.02%	0.08%	0.21%	17.29%	2.86%
WFG8	2.68%	2.36%	1.37%	0.56%	0.09%	0.56%	18.69%	2.43%
WFG9	12.04%	8.51%	6.84%	4.64%	0.48%	0.19%	13.86%	3.17%
VNT1	7.65%	2.13%	9.01%	1.38%	0.93%	**0.00%**	1.65%	6.64%
VNT2	6.44%	4.27%	4.39%	2.88%	3.07%	0.01%	2.41%	8.61%
VNT3	7.24%	10.18%	2.70%	3.94%	2.19%	0.02%	0.65%	11.85%
DTLZ1	1.25%	0.41%	1.49%	0.30%	0.55%	0.74%	2.78%	1.08%
DTLZ2	9.55%	7.45%	7.66%	2.77%	0.34%	0.08%	2.93%	3.03%
DTLZ3	1.22%	0.68%	1.10%	1.08%	0.97%	2.26%	6.44%	2.08%
DTLZ4	9.53%	7.85%	6.56%	2.52%	0.36%	0.14%	5.81%	3.57%
DTLZ5	9.11%	7.52%	5.84%	2.82%	0.17%	0.13%	5.10%	12.68%
DTLZ6	0.01%	0.50%	2.35%	0.31%	0.05%	0.66%	5.29%	2.50%
DTLZ7	10.07%	3.72%	5.77%	1.22%	0.13%	0.19%	12.14%	8.46%
CDTLZ2	9.88%	5.62%	9.34%	1.46%	0.35%	0.07%	2.63%	3.70%
IDTLZ1	1.22%	0.37%	0.68%	0.33%	0.86%	0.46%	6.13%	0.95%
IDTLZ2	9.96%	5.21%	7.11%	1.54%	0.55%	0.06%	1.38%	9.79%
SDTLZ1	1.25%	0.37%	1.38%	0.43%	0.56%	0.73%	8.81%	1.01%
SDTLZ2	9.45%	5.72%	7.62%	3.68%	0.33%	0.31%	11.58%	3.38%
MPDMP	1.17%	0.65%	1.27%	0.11%	0.01%	0.01%	0.58%	0.91%
MLDMP	0.11%	0.17%	0.28%	**0.00%**	0.08%	0.01%	0.78%	0.36%
DTLZ1-5	7.53%	0.22%	19.18%	0.05%	0.42%	0.11%	1.28%	1.47%
DTLZ2-5	12.55%	2.40%	18.23%	1.14%	0.17%	0.02%	1.13%	2.69%
Average	5.424%	3.586%	4.240%	1.561%	0.421%	0.461%	9.366%	3.176%

- Pareto-based archivers NSGA-II and SPEA2 generally perform poorly as the density-based criterion can lead to the dominated solutions frequently to enter the archive.
- Inserting convergence information into the density-based criterion of Pareto-based archivers can reduce the deteriorations. This has been shown in NSGA-II+ϵ and SPEA2+SDE.
- For indicator-based and decomposition-based archivers, Pareto dominance should still be necessary as the first criterion to select solutions. This can be inferred from the comparison between MOEA/D and NSGA-III.

4   Concluding Remarks

An archiver with theoretical quality guarantee is of high importance. It can improve search efficiency, prevent performance oscillation, and return a subset of the Pareto optimal solutions found so far. This paper has made a practical investigation of the optimality/monotonicity properties of eight representative archivers on 37 test instances. The results have shown that deterioration happens most of the time, and some archivers only return a population with less than half solutions being optimal.

It is worth pointing out that our investigation is based on the whole EMO algorithms rather than on archiving methods alone. That is, each EMO algorithm generates a different sequence of solutions that is presented to its archiving component. As such, the results (OR and DR) could be affected by the algorithm performance of producing solutions. An investigation of archiving methods under the same input sequence of solutions, independent of any EMO algorithm, can better tell their differences, which will be our next work.

Finally, note that we cannot say that a population consisting of a significantly large proportion of the current Pareto optimal solutions already well converges into the Pareto front, as it might be in the "middle" of the evolution, for example, for the Pareto-based algorithms on WFG1 and DTLZ3. But, an algorithm with a low percentage of the Pareto optimal solutions should have lots of room to be improved. In this regards, MOEA/D is an interesting example, in which good solutions can be easily generated but easily discarded as well. A combination of MOEA/D and SMS-EMOA could be potentially promising, in the sense that MOEA/D is responsible for generating solutions and updating population, while an extra archive based on the archiving method of SMS-EMOA is used to keep solutions. This would lead to different archiving methods for the different purposes in EMO—internal archiving for fostering exploration and external archiving for reducing deterioration, as suggested in [4,5]. It is worth mentioning that a similar algorithm framework, called bi-criterion evolution [30], has been presented recently, where MOEA/D can mainly be used to generate solutions and a Pareto-based archiving method is used to keep solutions. However, this cannot prevent the occurrence of deterioration as the Pareto dominance relation and individuals' crowding degree are used to maintain the archive just like in NSGA-II and SPEA2.

Acknowledgement. This work was supported by EPSRC (Grant Nos. EP/J017515/1 and EP/P005578/1) and Science and Technology Innovation Committee Foundation of Shenzhen (Grant Nos. ZDSYS201703031748284 and JCYJ20170307105521943).

References

1. Aguirre, H., Tanaka, K.: Working principles, behavior, and performance of MOEAs on MNK-landscapes. Eur. J. Oper. Res. **181**(3), 1670–1690 (2007)
2. Aguirre, H., Zapotecas, S., Liefooghe, A., Verel, S., Tanaka, K.: Approaches for many-objective optimization: analysis and comparison on MNK-landscapes. In: Bonnevay, S., Legrand, P., Monmarché, N., Lutton, E., Schoenauer, M. (eds.) EA 2015. LNCS, vol. 9554, pp. 14–28. Springer, Cham (2016). https://doi.org/10.1007/978-3-319-31471-6_2
3. Beume, N., Naujoks, B., Emmerich, M.: SMS-EMOA: multiobjective selection based on dominated hypervolume. Eur. J. Oper. Res. **181**(3), 1653–1669 (2007)
4. Bezerra, L.C.T., López-Ibáñez, M., Stützle, T.: Automatic component-wise design of multiobjective evolutionary algorithms. IEEE Trans. Evol. Comput. **20**(3), 403–417 (2016)
5. Bezerra, L.C.T., López-Ibáñez, M., Stützle, T.: A large-scale experimental evaluation of high-performing multi-and many-objective evolutionary algorithms. Evol. Comput. (2018, in press)
6. Corne, D., Knowles, J.: Some multiobjective optimizers are better than others. In: The 2003 Congress on Evolutionary Computation, vol. 4, pp. 2506–2512. IEEE (2003)
7. Deb, K., Jain, H.: An evolutionary many-objective optimization algorithm using reference-point-based nondominated sorting approach, part I: solving problems with box constraints. IEEE Trans. Evol. Comput. **18**(4), 577–601 (2014)
8. Deb, K., Pratap, A., Agarwal, S., Meyarivan, T.: A fast and elitist multiobjective genetic algorithm: NSGA-II. IEEE Trans. Evol. Comput. **6**(2), 182–197 (2002)
9. Deb, K., Thiele, L., Laumanns, M., Zitzler, E.: Scalable test problems for evolutionary multiobjective optimization. In: Abraham, A., Jain, L., Goldberg, R. (eds.) Evolutionary Multiobjective Optimization. Theoretical Advances and Applications, pp. 105–145. Springer, Berlin (2005). https://doi.org/10.1007/1-84628-137-7_6
10. Fieldsend, J.E.: University staff teaching allocation: formulating and optimising a many-objective problem. In: Proceedings of the Genetic and Evolutionary Computation Conference (GECCO), pp. 1097–1104. ACM (2017)
11. Fieldsend, J.E., Everson, R.M., Singh, S.: Using unconstrained elite archives for multiobjective optimization. IEEE Trans. Evol. Comput. **7**(3), 305–323 (2003)
12. Fonseca, C., Fleming, P.: An overview of evolutionary algorithms in multiobjective optimization. Evol. Comput. **3**(1), 1–16 (1995)
13. Fonseca, C.M., Paquete, L., López-Ibáñez, M.: An improved dimension-sweep algorithm for the hypervolume indicator. In: Proceedings of IEEE Congress Evolutionary Computation CEC 2006, pp. 1157–1163 (2006)
14. Hanne, T.: On the convergence of multiobjective evolutionary algorithms. Eur. J. Oper. Res. **117**(3), 553–564 (1999)
15. Huband, S., Hingston, P., Barone, L., While, L.: A review of multiobjective test problems and a scalable test problem toolkit. IEEE Trans. Evol. Comput. **10**(5), 477–506 (2006)
16. Ishibuchi, H., Hitotsuyanagi, Y., Tsukamoto, N., Nojima, Y.: Many-objective test problems to visually examine the behavior of multiobjective evolution in a decision space. In: Schaefer, R., Cotta, C., Kołodziej, J., Rudolph, G. (eds.) PPSN 2010. LNCS, vol. 6239, pp. 91–100. Springer, Heidelberg (2010). https://doi.org/10.1007/978-3-642-15871-1_10

17. Jain, H., Deb, K.: An evolutionary many-objective optimization algorithm using reference-point based nondominated sorting approach, part II: handling constraints and extending to an adaptive approach. IEEE Trans. Evol. Comput. **18**(4), 602–622 (2014)
18. Jin, H., Wong, M.-L.: Adaptive, convergent, and diversified archiving strategy for multiobjective evolutionary algorithms. Expert Syst. Appl. **37**(12), 8462–8470 (2010)
19. Judt, L., Mersmann, O., Naujoks, B.: Non-monotonicity of observed hypervolume in 1-Greedy S-Metric selection. J. Multi-Criteria Decis. Anal. **20**(5–6), 277–290 (2013)
20. Knowles, J., Corne, D.: Properties of an adaptive archiving algorithm for storing nondominated vectors. IEEE Trans. Evol. Comput. **7**(2), 100–116 (2003)
21. Knowles, J., Corne, D.: Bounded Pareto archiving: theory and practice. In: Gandibleux, X., Sevaux, M., Sörensen, K., T'kindt, V., et al. (eds.) LNE, vol. 535, pp. 39–64. Springer, Heidelberg (2004). https://doi.org/10.1007/978-3-642-17144-4_2
22. Knowles, J.D., Corne, D.W., Fleischer, M.: Bounded archiving using the Lebesgue measure. In: The 2003 Congress on Evolutionary Computation, vol. 4, pp. 2490–2497. IEEE (2003)
23. Köppen, M., Yoshida, K.: Substitute distance assignments in NSGA-II for handling many-objective optimization problems. In: Obayashi, S., Deb, K., Poloni, C., Hiroyasu, T., Murata, T. (eds.) EMO 2007. LNCS, vol. 4403, pp. 727–741. Springer, Heidelberg (2007). https://doi.org/10.1007/978-3-540-70928-2_55
24. Kursawe, F.: A variant of evolution strategies for vector optimization. In: Schwefel, H.-P., Männer, R. (eds.) PPSN 1990. LNCS, vol. 496, pp. 193–197. Springer, Heidelberg (1991). https://doi.org/10.1007/BFb0029752
25. Laumanns, M., Thiele, L., Deb, K., Zitzler, E.: Combining convergence and diversity in evolutionary multiobjective optimization. Evol. Comput. **10**(3), 263–282 (2002)
26. Laumanns, M., Zenklusen, R.: Stochastic convergence of random search methods to fixed size Pareto front approximations. Eur. J. Oper. Res. **213**(2), 414–421 (2011)
27. Li, M., Grosan, C., Yang, S., Liu, X., Yao, X.: Multi-line distance minimization: a visualized many-objective test problem suite. IEEE Trans. Evol. Comput. **22**(1), 61–78 (2018)
28. Li, M., Yang, S., Liu, X.: Shift-based density estimation for Pareto-based algorithms in many-objective optimization. IEEE Trans. Evol. Comput. **18**(3), 348–365 (2014)
29. Li, M., Yang, S., Liu, X.: A test problem for visual investigation of high-dimensional multi-objective search. In: Proceedings of the IEEE Congress on Evolutionary Computation (CEC), pp. 2140–2147 (2014)
30. Li, M., Yang, S., Liu, X.: Pareto or non-pareto: bi-criterion evolution in multiobjective optimization. IEEE Trans. Evol. Comput. **20**(5), 645–665 (2016)
31. López-Ibáñez, M., Knowles, J., Laumanns, M.: On sequential online archiving of objective vectors. In: Takahashi, R.H.C., Deb, K., Wanner, E.F., Greco, S. (eds.) EMO 2011. LNCS, vol. 6576, pp. 46–60. Springer, Heidelberg (2011). https://doi.org/10.1007/978-3-642-19893-9_4
32. Reed, P.M., Hadka, D., Herman, J.D., Kasprzyk, J.R., Kollat, J.B.: Evolutionary multiobjective optimization in water resources: the past, present, and future. Adv. Water Resour. **51**, 438–456 (2013)

33. Rudolph, G., Agapie, A.: Convergence properties of some multi-objective evolutionary algorithms. In: Proceedings of the 2000 Congress on Evolutionary Computation, vol. 2, pp. 1010–1016. IEEE (2000)
34. Schaffer, J.D.: Multiple objective optimization with vector evaluated genetic algorithms. In: Proceedings of the First International Conference on Genetic Algorithms and their Applications, pp. 93–100 (1985)
35. Schütze, O., Laumanns, M., Coello, C.A., Dellnitz, M., Talbi, E.G.: Convergence of stochastic search algorithms to finite size Pareto set approximations. J. Glob. Optim. **41**(4), 559–577 (2008)
36. Vlennet, R., Fonteix, C., Marc, I.: Multicriteria optimization using a genetic algorithm for determining a Pareto set. Int. J. Syst. Sci. **27**(2), 255–260 (1996)
37. Zhang, Q., Li, H.: MOEA/D: a multiobjective evolutionary algorithm based on decomposition. IEEE Trans. Evol. Comput. **11**(6), 712–731 (2007)
38. Zitzler, E., Deb, K., Thiele, L.: Comparison of multiobjective evolutionary algorithms: empirical results. Evol. Comput. **8**(2), 173–195 (2000)
39. Zitzler, E., Künzli, S.: Indicator-based selection in multiobjective search. In: Yao, X., et al. (eds.) PPSN 2004. LNCS, vol. 3242, pp. 832–842. Springer, Heidelberg (2004). https://doi.org/10.1007/978-3-540-30217-9_84
40. Zitzler, E., Laumanns, M., Bleuler, S.: A tutorial on evolutionary multiobjective optimization. In: Gandibleux, X., Sevaux, M., Sörensen, K., T'kindt, V., et al. (eds.) Metaheuristics for Multiobjective Optimisation. LNE, vol. 535, pp. 3–37. Springer, Heidelberg (2004). https://doi.org/10.1007/978-3-642-17144-4_1
41. Zitzler, E., Laumanns, M., Thiele, L.: SPEA2: improving the strength Pareto evolutionary algorithm for multiobjective optimization. In: Evolutionary Methods for Design, Optimisation and Control, pp. 95–100, Barcelona, Spain (2002)

Evolutionary Multi-objective Optimization Using Benson's Karush-Kuhn-Tucker Proximity Measure

Mohamed Abouhawwash[1](✉) and Mohamed A. Jameel[2]

[1] Department of Mathematics, Faculty of Science, Mansoura University,
Mansoura 35516, Egypt
`saleh1284@mans.edu.eg`
[2] Department of Mathematics, Faculty of Education, Arts and Sciences,
Sana'a University, Khawlan, Sana'a, Yemen
`Mohjameel555@gmail.com`

Abstract. Many Evolutionary Algorithms (EAs) have been proposed over the last decade aiming at solving multi- and many-objective optimization problems. Although EA literature is rich in performance metrics designed specifically to evaluate the convergence ability of these algorithms, most of these metrics require the knowledge of the true Pareto Optimal (PO) front. In this paper, we suggest a novel Karush-Kuhn-Tucker (KKT) based proximity measure using Benson's method (we call it B-KKTPM). B-KKTPM can determine the relative closeness of any point from the true PO front, without prior knowledge of this front. Finally, we integrate the proposed metric with two recent algorithms and apply it on several multi and many-objective optimization problems. Results show that B-KKTPM can be used as a termination condition for an Evolutionary Multi-objective Optimization (EMO) approach.

Keywords: Multi-objective optimization ·
Karush-Kuhn-Tucker conditions · Evolutionary optimization ·
Termination criterion · Benson's method

1 Introduction

Many EAs have proven to be effective in solving optimization problems with multiple (often conflicting) objectives. This made them widely popular over the last two decades. One of the most important reasons for its popularity is its ability to find multiple, wide-spread, trade-off solutions in a single simulation run, unlike classical methods, which use a point-to-point strategy to generate each non-dominated solution [1,2]. Among the first attempts in this field are MOGA [3], NPGA [4] and NSGA [5], where researchers devised different methods to deal with more than one goal by achieving a balance between convergence and diversity preservation. To accomplish convergence, these algorithms used the

© Springer Nature Switzerland AG 2019
K. Deb et al. (Eds.): EMO 2019, LNCS 11411, pp. 27–38, 2019.
https://doi.org/10.1007/978-3-030-12598-1_3

concept of Pareto dominance [6]. For diversity preservation, this group of algorithms used various diversity measures studied in the context of single-objective evolutionary computation techniques [7–9]. A following generation of algorithms achieved faster convergence while maintaining good diversity among the set of non-dominated solutions. This second wave is represented by a number of studies, most notably, NSGA-II [10], NSGA-III [11,12], PAES [13], and SPEA-II [14].

KKT conditions are one of the earliest and most popular sets of optimality conditions [15–17] and thus they play a vital role in constrained optimization theory [18,19]. KKT conditions is used to test whether a point (solution) is optimal or not. First order gradient of both objectives and constraints are required to check KKT conditions at a specific point. Other researcher expanded the idea of using KKT conditions to deal with non-smooth problems, using subdifferentials [19]. However these studies were only proposed for single-objective optimization problems. In 2015, Deb and Abouhawwash extended KKTPM to the realm of multi-objective optimization problems (MOPs) [20]. Their approach relies on one of the most common approaches to solve the generic MOPs, namely, *Achievement Scalarizing Function (ASF)* [21]. They extended this approach to use numerical differentiation when exact gradients are not available [22]. Abouhawwash et al. [23] integrated KKTPM with an efficient local search approach to enhance convergence of NSGA-III [24].

In this study we propose a novel KKT based metric using Benson's method, one of the classical methods of aggregating objectives in multi-objective optimization problems. The proposed metric (B-KKTPM) is intended to measure how far a solution is from being Pareto optimal in a multi-objective optimization context. We also show that the proximity of a solution from being optimal is perfectly correlated with the respective B-KKTPM of this solution. *Benson's method* is a well-known method for checking Pareto-optimality and for producing efficient solutions. One of the main goals is to verify that our proposed metric can be used as a termination criterion for EAs. One of the advantages of using Benson's method is that it- unlike ASF- does not need augmentation to avoid weakly dominated solutions. In the rest of the paper, a brief study of KKT proximity measure in Sect. 2. Our proposed metric for multi and many-objective optimization problems is described in Sect. 3. Simulation results on different optimization problems using NSGA-II and NSGA-III are introduced in Sect. 4. Lastly, Sect. 5 presents concludes and suggests potential future work.

2 KKT Based Proximity Measure

Dutta et al. [25] defined an approximate-KKT solution to calculate a KKTPM value for any iterate (solution) $\mathbf{x}^k$ for single-objective optimization problem of the following type:

$$
\begin{aligned}
&\text{Minimize}_{(\mathbf{x})}\ f(\mathbf{x}),\\
&\text{Subject to } g_j(\mathbf{x}) \leq 0, \quad j = 1, 2, \ldots, J,
\end{aligned}
\tag{1}
$$

after the theoretical computation of authors, they proposed a procedure for calculating KKTPM value ϵ_k for a solution $\mathbf{x}^k$ as follows:

$$
\begin{aligned}
&\text{Minimize}_{(\epsilon_k, \mathbf{u})} \ \epsilon_k, \\
&\text{Subject to } \|\nabla f(\mathbf{x}^k) + \sum_{j=1}^J u_j \nabla g_j(\mathbf{x}^k)\|^2 \leq \epsilon_k, \\
&\qquad\quad \sum_{j=1}^J u_j g_j(\mathbf{x}^k) \geq -\epsilon_k, \\
&\qquad\quad u_j \geq 0, \quad \forall j.
\end{aligned}
\tag{2}
$$

Here, the variable vector is $(\epsilon_k, \mathbf{u})$. The closer KKTPM value is to zero; it gives us thought about the closeness of the solution for the optimal value ϵ_k^* is called exact (or optimal) KKT proximity measure for the above problem. In the next section, we extend the KKT proximity metric concept for MOP based on the concept of *Benson's method*.

3   Proposed B-KKT Proximity Measure

A typical M-objective optimization problem with inequality constraints formulated as:

$$
\begin{aligned}
&\text{Minimize}_{(\mathbf{x})} \ \{f_1(\mathbf{x}), f_2(\mathbf{x}), \ldots, f_M(\mathbf{x})\}, \quad \mathbf{x} \in R^n \\
&\text{Subject to } g_j(\mathbf{x}) \leq 0, \quad j = 1, 2, \ldots, J.
\end{aligned}
\tag{3}
$$

There are many methods to solve the above problem. One of the most well-known methods of checking Pareto optimality (or efficiency) status and of generating efficient solutions for the above problem is *Benson's method*, which was originally suggested by Benson [26]. This method needs an initial solution $\mathbf{z}^0$ (parameter of the Benson's problem), this problem is given as follows:

$$
\begin{aligned}
&\text{Maximize}_{(\mathbf{x})} \sum_{m=1}^{M} \max\left(0, (z_m^0 - f_m(\mathbf{x}))\right), \\
&\text{Subject to } f_m(\mathbf{x}) \leq z_m^0, \qquad\qquad m = 1, 2, \ldots, M, \\
&\qquad\quad x_i^{(L)} \leq x_i \leq x_i^{(U)}, \qquad\quad i = 1, 2, \ldots, n.
\end{aligned}
\tag{4}
$$

The idea and the aim of Benson's method are on identifying the Pareto-optimality status, i.e., checking whether a given feasible solution $\mathbf{x}$ is Pareto-optimal or not. If the mentioned feasible solution $\mathbf{x}$ is not Pareto-optimal, this method finds PO solution of the multi-objective optimization problem which dominates the mentioned inefficient feasible solution. Figure 1 shows the chosen reference solution $\mathbf{z}^0 \in R^M$ which is any suitable solution in the M-dimensional objective space, note that use of nadir point, $\mathbf{z}^{nad}$, as the chosen point might be found suitable here [6]. The above problem maximizes the distance between the vector of objective functions corresponding to the under-assessment feasible solution $\mathbf{x}^0$ and the vectors of objective functions corresponding to all feasible solutions which dominate $\mathbf{z}^0 = f(\mathbf{x}^0)$. The distance function in Benson's method is $\|.\|_1$ ($L^1 - norm$). The objective function in the above problem is non-differentiable, thereby causing difficulties for derivative-based methods to solve this problem. To tackle this shortcoming, a modified formulation of the

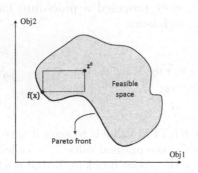
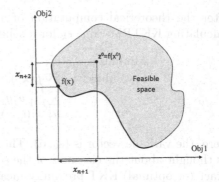

Fig. 1. Illustration of Benson's method.   **Fig. 2.** Illustration of Benson's problem (5).

optimization problem stated in Eq. 4 used, and which suggested in [27] for a smooth problem:

$$\text{Maximize}_{(\mathbf{x})} \sum_{i=1}^{M} x_{n+i},$$
$$\text{Subject to } f_i(\mathbf{x}) + x_{n+i} = f_i(\mathbf{x}^0), \tag{5}$$
$$x_{n+i} \geq 0 , \quad for \ all \quad i = 1, 2, \ldots, M,$$
$$\mathbf{x} \in X ,$$

where both $\mathbf{x} \in R^n$ and $x_{n+i} \in R_+^M$ are variables, $x_{n+i} = f_i(\mathbf{x}^0) - f_i(\mathbf{x})$ added as non-negative deviation variables. To clarify the idea in objective space (see Fig. 2). The initial feasible chosen, after that, the point $f_i(\mathbf{x}^0)$ has values greater than the PO solution $f_i(\mathbf{x})$. Maximizing the total deviation $x_{n+1} + x_{n+2}$, the purpose is to find a dominating solution, which is PO solution. For more details about the procedure of work for the above problem, see [27]. Note that sometimes equality constraints of the above problem cause computationally difficult. We must point out that problem stated in Eq. 5 is computationally badly conditioned because it has only one feasible solution ($x_{n+i} = 0$ for each i) if $\mathbf{x}^0$ is Pareto-optimal. Therefore, it is useful to note that equalities described in Eq. 5 could be replaced with inequalities $f_i(\mathbf{x}) + x_{n+i} \leq f_i(\mathbf{x}^0)$ for all $i = 1, 2, \ldots, M$ without affecting generality of the results presented [2,28].

Finally, we recount some of the advantages of Benson's method: (i) Its implementation is easy (ii) This method can be viewed as a hybridization of ε-constraint and weighted sum methods (iii) *Benson's method* constraints are more flexible than ε-constraint method (iv) If the point chosen appropriately then this method can also be used for non-convex MOPs. In this study, nadir point $\mathbf{z}^{nad}$ is selected as an appropriate initial point $\mathbf{z}^0$. According to the above explanation, to formulate the B-KKTPM proximity measure, a maximization problem described in Eq. 5 is transformed into a minimization problem as follows:

$$\text{Min.}_{(\mathbf{x},\, x_{n+i})}\ F(\mathbf{x}, x_{n+i}) = -\sum_{i=1}^{M} x_{n+i},$$

$$\text{Subject to } f_i(\mathbf{x}) + x_{n+i} - f_i(\mathbf{x}^0) \leq 0, i = 1, 2, \ldots, M,$$

$$g_j(\mathbf{x}) \leq 0, \quad j = 1, 2, \ldots, J. \tag{6}$$

Since the number of new variables x_{n+i} equals the number of objectives M. The above problem has $(n + M)$ variables: $\mathbf{y} = (\mathbf{x}, x_{n+i})$ for all $i = 1, 2, \ldots, M$. The idea of the above formulation of the optimization problem could be explained briefly as follows: Suppose that $\mathbf{x}^k$ a strict, efficient solution, then, Benson's optimization problem expected to produce the same solution, at the same time, KKT conditions expected to satisfy at this solution. But, if the solution $\mathbf{x}^k$ is away from the PO front, our new proposed KKTPM metric expected to give a monotonic metric value related to the nearness of any point from the corresponding Benson's optimal solution. For any suitable reference solution $\mathbf{z}^0 = f(\mathbf{x}^0)$, the optimal solution to the above optimization problem expected to make $x_{n+i}^* \geq 0$. Finally, there exist $M + J$ inequality constraints $G_j(y)$ for the above problem:

$$G_j(\mathbf{y}) = f_j(\mathbf{x}) + x_{n+j} - z_j^0 \leq 0, \quad j = 1, \ldots, M, \tag{7}$$

$$G_{M+j}(\mathbf{y}) = g_j(\mathbf{x}) \leq 0, \quad j = 1, 2, \ldots, J. \tag{8}$$

Now, B-KKT proximity measure for the above smooth objective function $\mathbf{y} = (\mathbf{x}, x_{n+i})$ stated in Eq. 6 computed as follows:

$$\text{Min.}_{(\epsilon_k,\, x_{n+j}, \mathbf{u})}\ \epsilon_k + \sum_{j=1}^{J} \left(u_{M+j} g_j(\mathbf{x}^k) \right)^2,$$

$$\text{Subject to } \|\nabla F(\mathbf{y}) + \sum_{j=1}^{M+J} u_j \nabla G_j(\mathbf{y})\|^2 \leq \epsilon_k,$$

$$\sum_{j=1}^{M+J} u_j G_j(\mathbf{y}) \geq -\epsilon_k, \tag{9}$$

$$u_j \geq 0, \quad j = 1, 2, \ldots, (M + J),$$

$$-x_{n+j} \leq 0, \quad j = 1, 2, \ldots, M.$$

The variables in this problem are ϵ_k, $x_{n+i} \in R^M$ for $i = 1, 2, \ldots, M$ and Lagrange multiplier vector $u_j \in R^{M+J}$. The first order gradient of F and G functions for the first constraint of the above problem required. If the complementary slackness condition value is zero at iterate $\mathbf{x}^k$, meaning that it is Pareto-optimal (by either $g_j(\mathbf{x}^k) = 0$ or $u_{M+j} = 0$) and hence the objective value for the above optimization should equal to zero. In the above optimization problem, the left side expressions of first and second constraints are not anticipated to equal zero for any other solution (not Pareto-optimal). The above problem has $(2M + J + 1)$ and $(2M + 2J + 1)$ variables and inequality constraints, respectively. To make B-KKTPM value has a larger (and worse) value for an infeasible solution, the second term in the objective function added as a penalty. Finally, to make sure that x_{n+j} $(\forall j, j = 1, 2, \ldots, M)$ have non-negative values in the above optimization problem, the final constraint added. An analysis of the optimization problem described in Eq. 9 and the subsequent simplifications uncover an essential outcome. The exact (or optimal) B-KKT proximity measure for feasible solutions is given as follows:

$$\epsilon_k^* = M - M \sum_{j=1}^{M} u_j^* - \sum_{j=1}^{J} \left(u_{M+j}^* g_j(\mathbf{x}^k) \right)^2, \tag{10}$$

where M is the number of objectives, $u_j^* \geq 0$ for all j and the third term is always non-negative. For any feasible solution $\mathbf{x}^k$, ϵ_k^* is always bounded in $[0, M]$. Therefore, B-KKTPM metric for any solution $\mathbf{x}^k$ is calculated as follows:

$$\text{B-KKTPM metric}(\mathbf{x}^k) = \begin{cases} \epsilon_k^*, & \text{if } \mathbf{x}^k \text{ is feasible,} \\ M + \sum_{j=1}^{J} \langle g_j(\mathbf{x}^k) \rangle^2, & \text{otherwise.} \end{cases} \tag{11}$$

where $\langle \alpha \rangle = \alpha$ if $\alpha > 0$; zero, otherwise. For PO solution $\mathbf{x}^k = \mathbf{x}^*$, the B-KKTPM value stated in Eq. 10 is always zero ($\epsilon_k^* = 0$), the second term is always M (in this term $\sum_{j=1}^{M} u_j^* = 1$ for all j) and the third term is always zero because that complementary slackness condition satisfied ($u_{M+j}^* g_j(\mathbf{x}^*) = 0$) for all constraints. To get a value greater than M for B-KKTPM metric in an infeasible solution case, M is added to ϵ_k^* as shown Eq. 11. Since the computation procedure for our proposed metric and KKTPM metric are same coincided, then we can find the computational cost for our proposed metric by using the same methods of KKTPM metric which mentioned in [29].

4   Results

Before discussing the performance of the proposed metric on multi- and many-objective optimization test problems, two of the most common MOEA algorithms used to solve all test problems. Bi-objective problems solved utilizing NSGA-II [10], while three or more objective problems solved utilizing NSGA-III [11]. These algorithms are used with standard parameter settings: Simulated Binary Crossover (SBX) probability = 0.9, SBX index = 30, polynomial mutation probability = $1/n$ (where n is the number of variables), and mutation index = 20 are applied. Description of SBX and polynomial mutation operations found in [30] and [6], respectively. In our metric, we suppose that if a solution (iterate $\mathbf{x}^k$) has B-KKTPM value which is less than or equal to 0.01, then we can say that this solution is the optimal solution. Some details of the test problems used in this work presented in Table 1.

4.1   Two-Objective Optimization Problems

Now, we apply B-KKTPM to the non-dominated solutions at the end of each generation in an NSGA-II run.

Two-Objective Test ZDT Problems. The simulation results for some two-objective ZDT test problems [31] are presented. The Variation for the smallest, first quartile, median, third quartile, and largest B-KKTPM values for NSGA-II

Table 1. Parameters - columns represent – left-to-right – test problem name, number of objectives, number of variables, bounds on the decision variables, population size, maximum number of generation.

Test problem	Num. of objectives (M)	Num. of variables (n)	Bounds x	Popsize (N)	MaxGen.
ZDT1	2	30	$x_i \in [0,1]$	40	200
ZDT3	2	30	$x_i \in [0,1]$	40	200
ZDT4	2	30	$x_1 \in [0,1]$ $x_{2:30} \in [-1,1]$	48	300
TNK	2	2	$x_i \in [0,\pi]$	40	250
BNH	2	2	$x_1 \in [0,5]$ $x_2 \in [0,3]$	200	500
DTLZ1	5	7	$x_i \in [0,1]$	212	1000
DTLZ2	3	8	$x_i \in [0,1]$	92	400
	5	14	$x_i \in [0,1]$	212	400
DTLZ5	3	12	$x_i \in [0,1]$	92	500
CAR	3	7	$x_{1,3,4} \in [0.5,1.5]$ $x_2 \in [0.45,1.35]$ $x_5 \in [0.875,2.625]$ $x_{6,7} \in [0.4,1.2]$	92	1000

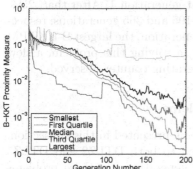
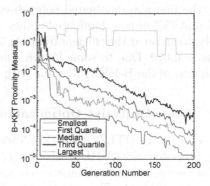

Fig. 3. B-KKT proximity measures versus generation numbers for non-dominated points on ZDT1 problem.

Fig. 4. B-KKT proximity measures versus generation numbers for non-dominated points on ZDT3 problem.

procedure on problems ZDT1, ZDT3 and ZDT4 illustrated in Figs. 3, 4 and 5, respectively. First, ZDT1 problem has convex PO front. The first population member arrived within B-KKTPM value (less than or equal 0.01) at 36 generations. Half and third quartile population members took about 105 and 110 generations to reach the true PO front. Note the decrease of B-KKTPM value to zero with the increasing number of generations, meaning that possibility of using B-KKTPM metric as a termination condition to the optimization algorithms. Next, ZDT3 problem has a convex discontinuous efficient front. The first strictly efficient solution was found at generation 11. Note that the convergence towards the efficient front was faster for about 75% of population members, as clear from Fig. 4. 27 and 62 of generations were necessary to half and third quartile of population members, respectively, until they arrive the efficient front. However, not all the population members (population size 40) were able to convergence to the efficient front at 200 generations. The ZDT4 function has about 100 local PO fronts and therefore is highly multi-modal. This problem solved utilizing the NSGA-II algorithm having 48 population members. Variation of B-KKTPM values with Benson's formulation for this problem shown in Fig. 5. Eighty-nine generations required for the first population member to reach the true PO front, while 202, and 225 generations required for both half and 75-th percentile population members, respectively, to reach the true PO front. To avoid points for which the gradient does not exist in this problem, we do not consider points had $x_1 = 0$.

Constrained Test Problems. Now, TNK and BNH constrained test problems [6] are presented, respectively. The variation for B-KKTPM values with Benson's formulation for problems TNK and BNH shown in Figs. 6 and 7, respectively. For the TNK test problem, At 175 generation, all Pareto-optimal solutions nearly converge (within 0.01 B-KKTPM value) to the efficient front. In the BNH problem, the first non-dominated point discovered at generation 1. After that, 25%, 50% and 75% of non-dominated points took 22, 105 and 365 generations, respectively, to arrive at the efficient front. At 500-th generation, the largest B-KKTPM value is 0.079. Due to some known complexity in solving this problem, a slow reduction of the B-KKTPM value with the generating counter observed.

4.2   Three-Objective Optimization Problems

Three-objective DTLZ2 and DTLZ5 test problems presented in this subsection. The proposed NSGA-III procedure [12], applied to both DTLZ2 and DTLZ5 problems (population members and maximum generations for these problems shown in Table 1). The variation for B-KKTPM values of both DTLZ2 and DTLZ5 problems illustrated in Figs. 8 and 9, respectively. For DTLZ2 problem, at generation 1 itself, the first non-dominated point was found. After that, 25%, 50% and 75% of the non-dominated points took 8, 41 and 72 generations, respectively, to arrive at the efficient front. For all 92 non-dominated points, 389 generations required to convergence all these points to the efficient front, while only 226 generations required to reach all non-dominated points to the efficient front with a minimal value of BKKTPM measure for DTLZ5 problem.

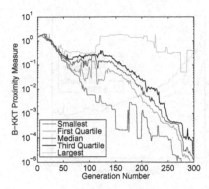

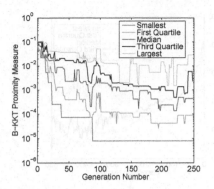

Fig. 5. B-KKT proximity measures versus generation numbers for non-dominated points on ZDT4 problem.

Fig. 6. B-KKT proximity measures versus generation numbers for non-dominated points on TNK problem.

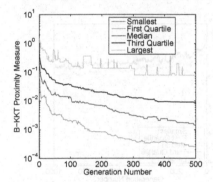

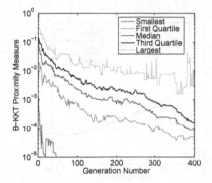

Fig. 7. B-KKT proximity measures versus generation numbers for non-dominated points on BNH problem.

Fig. 8. B-KKT proximity measures versus generation numbers for non-dominated points on 3-objective DTLZ2 problem.

4.3 Many-Objective Optimization Problems

For many objective optimization problems, EMO algorithms have difficulty in solving them. Here, DTLZ1 and DTLZ2 introduced using five objectives. With 212 population members for both DTLZ1 and DTLZ2, NSGA-III run until 1,000 and 400 maximum generations, respectively. In 5-objective DTLZ1 problem, the first strictly efficient solution is found at generations 84, while 298 generations were enough for half of the non-dominated solutions to reach to the PO front (see Fig. 10). At 1000-th generation, all population members used for this problem (see Table 1) could not reach to the efficient front. For 5-objective DTLZ2 problem, the convergence toward the Pareto-optimal front was faster than the DTLZ1 problem (see Fig. 11) because this problem is relatively easier to solve to Pareto-optimality. Since that all non-dominated points took only 215 generations

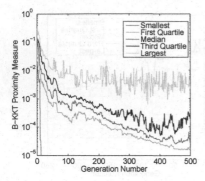

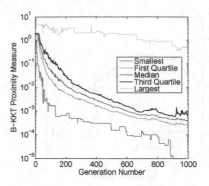

Fig. 9. B-KKT proximity measures versus generation numbers for non-dominated points on 3-objective DTLZ5 problem.

Fig. 10. B-KKT proximity measures versus generation numbers for non-dominated points on 5-objective DTLZ1 problem.

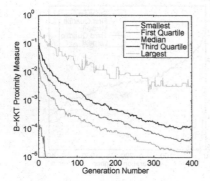

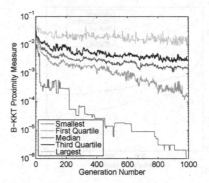

Fig. 11. B-KKT proximity measures versus generation numbers for non-dominated points on 5-objective DTLZ2 problem.

Fig. 12. B-KKT proximity measures versus generation numbers for non-dominated points on car side impact problem.

to arrive at the PO front. This figure indicated that the maximum of 215 generations was sufficient to all non-dominated points to reach close to the Pareto-optimal front.

4.4 Engineering Design Problem

Finally, the variation for B-KKTPM values with generations illustrated in Fig. 12 for 3-objective car-side-impact (CAR) [12] problem. NSGA-III run until 1000 generations with the number of special population members (see Table 1) for this problem. Although 75% of PO solutions are found fairly early (76 generation), but all PO solutions took more than 995 to reach to the efficient front.

5   Conclusions

In this paper, a novel KKT proximity measure for multi and many objective optimization problems is presented. A smooth decrease in the value of B-KKTPM is observed as solutions get closer to the PO front. Based on our results, the value of the new metric is consistently directly proportional to how far the solution is from being optimal. One essential and important result of this study is that the proposed B-KKTPM metric can be used as a termination condition in an optimization algorithm. At each point, the derivatives of both objective and constraint functions are required and thereby make B-KKTPM procedure computationally expensive and therefore time-consuming for some practical uses. We are working on reducing the computational cost of the proposed metric.

References

1. Chankong, V., Haimes, Y.Y.: Multiobjective Decision Making Theory and Methodology. North-Holland, New York (1983)
2. Miettinen, K.: Nonlinear multiobjective optimization, vol. 12. Springer, Heidelberg (2012)
3. Fonseca, M., Fleming, J.: Genetic algorithms for multiobjective optimization: formulation discussion and generalization. ICGA, pp. 416–423 (1993)
4. Horn, J., Nafpliotis, N., Goldberg, D.E.: A niched Pareto genetic algorithm for multiobjective optimization. In: Proceedings of the First IEEE Conference on Evolutionary Computation, IEEE World Congress on Computational Intelligence, vol. 1, pp. 82–87 (1994)
5. Srinivas, N., Deb, K.: Muiltiobjective optimization using nondominated sorting in genetic algorithms. Evolut. comput. **2**(3), 221–248 (1994)
6. Deb, K.: Multi-objective Optimization Using Evolutionary Algorithms. Wiley, New York (2001)
7. Goldberg, E., Richardson, J.: Genetic algorithms with sharing for multimodal function optimization, pp. 41–49 (1987)
8. DeJong, K.A.: An analysis of the behavior of a class of genetic adaptive systems. Ph.D. thesis, Ann Arbor, MI: University of Michigan (1975)
9. Deb, K.: An investigation of niche and species formation in genetic function optimizatio. In: ICGA vol. 89, pp. 42–50 (1989)
10. Deb, K., Pratap, A., Agarwal, S., Meyarivan, T.: A fast and elitist multiobjective genetic algorithm: NSGA-II. IEEE Trans. Evol. Comput. **6**(2), 182–197 (2002)
11. Deb, K., Jain, H.: An evolutionary many-objective optimization algorithm using reference-point-based nondominated sorting approach, part I: solving problems with box constraints. IEEE Trans. Evol. Comput. **18**(4), 577–601 (2014)
12. Jain, H., Deb, K.: An evolutionary many-objective optimization algorithm using reference-point based nondominated sorting approach, Part II: handling constraints and extending to an adaptive approach. IEEE Trans. Evol. Comput. **18**(4), 602–622 (2014)
13. Knowles, J., Corne, D.: The pareto archived evolution strategy: a new baseline algorithm for pareto multiobjective optimisation. In: Evolutionary Computation, pp. 98–105 (1999)
14. Zitzler, E., Laumanns, M., Thiele, L.: SPEA2: improving the strength Pareto evolutionary algorithm. TIK-Report 103, pp. 1–20 (2001)

15. Kuhn, H.W., Tucker, A.W.: Nonlinear programming. In: Proceedings of the Second Berkeley Symposium on Mathematical Statistics and Probability, vol. 1950, pp. 481–492 (1951)
16. Tulshyan, R., Arora, R., Deb, K., Dutta, J.: Investigating ea solutions for approximate KKT conditions for smooth problems. In: Proceedings of Genetic and Evolutionary Algorithms Conference (GECCO), pp. 689–696. ACM Press (2010)
17. Haeser, G., Schuverdt, M.L.: On approximate KKT condition and its extension to continuous variational inequalities. J. Optim. Theory Appl. 149(3), 528–539 (2011)
18. Reklaitis, G.V., Ravindran, A., Ragsdell, K.M.: Engineering Optimization Methods and Applications. Wiley, New York (2006)
19. Rockafellar, R.T.: Convex Analysis. Princeton University Press, Princeton (1996)
20. Deb, K., Abouhawwash, M.: An optimality theory based proximity measure for evolutionary multi-objective and many-objective optimization. IEEE Trans. Evol. Comput. 20(4), 515–52 (2016)
21. Wierzbicki, A.P.: The use of reference objectives in multiobjective optimization. In: Fandel, G., Gal, T. (eds.) Multiple Criteria Decision Making Theory and Application. Lecture Notes in Economics and Mathematical Systems, vol. 177, pp. 468–486. Springer, Heidelberg (1980). https://doi.org/10.1007/978-3-642-48782-8_32
22. Abouhawwash, M., Deb, K.: Karush-Kuhn-Tucker proximity measure for multi-objective optimization based on numerical gradients. In: Proceedings of the Genetic and Evolutionary Computation Conference, pp. 525–532 (2016)
23. Abouhawwash, M., Seada, H., Deb, K.: Towards faster convergence of evolutionary multi-criterion optimization algorithms using Karush Kuhn Tucker optimality based local search. Comput. Oper. Res. 79, 331–346 (2017)
24. Seada, H., Deb, K.: A unified evolutionary optimization procedure for single, multiple, and many objectives. IEEE Trans. Evol. Comput. 20(3), 358–369 (2016)
25. Dutta, J., Deb, K., Tulshyan, R., Arora, R.: Approximate KKT points and a proximity measure for termination. J. Global Optim. 56(4), 1463–1499 (2013)
26. Benson, H.P.: Existence of efficient solutions for vector maximization problems. J. Optim. Theory Appl. 26(4), 569–580 (1978)
27. Ehrgott, M.: Multicriteria Optimization, 2nd edn. Springer, Heidelberg (2005). https://doi.org/10.1007/3-540-27659-9
28. Branke, J.: Consideration of partial user preferences in evolutionary multiobjective optimization. In: Branke, J., Deb, K., Miettinen, K., Słowiński, R. (eds.) Multiobjective Optimization. LNCS, vol. 5252, pp. 157–178. Springer, Heidelberg (2008). https://doi.org/10.1007/978-3-540-88908-3_6
29. Seada, H., Abouhawwash, M., Deb, K.: Multi-phase balance of diversity and convergence in multiobjective optimization. IEEE Trans. Evol. Comput. (2018, in press)
30. Deb, K., Agrawal, R.B.: Simulated binary crossover for continuous search space. Complex Syst. 9(2), 115–148 (1995)
31. Zitzler, E., Deb, K., Thiele, L.: Comparison of multiobjective evolutionary algorithms: empirical results. Evol. Comput. 8(2), 173–195 (2000)

On the Convergence of Decomposition Algorithms in Many-Objective Problems

Ricardo H. C. Takahashi[(✉)]

Department of Mathematics, Universidade Federal de Minas Gerais,
Belo Horizonte, MG, Brazil
taka@mat.ufmg.br

Abstract. Decomposition methods have been considered for dealing with many-objective problems, since the Pareto-dominance selection was found to become ineffective as the number of objectives grow beyond four. As decomposition methods change the multiobjective problem into a set of single-objective problems, the difficulties found by evolutionary algorithms in many-objective optimization were expected to become alleviated. This paper studies the convergence properties of two decomposition schemes, respectively based on Euclidean norm and on Tchebyschev norm, in many-objective optimization. Numerical experiments show that the solution sequences obtained from Tchebyschev norm decomposition becomes stuck at a finite distance from the Pareto-set, while the sequences obtained from Euclidean norm decomposition may be adjusted such that an asymptotic convergence is achieved. Explanations for those different convergence behaviors are obtained from recently developed analytical tools.

Keywords: Many-objective problems · Decomposition methods ·
Tchebyschev decomposition

1 Introduction

The first mention to the difficulties that Evolutionary Multiobjective Optimization (EMO) algorithms would find in problems with a number of objectives higher than usual seems to appear in [1], in 1995. Some early experimental studies of convergence loss of EMO algorithms in those problems were published in 2001 [2,3]. In 2003, a study about the effect of increasing the number of objective functions on the performance of existing EMO algorithms was presented in [4]. In that work, it was verified empirically that for a number of objectives higher than three, the main EMO algorithms presented a significant degradation of performance. Those problems with four or more objectives were named the *many-objective* problems. In 2007, a detailed experimental evaluation of that phenomenon was presented in [6]. In the years that followed, the pursuit of the reasons behind that performance degradation and the search for new approaches that could alleviate this degradation have become a major theme of research [5].

© Springer Nature Switzerland AG 2019
K. Deb et al. (Eds.): EMO 2019, LNCS 11411, pp. 39–50, 2019.
https://doi.org/10.1007/978-3-030-12598-1_4

In 2018, an analytical description of the convergence loss of Pareto-dominance based algorithms in many-objective problems was presented in [7].

Decomposition-based EMO algorithms rely on the idea of defining a number of single-objective optimization problems which are solved simultaneously by the evolution of a single population. Even before the subject of many-objective optimization received so much attention, there were attempts to use decomposition methods which resulted in important evolutionary algorithms, such as the MOEA/D [8]. The empirical observation of the difficulties in the usage of dominance-based selection for many-objective problems motivated the employment of *decomposition-based algorithms*, which were expected to keep unaffected by the difficulties related to many-objective optimization. However, different decomposition methods have demonstrated very different performances in many-objective problems, and those differences are not well understood on this moment.

This paper studies the aggregation of objectives performed by the minimization of the distance to a reference point, with that distance defined by: (i) a Tchebyschev norm; and (ii) an Euclidean norm. The algorithm behavior under such decomposition schemes are examined here, considering the analytical tools developed in [7]. The paper is organized as follows. The Sect. 2 presents some numerical experiments that reveal some patterns of convergence loss in many-objective optimization. Specifically, the sequences of solutions obtained from Tchebyschev decomposition are found to become stuck at a finite distance from the Pareto set. Section 3 briefly presents the analytical tools developed in [7] and apply them to the analysis of the numerical results that were obtained in the former section. Section 4 presents some conclusions.

2   Numerical Experiments

In a finite-dimensional continuous-variable multi-criteria decision problem, a decision variable $\mathbf{x}$ should be chosen from a set $\Omega \subseteq \mathbb{R}^n$, according to m criteria functions $f_i : \Omega \mapsto \mathbb{R}$. Let $\mathbf{x}_1 \in \Omega$ and $\mathbf{x}_2 \in \Omega$. It is assumed, by convention, that $\mathbf{x}_1$ is better than $\mathbf{x}_2$ in criterion f_i if $f_i(\mathbf{x}_1) < f_i(\mathbf{x}_2)$. As the problem deals with m different criteria, the following relational operators are defined, in order to compare vectors. Let $\mathbf{u}, \mathbf{v} \in \mathbb{R}^n$, then:

$$\mathbf{u} \preceq \mathbf{v} \Leftrightarrow u_i \leq v_i, \, i = 1, \dots, m$$
$$\mathbf{u} \neq \mathbf{v} \Leftrightarrow \exists i \in \{1, \dots, m\} : u_i \neq v_i$$
$$\mathbf{u} \prec \mathbf{v} \Leftrightarrow \mathbf{u} \preceq \mathbf{v} \text{ and } \mathbf{u} \neq \mathbf{v}$$

Considering two candidate points $\mathbf{x}_1, \mathbf{x}_2 \in \Omega$, if $\mathbf{f}(\mathbf{x}_1) \prec \mathbf{f}(\mathbf{x}_2)$ then $\mathbf{x}_1$ *dominates* $\mathbf{x}_2$. The solutions $\mathbf{x} \in \Omega$ which are not dominated by any other solution in Ω are the *minimal* solutions in Ω, considering the strict partial order $\prec$. A *multi-objective optimization* problem is defined as the problem of finding such minimal solutions:

$$\min \mathbf{f}(\mathbf{x}) = (f_1(\mathbf{x}), f_2(\mathbf{x}), \cdots, f_m(\mathbf{x}))$$
$$\text{subject to: } \mathbf{x} \in \Omega \tag{1}$$

Formally, the solution set of this problem, denoted by $\mathcal{P}$, is defined by:

$$\mathcal{P} = \{\mathbf{x} \in \Omega \mid \nexists \bar{\mathbf{x}} \in \Omega \text{ such that } \mathbf{f}(\bar{\mathbf{x}}) \prec \mathbf{f}(\mathbf{x})\} \tag{2}$$

The set $\mathcal{P}$ is called the *Pareto-set* or the *non-dominated solution set* of the problem. The image of $\mathcal{P}$ in the space of objectives is called the *Pareto front* of the problem.

The problem of minimization of the following set of functions is considered here:

$$f_i(\mathbf{x}) = \|\mathbf{x} - \mathbf{e}_i\|_2^2 \quad , \quad i = 1, \ldots, m \tag{3}$$

where $\mathbf{x} \in \mathbb{R}^n$, $\mathbf{e}_i \in \mathbb{R}^n$, $i = 1, \cdots, m$ is the i-th canonical basis vector (the vector with all coordinates equal to zero except the i-th one, which is equal to 1), and $\|\cdot\|_2$ stands for the Euclidean norm of the argument vector. The Pareto-set of this problem is the $(m-1)$-dimensional simplex with vertices in $\{\mathbf{e}_1, \mathbf{e}_2, \ldots, \mathbf{e}_m\}$.

The analysis of the behavior of decomposition-based searches on this simple problem has the purpose of allowing the identification of the structural difficulties related only to the high number of objectives, removing other possibly interfering factors such as multi-modality, ill-conditioning, deceptive behavior, and so forth.

In the study conducted here, an initial point $\mathbf{x}_0$ is considered, and a sequence $[\mathbf{x}]_k$ is built in order to reach the reference point $\mathbf{x}_{ref} = (\frac{1}{n}, \frac{1}{n}, \ldots, \frac{1}{n})$, which is situated exactly on the center of the Pareto front. This sequence is generated according to:

1. The Euclidean distance d from the current point to the reference point is calculated:

$$d \leftarrow \|\mathbf{x}_k - \mathbf{x}_{ref}\|_2$$

2. A tentative step is generated from a Gaussian distribution, with standard deviation given by γd:

$$\zeta \leftarrow \mathcal{N}(0, \gamma d)$$

   with $\gamma = 0.02$.

3. The tentative new point is given by:

$$\mathbf{x}_a = \mathbf{x}_k + \zeta$$

4. If

$$\|\mathbf{x}_a - \mathbf{x}_{ref}\|_{(\cdot)} < \|\mathbf{x}_k - \mathbf{x}_{ref}\|_{(\cdot)}$$

   then

$$\mathbf{x}_{k+1} \leftarrow \mathbf{x}_a$$

   else keep trying with

$$\zeta \leftarrow \zeta/\alpha^j \text{ for } \alpha = 1.2 \text{ and } j = 1, 2, \ldots, 25$$

   If the acceptance condition is not reached, make $\mathbf{x}_{k+1} \leftarrow \mathbf{x}_k$.

5. Make $k \leftarrow k + 1$ and go to step 1.

The sequences generated by those steps are studied here for the cases of acceptance criterion (step 4) defined by the Euclidean norm, indicated by $\| \cdot \|_2$, and by the Tchebyschev norm, indicated by $\| \cdot \|_\infty$. Some experiments were performed with $n = 10$ and $m = 6$, and a random initial point defined by:

$$\mathbf{x}_0 = 2\mathbf{x}_{ref} + 0.1\epsilon$$

with $\epsilon \leftarrow \mathcal{N}(1, 0)$. The results are presented in Fig. 1.

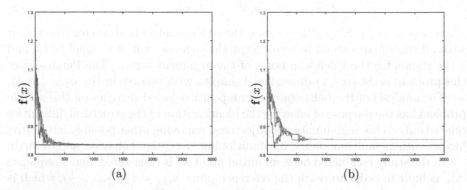

(a)   (b)

Fig. 1. (a) Evolution of $\mathbf{f}(\mathbf{x}_k)$ vector, for Euclidean norm decomposition. (b) Evolution of $\mathbf{f}(\mathbf{x}_k)$ vector, for Tchebyschev norm decomposition. In all cases, the reference values to be reached are represented by horizontal lines, and the abscissas represent the sequence index k.

Figure 1(a) shows that the sequence produced by the acceptance rule based on the Euclidean norm seems to converge to the reference point. On the other hand, Fig. 1(b) shows that the sequence produced by the acceptance rule based on the Tchebyschev norm becomes stuck at a distance from the reference.

Another run was performed for an initial point given by:

$$\mathbf{x}_0 = 1.1\mathbf{x}_{ref} + 0.1\epsilon$$

with all other simulation parameters kept constant. Figure 2(b) now shows a similar behavior for the sequence produced by the acceptance rule based on the Tchebyschev norm, which still becomes stuck at a distace from the reference. Figure 2(a) indicates that the sequence produced by the acceptance rule based on the Euclidean norm also becomes stuck now, although much nearer to the reference than the other sequence.

A new simulation is performed, on this time considering a reference point $\mathbf{x}_{ref} = 0.9(\frac{1}{n}, \frac{1}{n}, \ldots, \frac{1}{n})$, which is situated behind the Pareto-front, still considering the same initial condition from Fig. 2. The results are presented in Fig. 3. Now, the sequence produced by the acceptance rule based on the Euclidean norm converges to the Pareto set, while the sequence produced by the Tchebyschev norm rule still becomes stuck without converging.

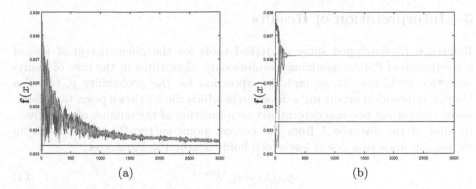

(a)   (b)

Fig. 2. Initial point close to the Pareto-set. (a) Evolution of $\mathbf{f}(\mathbf{x}_k)$ vector, for Euclidean norm decomposition. (b) Evolution of $\mathbf{f}(\mathbf{x}_k)$ vector, for the Tchebyschev norm decomposition. The reference values to be reached are represented by horizontal lines, and the abscissas represent the sequence index k.

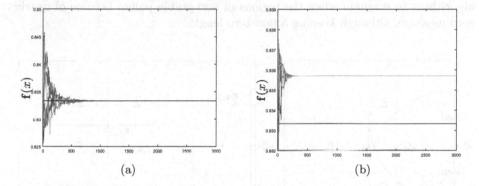

(a)   (b)

Fig. 3. Initial point close to the Pareto-set, reference point behind the Pareto front. (a) Evolution of $\mathbf{f}(\mathbf{x}_k)$ vector, for Euclidean norm decomposition. (b) Evolution of $\mathbf{f}(\mathbf{x}_k)$ vector, for the Tchebyschev norm decomposition. The reference values to be reached are represented by horizontal lines, and the abscissas represent the sequence index k.

The interpretation of those results is presented in the next section. Before continuing, it should be mentioned that, although more exhaustive simulations are not presented here, due to space limits, a much larger set of experiments was performed. In all cases of $m \geq 6$, results similar to the ones presented here were obtained. There was not, also, any significant difference for different distributions of the initial point. The only noticeable difference to the results presented here occurs for $m \leq 4$, when the final distance from the Tchebyschev sequence to the reference point becomes much smaller.

3   Interpretation of Results

Reference [7] developed some analytical tools for the phenomenon of loss of convergence of Pareto-dominance evolutionary algorithms in the case of many-objective problems. An asymptotic expression for the probability $p_m(\lambda)$ of a Gaussian mutation producing a direction in which there exists a point that dominates the current one was determined, as a function of the number of objectives, m, and of the distance λ from the current point to the Pareto-set. Such an expression leads to a power law which holds nearby the Pareto-set:

$$p_m(\lambda) = c_m \, \lambda^{m-1} \tag{4}$$

In this way, the cause of loss of convergence associated to Pareto-dominance selection was described as the fast decrease of the relative size of the space region which contains points that dominate current solutions. The probability of generation of any dominating point rapidly approaches zero, leading those algorithms to stagnate when the regions of acceptable points become of nearly zero measure, although keeping a non-zero length.

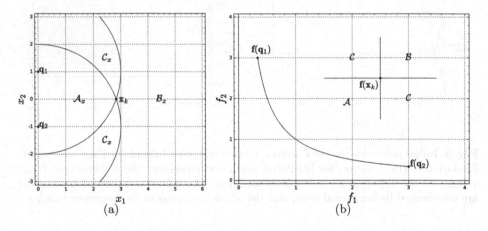

Fig. 4. Bi-objective problem with individual minima located in q_1 and q_2 and current point x_k. (a) Decision variable space: points in $\mathcal{A}_x$ dominate x_k, points in $\mathcal{B}_x$ are dominated by x_k and points in $\mathcal{C}_x$ neither dominate nor are dominated by x_k. (b) Space of objectives: the images of the same entities in the decision variable space, now including also the Pareto front (the curve from $f(q_1)$ to $f(q_2)$).

A pictorial explanation of the phenomenon which results in the collapse of the probability of enhancement of a current point x_k into a new point x_{k+1} which dominates the current one quite before reaching the Pareto-set is provided in Figs. 4, 5 and 6. Figure 4 shows the following regions: $\mathcal{A}_x$ (points which dominate x_k), $\mathcal{B}_x$ (points which are dominated by x_k), and $\mathcal{C}_x$ (points which neither are dominated by x_k nor dominate it). In the space of objectives, those

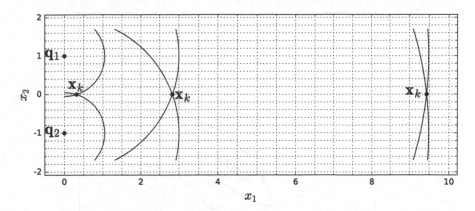

Fig. 5. Bi-objective problem with individual minima located in q_1 and q_2. As x_k approaches the Pareto-set, the relative size of $\mathcal{A}_x$ decreases, while the relative size of $\mathcal{C}_x$ increases.

regions correspond to quadrants symmetrically disposed around $f(x_k)$, respectively indicated by $\mathcal{A}$, $\mathcal{B}$ and $\mathcal{C}$.

The Fig. 5 shows what happens to the relative sizes of $\mathcal{A}_x$, $\mathcal{B}_x$ and $\mathcal{C}_x$ as x_k approaches the Pareto-set: for a farther x_k, the largest region is $\mathcal{B}_x$ and the smallest one is $\mathcal{C}_x$. For a nearer x_k, $\mathcal{C}_x$ becomes the largest region, while $\mathcal{A}_x$ becomes the smallest one.

The key illustration that explains why the search for new solutions that dominate the current solution eventually becomes stuck at a non-zero distance from the Pareto-set is provided in Fig. 6. This figure shows a sequence of contour curves that represent the boundary of set $\mathcal{A}_x$ (points which dominate x_k) for different distances from x_k to the Pareto-set. It can be seen that, instead of being similar to the outer contours, the inner contours become more elongated, eventually collapsing into a line segment before reaching the ultimate Pareto-set point x_{ref} to which the convergence was expected. Once x_k reaches that line segment, the probability of further finding a x_{k+1} inside the same line segment which is nearer to x_{ref} becomes virtually null.

It should be noticed that the phenomenon of convergence loss would occur for any number of objectives. In fact, from the Eq. (4) it can be inferred that for $m = 3$ or greater, the probability of solution enhancement would decay faster (by orders of magnitude) than the distance to the desired solution. This phenomenon has not been reported for $m = 3$ because it appears only at a very small distance to the solution set, which renders it irrelevant in that case. For larger numbers of objectives, this is the fundamental cause of the convergence difficulties that have been reported for "many-objective" problems.

The Tchebyschev decomposition has a geometry which is similar to the geometry of Pareto-dominance, with an important difference: In the case of an acceptance rule based on Pareto-dominance, the next point to be generated is constrained to dominate the current one. This leads to Eq. (4), with all

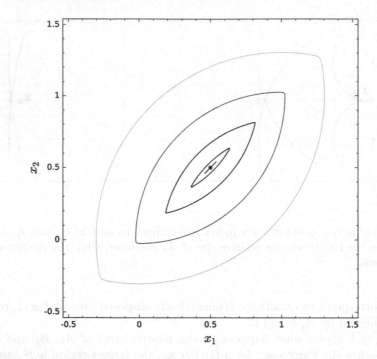

Fig. 6. Sequence of contour curves that represent the boundary of set $\mathcal{A}_x$ (the points which dominate $\mathbf{x}_k$). This sequence converges first to a line segment which contains the Pareto-optimal point $\mathbf{x}_{ref}$, before converging to $\mathbf{x}_{ref}$.

its consequences, as discussed above. In the case of an acceptance rule based on Tchebyschev decomposition, on the other hand, the constraint that the new point should have Tchebyschev norm smaller than the Tchebyschev norm of the current point means that, in addition to the points that dominate the current one, also some of the points which neither dominate nor are dominated by the current one are acceptable. This means that the convergence properties of sequences generated by Tchebyschev decomposition are expected to be better than in the case of Pareto-dominance acceptance rule.

However, the analysis to be presented next shows that the Tchebyschev acceptance rule eventually becomes equivalent to a Pareto-dominance acceptance rule, therefore leading to a similar loss of convergence. The explanation for this phenomenon is provided with the help of Fig. 7.

Figure 7 shows a representation, in the space of objectives, of the regions around a current point $\mathbf{f}(\mathbf{x}_k) = (p, q)$. The reference point is assumed to be on the origin. As $p > q$, the Tchebyschev norm of $\mathbf{f}(\mathbf{x}_k)$ is equal to p. Region $\mathcal{A}$ represents the points which dominate $\mathbf{f}(\mathbf{x}_k)$. Region $\mathcal{C}_1$ represents the points which neither dominate nor are dominated by $\mathbf{f}(\mathbf{x}_k)$ and which have Tchebyschev norm smaller than p. Region $\mathcal{C}_2$ contains the points which neither dominate nor are dominated by $\mathbf{f}(\mathbf{x}_k)$ and which have Tchebyschev norm greater than

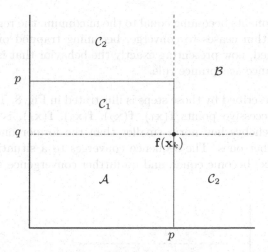

Fig. 7. Configuration of the regions of the space of objectives around a current point $f(x_k) = (p, q)$. The Tchebyschev norm of $f(x_k)$, considering the reference point on the origin, is equal to p. Region $\mathcal{A}$ represents the points which dominate $f(x_k)$. Region $\mathcal{C}_1$ represents the points which neither dominate nor are dominated by $f(x_k)$ and which have Tchebyschev norm smaller than p. Region $\mathcal{C}_2$ contain the points which neither dominate nor are dominated by $f(x_k)$ and which have Tchebyschev norm greater than p. Region $\mathcal{B}$ represents the points which are dominated by $f(x_k)$.

p. Region $\mathcal{B}$ represents the points which are dominated by $f(x_k)$. From this configuration, a next point $f(x_{k+1})$ is to be generated.

The relevant analysis of a situation like this will occur when the number of objectives grows. Although the figure becomes more complex, the five regions $\mathcal{A}$, $\mathcal{B}$, $\mathcal{C}_1$ and $\mathcal{C}_2$ will still appear, with the same definition above, for problem instances with any number of objectives, provided that there is one component of $f(x_k)$ which is equal to p and is greater than the other components. In such a case, the analysis tools provided earlier in this section will lead to the following consequences for a current point x_k sufficiently close to the Pareto-set:

1. The size of the preimage of set $\mathcal{A}$ (the set $\mathcal{A}_x$) becomes smaller as x_k approaches the Pareto-set. For a large number of objectives, when x_k is near the Pareto-set, the probability of x_{k+1} to fall on $\mathcal{A}_x$ approaches zero.
2. Considering the Tchebyschev acceptance rule, that disregards any x_{k+1} whose image $f(x_{k+1})$ falls on the sets $\mathcal{B}$ or $\mathcal{C}_2$, and also the consequence 1 above, the new point x_{k+1} will have its image $f(x_{k+1})$ in the set $\mathcal{C}_1$. This means that the new $f(x_{k+1})$ will have its greatest component smaller than p, while the other components, which were smaller than p, are likely to grow, being however bounded by p.
3. The successive application of such steps will lead the smaller components of $f(x_{k+j})$ to approach the maximum component. One component after the other will become approximately equal to the maximum component.

4. After all components becoming equal to the maximum, the region $\mathcal{C}_1$ vanishes, and the algorithm ceases to converge, becoming trapped on the last point that was reached, now presenting exactly the behavior that is predicted for a Pareto-dominance acceptance rule.

The process described by those steps is illustrated in Fig. 8. This figure shows a sequence of successive points $\mathbf{f}(\mathbf{x}_1)$, $\mathbf{f}(\mathbf{x}_2)$, $\mathbf{f}(\mathbf{x}_3)$, $\mathbf{f}(\mathbf{x}_4)$, is such that each new point has Tchebyschev norm smaller than the former one, and no point dominates the other ones. The sequence converges to a situation in which all components of $\mathbf{f}(\mathbf{x})$ become equal, and no further convergence will occur.

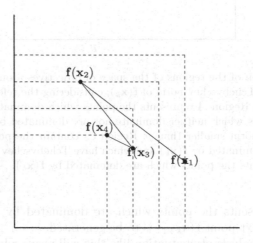

Fig. 8. A sequence of successive points $\mathbf{f}(\mathbf{x}_1)$, $\mathbf{f}(\mathbf{x}_2)$, $\mathbf{f}(\mathbf{x}_3)$, $\mathbf{f}(\mathbf{x}_4)$, is such that each new point has Tchebyschev norm smaller than the former one, and no point dominates the other ones.

In this way, it is expected that the minimization of a Tchebyschev norm goes farther than a pure Pareto-dominance search, because the algorithm presents some progress towards the Pareto-set while the sequence $\mathbf{f}(\mathbf{x}_k)$ approaches the situation in which all its components become equal. However, the search sequence eventually converges to the curve defined by $\{f_1(\mathbf{x}) = f_2(\mathbf{x}) = \ldots = f_m(\mathbf{x})\}$ before reaching the Pareto-set, becoming trapped on that curve.

The analysis presented in this section provides a full explanation for the behavior of the sequences produced by Tchebyschev-norm acceptance rule, as presented in Sect. 2. In this way: (i) the sequences cease to converge at a finite distance to the Pareto-set; (ii) this distance is about one order of magnitude smaller than the size of the Pareto-set (as predicted in [7]); (iii) the solution point in the space of objectives ($\mathbf{f}(\mathbf{x})$) has all its components on the same value when the convergence ceases; and (iv) this issue appears with a smaller intensity in Tchebyschev decomposition schemes than in Pareto-dominance algorithms, which is confirmed by the observation that a number of objectives $m = 4$ was enough for the full manifestation of the convergence loss in the Pareto-dominance case, while in the case of Tchebyschev decomposition the phenomenon appears clearly only for $m = 6$ or above.

The only remaining issue is concerned with the behavior of the sequences generated by Euclidean norm acceptance rule. The discussion about their behavior may be performed on the basis of Fig. 9.

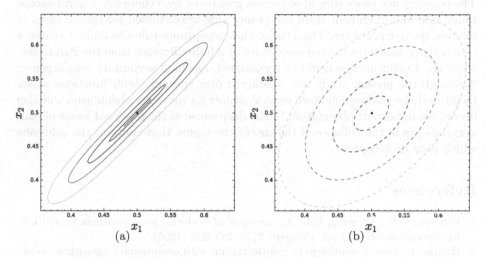

(a)   (b)

Fig. 9. Sequences of contour curves that represent the set of points which have Euclidean distance from $\mathbf{x}_{ref}$ smaller than $\mathbf{x}_k$, for the cases: (a) the reference point $\mathbf{x}_{ref}$ is on the Pareto-set; and (b) the reference point $\mathbf{x}_{ref}$ is behind the Pareto-set.

Figure 9(a) shows that, in the case of the reference point $\mathbf{x}_{ref}$ on the Pareto-set, the contour curves that enclose the acceptable points (the points which enhance the current Euclidean norm) are such that the inner curves become increasingly more elongated. The sequence of such curves collapse into a line segment that contains $\mathbf{x}_{ref}$, before converging to that point. Similarly to the analysis which was performed for the case of the Pareto-dominance contour curves, this geometric configuration causes the sequence $[\mathbf{x}]_k$ to converge to that line segment, becoming stuck on the point to which the convergence occurred, without further convergence to $\mathbf{x}_{ref}$. However, in the case of the reference point $\mathbf{y}_{ref}$ being placed behind the Pareto front, the contour curves that enclose the acceptable points become regularized, as shown in Fig. 9(b). In this case, the only accumulation point of the sequence $[\mathbf{x}]_k$ becomes the point $\mathbf{x}^*$, which corresponds to the point whose image $\mathbf{f}(\mathbf{x}^*)$ is the nearest point, in the space of objectives, to the reference point $\mathbf{y}_{ref}$ (which, of course does not have a pre-image in the space of decision variables). This fully explains the behavior of the sequences generated by the Euclidean norm acceptance rule, as presented in Sect. 2. In short: (i) when the reference point is defined on the Pareto front, the sequence becomes stuck at a finite distance to the Pareto-set (although at a distance which is smaller than in the case of Tchebyschev norm acceptance rule); (ii) when the reference point is established behind the Pareto front, the sequence seems to converge asymptotically to the Pareto-optimal point $\mathbf{x}^*$.

4   Conclusion

The convergence properties of sequences generated by Tchebyschev norm acceptance rule and Euclidean norm acceptance rule were studied for special cases of many-objective problems. The Tchebyschev acceptance rule was found to cause a convergence loss, with the sequence stuck at a finite distance from the Pareto-set, while the Euclidean rule could be regularized, reaching asymptotic convergence. Although the present study was conducted over specially built functions which facilitated the analysis, in more general situations the same phenomena will also occur. An important observation is: the deep cause of the observed losses of convergence lies in the collapse of the size of the region that contains the solutions which may be accepted.

References

1. Fonseca, C.M., Fleming, P.J.: An overview of evolutionary algorithms in multiobjective optimization. Evol. Comput. **7**(3), 205–230 (1995)
2. Hanne, T.: Global multiobjective optimization with evolutionary algorithms: selection mechanisms and mutation control. In: Zitzler, E., Thiele, L., Deb, K., Coello Coello, C.A., Corne, D. (eds.) EMO 2001. LNCS, vol. 1993, pp. 197–212. Springer, Heidelberg (2001). https://doi.org/10.1007/3-540-44719-9_14
3. Ikeda, K., Kita, H., Kobayashi, S.: Failure of Pareto-based MOEAs: does nondominated really mean near to optimal? In: Proceedings of the 2001 IEEE Congress on Evolutionary Computation (CEC 2001), pp. 957–962 (2001)
4. Khare, V., Yao, X., Deb, K.: Performance scaling of multi-objective evolutionary algorithms. In: Fonseca, C.M., Fleming, P.J., Zitzler, E., Thiele, L., Deb, K. (eds.) EMO 2003. LNCS, vol. 2632, pp. 376–390. Springer, Heidelberg (2003). https://doi.org/10.1007/3-540-36970-8_27
5. Li, B., Li, J., Tang, K., Yao, X.: Many-objective evolutionary algorithms: a survey. ACM Comput. Surv. **48**(1), Article 13 (2015)
6. Purshouse, R.C., Fleming, P.J.: On the evolutionary optimization of many conflicting objectives. IEEE Trans. Evol. Comput. **11**(6), 770–784 (2007)
7. Santos, T., Takahashi, R.H.C.: On the performance degradation of dominance-based evolutionary algorithms in many-objective optimization. IEEE Trans. Evol. Comput. **22**, 19–31 (2018)
8. Zhang, Q., Li, H.: MOEA/D: a multiobjective evolutionary algorithm based on decomposition. IEEE Trans. Evol. Comput. **11**(6), 712–731 (2007)

Algorithms

Algorithms

A New Hybrid Metaheuristic for Equality Constrained Bi-objective Optimization Problems

Oliver Cuate[1][(⊠)], Lourdes Uribe[2], Antonin Ponsich[3], Adriana Lara[2], Fernanda Beltran[1], Alberto Rodríguez Sánchez[3], and Oliver Schütze[4]

[1] Computer Science Department, Cinvestav-IPN,
Av. Instituto Politécnico Nacional No. 2508, Col. San Pedro Zacatenco, Mexico
{ocuate,mbeltran}@computacion.cs.cinvestav.mx
[2] ESFM, Instituto Politécnico Nacional,
Av. Instituto Politécnico Nacional Edif, 9, Unidad Profesional Adolfo López Mateos,
Zacatenco, Mexico
{lourdesur,adriana}@esfm.ipn.mx
[3] Universidad Autónoma Metropolitana Azcapotzalco,
Av. San Pablo No. 180, Col. Reynosa Tamaulipas, Azcapotzalco, Mexico
{aspo,al2161801914}@azc.uam.mx
[4] Department of Applied Mathematics and Systems,
UAM Cuajimalpa. Dr. Rodolfo Quintero Chair, Mexico City, Mexico
schuetze@cs.cinvestav.mx

Abstract. The recently proposed Pareto Tracer method is an effective numerical continuation technique which allows performing movements along the set of KKT points of a given multi-objective optimization problem. The nature of this predictor-corrector method leads to constructing solutions along the Pareto set/front numerically; it applies to higher dimensions and can handle box and equality constraints. We argue that the right hybridization of multi-objective evolutionary algorithms together with specific continuation methods leads to fast and reliable algorithms. Moreover, due to the continuation technique, the resulting hybrid algorithm could have a certain advantage when handling, in particular, equality constraints. In this paper, we make the first effort to hybridize NSGA-II with the Pareto Tracer. To support our claims, we present some numerical results on continuously differentiable equality constrained bi-objective optimization test problems, to show that the resulting hybrid NSGAII/PT is highly competitive against some state-of-the-art algorithms for constrained optimization. Finally, we stress that the chosen approach could be applied to a more significant number of objectives with some adaptations of the algorithm, leading to a very promising research topic.

Keywords: Multi-objective optimization · Evolutionary algorithms ·
Continuation methods · Hybrid algorithms

The authors acknowledge support for CONACyT project No. 285599 and IPN project SIP20181450.

© Springer Nature Switzerland AG 2019
K. Deb et al. (Eds.): EMO 2019, LNCS 11411, pp. 53–65, 2019.
https://doi.org/10.1007/978-3-030-12598-1_5

1   Introduction

In many different applications, the problem arises that several conflicting objectives have to be optimized concurrently [6,19]. Such problems are termed as multi-objective optimization problems (MOPs) in literature. One important characteristic of a MOP is that its solution set, the *Pareto set*, and its image, the *Pareto front*, typically form locally $(k - 1)$-dimensional manifolds, where k is the number of objectives of the MOP. Since these sets can not be computed analytically—apart from simple academic problems—the question arises how to compute suitable finite size approximations of them.

So far, many numerical methods for the treatment of a given MOP have been proposed. There exist, for instance, specialized evolutionary algorithms (EAs), called multi-objective evolutionary algorithms (MOEAs), that have caught the interest of many researchers in the recent past (e.g, [1,4,7,10,11,26]). Reasons for this include the global approach of these population-based methods, their relatively low assumptions on the model, their high robustness, and that they are capable of delivering a finite size approximation of the entire set of interest in one single run of the algorithm. The latter is because the whole population will evolve. These significant advantages, however, come at a certain price. It is widely accepted that MOEAs are very good in detecting promising regions and are highly efficient in computing rough approximations of the solution sets. However, they typically need quite a few function evaluations to obtain reasonable estimates of the Pareto set/front. This drawback gets even more significant in case the model contains complex constraints and/or the decision variable space is high-dimensional. As a possible remedy, researchers have hybridized evolutionary strategies with local search techniques mainly coming from mathematical programming leading to so-called memetic strategies. As local search engines, however, mainly only scalarization methods have been utilized (e.g., [2,3,12–14,18]). After a certain euphoria for such methods, this has led to many frustrations as the cost for the resulting hybrid is very high in many cases. The reason for this is the above mentioned missing possibility to fine tune the scalarization methods to obtain a suitable distribution along the solution sets. As a result, scalarization based memetic algorithms are highly competitive in case the given MOP is relatively easy, but this advantage vanishes with increasing complexity of the problem. For complex problems, scalarization based hybrids even perform worse than their base MOEAs due to the relative high cost induced by the local search.

The content of this study is the hybridization of a recently proposed multi-objective continuation method with MOEAs to obtain fast and reliable solvers for continuous MOPs. The Pareto Tracer (PT, [17]) is currently probably the more affordable algorithm (measured regarding the function calls needed to detect the next solution from a given starting point) that allows to perform a solution movement along the Pareto set/front; it is applicable to higher dimensions, and can handle box and equality constraints. In this initial study, we will make the first effort for such hybridization and will restrict ourselves to bi-objective equality constrained problems. We stress, however, that the chosen approach is in general not limited to bi-objective problems. As problems with a more significant number of objectives, however, will need a separate consideration and adaptations of the algorithm, we will leave this for future work. The results we present in this work already show the vast potential of the chosen hybrid and indicate the possibilities for deeper research in this direction. The first hybridization of a MOEA with a continuation method has been proposed in [21]. In this study, however, only unconstrained MOPs have been considered, and no extensions to constrained problems are discussed.

The remainder of this paper is organized as follows: in Sect. 2, we briefly state the required background for the understanding of this work. In Sect. 3, we describe the proposed algorithm that aims for the fast and reliable numerical treatment of continuous bi-objective optimization problems that contain equality constraints. In Sect. 4, we present some numerical results on selected academic benchmark problems including comparisons to the state-of-the-art. Finally, we draw our conclusions in Sect. 5 and state some possible paths for future research in this promising direction.

2   Background

In this paper we focus on continuously differentiable bi-objective optimization problems (BOPs) that contain equality constraints:

$$\min_{x \in \mathbb{R}^n} F(x)$$
$$\text{s.t. } h_i(x) = 0, \quad i = 1, \ldots, p, \tag{1}$$

where $F(x) = (f_1(x), f_2(x))^T$ is defined as the vector of the objective functions. Denote by Q the domain of (1), i.e., $Q := \{x \in \mathbb{R}^n : h_i(x) = 0, j = i, \ldots, p\}$.

Definition 1.

(a) Let $v, w \in \mathbb{R}^k$. Then the vector v is less than w ($v <_p w$), if $v_i < w_i$ for all $i \in \{1, \ldots, k\}$. The relation $\leq_p$ is defined analogously.
(b) A vector $y \in Q$ is dominated by a vector $x \in Q$ ($x \prec y$) with respect to (1) if $F(x) \leq_p F(y)$ and $F(x) \neq F(y)$, else y is called non-dominated by x.
(c) $x \in Q$ is called (Pareto) optimal if there is no $y \in Q$ which dominates x.
(d) The set P_Q of all Pareto optimal solutions is called the Pareto set and its image $F(P_Q)$ the Pareto front.

In the literature, some MOEAs can be found that have been designed to tackle such problems including NSGA-II [5], NSGA-II$_{MPP}$ [20], GDE3 [15] and MOEA/D/D [16]. NSGA-II uses feasibility rules in the selection process. In case the selection process considers two infeasible individuals, a penalization function is used to determine which individual violates more the constraints. NSGA-II$_{MPP}$ is a modification of the NSGA-II, which repairs a percent of the infeasible solutions via a sub-problem using *Sequential Quadratic Programming*. GDE3 is an extension of differential evolution for global multi-objective optimization. This method handles constraints using the same principles as NSGA-II.

Finally, MOEA/D/D combines dominance and decomposition-based approaches for solving many-objective optimization problems. MOEA/D/D also uses feasibility rules and the penalization function proposed in NSGA-II, but it is not only used in the selection process. It is present also in the update procedure of the algorithm, and this consideration helps with the preservation of diversity of the population.

PT is a predictor-corrector method that allows performing a movement along the set of KKT points of a given MOP. In the following we briefly recall the main steps for equality constrained BOPs, for more information and the general case the reader is referred to [17].

Given an initial solution x_i, a subsequent solution x_{i+1} is computed in two steps as follows: first, a tangent vector ν_μ of the set of KKT points at x_i is computed via solving

$$\begin{pmatrix} W_\alpha & H^T \\ H & 0 \end{pmatrix} \begin{pmatrix} \nu_\mu \\ \xi \end{pmatrix} = \begin{pmatrix} -J^T \mu \\ 0 \end{pmatrix}, \tag{2}$$

where $\alpha \in \mathbb{R}^k$ its associated Lagrange multiplier, J is the Jacobian of F at x, $\mu \in \mathbb{R}^k$, $\xi \in \mathbb{R}^p$ and

$$W_\alpha := \sum_{i=1}^{k} \alpha_i \nabla^2 f_i(x) \in \mathbb{R}^{n \times n} \tag{3}$$

and

$$H := \begin{pmatrix} \nabla h_1(x)^T \\ \vdots \\ \nabla h_p(x)^T \end{pmatrix} \in \mathbb{R}^{p \times n}. \tag{4}$$

If the rank of W_α is n and the rank of H is p, then the matrix on the right hand side of (2) is regular and hence ν_μ is determined uniquely. For $\mu \in \mathbb{R}^2$ there are two choices for the bi-objective case:

$$\mu^{(1)} = (-1, 1)^T \quad \text{and} \quad \mu^{(2)} = (1, -1)^T. \tag{5}$$

For $\mu^{(1)}$, a **right-down** movement along the Pareto front is performed, while the movement is *left-up* for $\mu^{(2)}$. Using the tangent vector ν_μ, the predictor p_i is computed as

$$p_i := x_i + t_i \nu_\mu, \tag{6}$$

where the step size t_i is chosen as

$$t_i := \frac{\tau}{\|J\nu_\mu\|} \tag{7}$$

to obtain $\|F(p_i) - F(x_i)\| \approx \tau$, where $\tau > 0$ is user specified value. By doing so, a uniform spread along the Pareto front can be obtained.

In the second step, the predicted point p_i is projected back to the set of KKT points via a multi-objective Newton method starting with p_i. Hereby, the Newton direction is chosen via solving

$$\min_{(\nu,\delta)\in\mathbb{R}^n\times\mathbb{R}} \quad \delta$$

$$\text{s.t.} \quad \nabla f_i(x)^T\nu + \tfrac{1}{2}\nu^T\nabla^2 f_i(x)\nu \le \delta, \quad i = 1,\ldots,k, \tag{8}$$

$$h_i(x) + \nabla h_i(x)^T\nu = 0, \quad i = 1,\ldots,p.$$

which is a modification of the Newton method from [9] adapted to the equality constrained case.

3 Proposed Algorithm (M-NSGA-II/PT)

In this section, we present a hybrid of a MOEA with the PT. As PT is very efficient in making a move along the Pareto set/front of a given problem, but is in turn of local nature, the task of the global counterpart is to feed PT with promising solutions that are well-spread along the set of interest. For this, we have decided to use a micro-GA that is based on NGSA-II in the first phase of the algorithm. In a second and last phase, PT takes over and refines the obtained solutions. In the following, we describe the two stages of the resulting hybrid (M-NSGA-II/PT) in detail.

3.1 First Stage: Rough Approximation via Micro-NSGA-II

The aim of the Multi-Objective Evolutionary Algorithm (MOEA) to be implemented is to generate an approximated Pareto set that contains only a few solutions since the construction of the real front will be subsequently performed by the PT. However, these solutions should be diverse enough to identify all the components of a possibly disconnected front. Finally, the MOEA should be able to handle equality constraints efficiently. Decomposition-based algorithm constitutes a viable option, actually, some preliminary experiments were performed with MOEA/D, but the use of small populations drastically increases the velocity of information transfer within the population. As a consequence, using too many neighbors (3 or more) led to very few different solutions in the final population. On the other hand, with too few neighbors, the algorithm is not able to converge to the real front. Finally, those convergence troubles encouraged for the use of a dominance-based algorithm, NSGA-II, which can efficiently handle a two-objective problem and allows the easy integration of constraint handling mechanism.

In order to balance the cost of the two stages, the algorithm handles a small population (preliminary tests showed that 20 individuals allow a sufficiently good convergence) and the crowding distance operator is used to maintain diversity. Regarding constraint handling, the Constraint Dominance Principle (CDP) is combined with ε-constraint [25], to avoid the risk of premature convergence towards the first feasible solutions found by the algorithm. CDP implements the standard feasibility rules: if two solutions are infeasible, that with the lower constraint violation is selected; if one solution is feasible and the other one is infeasible, the feasible wins; finally, in case that both solutions are feasible, the decision is taken according to the dominance criterion.

In addition, according to the ϵ-constraint strategy, constraints are first relaxed at the beginning of the run, so that a solution x such that $\Phi(x) \leq \epsilon$ is considered as feasible (where $\Phi(x)$ represents the total amount of constraint violation of x). Then, ϵ is gradually reduced, having slightly infeasible solutions competing with feasible ones and allowing diversity preservation during the run. A decreasing schedule of ϵ was proposed in the framework of single-objective optimization in [24], where ϵ decreases according to a polynomial function until a critical generation T_c is reached. Then, ϵ is set to 0 and the constraint handling technique reduces to the above-mentioned CDP:

$$\epsilon = \begin{cases} \epsilon(0)\,(1 - t/T_c)^{cp} & \text{if } 0 < t < T_c \\ 0 & \text{if } t \geq T_c, \end{cases} \tag{9}$$

where t is the generation number, cp is a parameter controlling the speed of the decrease and $\epsilon(0)$ is the constraint relaxation level at the first generation. This parameter is computed as the total constraint violation of x^θ, which is the θ-th solution in the first population, sorted in decreasing order of the total constraint violation Φ: $\epsilon(0) = \Phi(x^\theta)$. Notice that this technique was embedded in MOEA/D in [8]. Finally, an additional parameter was introduced in order to improve diversity: parent selection is performed with tournaments implementing the CDP extended with ϵ-constraint. However, the resulting winner of the tournament is considered only with probability p_f: in other cases (i.e., with probability $1 - p_f$), the winner individual is randomly chosen. The entire process is shortly described, for the reader convenience, in Algorithm 1.

Algorithm 1. Micro-NGSA-II

$P \leftarrow pop\_init()$
Evaluate each individual $x^i \in P$ to obtain $F(x^i)$
Compute $\epsilon(0)$
for $t \leftarrow 1$ **to** $MaxGen$ **do**
    $P' \leftarrow crossover(P)$ ▷ Parent selection through tournament, CDP and ϵ-constraint
    $P'' \leftarrow mutation(P')$
    $Q \leftarrow P \cup P''$
    $Q' \leftarrow Feasible(Q, \epsilon)$, $Q'' = Infeasible(Q, \epsilon)$
    $Q' \leftarrow FastNonDominatedSorting(Q')$, $Q'' = SortConstraintViolation(Q'')$
    Fill P with Q', using crowding distance if necessary
    if $|Q'| < PopSize$ **then**
        Complete P with Q''
    end if
    Update ϵ through equation 9
end forreturn P and $F(P)$

3.2   Second Stage: Refinement via PT

The main task on this stage is to process the resulting archive P, which is provided by the Micro-NSGA-II, appropriately. For instance, PT could spend a lot of additional function evaluations computing non-optimal KKT points. We can solve this and other issues for bi-objective problems as described below. After the first stage, and before using PT, we need a post-processing procedure for archive P as in the following steps:

1. Improve every point in P by the modification of the Newton method (8).
2. Remove all dominated points in the new archive (possible local fronts).
3. Sort the final archive P, according to f_1, in the way that $f_1(p^{(1)}) < \ldots < f_1(p^{(m)})$, where $p^{(i)} \in P$, $i = 1, \ldots, m$.

Now, we can use PT as follows: we take the first element $p^{(1)} \in P$ as starting point for PT with a **left-up** movement ($\mu^{(2)}$), we compute as many solutions as possible (until here there is no conflict with the other points of P). Then, we perform the **right-down** movement ($\mu^{(1)}$) starting at $p^{(1)}$, but from this case, we have to consider the value of the next element in P (and also the previous one in case we have) in order to avoid extra function evaluations.

In general, let x_d be the current solution for the **right-down** movement of PT starting from $p^{(i)}$, $i = 1, \ldots, m-1$, τ as in Eq. (7), and $\theta \in (0, \tau)$; then we have the following stopping criteria:

- $\|F(x_d) - F(p^{(i+1)})\|_2 < \theta$. That is, we reach the next point in P. In case $x_d \prec p^{(i+1)}$, we delete $p^{(i+1)}$ from P and we continue with the right-down movement (compute a new x_d). Otherwise, we stop and select $p^{(i+1)}$ as a new starting point.
- $f_2(x_d) \prec f_2(p^{(i+1)})$. When first condition is not satisfied, this condition means that $x_d \prec p^{(i+1)}$. If that is the case, we delete $p^{(i+1)}$ from P and continue with the **right-down** movement (compute a new x_d).
- No improvements in f_2 direction could be achieved (PT stopping condition). If the PT stops, then we select $p^{(i+1)}$ as a new starting point (we move to a different connected component of the Pareto front).

Additionally, for $p^{(i)}$, $i = 2, \ldots, m$, we have also to perform and verify the **left-up** movement. Here, we assume that a previous right-down movement was made. Let x_u be the current solution for the **left-up** movement of PT starting from $p^{(i)}$, $i = 2, \ldots, m$, x_d the last solution obtained by the right-down movement starting from $p^{(i-1)}$, τ as in Eq. (7), and $\theta \in (0, \tau)$; then we have the following stopping criteria:

- $\|F(x_u) - F(x_d)\|_2 < \theta$. This condition prevents the computation of extra solutions in previously considered regions of the Pareto front.
- $f_2(x_u) > f_2(x_d)$. When first condition is not satisfied, this condition means that $x_d \prec x_u$. If that is the case, then we stop and we continue with the right-down movement for $p^{(i+1)}$.
- No improvements in f_1 direction could be achieved (PT stop condition). If the PT stops, then we continue with the right-down movement for next point $p^{(i+1)}$.

4 Numerical Results

In this section, we present some numerical examples that compare the behavior of state-of-the-art MOEAs against the proposed M-NSGA-II/PT when applied to the CZDT test suite proposed in [20] and the modification of the Das and

Table 1. Parameters of the MOEAs.

MOEA	Parameter	Value
NSGA-II	Population size	100
	Crossover probability	0.8
	Mutation probability	$\frac{1}{n}$
	Distribution index for crossover	20
	Distribution index for mutation	20
NSGA-II/PT	Initial Population size	20
	Crossover probability	1.0
	Mutation probability	$\frac{1}{n}$
	Distribution index for crossover	20
	Distribution index for mutation	20
	τ for CZDT(1, 2, 4)	0.015
	τ for CZDT3	0.004
	τ for CZDT6	0.0005
$NSGA - II_{MPP}$	Population size	40
	Crossover probability	0.9
	Mutation probability	0.1
	Distribution index for crossover	15
	Distribution index for mutation	20
	$fmincon$ tol	1e−10
MOEA/D/D	Population size	100
	# weight vectors	100
	Crossover probability	1
	Mutation probability	$1/n$
	Distribution index for crossover	30
	Distribution index for mutation	20
	Penalty parameter of PBI	5
	Neighborhood size	20
	Probability used to select in the neighborhood	0.9
GDE3	Population size	100
	CR	0.2
	F	0.2
	Distribution index for mutation	20

Dennis problem (D&D) stated in [17]. We consider a solution x as feasible, i.e. $h(x) = 0$, when $\|h(x)\| < \varepsilon$. Here we consider $\varepsilon = 1e - 04$, which is a value regularly reported in the specialized literature. The selected MOEAs for comparison against are GDE3, MOEA/D/D, NSGA-II, and NSGA-II$_{MPP}$. We do not compare against classical Mathematical Programming (MP) methods because of the nature of the selected test suite, that is CZDT is multi-modal and the performance of MP techniques depends on the given initial solution.

For every experiment, we have performed 20 independent runs. From the experimental analysis, we observed that M-NSGA-II/PT needs between 15,000 and 17,000 function evaluations to obtain good results. Then to have a fair comparison, we established a final budget of 20,000 function calls for all the selected MOEAs. Finally, we split the budget for both stages of M-NSGA-II/PT as follows: 15,000 and 5,000. Table 1 contains the algorithm parameter values used for the experiment. The performance indicator Δ_2 (see [22]) is used to measure the algorithm effectiveness in this proposed benchmark. We apply the Wilcoxon test to validate the results; we consider $\alpha = 0.05$. Table 2 and Figs. 1, 2, 3 and 4 show the results for CZDT and D&D functions. The theoretical PF is marked with . while the MOEA approximation is marked with $\triangle$. For this table, Column # sol shows the average number of feasible solutions at the end of each run. An arrow up means the result is statistically reliable. Observe that the proposed hybrid algorithm wins significantly in five of six test functions. Finally,

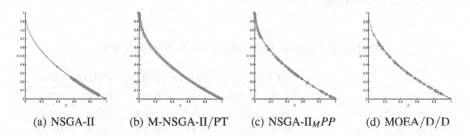

(a) NSGA-II   (b) M-NSGA-II/PT   (c) NSGA-II$_M$PP   (d) MOEA/D/D

Fig. 1. Pareto front approximation for the C-ZDT1 problem for the selected MOEAs with 20,000 function evaluations.

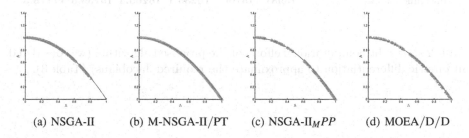

(a) NSGA-II   (b) M-NSGA-II/PT   (c) NSGA-II$_M$PP   (d) MOEA/D/D

Fig. 2. Pareto front approximation for the C-ZDT2 problem for the selected MOEAs with 20,000 function evaluations.

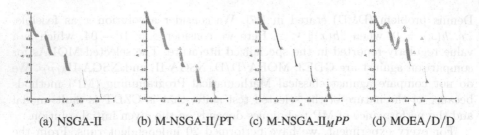

(a) NSGA-II   (b) M-NSGA-II/PT   (c) M-NSGA-II$_M$PP   (d) MOEA/D/D

Fig. 3. Pareto front approximation for the C-ZDT3 problem for the selected MOEAs with 20,000 function evaluations.

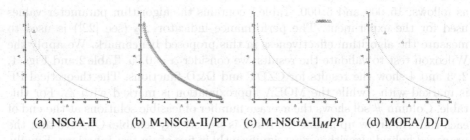

(a) NSGA-II   (b) M-NSGA-II/PT   (c) M-NSGA-II$_M$PP   (d) MOEA/D/D

Fig. 4. Pareto front approximation for the D&D problem for the selected MOEAs with 20,000 function evaluations.

Table 2. Computational effort for M-NSGA-II/PT.

	CZDT1	CZDT2	CZDT3	CZDT4	CZDT6	D&D
τ	0.015	0.015	0.004	0.015	0.0005	0.20
Function evaluations	196.2	150	280.3	367.4	147.9	511.9
Jacobian evaluations	784.8	600	954.8	954.8	591.6	1663.6
Total function evs PT	981	700	1235.1	1265.4	739.5	2175.5
Total function evs M-NSGA-II	15000	15000	15000	15000	15000	15000
Final function evs	**15981**	**15700**	**16235.1**	**16265.4**	**15739.5**	**17175.5**

Table 2 shows the computational efforts of the proposed algorithm (we considered automatic differentiation to approximate the required Jacobians) (Table 3).

Table 3. Results for CZDT and D&D functions with $k = 2$ for some MOEAs. (# is average of the number of feasible solutions at the end of each run).

	Method	20,000 Δ_p	# sol.		Method	20,000 Δ_p	# sol.
CZDT1	M-NSGA-II/PT (std.dev)	**0.0038** (1.8559e-04)	120.70	**CZDT4**	M-NSGA-II/PT (std.dev)	**0.0039** (1.8475e-04)	117.30
	NSGA-II ↑ (std.dev)	0.1214 (0.0560)	100		NSGA-II ↑ (std.dev)	1.1241 (0.5593)	1
	GDE3 ↑ (std.dev)	- (-)	0		GDE3 ↑ (std.dev)	- (-)	0
	MOEADD↑ (std.dev)	0.0166 (0.0101)	100		MOEADD ↑ (std.dev)	0.6620 (0.3253)	0.7
	NSGA-II$_{MPP}$ ↑ (std.dev)	0.0136 (7.6744e-04)	40		NSGA-II$_{MPP}$ ↑ (std.dev)	0.0150 (0.0011)	39.4
CZDT2	M-NSGA-II/PT (std.dev)	**0.0038** (1.9885e-04)	121.50	**CZDT6**	M-NSGA-II/PT (std.dev)	0.0884 (0.0180)	68.5
	NSGA-II ↑ (std.dev)	0.1147 (0.0721)	100		NSGA-II ↑ (std.dev)	0.4249 (0.0587)	100
	GDE3 ↑ (std.dev)	- (-)	0		GDE3 ↑ (std.dev)	- (-)	0
	MOEADD↑ (std.dev)	0.0197 (0.0210)	100		MOEADD ↑ (std.dev)	0.3518 (0.0749)	100
	NSGA-II$_{MPP}$ ↑ (std.dev)	0.0150 (0.0013)	40		NSGA-II$_{MPP}$ ↑ (std.dev)	**0.0100** (8.6138e-04)	40
CZDT3	M-NSGA-II/PT (std.dev)	**0.0164** (0.0158)	112.60	**D & D**	M-NSGA-II/PT (std.dev)	**0.5339** (0.8026)	70
	NSGA-II ↑ (std.dev)	0.2648 (0.1584)	100		NSGA-II↓ (std.dev)	6.6574 (0.6572)	69.8
	GDE3 ↑ (std.dev)	- (-)	0		GDE3 ↑ (std.dev)	- (-)	0
	MOEADD ↑ (std.dev)	0.1240 (0.0685)	100		MOEADD ↓ (std.dev)	3.9697 (0.8506)	10.8
	NSGA-II$_{MPP}$ ↓ (std.dev)	0.0396 (0.0583)	40		NSGA-II$_{MPP}$ ↑ (std.dev)	6.2986 (0.9729)	1.5

5 Conclusions and Future Work

In this paper, we have proposed M-NSGA-II/PT which is a two-phase hybrid algorithm combining the Micro-NSGA-II and the multi-objective continuation method named as Pareto Tracer. The task of Micro-NSGA-II is to primarily compute a rough, small sized but well-spread solution approximation of the given MOP. In a second stage, PT is used to refine this approximation by numerical continuation based on specific candidate solutions. In this study, we have focussed on bi-objective equality constrained problems. Numerical results and comparisons against four state-of-the-art MOEAs have shown that this new strategy is highly competitive and can lead to satisfying results with a moderate budget of function evaluations.

Though these results are strongly promising, more research has to be performed in this direction to obtain a new class of hybrid evolutionary algorithms for the fast and reliable numerical treatment of general MOPs. For this, it is mandatory to thoroughly discuss the cases of a general number of objectives and general constraints. Further, more comparisons have to be performed to demonstrate the benefit of the novel hybrid. Finally, it is intended to apply M-NSGA-II/PT to problems arising in real-world applications. For this task, it will likely be necessary to reduce the required derivative information, e.g. by utilizing approximation strategies like the one proposed in [23].

References

1. Beume, N., Naujoks, B., Emmerich, M.: SMS-EMOA: multiobjective selection based on dominated hypervolume. Eur. J. Oper. Res. **181**(3), 1653–1669 (2007)
2. Bosman, P.A.N., de Jong, E.D.: Exploiting gradient information in numerical multi-objective evolutionary optimization. In: Genetic and Evolutionary Computation Conference - GECCO 2005. ACM (2005)
3. Brown, M., Smith, R.E.: Directed multi-objective optimization. Int. J. Comput. Syst. Sig. **6**(1), 3–17 (2005)
4. Deb, K.: Multi-Objective Optimization Using Evolutionary Algorithms. Wiley, New York (2001)
5. Deb, K., Pratap, A., Agarwal, S., Meyarivan, T.: A fast and elitist multiobjective genetic algorithm: NSGA-II. IEEE Trans. Evol. Comput. **6**(2), 182–197 (2002)
6. Dilettoso, E., Rizzo, S.A., Salerno, N.: A weakly Pareto compliant quality indicator. Math. Comput. Appl. **22**(1) (2017)
7. Dominguez-Isidro, S., Mezura-Montes, E.: The baldwin effect on a memetic differential evolution for constrained numerical optimization problems. In: Proceedings of the Genetic and Evolutionary Optimization Conference, pp. 1–8 (2017)
8. Fan, Z., et al.: An improved epsilon constraint handling method embedded in MOEA/D for constrained multi-objective optimization problems. In: 2016 IEEE Symposium Series on Computational Intelligence (SSCI), pp. 1–8 (2016)
9. Fliege, J., Drummond, L.M.G., Svaiter, B.F.: Newton's method for multiobjective optimization. SIAM J. Optim. **20**, 602–626 (2009)
10. Gerstl, K., Rudolph, G., Schütze, O., Trautmann, H.: Finding evenly spaced fronts for multiobjective control via averaging Hausdorff-measure. In: Proceedings of 8th International Conference on Electrical Engineering, Computing Science and Automatic Control (CCE), pp. 1–6. IEEE Press (2011)
11. Hernández-Ocana, B., Mezura-Montes, E., del Pilar Pozos-Parra, M.: Evolutionary bacterial foraging algorithm to solve constrained numerical optimization problems. In: New Tendencies in Logic, Languages, Algorithms, and New Methods of Reasoning, pp. 29–42 (2016)
12. Hu, X., Huang, Z., Wang, Z.: Hybridization of the multi-objective evolutionary algorithms and the gradient-based algorithms. In: Proceedings of the IEEE Congress on Evolutionary Computation, pp. 870–877 (2003)

13. Knowles, J., Corne, D.: Memetic algorithms for multiojective optimization: issues, methods and prospects. In: Hart, W.E., Smith, J.E., Krasnogor, N. (eds.) Recent Advances in Memetic Algorithms. STUDFUZZ, vol. 166, pp. 313–352. Springer, Heidelberg (2005). https://doi.org/10.1007/3-540-32363-5_14

14. Knowles, J.D., Corne, D.W.: M-PAES: a memetic algorithm for multiobjective optimization. In: Proceedings of the IEEE Congress on Evolutionary Computation, Piscataway, New Jersey, pp. 325–332 (2000)

15. Kukkonen, S., Lampinen, J.: GDE3: the third evolution step of generalized differential evolution. In: The 2005 IEEE Congress on Evolutionary Computation, vol. 1, pp. 443–450. IEEE (2005)

16. Li, K., Deb, K., Zhang, Q., Kwong, S.: An evolutionary many-objective optimization algorithm based on dominance and decomposition. IEEE Trans. Evol. Comput. **19**(5), 694–716 (2015)

17. Martín, A., Schütze, O.: Pareto tracer: a predictor-corrector method for multiobjective optimization problems. Eng. Optim. **50**(3), 516–536 (2018)

18. Ong, Y.W., Keane, A.J.: Meta-Lamarckian learning in memetic algorithms. IEEE Trans. Evol. Comput. **8**(2), 99–110 (2004)

19. Peitz, S., Dellnitz, M.: A survey of recent trends in multiobjective optimal control-surrogate models, feedback control and objective reduction. Math. Comput. Appl. **23**(2) (2018)

20. Saha, A., Ray, T.: Equality constrained multi-objective optimization. In: 2012 IEEE Congress on Evolutionary Computation, CEC 2012, pp. 1–7, June 2012

21. Schütze, O., Coello Coello, C.A., Mostaghim, S., Talbi, E.-G., Dellnitz, M.: Hybridizing evolutionary strategies with continuation methods for solving multiobjective problems. Eng. Optim. **40**(5), 383–402 (2008)

22. Schütze, O., Esquivel, X., Lara, A., Coello Coello, C.A.: Using the averaged Hausdorff distance as a performance measure in evolutionary multiobjective optimization. IEEE Trans. Evol. Comput. **16**(4), 504–522 (2012)

23. Schütze, O., Alvarado, S., Segura, C., Landa, R.: Gradient subspace approximation: a direct search method for memetic computing. Soft Comput. **21**(21), 6331–6350 (2017)

24. Takahama, T., Sakai, S.: Constrained optimization by the ϵ-constrained differential evolution with gradient-based mutation and feasible elites (2006)

25. Takahama, T., Sakai, S., Iwane, N.: Solving nonlinear constrained optimization problems by the ε constrained differential evolution. In: IEEE 2006, vol. 3, pp. 2322–2327. IEEE (2006)

26. Trautmann, H., Wagner, T., Brockhoff, D.: R2-EMOA: focused multiobjective search using R2-indicator-based selection. In: Nicosia, G., Pardalos, P. (eds.) LION 2013. LNCS, vol. 7997, pp. 70–74. Springer, Heidelberg (2013). https://doi.org/10.1007/978-3-642-44973-4_8

Make Evolutionary Multiobjective Algorithms Scale Better with Advanced Data Structures: Van Emde Boas Tree for Non-dominated Sorting

Maxim Buzdalov[✉]

ITMO University, 49 Kronverkskiy Avenue, Saint Petersburg 197101, Russia
mbuzdalov@gmail.com

Abstract. We improve the worst-case time complexity of non-dominated sorting, an operation frequently used in evolutionary multiobjective algorithms, to $O(n \cdot (\log n)^{k-2} \log \log n)$, where n is the number of solutions, k is the number of objectives, and the random-access memory computation model is assumed. This improvement was possible thanks to the van Emde Boas tree, an "advanced" data structure which stores a set of non-negative integers less than n and supports many queries in $O(\log \log n)$. This is not only a theoretical improvement, as we also provide an efficient implementation of the van Emde Boas tree, which resulted in a competitive algorithm that scales better than other algorithms when n grows, at least for small numbers of objectives greater than two.

Keywords: Non-dominated sorting · Large-scale optimization · vEB tree

1 Introduction

The world is making its first steps through the era of Big Data. As the amount of data grows faster than the available computing power, researchers need algorithms that scale better than ever. The scalability can be improved in various directions, including scalability with the number of cores, scalability with the problem size, as well as scalability with the complexity of the problem, which makes sense for problem-agnostic solvers such as evolutionary algorithms.

There is an increasing trend among researchers in evolutionary computation to tackle problems with millions of decision variables, continuous [1] and discrete [6], or even with billions of variables [9]. All mentioned papers have been nominated for best paper awards at the corresponding conferences, and some have won, which signifies the interest of the community to such approaches.

The number of decision variables is only one of the critical measures for how difficult it is to solve the problem. The problem's complexity may influence the choices of parameters for evolutionary algorithms, which, in turn, can influence

© Springer Nature Switzerland AG 2019
K. Deb et al. (Eds.): EMO 2019, LNCS 11411, pp. 66–77, 2019.
https://doi.org/10.1007/978-3-030-12598-1_6

the overall running time through the computational complexity of the internal steps of the algorithm. As an example one can consider multimodal problems, which typically require larger population sizes to find good enough local optima. If the computational complexity of the algorithm is superlinear with regards to the population size, any increase of the multimodality results in a much larger increase in the running time. This means that the computational complexity of the internal steps of an evolutionary algorithm can determine its scalability with regards to the complexity of the solved problem.

This is especially true for multiobjective optimization. Since many evolutionary multiobjective algorithms contain subroutines that are basically computational geometry algorithms, sometimes in high dimensions, it is often the case that an iteration of such an evolutionary algorithm has at least quadratic complexity with regards to the population size. While for small population sizes this effect does not show up, since all the internals are dominated by fitness evaluation, it affects running times for large enough population sizes to the extent where parallelization of the update procedures seems to be necessary [16].

The hypervolume indicator [26], or S-metric, is probably the most well-known example of a computationally hard subroutine which is a part of several evolutionary multiobjective algorithms. Despite its highly desirable properties, it is #P-complete [2] and also NP-hard to approximate. While for small dimensions there do exist efficient algorithms, it is very unlikely that hypervolume indicator will ever scale well for the number of objectives starting with, for instance, six.

Non-dominated sorting is maybe the second most popular subroutine among those that can be bottlenecks in evolutionary multiobjective algorithms. The problem of non-dominated sorting was introduced along with the original NSGA algorithm [24], where an algorithm with time complexity $O(n^3k)$ was proposed to solve this problem. Here and onwards, n is the number of solutions (also "the number of points"), and k is the number of objectives (also "dimension"). Among other improvements, NSGA-II [10] brought a faster algorithm for non-dominated sorting with time and memory complexity of $O(n^2k)$. Non-dominated sorting remains one of the bottlenecks in NSGA-III [8] and many other algorithms.

Jensen [18] applied the divide-and-conquer paradigm to this problem, following the guidelines of Kung [19] for solving similar problems, and achieved the worst-case time complexity of $O(n \cdot (\log n)^{k-1})$ and linear memory complexity under certain conditions (e.g. no two objective vectors coincide in any of the objectives). It took more than ten years to settle down the same worst-case time complexity for arbitrary inputs [5,14]. The practical performance has also been improved since then, the paper [20] features the currently fastest variation of this idea. In the last ten years a number of algorithms were published which focused on improving the practical performance while having the $\Omega(n^2k)$ worst-case performance, among which Best Order Sort [22,23] and ENS-NDT [17] currently seem to be the fastest on average. In particular, the algorithm from [20] is a hybrid algorithm that joins the divide-and-conquer paradigm of [5,14,18] and a specially tailored version of ENS-NDT [17].

The worst-case time complexity bound of $O(n \cdot (\log n)^{k-1})$ was unbeatable since 2003, except for constant small dimensions. Some works exist for $k = 3$ (e.g. an $O(n \log n)$ algorithm in $O(n \log n/\log \log n)$ space [3], or an $O(n \cdot (\log \log n)^2)$ algorithm in the RAM machine model [21]), however, no implementations and no experimental evaluation are available, which makes a direct comparison difficult. There is currently no known lower bound except for the trivial $\Omega(nk)$, which can be slightly refined to $\Omega(n \log n + nk)$ if the pointer machine model is assumed.

We propose a *theoretically faster algorithm* with the worst-case time complexity of $O(n \cdot (\log n)^{k-2} \log \log n)$ which uses the RAM machine model. The speedup is due to the use of the van Emde Boas tree [12,13] instead of the binary search tree. The van Emde Boas tree is a data structure that stores a set of integers belonging to the interval $[0; M-1)$ and offers many operations in $O(\log \log M)$, including navigation to the next or to the previous element of the set. It is widely believed to be efficient only when the number of elements stored in it is large (of the same order as M). Despite this, we have written an *efficient implementation* of the van Emde Boas tree which resulted in competitive running times of the entire enclosing non-dominated sorting algorithm, which is comparable to the best available algorithms. The new implementation *eventually overcomes* other algorithms as n grows, which was observed experimentally at $n \approx 10^6 \ldots 10^7$ and $3 \leq k \leq 5$, and is not much slower at smaller n.

The rest of the paper is structured as follows. The necessary notation is introduced in Sect. 2. The structure of the divide-and-conquer algorithm from [5,14,18] is outlined in Sect. 3, along with the hybridization ideas successfully applied in [20]. The concept of the van Emde Boas tree is explained in Sect. 4. The details of our efficient implementation[1] are given in Sect. 5 for the tree and in Sect. 6 for the whole algorithm. We give our experimental setup, results and their discussion in Sect. 7. Section 8 summarizes our work.

2   Preliminaries

Since the algorithms studied in this paper operate only on the level of objective vectors, we ignore various problems of genotype-to-phenotype mapping, use interchangeably individuals and vectors of their objective values, and hence often call individuals the *points*. The letter n typically denotes the *number of points*, and the letter k denotes the number of objectives, or the *dimension*.

We consider points coming from the $\mathbb{R}^k$ space without loss of generality, as otherwise the values for each objective can be sorted and then compressed into integers from the range $[1; n]$. Also without loss of generality we consider multiobjective problems that require minimization of every objective. For two points p, q from the k-dimensional space we say that p *dominates* q, and write $p \prec q$, if $\forall i, 1 \leq i \leq k, p_i \leq q_i$ and $\exists j, 1 \leq j \leq k, p_j < q_j$.

Note that it follows from this definition that $p \nprec p$ for any p. Some implementations of non-dominated sorting assume that $p \prec p$, which introduces inconsistency to the results but can be desirable to punish multiple equal solutions. For

---

[1] Available at https://github.com/mbuzdalov/non-dominated-sorting/tree/v0.2.

adapting any divide-and-conquer algorithm for non-dominated sorting to this property we direct the reader to the recent work [4].

The *non-dominated sorting* problem is, given a (multi)set of points P, to assign each point a *rank* such that:

- for any point p such that $\forall q \in P$ it holds that $q \not\prec p$ the rank $r(p) \leftarrow 0$;
- for any other point $r(p) \leftarrow 1 + \max_{q \in P, q \prec p} r(q)$.

This definition assumes that ranks are assigned in an arbitrary order such that, if $p \prec q$, the rank is first assigned to p. It is not difficult to show that such an order exists, e.g. the lexicographical order of points has this property for an arbitrary set of points, so the definition is valid.

3   The Divide-and-Conquer Algorithm for Non-dominated Sorting

In this section we outline the divide-and-conquer algorithm initially proposed by Jensen [18] and then subsequently refined by a number of researchers [5,14,20]. This algorithm divides the problem into smaller subproblems of two types, which have somewhat more general formulation than just non-dominated sorting, and reduces the number of used objectives when possible. The algorithms based on the so-called sweep line solve the special cases of these subproblems that have only two objectives to consider. We will focus on them, since in this paper we aim at replacing these algorithms with their more efficient counterparts based on the van Emde Boas tree, and the rest of the algorithm will be described briefly.

3.1   The General Plan

The first step of the algorithm, similar to the ENS algorithm family [25], is to sort the individuals lexicographically (by first comparing them in the first objective, if equal move on to the second one, and so on), which can be done in $O(n \log n + nk)$ by an appropriate modification of the quicksort. While doing this, the algorithm retains a single representative of multiple equal points, if needed, so the rest of the algorithm can assume that no two points are equal; after the main work is done, all equal points are assigned the same rank as their representative. The resulting list of points $P = \{p_1, p_2, \ldots, p_n\}$ obviously has the following property: for any two $1 \leq i < j \leq n$ it holds that $p_j \not\prec p_i$.

When $k > 2$, the algorithm uses the divide-and-conquer technique to solve the problem. To do this, it defines two auxiliary subproblems called A and B:

A: A set of points $S \subset P$ is given, along with the number of meaningful objectives m and the *lower bounds* on the rank $r(s)$ for each point $s \in S$. It is assumed that all points from S have equal objective values for all objectives $m < i \leq k$. Assign the ranks according to the definition of non-dominated sorting and taking into account these lower bounds. It is assumed that all comparisons with points dominating all points from S are already performed, so the resulting ranks will be final.

B: Two sets of points L and R are given, along with the number of meaningful objectives m. It is known that $L \subset P$, $R \subset P$, $L \cap R = \emptyset$, all the ranks are final for all points in L, lower bounds are known for all points in R. For any point $l \in L$, $r \in R$ and objective $m < i \le k$ it holds that $l_i < r_i$. Update the lower bounds for the points in R taking into account the dominance relations between points from L and R, as well as the ranks of points in L.

Note that the non-dominated sorting problem itself is a special case of the subproblem A when all lower bounds on the ranks are zeros and $m = k$.

The subproblem B can be trivially solved if $|L| = 1$ or $|R| = 1$. If $m = 2$, a special algorithm is used, which is outlined in the following subsection. For performance reasons, as proposed in [20], the subproblem is solved by a tailored version of an efficient quadratic algorithm, such as ENS-SS [25] or ENS-NDT [17], when $|L| + |R| \le C$ for some (constant) value C. Otherwise, the algorithm finds a median M of the m-th objective across L and R. Then it splits L into three sets L_L, L_M, L_H, as well as R into R_L, R_M, R_H, such that all $\cdot_L$ sets feature the m-th objective less than M, for all $\cdot_M$ sets it is equal to M, and for all $\cdot_H$ sets it is greater than M. Then the instances of the subproblem B are solved for the pairs $(L_L; R_M)$, $(L_L; R_H)$, $(L_M; R_M)$, (L_M, R_H) with the meaningful objective value of $m - 1$, as well for the pairs $(L_L; R_L)$ and $(L_H; R_H)$ with the unchanged meaningful objective value. The use of the median guarantees that $\max(|L_L| + |R_L|, |L_H| + |R_H|) \le \frac{1}{2}(|L| + |R|)$. If an algorithm for $m = 2$ works in $O((|L| + |R|) \log(|L| + |R|))$, this results in the runtime bound of $O((|L| + |R|)(\log(|L| + |R|))^{m-1})$ thanks to the Master theorem [7].

Very similarly, the subproblem A is trivially solved if $|S| = 2$, delegated to the two-dimensional case if $m = 2$ and to a quadratic algorithm if $|S| \le C$. Otherwise, the set S is split into S_L, S_M, S_H around the median of the m-th objectives. Next, the instance of the subproblem A is solved for S_L, since no other points can dominate points from this set. Once it is solved, the ranks for S_L are final. After that, the instance of the subproblem B is solved for $L \leftarrow S_L$ and $R \leftarrow S_M$. Then the subproblem A is solved for S_M, next the subproblem B is solved for $L \leftarrow S_L \cup S_M$ and $R \leftarrow S_H$, and finally another instance of the subproblem A is solved for S_H. Then the Master theorem proves the $O(|S|(\log |S|)^{m-1})$ runtime if the algorithm for $m = 2$ works in $O(|S| \log |S|)$.

3.2   Sweep Line Algorithms for $m = 2$

The special algorithms for two-dimensional cases of the above problems use the concept of the *sweep line*. Such algorithms visit points as if they are hit by a line that is parallel to the ordinate axis and moves through the plane from small to large abscissas. They also maintain a data structure which is updated and/or queried when a point is visited.

The data structure used in the algorithms for non-dominated sorting is a sorted set which maps ordinate values to ranks. More precisely, for each last processed point p of rank r the mapping $p_y \to r$ is stored in the set. Additionally, if there exist two points p and q, such that $p_y \le q_y$ and $r(p) > r(q)$, the mapping for the point q need not be stored, since any point z, that comes lexicographically after both p and q and is dominated by q in first two objectives, is also dominated by p. This means that we can restrict the data structure to a *monotone sorted map*, where for any two mappings $k_1 \to v_1$ and $k_2 \to v_2$ it holds that $(k_1 < k_2) \leftrightarrow (v_1 < v_2)$.

With such a sorted map as a data structure on the sweep line, one can solve the subproblem B as follows. The points from $L \cup R$ are jointly traversed in lexicographical order. If the next point comes from L, the corresponding mapping (the ordinate of the point to the rank of this point) is added to the map, while preserving monotonicity. If the next point is $r \in R$, the smallest mapping with the not-exceeding ordinate is queried from the map, and the rank lower bound of r is updated. The subproblem A is solved similarly, however, both operations are performed on each point (first query then insertion).

The existing implementations of the divide-and-conquer non-dominated sorting algorithms use a binary search tree to implement the monotone sorted map, which supports insertions and queries in $O(\log n)$ time (amortized for insertions). This ensures $O(n \log n)$ worst-case running time for both subproblems on n points, more precisely, $O(|S| \log |S|)$ for the subproblem A and $O((|L| + |R|) \log |L|)$ for the subproblem B.

4   The Van Emde Boas Tree

The van Emde Boas tree [12,13] is a data structure for storing non-negative integers less than $D = 2^d$ and supporting queries common for sorted sets in time $O(\log \log D) = O(\log d)$. These operations include testing whether an element belongs to a set, querying the closest element not smaller than (the *next-query*), or not greater than (the *prev-query*) the given one, as well as insertion and removal. It additionally supports querying the minimum and the maximum among the stored elements in $O(1)$ time. It requires the random-access memory model and, in particular, the ability of indexing the elements of an array by the results of bitwise operations on keys in $O(1)$ time.

This data structure is parameterized by d (we will refer to a van Emde Boas tree with the parameter d as the *d-tree*) and is constructed recursively. It always stores the minimum and the maximum elements explicitly. For $d = 1$ there can be only two elements, so the implementation is trivial. For $d > 1$, the elements other than the minimum and the maximum are stored in subordinate d_L-trees, where $d_L = \lfloor \frac{d}{2} \rfloor$, which are stored in an array of size $D_H = 2^{\lceil \frac{d}{2} \rceil}$. We also denote $D_L = 2^{d_L}$ and $d_H = \lceil \frac{d}{2} \rceil$. A key x is stored in a subordinate tree as follows: the $(x \text{ div } D_L)$-th subordinate tree stores the value $x \mod D_L$. To speed up the next-queries and prev-queries, a d-tree also contains a d_H-tree that serves as an index tree: if the x-th subordinate tree is not empty, the index tree contains x.

Note that operations such as $x \text{ div } D_L$ and $x \mod D_L$ can be efficiently implemented in most modern computer architectures with the use of bitwise arithmetics (e.g. the former is implemented by bit shift, and the latter by bit masking), since D_L is a power of two. We assume this knowledge in the following sections.

Testing whether an element x belongs to a d-tree, $d > 1$, is straightforward: first it is compared with the minimum and the maximum; in a non-trivial case when it is between them, the $(x \operatorname{div} D_L)$-th subordinate tree is asked for whether it contains $x \bmod D_L$. Since the index tree and each subordinate tree have almost the same parameter value $\approx d/2$, each query in a d-tree amounts to $O(1)$ plus one query to a $(d/2)$-tree, which results in $O(\log d) = O(\log \log D)$ runtime.

A similar recursive construction enables efficient prev-queries and next-queries. For instance, for a next-query on a d-tree, $d > 1$, a key x is first compared with the maximum. If it is smaller (otherwise the query is trivial and $O(1)$), the $(x \operatorname{div} D_L)$-th tree is queried. If it returned that there is no next key, which happens in $O(1)$, then the index tree is queried for the next non-empty subordinate tree, and if any, its minimum is returned. In total, in every query to a d-tree, at most one query to a $(d/2)$-tree is performed, which results in $O(\log d)$ runtime.

The update procedures require considering a number of corner cases and an accurate implementation, so we will not go into details in this paper, however, recursion patterns similar to the ones above ensure the same $O(\log d)$ runtime bound for both insertion and removal.

5   Efficient Implementation of the Van Emde Boas Tree

This section is dedicated to an efficient implementation of a monotone sorted map, based on the van Emde Boas tree, to be used for a more efficient implementation of algorithms described in Sect. 3.2. Note that the actually used code does not support the entire interface of a van Emde Boas tree, but rather an interface required for non-dominated sorting. The Java classes mentioned in this section can be accessed on GitHub.[2]

We mention first that, in order to implement a monotone sorted map over a sorted set, we shall store also the values, and we shall also be able to "clean up" when a mapping is inserted that forces some existing mappings to be removed in order to preserve monotonicity. We store the values in a plain auxiliary array indexed by the unmodified keys, and some additional information about that array is passed around a few methods that should be able to clean up.

For performance reasons, we implement, aside from the generic van Emde Boas tree that contains other van Emde Boas trees as either subordinate trees or an index tree, six other implementations of a van Emde Boas tree for specific ranges of the parameter d, and a special implementation representing an empty tree. This is done for performance reasons, since for small values of d the corresponding operations can be implemented more efficiently with the use of bitwise arithmetics. More precisely, we have the following implementations:

- `AnyAnyBitSet`: the general case used for $d \geq 14$;
- `EmptyBitSet`: a singleton implementation of an always empty set, used for empty subordinate sets of an `AnyAnyBitSet` to save time and memory;

---

[2] https://github.com/mbuzdalov/non-dominated-sorting/tree/v0.2/
implementations/src/main/java/ru/ifmo/nds/util/veb.

- `IntBitSet`: a wrapper around a 32-bit integer, used for $0 \le d \le 5$;
- `LongBitSet`: a wrapper around a 64-bit integer, used for $d = 6$;
- `IntIntBitSet`: an implementation with both subordinate sets and the index set represented directly as 32-bit integers, used for $7 \le d \le 10$;
- `IntLongBitSet`: same as above, but the index set is a 64-bit integer, used for $d = 11$;
- `LongLongBitSet`: same as above, but the subordinate sets are also 64-bit integers, used for $d = 12$;
- `LongAnyBitSet`: same as above, but the index set is an `IntIntBitSet`, used for $d = 13$;

Note that in languages which allow integer parameters for classes and compile-time specialization of classes based on these parameters, such as C++, the number of implementations can be reduced in order to achieve the same performance effect, but would still require writing at least half the size of the code above.

Each of these classes contains implementations of the following methods:

- `isEmpty()`, `min()`, `max()` are implemented straightforwardly except for the cases of `IntBitSet` and `LongBitSet`, where the two latter are implemented through counting the number of leading/trailing zeros;
- `clear()` removes all elements in time linear to the number of elements;
- `next(int)`, `prev(int)`, `prevInclusively(int)` perform the strict next-query, the strict prev-query and the non-strict prev-query, correspondingly;
- `add(int)` and `remove(int)` to add and remove an element without checking the monotonicity, which are used internally;
- `setEnsuringMonotonicity` that takes a key, a value, an array for values and an offset in this array to be used with the key, and performs insertion of the mapping if necessary and cleaning up the mappings which are no longer needed;
- `cleanupUpwards` that takes a value, an array for values and an offset in this array, and cleans up the mappings which are no longer needed, starting from the minimum element.

In particular, the two latter methods can perform bulk removal of the stale mappings, which is typically faster than using `next` and `remove` iteratively. Two different methods for strict and non-strict prev-queries implement strictness-specific shortcuts to save more computation time. As a result of all these precautions, we ensure roughly $O(\log \log D)$ performance for queries on our monotone sorted map implementation, whereas we save sufficient resources, both time and memory, at the smaller end of the size range, to make it competitive.

6   Implementation and Analysis of the Whole Algorithm

Both algorithms given in Sect. 3.2 run in time $O(n \log \log n)$ with the use of the implementation of the monotone sorted map based on the van Emde Boas tree, assuming ordinates are non-negative integers less than n. However, in practice they are floating point values from an arbitrary range. To enable using our new implementation, we first compress these values in $O(n \log n)$ time using sorting and a linear scan. Since for $k > 2$ the resulting running times are still $\omega(n \log n)$, this does not impact the theoretical performance, however, it has a small but noticeable impact on the overall performance.

Repeating the analysis given in Sect. 3.1 with the performance for two-dimensional algorithms to solve subproblems A and B to be $O(|S| \log \log |S|)$ and $O((|L| + |H|) \log \log |L|)$ correspondingly, we get the overall running time of $O(n \cdot (\log n)^{k-2} \log \log n)$ for $k > 2$.

7 Experiments

We conducted experiments on uniform hypercube datasets (also known as "cloud" datasets) with dimension $k \in \{3, 4, 5, 7, 10\}$ and the numbers of points $n = \lfloor 10^{i/2} \rfloor$ for $1 \le i \le 14$. For these n and k, we randomly generated 10 datasets.

We did not use datasets with only non-dominated points, as in these conditions it does not matter which data structure is used for the two-dimensional case, since at most one point is stored in it during the entire run, and the theoretical performance becomes $O(n \cdot (\log n)^{k-2})$. We did not use datasets from the runs of the real evolutionary multiobjective algorithms on benchmark problems, since it is not clear yet which algorithms, on which problems and with which budgets will form a benchmark set that adequately represents the possible cases from the point of view of non-dominated sorting, especially for very large n.

The following algorithms were used in our comparison:

- Best Order Sort as in [23];
- ENS-NDT as in [17] with the split threshold of 4;
- two configurations of the divide-and-conquer algorithm based on a binary search tree hybridized as in [20] with ENS-SS [25] and with ENS-NDT, and two similar configurations based on the van Emde Boas tree.

We have executed each algorithm on each dataset for five times (which seems fair as runtimes are well concentrated). We did not execute Best Order Sort on datasets with $n > 10^5$ as it appeared to scale worse than other algorithms.

The results are given in Table 1. One can see that the algorithm based on the van Emde Boas tree hybridized with ENS-SS wins for $k = 3$ at $n \ge 3 \cdot 10^5$. The modification of [20] with the van Emde Boas tree starts winning for $4 \le k \le 5$ at $n = 10^7$. For other dimensions, the van Emde Boas based algorithms do not yet win for $n \le 10^7$, however they get closer to the best algorithms while n grows. It is slightly more difficult for a van Emde Boas based algorithm to overcome other algorithms with greater k and cloud datasets, as the maximum rank grows slower with n when k is large.

One can also see that our implementation of the van Emde Boas tree does not impose significant penalties on the running times. Indeed, for both hybrid

Table 1. Running times of algorithms, in seconds. The values of n are ordered in a column-first order. For every k, the best result for every n is highlighted dark gray. Results within 5% of the best are highlighted light gray.

Algorithm	$n=10^{\frac{1}{2}}$ / 10^1	$10^{\frac{3}{2}}$ / 10^2	$10^{\frac{5}{2}}$ / 10^3	$10^{\frac{7}{2}}$ / 10^4	$10^{\frac{9}{2}}$ / 10^5	$10^{\frac{11}{2}}$ / 10^6	$10^{\frac{13}{2}}$ / 10^7
$k = 3$							
BOS	$2.015\cdot10^{-7}$	$4.905\cdot10^{-6}$	$7.645\cdot10^{-5}$	$1.262\cdot10^{-3}$	$3.961\cdot10^{-2}$	—	—
[23]	$1.187\cdot10^{-6}$	$2.033\cdot10^{-5}$	$3.049\cdot10^{-4}$	$6.128\cdot10^{-3}$	$2.661\cdot10^{-1}$	—	—
ENS-NDT	$1.254\cdot10^{-7}$	$5.506\cdot10^{-6}$	$1.235\cdot10^{-4}$	$1.904\cdot10^{-3}$	$3.197\cdot10^{-2}$	$5.675\cdot10^{-1}$	$1.167\cdot10^{1}$
[17]	$8.634\cdot10^{-7}$	$3.068\cdot10^{-5}$	$4.837\cdot10^{-4}$	$7.755\cdot10^{-3}$	$1.358\cdot10^{-1}$	$2.444\cdot10^{0}$	$5.186\cdot10^{2}$
RBTree+ENS-SS	$1.124\cdot10^{-7}$	$2.840\cdot10^{-6}$	$7.557\cdot10^{-5}$	$1.272\cdot10^{-3}$	$1.950\cdot10^{-2}$	$2.910\cdot10^{-1}$	$4.171\cdot10^{0}$
	$5.472\cdot10^{-7}$	$1.722\cdot10^{-5}$	$3.274\cdot10^{-4}$	$5.095\cdot10^{-3}$	$7.287\cdot10^{-2}$	$1.141\cdot10^{0}$	$1.555\cdot10^{1}$
RBTree+ENS-NDT	$1.701\cdot10^{-7}$	$5.981\cdot10^{-6}$	$1.274\cdot10^{-4}$	$1.719\cdot10^{-3}$	$2.369\cdot10^{-2}$	$3.282\cdot10^{-1}$	$4.598\cdot10^{0}$
[20]	$1.088\cdot10^{-6}$	$3.203\cdot10^{-5}$	$4.572\cdot10^{-4}$	$6.488\cdot10^{-3}$	$8.584\cdot10^{-2}$	$1.240\cdot10^{0}$	$1.725\cdot10^{1}$
vEB+ENS-SS	$1.536\cdot10^{-7}$	$4.347\cdot10^{-6}$	$8.990\cdot10^{-5}$	$1.330\cdot10^{-3}$	$2.092\cdot10^{-2}$	$2.828\cdot10^{-1}$	$3.903\cdot10^{0}$
	$8.937\cdot10^{-7}$	$2.171\cdot10^{-5}$	$3.551\cdot10^{-4}$	$5.514\cdot10^{-3}$	$7.479\cdot10^{-2}$	$1.087\cdot10^{0}$	$1.481\cdot10^{1}$
vEB+ENS-NDT	$2.106\cdot10^{-7}$	$7.609\cdot10^{-6}$	$1.413\cdot10^{-4}$	$1.776\cdot10^{-3}$	$2.495\cdot10^{-2}$	$3.257\cdot10^{-1}$	$4.416\cdot10^{0}$
	$1.427\cdot10^{-6}$	$3.657\cdot10^{-5}$	$4.805\cdot10^{-4}$	$7.017\cdot10^{-3}$	$8.802\cdot10^{-2}$	$1.227\cdot10^{0}$	$1.617\cdot10^{1}$
$k = 4$							
BOS	$2.545\cdot10^{-7}$	$6.476\cdot10^{-6}$	$1.026\cdot10^{-4}$	$1.944\cdot10^{-3}$	$7.825\cdot10^{-2}$	—	—
[23]	$1.561\cdot10^{-6}$	$2.603\cdot10^{-5}$	$4.271\cdot10^{-4}$	$1.131\cdot10^{-2}$	$5.773\cdot10^{-1}$	—	—
ENS-NDT	$1.335\cdot10^{-7}$	$5.941\cdot10^{-6}$	$1.418\cdot10^{-4}$	$2.395\cdot10^{-3}$	$5.108\cdot10^{-2}$	$1.165\cdot10^{0}$	$3.636\cdot10^{1}$
[17]	$8.689\cdot10^{-7}$	$3.316\cdot10^{-5}$	$5.548\cdot10^{-4}$	$1.107\cdot10^{-2}$	$2.343\cdot10^{-1}$	$6.181\cdot10^{0}$	$1.735\cdot10^{2}$
RBTree+ENS-SS	$1.158\cdot10^{-7}$	$3.388\cdot10^{-6}$	$1.304\cdot10^{-4}$	$3.032\cdot10^{-3}$	$6.782\cdot10^{-2}$	$1.200\cdot10^{0}$	$2.121\cdot10^{1}$
	$5.723\cdot10^{-7}$	$2.194\cdot10^{-5}$	$6.711\cdot10^{-4}$	$1.462\cdot10^{-2}$	$2.692\cdot10^{-1}$	$5.327\cdot10^{0}$	$8.797\cdot10^{1}$
RBTree+ENS-NDT	$1.726\cdot10^{-7}$	$6.721\cdot10^{-6}$	$1.514\cdot10^{-4}$	$2.349\cdot10^{-3}$	$4.441\cdot10^{-2}$	$9.704\cdot10^{-1}$	$1.942\cdot10^{1}$
[20]	$1.105\cdot10^{-6}$	$3.524\cdot10^{-5}$	$5.643\cdot10^{-4}$	$1.020\cdot10^{-2}$	$2.196\cdot10^{-1}$	$4.502\cdot10^{0}$	$7.988\cdot10^{1}$
vEB+ENS-SS	$1.591\cdot10^{-7}$	$4.958\cdot10^{-6}$	$1.457\cdot10^{-4}$	$3.111\cdot10^{-3}$	$7.000\cdot10^{-2}$	$1.191\cdot10^{0}$	$2.080\cdot10^{1}$
	$9.067\cdot10^{-7}$	$2.646\cdot10^{-5}$	$7.117\cdot10^{-4}$	$1.546\cdot10^{-2}$	$2.730\cdot10^{-1}$	$5.160\cdot10^{0}$	$8.357\cdot10^{1}$
vEB+ENS-NDT	$2.099\cdot10^{-7}$	$8.262\cdot10^{-6}$	$1.682\cdot10^{-4}$	$2.527\cdot10^{-3}$	$4.694\cdot10^{-2}$	$9.893\cdot10^{-1}$	$1.995\cdot10^{1}$
	$1.428\cdot10^{-6}$	$3.994\cdot10^{-5}$	$6.219\cdot10^{-4}$	$1.061\cdot10^{-2}$	$2.262\cdot10^{-1}$	$4.505\cdot10^{0}$	$7.872\cdot10^{1}$
$k = 5$							
BOS	$3.079\cdot10^{-7}$	$7.861\cdot10^{-6}$	$1.264\cdot10^{-4}$	$2.594\cdot10^{-3}$	$1.272\cdot10^{-1}$	—	—
[23]	$1.870\cdot10^{-6}$	$3.109\cdot10^{-5}$	$5.526\cdot10^{-4}$	$1.698\cdot10^{-2}$	$9.433\cdot10^{-1}$	—	—
ENS-NDT	$1.341\cdot10^{-7}$	$6.086\cdot10^{-6}$	$1.479\cdot10^{-4}$	$2.843\cdot10^{-3}$	$6.741\cdot10^{-2}$	$1.883\cdot10^{0}$	$7.679\cdot10^{1}$
[17]	$9.312\cdot10^{-7}$	$3.459\cdot10^{-5}$	$6.281\cdot10^{-4}$	$1.397\cdot10^{-2}$	$3.201\cdot10^{-1}$	$1.203\cdot10^{1}$	$4.162\cdot10^{2}$
RBTree+ENS-SS	$1.210\cdot10^{-7}$	$3.912\cdot10^{-6}$	$1.502\cdot10^{-4}$	$4.153\cdot10^{-3}$	$1.241\cdot10^{-1}$	$2.693\cdot10^{0}$	$5.615\cdot10^{1}$
	$5.924\cdot10^{-7}$	$2.643\cdot10^{-5}$	$8.300\cdot10^{-4}$	$2.291\cdot10^{-2}$	$5.253\cdot10^{-1}$	$1.345\cdot10^{1}$	$2.547\cdot10^{2}$
RBTree+ENS-NDT	$1.747\cdot10^{-7}$	$6.808\cdot10^{-6}$	$1.667\cdot10^{-4}$	$3.000\cdot10^{-3}$	$6.219\cdot10^{-2}$	$1.688\cdot10^{0}$	$4.422\cdot10^{1}$
[20]	$1.146\cdot10^{-6}$	$3.846\cdot10^{-5}$	$6.699\cdot10^{-4}$	$1.397\cdot10^{-2}$	$3.388\cdot10^{-1}$	$8.718\cdot10^{0}$	$1.979\cdot10^{2}$
vEB+ENS-SS	$1.631\cdot10^{-7}$	$5.488\cdot10^{-6}$	$1.673\cdot10^{-4}$	$4.300\cdot10^{-3}$	$1.287\cdot10^{-1}$	$2.749\cdot10^{0}$	$5.632\cdot10^{1}$
	$9.300\cdot10^{-7}$	$3.097\cdot10^{-5}$	$8.858\cdot10^{-4}$	$2.378\cdot10^{-2}$	$5.339\cdot10^{-1}$	$1.364\cdot10^{1}$	$2.545\cdot10^{2}$
vEB+ENS-NDT	$2.157\cdot10^{-7}$	$8.271\cdot10^{-6}$	$1.840\cdot10^{-4}$	$3.170\cdot10^{-3}$	$6.448\cdot10^{-2}$	$1.716\cdot10^{0}$	$4.501\cdot10^{1}$
	$1.481\cdot10^{-6}$	$4.334\cdot10^{-5}$	$7.256\cdot10^{-4}$	$1.453\cdot10^{-2}$	$3.523\cdot10^{-1}$	$8.926\cdot10^{0}$	$1.973\cdot10^{2}$
$k = 7$							
BOS	$4.095\cdot10^{-7}$	$1.045\cdot10^{-5}$	$1.704\cdot10^{-4}$	$4.088\cdot10^{-3}$	$2.206\cdot10^{-1}$	—	—
[23]	$2.410\cdot10^{-6}$	$4.108\cdot10^{-5}$	$7.723\cdot10^{-4}$	$2.858\cdot10^{-2}$	$1.881\cdot10^{0}$	—	—
ENS-NDT	$1.356\cdot10^{-7}$	$6.729\cdot10^{-6}$	$1.604\cdot10^{-4}$	$3.502\cdot10^{-3}$	$9.785\cdot10^{-2}$	$3.519\cdot10^{0}$	$1.591\cdot10^{2}$
[17]	$9.644\cdot10^{-7}$	$3.643\cdot10^{-5}$	$7.237\cdot10^{-4}$	$1.830\cdot10^{-2}$	$5.160\cdot10^{-1}$	$2.458\cdot10^{1}$	$7.900\cdot10^{2}$
RBTree+ENS-SS	$1.306\cdot10^{-7}$	$5.249\cdot10^{-6}$	$1.894\cdot10^{-4}$	$5.146\cdot10^{-3}$	$1.854\cdot10^{-1}$	$5.649\cdot10^{0}$	$1.614\cdot10^{2}$
	$6.322\cdot10^{-7}$	$3.492\cdot10^{-5}$	$9.543\cdot10^{-4}$	$2.945\cdot10^{-2}$	$9.267\cdot10^{-1}$	$3.445\cdot10^{1}$	$8.643\cdot10^{2}$
RBTree+ENS-NDT	$1.863\cdot10^{-7}$	$7.395\cdot10^{-6}$	$1.905\cdot10^{-4}$	$4.199\cdot10^{-3}$	$1.024\cdot10^{-1}$	$2.745\cdot10^{0}$	$9.130\cdot10^{1}$
[20]	$1.192\cdot10^{-6}$	$4.078\cdot10^{-5}$	$8.392\cdot10^{-4}$	$2.182\cdot10^{-2}$	$5.310\cdot10^{-1}$	$1.514\cdot10^{1}$	$4.390\cdot10^{2}$
vEB+ENS-SS	$1.724\cdot10^{-7}$	$6.536\cdot10^{-6}$	$2.060\cdot10^{-4}$	$5.363\cdot10^{-3}$	$1.887\cdot10^{-1}$	$5.789\cdot10^{0}$	$1.621\cdot10^{2}$
	$9.621\cdot10^{-7}$	$3.965\cdot10^{-5}$	$1.013\cdot10^{-3}$	$3.012\cdot10^{-2}$	$9.433\cdot10^{-1}$	$3.485\cdot10^{1}$	$8.861\cdot10^{2}$
vEB+ENS-NDT	$2.334\cdot10^{-7}$	$8.924\cdot10^{-6}$	$2.082\cdot10^{-4}$	$4.434\cdot10^{-3}$	$1.053\cdot10^{-1}$	$2.796\cdot10^{0}$	$9.279\cdot10^{1}$
	$1.521\cdot10^{-6}$	$4.539\cdot10^{-5}$	$9.008\cdot10^{-4}$	$2.248\cdot10^{-2}$	$5.406\cdot10^{-1}$	$1.529\cdot10^{1}$	$4.537\cdot10^{2}$
$k = 10$							
BOS	$6.174\cdot10^{-7}$	$1.409\cdot10^{-5}$	$2.251\cdot10^{-4}$	$6.047\cdot10^{-3}$	$3.518\cdot10^{-1}$	—	—
[23]	$3.500\cdot10^{-6}$	$5.536\cdot10^{-5}$	$1.056\cdot10^{-3}$	$4.363\cdot10^{-2}$	$3.797\cdot10^{0}$	—	—
ENS-NDT	$1.550\cdot10^{-7}$	$7.055\cdot10^{-6}$	$1.790\cdot10^{-4}$	$4.209\cdot10^{-3}$	$1.366\cdot10^{-1}$	$5.645\cdot10^{0}$	$2.478\cdot10^{2}$
[17]	$1.043\cdot10^{-7}$	$3.969\cdot10^{-5}$	$8.124\cdot10^{-4}$	$2.402\cdot10^{-2}$	$7.600\cdot10^{-1}$	$4.098\cdot10^{1}$	$1.599\cdot10^{3}$
RBTree+ENS-SS	$1.481\cdot10^{-7}$	$5.541\cdot10^{-6}$	$2.449\cdot10^{-4}$	$6.954\cdot10^{-3}$	$2.188\cdot10^{-1}$	$7.255\cdot10^{0}$	$2.540\cdot10^{2}$
	$6.785\cdot10^{-7}$	$4.178\cdot10^{-5}$	$1.244\cdot10^{-3}$	$3.738\cdot10^{-2}$	$1.197\cdot10^{0}$	$4.850\cdot10^{1}$	$1.589\cdot10^{3}$
RBTree+ENS-NDT	$2.047\cdot10^{-7}$	$7.828\cdot10^{-6}$	$1.978\cdot10^{-4}$	$5.312\cdot10^{-3}$	$1.512\cdot10^{-1}$	$4.418\cdot10^{0}$	$1.400\cdot10^{2}$
[20]	$1.262\cdot10^{-6}$	$4.253\cdot10^{-5}$	$9.498\cdot10^{-4}$	$3.061\cdot10^{-2}$	$8.340\cdot10^{-1}$	$2.394\cdot10^{1}$	$7.132\cdot10^{2}$
vEB+ENS-SS	$1.927\cdot10^{-7}$	$6.845\cdot10^{-6}$	$2.618\cdot10^{-4}$	$7.159\cdot10^{-3}$	$2.210\cdot10^{-1}$	$7.328\cdot10^{0}$	$2.520\cdot10^{2}$
	$1.021\cdot10^{-6}$	$4.671\cdot10^{-5}$	$1.297\cdot10^{-3}$	$3.806\cdot10^{-2}$	$1.194\cdot10^{0}$	$4.871\cdot10^{1}$	$1.597\cdot10^{3}$
vEB+ENS-NDT	$2.508\cdot10^{-7}$	$9.324\cdot10^{-6}$	$2.137\cdot10^{-4}$	$5.513\cdot10^{-3}$	$1.539\cdot10^{-1}$	$4.459\cdot10^{0}$	$1.396\cdot10^{2}$
	$1.595\cdot10^{-6}$	$4.698\cdot10^{-5}$	$1.009\cdot10^{-3}$	$3.122\cdot10^{-2}$	$8.368\cdot10^{-1}$	$2.405\cdot10^{1}$	$7.156\cdot10^{2}$

variants the running time of the new implementations does not exceed 1.5 times the running time of the old implementations even for the smallest values of n, and this ratio tends to decrease towards 1.0 and even below while n increases.

8 Conclusion

We proposed a theoretic improvement over the existing runtime complexity of non-dominated sorting. Our algorithm is based on the van Emde Boas tree and works in time $O(n \cdot (\log n)^{k-2} \log \log n)$, a factor of $O(\log n/\log \log n)$ faster than any previous approach for arbitrary k. We also provided its implementation able to outperform the existing algorithms on large enough n and small enough k.

This paper evaluated the algorithms only on artificial uniform hypercube datasets. We expect to perform a more thorough comparison of all the available non-dominated sorting algorithms on real-world datasets and on various scales. Apart from implementation of the algorithms from [3,21] we also plan to test Pareto archiving algorithms [11,15] wrapped in the ENS framework [25].

Acknowledgment. The research is financially supported by The Russian Science Foundation, Agreement No. 17-71-30029 with co-financing of Bank Saint Petersburg.

References

1. Bouter, A., Alderliesten, T., Witteveen, C., Bosman, P.A.N.: Exploiting linkage information in real-valued optimization with the real-valued gene-pool optimal mixing evolutionary algorithm. In: Proceedings of Genetic and Evolutionary Computation Conference, pp. 705–712 (2017)
2. Bringmann, K., Friedrich, T.: Approximating the least hypervolume contributor: NP-hard in general, but fast in practice. In: Ehrgott, M., Fonseca, C.M., Gandibleux, X., Hao, J.-K., Sevaux, M. (eds.) EMO 2009. LNCS, vol. 5467, pp. 6–20. Springer, Heidelberg (2009). https://doi.org/10.1007/978-3-642-01020-0_6
3. Buchsbaum, A.L., Goodrich, M.T.: Three-dimensional layers of maxima. Algorithmica **39**, 275–286 (2004)
4. Buzdalov, M.: Generalized offline orthant search: one code for many problems in multiobjective optimization. In: Proceedings of Genetic and Evolutionary Computation Conference, pp. 593–600 (2018)
5. Buzdalov, M., Shalyto, A.: A provably asymptotically fast version of the generalized jensen algorithm for non-dominated sorting. In: Bartz-Beielstein, T., Branke, J., Filipič, B., Smith, J. (eds.) PPSN 2014. LNCS, vol. 8672, pp. 528–537. Springer, Cham (2014). https://doi.org/10.1007/978-3-319-10762-2_52
6. Chicano, F., Whitley, D., Ochoa, G., Tinós, R.: Optimizing one million variable NK landscapes by hybridizing deterministic recombination and local search. In: Proceedings of Genetic and Evolutionary Computation Conference, pp. 753–760 (2017)
7. Cormen, T.H., Leiserson, C.E., Rivest, R.L., Stein, C.: Introduction to Algorithms, 2nd edn. MIT Press, Cambridge (2001)
8. Deb, K., Jain, H.: An evolutionary many-objective optimization algorithm using reference-point-based nondominated sorting approach, part I: solving problems with box constraints. IEEE Trans. Evol. Comput. **18**(4), 577–601 (2013)

9. Deb, K., Myburgh, C.: A population-based fast algorithm for a billion-dimensional resource allocation problem with integer variables. Eur. J. Oper. Res. **261**(2), 460–474 (2017)
10. Deb, K., Pratap, A., Agarwal, S., Meyarivan, T.: A fast and elitist multi-objective genetic algorithm: NSGA-II. IEEE Trans. Evol. Comput. **6**(2), 182–197 (2002)
11. Drozdik, M., Akimoto, Y., Aguirre, H., Tanaka, K.: Computational cost reduction of nondominated sorting using the m-front. IEEE Trans. Evol. Comput. **19**(5), 659–678 (2015)
12. van Emde Boas, P., Kaas, R., Zijlstra, E.: Design and implementation of an efficient priority queue. Math. Syst. Theory **10**, 99–127 (1976)
13. van Emde Boas, P.: Preserving order in a forest in less than logarithmic time. In: Proceedings of the Annual Symposium on Foundations of Computer Science, pp. 75–84 (1975)
14. Fortin, F.A., Grenier, S., Parizeau, M.: Generalizing the improved run-time complexity algorithm for non-dominated sorting. In: Proceedings of Genetic and Evolutionary Computation Conference, pp. 615–622. ACM (2013)
15. Glasmachers, T.: A fast incremental BSP tree archive for non-dominated points. In: Trautmann, H., et al. (eds.) EMO 2017. LNCS, vol. 10173, pp. 252–266. Springer, Cham (2017). https://doi.org/10.1007/978-3-319-54157-0_18
16. Gupta, S., Tan, G.: A scalable parallel implementation of evolutionary algorithms for multi-objective optimization on GPUs. In: Proceedings of IEEE Congress on Evolutionary Computation, pp. 1567–1574 (2015)
17. Gustavsson, P., Syberfeldt, A.: A new algorithm using the non-dominated tree to improve non-dominated sorting. Evol. Comput. **26**(1), 89–116 (2018)
18. Jensen, M.T.: Reducing the run-time complexity of multiobjective EAs: the NSGA-II and other algorithms. IEEE Trans. Evol. Comput. **7**(5), 503–515 (2003)
19. Kung, H.T., Luccio, F., Preparata, F.P.: On finding the maxima of a set of vectors. J. ACM **22**(4), 469–476 (1975)
20. Markina, M., Buzdalov, M.: Towards large-scale multiobjective optimisation with a hybrid algorithm for non-dominated sorting. In: Auger, A., Fonseca, C.M., Lourenço, N., Machado, P., Paquete, L., Whitley, D. (eds.) PPSN 2018. LNCS, vol. 11101, pp. 347–358. Springer, Cham (2018). https://doi.org/10.1007/978-3-319-99253-2_28
21. Nekrich, Y.: A fast algorithm for three-dimensional layers of maxima problem. In: Dehne, F., Iacono, J., Sack, J.-R. (eds.) WADS 2011. LNCS, vol. 6844, pp. 607–618. Springer, Heidelberg (2011). https://doi.org/10.1007/978-3-642-22300-6_51
22. Roy, P.C., Deb, K., Islam, M.M.: An efficient nondominated sorting algorithm for large number of fronts. IEEE Trans. Cybern. (2018). https://doi.org/10.1109/TCYB.2017.2789158
23. Roy, P.C., Islam, M.M., Deb, K.: Best order sort: a new algorithm to non-dominated sorting for evolutionary multi-objective optimization. In: Proceedings of Genetic and Evolutionary Computation Conference Companion, pp. 1113–1120 (2016)
24. Srinivas, N., Deb, K.: Multiobjective optimization using nondominated sorting in genetic algorithms. Evol. Comput. **2**(3), 221–248 (1994)
25. Zhang, X., Tian, Y., Cheng, R., Jin, Y.: An efficient approach to nondominated sorting for evolutionary multiobjective optimization. IEEE Trans. Evol. Comput. **19**(2), 201–213 (2015)
26. Zitzler, E., Thiele, L.: Multiobjective evolutionary algorithms: a comparative case study and the Strength Pareto approach. IEEE Trans. Evol. Comput. **3**(4), 257–271 (1999)

Toward a New Family of Hybrid Evolutionary Algorithms

Lourdes Uribe[1,2]([✉]), Oliver Schütze[1,2], and Adriana Lara[1,2]

[1] ESFM, Instituto Politécnico Nacional,
Av. Instituto Politécnico Nacional Edif, 9, Unidad Profesional Adolfo López Mateos,
Zacatenco, Mexico
{lourdesur,adriana}@esfm.ipn.mx, schuetze@cs.cinvestav.mx
[2] Department of Applied Mathematics and Systems,
UAM Cuajimalpa. Dr. Rodolfo Quintero Chair, Mexico City, Mexico

Abstract. Multi-objective optimization problems (MOPs) arise in a natural way in diverse knowledge areas. Multi-objective evolutionary algorithms (MOEAs) have been applied successfully to solve this type of optimization problems over the last two decades. However, until now MOEAs need quite a few resources in order to obtain acceptable Pareto set/front approximations. Even more, in certain cases when the search space is highly constrained, MOEAs may have troubles when approximating the solution set. When dealing with constrained MOPs (CMOPs), MOEAs usually apply penalization methods. One possibility to overcome these situations is the hybridization of MOEAs with local search operators. If the local search operator is based on classical mathematical programming, gradient information is used, leading to a relatively high computational cost. In this work, we give an overview of our recently proposed constraint handling methods and their corresponding hybrid algorithms. These methods have specific mechanisms that deal with the constraints in a wiser way without increasing their cost. Both methods do not explicitly compute the gradients but extract this information in the best manner out of the current population of the MOEAs. We conjecture that these techniques will allow for the fast and reliable treatment of CMOPs in the near future. Numerical results indicate that these ideas already yield competitive results in many cases.

Keywords: Multi-objective optimization ·
Evolutionary computation · Mathematical programming ·
Hybrid meta-heuristics

1 Introduction

In many engineering applications one is faced with the problem that several objectives have to be optimized concurrently resulting in a multi-objective optimization problem (MOP). For the treatment of MOPs, traditional optimization

---

The authors acknowledge support for CONACyT project No. 285599 and IPN project SIP20181450.

© Springer Nature Switzerland AG 2019
K. Deb et al. (Eds.): EMO 2019, LNCS 11411, pp. 78–90, 2019.
https://doi.org/10.1007/978-3-030-12598-1_7

techniques establish effective search directions by using differentiability properties of the objective functions. These directions should be able to (at least locally) lead toward better solutions with respect to the objective value. The computation of these proper multi-objective search directions requires, in general, the numerical approximation of the derivatives. These search directions produce a sequence of trial points which eventually converge to one single local optimum point of the problem. It is worth to notice that the solution of a MOP, the Pareto set (PS), typically forms a $(k - 1)$-dimensional object where k is the number of objectives.

Multi-objective evolutionary algorithms (MOEAs) have caught the interest of many researchers (see, e.g. [4,6,30]) over the last two decades. Some reasons for this include that MOEAs are of global nature, and hence, they do not depend on the initial population. Further, due to their set based approach they compute a finite size approximation of the entire PS in one single run of the algorithm. Also, MOEAs do not require gradient information. Recently, hybrid algorithms have gained popularity. They combine gradient-based local search with MOEAs. In particular, for unconstrained MOPs we refer to [13,20,23,24], and to [18, 25] for constrained MOPs. Also, hybrid MOEAs with non-gradient based local search can be found (e.g., [28] and [29]). Designing these hybrid algorithms is not a direct process since MOEAs are stochastic by nature, and making solution improvements in a deterministic way can affect the convergence of the set based algorithm. Besides, the cost of performing gradient-based local search could be excessive, considering the particular improvement. For this reason, designing effective local search procedures is highly relevant. For the case of constrained MOPs, hybrid algorithms mostly rely on their evolutionary part to manage the feasibility of the solutions, and do not involve constraint function information during the evolutionary process.

In [20], an analysis about the behavior of multi-objective stochastic local search (MOSLS) was presented. This analysis showed that a pressure both toward and along the Pareto front (PF) is already inherent for unconstrained problems. **For the constrained case however, this behavior is not preserved**. Although, based on the Karush-Kuhn Tucker (KKT) equations for optimality, one can identify subspaces that allow a movement along the Pareto front for points that are near to the solution set. Since gradients are required to obtain these subspaces, an increment in the computational cost is expected. In this work, we present an overview of our recently proposed constraint handling methods and their corresponding hybrid algorithms. These methods have specific mechanisms, that take advantage of these subspaces, i.e., they move along the PF. In addition, these methods extract information in a best manner out of the current population of the MOEAs. In this way, we are able to move along the Pareto front and maintain a low computational cost.

We first present the Subspace Mutation Operator (SPM) as an alternative variation operator for MOEAs. Classically, mutation has been guided for "small-moves heuristics" [22] or differences on the objective values [15]. For the constrained case, the constraints management is left to the selection process in the

MOEA. On the contrary, SPM takes advantage of the studied neighborhood sampling to perform promising mutations when dealing with CMOPs. Second, we extend the formula introduced in [16] for computing a descent direction to the case of optimizing two objective functions under constraints. We also propose an efficient gradient-free computation of this descent direction for CMOPs, providing support with theoretical results. Both techniques are coupled with a state-of-the-art MOEAs to empirically show the practical potential of the presented theory. We conjecture that these new insights will help with the fast and reliable treatment of CMOPs. Finally, the mention to a real world application, in [27], is included. For the solution of this problem, a hybrid metaheuristic for constrained optimization was developed.

2   Background

In the following we consider continuous MOPs that can be expressed as

$$\min_{x \in Q} F(x),$$

where $x \in Q \subset \mathbb{R}^n$ contains the decision variables and $F : Q \to \mathbb{R}^k$ is the vector of objective functions $F(x) = [f_1(x), \ldots, f_k(x)]^T$. Each objective $f_i : Q \to \mathbb{R}$ is assumed for simplicity to be continuously differentiable. We stress, however, that the sampling algorithm used in this work does not use any explicit gradient information. Traditional optimization techniques use differentiability properties of the functions. The idea of using the gradient-based information to find descent directions, has been extensively exploited for single objective optimization. For the case of MOPs, moving an individual solution toward a particular improvement direction is also wanted. Multi-objective descent directions (MODDs) should be able to (at least locally) lead toward better solutions regarding all the functions simultaneously. So far, some proposals to compute MODDs using first or second order information are available in [2,3,9,11]. Another point-wise iterative search procedure (called Directed Search Method) [20] has the feature of steering the search in any direction given in objective function space.

Gradient Subspace Approximation (GSA) [21] is a method that seeks to compute the most greedy descent direction of an objective function f, by exploiting the neighborhood information available from the population of a MOEA. Consider a point x_0 chosen to start local search, as well as a point x_i from a neighborhood of x_0. Notice that both points come from the population of the evolutionary algorithm; therefore their objective value is already known. Then, with the available information it is possible to approximate (without any additional cost regarding function evaluations) the directional derivative in the direction

$$\nu_i = \frac{x_i - x_0}{\|x_i - x_0\|}. \tag{1}$$

To be more precise, it holds that

$$f'_{\nu_i} = \langle \nabla f(x_0), \nu_i \rangle = \frac{f(x_i) - f(x_0)}{\|x_i - x_0\|} + O(\|x_i - x_0\|), \tag{2}$$

where O denotes the Landau symbol.

Assume we are given $\nu_1, \ldots, \nu_r \in \mathbb{R}^n$. The best approximation of the gradient, $\nu \in span\{\nu_1, \ldots, \nu_r\}$, can be obtained by solving the following problem:

$$\min_{\lambda \in \mathbb{R}^k} \sum_{i=1}^r \lambda_i \langle \nabla f(x_0), \nu_i \rangle \tag{3}$$

$$s.t. \ \lambda^T V^T V \lambda - 1 = 0,$$

where $V = (\nu_1, \ldots, \nu_r) \in \mathbb{R}^{n \times r}$. Then by solving (3) we get that $\lambda^* := \frac{\tilde{\lambda}^*}{\|V\tilde{\lambda}^*\|}$, is the unique solution of (3), where $\tilde{\lambda}^* = -(V^T V)^{-1} V^T \nabla f(x_0)$. Therefore, $\nu^* := V\lambda^*$, is the most greedy search direction in $span\{\nu_1, \ldots, \nu_r\}$ [21]. This approach has a gradient free realization. For this, x_0 and $x_1, x_2, \ldots, x_r$ in the neighborhood of x_0 are given and whose objective values $f(x_i)$, $i = 1, 2, \ldots, r$, are known. Define,

$$\nu_i := \frac{x_i - x_0}{\|x_i - x_0\|_2}, \quad d_i := \frac{f(x_i) - f(x_0)}{\|x_i - x_0\|_2}, i \in \{1, \ldots, r\}. \tag{4}$$

We finally obtain

$$\tilde{\nu}^* = -\frac{1}{\|\tilde{V}\tilde{\lambda}^*\|} \tilde{V}(\tilde{V}^T \tilde{V})^{-1} d. \tag{5}$$

where $\tilde{V} := (\nu_1, \ldots, \nu_r)$. Note that the computation of $\tilde{\nu}^*$ is gradient-free. This approach can be extended both to equality and inequality constraints, see [21] for details.

3   Subspace Polynomial Mutation Operator

In this section, we present the Subspace Polynomial Mutation (SPM) operator; that identifies subspaces that allow to perform a movement along the Pareto front for points near to the solution set. We present the case for general inequalities. For this, assume for a moment that we are given the search directions $\nu_1, \ldots, \nu_r \in \mathbb{R}^n$ as well as the directional derivatives

$$\langle \nabla g_{j_i}(x), \nu_s \rangle, \quad i = 1, \ldots, l, \ s = 1, \ldots, r, \tag{6}$$

where $g_{j_1}, \ldots, g_{j_l}$, $1 \leq l$, are the active inequality constraints at x and further, that we are interested in directions ν within the subspace $span\{\nu_1, \ldots, \nu_r\}$ such that

$$0 = \langle \nabla g_{j_i}(x), \nu \rangle = \langle \nabla g_{j_i}(x), \sum_{s=1}^r \xi_s \nu_s \rangle = \sum_{s=1}^r \xi_s \langle \nabla g_{j_i}(x), \nu_s \rangle, \quad i = 1, \ldots, l, \tag{7}$$

where $\xi \in \mathbb{R}^{r-l}$ at random. Define V as above and then,

$$GV = (\langle \nabla g_{j_i}(x), \nu_s \rangle)_{\substack{i=1,\ldots,l \\ s=1,\ldots,r}} \in \mathbb{R}^{l \times r} \tag{8}$$

is composed of the directional derivatives from (6). Here we assume that $r > l$. Note that (7) is equivalent to $GV\xi = 0$, i.e., we are interested in the kernel of

GV. To find an orthonormal basis of $ker(GV)$ we can utilize a QR factorization: $(GV)^T = QR = (q_1, \ldots, q_l, q_{l+1}, \ldots, q_r)R$ and define:

$$Q_1 := (q_{l+1}, \ldots, q_r) \in \mathbb{R}^{r \times (r-l)}. \tag{9}$$

The columns of Q_1 build the desired orthonormal basis of the kernel of GV. This is used for the construction of the suitable subspace that SPM needs, see Algorithm 1.

Algorithm 1. $y = SPM(x)$ - Subspace Polynomial Mutation

Require: x : solution for mutation.
Ensure: y : mutated solution.
1: $y := x$;
2: Compute Q_1 as in Eq. (9).
3: Compute $S_x := V * Q_1 \in \mathbb{R}^{n \times (r-l)}$
4: **for** $i = 1, \cdots, (r-l)$ **do**
5:     Compute step size t_i.
6:     $y := y + t_i s_i$;
7: **end for**

Figures 1 and 2 show the different behaviors of uniform sampling and SPM sampling in decision and objective space for CMOPS. In Fig. 2, observe how SPM is able to perform movements along the feasible subspace.

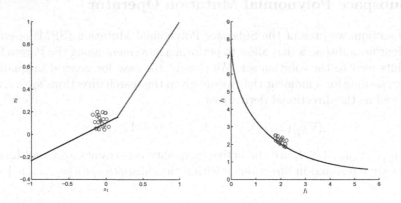

(a) Uniform sampling in decision space   (b) Uniform sampling in objective space

Fig. 1. Example of uniform sampling in decision and objective space for a CMOP.

Given a particular individual of the population, SPM stochastically generates a new individual considering a suitable subspace which promotes feasibility. Therefore, the survival rate of the mutated individual increases, i.e., represents a successful mutation. Also, the exploitation of this suitable subspace is very

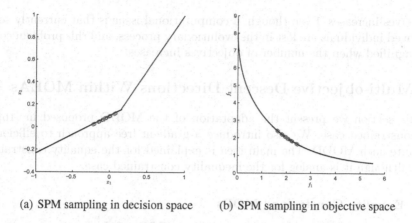

(a) SPM sampling in decision space   (b) SPM sampling in objective space

Fig. 2. Example of SPM sampling in decision and objective space for a CMOP.

convenient due to the fact that the mutated candidate probably lies along the Pareto set of the constrained MOP. SPM adapts polynomial mutation (PM) to effectively deal with constrained MOPs. SPM performs movements over the basic vectors of a suitable subspace for mutation instead if moving on each canonical coordinate variable. The main idea is that given a parent solution x, we compute the mutation subspace. Then, a perturbation is performed along each basic direction of the mutation subspace. SPM is applied just when it is detected that can be useful. Otherwise, polynomial mutation is applied. It is worth to notice that SPM is a free-cost operator (in terms of function evaluations) but its application is controlled by the conditions above, in order to preserve diversity in the evolutionary process.

Numerical Results. In order to test SPM and to investigate its advantages over polynomial mutation when dealing with constrained MOPs, we replaced the mutation module of the algorithms NSGA-II [8] and GDE3 [12]. The performance of the algorithms (i) NSGA-II + SPM and (ii) GDE3 + SPM are compared against their corresponding base MOEAs, NSGA-II and GDE, that uses traditional polynomial mutation. It is worth to notice that the intention of this experiment is to show that the proposed neighborhood sampling represents wide open possibilities for the design of specialized operators for MOEAs and multi-objective hybrid algorithms. The comparison of the two mutation operators, for each test problem (CTP suite), is presented in [14]. Although the results of the SPM are already very promising, there are some aspects that have to be taken care of in the future. For instance, the extension for handling equality constraints. Also, other hybrid aspects inherent to the inclusion of neighborhood sampled operators into population strategies should be investigated, and are subject to ongoing work. For example, our approach is (theoretically) not limited to a few objectives. Since the computational cost is low due to the information extracted from the population, the efficiency is not affected when the number of objectives increases. Even though, a computational issue is that currently some improved individuals are lost in the evolutionary process, and this problem could be magnified when the number of objectives increases.

4 Multi-objective Descent Directions Within MOEAs

In this section we present the adaptation of the MODD proposed in [16] to the constrained case. We also introduce a gradient-free approach to efficiently compute such MODD. The main idea is explained for the equality constrained case, although it is analog for the inequality constrained case.

4.1 Equality Constrained MOPs

Consider a bi-objective optimization problem (BOP) of the form

$$\min_{x \in \mathbb{R}^n} F(x) := [f_1(x), f_2(x)]^T, \tag{10}$$

$$\text{s.t} \qquad h_j(x) = 0, \qquad j \in \{1, \ldots, m\},$$

where $F : \mathbb{R}^n \to \mathbb{R}^2$ is the objective function. Consider the scenario that x is a feasible initial point for MOP (10); i.e., that $h_j(x) = 0$ for $j \in \{1, \ldots, m\}$. In this case $H = \left(\nabla h_1(x)^T, \ldots, \nabla h_m(x)^T\right)^T \in \mathbb{R}^{m \times n}$, is the matrix formed by the gradients of the equality constraints at x. Assuming that $rank(H) = m$, we decompose the matrix as $H^T = QR = (q_1, \ldots, q_m, q_{m+1}, \ldots, q_n) R$, where $Q \in \mathbb{R}^{n \times n}$ is orthogonal, and $R \in \mathbb{R}^{n \times m}$ is right upper triangular. It is worth to notice that the last $n - m$ column vectors of Q form an orthonormal basis of the tangent space of the feasible set $h^{-1}(0)$ at x, where $\nabla f_i(x) \neq 0$ and $h(x) = 0$. We denote this submatrix by $\tilde{Q} := (q_{m+1}, \ldots, q_n)$.

Then, we consider ν_L which is the selected MODD. Then, for constrained BOP.

$$\nu_p = \tilde{Q}\tilde{Q}^T \nu_L \tag{11}$$

is the orthogonal projection of ν_L onto the set of feasible directions. The following proposition establishes criteria for the practical use of this proposal.

Proposition 1. [26] *For a BOP of the form* (10), *with* $k = 2$, *suppose* $\nabla f_i(x) \neq 0$ *for* $i \in \{1, 2\}$. *Assume* ν_p, *given by Eq.* (11), *such that* $\langle \nu_p, h_j(x) \rangle \neq 0$ *for* $j \in \{1, \ldots, m\}$ *and let* $x \in \mathbb{R}^n$ *such that* $h_j(x) = 0$ *for* $j \in \{1, \ldots, m\}$. *Then the following holds:*

(a) If $\nabla f_1(x)^T \tilde{Q}\tilde{Q}^T \nabla f_2(x) > 0$, *then* ν_p *is a MODD of* F *at* x.
(b) If $\nabla f_1(x)^T \tilde{Q}\tilde{Q}^T \nabla f_2(x) = 0$ *and* $\tilde{Q}^T \nabla f_i(x) \neq 0$ *for an index* $i \in \{1, 2\}$, *then* ν_p *is a MODD of* F *at* x.
(c) If $\nabla f_1(x)^T \tilde{Q}\tilde{Q}^T \nabla f_2(x) < 0$, *then* ν_p *is no descent direction of* F *at* x.

In particular, ν_p is a descent direction of F at x if $\nabla f_1(x)$ and $\tilde{Q}\tilde{Q}^T \nabla f_2(x)$ point to the same direction. In case $\nabla f_1(x)$ and $\tilde{Q}\tilde{Q}^T \nabla f_2(x)$ point to the different directions, ν_p is not a descent direction. This can be extended to the inequality constrained case.

4.2   Gradient-Free Descent Direction

In a next step we utilize GSA in order to obtain the MODD from Proposition 1 gradient free within the use of MOEAs. Assuming that x_0 is a feasible solution, first we apply GSA gradient-free realization similar to the unconstrained case to approximate ν_L. Then, we proceed analogously to the equality constrained case, in order to compute $\tilde{\nu}_p$, we apply a QR decomposition to $\tilde{H}^T := \tilde{V}(\tilde{V}^T\tilde{V})^{-1}M$, where $M \in \mathbb{R}^{m \times r}$ is approximated via GSA. Define $O := (\tilde{q}_{m+1}, \ldots, \tilde{q}_n)$, where $\tilde{q}_i$ from $i \in \{m+1, \ldots, n\}$ are the last $n - m$ column vectors of the orthogonal matrix Q obtain by the QR-decomposition of $\tilde{H}^T$. Then

$$\tilde{\nu}_p := OO^T\tilde{\nu}_L, \tag{12}$$

is the orthogonal projection of $\tilde{\nu}_p$ onto the set of feasible directions. Finally, assuming the notation for $\tilde{V}$ and d_i^j as in Eq. (4), the following proposition states the criteria for the application of our gradient-free proposal:

Proposition 2. [26] *For a MOP of the form* (10), *suppose* $\nabla f_i(x) \neq 0$ *for* $i \in \{1, 2\}$. *Assume* $\tilde{\nu}_p$, *given by Eq.* (12) *such that* $\langle \tilde{\nu}_p, h_j(x) \rangle \neq 0$ *for* $j \in \{1, \ldots, m\}$ *and let* $x \in \mathbb{R}^n$ *such that* $h_j(x) = 0$ *for* $j \in \{1, \ldots, m\}$. *Then the following statements hold:*

(a) *If* $d^{1T}(\tilde{V}^T\tilde{V})^{-1}\tilde{V}^TOO^T\tilde{V}(\tilde{V}^T\tilde{V})^{-1}d^2 > 0$, *then* $\tilde{\nu}_p$ *is a MODD of F at x.*
(b) *If* $d^{1T}(\tilde{V}^T\tilde{V})^{-1}\tilde{V}^TOO^T\tilde{V}(\tilde{V}^T\tilde{V})^{-1}d^2 = 0$ *and* $d^{iT}(\tilde{V}^T\tilde{V})^{-1}\tilde{V}^TOO^T\tilde{V}(\tilde{V}^T\tilde{V})^{-1}d^i \neq 0$ *for an index* $i \in \{1, 2\}$, *then* $\tilde{\nu}_p$ *is a MODD of F at x.*
(c) *If* $d^{1T}(\tilde{V}^T\tilde{V})^{-1}\tilde{V}^TOO^T\tilde{V}(\tilde{V}^T\tilde{V})^{-1}d^2 < 0$, *then* $\tilde{\nu}_p$ *is not a MODD F at x.*

Note that the above propositions allow us to know whether the computed direction is a descent direction without any additional cost. Thus, if it is a descent direction we compute the new iterative point x_i as follows: $x_i := x_0 + t\tilde{\nu}_p$, where t is a suitable step length. For the interleaving of this proposal into a MOEA, we have decided to couple the proposed gradient-free MODD with NSGA-III [7]. The reasons for this choice are (i) that the *niching process* imposed to the population of NSGA-III induces the neighborhood selection needed for the local search process, and (ii) that the ideas presented here also apply to MOPs with more than two objectives when considering other MODDs like those referred in Sect. 2, instead of the one in [16]. It is worth to notice that the niching procedure is used to preserve solutions and also to define a neighborhood around the given reference points. Then these r neighbor solutions allow the computation of the proposed gradient-free MODD. Doing this, we guarantee keeping the best solutions at every generation; also we anticipated an accelerated convergence toward the solution by refining the suitable individuals.

Numerical Results. In the following, we compare the performance of the hybrid algorithm against the base MOEA to assess the benefits of our proposal. We tested both algorithms on equality constrained test BOPs defined in [17]; also, on the well known CTP test problems [6]. For all experiments we have executed 30 independent runs using 50,000 and 15,000 function calls respectively for each benchmark. For equality test functions (2 to 4) we use one hundred decision variables. The comparison of the two algorithms, for each test problem, is presented in Table 1. The performance indicators Δ_2 [19] and the hypervolume HV are used to measure algorithm effectiveness. We apply the Wilcoxon test to validate the results; the obtained p–value for all test problems appears in both tables; we consider $\alpha = 0.05$. Table 1 shows that $NSGA - III/GFDD$ returns better values in 7 out of 10 test problems. Also, almost all the results are statistically significant. Note that the difference between both algorithms is more evident when dealing with inequality constrained test problems; we expected this from the design of the refinement mechanism. Although the results of this new hybrid algorithm are very promising, some aspects are still pending for exploration. For example, we focused on the bi-objective case as the first attempt of this approach, but we claim that the methods developed in this work can lead to proposals that are able to deal with more objective functions. For example, we can use the MODD defined in [9] instead of the one given by [16]. See [26] for more details.

Table 1. Values obtained for Δ_2 and HV performance indicators. These results are averaged over 30 independent runs.

Problem	Δ_2			HV		
	NSGA-III	NSGA-III/GFDD	p-value	NSGA-III	NSGA-III/GFDD	p-value
MOP-EQ1 (std.dev)	1.3993 (0.9608)	**0.112** (0.1021)	7.39E-011	34.0588 (5.5815)	**40.6781** (0.7866)	6.68E-011
MOP-EQ2 (std.dev)	1.1048 (0.1184)	**1.0767** (0.0789)	0.0002	1.0043 (0.2142)	**2.1877** (0.1655)	3.02E-011
MOP-EQ3 (std.dev)	1.0703 (0.0754)	**0.9011** (0.0972)	6.53E-08	1.0073 (0.1413)	**1.7256** (0.3642)	2.03E-09
CTP1 (std.dev)	0.0108 (0.0043)	**0.0095** (0.0007)	3.51E-02	0.4675 (0.0014)	**0.4677** (0.0012)	4.36E-02
CTP2 (std.dev)	**0.0038** (0.0003)	0.0045 (0.0003)	3.50E-09	**0.5137** (0.0004)	0.5129 (0.0003)	1.69E-09
CTP3 (std.dev)	**0.0161** (0.0107)	0.0273 (0.0112)	0.0001	0.6069 (0.0025)	**0.6098** (0.0145)	0.5392
CTP4 (std.dev)	0.1974 (0.1258)	**0.1865** (0.1141)	0.5742	**0.4463** (0.0464)	0.442 (0.0563)	0.9058
CTP5 (std.dev)	0.0113 (0.0044)	**0.0109** (0.003)	0.0224	0.4814 (0.0035)	**0.4836** (0.012)	0.0108
CTP6 (std.dev)	**0.012** (0.0009)	0.022 (0.0007)	3.02E-011	**2.078** (0.0027)	2.0669 (0.0014)	3.02E-011
CTP7 (std.dev)	0.0122 (0.0149)	**0.0098** (0.0151)	0.0150	0.8759 (0.0035)	**0.8763** (0.0042)	0.0351

5   Application: Hybrid Algorithm for Constrained Optimization

In this section we present how to solve a real world application (Garch with trend model) using a hybrid algorithm. This model was proposed in [10]. Here, the authors introduce the Garch with trend model in order to analyze the behavior of several international and Mexican commodities. In particular, this work we extend the Garch model [1] to test for a linear trend in the volatility. The main aim of this work is to efficiently solve scalar optimization problems (SOPs) that are related to Garch with trend models. The Newton method, which is usually taken as solver for such problems, is not reliable. Reasons for this include that: (a) the objective function is highly multi-modal resulting in many local minima; (b) the set of feasible points is in many cases disconnected, in particular near to the global solutions, due to the existence of the inequality constraints; and (c) the performance of the Newton method on constrained problems is rather slow if the initial starting point is not near to the solution. Figure 3 shows a particular time series. The trend is shown by the red line. In addition, a spurious trend is shown in blue which results from a local minimum of the related SOP. As such spurious trends cannot be detected visually, it is hence desired to obtain the global optimum of the given SOP.

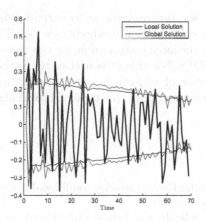

Fig. 3. Optimal Volatility (red line) and spurious Volatility (blue line) of a given time series. The *spurious* volatility attains a local maximum, on the other hand the *optimal* volatility attains the global maximum of the objective function (taken from [27]). (Color figure online)

We propose in [27] the algorithm DE–N which is a hybrid of Differential Evolution and the Newton method. We show the strength of the proposal on a benchmark suite consisting of 44 monthly CPI series of agricultural commodities and two different series related to international prices[1]. Numerical results indicate that DE–N accomplishes its task within a reasonable effort. Further, the algorithm is highly competitive to the state-of-the-art. We test DE–N against two global methods that are used to solve Garch related SOPs which are, however, not designed for the particular problem class (DE, *Fuzzy Genetic Algorithm (F-GA)* and *Monte Carlo Method (MC)*). For all experiments we have executed 30 independent runs using 60,000 and 100,000 function calls. The success rate of all tested algorithms for 60,000 function evaluations is: *DE-N: 86.36%, DE: 63.64%, F-GA: 0% and MC: 0%.* and for 100,000 function evaluations is: *DE-N: 100%, DE: 100%, F-GA: 20.45% and MC: 25%.* This proposal is published in [27].

6   Conclusions and Future Work

We present an overview of our recently proposed constraint handling methods and their corresponding hybrid algorithms. These methods have specific mechanisms that deal with the constraints in a wiser way without increasing their cost. That is, we take advantage of suitable subspaces that allow a movement along the PF. We notice that by utilizing the approximation strategies proposed in GSA, we are capable of building these low-cost operators (SPM and gradient-free MODD).

Another possibility for future work is to extend this approach to hybrid algorithms that employ continuation methods (see [5]). These ideas guide us to a new class of hybrid evolutionary algorithms for the fast and reliable numerical treatment of general MOPs. For this, it is necessary to study more general cases such as a general number of objectives, and the treatment of both types of constraints. Also, we desire to apply this type of hybrid algorithms to real-world problems.

References

1. Bollerslev, T.: Generalized autoregressive conditional heteroskedasticity. J. Econometrics **31**(3), 307–327 (1986)
2. Bosman, P.A.: On gradients and hybrid evolutionary algorithms for real-valued multiobjective optimization. IEEE Trans. Evol. Comput. **16**(1), 51–69 (2012)
3. Brown, M., Smith, R.E.: Effective use of directional information in multi-objective evolutionary computation. In: Cantú-Paz, E., et al. (eds.) GECCO 2003. LNCS, vol. 2723, pp. 778–789. Springer, Heidelberg (2003). https://doi.org/10.1007/3-540-45105-6_92
4. Coello, C.A.C., Van Veldhuizen, D.A., Lamont, G.B.: Evolutionary Algorithms for Solving Multi-Objective Problems, vol. 242. Springer, Heidelberg (2002). https://doi.org/10.1007/978-1-4757-5184-0

---

[1] Time series of domestic price indexes were obtained from the National Institute of Statistics and Geography of Mexico (INEGI, by its Spanish acronym).

5. Cuate, O., et al.: A new hybrid metaheuristic for equality constrained bi-objective optimization problems. In: EMO 2019 (2019)
6. Deb, K.: Multi-Objective Optimization Using Evolutionary Algorithms. Wiley, New York (2001)
7. Deb, K., Jain, H.: An evolutionary many-objective optimization algorithm using reference-point-based nondominated sorting approach, part I: solving problems with box constraints. IEEE Trans. Evol. Comput. **18**(4), 577–601 (2014)
8. Deb, K., Pratap, A., Agarwal, S., Meyarivan, T.: A fast and elitist multiobjective genetic algorithm: NSGA-II. IEEE Trans. Evol. Comput. **6**(2), 182–197 (2002)
9. Fliege, J., Svaiter, B.F.: Steepest descent methods for multicriteria optimization. Math. Methods Oper. Res. **51**(3), 479–494 (2000)
10. Guerrero, S., Hernandez-del-Valle, G., Juárez-Torres, M.: A functional approach to test trending volatility: evidence of trending volatility in the price of Mexican and international agricultural products. Agricultural Economics
11. Harada, K., Sakuma, J., Kobayashi, S.: Local search for multiobjective function optimization: pareto descent method. In: Proceedings of the 8th Annual Conference on Genetic and Evolutionary Computation, pp. 659–666. ACM (2006)
12. Kukkonen, S., Lampinen, J.: GDE3: the third evolution step of generalized differential evolution. In: CEC 2005, vol. 1, pp. 443–450. IEEE (2005)
13. Lara, A., Sanchez, G., Coello, C.A.C., Schütze, O.: HCS: a new local search strategy for memetic multiobjective evolutionary algorithms. IEEE Trans. Evol. Comput. **14**(1), 112–132 (2010)
14. Lara, A., Uribe, L., Alvarado, S., Sosa, V.A., Wang, H., Schütze, O.: On the choice of neighborhood sampling to build effective search operators for constrained MOPs. Memet. Comput. 1–19 (2018)
15. Li, J., Tan, Y.: Orienting mutation based fireworks algorithm. In: CEC 2015, pp. 1265–1271. IEEE (2015)
16. López, A.L., Coello, C.A.C., Schütze, O.: A painless gradient-assisted multi-objective memetic mechanism for solving continuous bi-objective optimization problems. In: CEC 2010, pp. 1–8. IEEE (2010)
17. Martín, A., Schütze, O.: Pareto tracer: a predictor-corrector method for multi-objective optimization problems. Eng. Optim. **50**(3), 516–536 (2018)
18. Saha, A., Ray, T.: Equality constrained multi-objective optimization, pp. 1–7, June 2012
19. Schütze, O., Esquivel, X., Lara, A., Coello Coello, C.A.: Using the averaged Hausdorff distance as a performance measure in evolutionary multiobjective optimization. IEEE Trans. Evol. Comput. **16**(4), 504–522 (2012)
20. Schütze, O., Martín, A., Lara, A., Alvarado, S., Salinas, E., Coello Coello, C.A.: The directed search method for multi-objective memetic algorithms. Comput. Optim. Appl. 1–28 (2015)
21. Schütze, O., Alvarado, S., Segura, C., Landa, R.: Gradient subspace approximation: a direct search method for memetic computing. Soft Comput. **21**(21), 6331–6350 (2017)
22. Shalamov, V., Filchenkov, A., Chivilikhin, D.: Small-moves based mutation for pick-up and delivery problem. In: Proceedings of the 2016 on Genetic and Evolutionary Computation Conference Companion, pp. 1027–1030. ACM (2016)
23. Shukla, P.K.: On gradient based local search methods in unconstrained evolutionary multi-objective optimization. In: Obayashi, S., Deb, K., Poloni, C., Hiroyasu, T., Murata, T. (eds.) EMO 2007. LNCS, vol. 4403, pp. 96–110. Springer, Heidelberg (2007). https://doi.org/10.1007/978-3-540-70928-2_11

24. Sun, J.Q., Xiong, F.R., Schütze, O., Hernández, C.: Cell Mapping Methods—Algorithmic Approaches and Applications. Springer, Singapore (2018). https://doi.org/10.1007/978-981-13-0457-6
25. Takahama, T., Sakai, S.: Constrained optimization by the ε constrained differential evolution with an archive and gradient-based mutation. In: CEC 2010, pp. 1–9. IEEE (2010)
26. Uribe, L., Lara, A., Schütze, O.: On the efficient computation and use of multiobjective descent directions within MOEAs. Technical report (2018)
27. Uribe, L., Perea, B., Hernández-del Valle, G., Schütze, O.: A hybrid metaheuristic for the efficient solution of garch with trend models. Comput. Econ. **52**(1), 145–166 (2018)
28. Zapotecas-Martínez, S., Coello Coello, C.A.: A hybridization of MOEA/D with the nonlinear simplex search algorithm. In: 2013 IEEE Symposium on Computational Intelligence in Multi-Criteria Decision-Making, pp. 48–55. IEEE (2013)
29. Zapotecas-Martínez, S., Coello Coello, C.A.: MONSS: a multi-objective nonlinear simplex search approach. Eng. Optim. **48**(1), 16–38 (2016)
30. Zhang, Q., Li, H.: MOEA/D: a multiobjective evolutionary algorithm based on decomposition. IEEE Trans. Evol. Comput. **11**(6), 712–731 (2007)

Adjustment of Weight Vectors
of Penalty-Based Boundary Intersection
Method in MOEA/D

Hui Li[1]($\boxtimes$), Jianyong Sun[1], Qingfu Zhang[2], and Yuxiang Shui[1]

[1] Xi'an Jiaotong University, Xi'an, Shaanxi, China
{lihui10,jy.sun}@xjtu.edu.cn
[2] City University of Hong Kong, Kowloon Tong, Hong Kong
qingfu.zhang@cs.cityu.edu.hk

Abstract. Multi-objective Evolutionary Algorithm Based on Decomposition (MOEA/D) is one of the dominant algorithmic frameworks for multi-objective optimization in the area of evolutionary computation. The performance of multi-objective algorithms based on MOEA/D framework highly depends on how a diverse set of single objective subproblems are generated. Among all decomposition methods, the Penalty-based Boundary Intersection (PBI) method has received particular research interest in MOEA/D due to its ability for controlling the diversity of population for many-objective optimization. However, optimizing multiple PBI subproblems defined via a set of uniformly-distributed weight vectors may not be able to produce a good approximation of Pareto-optimal front when objectives have different scales. To overcome this weakness, we suggest a new strategy for adjusting weight vectors of PBI-based subproblems in this paper. Our experimental results have shown that the performance of MOEA/D-PBI with adjusted weight vectors is competitive to NSGA-III in diversity when dealing with the scaled version of some benchmark multi-objective test problems.

Keywords: MOEA/D · Penalty-based Boundary Intersection (PBI) ·
Objective normalization

1 Introduction

Over the past twenty years, evolutionary algorithms (EAs) have became a class of popular methodologies for solving multi-objective optimization problems (MOPs). This is because the population-based mechanism enables EAs to find multiple Pareto-optimal solutions in a single run. Unlike the fitness assignment in single objective EAs, the fitness values of individuals in multi-objective evolutionary algorithms (MOEAs) are often assigned in terms of Pareto dominance, or decomposition (i.e., scalarization), or performance indicator. The representative MOEAs with above three schemes for fitness assignment are NSGA-II [1], MOEA/D [2], and IBEA [3], respectively. In recent a few years, there has been

© Springer Nature Switzerland AG 2019
K. Deb et al. (Eds.): EMO 2019, LNCS 11411, pp. 91–100, 2019.
https://doi.org/10.1007/978-3-030-12598-1_8

an increasing research interests on MOEA/D in the area of MOEAs due to its ability to deal with various problem difficulties. For example, the combination of MOEA/D with differential evolution (DE) is an efficient optimization strategy when solving MOPs with complicated Pareto sets [4]. Compared with Pareto-based MOEAs, MOEA/D is more suitable for solving many-objective optimization problems since the selection pressure of population can be guaranteed by optimizing single objective subproblems. So far, a large number of MOEAs for many-objective optimization were developed on the basis of MOEA/D framework, which employed the idea of decomposition for fitness assignment or diversity control.

To find a set of nondominated solutions with good spread, diversity maintenance plays a very crucial role in MOEAs. In Pareto-based MOEAs, the diversity of population is often controlled by calculating the density values of all population members. The well-known examples of these density strategies include crowding distance in NSGA-II, nearest neighboring method in SPEA2 [5], grid-based density estimation in PAES [6]. In contrast, MOEA/D maintains the diversity of population by optimizing multiple single objective subproblems with no density estimation. It should be pointed out that the original version of MOEA/D made an assumption on decomposition that optimizing multiple subproblems can produce a set of well-distributed Pareto-optimal solutions if the corresponding weight vectors for subproblems are appropriately chosen. This assumption is reasonable when all objectives have similar scales and the shape of Pareto-optimal front is relatively simple. In fact, some efforts have also been made on improving the performance of MOEA/D in diversity. The most commonly-used strategies on the improvement of diversity in MOEA/D include objective normalization [2] and adjustment of weight vectors [7,8].

When solving the MOPs with continuous objective space, the weighted Tchebycheff method is more preferred than other decomposition methods. This is mainly because: (i) it is easy to understand and implement, and (ii) it is effective for both convex Pareto-optimal front and concave Pareto-optimal front. However, the performance of weighted Tchebycheff method in diversity is not very satisfactory on some benchmark multi-objective test problems, such as DTLZ1 and DTLZ2 [9], when a number of uniformly-distributed weight vectors are used in decomposition. To overcome this weakness, MOEA/D is combined with the PBI method, which calculates the projected distance for convergence and the perpendicular distance for diversity. Very interestingly, the idea of PBI method has been widely adopted to associate population members with search directions in many MOEAs for many-objective optimization. However, the PBI method has very poor performance on the MOPs with different objective scales, which was reported in the paper on NSGA-III [10]. It is well-known that the weighted Tchebycheff method with objective normalization can deal with the MOPs the different objective scales [2]. Similarly, a modified PBI method with objective normalization was studied in [11]. It should be mentioned that this modified PBI method is sensitive to one extra parameter apart from the penalty factor. In this paper, we investigate a new strategy to improve the performance of the

PBI method in diversity without using objective normalization. The basic idea in our proposed strategy is to adjust the weight vector of each PBI subproblem based on the objective ranges of extreme solutions, which are optimal to one of the objective functions. Our experimental results have shown that the proposed PBI method is effective when solving the MOPs with different objective scales.

The rest of this paper is organized as follows. Section 2 introduces the basic concepts in multi-objective optimization and provides a brief survey on the diversity strategies in MOEA/D. In Sect. 3, the PBI method with adjusted weight vectors is presented. In the following section, some experimental results are reported to show the effectiveness of our proposed strategy. The final section concludes this paper.

2 Related Works

In many real-world optimization problems, multiple conflicting objectives are often involved. The mathematical formulation of an MOP can be written as follows:

$$\text{minimize} \quad F(x) = (f_1(x), \dots, f_m(x)) \qquad \text{subject to} \quad x \in \Omega \qquad (1)$$

where x is a decision vector, and Ω is a feasible region. The optimality of (1) can be defined in terms of Pareto dominance [12]. The set of all Pareto solutions is called Pareto set (PS). The set of their objective vectors is called Pareto front (PF).

According to the relationship between decision and search, multi-objective methods can be prior (decision before search), or posterior (decision after search), or interactive (decision during search). Traditional multi-objective optimization methods, such as weighted sum method and weighted Tchebycheff method, often belong to prior methods. In each run of these methods, only one preferred solution is found. In contrast, posterior methods, such as MOEAs, aim at obtaining a set of Pareto solutions with good distribution along Pareto front. To achieve this goal, two major research issues, i.e., fitness assignment and diversity, must be highly addressed. Among all MOEAs, the decomposition-based MOEAs, such as MOEA/D, have received much attention over the past ten years. In MOEA/D, a family of subproblems are defined via scalarization functions or subregions with prior weight vectors, and optimized by evolving a population of individuals. When subproblems are single objective, a set of uniformly-distributed weight vectors are often considered. This naive strategy for the settings of weight vectors used in MOEA/D can work well on some normalized benchmark multi-objective test problems, such as ZDT test problems.

In fact, the performance of MOEA/D in diversity depends on both the selection of decomposition methods and the settings of weight vectors. For example, the weighted Tchebycheff method needs to consider the following optimization problem.

$$\text{minimize} \quad g^{(tch)}(x|\lambda) = \max_{i \in \{1,\dots,m\}} \lambda_i |f_i(x) - f_i^*| \qquad (2)$$

where

- $\lambda = (\lambda_1, \ldots, \lambda_m)$ is a normalized weight vector with $\lambda_i \geq 0, i = 1, \ldots, m$ and $\sum_{i=1}^{m} \lambda_i = 1$;
- $(f_1^*, \ldots, f_m^*)$ is an ideal point satisfying $f_i^* = \min_{x \in \Omega} f_i(x), i = 1, \ldots, m$.

The performance of Tchebycheff approach in diversity is highly related to the scales of objectives. When all objectives have similar scales, a set of uniformly distributed weight vectors are often generated by the simplex-lattice method, in which each component of normalized weight vectors is taken the value from the following set:

$$\left\{ 0, \frac{1}{H}, \frac{2}{H}, \ldots, \frac{H-1}{H}, 1 \right\} \tag{3}$$

where H is a positive integer number. The total number of normalized weight vectors determined in the above method is C_{m-1}^{m+H-1}. The performance of MOEA/D with Tchebycheff decomposition on the 3-objective normalized test problems, such as DTLZ1 and DTLZ2, can be further improved by transforming weight vectors in the following way:

$$\bar{\lambda}_i = \frac{\frac{1}{\lambda_i + \delta}}{\frac{1}{\lambda_1 + \delta} + \frac{1}{\lambda_2 + \delta} + \cdots + \frac{1}{\lambda_m + \delta}} \tag{4}$$

where δ is a very small positive number.

When dealing with different scales of objectives, the performance of the weighted Tchebycheff method in MOEA/D can be improved by objective normalization, which transforms the value of each objective function as follows:

$$\bar{f}_i(x) = \frac{f_i(x) - f_i^{min}}{f_i^{max} - f_i^{min}}, i = 1, \ldots, m \tag{5}$$

with

$$f_i^{min} = \min_{x \in PS} f_i(x) \quad \text{and} \quad f_i^{max} = \max_{x \in PS} f_i(x),$$

Consequently, the values of normalized objective function $\bar{f}_i, i = 1, \ldots, m$, belong to [0,1].

The penalty boundary intersection method is the other commonly-used decomposition method for continuous MOPs. It aims at minimizing the combination of two distance functions formulated as follows:

$$g^{(pbi)}(x, \lambda, z) = d_1(F(x), \lambda, z) + \theta \cdot d_2(F(x), \lambda, z) \tag{6}$$

where

- d_1 is the projected distance of $F(x)$ along the reference line L as shown in Fig. 1, and d_2 is the perpendicular distance from $F(x)$ to the reference line L.

$$d_1(F(x), \lambda, z) = \frac{|(F(x) - z)^T \lambda|}{\|\lambda\|} \quad \text{and} \quad d_2(F(x), \lambda, z) = \left\| F(x) - z - d_1 \frac{\lambda}{\|\lambda\|} \right\|$$

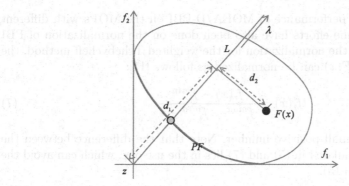

Fig. 1. Penalty boundary intersection method for decomposition in MOEA/D.

- θ is a positive penalty factor. It plays an important role in striking the balance between convergence and diversity in MOEA/D.

The existing experimental results have shown that MOEA/D with PBI performs better than that with Tchebycheff in diversity when handling the MOPs with simplex or spherical PF shapes, such as DTLZ1 and DTLZ2.

3 MOEA/D-PBI with Adjusted Weight Vectors

In this paper, we investigate both objective normalization and adjusted weight vectors in MOEA/D with PBI, which uses the following baseline MOEA/D framework:

- **Step 0: Initialization:**
  - Initialize a set of N weight vectors $\{\lambda^1, \ldots, \lambda^N\}$, and calculate the neighborhoods of subproblems $\{B_1, \ldots, B_N\}$;
  - Initialize a population of N solutions $\{x^1, \ldots, x^N\}$ randomly, and calculate their objective values $\{F(x^1), \ldots, F(x^N)\}$;
  - Compute the reference point $z_j = \min_{i \in \{1, \ldots, N\}} f_j(x^i), j = 1, \ldots, m$.
- **Step 1: Reproduction and Update:**
  - Select two indexes i_1 and i_2 from one neighborhood B_c with $c \in \{1, \ldots, N\}$ randomly;
  - Generate an offspring solution y by recombining x^{i_1} and x^{i_2} via simulated binary crossover and disturb it via polynomial mutation;
  - Update the current solutions of subproblems in B_c with y in terms of PBI-based criteria.
- **Step 2: Stopping Criteria:**
  - If the total number of function evaluations reach the computational budget, then output the population and stop; otherwise, go to **Step 1** for further iterations.

To improve the performance of MOEA/D-PBI on the MOPs with different objective scales, some efforts have also been done on the normalization of PBI method. Similar to the normalization for the weighted Tchebycheff method, the objective function $F(x)$ can be normalized as follows [11]:

$$\bar{f}_i(x) = \frac{f_i(x) - f_i^{min}}{f_i^{max} - f_i^{min} + \epsilon} \qquad (7)$$

where ϵ is a very small positive number. Note that the difference between the normalizations formulated in (5) and (7) lies in the use of ϵ, which can avoid the zero denominator in (5).

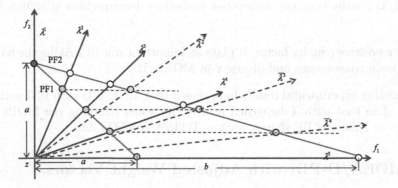

Fig. 2. The distribution of intersection points between five weight vectors with even spread and two PFs (i.e., PF1 with same scales and PF2 with different scales)

In this paper, we propose a new strategy to improve the performance of PBI on the MOPs with different objective scales without objective normalization. The main idea of our method is illustrated in Fig. 2. When dealing with the linear PF with same scales in all objectives, such as PF1 in Fig. 2, PBI with a set of uniformly-distributed weight vectors can obtain a set of Pareto solutions with good spread. However, this is not the case for the PF2 with different scales in two objectives. Assume that five points obtained by PBI with five weight vectors $\lambda^1, \ldots, \lambda^5$ are equally spaced in PF1, the corresponding five equally-spaced solutions in PF2 can be obtained by optimizing the PBI subproblems with five new weight vectors $\bar{\lambda}^1, \ldots, \bar{\lambda}^5$ satisfying:

$$\begin{bmatrix} \bar{\lambda}_1^i \\ \bar{\lambda}_2^i \end{bmatrix} = \begin{bmatrix} b \times \lambda_1^i \\ a \times \lambda_2^i \end{bmatrix} \qquad (8)$$

Based on our discussions above, a set of N weight vectors $\bar{\lambda}^1, \ldots, \bar{\lambda}^N$ used in MOEA/D-PBI need to be transformed in the following way:

$$\bar{\lambda}_k^i = \frac{\lambda_k^i \times (f_k^{max} - f_k^{min})}{\sum_{j=1}^m (\lambda_j^i \times (f_j^{max} - f_j^{min}))}, k = 1, \ldots, m \qquad (9)$$

In this work, the values of $f_i^{min}, i = 1, \ldots, m$, are computed by:

$$f_k^{min} = \min_{x \in \{x^1, \ldots, x^N\}} f_k(x), k = 1, \ldots, m.$$

The values of $f_k^{max}, k = 1, \ldots, m$, are computed by finding intercepts of extreme points as in the objective normalization of NSGA-III.

4 Computational Experiments

In this section, the settings of our experiments are first introduced. Then, the experimental results are reported and discussed.

4.1 Experimental Settings

To verify the effectiveness of our proposed strategy for adjusting weight vectors in MOEA/D-PBI, the rescaled versions of two benchmark multiobjective test problems, i.e., DTLZ1 and DTLZ2, are considered in our experiments. The changes of these MOPs are summarized as follows (Table 1):

Table 1. The changes in four rescaled benchmark multiobjective test problems. The distance functions g are the same as those in the original benchmark MOPs.

Rescaled instance	Formulation	Dimensionality
DTLZ1/R	$f_1(x) = (1 + g(x)) \times x_1 x_2$ $f_2(x) = (1 + g(x)) \times 2 \times x_1(1 - x_2)$ $f_3(x) = (1 + g(x)) \times 10 \times (1 - x_1)$	$n = 10$
DTLZ2/R	$f_1(x) = (1 + g(x)) \times \cos(0.5x_1\pi) \cos(0.5x_2\pi)$ $f_2(x) = (1 + g(x)) \times 2 \times \cos(0.5x_1\pi) \sin(0.5x_2\pi)$ $f_3(x) = (1 + g(x)) \times 10 \times \sin(0.5x_1\pi)$	$n = 10$

In this work, three versions of MOEA/D-PBI and NSGA-III are considered in performance comparison. The configurations of three MOEA/D-PBI variants are as follows:

- MOEA/D-PBI-V1: No transformation on objective functions and no adjustment of weight vectors;
- MOEA/D-PBI-V2: objective normalization in (7);
- MOEA/D-PBI-V3: adjustment of weight vectors in (9).

The population size is set to 300 for DTLZ1/R and DTLZ2/R. The penalty factor θ in MOEA/D-PBI-V1 and MOEA/D-PBI-V2 is set to 5 while that in MOEA/D-PBI-V3 is set to 20. The neighborhood size in all three MOEA/D-PBI variants is set to $0.1 \times N$. The total number of generations is set to 500.

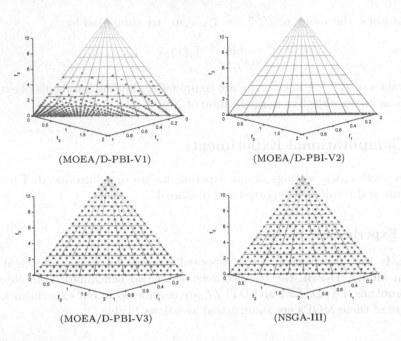

(MOEA/D-PBI-V1)   (MOEA/D-PBI-V2)

(MOEA/D-PBI-V3)   (NSGA-III)

Fig. 3. The final populations found by three MOEA/D-PBI variants and NSGA-III on DTLZ1/R in one run.

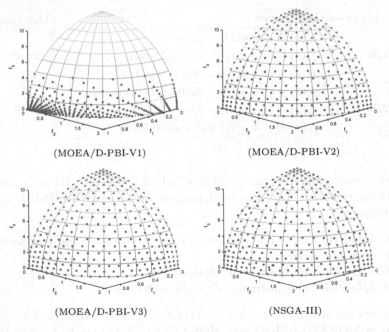

(MOEA/D-PBI-V1)   (MOEA/D-PBI-V2)

(MOEA/D-PBI-V3)   (NSGA-III)

Fig. 4. The final populations found by three MOEA/D-PBI variants and NSGA-III on DTLZ2/R in one run.

4.2   Experimental Results

The final populations found by four algorithms, i.e., MOEA/D-PBI-V1, MOEA/D-PBI-V2, MOEA/D-PBI-V3, and NSGA-III, on DTLZ1/R and DTLZ2/R are plotted in Figs. 3 and 4. From the results reported in these two figures, we can have the following observations:

- On DTLZ1/R, both MOEA/D-PBI-V1, i.e., the original MOEA/D with PBI, and MOEA/D-PBI-V2 with objective normalization obtained part of Pareto solutions in the PF. Neither of these two MOEA/D-PBI variants found the whole PF. However, MOEA/D-PBI-V1 performed better than MOEA/D-PBI-V2 in diversity on this instance. In this case, the use of objective normalization in MOEA/D-PBI didn't play a positive role in enhancing the performance in diversity. The reason might be due to the difficulty in minimizing the multimodal distance function in DTLZ1/R. As a result, the range of function values in one objective could be very tiny. In contrast, both MOEA/D-PBI-V3 and NSGA-III found a very good approximation of the whole PF on this instance.
- On DTLZ2/R, the final populations found by all four algorithms except MOEA/D-PBI-V1 approximate the whole PF very well both in convergence and in diversity. Again, MOEA/D-PBI-V1 with no objective normalization and no adjustment of weight vectors failed to find the whole PF. The good performance of MOEA/D-PBI-V2 on this instance is due to the success in minimizing the unimodal distance function in DTLZ2/R. However, MOEA/D-PBI-V3 is clearly superior to MOEA/D-PBI-V1 on this instance. Both MOEA/D-PBI-V3 and NSGA-III are competitive in approximating the whole PF of DTLZ2/R.

The above experimental results indicate that the performance of MOEA/D-PBI with objective normalization in diversity is less robust than that of MOEA/D-PBI with our proposed adjusted weighting method. It should also be pointed out that a large value of the penalty factor θ must be considered in MOEA/D-PBI with adjusted weight vectors for guaranteeing good diversity of final population.

5   Conclusions

This paper studied a new strategy for adjusting weight vectors in MOEA/D with PBI decomposition. The main idea of our proposed strategy is to modify the component of each weight vector by multiplying a factor, which corresponds to the range of associated objective values of solutions in current population. Our experimental results have shown MOEA/D-PBI with adjusted weight vectors has the ability to deal with the MOPs with different objective scales, and its performance is competitive to that of NSGA-III on the rescaled 3-objective benchmark test problems. In our future work, the application of MOEA/D-PBI with our proposed weighting method for many-objective optimization will be investigated.

Acknowledgment. The authors would like to thank the anonymous reviewers for their insightful comments. This work was supported by National Natural Science Foundation of China (NSFC) grants 61573279, 61175063 and 11131006.

References

1. Deb, K., Pratap, A., Agarwal, S., Meyarivan, T.: A fast and elitist multiobjective genetic algorithm: NSGA-II. IEEE Trans. Evol. Comput. **6**(2), 182–197 (2002)
2. Zhang, Q., Li, H.: MOEA/D: a multiobjective evolutionary algorithm based on decomposition. IEEE Trans. Evol. Comput. **11**(6), 712–731 (2007)
3. Zitzler, E., Künzli, S.: Indicator-based selection in multiobjective search. In: Yao, X., et al. (eds.) PPSN 2004. LNCS, vol. 3242, pp. 832–842. Springer, Heidelberg (2004). https://doi.org/10.1007/978-3-540-30217-9_84
4. Li, H., Zhang, Q.: Multiobjective optimization problems with complicated Pareto sets, MOEA/D and NSGA-II. IEEE Trans. Evol. Comput. **12**(2), 284–302 (2009)
5. Zitzler, E., Laumanns, M., Thiele, L.: SPEA2: improving the strength pareto evolutionary algorithm for multiobjective optimization. In: Proceedings of Evolutionary Methods Design, Optimisation and Control With Applications to Industrial Problems, pp. 95–100 (2001)
6. Knowles, J., Corne, D.: The pareto archived evolution strategy: a new baseline algorithm for pareto multiobjective optimisation. In: Proceedings of the 1999 Congress on Evolutionary Computation, CEC 1999, vol. 1, January 1999
7. Li, H., Landa-Silva, D.: An adaptive evolutionary multi-objective approach based on simulated annealing. Evol. Comput. **19**(4), 561–595 (2011)
8. Qi, Y., et al.: MOEA/D with adaptive weight adjustment. Evol. Comput. **22**(2), 231–264 (2014)
9. Deb, K., Thiele, L., Laumanns, M., Zitzler, E.: Scalable test problems for evolutionary multi-objective optimization. Technical report 112, Computer Engineering and Networks Laboratory, ETH Zurich (2001)
10. Deb, K., Jain, H.: An evolutionary many-objective optimization algorithm using reference-point-based nondominated sorting approach, part I: solving problems with box constraints. IEEE Trans. Evol. Comput. **18**(4), 577–601 (2014)
11. Ishibuchi, H., Doi, K., Nojima, Y.: On the effect of normalization in MOEA/D for multi-objective and many-objective optimization. Complex Intell. Syst. **3**, 279–294 (2017)
12. Miettinen, K.: Nonlinear Multiobjective Optimization. Springer, Heidelberg (1998)

GDE4: The Generalized Differential Evolution with Ordered Mutation

Azam Asilian Bidgoli[1,2]($\boxtimes$) (iD), Sedigheh Mahdavi[2] (iD),
Shahryar Rahnamayan[2] (iD), and Hessein Ebrahimpour-Komleh[1] (iD)

[1] Department of Computer and Electrical Engineering, University of Kashan,
Kashan, Iran
ebrahimpour@kashanu.ac.ir

[2] Nature Inspired Computational Intelligence (NICI) Lab,
Department of Electrical, Computer, and Software Engineering,
University of Ontario Institute of Technology (UOIT), Oshawa, Canada
{azam.asilianbidgoli,sedigheh.mahdavi,shahryar.rahnamayan}@uoit.ca

Abstract. Differential Evolution (DE) is one the most popular evolutionary algorithm (EA) to handle optimization problems with an efficient performance. Due to its success and popularity, it has been utilized by researchers in multi-objective optimization, so there are various multi-objective versions of DE. Similar to other population-based algorithms, DE uses a mutation operator to produce the new individual for the next generation. Although the original version of DE randomly selects three candidate solutions from the population without considering any ordering in its mutation scheme, this paper proposes ordering strategy of individuals which influences the performance of the algorithm. An enhanced version (GDE4) of Generalized Differential Evolution (GDE) with ordered mutation operator is designed. GDE is a multi-objective evolutionary algorithm based on DE. The proposed approach orders candidate individuals using popular ranking measures of multi-objective optimization problems to utilize the ordered solutions in mutation operator. The best one of three randomly selected solutions is considered as the parent, and two others are applied as second and third candidate solutions in DE mutation, respectively. Unlike most of the multi-objective methods which consider multi-objectiveness during the selection process, the proposed method improves the performance using a modification on the genetic operator. The standard benchmark functions and measures are adopted to evaluate the performance of GDE4. The conducted experiments are on 5, 10, and 15 objectives for the utilized benchmark set. The comparison results reveal that GDE4 algorithm outperforms GDE3, the last version of GDE.

Keywords: Evolutionary computation ·
Multi-objective optimization · Generalized differential evolution ·
Ordered mutation

© Springer Nature Switzerland AG 2019
K. Deb et al. (Eds.): EMO 2019, LNCS 11411, pp. 101–113, 2019.
https://doi.org/10.1007/978-3-030-12598-1_9

1 Introduction

Since many real-world problems involve more than one objective, solving multi-objective optimization problems is considered as an important subject in many fields of science and engineering. The main issue that makes such problems harder than single objective problems is that how it is possible to compare solutions with two or more conflicting objectives. Evolutionary computation (EC) as a powerful method has been used to solve multi-objective optimization problems. There are a wide variety of single objective evolutionary algorithms (EA's) which have been adapted for multi-objective schemes [1,2]. Differential evolution (DE) is one of them which its simplicity offers a great characteristic to apply it in single- and multi-objective optimization. Generalized differential evolution (GDE) [3] is a multi-objective version of DE. There are some research works to improve GDE to be more successful in the optimization. The third version (GDE3) [4] was proposed to handle all types of multi-objective optimization problems including non-constrained and constrained ones.

Creation of a new individual in population-based algorithms, is one of the most important steps to make the generation more progressive. So selecting parents can influence producing better population and increasing elitism in the next generations. DE mutation which uses three randomly selected individuals from the population to create a new offspring. For single objective DE, there are some designed schemes of ordered mutation based on objective function value which have shown significant improvement in obtained results [5–7].

GDE3 also uses DE mutation operator, so ordered selection can improve the results of multi-objective optimization. The difficulty of ordered selection in multi-objective optimization case compared to the single objective one is the defining strategy of ranking of three selected individuals. Since there are two or more conflicting objective values, decision making in which candidate solutions are better, is sophisticated. This paper presents a version of GDE3 with ordered mutation (GDE4) for multi-objective optimization problems. Three selected candidate solutions are sorted based on two known measures, non-dominating sorting and crowding distance [8]. The best one is used as a base vector, and two other ranked candidate solutions are considered as the second and third individuals in DE mutation of GDE3. Since optimality doesn't have a straightforward definition, most of the multi-objective algorithms consider multi-objectiveness during their selection process. They concentrate on the proposing a method to rank candidate solutions while the proposed method improves multi-objectiveness in generative operator. Experiments show an enhancement of results in GDE4 compared to GDE3 in standard benchmarks. The organization of the rest of this paper is as follows. Section 2 gives a brief background review of GDE3 algorithms. Section 3 describes the proposed scheme in detail. Section 4 presents a simple algorithm and the experimental results to support discussion on the proposed scheme. Finally, the paper is concluded in Sect. 5.

2   Background Review

Many real-world optimization problems have more than one conflicting objectives to be optimized. The definition of the optimality is not as simple as the single-objective optimization. Therefore it is necessary to make a tradeoff between objective values. There are some well-known concepts to compare two candidate solutions in the multi-objective problem space. Since this paper utilizes non-dominated sorting and crowding distance to order candidate solutions for DE mutation scheme, in this section, we define these measures in detail. A minimization multi-objective optimization problem is defined as follows:

$$Minimize\ F(x) = [f_1(x), f_2(x), ..., f_M(x)] \quad L_i \le x_i \le U_i, i = 1, 2, ..., d \quad (1)$$

where M is the number of objectives, d is the number of variables (dimension) of solution vector, x_i is in interval $[L_i, U_i]$. f_i represent the objective function which should be minimized.

If $x = (x_1, x_2, ..., x_d)$ and $\acute{x} = (\acute{x}_1, \acute{x}_2, ..., \acute{x}_d)$ are two vectors in search space, x dominates $\acute{x}$ ($x \succ \acute{x}$) if and only if:

$$\forall i \in \{1, 2, ..., d\}, f(x_i) \le f(\acute{x}_i) \land \exists i \in \{1, 2, ..., d\} : f(x_i) < f(\acute{x}_i) \quad (2)$$

It defines optimality for solutions in objective space. Candidate solution x is better than $\acute{x}$ if it is not bigger than $\acute{x}$ in any of objectives and at least it has a smaller value in one of the objectives. All solutions that are not dominated using none of other solutions in the population called non-dominated solutions and they create the Pareto front set.

Non-dominated sorting is an algorithm to rank obtained solutions to different levels in the processing of multi-objective optimization. All non-dominated solutions are in the first rank and then the second rank is made of solutions which are non-dominated by removing the first rank from the population. This process is repeated until all solutions are ranked using this concept.

Crowding distance is another measure which usually completes comparison of solutions along with non-dominating sorting. It is a measure to compute the diversity of obtained solutions by calculating the distance between adjacent solutions. In the beginning, the set of solutions in the same rank are sorted according to each objective function value in ascending order. To get crowding distance, the difference between neighbors objective values of each solution is computed. This computation is done for all objectives, then the sum of individual distance values corresponding to each objective is considered as overall crowding distance. The bigger value of crowding distance for a vector in population shows less diversity around that vector.

2.1   Generalized Differential Evolution

The DE is an evolutionary algorithm originally for solving continuous optimization problems which improves initial population using the crossover and mutation operations. Creation of new generation is done by a mutation and a crossover operator. The mutation operator for a gene, j, is defined as follows:

$$v_{j,i} = x_{j,i_1} + F \cdot (x_{j,i_2} - x_{j,i_3}) \tag{3}$$

Applying this operator generates a new D dimensional vector, v_i, using three randomly selected individuals, x_{j,i_1}, x_{j,i_2}, and x_{j,i_2} from the current population. Parameter F, mutation factor, scales difference between two vectors. The crossover operator changes some or all of the genes of parent solution based on Crossover Rate (CR). Similar to other population-based algorithms, the single objective version of DE starts with a uniform randomly generated population. Next generation is created using mentioned mutation and crossover operations; then best individual (between parent and new individual) is selected based on their objective values; which is called a greedy selection. It iterates until meeting stopping criterion such as a predefined number of generations.

There are also several variants of DE algorithms for multi-objective optimization. The first version of Generalized Differential Evolution (GDE) [3] changed the DE selection mechanism for producing the next generation. The idea in the selection was based on constraint-domination. The new vector is selected if it dominates the old vector. GDE2 [9], the next version of multi-objective DE algorithm, added the crowding distance measure to its selection scheme. If both vectors are non-dominating each other, the vector with a higher crowding distance will be selected.

The third version of GDE (GDE3) extends DE algorithm for multi-objective optimization problems with M objectives and K constraints. DE operators are applied using three randomly selected vectors to produce an offspring per parent in each generation. The selection strategy is similar to the GDE2 except in two parts: 1. Applying constraints during selection process. 2. The non-dominating case of two candidate solutions. Selection rules in GDE3 are as follows: when old and new vectors are infeasible solutions, each solution that dominates other in constraint violation space is selected. In the case that one of them is feasible vector, feasible vector is selected. If both vectors are feasible, then one is selected for the next generation that dominates other. In non-dominating case, both vectors are selected. Therefore, the size of the population generated may be larger than the population of the previous generation. If this is the case, it is then decreased back to the original size. Selection strategy for this step is similar to NSGA-II algorithm [10]; it sorts individuals in the population, based on the non-dominated sorting algorithm and crowding distance measure. Similar to other population-based multi-objective algorithms, the selected individuals go to the next generation to continue optimization processing. The common point about all of these versions is the utilizing randomly selected individuals to produce a new vector using the main mutation operator of DE which will be modified in our proposed algorithm in this paper. So, even the mutation scheme would be tailored to support multi-objective optimization strategy.

2.2 Existing Single Objective Differential Evolution with Ordered Mutation

In some versions of DE algorithm, the ordering of the candidate solutions is utilized for the mutation operator to enhance the performance of DE algorithm for solving the single objective optimization problems. A new scheme of mutation operator, DE/2-Opt, was defined in [5] which sorts two first candidate solutions in the mutation operator according (for minimization case) to their objective function value in ascending order to place as x_{i_1} and x_{i_2} in the mutation operator as:

'DE/2-Opt/1':

$$v_i = \begin{cases} x_{i_1} + F \cdot (x_{i_2} - x_{i_3}) & \text{if } f(x_{i_1}) <= f(x_{i_2}) \\ x_{i_2} + F \cdot (x_{i_1} - x_{i_3}) & \text{if } f(x_{i_2}) < f(x_{i_1}) \end{cases} \tag{4}$$

'DE/2-Opt/2':

$$v_i = \begin{cases} x_{i_1} + F \cdot (x_{i_2} - x_{i_3} + x_{i_4} - x_{i_5}) & \text{if } f(x_{i_1}) <= f(x_{i_2}) \\ x_{i_2} + F \cdot (x_{i_1} - x_{i_3} + x_{i_4} - x_{i_5}) & \text{if } f(x_{i_2}) < f(x_{i_1}) \end{cases} \tag{5}$$

In [6], the winner mutation (DE/win) was proposed which uses the best candidate of three selected random candidate solutions for the base vector as follows:

'DE/win/1':

$$v_i = \begin{cases} x_{i_1} + F \cdot (x_{i_2} - x_{i_3}) & \text{if } f(x_{i_1}) <= f(x_{i_2}), f(x_{i_3}) \\ x_{i_2} + F \cdot (x_{i_1} - x_{i_3}) & \text{if } f(x_{i_2}) < f(x_{i_1}), f(x_{i_3}) \\ x_{i_3} + F \cdot (x_{i_2} - x_{i_1}) & \text{if } f(x_{i_3}) < f(x_{i_2}), f(x_{i_1}) \end{cases} \tag{6}$$

In [7], a modified DE algorithm with the order mutation scheme was proposed which three selected random solutions are sorted in ascending order according to their fitness values for placing as vectors (x_{i_1}, x_{i_2}, and x_{i_3}) in the mutation operator.

'DE/order/1':

$$v_i = x_{i_1} + F \cdot (x_{i_2} - x_{i_3}) \, s.t. \ f(x_{i_1}) <= f(x_{i_2}) <= f(x_{i_3}) \tag{7}$$

Where $f(x)$ indicates the objective function. This method outperforms previous mentioned DE schemes.

3 Proposed Algorithm: The Generalized Differential Evolution with the Ordered Mutation (GDE4)

The proposed enhanced version of GDE3 method has the same components of GDE3 method except for the mutation operator in the DE algorithm. In this paper, a new mutation scheme is proposed according to a defined order for the candidate solutions involved in the mutation of the DE algorithm. GDE3 uses the

DE/rand/1/bin method to solve problems with M objectives and K constraints. The basic mutation, in the classical DE (DE/rand/1/bin) generates the mutant vector as a linear combination of three selected individual candidate solutions from the current population as follows:

$$v_i = x_{i_1} + F \cdot (x_{i_2} - x_{i_3}) \tag{8}$$

Where i_1, i_2, i_3 are different random integer numbers within $[1, NP]$ and NP is the population size. In [7], an ordered mutation scheme was proposed to improve the performance of DE algorithm and we change this mutation scheme by defining a new order of the randomly selected candidate solutions for the problems with M objectives. In the GDE4, we propose an order mutation scheme which uses non-dominance and crowding distance measures to sort three different random candidate solutions to set as vectors in the mutation scheme. The sorted candidate solutions can be called as the best (x_b), the second best (x_{sb}), and the worst candidate (x_w) solutions.

In the following, we explain how the three randomly selected candidate solutions are sorted. First, all candidate solutions are sorted by non-dominated sorting method [10] and they are associated with their corresponding non-dominated ranks $(Rank_d)$ obtained from non-dominated sorting. Random candidate solutions can be faced with four possible cases based on their non-dominance ranks:

1. In the first case, all three candidate solutions are in different Pareto fronts; therefore, they are set to x_b, x_{sb}, and x_w to their non-dominated ranks. The ordered mutation scheme $(DE/order/1)$ is defined as follows: 'DE/order/1':

$$v_i = x_{i_b} + F \cdot (x_{i_{sb}} - x_{i_w})$$
$$s.t. \ Rank_d(x_{i_b}) < Rank_d(x_{i_{sb}}) < Rank_d(x_{i_w})$$

2. In this case, two candidate solutions are in the same Pareto front, so we compute crowding distance (CD) measure to sort these solutions. The ordered mutation scheme $(DE/order/1)$ is defined as two possible cases:
   (a) 'DE/order/1':

$$v_i = x_{i_b} + F \cdot (x_{i_{sb}} - x_{i_w})$$
$$s.t. \ Rank_d(x_{i_b}) = Rank_d(x_{i_{sb}}) < Rank_d(x_{i_w})$$
$$CD(x_{i_b}) > CD(x_{i_{sb}})$$

   (b) 'DE/order/1':

$$v_i = x_{i_b} + F \cdot (x_{i_{sb}} - x_{i_w})$$
$$s.t. \ Rank_d(x_{i_b}) < Rank_d(x_{i_w}) = Rank_d(x_{i_{sb}})$$
$$and \ CD(x_{i_{sb}}) > CD(x_{i_w})$$

3. If all three random candidate solutions are in the same Pareto front, they are sorted based on their crowding distance (CD) to place in the mutation scheme. The ordered mutation scheme $(DE/order/1)$ is defined as follows: 'DE/order/1':

$$v_i = x_{i_b} + F \cdot (x_{i_{sb}} - x_{i_w})$$
$$s.t. \ CD(x_{i_b}) > CD(x_{i_{sb}}) > CD(x_{i_w})$$

The proposed method uses the order mutation scheme for DE algorithm in GDE3, and other components remain untouched. Generalized Differential Evolution with the ordered mutation (GDE4) suggests that placing the best solution of three selected candidate solutions according to two measures, non-dominance and crowding distance, as the base vector causes to generate more promising trial solutions. Also, we use the worst candidate solution of three candidate solutions as the third vector in the mutation which causes the new trial candidate solution to get away from the worst candidate and move toward the second best candidate solution.

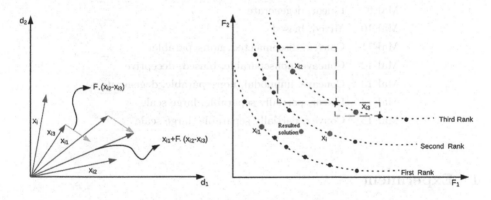

Fig. 1. An example of variable and objective spaces for ordered DE mutation.

Figure 1 presents variable and objective spaces in a case of ordered mutation and clarifies the benefits of this strategy in creating a promising new solution. As it is shown, for a parent solution, x_i, three randomly selected candidate solutions are ordered based on the proposed strategy in GDE4 algorithm. In this case, the first candidate solution, x_{i1} is in the first rank of non-dominated sorting, so it is considered as the base vector (best). x_{i2} and x_{i3} are in the same rank therefore they ordered according to crowding distance. x_{i2} has a bigger crowding distance comparing to x_{i3}, so they are ordered as second (better) and third vector (worst) in mutation operator. Right sub-figure in Fig. 1 shows the operation of mutation on selected vectors. $F \cdot (x_{i2} - x_{i3})$ leads new vector moves toward to better solution and gets away from the worst while F is considered 1. In this example, better solution is one with a bigger crowding distance. Moving toward this solution causes the creation of a vector in a less crowded region to have a well-distributed Pareto front. Then summation operation on x_{i1} and $F.(x_{i2} - x_{i3})$ causes the final resulted vector goes toward the best candidate solution. So it is expected to generate a more promising candidate solution.

Table 1. Main properties of the test functions [11].

Problem	Properties
MaF1	Linear
MaF2	Concave
MaF3	Convex, multimodal
MaF4	Concave, multimodal
MaF5	Convex, biased
MaF6	Concave, degenerate
MaF7	Mixed, disconnected, multimodal
MaF8	Linear, degenerate
MaF9	Linear, degenerate
MaF10	Mixed, biased
MaF11	Convex, disconnected, nonseparable
MaF12	Concave, nonseparable, biased, deceptive
MaF13	Concave, unimodal, nonseparable, degenerate
MaF14	Linear, partially separable, large scale
MaF15	Convex, partially separable, large scale

4 Experiment

GDE4 is evaluated with a set of test problems and compared to GDE3 regarding multi-objective evaluation measures. The same settings are considered for two algorithms. The mutation amplification factor (F) and crossover rate (CR) are set to 0.5 and 1, respectively. For population size and maximum evaluation number, value 100 and $3000 * D$ are considered. To evaluate the performance of the proposed algorithm, we use the inverse generational distance (IGD) metric [12–14], which measures the convergence and the diversity of the obtained Pareto-optimal solutions at the same time. The IGD metric measures the distances between each solution composing the Pareto-optimal front and the obtained solution. The IGD metric is defined as follow:

$$IGD = \frac{\sqrt{\sum_{i=1}^{n} d_i}}{n} \tag{9}$$

Where n is the number of solutions in the Pareto-optimal front, and d_i is the Euclidean distance (measured in the objective space) between each point of the Pareto-optimal front (reference Pareto front) and the nearest member of obtained solution. Also, all algorithms were executed 51 times independently, and the best, the worst, the median, and the average results of each algorithm are reported. Additionally, the Wilcoxon's signed rank statistical test with a confidence interval of 95% is conducted to evaluate the statistical significance of the obtained results. We have utilized GDE3 algorithm in the MATLAB based MOEA platform (PlatEMO) [15] and it was modified by changing its mutation operator to the order mutation as explained for the GDE4.

In the experiments, fifteen test problems are used to evaluate the performance of the proposed algorithm from the MaF test suite which is designed for the assessment of MOEAs in the CEC 2017 competition on evolutionary many-objective optimization [11]. These benchmark functions have many properties to resemble various real-world scenarios such as multi-modal, disconnected, degenerate, and/or nonseparable, and having an irregular Pareto front shape, a complex Pareto set or a large number of decision variables. The main properties of functions are detailed in Table 1. Experiments are performed on 5, 10 and 15 objective functions.

Figure 2 illustrates the distribution of obtained solutions by GDE4 and GDE3 for MFa11 test problem in different number of objectives. The diagrams are resulted based on median value of IGD. As the figure shows both algorithms are able to find distributed solutions with same performance when the number of objectives is 5. However, as the number of objectives of the test problem increases, GDE4 performs significantly better to find well-distributed solutions. The difference between diversity of obtained solutions using GDE3 and GFE4 is more remarkable with 15 objectives.

The results of IGD metric for two comparing methods are summarized in Table 2. Better mean of IGDs are highlighted based on Wilcoxon's signed rank statistical test. It can be seen from the tables, on functions with five objectives, GDE4 can achieve the better results than GDE3 on seven functions while GDE3 is better than GDE4 on five functions, and they are similar results on three functions. On functions with ten objectives, GDE4 can achieve the better results than GDE3 on ten functions while GDE3 can obtain better results than GDE4 on three functions; and they are similar results on two functions. On functions with fifteen objectives, GDE4 outperforms GDE3 on nine functions while GDE3 can obtain better results than GDE4 on five functions; and they are similar results on one functions. Results show that by increasing the number of objectives in the many-objective functions, GDE4 preforms significantly better than GDE3 regarding statistical test. Furthermore, comparing results according to median and best IGD confirms better performance of GDE4. Median IGDs of GDE4 are better in 9, 11 and 9 out of 15 functions for 5, 10, and 15 objective problems respectively comparing to GDE3.

According to best IGDs, GDE4 achieve better results in 7, 9, and 8 out of 15 functions for 5, 10, and 15 objective problems respectively. So the order of

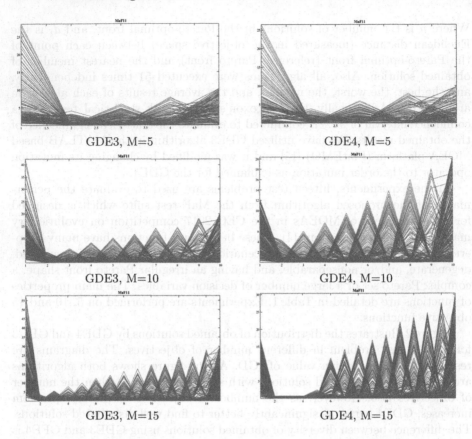

Fig. 2. Comparison of obtained Pareto fronts by GDE3 and GDE4 for MFa11 test problem in different dimensions.

solutions in DE mutation operator improves the search processing in many-objective optimization problems using generating better (non-dominated) solutions. The generated solution is expected to create in place close to the best solution in term of the rank of non-dominated sorting and the less crowded region. As another advantage of the proposed method, it can be clarified that this improvement is achieved without any extra objective function evaluation. The method needs only ordering of three existing solutions, so there isn't overhead computation for applying mutation comparing to previous version.

Table 2. Results of GDE3 and GDE4 algorithms for the functions MaF_1-MaF_{15}. The highlighted entries are significantly better.

Function		#Objectives=5		#Objectives=10		#Objectives=15	
		GDE3	GDE4	GDE3	GDE4	GDE3	GDE4
MaF_1	Mean	0.2052	**0.1696**	0.3529	**0.3015**	0.3718	**0.3220**
	Median	0.2049	0.1700	0.3542	0.3006	0.3676	0.3207
	Worst	0.2244	0.1843	0.3687	0.3096	0.4312	0.3489
	Best	0.1923	0.1570	0.3346	0.2894	0.3510	0.3038
MaF_2	Mean	**0.1502**	0.1414	**0.1691**	0.1717	0.1960	**0.1639**
	Median	0.1448	0.1393	0.1691	0.1718	0.1962	0.1638
	Worst	0.2016	0.2059	0.1783	0.1802	0.2159	0.1698
	Best	0.0983	0.1006	0.1619	0.1630	0.1774	0.1580
MaF_3	Mean	2.9848e+4	**1.1454e+4**	8.9508e+4	**4.7488e+4**	6.3820e+6	**7.6529e+4**
	Median	3.0976e+4	9.6593e+3	6.5059e+04	3.8852e+4	1.2809e+5	6.0298e+4
	Worst	5.1531e+4	2.9545e+4	2.6107e+5	1.2766e+5	1.9288e+8	2.5940e+5
	Best	303.1595	4.0487e+3	3.3775e+4	1.2146e+4	5.5970e+4	9.7001e+3
MaF_4	Mean	**185.6015**	154.0321	1.8330e+4	**6.5089e+3**	6.1292e+5	**1.7718e+5**
	Median	132.2845	157.4819	1.7994e+4	5.7372e+3	5.6660e+5	1.6946e+5
	Worst	558.9949	272.2185	3.8579e+4	1.6750e+4	1.2662e+6	4.3676e+5
	Best	2.8850	56.2430	73.5666	1.3418e+3	5.1928e+3	4.2660e+4
MaF_5	Mean	3.4354	**2.4941**	81.7945	**52.6467**	1.6891e+3	**1.2131e+3**
	Median	3.4658	2.5029	78.1735	52.7151	1.7065e+3	1.2137e+3
	Worst	4.1599	2.9408	130.9308	66.7699	2.2733e+3	1.6161e+3
	Best	2.9288	2.0434	58.0378	43.6443	1.3950e+3	1.0154e+3
MaF_6	Mean	0.0043	**0.0042**	0.5219	**0.2241**	**0.3858**	0.3430
	Median	0.0043	0.0041	0.4389	0.2496	0.3418	0.3425
	Worst	0.0045	0.0049	1.2602	0.3195	0.7446	0.3474
	Best	0.0039	0.0038	0.3101	0.0025	0.3415	0.3415
MaF_7	Mean	0.5699	**0.4674**	1.8627	**1.6517**	2.0686	3.2783
	Median	0.5701	0.4653	1.8428	1.5986	2.0761	3.0007
	Worst	0.6658	0.5433	2.0059	2.4190	2.1438	5.5710
	Best	0.4885	0.3954	1.7207	1.4164	1.9249	2.1578
MaF_8	Mean	**0.1352**	0.5757	**0.1420**	1.3324	**0.1414**	2.5921
	Median	0.1342	0.5621	0.1417	1.2636	0.1417	2.2632
	Worst	0.1618	0.9057	0.1505	2.5778	0.1463	7.5285
	Best	0.1214	0.3688	0.1363	0.7950	0.1349	1.2687
MaF_9	Mean	**0.7077**	1.1437	**64.3750**	53.1231	**0.8668**	9.3834
	Median	0.7029	1.1106	46.9070	45.8173	0.8606	12.0917
	Worst	0.7417	1.7796	173.4796	155.0931	1.0069	15.3572
	Best	0.6873	0.8050	12.3038	2.4595	0.7851	1.7868
MaF_{10}	Mean	2.3053	**1.9520**	4.0770	**3.0795**	4.8911	**4.0115**
	Median	2.2896	1.9442	4.1003	3.0895	4.9108	4.0144
	Worst	2.5162	2.0167	4.2786	3.1721	5.0849	4.1371
	Best	2.1792	1.9116	3.7678	3.0070	4.7297	3.8524
MaF_{11}	Mean	0.9947	**0.6098**	1.6504	**0.8708**	1.9490	**1.4747**
	Median	0.9777	0.5826	1.7212	1.0780	2.2806	1.8268
	Worst	1.1922	0.9363	2.2222	1.6269	3.0181	2.2427
	Best	0.8443	0.4961	0.5462	0.1776	0.7014	0.2508
MaF_{12}	Mean	**1.5934**	1.6983	5.7623	**5.4441**	8.6395	**7.7793**
	Median	1.5959	1.7319	5.7621	5.4454	8.6360	7.8252
	Worst	1.7707	1.8636	5.8966	5.6727	8.9330	8.0114
	Best	1.4322	1.5049	5.5960	4.9936	8.3844	7.3945
MaF_{13}	Mean	0.1869	**0.1209**	0.1232	**0.1071**	0.1045	**0.0953**
	Median	0.1748	0.1219	0.1204	0.1063	0.1017	0.0945
	Worst	0.2502	0.1308	0.1647	0.1235	0.1500	0.1129
	Best	0.1372	0.1045	0.1089	0.0928	0.0921	0.0845
MaF_{14}	Mean	**0.9794**	25.9004	**8.0794**	25.5172	**3.1449**	41.3429
	Median	0.9796	28.4090	8.3299	23.0875	1.0996	39.7949
	Worst	0.9796	45.2110	18.2883	48.8730	12.0864	59.5061
	Best	0.9774	8.0826	1.9220	11.4911	1.0963	27.2051
MaF_{15}	Mean	**9.9889**	11.2497	**56.4160**	53.1113	**50.6759**	72.4516
	Median	9.2002	11.2324	49.1652	51.9795	53.6135	72.7500
	Worst	15.1699	16.8106	108.9222	73.1437	90.7196	84.7248
	Best	6.5229	6.6440	32.3527	40.5497	23.5498	59.5537

5   Conclusion Remarks

This paper proposes GDE4, a new version of Generalized Differential Evolution algorithm for multi-objective optimization problems. The ordering of randomly selected candidate solutions for DE mutation operator is investigated. Method sorts three solutions at first, based on non-dominated sorting approach and then crowding distance measure to utilize as first, second and best solutions in DE mutation to generate a new individual exhibiting better fitness. DE summation and subtraction operators cause moving of new solution toward the first and second vectors and getting away from the third vector. So ordered vectors has inherited the quality of best and better candidate solutions. The performance of the method is evaluated using standard benchmark functions of CEC 2017 competition on evolutionary many-objective optimization problems. The results indicate that the proposed algorithm outperforms GDE3 which puts solutions in mutation operator randomly in most test problems. In the future, it is intended to investigate new strategies to order candidate solutions, such as the distance of each vector from an ideal point.

References

1. Ali, M., Siarry, P., Pant, M.: An efficient differential evolution based algorithm for solving multi-objective optimization problems. Eur. J. Oper. Res. **217**(2), 404–416 (2012)
2. Deb, K., Agrawal, S., Pratap, A., Meyarivan, T.: A fast elitist non-dominated sorting genetic algorithm for multi-objective optimization: NSGA-II. In: Schoenauer, M., et al. (eds.) PPSN 2000. LNCS, vol. 1917, pp. 849–858. Springer, Heidelberg (2000). https://doi.org/10.1007/3-540-45356-3_83
3. Lampinen, J.: DEs selection rule for multiobjective optimization. Technical report, Lappeenranta University of Technology, Department of Information Technology, pp. 03–04 (2001)
4. Kukkonen, S., Lampinen, J.: GDE3: the third evolution step of generalized differential evolution. In: The 2005 IEEE Congress on Evolutionary Computation, vol. 1, pp. 443–450. IEEE (2005)
5. Chiang, C.W., Lee, W.P., Heh, J.S.: A 2-opt based differential evolution for global optimization. Appl. Soft Comput. **10**(4), 1200–1207 (2010)
6. Yeh, M.F., Lu, H.C., Chen, T.H., Huang, P.J.: System identification using differential evolution with winner mutation strategy. In: 2014 International Conference on Machine Learning and Cybernetics (ICMLC), vol. 1, pp. 77–81. IEEE (2014)
7. Mahdavi, S., Rahnamayan, S., Karia, C.: Analyzing effects of ordering vectors in mutation schemes on performance of differential evolution. In: 2017 IEEE Congress on Evolutionary Computation (CEC), pp. 2290–2298 (2017). https://doi.org/10.1109/CEC.2017.7969582
8. Seada, H., Deb, K.: Non-dominated sorting based multi/many-objective optimization: two decades of research and application. In: Mandal, J.K., Mukhopadhyay, S., Dutta, P. (eds.) Multi-Objective Optimization, pp. 1–24. Springer, Singapore (2018). https://doi.org/10.1007/978-981-13-1471-1_1
9. Kukkonen, S., Lampinen, J.: An extension of generalized differential evolution for multi-objective optimization with constraints. In: Yao, X., et al. (eds.) PPSN 2004. LNCS, vol. 3242, pp. 752–761. Springer, Heidelberg (2004). https://doi.org/10.1007/978-3-540-30217-9_76
10. Deb, K., Pratap, A., Agarwal, S., Meyarivan, T.: A fast and elitist multiobjective genetic algorithm: NSGA-II. IEEE Trans. Evol. Comput. **6**(2), 182–197 (2002)
11. Cheng, R., et al.: A benchmark test suite for evolutionary many-objective optimization. Complex Intell. Syst. **3**(1), 67–81 (2017)
12. Hansen, N., Ostermeier, A.: Adapting arbitrary normal mutation distributions in evolution strategies: the covariance matrix adaptation. In: 1996 Proceedings of IEEE International Conference on Evolutionary Computation, pp. 312–317. IEEE (1996)

13. Iorio, A.W., Li, X.: Solving rotated multi-objective optimization problems using differential evolution. In: Webb, G.I., Yu, X. (eds.) AI 2004. LNCS (LNAI), vol. 3339, pp. 861–872. Springer, Heidelberg (2004). https://doi.org/10.1007/978-3-540-30549-1_74
14. Deb, K., Jain, H.: An evolutionary many-objective optimization algorithm using reference-point-based nondominated sorting approach, part I: solving problems with box constraints. IEEE Trans. Evol. Comput. 18(4), 577–601 (2014)
15. Tian, Y., Cheng, R., Zhang, X., Jin, Y.: PlatEMO: a MATLAB platform for evolutionary multi-objective optimization [educational forum]. IEEE Comput. Intell. Mag. 12(4), 73–87 (2017)

Multi-objective Techniques
for Single-Objective Local Search:
A Case Study on Traveling
Salesman Problem

Jialong Shi[1]([✉]), Jianyong Sun[1], and Qingfu Zhang[2,3]

[1] School of Mathematics and Statistics, Xi'an Jiaotong University, Xi'an, China
{jialong.shi,jy.sun}@xjtu.edu.cn
[2] Department of Computer Science, City University of Hong Kong,
Hong Kong, Hong Kong
[3] Shenzhen Research Institute, City University of Hong Kong, Shenzhen, China
qingfu.zhang@cityu.edu.hk

Abstract. In this paper, we show that the techniques widely used in multi-objective optimization can help a single-objective local search procedure escape from local optima and find better solutions. The Traveling Salesman Problem (TSP) is selected as a case study. Firstly the original TSP f_0 is decomposed into two TSPs f_1 and f_2 such that $f_0 = f_1 + f_2$. Then we propose the Non-Dominance Search (NDS) method which applies the non-domination concept on (f_1, f_2) to guide a local search out of the local optima of f_0. In the experimental study, NDS is combined with Iterated Local Search (ILS), a well-known metaheuristic for the TSP. Experimental results on some selected TSPLIB instances show that the proposed NDS can significantly improve the performance of ILS.

Keywords: Multi-objective optimization ·
Traveling Salesman Problem · Local search · Metaheuristic ·
Iterated Local Search

1 Introduction

A single-objective problem is defined as follows:

$$\begin{aligned} \text{minimize} \quad & f(x) \\ \text{subject to} \quad & x \in \mathcal{S}, \end{aligned} \tag{1}$$

where $f : \mathcal{S} \to \mathbb{R}$ is the objective function and $\mathcal{S}$ is the solution space. When $\mathcal{S}$ is a finite set, we face a Combinatorial Optimization Problem (COP). For COPs, many state-of-the-art algorithms use local search as a basic building block. However, a local search procedure stops at locally optimal solutions. When the objective function in Eq. (1) becomes an objective vector F which contains more than

© Springer Nature Switzerland AG 2019
K. Deb et al. (Eds.): EMO 2019, LNCS 11411, pp. 114–125, 2019.
https://doi.org/10.1007/978-3-030-12598-1_10

one objective, we face a multi-objective problem. Usually, in a multi-objective problem there exists a trade-off between different objectives and no single solution can optimize all the objectives. Most existing multi-objective algorithms are based on the non-dominance concept. They intend to find the solutions that are not dominated by any member of the current population and remove the population members that are dominated by the newly added solutions. A formal definition of the dominance/non-dominance relationship can be found in Sect. 2.

In this paper, we show that the multi-objective optimization techniques can help a single-objective local search procedure escape from local optima. We investigate a single case where local search for the Traveling Salesman Problem (TSP) can be improved by adding a simple Non-Dominance Search (NDS) procedure. Firstly, the original problem f_0 is decomposed into two problems f_1 and f_2 such that $f_0(x) = f_1(x) + f_2(x)$ for any $x \in S$. When the local search on f_0 falls in a local optimum x_*, the NDS procedure searches the neighborhood of x_* and tries to find a neighboring solution x' which is non-dominated by x_* on the two objectives (f_1, f_2). If such a solution is found, NDS searches the neighborhood of x'. If in the neighborhood of x' NDS finds a solution x'' that satisfies $f_0(x'') < f_0(x)$. Then the local optimum x_* is replaced by x'' and a new round of local search is started from x''. Otherwise, a random perturbation will be executed to jump out of the local optimum x_*. Experiments on some TSP instances have shown that NDS can significantly improve the performance of a widely used TSP algorithm, Iterated Local Search (ILS). Note here that the goal of this paper is to show the possibility of using multi-objective techniques to escape from local optima, not to propose a competitive algorithm for the TSP.

The rest of this paper is organized as follows. Section 2 introduces the multi-objective problems and the corresponding optimization techniques. Section 3 introduces the TSP and the method to decompose a TSP into two TSPs. Section 4 presents the proposed NDS method. In Sect. 5 the experimental studies have been conducted to show that NDS can improve the performance of ILS on the test TSP instances. Section 6 concludes this paper.

2   Multi-objective Optimization

A multi-objective problem is defined as:

$$\begin{aligned} \text{minimize} \quad & F(x) = (f_1(x), \ldots, f_m(x)) \\ \text{subject to} \quad & x \in S, \end{aligned} \tag{2}$$

where $F : S \to \mathbb{R}^m$ is the objective vector function which contains m objectives. When the solution space S is a finite set, it is referred to a Multi-objective COP (MCOP). Usually, there exists a trade-off between objectives and no single solution can optimize all objectives simultaneously. The goal of a multi-objective algorithm is to approximate the Pareto Set(PS). Some of the key concepts in multi-objective optimization are defined as follows.

- **Definition 1 (Dominance).** A vector $u = (u_1, \ldots, u_m)$ is said to *dominate* a vector $v = (v_1, \ldots, v_m)$, if and only if $u_k \le v_k, \forall k \in \{1, \ldots, m\} \wedge \exists k \in \{1, \ldots, m\} : u_k < v_k$, denoted as $u \prec v$.

- **Definition 2 (Non-dominance).** If u is not dominated by v and v is not dominated by u, we say that u and v are *non-dominated* by each other, denoted as $u \not\prec v$ or $v \not\prec u$.
- **Definition 3 (Pareto Optimal Solution).** A feasible solution $x \in S$ is called a *Pareto optimal solution*, if and only if $\nexists y \in S$ such that $F(y) \prec F(x)$.
- **Definition 4 (Pareto Set and Pareto Front).** For a multi-objective problem, the set of all the Pareto optimal solutions is called the *Pareto Set* (PS), denoted as $PS = \{x \in S | \nexists y \in S, F(y) \prec F(x)\}$ and the Pareto front (PF) is defined as $PF = \{F(x) | x \in PS\}$.

For convenience, in the following when we state that a solution x is dominated/non-dominated by a solution y, we mean that $F(x)$ is dominated/non-dominated by $F(y)$.

In the well known Multi-objective Evolutionary Algorithm based on Decomposition (MOEA/D) [1], the single-objective optimization techniques is employed to help solve multi-objective problems. The original multi-objective problem is decomposed into L scalar optimization subproblems using the predefined weight vectors $\{\lambda_1, \ldots, \lambda_L\}$. For example, given a bi-objective problem (f_1, f_2) and a weight vector $\lambda = (0.5, 0.5)$, the weighted sum method is used to generate a scalar optimization subproblem:

$$\text{minimize} \quad f^{ws}(x|\lambda) = 0.5 * f_1(x) + 0.5 * f_2(x). \tag{3}$$

In the population of MOEA/D, each solution is associated with a subproblem and different subproblems are optimized in a collaborative manner. Specifically, each subproblem has a number of neighboring subproblems. In the reproduction procedure of each subproblem, two neighboring subproblems are selected randomly and a new solution is generated from the associated solutions of the two neighboring subproblems using evolutionary operators like crossover and mutation. Then, the newly generated solution is used to update the population. The idea of using single-objective problem to guide multi-objective optimization also can be found in the existing multi-objective COP algorithms. For example Shi et al. [2,3] proposed to use the single-objective subproblems to guide the Pareto Local Search (PLS).

In the literature several studies have proposed to use a multi-objective evolutionary algorithm to solve a single-objective problem, which are termed as *multi-objectivization* [4-7]. In these studies, the original objective is converted into a bi-objective problem and a multi-objective evolutionary algorithm (e.g. NSGA-II) is executed on it. There are also other existing studies which use multi-objective techniques to benefit a single-objective optimization procedure. For example, Deb and Saha [8] convert a multi-modal problem into a bi-objective problem by considering the gradient or neighborhood information as the second objective. Then they use NSGA-II to address the Pareto front such that all the local optima of the original problem can be found. Some studies [9-11] transform a constrained single-objective optimization problem into an unconstrained multi-objective optimization problem. Bui et al. [12] considers the diversity of an evolutionary algorithm as the second objective to avoid falling in local optima.

Alsheddy [13] constructs the second objective by a penalty-based approach and run Pareto local search on the generated bi-objective problem. This paper differs from these existing studies in the following aspects.

- This paper focuses on helping a local search procedure escape from local optima and find better solutions. Based on the proposed method, an enhanced ILS is proposed.
- Instead of maintaining a population of non-dominated solutions, the method proposed in this paper only maintains one solution in its memory, which is easy to be operated.
- A new TSP decomposition method is introduced.

3 Traveling Salesman Problem

In the Traveling Salesman Problem (TSP), a salesman wants to find the shortest tour of n cities, starting and ending at the same city and visiting each of the other cities exactly once. Let $G = (V, E)$ be a fully connected graph where V is its node set and E the edge set. Let $c_e > 0$ be the cost of $e \in E$. A feasible solution x is a cycle passing through every node in V exactly once. Thus the TSP is defined as follows:

$$\text{minimize} \quad f(x) = \sum_{e \in x} c_e. \tag{4}$$

A node in G can be interpreted as a city and c_e as the travel cost from the source node of edge e to its destination node. TSP is NP-hard and is one of the most widely considered test problems in the area of combinatorial optimization. In this paper we focus on the symmetric TSPs. In a symmetric TSP, G is undirected, i.e., the cost of travel from node A to node B is the same as that from B to A.

The *distance matrix* of a TSP lists the costs (distances) of all edges in this TSP. A TSP instance is uniquely determined by its distance matrix. For example, Fig. 1 shows a 5-city TSP instance and its distance matrix. Since the TSP instance in Fig. 1 is a symmetric TSP, its distance matrix is a symmetric matrix.

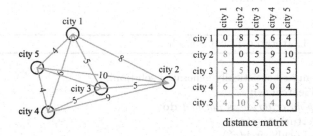

Fig. 1. An example of a 5-city TSP and its distance matrix

3.1   Iterated Local Search for TSP

Local search is a basic heuristic for COPs. It is a basic building block in many existing metaheuristics. Local search defines a neighborhood for every candidate solution in the search space. It maintains one candidate solution and iteratively improves it. It searches the neighborhood of the current solution and moves to a neighboring solution which has a better objective function value. Since the neighborhood size is limited, local search usually stops at solutions that are not worse than their neighbors but not necessarily all other solutions in the search space, i.e. the locally optimal solutions.

Here we use the 2-Opt local search as an example to show the mechanism of local search. The 2-Opt local search is a widely used local search method for the TSP. In a 2-Opt move, two non-adjacent edges are replaced by the other two edges if the resulting solution is better than the original one. As illustrated in Fig. 2, edges AB and CD are replaced by edges AC and BD, so that the cost of the solution is reduced.

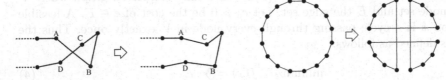

Fig. 2. An example of 2-Opt move on the TSP

Fig. 3. An example of the double bridge perturbation on the TSP

There are many efficient algorithms for the TSP [14–16], among which Iterated Local Search (ILS) [17] is a well-known metaheuristic based on local search. At each iteration of ILS, a local search procedure is performed. When the local search procedure is trapped in a local optimum, ILS executes the *perturbation* operator to escape from the attraction region of the local optimum. In the next iteration, a new local search procedure will start from the perturbed solution. During the search, ILS keeps recording the historical best solution. Algorithm 1 shows the procedure of ILS.

Algorithm 1. Iterated Local Search

1 $j \leftarrow 0$;
2 $x_0 \leftarrow$ random or heuristically generated solution.;
3 $x_0 \leftarrow$ LocalSearch(x_0);
4 **while** *stopping criterion is not met* **do**
5 $\quad$ $x'_j \leftarrow$ Perturbation(x_j);
6 $\quad$ $x_{j+1} \leftarrow$ LocalSearch(x'_j);
7 $\quad$ $j \leftarrow j + 1$;
8 **return** *the historical best solution* x_{best}

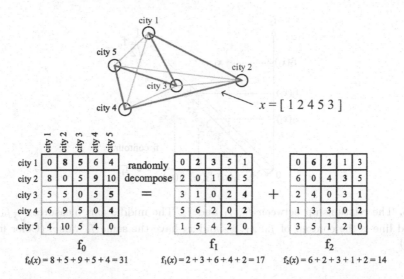

Fig. 4. A possible decomposition result of the 5-city TSP instance. The red lines marks an example solution $x = [1\ 2\ 4\ 5\ 3]$. After the decomposition it satisfies that $f_0(x) = f_1(x) + f_2(x)$ (Color figure online)

For the TSP, the *double bridge* perturbation is a widely used perturbation method. Figure 3 shows an example of the double bridge perturbation.

3.2   TSP Decomposition

In this section, we introduce a method to decompose a TSP f_0 into two different TSPs f_1 and f_2. The decomposition method is very simple. Since a TSP is uniquely determined by its distance matrix and its objective function f_0 is the linear addition of the elements in the distance matrix, we just randomly divide the original distance matrix into two positive symmetric matrices. Each matrix defines a new TSP instance. For example, Fig. 4 shows a possible decomposition result of the 5-city TSP instance. In Fig. 4, a solution $x = [1\ 2\ 4\ 5\ 3]$ is also shown and we can see that after the decomposition it satisfies that $f_0(x) = f_1(x) + f_2(x)$.

4   Non-Dominance Search

In this paper, we propose the *Non-Dominance Search (NDS)* method which can help a local search procedure escape from local optimum and finds better solutions. Firstly the original TSP f_0 is decomposed into two TSPs f_1 and f_2 using the method introduced in Sect. 3. Since $f_0 = f_1 + f_2$, the relationship between f_0 and the bi-objective problem (f_1, f_2) can be illustrated in Fig. 5. In Fig. 5, f_1 and f_2 form a bi-objective space. If we put an extra axis in the middle of the f_1 axis and f_2 axis, then this new axis measures $\frac{1}{\sqrt{2}} f_0$. Minimizing $\frac{1}{\sqrt{2}} f_0$ is minimizing f_0 itself. The contour of f_0 is perpendicular to the $\frac{1}{\sqrt{2}} f_0$ axis. For example, in Fig. 5, x_1, x_2 and x_3 have the same f_0 value.

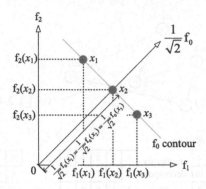

Fig. 5. The relationship between f_0, f_1 and f_2. The middle axis measures $\frac{1}{\sqrt{2}} f_0$ and the red line is the contour of f_0. x_1, x_2 and x_3 have the same f_0 value. (Color figure online)

4.1 The Idea

To illustrate the idea of NDS, we assume that the local optimum we want to escape is x_* and x_* has six neighboring solutions $\{x_1', \ldots, x_6'\}$, as shown in Fig. 6. Since x_* is locally optimal on f_0, all of $\{x_1', \ldots, x_6'\}$ are located above the $f_0(x_*)$ contour. Our algorithm intend to find a neighboring solution whose neighborhood can break through the contour of $f_0(x_*)$. From Fig. 6 we can see that the neighboring solution that are not dominated by x_* (e.g. x_6') are more likely to be close to the contour of $f_0(x_*)$, compared to the solutions that are dominated by x_* (e.g. x_3'). Hence the neighborhood of x_6' are more likely to contain a solution that can break through the $f_0(x_*)$ contour than the neighborhood of x_3'. This is the hypothesis that the proposed NDS method is based on.

Starting from a local optimum x_* of f_0, the procedure of NDS is shown in Algorithm 2. NDS follows the first-improving strategy. If it finds a candidate solution x'' that satisfies $f_0(x'') < f_0(x_*)$, it stops evaluating the rest candidate solutions and outputs x'' immediately. In NDS, the neighborhood of the neighborhood of x_* is explored, hence NDS can be seen as a local search on an enlarged neighborhood structure. The novelty of NDS is that it uses multi-objective techniques (i.e. the non-dominance concept) to guide the search on the enlarged neighborhood.

4.2 Improving ILS by NDS

NDS is a method to escape from local optima and find better solutions. It can be employed by different algorithms. In this section we show that the proposed NDS method can improve the performance of ILS. The resulting ILS variant is called ILS+NDS. Roughly speaking, ILS+NDS alternately executed a local search procedure and an NDS procedure. If NDS fails to escape from the current local optimum, a perturbation operator will be applied to this local optimum. Since NDS can be seen as an enlarged neighborhood search, if the current local optimum

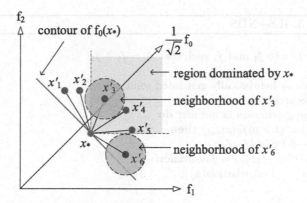

Fig. 6. Assume the local optimum x_* has six neighboring solutions $\{x'_1, \ldots, x'_6\}$. The neighborhood of x'_6 are more likely to break through the contour of $f_0(x_*)$ than the neighborhood of x'_3 since x'_6 is non-dominated by x_*.

Algorithm 2. Non-Dominance Search

Input: x_*, f_0, f_1, f_2
1 **for** each $x' \in Neighborhood(x_*)$ **do**
2   **if** $(f_1(x'), f_2(x')) \not\prec (f_1(x_*), f_2(x_*))$ **then**
3   **for** each $x'' \in Neighborhood(x')$ **do**
4   **if** $f_0(x'') < f_0(x_*)$ **then**
5             $x_{output} \leftarrow x''$;
6   go to step 7;

7 **return** x_{output}

---

has relatively low quality, then there is no need to explore the neighborhood of this local optimum. Hence in ILS+NDS, we design a mechanism to skip the low-quality local optima. Assuming the current best solution is x_{best}, ILS+NDS only applies NDS on the local optimum x_* that satisfies $f_0(x_*) < (1 + p)f_0(x_{best})$. Here $p > 0$ is a pre-defined parameter. For the local optima that do not meet this criterion, ILS+NDS directly executes perturbation on them. Algorithm 3 shows the procedure of ILS-NDS.

5 Experimental Study

To show that NDS can truly improve the performance of ILS. In the experimental study we compare ILS+NDS against the original ILS on ten TSP instances from TSPLIB [18]. The goal of this paper is to show the possibility of using multi-objective techniques in single-objective local search, not to propose a competitive algorithm for the TSP, hence we do not compare ILS+NDS against the state-of-the-art TSP algorithms. Both ILS and ILS+NDS are implemented in C++.

Algorithm 3. ILS+NDS

Input: f_0
1 Decompose f_0 into f_1 and f_2 such that $f_0 = f_1 + f_2$;
2 $j \leftarrow 0$;
3 $x_0 \leftarrow$ random or heuristically generated solution;
4 $x_0 \leftarrow$ LocalSearch(x_0);
5 **while** *stopping criterion is not met* **do**
6   **if** $f_0(x_j) < (1+p)f_0(x_{best})$ **then**
7         $x'_j \leftarrow$ NDS($x_j|f_1, f_2$);
8   **if** *NDS($x_j|f_1, f_2$) is failed* **then**
9             $x'_j \leftarrow$ Perturbation(x_j);

10   **else**
11         $x'_j \leftarrow$ Perturbation(x_j);
12     $x_{j+1} \leftarrow$ LocalSearch(x'_j);
13     $j \leftarrow j + 1$;

14 **return** *the historical best solution* x_{best}

The experimental platform is two 6-core 2.00GHz Intel Xeon E5-2620 CPUs (24 Logical Processors) under Ubuntu OS.

In our experiment, the TSP instances {eil51, pr76, rd100, bier127, kroA150, d198, ts225, pr264, pr299, lin318} are selected from the TSPLIB as the test instances. The numbers in the instances' names indicate their city numbers. In the ILS+NDS implementation, the 2-opt local search and the double bridge perturbation are employed. To get the proper p value for ILS+NDS, we test $p = 0.005, 0, 01, 0.015, \ldots, 0.05$. For each setting of p, we run the ILS+NDS 50 times on each test instances. Each run starts from a randomly generated solution and ends when the global optimum is reached or when the function evaluation times reaches 10^{10}. For comparison, we also run ILS 50 times on each instances. After the experiment, we calculate the *excess* of the best solution found in each run. Here the excess is defined by:

$$\text{excess} = \frac{f_0(x_{best}) - f_{0,opt}}{f_{0,opt}}, \tag{5}$$

where $f_{0,opt}$ is the globally optimal value of f_0. It is obvious that the lower excess value, the better performance of an algorithm.

Figure 7 shows the boxplot of the resulting excess data. From Figs. 7a and b we can see that, on the instance eil51 and pr76, all algorithms get a zero excess, which means that they all find the global optimum within 10^{10} function evaluations. Hence in Fig. 8 we show the function evaluations that each algorithm takes to find the global optimum of eil51 and pr76. From Figs. 7 and 8 we can conclude that, ILS+NDS performs better than ILS on all the ten test instances when the parameter p is properly set. In addition, we can find that the performance of ILS+NDS deteriorates when p is too small or too large. When p is too small, the criterion of executing NDS (i.e. $f_0(x_*) < (1+p)f_0(x_{best})$) becomes hard

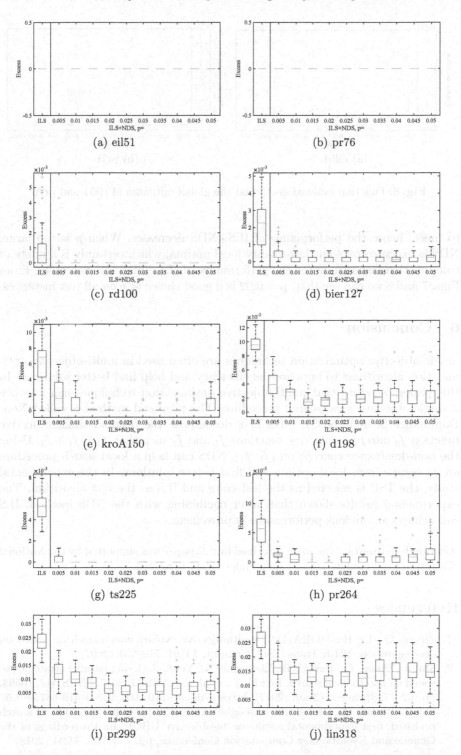

Fig. 7. The final excess achieved by ILS and ILS+NDS with different p values

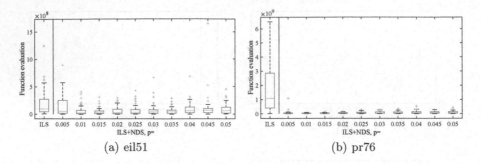

(a) eil51   (b) pr76

Fig. 8. Function evaluations to find the global optimum of eil51 and pr76

to meet, hence the performance of ILS+NDS decreases. When p is too large, NDS is executed on some low-quality local optima, which certainly is a waste of computation resource, hence the performance of ILS+NDS also decreases. From Figs. 7 and 8 we can see that, $p = 0.02$ is a good choice of p on all test instances.

6   Conclusion

Single-objective optimization techniques are often used in multi-objective optimization algorithms to improve the efficiency and help find better solutions. In this paper, we state that multi-objective optimization techniques can improve a single-objective local search procedure. The proposed method is called Non-Dominance Search (NDS). NDS first decompose the original single-objective function f_0 into two objective functions f_1 and f_2 such that $f_0 = f_1 + f_2$. Using the non-dominance concept on (f_1, f_2), NDS can help a local search procedure on f_0 escape from local optima and find better solutions. In the experimental study, the TSP is selected as the test suite and ILS as the test algorithm. The experimental results shown that, after combining with the NDS method, ILS can achieve an obvious performance improvement.

Acknowledgments. The work described in this paper was supported by the National Natural Science Foundation of China under Grant 61876163.

References

1. Zhang, Q., Li, H.: MOEA/D: a multiobjective evolutionary algorithm based on decomposition. IEEE Trans. Evol. Comput. **11**(6), 712–731 (2007)
2. Shi, J., Zhang, Q., Derbel, B., Liefooghe, A., Verel, S.: Using parallel strategies to speed up pareto local search. In: Shi, Y., et al. (eds.) SEAL 2017. LNCS, vol. 10593, pp. 62–74. Springer, Cham (2017). https://doi.org/10.1007/978-3-319-68759-9_6
3. Shi, J., Zhang, Q., Derbel, B., Liefooghe, A., Sun, J.: Parallel pareto local search revisited: first experimental results on bi-objective UBQP. In: Proceedings of the Genetic and Evolutionary Computation Conference, pp. 753–760. ACM (2018)

4. Knowles, J.D., Watson, R.A., Corne, D.W.: Reducing local optima in single-objective problems by multi-objectivization. In: Zitzler, E., Thiele, L., Deb, K., Coello Coello, C.A., Corne, D. (eds.) EMO 2001. LNCS, vol. 1993, pp. 269–283. Springer, Heidelberg (2001). https://doi.org/10.1007/3-540-44719-9_19
5. Jensen, M.T.: Helper-objectives: using multi-objective evolutionary algorithms for single-objective optimisation. J. Math. Model. Algorithms 3(4), 323–347 (2004)
6. Ishibuchi, H., Nojima, Y.: Optimization of scalarizing functions through evolutionary multiobjective optimization. In: Obayashi, S., Deb, K., Poloni, C., Hiroyasu, T., Murata, T. (eds.) EMO 2007. LNCS, vol. 4403, pp. 51–65. Springer, Heidelberg (2007). https://doi.org/10.1007/978-3-540-70928-2_8
7. Jähne, M., Li, X., Branke, J.: Evolutionary algorithms and multi-objectivization for the travelling salesman problem. In: Proceedings of the 11th Annual Conference on Genetic and Evolutionary Computation, pp. 595–602. ACM (2009)
8. Deb, K., Saha, A.: Multimodal optimization using a bi-objective evolutionary algorithm. Evol. Comput. 20(1), 27–62 (2012)
9. Coello, C.A.C.: Treating constraints as objectives for single-objective evolutionary optimization. Eng. Optim.+ A35 32(3), 275–308 (2000)
10. Mezura-Montes, E., Coello, C.A.C.: Constrained optimization via multiobjective evolutionary algorithms. In: Knowles, J., Corne, D., Deb, K., Chair, D.R. (eds.) Multiobjective Problem Solving from Nature. Natural Computing Series, pp. 53–75. Springer, Heidelberg (2008). https://doi.org/10.1007/978-3-540-72964-8_3
11. Singh, H.K., Isaacs, A., Nguyen, T.T., Ray, T., Yao, X.: Performance of infeasibility driven evolutionary algorithm (IDEA) on constrained dynamic single objective optimization problems. In: The 2009 IEEE Congress on Evolutionary Computation, pp. 3127–3134. IEEE (2009)
12. Bui, L.T., Abbass, H.A., Branke, J.: Multiobjective optimization for dynamic environments. In: The 2005 IEEE Congress on Evolutionary Computation, vol. 3, pp. 2349–2356. IEEE (2005)
13. Alsheddy, A.: A penalty-based multi-objectivization approach for single objective optimization. Inf. Sci. 442, 1–17 (2018)
14. Helsgaun, K.: An effective implementation of the lin-kernighan traveling salesman heuristic. Eur. J. Oper. Res. 126(1), 106–130 (2000)
15. Dorigo, M., Birattari, M., Stutzle, T.: Ant colony optimization. IEEE Comput. Intell. Mag. 1(4), 28–39 (2006)
16. Shi, J., Zhang, Q., Tsang, E.: EB-GLS: an improved guided local search based on the big valley structure. Memetic Comput. 10, 1–18 (2017)
17. Lourenço, H.R., Martin, O.C., Stützle, T.: Iterated local search: framework and applications. In: Gendreau, M., Potvin, J.Y. (eds.) Handbook of Metaheuristics. International Series in Operations Research & Management Science, pp. 363–397. Springer, Boston (2010). https://doi.org/10.1007/978-1-4419-1665-5_12
18. Reinelt, G.: TSPLIB-a traveling salesman problem library. ORSA J. Comput. 3(4), 376–384 (1991)

Multimodality in Multi-objective Optimization – More Boon than Bane?

Christian Grimme(✉), Pascal Kerschke, and Heike Trautmann

Information Systems and Statistics, ERCIS, University of Münster,
Leonardo-Campus 3, 48149 Münster, Germany
{christian.grimme,kerschke,trautmann}@uni-muenster.de

Abstract. This paper addresses multimodality of multi-objective (MO) optimization landscapes. Contrary to common perception of local optima, according to which they are hindering the progress of optimization algorithms, it will be shown that local efficient sets in a multi-objective setting can assist optimizers in finding global efficient sets. We use sophisticated visualization techniques, which rely on gradient field heatmaps, to highlight those insights into landscape characteristics. Finally, the MO local optimizer MOGSA is introduced, which exploits those observations by sliding down the multi-objective gradient hill and moving along the local efficient sets.

Keywords: Multi-objective optimization · Multimodality ·
Fitness landscapes · Basins of attraction · Local search · Gradients

1 Introduction

In single-objective (SO) continuous optimization, multimodality of the problem landscape is a crucial factor determining problem hardness. It is well-known that solvers might get trapped in local optima or at least require a large computational budget to (repeatedly) escape from the latter [22]. We will show that, counter-intuitively, MO optimizers do not necessarily face the same challenges. Contrarily, the existence of local efficient sets is potentially beneficial for sliding towards the global optimum along them. For this purpose a sophisticated visualization technique based on gradient field heatmaps, using the cumulated lengths of the normalized (approximated) gradients of both objectives towards the respective attracting local efficient set, is proposed. Respective figures reveal ridges and basins of attractions of local efficient sets - both in decision, as well as in objective space.

Basically, the gained insights can be exploited in two different ways: First, we pave the ground for retrieving as much information about the problems' landscape characteristics as possible with the potential of deriving informative exploratory landscape features for MO optimization problems, which is a rather new research field with only few results so far. Secondly, we introduce MOGSA, a (local) MO optimizer (MOO), which builds upon the straightforward idea of

© Springer Nature Switzerland AG 2019
K. Deb et al. (Eds.): EMO 2019, LNCS 11411, pp. 126–138, 2019.
https://doi.org/10.1007/978-3-030-12598-1_11

sliding down the MO gradient hill towards the global efficient set by exploiting properties of local efficient sets. Experiments show that common benchmark sets almost exclusively show similarities in enabling the algorithm to exploit the multimodal problem nature. MOGSA has the potential to even outperform competitive state-of-the-art MO optimizers and to be efficiently hybridized with other MOO approaches. Section 2 gives an overview of related work followed by a detailed description of our proposed visualization approach in Sect. 3. MOGSA is conceptually introduced in Sect. 4 together with preliminary experimental results. Conclusions are drawn in Sect. 5.

2 Related Work

SO continuous optimization insights are often directly transferred to MOO: *Multimodality implies the existence of traps for local search methods in SO, which prevent global convergence. Thus, multimodality in MO must be challenging for finding the global efficient set.* This assumption is the more astonishing, as almost no insights into the landscapes of continuous MO problems exist. It is restricted to very few general visualization techniques [29] and an early approach of [11]. These, however, only provide limited information on locality or landscape features. To the authors' best knowledge, only recent own work [18,20] provides insights into MO landscapes. Theoretical and empirical results [13,16,21] imply, that local optima in multimodal MOO are not necessarily traps for optimizers but following combined gradient directions can strongly support MOO. This is supported by many works (e.g., [14,24–26]), in which gradient-based directed search methods are applied to MOO problems. Note that for combinatorial MOO, analogies from SO landscape analysis (modality, ruggedness, correlation, and plateaus) are often used for making abstract problem features accessible [6,9,23,32]. However, due to the high dimensionality of combinatorial problems, a visual landscape representation is usually not possible or helpful.

3 A Gradient-Based Methodology for Visualizing Multi-objective Landscapes

Formal Preliminaries. Let $f : \mathcal{X} \to \mathbb{R}^p$ a vector valued function and define $x \in \mathcal{X}$ dominates $x' \in \mathcal{X} \Leftrightarrow f(x) \leq f(x') \wedge f(x) \neq f(x')$. The challenge of a MOO problem (MOP) is to find all points in $\mathcal{X}$ that are not dominated by other points in $\mathcal{X}$. This so-called *efficient set* $\mathcal{X}^*$ has an image $f(\mathcal{X}^*)$ w.r.t. all objectives which is called the *Pareto front*. In the following, we are considering unconstrained continuous minimization MOPs, that is $\mathcal{X} = \mathbb{R}^d$ for some dimension d.

We define two cases of MOP local optimality [13]. First, the usual definition of locality in neighborhood $B_\varepsilon(x) = \{x' \in \mathbb{R}^d | \|x - x'\| \leq \varepsilon\}$: A *local efficient point* $x \in \mathbb{R}^d$ is a point for which exists $\varepsilon > 0$ such that no $x' \in B_\varepsilon(x)$ dominates x. Further, $x \in \mathbb{R}^d$ is called a *strictly local efficient point*, if there exists $\varepsilon > 0$

such that x dominates all $x' \in B_\varepsilon(x) \setminus \{x\}$. We consider a set $A \subseteq \mathbb{R}^d$ to be *connected* iff there do not exist two open and *disjoint* subsets $U_1, U_2 \subseteq \mathbb{R}^d$ such that $A \subseteq (U_1 \cup U_2)$, $(U_1 \cap A) \neq \emptyset$, and $(U_2 \cap A) \neq \emptyset$. Further, let $B \subseteq \mathbb{R}^d$. A subset $C \subseteq B$ is a *connected component* of B iff $C \neq \emptyset$ is connected, and there exists no connected set D with $D \subseteq B$ such that $C \subset D$. Consequently, each connected component consisting of locally efficient points is called *local efficient set*.

The transfer of strict local efficiency to sets needs another definition. A local efficient set $\mathcal{X}_L$ is called *strictly local efficient set*, iff there exists an $\varepsilon > 0$ such that the environment set $E_\varepsilon(\mathcal{X}_L) = \{x' \in \mathbb{R}^d \setminus \mathcal{X}_L \mid \exists x \in \mathcal{X}_L : ||x - x'|| \leq \varepsilon\}$ is not empty and each point in $E_\varepsilon(\mathcal{X}_L)$ is dominated by at least one point in $\mathcal{X}_L$, see Fig. 1. The distinction between local and strict local efficient sets allows for a fine-grained analysis, where only the latter describes (rare) local traps for MOO algorithms.

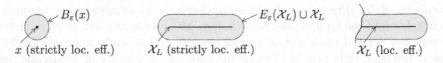

x (strictly loc. eff.) $\mathcal{X}_L$ (strictly loc. eff.) $\mathcal{X}_L$ (loc. eff.)

Fig. 1. Schematic examples illustrating the definitions of a strictly local efficient point (left) and set (center), as well as of a non-strictly local efficient set (right).

Idea of Gradient-Based Visualization for Multi-objective Landscapes.

For visually relating decision and objective space in MOO, a utility function based on the aggregated MO gradient is proposed in [18] and a necessary condition in [17] leading to the following concept:

Let the objectives be continuously differentiable in $\mathbb{R}^d$ and $x^* \in \mathcal{X}$ be a local efficient point of $\mathcal{X}$. Then, there exists a vector $\nu \in \mathbb{R}^p$ with $0 \leq \nu_i$, $i = 1, \ldots, p$, and $\sum_{i=1}^p \nu_i = 1$, such that $\sum_{i=1}^p \nu_i \nabla f_i(x^*) = 0$. For a local efficient point and a suitable weighting vector, gradients for all objectives cancel each other out. In the special case of a bi-objective problem, gradients become anti-parallel and only differ in length. By normalizing the SO gradients, the then *normalized multi-objective gradient* becomes zero when a local efficient point is reached. Otherwise, the length and the direction of the normalized multi-objective gradient provides information on the attraction area and closeness of a local efficient point or set. For visualization purposes, we compute the discretized path of a given point to a (local) efficient point following the MO gradient direction. Therefore, the search space is divided into a grid of discrete points and then the combined bi-objective gradient is computed for each of the grid points. The accumulated length of the path towards the local efficient point is considered as utility value that determines the "height" of the respective decision vector.

If the MO gradient ∇f is unknown, it can be approximated as g by means of its SO gradients ∇f_j, $j = 1, \ldots, p$ via $g_j = \nabla f_j^{(t)} =$

$\sum_{i=1}^{d} \frac{f_j(x^{(t)}+\delta \cdot e_i) - f_j(x^{(t)} - \delta \cdot e_i)}{2 \cdot \delta}$ $(*)$, where f_j is the j-th (single) objective, δ is
a (small) step-size and e_i is the d-dimensional unit vector. Using MO gradient
paths on all grid points, the MO landscape can be visualized in a two-dimensional
heatmap as exemplarily shown in Fig. 2.

Visual Inspection of State-of-the-Art Benchmarks for MOP. First, we
illustrate the concept for a highly multimodal MOP of the very recent *bi-objective
black-box optimization benchmark* (BBOB) [4,30]: bi-objective problems are constructed by combining multimodal SO functions of the BBOB set [15].

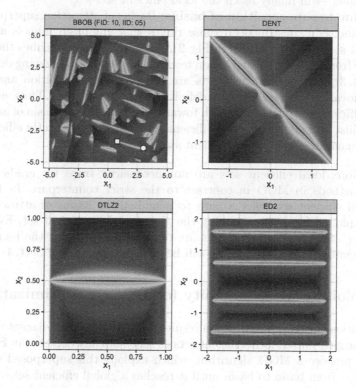

Fig. 2. Visualization of some exemplary and representative MOPs. The box and circle
(top left sub-figure) denote the global SO optima for the SO functions the BBOB
instance consists of. (Color figure online)

We also consider DENT [12] with a partly convex and concave Pareto-front,
while the global efficient set has a rather simple structure. DTLZ2 [8], which
has a completely concave Pareto-front, is focused followed by ED2 [10] similarly
constructed as DTLZ2 but allowing specific adjustment of concavity and locality.

Figure 2 depicts the respective 2D decision space of the considered problems. Red colored areas denote a "long" distance (w.r.t. gradient descent) to
the respective local efficient set. (Strictly) Local efficient sets are colored in blue
and usually surrounded by green to yellow areas which denote "small" distances.

We observe known properties of the problems in decision space, like strong multimodality for the BBOB problem or rather simple structures of efficient sets for the remaining problem instances. Based on comprehensive visual inspections of all well-known MO benchmark sets, two crucial observations could be made:

Basins of attraction: We can visually inspect the basins of attraction for all local efficient sets based on the MO gradient on the discretized decision space. Each basin can be considered as funnel towards its local efficient set. We may imagine a "MO ball" following the MO gradient: then, the ball - rolling down this funnel - will finally reach the local efficient set.

Ridges due to superposition of basins: Basins of attraction superpose each other and as a consequence expose ridges that cut basin funnels and local efficient sets (see e.g., BBOB in Fig. 2). A superposition describes the abrupt change from one basin of attraction towards another one. Following definitions in Sect. 3, only local efficient sets are subject to superposition and ridges. Strict local efficient sets are - by definition - not superposed. Thus, non-strict local efficient sets offer a sure path towards the superposing basin of attraction (if we just walk along the local efficient set itself), while strictly efficient sets can be considered traps for gradient steered optimization.

Therefore, local efficient sets are not necessarily traps for gradient-based descent methods in MOO in contrast to the strict counterpart. In fact, not strict local efficient sets offer a path to neighbouring basins of attraction and can be exploited for finally reaching the global (strict) efficient set. Even more interesting, strict local efficient sets are (empirically) rare for the benchmarks we investigated so far (an example will be shown at the end of Sect. 4).

4   Exploiting Multimodality for Efficient Optimization

Superposition of basins of attraction results in ridges between adjacent basins, as well as abruptly cutted local efficient sets as schematically depicted in Fig. 3. We therefore propose a MOO algorithm, which exploits this superposed structure by "sliding" from basin to basin until it reaches a global efficient set.

An Optimization Algorithm that Slides Through Local Optima. MOGSA, a *multi-objective gradient sliding algorithm*, consists of (multiple repetitions of) the following two phases detailed below: (1) follow the MOP's MO gradient until a (locally or globally) efficient point was found, (2) explore the corresponding efficient set by following the gradients of the MOP's SO components.

Find a local (or global) efficient point: At first, MOGSA performs a local search by sliding down the MO gradient landscape as described in Algorithm 1. Given an initial individual[1], the MOP's SO gradients are approximated (line 2). Next,

---

[1] If no initial point is given, it will be sampled randomly within the search space.

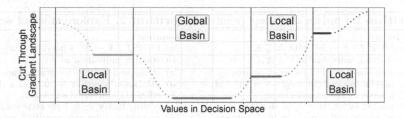

Fig. 3. Schematic view on the superimposed structure of basins of attraction. The vertical red lines represent the ridges distinguishing adjacent basins from each other, solid horizontal lines (within each basin) illustrate the respective efficient sets and the dotted lines represent the path towards the attracting efficient set. (Color figure online)

the MO gradient[2] at the current position is computed by summing up the normalized SO gradients (line 3). Note that the length of the individual's combined gradient already provides information on its location w.r.t. its attracting efficient set. If it is large (i.e., its length is close to two), both SO gradients point into similar directions, whereas a length of zero indicates opposing and hence offsetting gradients. In the latter case, the respective individual is regarded as being locally efficient. In order to account for numerical imprecision, MOGSA considers all individuals whose MO gradients have a length less than $\gamma = 10^{-6}$ as locally efficient and stops its downhill search once such a point was found (lines 4–6). Otherwise, it performs a gradient-descent step using the gradient length scaled by σ_1 (line 7). If it was too short (i.e., less than $\varepsilon = 10^{-6}$), the algorithm reached a dead end. Such a dead end could for instance occur, if MOGSA evaluates a point on the boundary of the feasible space and its gradient is pointing towards the infeasible area. In such a scenario, MOGSA leaves this dead end by restarting from an unexplored region of the MOP's feasible space (lines 8–10).

Once the algorithm has successfully performed three consecutive gradient steps (lines 11–18), it computes the angle between the three "youngest" individuals to detect whether the latter two are located on the same or opposite side of the attracting efficient set. In the former case, MOGSA continues its downhill search. However, in the latter scenario, i.e., $x^{(t)}$ and $x^{(t+1)}$ are located on opposite sides, it performs an interval bisection procedure, which exploits the individual's closeness to the efficient set in order to quickly converge to the efficient point. For this purpose, we modified the classical bisection method [5] such that the interval will be split according to the ratio of the lengths of the gradients, i.e., $x^{(t+2)} = x^{(t)} + (x^{(t+1)} - x^{(t)}) \cdot \|g^{(t)}\|/\|g^{(t+1)}\|$, rather than simply at its center.

Explore Efficient Set: Once a locally efficient individual was found by Algorithm 1, it starts its exploration phase from there (see Algorithm 2). MOGSA computes the SO gradients (line 2) and follows the (normalized) gradient of the first objective scaled by σ_2 (line 4) as long as the step size is at least $\varepsilon = 10^{-6}$

---

[2] Note that the current implementation of MOGSA only enables the optimization of bi-objective problems.

Algorithm 1. Find local efficient point

1: **Require:**
   a) MOP $f : \mathcal{X} \to \mathbb{R}^p$ with $\mathcal{X} \subseteq \mathbb{R}^d$,
   b) starting individual $x^{(t)} \in \mathcal{X}$,
   c) step-size $\delta = 10^{-6}$ for grad.-approx.,
   d) maximum gradient length $\gamma = 10^{-6}$ of a locally efficient individual,
   e) scaling factor $\sigma_1 = 1$ for step-size,
   f) maximum difference $\varepsilon = 10^{-6}$ between individuals to be considered identical.
2: Approximate single-objective gradients $g_j^{(t)}$, currently only for $j \in \{1, 2\}$, using Eqn. (∗)
3: Combine normalized single-obj. gradients:
$$g^{(t)} = g_1^{(t)}/\|g_1^{(t)}\| + g_2^{(t)}/\|g_2^{(t)}\|$$
4: **If** $\|g^{(t)}\| < \gamma$ **then**
5:    $x^{(t)}$ is locally efficient ⤳ exit algorithm
6: **end if**
7: Do gradient step: $x^{(t+1)} = x^{(t)} + \sigma_1 \cdot g^{(t)}$
   (if $x^{(t+1)} \notin \mathcal{X}$, place it on boundary)
8: **If** $\|x^{(t)} - x^{(t+1)}\| \le \varepsilon$ **then**
9:    restart, i.e., draw an alternative $x^{(t+1)}$
   (using *Optimal Augmented Latin Hypercube Sampling*) and proceed to step 2
10: **end if**
11: **If** $x^{(t+1)}$ is at least 3rd element since last restart **then**
12:   comp. $\omega = \angle(x^{(t+1)} - x^{(t)}, x^{(t)} - x^{(t-1)})$
13:   **If** $\omega \le 90°$ **then**
14:     individuals are approaching efficient set from same side ⤳ proceed to step 2
15:   **else**
16:     $x^{(t)}$ and $x^{(t+1)}$ are located on opposite sides of efficient set ⤳ perform weighted interval bisection between them
17:   **end if**
18: **end if**
19: **Return** archive of visited points

Algorithm 2. Explore efficient set

1: **Require:**
   a) MOP $f : \mathcal{X} \to \mathbb{R}^p$ with $\mathcal{X} \subseteq \mathbb{R}^d$,
   b) starting individual $x^{(0)} \in \mathcal{X}$,
   c) step-size $\delta = 10^{-6}$ for grad.-approx.,
   d) maximum length $\gamma = 10^{-6}$ of a (local efficient) individual's gradient,
   e) scaling factor $\sigma_2 = 1$ for step-size,
   f) maximum difference $\varepsilon = 10^{-6}$ between individuals to be considered identical.
2: **Initialize**, i.e., set $x^{(t)} = x^{(0)}$ and approx. single-obj. gradients $g_j^{(t)}$ (for $j \in \{1, 2\}$)
3: **Explore set from $x^{(0)}$ in direction of g_1:**
4:   $x^{(t+1)} = x^{(t)} + \sigma_2 \cdot (g_1^{(t)}/\|g_1^{(t)}\|)$
   (if $x^{(t+1)} \notin \mathcal{X}$, place it on boundary)
5: **If** $\|x^{(t)} - x^{(t+1)}\| \le \varepsilon$ **then**
6:    no step performed ⤳ proceed to step 19
7: **end if**
8: Approx. $g_1^{(t+1)}$ and $g_2^{(t+1)}$ using Eqn. (∗)
9: **If** $\left(\|g_1^{(t+1)}\| \le \gamma\right)$ or $\left(\|g_2^{(t+1)}\| \le \gamma\right)$
10: **then**
11:    found single-objective optimum ⤳ proceed to step 19
12: **end if**
13: compute angle $\alpha = \angle(g_1^{(t)}, g_1^{(t+1)})$
14: compute angle $\beta = \angle(g_1^{(t+1)}, g_2^{(t+1)})$
15: **If** $(\alpha > 90°)$ or $(\beta < 90°)$ **then**
16:    left efficient set ⤳ proceed to step 19
17: **end if**
18: still in efficient set ⤳ proceed to step 4
19: **Explore set from $x^{(0)}$ in direction of g_2:**
20:    analog to steps 4-18, but using exchanged gradients g_1 and g_2
21: **Return** archive of visited points

(lines 5–7). These steps are repeated until MOGSA has reached the local optimum of the first objective (lines 9–12) or even left the efficient set (lines 13–17). The latter can have two reasons: (i) it left the efficient set, but remains in the same basin of attraction (indicated by an angle of more than 90° between two consecutive gradients of the first objective), or (ii) it left the basin of attraction and crossed the ridge to an adjacent basin (indicated by an angle of less than 90° between the two single-objective gradients in the current individual). Once MOGSA finished exploring one part of the efficient set (by following the first objective), it explores the set once more (starting in the same initial individual), but this time follows the second objective (lines 19–20). If at least one of the two exploration phases stopped because of a crossed ridge, the respective efficient set can not be globally efficient as it is apparently superimposed by another basin of attraction. In such a case, MOGSA again executes Algorithms 1 and 2 - starting from the individual belonging to the adjacent, and thus more promising basin of attraction. However, in case neither end of the efficient set has been cut by a ridge, a strictly local efficient set - and thus likely a globally efficient Pareto set - was found.

Comparison of MOGSA and State-of-the-Art MOO Algorithms. Figure 4 illustrates the different search behaviors of MOGSA (top row), NSGA-II [7] (middle row) and SMS-EMOA [1] (bottom) with default parameter settings. The traces of their optimization paths are shown in the decision spaces of two exemplary MOPs: an instance of the bi-objective BBOB (left column) and DTLZ2 (right). For both problems, MOGSA was executed until it terminated

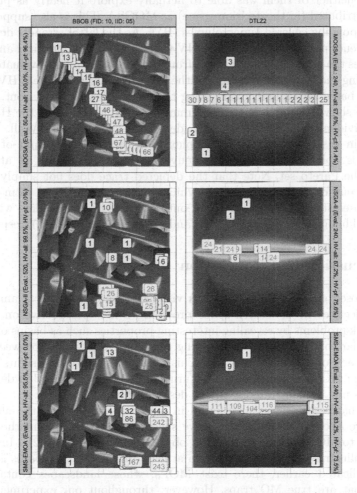

Fig. 4. Exemplary comparison of the search behavior of MOGSA (top row), NSGA-II (middle) and SMS-EMOA (bottom) in the search space of two popular MOPs: an instance from the bi-objective BBOB (FID 10, IID 5; left column) and DTLZ 2 (right).

successfully, i.e., after only 504 resp. 240 function evaluations including gradient approximation while its two contenders were then executed with the same budget. Note that the MOPs were created using the R-package smoof [3], and for the competing MOEAs (i.e., NSGA-II and SMS-EMOA) ecr2 [2] was used.

The deceptive structure of DTLZ2 initially lured MOGSA towards its boundaries (see top right image of Fig. 4). However, once it reached the boundaries, it immediately restarted and quickly converged towards the global efficient set, which it then explored very efficiently. Given the rather small amount of function evaluations needed by MOGSA, we restricted the population sizes of NSGA-II and SMS-EMOA to $\mu = 5$ individuals such that they were able to run for a reasonable number of generations. Although both solvers approached the efficient set, neither of them was able to actually explore it nearly as precise or evenly distributed (in the decision space) as MOGSA. This is also supported by the corresponding covered hypervolumes (HV). As the latter strongly depend on their reference points, we provide two HV values per pair of optimizer and MOP. HV-all uses the nadir of *all* individuals from the archive of points evaluated when constructing the heatmap or by any of the three optimizers, whereas HV-pf uses the nadir based on all individuals along the (theoretical) Pareto front. All HV values are shown as ratios of the maximum achievable HV - i.e., the HV based on all individuals, which were used for identifying the nadir of HV-all.

Even in case of much more complex landscapes (left column of Fig. 4), MOGSA is able to successfully maneuver through the basins of attraction towards the Pareto set. Note that the depicted trace does not simply display a positive outlier. Out of ten runs, in which we executed MOGSA from ten randomly chosen starting points, MOGSA outperformed its contenders - within the considered budget - in nine (NSGA-II) and ten runs (SMS-EMOA), respectively.

5   Discussion and Conclusion

Our proposed gradient field heatmaps visualize interaction effects among different objectives in the search space providing a new perspective on popular benchmark problems by revealing interesting properties such as basins of attraction (similar to the idea of cell mapping, see, e.g., [19]), ridges between them, local and/or global efficient sets, etc. This results in important insights into the structure of MO landscapes, which in turn can be used for algorithm design - as we successfully demonstrated with the design of MOGSA in Sect. 4.

Note that certain structures found in search space (such as the attraction basins) are also visible in the objective space. As a result, our method is not restricted to MOPs, whose search *and* objective space both are 2D. Instead, it allows to illustrate any MOP for which at least *one* of the two spaces is 2D.

As indicated earlier, there exist MOPs, whose landscapes contain local optima that are true MO traps. However, throughout our experiments with more than a hundred of different MOPs, we only encountered one single landscape (displayed in Fig. 5), which displayed such a "malign" strictly local efficient set. Despite possessing only *one* Pareto front in the objective space, it contains *two* cut-free efficient sets in the decision space. The *local* front - spanning from the pink square to the cyan circle - is entirely dominated by the Pareto front. However, as both ends of the local front are nondominated in its close neighborhood, MOGSA is unable to leave it towards an improving attractor. In future work, we intend to analyze (a) the causes for such traps, and (b) whether real-world applications (in contrast to artificial problems) possess such traps more often.

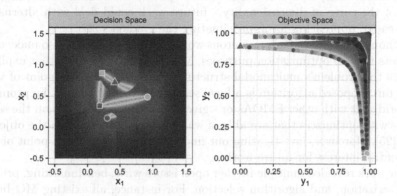

Fig. 5. Example of a MOP with a true MO trap (ranging from pink square to cyan circle). The problem combines two multi-sphere problems created using MPM2 [33]. (Color figure online)

Within our exploratory proof-of-concept experiments, MOGSA reached the global Pareto front much faster in terms of function evaluations - despite the costs for *approximating* the gradient - than NSGA-II and SMS-EMOA. This effect would increase even further, if the exact gradient was accessible. Without a doubt MOGSA's performance strongly depends on its parametrization - especially on its step-size related factors. If they are set too high, the algorithm jumps across entire basins, whereas short steps increase the number of iterations and thus function evaluations. Therefore, a future aspect is the development of sophisticated step-size adaptation mechanisms, which ideally adjust the parameters automatically to the landscape at hand.

Given a reasonable parametrization, MOGSA converges rather quickly and hence only consumes a small part of the budget. To further improve the algorithm's performance, one could invest additional budget into further runs of the optimizer as these allow to (a) avoid running into MO traps, and (b) detect different parts of the global Pareto front (if existent). Given the deterministic search behavior, one should restart from unexplored regions as MOGSA would otherwise quickly move into already visited basins. The chances of starting "far away" from previously seen areas, can be increased by applying sophisticated sampling mechanisms such as *optimal augmented latin hypercube sampling* [28].

Currently, the results were only shown for 2D problems, however, the algorithm's concept is transferable to higher dimensional search and objective spaces.

Adapting it to larger search spaces is straightforward as this simply requires the gradient approximation in further dimensions - resulting in two additional function evaluations per individual and search space dimension. In case of larger objective spaces, it might be beneficial to first optimize w.r.t. two objectives and once a strictly local efficient set was found, MOGSA could travel along the adjacent fronts (by following different objectives) until a cut-free simplex of global efficient points (comprising the global optimum) is found. In order to keep the optimizer competitive in larger search spaces - and hence diminishing the effect of the curse of dimensionality - future work could deal with alternative, i.e., cheaper, approaches for approximating the gradients [27].

Although there have been numerous works in the past, which also make use of gradients for MO optimization purposes, MOGSA is the first one that explicitly exploits the problem's multimodal structure from a search space point of view.

As our proposed algorithm is a local search algorithm, it might be promising to hybridize it with other EMOAs or - given that MOGSA focuses on the search space - with optimizers that are able of walking along the fronts in the objective space [25]. Moreover, investigating our findings from a theoretical point of view is of central interest for future work.

Our work revealed multiple further open issues w.r.t. benchmarking, problem characterization, and algorithm selection. For instance, all existing MO benchmarks should be compared thoroughly (e.g., using our visualization approach). The gained insights can then be used to group the MOPs in an appropriate way, similar to the five groups of BBOB [15]. Once important properties of MO landscapes have been identified, landscape features (see, e.g., [31] or [20]), which (a) 'measure' these different properties, and (b) can later on be used to perform algorithm selection on a portfolio of complementary, powerful MO algorithms, can be designed. Such a complementary portfolio requires an extensive benchmark of competitive state-of-the-art algorithms on a manifold of MOPs.

References

1. Beume, N., Naujoks, B., Emmerich, M.T.M.: SMS-EMOA: multiobjective selection based on dominated hypervolume. EJOR **181**(3), 1653–1669 (2007)
2. Bossek, J.: ecr 2.0: a modular framework for evolutionary computation in R. In: Proceedings of GECCO Companion, pp. 1187–1193. ACM (2017)
3. Bossek, J.: smoof: single- and multi-objective optimization test functions. R J. (2017). https://journal.r-project.org/archive/2017/RJ-2017-004/
4. Brockhoff, D., Tran, T.D., Hansen, N.: Benchmarking numerical multiobjective optimizers revisited. In: Proceedings of GECCO, pp. 639–646. ACM (2015)
5. Burden, R.L., Faires, D.J.: Numeric Analysis, 3rd edn. Prindle, Weber & Schmidt Publishing Company, Boston (1985)
6. Daolio, F., Liefooghe, A., Verel, S., Aguirre, H.E., Tanaka, K.: Global vs local search on multi-objective NK-landscapes: contrasting the impact of problem features. In: Proceedings of GECCO, pp. 369–376. ACM (2015)
7. Deb, K., Pratap, A., Agarwal, S., Meyarivan, T.: A fast and elitist multiobjective genetic algorithm: NSGA-II. IEEE TEVC **6**(2), 182–197 (2002)

8. Deb, K., Thiele, L., Laumanns, M., Zitzler, E.: Scalable test problems for evolutionary multiobjective optimization. In: Abraham, A., Jain, L., Goldberg, R. (eds.) Evolutionary Multiobjective Optimization, pp. 105–145. Springer, London (2005)
9. Ehrgott, M., Klamroth, K.: Connectedness of efficient solutions in multiple criteria combinatorial optimization. EJOR **97**(1), 159–166 (1997)
10. Emmerich, M.T.M., Deutz, A.H.: Test problems based on Lamé superspheres. In: Obayashi, S., Deb, K., Poloni, C., Hiroyasu, T., Murata, T. (eds.) EMO 2007. LNCS, vol. 4403, pp. 922–936. Springer, Heidelberg (2007). https://doi.org/10. 1007/978-3-540-70928-2_68
11. da Fonseca, C.M.M.: Multiobjective genetic algorithms with application to control engineering problems. Ph.D. thesis, University of Sheffield (1995)
12. Gerstl, K., Rudolph, G., Schtze, O., Trautmann, H.: Finding evenly spaced fronts for multiobjective control via averaging Hausdorff-measure. In: 2011 8th International Conference on Electrical Engineering, Computing Science and Automatic Control, pp. 1–6 (2011). https://doi.org/10.1109/ICEEE.2011.6106656
13. Grimme, C., Kerschke, P., Emmerich, M.T.M., Preuss, M., Deutz, A.H., Trautmann, H.: Sliding to the global optimum: how to benefit from non-global optima in multimodal multi-objective optimization. In: Proceedings of LeGO (2018, accepted)
14. Grimme, C., Lepping, J., Papaspyrou, A.: Adapting to the habitat: on the integration of local search into the predator-prey model. In: Ehrgott, M., Fonseca, C.M., Gandibleux, X., Hao, J.-K., Sevaux, M. (eds.) EMO 2009. LNCS, vol. 5467, pp. 510–524. Springer, Heidelberg (2009). https://doi.org/10.1007/978-3-642-01020-0_40
15. Hansen, N., Finck, S., Ros, R., Auger, A.: Real-parameter black-box optimization benchmarking 2009: noiseless functions definitions. Technical report, INRIA (2009)
16. Jin, Y., Sendhoff, B.: Connectedness, regularity and the success of local search in evolutionary multi-objective optimization. In: Proceedings of the IEEE CEC, vol. 3, pp. 1910–1917. IEEE (2003)
17. John, F.: Extremum problems with inequalities as subsidiary conditions. In: Studies and Essays, Courant Anniversary Volume, pp. 187–204. Interscience (1948)
18. Kerschke, P., Grimme, C.: An expedition to multimodal multi-objective optimization landscapes. In: Trautmann, H., et al. (eds.) EMO 2017. LNCS, vol. 10173, pp. 329–343. Springer, Cham (2017). https://doi.org/10.1007/978-3-319-54157-0_23
19. Kerschke, P., et al.: Cell mapping techniques for exploratory landscape analysis. In: Tantar, A.-A., et al. (eds.) EVOLVE - A Bridge between Probability, Set Oriented Numerics, and Evolutionary Computation V. AISC, vol. 288, pp. 115–131. Springer, Cham (2014). https://doi.org/10.1007/978-3-319-07494-8_9
20. Kerschke, P., et al.: Towards analyzing multimodality of continuous multiobjective landscapes. In: Handl, J., Hart, E., Lewis, P.R., López-Ibáñez, M., Ochoa, G., Paechter, B. (eds.) PPSN 2016. LNCS, vol. 9921, pp. 962–972. Springer, Cham (2016). https://doi.org/10.1007/978-3-319-45823-6_90
21. Kerschke, P., et al.: Search dynamics on multimodal multi-objective problems. Evol. Comput. 1–33 (2018). https://doi.org/10.1162/evco_a_00234
22. Preuss, M.: Multimodal Optimization by Means of Evolutionary Algorithms. Springer, Cham (2015). https://doi.org/10.1007/978-3-319-07407-8. https://www.springer.com/de/book/9783319074061
23. Rosenthal, S., Borschbach, M.: A concept for real-valued multi-objective landscape analysis characterizing two biochemical optimization problems. In: Mora, A.M., Squillero, G. (eds.) EvoApplications 2015. LNCS, vol. 9028, pp. 897–909. Springer, Cham (2015). https://doi.org/10.1007/978-3-319-16549-3_72

24. Schütze, O., Hernández, V.A., Trautmann, H., Rudolph, G.: The hypervolume based directed search method for multi-objective optimization problems. J. Heuristics **22**(3), 273–300 (2016)
25. Schütze, O., Martín, A., Lara, A., Alvarado, S., Salinas, E., Coello, C.A.: The directed search method for multi-objective memetic algorithms. Comput. Optim. Appl. **63**(2), 305–332 (2016)
26. Schütze, O., Sanchez, G., Coello Coello, C.A.: A new memetic strategy for the numerical treatment of multi-objective optimization problems. In: Proceedings of GECCO, pp. 705–712. ACM (2008)
27. Spall, J.C.: Multivariate stochastic approximation using a simultaneous perturbation gradient approximation. IEEE Trans. Autom. Control **37**(3), 332–341 (1992)
28. Stein, M.: Large sample properties of simulations using latin hypercube sampling. Technometrics **29**, 143–151 (1987)
29. Tušar, T., Filipič, B.: Visualization of Pareto front approximations in evolutionary multiobjective optimization: a critical review and the prosection method. IEEE TEVC **19**(2), 225–245 (2015)
30. Tušar, T., Brockhoff, D., Hansen, N., Auger, A.: COCO: the bi-objective black box optimization benchmarking (bbob-biobj) test suite. arXiv preprint (2016)
31. Ulrich, T., Bader, J., Thiele, L.: Defining and optimizing indicator-based diversity measures in multiobjective search. In: Schaefer, R., Cotta, C., Kołodziej, J., Rudolph, G. (eds.) PPSN 2010. LNCS, vol. 6238, pp. 707–717. Springer, Heidelberg (2010). https://doi.org/10.1007/978-3-642-15844-5_71
32. Verel, S., Liefooghe, A., Jourdan, L., Dhaenens, C.: On the structure of multiobjective combinatorial search space: MNK-landscapes with correlated objectives. Eur. J. Oper. Res. **227**(2), 331–342 (2013)
33. Wessing, S.: Two-stage methods for multimodal optimization. Ph.D. thesis, Technische Universität Dortmund (2015). http://hdl.handle.net/2003/34148

Solving Nonlinear Equation Systems Using Multiobjective Differential Evolution

Jing-Yu Ji[1], Wei-Jie Yu[1(✉)], and Jun Zhang[2]

[1] Sun Yat-sen University, Guangzhou 510275, China
ywj21c@163.com
[2] South China University of Technology, Guangzhou 510641, China

Abstract. Nonlinear equation systems (NESs) usually have more than one optimal solution. However, locating all the optimal solutions in a single run, is one of the most challenging issues for evolutionary optimization. In this paper, we address this issue by transforming all the optimal solutions of an NES to the nondominated solutions of a constructed multiobjective optimization problem (MOP). In the general case, we prove that the proposed transformation fully matches the requirement of multiobjective optimization. That is, the multiple objectives always conflict with each other. In this way, multiobjective optimization techniques can be used to locate these multiple optimal solutions simultaneously as they locate the nondominated solutions of the MOPs. Our proposed approach is evaluated on 22 NESs with different features, such as linear and nonlinear equations, different numbers of optimal solutions, and infinite optimal solutions. Experimental results reveal that the proposed approach is highly competitive with some other state-of-the-art algorithms for NES.

Keywords: Nonlinear equation systems ·
Multiobjective optimization · Differential evolution

1 Introduction

Nonlinear equation systems (NESs) arise in many engineering and scientific domains [1,13,19,20]. In many practical optimization problems, solving NESs is time consuming, since the calculation in NESs is computationally expensive and solutions are obtained in real-time due to several numerical issues [21]. Hence, an efficient way to solve NESs is of great practical significance.

An NES can be defined as follows:

$$f(\mathbf{x}) = \begin{bmatrix} f_1(\mathbf{x}) \\ f_2(\mathbf{x}) \\ \vdots \\ f_n(\mathbf{x}) \end{bmatrix} \qquad (1)$$

© Springer Nature Switzerland AG 2019
K. Deb et al. (Eds.): EMO 2019, LNCS 11411, pp. 139–150, 2019.
https://doi.org/10.1007/978-3-030-12598-1_12

where $\mathbf{x} = (x_1, x_2, \ldots, x_D) \in S$ denotes D variables for n equations, $S = \prod_{i=1}^{D} [\underline{x}_i, \overline{x}_i]$ is the decision space in which $\underline{x}_i \leq x_i \leq \overline{x}_i$, $f(\mathbf{x}) = (f_1(\mathbf{x}), f_2(\mathbf{x}), \ldots, f_n(\mathbf{x})) \in \Re^n$ refers to n nonlinear functions, and $\Re^n$ is the objective space mapped from S to $\Re^n$. If a solution is the optimal solution of the NES, it satisfies the condition that every equation in (1) is equal to zero, i.e.,

$$\begin{cases} f_1(x_1, x_2, \ldots, x_D) = 0 \\ f_2(x_1, x_2, \ldots, x_D) = 0 \\ \quad \vdots \\ f_n(x_1, x_2, \ldots, x_D) = 0 \end{cases} \tag{2}$$

Solving NESs often involves finding multiple optimal solutions, especially in the case of $D > n$. Since each optimal solution may represent the equal importance in practical applications, and locating them may require computationally expensive experiments, it is desirable to locate them as many as possible in a single run. the computational budget can be saved, and decision makers can also have multiple choices.

To obtain multiple optima of an NES simultaneously, the population-based evolutionary algorithms (EAs) [2] have been gaining increasing attention recently. A variety of EAs has been proposed and developed [11,14,15,17] for solving NESs. In general, an EA for locating multiple solutions includes two significant components: (1) search algorithm and (2) multimodality handling technique. As a search engine, different search algorithms and their variants use different strategies to generate new offspring, and hence have different impacts on the performance of solving NES. Among the various search algorithms, such as genetic algorithm [22], particle swarm optimization [4,10], ant colony optimization [9], and differential evolution [18], most of them originally aim at solving numerical optimization with single optimum. Therefore, when they are applied to locating multiple optimal solutions, the techniques that can aid the search algorithms in handling the multimodality of NESs are required.

Recently, multiobjective optimization techniques have been widely applied to solving NESs [8,16,23]. In this way, all the optimal solutions are located at the whole population level. Hence, the fact that the search is sensitive to the user-defined parameters does not exist in this kind of techniques. By some specific transformations, the NES first is transformed into a multiobjective optimization problem (MOP). Meanwhile, the multiple optimal solutions of the NES are transformed into the nondominated solutions of the new constructed MOP. Then, these nondominated solutions can be located simultaneously by multiobjective optimization techniques [8,16]. However, the feasibility of solving NESs by multiobjective optimization techniques depends on whether the multiple objectives always conflict in the constructed MOP. If not, the multiple optimal solutions converted as nondominated solutions does not exist with regard to the transformed MOP.

In principle, the MOP is constructed by the objective function itself and the additional information from variables, gradients, and individual distances [3,7,22], etc. In [7,22], the gradient of the objective function is utilized to recast a bi-objective optimization problem. However, if the objective function is non-differentiable or discontinuous, the second objective would be difficult to construct, which may infeasible in practical applications. In [3], the distance information of each individual is considered as an additional objective value. However, the performance of this algorithm may suffer from degradation in particular landscapes, since the two objectives do not conflict with each other.

The above mentioned issues drive the motivation of proposing a stable and feasible transformation from NES to MOP. On one hand, the additional information adopted to construct the objectives of an MOP should be easy to obtained. On the other hand, the multiple objectives should always conflict. Based on these two aspects, a bi-objective transformation combined with DE search algorithm, named BiTDE, is proposed in this paper. Specifically, the characteristics of BiTDE are stated as follows:

- The two constructed objectives always conflict with each other, which are fully suitable to the requirement of multiobjective optimization.
- We theoretically prove that all the optimal solutions of an NES are mapped into the nondominated solutions of the transformed MOP.
- As for the NESs with infinite optimal solutions, BiTDE can obtain the optimal solutions evenly distributed for decision makers.

To demonstrate the feasibility of our proposed bi-objective transformation, and verify the performance of the BiTDE, extensive experiments are conducted on 22 NESs collected from practical applications. The experiments on these 22 NESs show that our proposed approach is better or at least competitive against the compared state-of-the-art algorithms.

The reminder of this paper is organized as follows. The next section introduces the preliminary knowledge. Section 3 details our proposed BiTDE. Section 4 presents the experimental results and analyses. Finally, Sect. 5 draws the conclusion.

2 Preliminary Knowledge

2.1 Multiobjective Optimization Techniques

Generally, an MOP can be formulated as follows:

$$\text{Minimize} \quad F(\mathbf{x}) = (F_1(\mathbf{x}), \ldots, F_m(\mathbf{x}))$$
$$\text{subeject to } \mathbf{x} = (x_1, \ldots, x_D) \in S \tag{3}$$

where $\mathbf{x}$ is a decision vector with n dimensions, and $F(\mathbf{x})$ consists of m objectives that always conflict.

With regard to multiobjective optimization, four essential definitions are involved in our proposed approach.

Definition 1. *Given two vectors u and v, u Pareto Dominates v, denoted by $\prec$, if*

$$\forall i \in \{1, \ldots, m\}, \ F_i(u) \leq F_i(v) \ and \ \exists i \in \{1, \ldots, m\}, F_i(u) < F_i(v).$$

Definition 2. *The vector u is Pareto Optimal, if and only if*

$$\neg \exists v \in S, \ s.t. \ v \prec u.$$

Definition 3. *The set of all Pareto optimal vectors (i.e., Pareto set), denoted by PS, is defined as*

$$PS = \{u \in S | \neg \exists v \in S, v \prec u\}.$$

Definition 4. *The set of all Pareto optimal vectors mapped in objective space (i.e., Pareto front), denoted by PF, is defined as*

$$PF = \{F(u) | u \in PS\}.$$

In an MOP, the m objectives always conflict, and thus a set of tradeoff solutions would be finally obtained by multiobjective optimization techniques.

2.2 Differential Evolution

Differential evolution (DE) is a simple and efficient evolutionary algorithm, and originally proposed for single objective numerical optimization. Recently, it has been widely employed as a search engine to generate new candidate vectors by its mutation and crossover operators [5].

Mutation Operator. When generating new vectors or individuals of DE, differential information is utilized to create the mutant vectors first. Specifically, the ith mutant vector **v** generated by mutation operator (DE/rand/1) can be expressed as follows:

$$v_i = x_{r_1} + F_s \cdot (x_{r_2} - x_{r_3}) \tag{4}$$

where F_s is the scale factor, and x_{r_1}, x_{r_2}, x_{r_3} are three distinct vectors that randomly selected in the population.

Crossover Operator. Crossover operator is applied to produce a trial vector u_i based on target vector x_i and mutant vector v_i:

$$u_{i,j} = \begin{cases} v_{i,j}, \text{ if } rand_j \leq C_r, \text{ or } j == j_{rand} \\ x_{i,j}, \text{ otherwise} \end{cases} \tag{5}$$

where $rand_j$ is a random number uniformly generated from $[1,0]$, C_r is the crossover rate, and j_{rand} is an integer randomly selected from $\{1, \ldots, D\}$.

3 The Proposed BiTDE

3.1 Bi-objective Transformation

In our proposed transformation, two conflicting objectives are designed by the original objective function of NESs and additional information from the population.

As shown in (1) and (2), the objective function $f(\mathbf{x})$ has n equations, but these equations cannot be directly solved as numerical functions in evolutionary optimization. Hence, we modify the $f(\mathbf{x})$ to the following formulation:

$$\text{Minimize } f(\mathbf{x}) = \sum_{j=1}^{n} |f_j(\mathbf{x})| \tag{6}$$

The additional information for each individual is computing from a number of fixed reference points initiated randomly at the beginning of search. More specifically, the additional information ad for a given individual $\mathbf{x}_i$ is calculated as follows:

$$ad(\mathbf{x}_i) = \sum_{j=1}^{R_p} \sqrt{\sum_{t=1}^{D}(x_i^t - x_j^t)^2} \tag{7}$$

where R_p denotes the number of reference points.

Since multiobjective optimization also involves multiple optimal solutions, it has the technical feasibility to deal with NESs. However, the requirement of multiobjective optimization is that the multiple objectives have to conflict with each other. Hence, how to establish an effective transformation matching the requirement is the key to solve NESs by multiobjective optimization. To address this issue, a bi-objective transformation is proposed to construct **two totally conflicting objectives**:

$$\begin{cases} F_1(\mathbf{x}) = ad(\mathbf{x})_{norm} + \xi * f(\mathbf{x})_{norm} \\ F_2(\mathbf{x}) = 1.0 - ad(\mathbf{x})_{norm} + \xi * f(\mathbf{x})_{norm} \end{cases} \tag{8}$$

where $ad(\mathbf{x})_{norm}$ and $f(\mathbf{x})_{norm}$ represent the values obtained after min-max normalization, and ξ is a scaling factor controlled as follows:

$$\xi = \xi_o * (t/MaxGen)^2 \tag{9}$$

where ξ_o is an initial value, t is the current generation, and $MaxGen$ denotes the maximum evolution generation.

In (8), if $F_1(\mathbf{x})$ increases with the increasing of $\mathbf{x}$, $F_2(\mathbf{x})$ goes to decrease, and vice versa. Hence, $F_1(\mathbf{x})$ always conflict with $F_2(\mathbf{x})$. As a result, if an individual is one of the multiple optimal solutions of an NES, **it must be** the nondominated solution in the bi-objective problem described by (8).

Proof. If an individual of $\mathbf{u}$ is an optimal solution of an NES but not the nondominated solution of (8), then there must be an individual $\mathbf{v}$ that Pareto dominates $\mathbf{u}$, which is as follows:

$$\begin{cases} F_1(\mathbf{v}) < F_1(\mathbf{u}) \\ F_2(\mathbf{v}) \le F_2(\mathbf{u}) \end{cases} \text{(10a) or} \quad \begin{cases} F_1(\mathbf{v}) \le F_1(\mathbf{u}) \\ F_2(\mathbf{v}) < F_2(\mathbf{u}) \end{cases} \text{(10b)}$$

As for (10a), because $\mathbf{u}$ is an optimal solution, the value of $f(\mathbf{u})_{norm}$ is 0.0.

Thus, the expression can be simplified as follows:

$$\begin{cases} \xi * f(\mathbf{v})_{norm} \le ad(\mathbf{u})_{norm} - ad(\mathbf{v})_{norm} \\ \xi * f(\mathbf{v})_{norm} < -[ad(\mathbf{u})_{norm} - ad(\mathbf{v})_{norm}] \end{cases} \qquad (11)$$

since ξ and $f(\mathbf{x})_{norm}$ are positive, the two inequalities in (11) are contradictory. This means that $\mathbf{u}$ is a nondominated solution in (8). The situation in (10b) is the same.

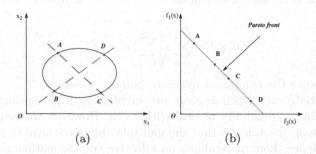

(a)   (b)

Fig. 1. The principle of bi-objective transformation. (a) Four optimal solutions A, B, C and D of an NES in decision space. (b) Converting the four optimal solutions into nondominated solutions on the Pareto front.

Figure 1 gives an example of the mapping relationship between an NES and an MOP. The four solid points are multiple optima of an NES, as shown in the decision space S in Fig. 1(a). Through the bi-objective transformation, they are converted into the nondominated solutions on the Pareto front, as shown in objective space $\Re^2$ in Fig. 1(b).

3.2   Solving the Transformed Problem

In essence, the existing multiobjective optimization techniques can cope with the transformed bi-objective problem straightforwardly. In this paper, the nondominated and crowding distance sorting technique [6], and DE search algorithm are adopted in BiTDE. The implementation of BiTDE is stated as follows:

Step 1. Set $t = 1$.
Step 2. Randomly generate N individuals as the initial population (P) and R_p reference points from the decision space S.
Step 3. Calculate the additional information ad for each individual in P based on (7).
Step 4. Calculate the current value of ξ.
Step 5. Evaluate each individual in P based on (8).
Step 6. Using the mutation and crossover operators to produce N new individuals as offspring population Q.

Step 7. Implement the fast nondominated and crowding distance sorting approaches sequentially to select N best individuals from the combined population $P \bigcup Q$.

Step 8. If $t < MaxGen$, update t to $t+1$ and go to **Step 4**, otherwise, output the nondominated solutions in P.

As for the **Step 7**, the implementations of the fast nondominated and crowding distance sorting approaches can refer to [6]. It is noteworthy that BiTDE does not introduce any serious burden on time complexity. As for calculating the additional information, the computational time complexity is $O(N)^2$ as well as the fast nondominated sorting, and hence, the total time complexity of BiTDE is still $O(N)^2$.

4 Experiments

Twenty-two NESs [8] are adopted to evaluate the performance of our proposed algorithm. The properties of these 22 NESs are given in Table 1, where LE and NE denote the linear and nonlinear equations respectively, NOS denotes the number of optimal solutions, and the maximum times of fitness evaluations ($MaxFEs$) is $5.0E + 04$ for each instance.

The proposed BiTDE is compared with six state-of-the-art EAs for NES. The six algorithms are A-MONES and A-MOMMOP [8] which also utilize multiobjective optimization techniques to cope with NESs, NCDE and NSDE [15] which are niching-method-based algorithms, and Rep-SHADE and Rep-CLPSO [8] which combine the repulsion strategy with two developed variants of DE and particle swarm optimization search algorithms. As for these six algorithms, the parameters are set same according to the original papers.

Table 1. Properties of 22 test NESs

Ins.	D	S	LE	NE	NOS	Ins.	D	S	LE	NE	NOS
NES01	2	$[-1,1]^2$	1	1	2	NES02	2	$[-1,1]^2$	1	1	11
NES03	2	$[-10,10]^2$	0	2	13	NES04	10	$[-2,2]^{10}$	0	10	1
NES05	5	$[-10,10]^5$	4	1	3	NES06	2	$[-2,2]^2$	0	2	10
NES07	2	$[-5,5]^2$	0	2	9	NES08	2	$[0,2\pi]^2$	0	2	13
NES09	2	$[-2,2]^2$	0	2	6	NES10	2	$[-2,2]^2$	0	2	4
NES11	2	$[-2,2]^2$	0	2	6	NES12	3	$[-5,5] \times [-1,3] \times [-5,5]$	0	3	2
NES13	2	$[0,1] \times [-10,0]$	0	2	2	NES14	2	$[0,2.5] \times [-4,6]$	0	2	4
NES15	2	$[-1,1] \times [-10,10]$	0	2	4	NES16	2	$[0.25,1] \times [1.5,2\pi]$	0	2	2
NES17	3	$[3,5] \times [2,4] \times [0.5,2]$	0	3	1	NES18	3	$[-1,-0.1] \times [-2,2]$	0	2	2
NES19	2	$[-5,1.5] \times [0,5]$	0	2	3	NES20	2	$[0,2] \times [10,30]$	0	2	2
NES21	3	$[-2,2]^3$	1	1	$Inf.$	NES22	20	$[-1,1]^{20}$	1	19	$Inf.$

4.1   Evaluation Criterion

Two commonly used evaluation criteria [12], i.e., the peak ratio (PR) and the successful rate (SR), are utilized to evaluate the performance of different algorithms under a given accuracy level ε. The PR and SR are computed as follows:

$$PR = \frac{\sum_{i=1}^{NR} NPF_i}{NKP \times NOS}, \quad SR = \frac{NSR}{NR} \tag{12}$$

where NPF_i is the number of optimal solutions found in the ith run, NR is the number of runs, and NSR denotes the number of successful runs where all optimal solutions are found.

4.2   Experimental Setup

As for the parameter settings, the population size N is set to 100 for BiTDE, ξ_o is set to 999, and $MaxGen = MaxFes/N$. When generating the offspring population by DE, the scale factor F_s and crossover rate C_r are randomly selected from the scaling factor pool $[0.6, 0.8, 1.0]$ and the crossover rate pool $[0.1, 0.5, 1.0]$ for every individual, respectively. All the compared algorithms run 50 times for each test instance of the 22 NESs.

4.3   Experimental Results and Comparisons

Table 2 summarizes the statistical results of PR and SR for the first 20 NESs with finite optimal solutions. It can be seen that the results obtained by BiTDE are very competitive. The BiTDE can successfully solve 16 test instances of the first 20 NESs where the PR and SR values are 100% for these test instances. However, all of the other six algorithms cannot cope with the $NES03$ and $NES08$ consecutively on 50 independent runs. A-MONES, A-MOMMOP, NCDE, NSDE, Rep-SHADE and Rep-CLPSO can solve 7, 9, 10, 10, 13 and 10 test instances, respectively. As for the rest of the four NESs, BiTDE still outperforms the other six algorithms on $NES05$ and $NES06$. Only on $NES07$ and $NES12$, BiTDE performs worse than the other five algorithms, i.e., A-MONES, A-MOMMOP, NCDE, NSDE, Rep-SHADE.

Moreover, the nonparametric Wilcoxon and the Friedman test have been adopted to test the statistical differences between BiTDE and the other six algorithms. Table 3 and Fig. 2 present the experimental results. Table 3 reveals that BiTDE has significant improvements over A-MONES, A-MOMMOP, NCDE, NSDE and Rep-CLPSO, respectively. Moreover, Fig. 2 shows that BiTDE ranks the first place among these seven compared algorithms based on Friedman's test values.

4.4   Further Study

When solving the first 20 NESs with finite optimal solutions, BiTDE achieves highly competitive performance. However, there is another kind of NESs which

Table 2. Comparison of the mean PR and SR obtained by the algorithms

PR/SR	BiTDE	A-MONES	A-MOMMOP	NCDE	NSDE	Rep-SHADE	Rep-CLPSO
NES01	1.00/1.00	1.00/1.00	1.00/1.00	1.00/1.00	1.00/1.00	1.00/1.00	1.00/1.00
NES02	1.00/1.00	1.00/1.00	1.00/1.00	0.98/0.88	0.96/0.66	0.98/0.86	0.94/0.62
NES03	1.00/1.00	0.97/0.86	0.50/0.00	0.64/0.00	0.81/0.04	0.77/0.08	0.50/0.00
NES04	1.00/1.00	1.00/1.00	1.0000/1.00	1.00/1.00	1.00/1.00	1.00/1.00	1.00/1.00
NES05	**0.89/0.72**	0.73/0.50	0.78/0.42	0.00/0.00	0.07/0.00	0.29/0.04	0.00/0.00
NES06	**0.98/0.92**	0.71/0.50	0.88/0.24	0.65/0.00	0.86/0.28	0.92/0.42	0.84/0.22
NES07	0.88/0.36	**0.99/0.96**	0.98/0.90	0.98/0.84	0.98/0.88	0.97/0.80	0.86/0.28
NES08	1.00/1.00	0.44/0.00	0.99/0.98	0.85/0.04	0.85/0.02	0.88/0.10	0.92/0.36
NES09	1.00/1.00	0.75/0.50	0.86/0.48	1.00/1.00	1.00/1.00	1.00/1.00	1.00/1.00
NES10	1.00/1.00	0.51/0.00	0.57/0.14	1.00/1.00	1.00/1.00	1.00/1.00	1.00/1.00
NES11	1.00/1.00	0.76/0.50	**1.00/1.00**	0.99/0.98	0.99/0.94	1.00/1.00	1.00/1.00
NES12	0.92/0.84	0.97/0.94	0.98/0.96	**0.99/0.98**	**0.99/0.98**	0.98/0.96	0.81/0.62
NES13	1.00/1.00	1.00/1.00	1.00/1.00	1.00/1.00	1.00/1.00	1.00/1.00	1.00/1.00
NES14	1.00/1.00	1.00/1.00	1.00/1.00	1.00/1.00	1.00/1.00	1.00/1.00	1.00/1.00
NES15	1.00/1.00	0.75/0.50	1.00/1.00	0.97/0.88	1.00/1.00	1.00/1.00	0.99/0.96
NES16	1.00/1.00	0.86/0.74	1.00/1.00	1.00/1.00	1.00/1.00	1.00/1.00	0.99/0.98
NES17	1.00/1.00	1.00/1.00	1.00/1.00	0.40/0.40	0.84/0.84	1.00/1.00	1.00/1.00
NES18	1.00/1.00	1.00/1.00	1.00/1.00	1.00/1.00	0.98/0.96	1.00/1.00	1.00/1.00
NES19	1.00/1.00	0.94/0.86	0.99/0.98	0.98/0.96	1.00/1.00	1.00/1.00	1.00/1.00
NES20	1.00/1.00	0.91/0.84	0.94/0.88	1.00/1.00	1.00/1.00	1.00/1.00	0.92/0.84
Total	**0.98/0.94**	0.86/0.73	0.92/0.79	0.87/0.75	0.91/0.78	0.94/0.81	0.88/0.74

Table 3. Wilcoxon signed ranks test results for BiTDE versus the selected algorithms.

Comparison	PR			SR		
BiTDE	R^+	R^-	p-value	R^+	R^-	p-value
vs. A-MONES	161.5	28.5	5.75E−03	159.5	30.5	8.00E−05
vs. A-MOMMOP	152.5	57.5	6.14E−02	154.0	56.0	5.14E−02
vs. NCDE	148.5	41.5	1.87E−02	149.0	41.0	1.76E−02
vs. NSDE	152.0	58.0	6.41E−02	154.5	55.5	3.45E−02
vs. Rep-SHADE	120.5	69.5	2.42E−01	121.0	69.0	2.73E−01
vs. Rep-CLPSO	182.5	27.5	2.61E−03	182.5	27.5	2.03E−03

have infinite optimal solutions in real-world applications. To investigate the capability of BiTDE to solve this kind of NESs, the performance of BiTDE on $NES21$ and $NES22$ with infinite optimal solutions are studied and compared in this section.

The PR and SR evaluation criterion are not suitable for measuring the algorithm performance on NESs with infinite optimal solutions. Thus, we plot the nondominated solutions in the final population obtained by BiTDE in the objective spaces straightforwardly.

Figure 3 shows the nondominated solutions obtained by BiTDE for $NES21$ and $NES22$, respectively. It can be seen that all the soft red points are on the line $f_2(x) = 1 - f_1(x)$, which means that these solutions are the optimal

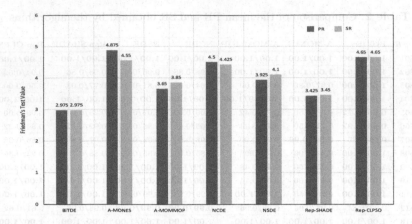

Fig. 2. Friedman's test for BiTDE, A-MONES, A-MOMMOP, NCDE, NSDE, Rep-SHADE and Rep-CLPSO.

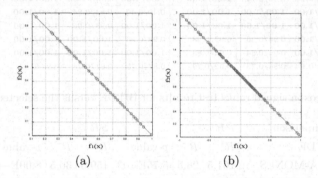

Fig. 3. The nondominated solutions obtained by BiTDE plotted in bi-objective space. (a) The NES21 test instance. (b) The NES22 test instance. (Color figure online)

solutions of the NESs. Usually the quality of the nondominated solutions is indicated by the hypervolume value [24]. To compare the nondominated solutions obtained from different algorithms, Table 4 records the hypervolume values of all the compared algorithms. For each algorithm, the nondominated solution obtained in the final population are preprocessed by the weighted bi-objective transformation technique [8] to have a fair comparison.

The hypervolume values obtained by BiTDE are larger than those of the other six algorithms on $NES21$ and $NES22$, respectively. This implies that BiTDE has a better capability to solve the NESs with infinite solutions than the other compared algorithms.

Table 4. Wilcoxon signed ranks test results for BiTDE versus the selected algorithms.

Ins./HV	BiTDE	A-MONES	A-MOMMOP	NCDE	NSDE	Rep-SHADE	Rep-CLPSO
NES21	0.475833	0.288931	0.298077	0.291552	0.291554	0.296955	0.300833
NES22	0.044486	0.027886	0.039333	0.031114	0.031185	0.034823	0.034317

5 Conclusion

In this paper, we have first proposed a bi-objective transformation for NESs. Since the two objectives strictly conflict with each other, the transformation is fully suitable for the requirement of multiobjective optimization, and thus the transformed problem is feasible to be solved by multiobjective optimization techniques. After the transformation, the nondominated solutions of the transformed bi-objective optimization problem contains all the optimal solutions of the original NES. Sequentially, the fast nondominated sorting approach and DE search algorithm are naturally implemented in our proposed transformation to locate the multiple optimal solutions of the original NES simultaneously.

The performance of BiTDE has been evaluated on 22 NESs test instances from real-world applications. Six state-of-the-art EAs with different strategies, such as multiobjective optimization techniques, niching methods and repulsion, are adopted to compare with our proposed algorithm. The experimental results show that BiTDE has high capabilities to solve most of the tested instances, and it is competitive with the six compared algorithms, especially on the NESs with infinite optimal solutions. As for future work, we plan to improve our proposed algorithm on more complex NESs from practical applications, and develop it to be more robust.

Acknowledgement. This work was supported by the Science and Technology Planning Project of Guangdong Province, China (Grant No. 2014B050504005).

References

1. Aleenejad, M., Ahmadi, R., Moamaei, P.: Selective harmonic elimination for cascaded multicell multilevel power converters with higher number of H-bridge modules. In: Power and Energy Conference at Illinois (PECI), pp. 1–5. IEEE (2014)
2. Bäck, T., Fogel, D.B., Michalewicz, Z.: Handbook of Evolutionary Computation. CRC Press, Boca Raton (1997)
3. Basak, A., Das, S., Tan, K.C.: Multimodal optimization using a biobjective differential evolution algorithm enhanced with mean distance-based selection. IEEE Trans. Evol. Comput. **17**(5), 666–685 (2013)
4. Campos, M., Krohling, R.A., Enriquez, I.: Bare bones particle swarm optimization with scale matrix adaptation. IEEE Trans. Cybern. **44**(9), 1567–1578 (2014)
5. Das, S., Mullick, S.S., Suganthan, P.N.: Recent advances in differential evolution-an updated survey. Swarm Evol. Comput. **27**, 1–30 (2016)
6. Deb, K., Pratap, A., Agarwal, S., Meyarivan, T.: A fast and elitist multiobjective genetic algorithm: NSGA-II. IEEE Trans. Evol. Comput. **6**(2), 182–197 (2002)

7. Deb, K., Saha, A.: Multimodal optimization using a bi-objective evolutionary algorithm. Evol. Comput. **20**(1), 27–62 (2012)
8. Gong, W., Wang, Y., Cai, Z., Yang, S.: A weighted biobjective transformation technique for locating multiple optimal solutions of nonlinear equation systems. IEEE Trans. Evol. Comput. **21**(5), 697–713 (2017)
9. Hu, X.M., Zhang, J., Chung, H.S.H., Li, Y., Liu, O.: Samaco: variable sampling ant colony optimization algorithm for continuous optimization. IEEE Trans. Syst. Man Cybern. Part B (Cybern.) **40**(6), 1555–1566 (2010)
10. Kennedy, J.: Particle swarm optimization. In: Sammut, C., Webb, G.I. (eds.) Encyclopedia of Machine Learning, pp. 760–766. Springer, Heidelberg (2011). https://doi.org/10.1007/978-0-387-30164-8_630
11. Li, X.: Efficient differential evolution using speciation for multimodal function optimization. In: Proceedings of the 7th Annual Conference on Genetic and Evolutionary Computation, pp. 873–880. ACM (2005)
12. Li, X., Engelbrecht, A., Epitropakis, M.G.: Benchmark functions for CEC 2013 special session and competition on niching methods for multimodal function optimization. Technical report, RMIT University, Evolutionary Computation and Machine Learning Group, Australia (2013)
13. Long, J., Szeto, W., Gao, Z., Huang, H.J., Shi, Q.: The nonlinear equation system approach to solving dynamic user optimal simultaneous route and departure time choice problems. Transp. Res. Part B: Methodol. **83**, 179–206 (2016)
14. Qu, B.Y., Suganthan, P.N., Das, S.: A distance-based locally informed particle swarm model for multimodal optimization. IEEE Trans. Evol. Comput. **17**(3), 387–402 (2013)
15. Qu, B.Y., Suganthan, P.N., Liang, J.J.: Differential evolution with neighborhood mutation for multimodal optimization. IEEE Trans. Evol. Comput. **16**(5), 601–614 (2012)
16. Song, W., Wang, Y., Li, H.X., Cai, Z.: Locating multiple optimal solutions of nonlinear equation systems based on multiobjective optimization. IEEE Trans. Evol. Comput. **19**(3), 414–431 (2015)
17. Stoean, C., Preuss, M., Stoean, R., Dumitrescu, D.: Multimodal optimization by means of a topological species conservation algorithm. IEEE Trans. Evol. Comput. **14**(6), 842–864 (2010)
18. Storn, R., Price, K.: Differential evolution-a simple and efficient heuristic for global optimization over continuous spaces. J. Global Optim. **11**(4), 341–359 (1997)
19. Turgut, O.E., Turgut, M.S., Coban, M.T.: Chaotic quantum behaved particle swarm optimization algorithm for solving nonlinear system of equations. Comput. Math. Appl. **68**(4), 508–530 (2014)
20. Turkyilmazoglu, M.: Determination of the correct range of physical parameters in the approximate analytical solutions of nonlinear equations using the Adomian decomposition method. Mediterr. J. Math. **13**(6), 4019–4037 (2016)
21. Van Hentenryck, P., McAllester, D., Kapur, D.: Solving polynomial systems using a branch and prune approach. SIAM J. Numer. Anal. **34**(2), 797–827 (1997)
22. Yao, J., Kharma, N., Grogono, P.: Bi-objective multipopulation genetic algorithm for multimodal function optimization. IEEE Trans. Evol. Comput. **14**(1), 80–102 (2010)
23. Yu, W.J., Ji, J.Y., Gong, Y.J., Yang, Q., Zhang, J.: A tri-objective differential evolution approach for multimodal optimization. Inf. Sci. **423**, 1–23 (2018)
24. Zitzler, E., Thiele, L.: Multiobjective evolutionary algorithms: a comparative case study and the strength Pareto approach. IEEE Trans. Evol. Comput. **3**(4), 257–271 (1999)

Process-Monitoring-for-Quality—
A Model Selection Criterion
for Genetic Programming

Carlos A. Escobar[1,3(✉)], Diana M. Wegner[1], Abhinav Gaur[2],
and Ruben Morales-Menendez[3]

[1] General Motors, Research and Development, Warren, MI 48092, USA
carlos.1.escobar@gm.com
[2] Michigan State University, East Lansing, MI 48823, USA
[3] Tecnológico de Monterrey, 64849 Monterrey, NL, Mexico

Abstract. *Process Monitoring for Quality* is a manufacturing quality
philosophy aimed at defect detection through binary classification that
is founded on *big data* and *big models*. *Genetic Programming (GP)* algorithms have been successfully applied by following the *big models* learning paradigm for rare quality event detection (classification). Since it is
a bias-free technique unmarred by human preconceptions, it can potentially generate better solutions (models) compared with the best human
efforts. However, since *GP* uses random search methods based on Dar-
winian philosophy of "survival of the fittest", hundreds, or even thousands of models need to be created to find a good solution. In this context,
model selection becomes a critical step in the process of finding the *final
model* to be deployed at the plant. A three-objective optimization *model
selection* criterion $(3D - GP)$ is introduced for analyzing highly/ultra
unbalanced data structures. It uses three competing attributes – prediction, separability, complexity – to project *candidate models* into a
three-dimensional space to select the *final model* that solves the posed
tradeoff between them the best.

Keywords: Genetic programming · Separability index ·
Model selection · Binary classification ·
Highly unbalanced data structures · Manufacturing

1 Introduction

Process Monitoring for Quality (PMQ) is a *big data*-driven quality philosophy
aimed at defect detection through binary classification [1]. It is founded on *Big
Models*, a predictive modeling paradigm based on machine learning, statistics
and optimization aimed at developing a manufacturing-functional model [12]. We
are living in an era of high conformance manufacturing environment where most
mature organizations generate only a few *Defects Per Million of Opportunities
(DPMO)*, therefore, manufacturing-derived data sets for binary classification of

© Springer Nature Switzerland AG 2019
K. Deb et al. (Eds.): EMO 2019, LNCS 11411, pp. 151–164, 2019.
https://doi.org/10.1007/978-3-030-12598-1_13

quality tend to be highly/ultra (minority class count < 1%) unbalanced. *PMQ* has the potential to solve a new range of hitherto intractable problems [14], where detecting these few *DPMO* is one of them. *Genetic Programming (GP)* has been successfully applied following the *big models* modeling paradigm to perform this task [1].

Inspired by biological evolution and its fundamental mechanisms, *GP* algorithm uses random mutation, crossover, a fitness function, and multiple generations of evolution to optimize a user-defined fitness-function. In classification, the input features are used to generate a random population of models, then the performance of each of them is evaluated based on the fitness-function. Models that perform better have a higher probability of being included in the breeding of the next generation [24]. Finally, the fittest models become *candidate models*, where oftentimes multi-attribute optimization or Pareto optimization techniques [7] are used to develop a Pareto frontier, a set of *Pareto Optimal (PO)* solutions that are not dominated by any other feasible solution. Model complexity and generalization ability (prediction on unseen data) are the most common competing attributes considered for *Model Selection (MS)*.

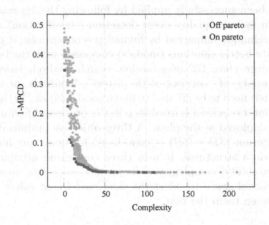

Fig. 1. Pareto optimization (source [1]), *PO* in green squares. (Color figure online)

GP is a technique free of human preconceptions or biases, that relies on computer power to develop a good predictive model. In addition to optimizing model parameters, the model structure is also optimized. This broad approach, stochastic approach allows better search and optimization as well as a better understanding of the natural physical processes which are being modeled [8]. However, the outcome can result in hundreds, thousands, or even tens of thousands of models are developed in this process. Although the Pareto frontier and pruning techniques help to downsize the list by identifying the non-dominated models (as shown in Fig. 1), the number of options in *PO* may still be huge. In such cases, engineers may be stuck with many alternatives and little guidance as to how to choose between them.

In this context, the separation distance between the two classes (separability) can be used to further discriminate from two competing models. This problem representation, highlights the importance of developing a three-objective optimization MS criterion.

A three-dimensional MS criterion $(3D - GP)$ based on prediction, separability, and complexity is developed. The three attributes are used to project each *candidate model* into a three dimensional space to select the *final model* that solves the posed tradeoff between them the best. Proposed criterion is supported by a novel *Separability Index (SI)*, which is aimed at analyzing highly/ultra unbalanced data structures. MS is the main focus of this paper.

The rest of the paper is organized as follows. Acronyms in Table 1, a brief theoretical background is in Sect. 2. The proposed SI is described in Sect. 3. Section 4 describes the MS criterion. Section 5 shows the conclusions and future research.

Table 1. Acronyms definition

Acronyms	Definition
CM	Candidate Model
CT	Classification Threshold
DPMO	Defect Per Million of Opportunities
EMO	Evolutionary Multi-Objective Optimization
FN	False Negative(s)
FP	False Positive(s)
GA	Genetic Algorithm(s)
GP	Genetic Programming
MM	Min-Max
MPCD	Maximum Probability of Correct Decision
MS	Model Selection
PMQ	Process Monitoring for Quality
PO	Pareto Optimal
SI	Separability Index
SVM	Support Vector Machine
TN	True Negative(s)
TP	True Positive(s)
UMW	Ultrasonic Metal Welding
VS	Validation Set
3D	Three Dimension

2   Theoretical Background

2.1   Genetic Programming Algorithm – Model Development Process

GP [18] is an evolutionary computational method to learn predictive models. In this section, the model development method based on a generic multi-objective *GP* is described, with the following remarks:

- *Evolutionary Multi-Objective Optimization (EMO)* is an efficient way to develop classification models and
- once a set of *candidate models* have been developed by the algorithm, $3D-GP$ criterion can be applied to select the *final model*.

2.2   A Primer on Genetic Programming

A *GP* can be considered an application of *Genetic Algorithms (GA)* [17] when the space of solutions to search consists of programs or equations for solving a task [4]. Figure 2 shows a flowchart of a generic *GA*. Instead of decision variables, an individual is a program or an equation to solve a task. In order to create an initial population, a *terminal* set $\mathcal{T}$ and a *function* set $\mathcal{F}$ must be pre-specified based on the application. A terminal set consists of constants and variables where as a function set consists of operators or basic functions. $\mathcal{F} = \{+, -, \times, \div\}$ and $\mathcal{T} = \{x_i \; \forall i \in \{1, 2, \ldots, n_x\}\}$ where x_i's are problem features, are examples of function set and terminal sets. Based on the application, an extended operator set (i.e. square root, power, log, min, max) may be needed to better model a physical process. An individual in *GP* can be represented using different data structures such as string of words [6], or trees [28] or graphs [26]. In this work, the tree data structure is used to represent a solution/equation, hence a few important concepts in the context of tree-based *GP*s will be discussed.

Consider a *GP* with terminal set $\mathcal{T} = \{1, 2, x\}$ and function set $\mathcal{F} = \{+, -, *\}$. The values '1' and '2' in terminal set represent the set ephemeral constants allowed in the classification rules and let x be a feature. Then, Fig. 3 shows two candidate solutions that belong to the set of valid *GP* individuals for such a *GP*. Furthermore, *sub-tree swap crossover* [3] is a popular crossover mechanism used in tree based *GP*s. A sub-tree to be exchanged between two *GP* individuals (parents) is first chosen at random in each parent. Then, the subtree crossover operation is completed by exchanging the chosen subtrees between the two parents. A Koza-style subtree mutation [18] involves swapping either a terminal with another element from the terminal set or a function with another element from the function set. When swapping functions, care must be taken to maintain the arity of functions being swapped are the same. Bloat is known issue with single objective *GP* algorithms [20]. Bloating of *GP* trees causes the *GP* tree sizes to grow very fast without significant improvement in the model prediction ability. One of the ways of keeping bloat in check is to use multi-objective optimization [5]. Model error and model complexity are two most common objectives to minimize in the model development process. Examples of additional modeling objectives which can be added include model dimensionality and model age.

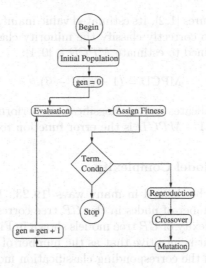

Fig. 2. A flowchart of the working principle of a genetic algorithm (Source [9]).

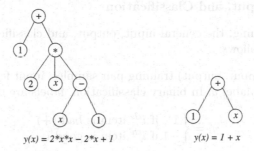

Fig. 3. Example of two *GP* solutions using tree representation.

2.3 Objective 1: Maximum Probability of Correct Decision

In predictive analytics, a confusion matrix (Table 2) [15] is a table with two rows and two columns that reports the performance of a classifier function by including the number of *False Positives (FP)*, *False Negatives (FN)*, *True Positives (TP)*, and *True Negatives (TN)*. This allows more detailed analysis than just the proportion of correct guesses (accuracy). A type-I error (α) may be compared with a *FP* prediction; a type-II (β) error may be compared with a false *FN* [10]. They are defined as:

$$\alpha = \frac{\text{FP}}{\text{FP} + \text{TN}}, \quad \beta = \frac{\text{FN}}{\text{FN} + \text{TP}}. \tag{1}$$

where α is the incorrect rejection of a true null hypothesis, and β is the incorrect retain of a false null hypothesis.

The *Maximum Probability of Correct Decision (MPCD)* is a probabilistic-based measure of classification performance aimed at analyzing highly/ultra unbalanced data structures [1,2]. Its estimated value mainly describes the ability of a *candidate model* to correctly classify the minority class (detection). The α and β values are combined to estimate $MPCD \in [0,1]$:

$$MPCD = (1 - \alpha)(1 - \beta). \tag{2}$$

where higher score indicates better classification performance. In the multi-objective GP method, $1 - MPCD$ is the error function to be minimized.

2.4 Objective 2: Model Complexity

Model complexity can be defined in many ways [19,23]. In this research, it is defined as the total number of nodes in the GP tree corresponding to a model. For example, the complexity of GP tree models shown in Fig. 3 is eight and three respectively. However, it is intuitive that as the number of nodes in a tree rises, the number of terms in the corresponding classification model also rises.

2.5 Input, Output, and Classification

In supervised learning, the general input, output, and classification process for GP is defined as follows:

- **Input:** set of (input, output) training pair samples; input features x, and the associated class label y. In binary classification, labels are defined as:

$$y_i = \begin{cases} 1 & \text{if } i^{th} \text{ item is bad } (+) \\ -1 & \text{if } i^{th} \text{ item is good } (-) \end{cases} \tag{3}$$

- **Output:** a discriminative function $z_i = f(x_i)$ and its associated *classification threshold (CT)*.
- **Classification:** the discriminative function value z_i is compared to the CT to assign the predicted label $(\hat{y})$ to an item.

$$\hat{y}_i = \begin{cases} 1 & \text{if } z_i \geq CT \Rightarrow i^{th} \text{ item is predicted bad } (+) \\ -1 & \text{if } z_i < CT \Rightarrow i^{th} \text{ item is predicted good } (-). \end{cases} \tag{4}$$

Although the multi-objective GP is effective in controlling bloat and producing a diverse set of *candidate models*, most of the time engineers are left with the problem of choosing the *final model*. This is where the idea of $3D - GP$ is helpful as explained in coming sections.

3 Separability Index

The large margin theory, which was originally applied to explain the success of boosting [25] and to develop the *Support Vector Machine (SVM)* algorithm [22,29] plays a crucial role in modern machine learning research. Based on this concept, a larger gap between classes is preferred to avoid the negative effects of redundant and noisy examples. Several studies have shown that the generalization performance of a classifier is related to the distribution of its margins [16].

To evaluate the robustness of the predictions of a classifier, the third model attribute is the *SI* introduced here, which is aimed at analyzing highly/ultra unbalanced data structures. Its score is computed based on the confusion matrix (Table 2) and the absolute difference of each predicted value to the *CT*. First, $\hat{y}_i$ is used to populate the confusion matrix, then the *SI* uses the absolute difference to reward for correct classifications (*TN*, *TP*) and to penalize for miss-classifications (*FN*, *FP*) in the *Validation Set (VS)*.

Table 2. Confusion matrix.

	Declare good	Declare bad
Good	True Negative (TN)	False Positive (FP)
Bad	False Negative (FN)	True Positive (TP)

Proposed formulation is broken down into two terms; the former (SI_{t1}) analyzes the majority class (e.g., good) and the later (SI_{t2}) the minority class (e.g., bad). Both terms are scaled (divided) by their total count, therefore they represent the average distance to the threshold. Finally, to make the *SI* score highly sensitive to *FN* – fail to detect – both terms are multiplied:

$$SI = \frac{(\sum_{tn=1}^{TN} |z_{tn} - CT| - \sum_{fp=1}^{FP} |z_{fp} - CT|)}{TN + FP} \times \frac{(\sum_{tp=1}^{TP} |z_{tp} - CT| - \sum_{fn=1}^{FN} |z_{fn} - CT|)}{TP + FN}. \tag{5}$$

where CT = classification threshold, TN = # of TN in the VS, FP = # of FP in the VS, TP = # of TP in the VS, FN = # of FN in the VS, $z_t n$ = predicted value of the TN_{tn}, $z_f p$ = predicted value of the FP_{fp}, $z_t p$ = predicted value of the TP_{tp}, $z_f n$ = predicted value of the FN_{fn}.

Virtual scenarios described in Fig. 4 are used to illustrate the properties of the *SI*. To perform this analysis, the confusion matrix is used to compute *MPCD* and the *SI* of each scenario, Table 3. With a virtual validation sample size of 3003 (3000 - *good*, 3 - *bad*), triangles are used to denote *good* units and circles for *bad* units, negative predicted values are classified as *bad*, and positive predictive values as *good*, and $CT = 0$.

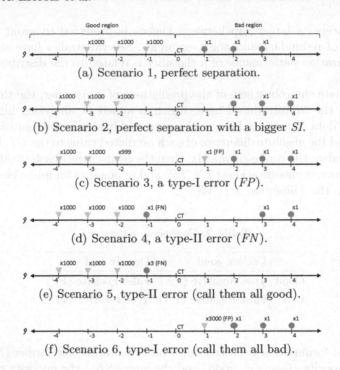

(a) Scenario 1, perfect separation.

(b) Scenario 2, perfect separation with a bigger *SI*.

(c) Scenario 3, a type-I error (*FP*).

(d) Scenario 4, a type-II error (*FN*).

(e) Scenario 5, type-II error (call them all good).

(f) Scenario 6, type-I error (call them all bad).

Fig. 4. Virtual scenarios.

Table 3. Associated *SI* scores by scenarios.

Scenario	TN	TP	FN	FP	α	β	MPCD	SI_{t1}	SI_{t2}	SI
S1	3000	3	0	0	0	0	1	2	2	4
S2	3000	3	0	0	0	0	1	3	3	9
S3	2999	3	0	1	3.33e−4	0	0.9997	2.999	3	8.997
S4	3000	2	1	0	0	0.3333	0.6667	3	2	6
S5	3000	0	3	0	0	1	0	3	−1	−3
S6	0	3	0	3000	1	0	0	−1	3	−3

Analysis 1. Scenario 1 (Fig. 4(a)) shows a classifier that perfectly separates the data, $MPCD = 1$ and $SI_{S1} = 4$. Scenario 2 (Fig. 4(b)) also shows a classifier that perfectly separates the data, but the $SI_{S2} = 9$ helps to further discriminate between S1 and S2. With the later being the best option.

Analysis 2. To demonstrate the sensibility of the proposed index with respect to *FN*, virtual Scenarios 3–4 (Fig. 4(c), (d) respectively) are also presented and analyzed. Using Scenario 2 as a reference, one type-I (*FP*), one type-II (*FN*) and two type-I errors are committed:

- if one type-I error (Scenario 3) is committed, $MPCD = 0.9997$ and $SI_{S3} = 8.997$, both scores show a decrease of 0.03% compared to SI_{S2}.
- if one type-II error (Scenario 4) is committed, $MPCD = 0.6667$ and $SI_{S4} = 5$, both scores show a decrease of 33.33% compared to SI_{S2}.

Analysis 3. Scenario 5 shows a common situation when analyzing highly/ultra unbalanced data, where the classifier fails to detect, this is an illustrative example where common measures of classification performance (e.g., accuracy) fail in evaluating a classifier under these conditions, since $accuracy = 0.9997$ is misleading. On the other hand, scenario 6 (less common) classified all the examples as bad, in both cases $MPCD = 0$ and $SI = -3$.

Based on analyzes 1–2 it is illustrated how the $MPCD$ and SI can work together in selecting the *final model*. If only the measure of classification performance is used, there is no difference between scenarios 1–2 (if same complexity is assumed), this situation highlights the importance of including the SI in the MS process. On the other hand, analysis 2 shows how $MPCD$ and SI would penalize a classifier for failing to detect (FN). Analysis 3 shows that the SI can take negative values, in such cases *candidate models* are not good, and should not be included in the model selection process, since the $3D - GP$ criterion is based on the *Euclidean* distance.

The SI index is a relative measure of classification robustness, that requires a measure of classification performance (e.g., $MPCD$) to perform a comprehensive classifier evaluation.

4   The *MS* Criterion

The $3D - GP$ MS criterion uses the three attributes (prediction, separability, complexity) to project each *candidate model* into a three dimensional space to select the *final model* that solves the posed tradeoff between them the best. The first two attributes work together in rewarding a model with high prediction and high separability abilities. Although these values tend to be highly correlated, adding the SI in the MS process, helps to further discriminate between competing models with similar prediction and complexity, as shown in the virtual case in Sect. 3. Since over-complex models are usually not trusted by engineers (and therefore never deployed), proposed criterion, also penalizes for model complexity.

4.1   Euclidean Distance

The *Euclidean* distance is a straight forward way of representing distance between two points in the *Euclidean* space [11]. In *Euclidean* three-dimensional space (attributes - x, y, z), the distance between points (x_1, y_1, z_1) and (x_2, y_2, z_2) is given by:

$$E = \sqrt{(x_2 - x_1)^2 + (y_2 - y_1)^2 + (z_2 - z_1)^2}. \tag{6}$$

If the relative importance of each attribute is known, weighted *Euclidean* distance can be used:

$$E_w = \sqrt{w_x(x_2 - x_1)^2 + w_y(y_2 - y_1)^2 + w_z(z_2 - z_1)^2}. \qquad (7)$$

The range of all attributes should be normalized [21] to compute the final distance. Since the measure of classification performance, $MPCD$, is scaled to $[0, 1]$, $Min - Max(MM)$ normalization was the rescaling used method, given attribute x, its normalized value is computed by:

$$MM(x_i) = \frac{x_i - min(x)}{max(x) - min(x)}. \qquad (8)$$

4.2   Attributes

1. **Prediction (detection):** a rewarding attribute based on generalization (validation $MPCD$), $p = 1 - MPCD$ (smaller better). It is aimed at selecting a model with a high detection ability. Normalization is not needed.
2. **Separability (robustness):** a rewarding attribute based on the SI, $s = 1 - SI$ (smaller better). It is aimed at selecting a model with a high separability distance between classes. Normalization is needed.
3. **Complexity:** a penalizing attribute based on the complexity defined by the classifier, c (smaller better). It is aimed at preventing over-complexity. Normalization is needed.

For each *candidate model*, (CM_i^m), $i = 1, ..., m$, – where m is the number of models – the tree associated attribute values are mapped into a three-dimensional space and the weighted *Euclidean* (E_i) distance to the *utopian point* $(0, 0, 0)$ is computed, Eq. 9. Then, the closest model $(3D - GP^*)$ is selected, Eq. 10. In this context, the *utopian point*, is an ideal model that optimizes the three attribute-functions simultaneously; however, most of the times, a model cannot be improved in any of the attributes without degrading at least one of the others.

$$E_{wi} = \sqrt{(p_i - 0)^2 + (s_i - 0)^2 + 0.01 \times (c_i - 0)^2}. \qquad (9)$$

$$3D - GP^* = min(E_{wi})_i^m. \qquad (10)$$

The basic idea of keeping the complexity weight $(w_c = 0.01)$ relatively low, is to maintain prediction and separability as the main drivers; thus, the criterion is not hampered from selecting a model with high predictive capacity. However, this weight can be increased to induce lesser complexity. Penalization for complexity helps to eliminate over-complexity.

4.3 The *MS* Process

To illustrate the *MS* process, a set of 25 *GP candidate models* are developed using the two objective functions described in Sect. 2.1 with a data set derived from *Ultrasonic Metal Welding (UMW)* [27] of battery tabs for the Chevrolet Volt [1]. A very stable process, that only generates a few defective welds per million of opportunities. The data set has 54 features and a binary outcome (*good/bad*). It is highly unbalanced, since it contains only 36 bad batteries out of 30,731 examples (0.09%). Because manufacturing systems tend to be time-dependent, the data set is partitioned following the time-ordered hold-out validation scheme [12]: training set (18,495 - including 20 *bads*), validation set (12,236 - 9 *bads*).

Candidate model information is summarized in Fig. 5. The *MS* process is illustrated in Fig. 6, first, the three attributes of each CM_i are mapped into a three dimensional space, Fig. 6(a), then, the weighted *Euclidean* distance of each CM_i is computed, and the one closest to the *utopian point* is selected, Fig. 6(b).

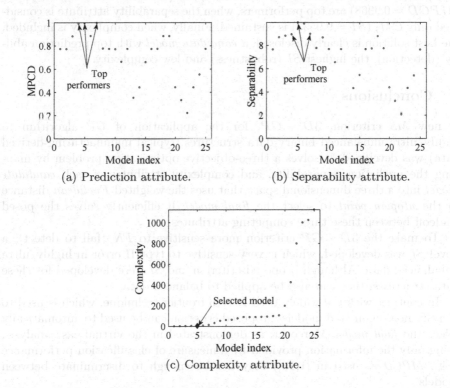

(a) Prediction attribute.   (b) Separability attribute.

(c) Complexity attribute.

Fig. 5. Candidate model information.

According to the *MS* criterion, the model that optimizes the three attributes the best is CM_5. With an estimated $E_5 = 0.0062$, and attribute values of

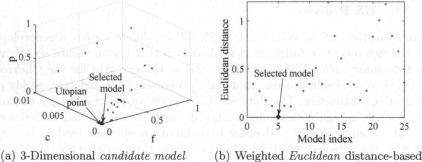

(a) 3-Dimensional *candidate model*
location.

(b) Weighted *Euclidean* distance-based
MS.

Fig. 6. *MS* process based on the weighted *Euclidean* distance.

$MPCD = 0.9937$, $SI = 9.9533$ and $c = 5$. As shown in Fig. 5, from prediction perspective CM_4 ($MPCD = 0.9906$), CM_5 ($MPCD = 0.9937$) and CM_7 ($MPCD = 0.9968$) are top performers, when the separability attribute is considered only CM_5 ($SI = 9.9533$) is sustained. Finally, when complexity is included, the best solution is clearly disclosed, a *candidate model* with top prediction ability (detection), the highest SI (robustness) and low complexity.

5   Conclusions

A new MS criterion, $3D - GP$, for the application of GP algorithm to highly/ultra unbalanced binary data structures (typical manufacturing-derived data) was developed. It solves a three-objective optimization problem by mapping the prediction, separability, and complexity attributes of each *candidate model* into a three dimensional space that uses the weighted *Euclidean* distance to the *utopian point* to select the *final model*. It efficiently solves the posed tradeoff between these three competing attributes.

To make the $3D - GP$ criterion more sensitive to FN (fail to detect), a novel SI was developed, which is very sensitive to type-II error in highly/ultra unbalanced data. Although proposed criterion and SI were developed for these data structures, they can also be applied to balanced data.

In contrast with the widely used Pareto frontier technique, which is used to identify non-dominated models, proposed criterion can be used to automatically select the *final model*. Moreover, as demonstrated in the virtual case analysis, using only the information provided by a measure of classification performance (e.g., $MPCD$ as used in [1,13]) may not be enough to discriminate between models.

Rare quality event detection is one of the main applications of PMQ, this task is broken down into classifier creation and MS. Proposed criterion addresses the research challenges of selecting the *final model* when GP is applied for this task.

Future research can focus on extending the application of the proposed criterion to the support vector machine algorithm. Another future research along this path, would be to develop a new classifier that uses the proposed *SI* as the target function.

References

1. Abell, J.A., Chakraborty, D., Escobar, C.A., Im, K.H., Wegner, D.M., Wincek, M.A.: Big data driven manufacturing—process-monitoring-for-quality philosophy. ASME J. Manuf. Sci. Eng. Data Sci.-Enhanced Manuf. **139**(10) (2017)
2. Abell, J.A., Spicer, J.P., Wincek, M.A., Wang, H., Chakraborty, D.: Binary Classification of Items of Interest in a Repeatable Process. US Patent (US8757469B2), June 2014. www.google.com/patents/US20130105556
3. Angeline, P.J.: Subtree crossover: building block engine or macromutation. Genet. Program. **97**, 9–17 (1997)
4. Banzhaf, W., Nordin, P., Keller, R.E., Francone, F.D.: Genetic Programming: An Introduction, vol. 1. Morgan Kaufmann, San Francisco (1998)
5. Bleuler, S., Brack, M., Thiele, L., Zitzler, E.: Multiobjective genetic programming: reducing bloat using SPEA2. In: Proceedings of the 2001 Congress on Evolutionary Computation, pp. 536–543 (2001)
6. Brameier, M.F., Banzhaf, W.: Linear Genetic Programming. Springer, Heidelberg (2007). https://doi.org/10.1007/978-0-387-31030-5
7. Branke, J., Deb, K., Miettinen, K., Słowiński, R. (eds.): Multiobjective Optimization: Interactive and Evolutionary Approaches. LNCS, vol. 5252. Springer, Heidelberg (2008). https://doi.org/10.1007/978-3-540-88908-3
8. Chaudhari, N.S., Tiwari, A., Purohit, A.: Genetic programming for classification. Int. J. Comput. Electron. Eng. IJCEE **1**, 69–76 (2009)
9. Deb, K.: Multi-objective Optimization Using Evolutionary Algorithms, vol. 16. Wiley, Hoboken (2001)
10. Devore, J.: Probability and Statistics for Engineering and the Sciences. Cengage Learning, Boston (2015)
11. Deza, M.M., Deza, E.: Encyclopedia of distances. In: Deza, M.M., Deza, E. (eds.) Encyclopedia of Distances, pp. 1–583. Springer, Heidelberg (2009). https://doi.org/10.1007/978-3-642-00234-2_1
12. Escobar, C.A., Abell, J.A., Hernández-de Menéndez, M., Morales-Menendez, R.: Process-monitoring-for-quality—big models. Procedia Manuf. **26**, 1167–1179 (2018)
13. Escobar, C.A., Morales-Menendez, R.: Process-monitoring-for-quality—a model selection criterion. Manuf. Lett. **15 Part A**, 55–58 (2018)
14. Escobar, C.A., Wincek, M.A., Chakraborty, D., Morales-Menendez, R.: Process-monitoring-for-quality—applications. Manuf. Lett. **16**, 14–17 (2018)
15. Fawcett, T.: An introduction to ROC analysis. Pattern Recogn. Lett. **27**(8), 861–874 (2006)
16. Feng, W., Huang, W., Ren, J.: Class imbalance ensemble learning based on the margin theory. Appl. Sci. **8**(5), 815 (2018)
17. Goldberg, D.E.: Genetic Algorithms. Pearson Education India (2006)
18. Koza, J.R.: Genetic Programming II, Automatic Discovery of Reusable Subprograms. MIT Press, Cambridge (1992)
19. Le, N., Xuan, H.N., Brabazon, A., Thi, T.P.: Complexity measures in genetic programming learning: a brief review. In: 2016 IEEE Congress on Evolutionary Computation, pp. 2409–2416 (2016)

20. Luke, S.: Issues in scaling genetic programming: breeding strategies, tree generation, and code bloat. Ph.D. thesis, Research Directed by Department of Computer Science, University of Maryland, College Park (2000)

21. Mohamad, I.B., Usman, D.: Standardization and its effects on K-means clustering algorithm. Res. J. Appl. Sci. Eng. Technol. **6**(17), 3299–3303 (2013)

22. Murphy, K.: Machine Learning: A Probabilistic Perspective. MIT Press, Cambridge (2012)

23. Ni, J., Rockett, P.: Tikhonov regularization as a complexity measure in multiobjective genetic programming. IEEE Trans. Evol. Comput. **19**(2), 157–166 (2015)

24. Poli, R., Langdon, W.B., McPhee, N.F., Koza, J.R.: A Field Guide to Genetic Programming. Lulu.com, Morrisville (2008)

25. Schapire, R.E., Freund, Y., Bartlett, P., Lee, W.S., et al.: Boosting the margin: a new explanation for the effectiveness of voting methods. Ann. Stat. **26**(5), 1651–1686 (1998)

26. Schmidt, M., Lipson, H.: Distilling free-form natural laws from experimental data. Science **324**(5923), 81–85 (2009)

27. Shao, C., et al.: Feature selection for manufacturing process monitoring using cross-validation. J. Manuf. Syst. **10** (2013)

28. Silva, S., Almeida, J.: GPLAB-A genetic programming toolbox for MATLAB. In: Proceedings of the Nordic MATLAB Conference, pp. 273–278. Citeseer (2003)

29. Vapnik, V.: The Nature of Statistical Learning Theory. Springer, Heidelberg (2013). https://doi.org/10.1007/978-1-4757-3264-1

Evolutionary Many-Constraint Optimization: An Exploratory Analysis

Mengjun Ming, Rui Wang[✉], and Tao Zhang

College of System Engineering, National University of Defense Technology,
Changsha 410073, Hunan, People's Republic of China
ruiwangnudt@gmail.com

Abstract. Many engineering optimization problems involve handling constraints. Existing constraint-handling methods, dealing with all constraints simultaneously as a whole, may become less effective when the number of constraints is large, termed many-constraint optimization problems (MCOPs). Since different constraints usually pose different degrees of difficulty to optimization problems (the constraint satisfying order may also be defined by a decision-maker), intuitively, a potential way is to progressively introduce each constraint into the search wherein the constraint-handling order becomes crucial. However, MCOPs and the problem-solver are far from being well investigated. This study therefore fills in this research gap. First, MCOPs are formulated, followed by an analysis of the difficulty of MCOPs. Then the concept of constraint-ranking is introduced. Based on the ranking results, a novel framework, i.e., cascaded constraint-handling (CCH), that follows "the most interesting first" principle is proposed to solve MCOPs. This implicitly enables the search to start from both interior and exterior of the feasible region. To demonstrate the effectiveness of the CCH framework, first an MCOP benchmark suite is designed. Then the penalty function based constraint-handling technique with and without the CCH is compared. Experimental results clearly show the superiority of the CCH framework.

Keywords: Many-constraint optimization problem (MCOP) ·
Constrained optimization ·
Cascaded constraint-handling (CCH) framework ·
Evolutionary algorithms

1 Introduction

Constrained optimization problems (COPs) arise regularly in many engineering applications. In COPs the goal is to find a solution that satisfies all constraints and also optimizes the objective function. In the last two decades, evolutionary algorithms (EAs), integrated with constraint-handling techniques, have attracted a great deal of attention to handle COPs [4,14]. Some of representative methods include, for example, penalty function based methods [1,8,11], multi-objective approaches [2,3,22], feasibility-rule based methods [7,12], stochastic ranking based methods [16] and ϵ-constraint method [19].

© Springer Nature Switzerland AG 2019
K. Deb et al. (Eds.): EMO 2019, LNCS 11411, pp. 165–176, 2019.
https://doi.org/10.1007/978-3-030-12598-1_14

The literature has revealed that these methods have their own advantages and disadvantages, thus, exhibiting differently on various problems [17]. However, the above methods share a common ground, i.e., dealing with all constraints simultaneously as a whole. They may be less effective on many-constraint optimization problems (MCOPs), for the reasons that—(i) searching a feasible solution in an MCOP is more difficult than in a general COP. This is because the feasible search space becomes even more complex; and (ii) the *handling-difficulty* of different constraints is not considered. That is, a solution that violates an easy constraint or a hard constraint is treated equally. Additionally, when the number of constraints is large, decision-makers may have their priorities, i.e., some of constraints should be satisfied first. The above institutive observation motivates us to re-investigate constrained optimization problems, especially when the number of constraints is large.

This study therefore presents a systematic analysis of MCOPs. In order to effectively handle MCOPs, a methodology adopting the "divide-and-conquer" concept is proposed. It first divides all constraints into different levels according to their *handling-difficulty*. Then the constraints are sequentially added into the evolutionary search. In this study "the most interesting first" principle is followed which aims to maximize the degree of constraint satisfaction of decision-makers. It is worth noting that when decision-makers have no priority on constraints "the most interesting one" is interpreted as "the most difficult one" so as to distribute as much search effort as possible to handle those challenging constraints.

Overall the main contributions of this study are as follows. (i) Limitations of constraint-handling techniques in literature are discussed. Correspondingly, the challenge of many-constraint optimization is highlighted. (ii) Based on the scalable many-objective test problems, DTLZ test suite [6], a set of MCOP benchmarks is designed. (iii) A simple yet effective framework, namely, the cascaded constraint-handling (CCH) is proposed. The initial results show that the CCH framework can significantly improve the effectiveness of penalty function based constraint-handling technique on MCOPs.

The rest of this paper is organized as follows. Section 2 formulates MCOPs and presents the motivation. Section 3 elaborates the proposed CCH framework, following the introduction of constraint ranking. In Sect. 4, the performance of the CCH framework is examined by comparing the only use of penalty function and the CCH integrated version on a set of newly designed benchmarks. Section 5 concludes the paper and identifies future studies.

2   MCOP: Formulation and Motivation

Without loss of generality, a constrained optimization problem can be formulated as follows [2]:

$$
\begin{aligned}
&\min f(\mathbf{x}), \mathbf{x} = (x_1, x_2, \ldots, x_n) \\
&\underline{x}_i \leq x_i \leq \overline{x}_i, i = 1, 2, \ldots, n \\
&g_j(\mathbf{x}) \leq 0, j = 1, \ldots, q \\
&h_j(\mathbf{x}) = 0, j = q + 1, \ldots, m
\end{aligned}
\tag{1}
$$

where $\underline{x}_i$ and $\overline{x}_i$ are the lower and the upper bound of x_i, respectively, and q is the number of inequality constraints and $m - q$ is the number of equality constraints. Generally, an equality constraint is handled by converting it into an inequality one. That is, $h_j(\mathbf{x}) = 0$ is converted into $|h_j(\mathbf{x})| - \delta \leq 0$ with δ being a positive close-to-zero number [15]. Thus the feasible region (F) can be generalized as:

$$g_j(\mathbf{x}) \leq 0, \forall j \in \{1, 2, \ldots, m\} \tag{2}$$

When the number of constraints is large, the general COP is termed **many-constraint optimization problem (MCOP)**.

We argue that MCOPs should be paid more attention since many engineering problems involve hundreds of or even more constraints. These constraints usually pose different degrees of difficulty to the problem. However, the literature so far often deals with constraints simultaneously as a whole, which, in other words, has not considered the difference amongst constraints. Thus, the existing constraint-handling methods need to be re-examined when dealing with MCOPs.

Given a large number of constraints, locating feasible solutions is already difficult, not to mention finding the optimum. Thus, eliciting information from infeasible solutions becomes crucial and helpful. In this sense, the priorities over different constraints in MCOPs could serve as a baseline for selecting potential infeasible solutions. More specifically, considering two infeasible solutions, the one that satisfies high-priority constraints is preferred. Certainly, this requires to rank constraints appropriately which will be introduced in the next section.

3 MCOP: Methodology

3.1 Constraint Ranking

As previously mentioned, constraints in an MCOP feature different degrees of difficulty, that is, some are easy to be satisfied while others are difficult. In addition to the difficulty posed by constraints themselves, decision-makers may also have their priorities on the constraint satisfaction. In view of the difference amongst constraints, it is natural to consider ranking them, which can therefore provide a standard when maximizing the degree of constraint satisfaction.

In accordance with the above cognition, "the most interesting first" principle is adopted. Specifically, the interest level (L) of a constraint (e.g., g_j) can be determined either by the preference information from decision-makers or the degree of *handling-difficulty*.

Here, the *handling-difficulty* of g_j is measured by the proportion of solutions that dissatisfy it, as shown in Eq. (3).

$$L(g_j) = \frac{N_j^S}{N}, \quad j = 1, 2, \ldots, m \tag{3}$$

where N_j^S is the number of solutions violating against the j-th constraint in the initially-produced population and N is the population size. Note that in addition to this naive strategy, other advanced indicators can be employed which also deserve further study. And if decision-makers have priorities for some constraints, the corresponding interest levels can be directly determined by the decision-makers.

Assuming that the interest level order of m constraints in an MCOP is as follows.

$$L(g_{j_1}) \geqslant L(g_{j_2}) \geqslant \ldots \geqslant L(g_{j_m}) \tag{4}$$

where $j_1, j_2, \ldots, j_m \in \{1, 2, \ldots, m\}$ and $j_1 \neq j_2 \neq \ldots \neq j_m$. Thus, these constraints can be ranked as $g_{j_1}, g_{j_2}, \ldots, g_{j_m}$ which means that g_{j_1} is the most interesting one, followed by g_{j_2} and so on.

3.2   Cascaded Constraint-Handling Framework

Observing the difference of constraints, this section elaborates a simple yet effective algorithmic framework, namely, cascaded constraint-handling (CCH). The CCH, as the name says, handles the constraints in a sequential way. That is, constraints are progressively introduced into the search.

Provided that there is no preference from decision-makers, all the constraints can be grouped into different levels in terms of the interest level, i.e., the *handling-difficulty*. To describe the CCH framework, we assume that the rank of constraints has been determined as $g_{j_1}, g_{j_2}, \ldots, g_{j_m}$.

The evolutionary process is then divided into multiple stages whose number is the same as the quantity of different interest levels, and the number of generation assigned to each stage is roughly equivalent to the division quotient. Then the constraints are successively embraced by the CCH. In detail, at the first stage, only the most difficult constraint is taken into account. The population keeps evolving until it enters the next stage in which the second difficult constraint is added. Sequentially, the remaining constraints are re-included into the problem following their own rank, see Eqs. (5) to (7).

$$\min f(\mathbf{x}) \quad \text{s.t.} \quad g_{j_1}(\mathbf{x}) \leq 0 \tag{5}$$

$$\Downarrow$$

$$\min f(\mathbf{x}) \quad \text{s.t.} \quad \begin{cases} g_{j_1}(\mathbf{x}) \leq 0 \\ g_{j_2}(\mathbf{x}) \leq 0 \end{cases} \tag{6}$$

$$\Downarrow$$

$$\ldots$$

$$\Downarrow$$

$$\min f(\mathbf{x}) \quad \text{s.t.} \quad \begin{cases} g_{j_1}(\mathbf{x}) \leq 0 \\ g_{j_2}(\mathbf{x}) \leq 0 \\ \ldots \\ g_{j_m}(\mathbf{x}) \leq 0 \end{cases} \tag{7}$$

It is worth mentioning that (i) at the initialization stage, expect for the box constraints, none of constraints ($g_j(\mathbf{x})$) is considered. Subsequently, constraints are added in a stepwise manner such that the feasible region shrinks progressively. Also, during this process some infeasible solutions are provisionally acceptable (i.e., considered as feasible in early stages). This enables the search to effectively work from both interior and exterior of the feasible region; (ii) Priority is assigned to those difficult constraints, that is, being handled in advance. This leads more search effort to be distributed to handle the most difficult ones. Note that the study [13] also proposed to handle constraints in a sequential manner. However, its basic idea focuses on finding feasible solutions first then searching for the optimal one, which differs from our method as described above.

In principle most of constraint-handling methods, e.g., penalty function [11], stochastic ranking [16], can be integrated into the CCH framework. To demonstrate the effectiveness of the CCH framework, the penalty function is taken as an example. The derived algorithm is then called penalty function based cascaded constraint-handling method, denoted as CCH-PF. Moreover, essentially, the CCH-PF utilizes the "divide-and-conquer" concept, that is, first constraints are divided into different levels and assigned to different evolutionary stages by the CCH, then those constraints are conquered by the PF method individually.

The use of penalty function is to penalize infeasible solutions by adding a positive value to the objective function f [20]:

$$eval(\mathbf{x}) = \begin{cases} f(\mathbf{x}) & \text{if } \mathbf{x} \in F \\ f(\mathbf{x}) + \alpha(t) \times \sum_{r=1}^{N_r} G_r(\mathbf{x}) & \text{otherwise} \end{cases} \tag{8}$$

where $eval(\mathbf{x})$ is the composite evaluation value used for comparison between individuals. The penalty coefficient α can either be static [21] or adaptively adjusted based on the generation counter and/or the population information [9]. N_r is the number of constraints involved in the current stage. Correspondingly, the violation against current constraints is summed as the total violation degree. Amongst them, the violation of the individual $\mathbf{x}$ against the r-th ranked constraint can be constructed as $G_r(\mathbf{x}) = \max\{0, g_{j_r}(\mathbf{x})\}$.

The pseudo-code of the CCH-PF is shown in Algorithm 1. First, interest levels of all constraints are calculated. Constraints are then ranked based on their interest levels (lines 3–4), and are grouped into corresponding stages which is of the same number as the interest levels (line 5). In the subsequent generations, the evolution stage is identified based on the generation that the evolution is in (line 12). If it enters a new stage, the constraints that are ranked in the next level are added, leading to the change of the constraint violation function (line 13). Note that the penalty function can be replaced by other constraint-handling techniques. If so, the pseudo-code in lines 13–14 of Algorithm 1 should be changed accordingly.

To further explain the scheme of CCH-PF, a minimization problem with two variables, as shown in Fig. 1, is presented. The surface in Fig. 1(a) shows the original objective values of all solutions in the search space with the

---

Algorithm 1. Penalty function based cascaded constraint-handling method (CCH-PF)

Input: N: population size; m: the number of constraints; $f(\cdot)$: objective function; C: constraints; $maxGen$: maximum generation; α: the initial penalty factor.

Output: the optimal solution.

1 Initialize population $S \leftarrow \{\mathbf{x}_1, \mathbf{x}_2, \ldots, \mathbf{x}_N\}$;
2 Calculate objective values $F = f(S)$;
3 Calculate the interest level (L) of the constraints in C using Eq. (3);
4 Rank m constraints according to L with the largest ranking first, obtaining C';
5 Group the constraints into different stages of the evolutionary process ;
6 Initialize the problem as an unconstrained problem;
7 **for** $t \leftarrow 1$ **to** $maxGen$ **do**
8 $\quad$ Generate offspring S' by differential evolution (DE) [18] and polynomial mutation (PM) [5] of parent solutions in S;
9 $\quad$ $F' = f(S')$;
10 $\quad$ $S_{joint} = S \uplus S'$;
11 $\quad$ $F_{joint} = F \uplus F'$;
12 $\quad$ Identify the current stage and the N_r constraints ranked forward in C';
13 $\quad$ Add the corresponding new constraints and update the constraint violation function if the evolution enters a new stage;
14 $\quad$ Obtaining the aggregating value by Eq. (8):
$\quad$ $eval(S_{joint}) = F_{joint} + \alpha \times \sum_{r=1}^{N_r} G_r(S_{joint})$;
15 $\quad$ Replace solutions in S with better solutions in S_{joint};
16 $\quad$ $\mathbf{x}^* \leftarrow \arg \min_{\mathbf{x}_k \in S_{joint}} eval_k, k = 1, 2, \ldots, 2N$;
17 **end**
18 **return** $\mathbf{x}^*$ and its objective values.

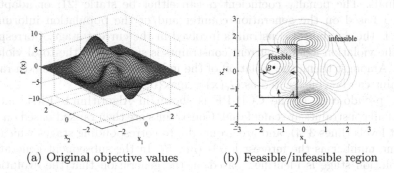

(a) Original objective values   (b) Feasible/infeasible region

Fig. 1. Illustration of a simple constrained optimization problem

contours plotted. Figure 1(b) illustrates the feasible region (see the black rectangle) bounded by four constraints. In this problem, points A and B correspond to the best and the second-best solutions, respectively.

Assuming that the rank of constraints is "the right>the upper>the left>the lower", by applying the CCH-PF method, these constraints are subsequently added and handled in four stages, as presented in Fig. 2. The solutions obtained by the CCH-PF at the beginning, the mid-term and the end of the evolutionary process are demonstrated in Fig. 3, compared with the penalty function without the CCH framework.

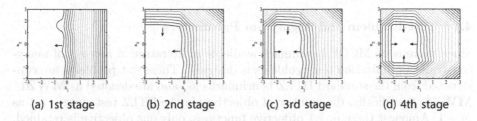

(a) 1st stage   (b) 2nd stage   (c) 3rd stage   (d) 4th stage

Fig. 2. Operational process of the cascaded constraint-handling (CCH) framework

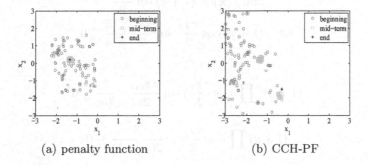

(a) penalty function   (b) CCH-PF

Fig. 3. Solutions obtained by penalty function with and without the CCH framework under different evolutionary phases

From Fig. 3, it can be observed that,

- by the CCH-PF some infeasible solutions are retained at the early stages and then move towards the boundary from the exterior of the feasible region. Since starting from both sides of the boundary can provide more information for the search, it is more likely to find the global optimal solution, especially when it is located on the boundaries, e.g., A in Fig. 1(b).
- in comparison, the solutions in Fig. 3(a) are trapped into local minima. The reason may be that, without CCH framework, the emphasis is laid on the solutions satisfying all constraints, information from infeasible ones ignored. Thus, B is of high probability to be obtained for its central location in the feasible region. Once B is found, the search would be easily restricted within its vicinity, resulting in the premature convergence.

4 Experiments

This section examines the performance of the proposed CCH framework when integrated with the penalty function method. Moreover, in order to eliminate the possible interference caused by the penalty factor (α), in addition to the static penalty function used above, the adaptive penalty function [9] is also adopted, abbreviated as SPF and APF, respectively.

4.1 Test Problem and Algorithm Parameters

Since there is no MCOP benchmark available in literature, a new set of many-constraint optimization test problems is designed. These test problems are constructed from the standard DTLZ benchmarks [6], and are denoted as **MWZ1–MWZ6**. Specifically, the number of objectives in the DTLZ test suite is set as $m + 1$. Amongst these $m + 1$ objective functions, only one objective is retained. All the other m objectives are converted into constraints. For example, **MWZ2** is described as follows.

$$\min f(\mathbf{x}) = (1 + g) \sin \frac{\pi x_1}{2}$$

$$s.t. \quad (1 + g) \cos \frac{\pi x_1}{2} \sin \frac{\pi x_2}{2} - \frac{3}{4} \leq 0$$

$$\dots$$

$$(1 + g)(\prod_{j=1}^{m-1} \cos \frac{\pi x_j}{2}) \sin \frac{\pi x_m}{2} - \frac{3}{2m} \leq 0 \tag{9}$$

$$(1 + g) \prod_{j=1}^{m} \cos \frac{\pi x_j}{2} - \frac{3}{2(m + 1)} \leq 0$$

with

$$g = \sum_{i=m+1}^{n} (x_i - 0.5)^2 \tag{10}$$

General parameters of the test algorithms are set as follows: The population size N, the variable dimension n and the maximal generation number $maxGen$ are fixed as 100, 50 and 500, respectively. In these test problems, the number of constraints is set as $m = 39$. And the penalty factor α is initialized as 50. All the instances are independently run 31 times.

4.2 Experimental Results

In this section the performances of SPF and CCH-SPF, APF and CCH-APF on the six benchmark problems are examined. The best, mean, worst value among the obtained optimal objectives, as well as the standard deviation, across 31 independent runs of these algorithms are summarized in Table 1. Moreover, the non-parametric Wilcoxon-ranksum two-sided comparison [10] procedure at the

95% confidence level is employed to examine whether the results are significantly different or not. The symbol '+', '−' or '=' in Table 1 means that the proposed methodology is better than, worse than or comparable to the technique without the CCH framework.

Table 1. The best, mean, worst value and the standard deviation of the minimum objectives obtained by SPF/ APF with and without the CCH framework on all test problems. The statistical test using non-parametric Wilcoxon-ranksum two-sided comparison is performed. The symbol '+', '−' or '=' means that the technique integrated with the CCH is statistically better than, worse than or comparable to the counterpart.

	Best results		Mean results		Worst results		Std. dev.		+/−/=
	SPF	CCH-SPF	SPF	CCH-SPF	SPF	CCH-SPF	SPF	CCH-SPF	
MWZ1	0.0006	0.0000	0.0029	0.0000	0.0084	0.0000	0.0021	0.0000	+
MWZ2	0.0008	0.0000	0.1246	0.0019	0.4302	0.0295	0.1385	0.0054	+
MWZ3	0.0050	0.0000	0.0244	0.0005	0.0582	0.0075	0.0124	0.0014	+
MWZ4	0.9933	0.9927	0.9976	0.9956	0.9999	0.9992	0.0023	0.0025	+
MWZ5	0.0001	0.0000	0.0004	0.0003	0.0008	0.0009	0.0002	0.0002	+
MWZ6	0.0004	0.0000	0.0016	0.0000	0.0049	0.0003	0.0011	0.0001	+
	Best results		Mean results		Worst results		Std. dev.		+/−/=
	APF	CCH-APF	APF	CCH-APF	APF	CCH-APF	APF	CCH-APF	
MWZ1	0.0000	0.0000	0.0028	0.0002	0.0074	0.0011	0.0019	0.0003	+
MWZ2	0.0007	0.0000	0.2488	0.0131	0.4704	0.2142	0.1361	0.0431	+
MWZ3	0.0043	0.0000	0.0229	0.0003	0.0821	0.0099	0.0148	0.0018	+
MWZ4	0.9929	0.9929	0.9975	0.9955	0.9999	0.9988	0.0025	0.0022	+
MWZ5	0.0001	0.0000	0.0005	0.0002	0.0011	0.0004	0.0002	0.0001	+
MWZ6	0.0004	0.0000	0.0012	0.0000	0.0030	0.0002	0.0006	0.0001	+

In addition, the comparison results obtained by two pairs of algorithms are box-plotted in Fig. 4, allowing a visual inspection.

According to the experimental results, it is observed that the superiority of the proposed methodology is evident. To be more precise, amongst the six test problems, CCH-SPF and CCH-APF are superior to their counterpart, presented in Table 1 and Fig. 4. Additionally, both of CCH-SPF and CCH-APF obtain comparably good performances.

In order to demonstrate the effectiveness of the CCH framework even further, the SPF and the APF is individually compared against CCH-SPF and CCH-APF on the **MWZ2** under an incremental number of constraints. The comparison results are shown in Fig. 5. It can be observed that, regardless of the number of the constraints, the CCH integrated version can approximately converge to the optimal objective value with little error while the competitors cannot. Moreover, the performances of SPF and APF fluctuate to some extent.

Combining the above results, we therefore can tentatively conclude that the proposed CCH framework is helpful in improving the effectiveness of the penalty function method when dealing with MCOPs.

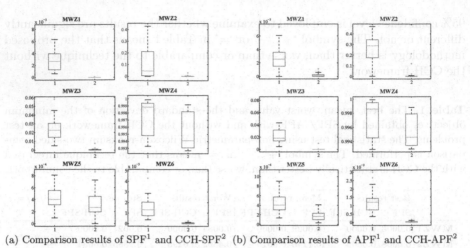

(a) Comparison results of SPF[1] and CCH-SPF[2]   (b) Comparison results of APF[1] and CCH-APF[2]

Fig. 4. Comparison results of the constraint-handling technique without and with the CCH framework on different test problems

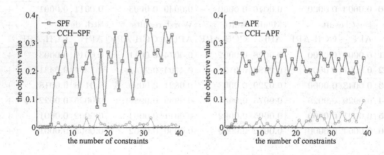

Fig. 5. The change of optimal objective values obtained by SPF and APF, without and with the CCH framework as the number of constraints increases.

5   Conclusion

Many engineering problems involve handling a large number of constraints. However, few studies in literature have explicitly investigated how to effectively handle many-constraint optimization problems (MCOPs). This study, to the best of the authors' knowledge, provides the first initial analysis on many-constraint optimization. First, the MCOPs are formulated, and a new set of benchmarks is proposed. Then, a simple yet effective algorithmic framework, the cascaded constraint-handling (CCH), is proposed. The CCH framework handles constraints progressively during the search process. That is, constraints are added sequentially. Regarding the order of adding constraints, "the most interesting one first" principle is adopted. Accordingly, a method for ranking the interest level of constraints is proposed. Experimental results show that the

CCH integrated penalty function is much more effective than the individual use of penalty function method on MCOPs.

This study is limited in a certain number of ways. First, the effectiveness of the CCH framework is only examined for the penalty function method. Therefore caution is advised in generalizing the CCH to other constraint-handling techniques. Second, the adopted MWZ benchmarks are constructed based on the DTLZ many-objective optimization problems. These problems are known for the conflicting relationship amongst objectives. That is to say, the MWZ also holds the conflicting relationship which however is not usual in real constrained optimization problems. Thus, more advanced benchmarks are highly required. A final point to consider is the issue of equality constraints. According to the current constraint-ranking method, equality constraints would be very likely to be the most difficult constraints, and thus make the search of feasible solutions extremely difficult. In this sense, some other strategies to handle equality constraints can be considered, e.g., the variable reduction method [23, 24].

In terms of future research, since the CCH framework examined in this paper is very simplified, more effective strategies within this framework are encouraged which include, e.g., advanced constraint-ranking method, appropriate partition of constraints. Consideration may also be given to scaling up the framework for being applicable to other constrained optimization algorithms.

In summary, this study has highlighted many-constraint optimization problems. However, both the MCOP benchmarks and the methodologies, e.g., the CCH framework, are still in their infancy. More research along this direction is needed.

Acknowledgement. This work is supported by the National Natural Science Foundation of China (Nos. 61773390 and 71571187) and the Outstanding Natural Science Foundation of Hunan Province (2017JJ1001).

References

1. Back, T., Hoffmeister, F., Schwefel, H.P.: A survey of evolution strategies. In: International Conference on Genetic Algorithms, pp. 2–9 (1991)
2. Cai, Z., Wang, Y.: A multiobjective optimization-based evolutionary algorithm for constrained optimization. IEEE Trans. Evol. Comput. **10**(6), 658–675 (2006)
3. Coello, C.A.C.: Treating constraints as objectives for single-objective evolutionary optimization. Eng. Optim. **32**(3), 275–308 (2000)
4. Coello, C.A.C.: Theoretical and numerical constraint-handling techniques used with evolutionary algorithms: a survey of the state of the art. Comput. Methods Appl. Mech. Eng. **191**(11C12), 1245–1287 (2002)
5. Deb, K.: Multi-Objective Optimization Using Evolutionary Algorithms, vol. 16. Wiley, Weinheim (2001)
6. Deb, K., Thiele, L., Laumanns, M., Zitzler, E.: Scalable multi-objective optimization test problems. In: IEEE Congress on Evolutionary Computation, pp. 825–830 (2002)
7. Deb, K.: An efficient constraint handling method for genetic algorithms. Comput. Methods Appl. Mech. Eng. **186**(2), 311–338 (2000)

8. Farmani, R., Wright, J.A.: Self-adaptive fitness formulation for constrained optimization. IEEE Trans. Evol. Comput. **7**(5), 445–455 (2003)
9. Hamida, S.B., Schoenauer, M.: ASCHEA: new results using adaptive segregational constraint handling. In: IEEE Congress on Evolutionary Computation, pp. 884–889 (2002)
10. Hollander, M., Wolfe, D.: Nonparametric Statistical Methods. Wiley-Interscience, Boston (1999)
11. Joines, J.A., Houck, C.R.: On the use of non-stationary penalty functions to solve nonlinear constrained optimization problems with GA's. In: First IEEE Conference on Evolutionary Computation, IEEE World Congress on Computational Intelligence, pp. 579–584 (1994)
12. Mallipeddi, R., Suganthan, P.N.: Differential evolution with ensemble of constraint handling techniques for solving CEC 2010 benchmark problems. In: Evolutionary Computation, pp. 1–8 (2010)
13. Marc, S., Spyros, X.: Constrained GA optimization. In: International Conference on Genetic Algorithms, pp. 573–580 (1993)
14. Michalewicz, Z., Schoenauer, M.: Evolutionary algorithms for constrained parameter optimization problems. Evol. Comput. **4**(1), 1–32 (2014)
15. Okamoto, T., Hirata, H.: Constrained optimization using the quasi-chaotic optimization method with the exact penalty function and the sequential quadratic programming. In: IEEE International Conference on Systems, Man, and Cybernetics, pp. 1765–1770 (2011)
16. Runarsson, T.P., Yao, X.: Stochastic ranking for constrained evolutionary optimization. IEEE Trans. Evol. Comput. **4**(3), 284–294 (2000)
17. Sheth, P.D., Umbarkar, A.J.: Constrained optimization problems solving using evolutionary algorithms: a review. In: International Conference on Computational Intelligence and Communication Networks, pp. 1251–1257 (2016)
18. Storn, R., Price, K.: Differential evolution–a simple and efficient heuristic for global optimization over continuous spaces. J. Glob. Optim. **11**(4), 341–359 (1997)
19. Takahama, T., Sakai, S.: Constrained optimization by the ϵ constrained differential evolution with an archive and gradient-based mutation. In: IEEE Congress on Evolutionary Computation, pp. 1–9 (2010)
20. Tessema, B., Yen, G.G.: A self adaptive penalty function based algorithm for constrained optimization. In: IEEE Congress on Evolutionary Computation, pp. 246–253 (2006)
21. Wang, Y., Ma, W.: A penalty-based evolutionary algorithm for constrained optimization. In: Jiao, L., Wang, L., Gao, X., Liu, J., Wu, F. (eds.) ICNC 2006. LNCS, vol. 4221, pp. 740–748. Springer, Heidelberg (2006). https://doi.org/10.1007/11881070_99
22. Wei, W., Wang, J., Tao, M.: Constrained differential evolution with multiobjective sorting mutation operators for constrained optimization. Appl. Soft Comput. **33**(C), 207–222 (2015)
23. Wu, G., Pedrycz, W., Suganthan, P., Li, H.: Using variable reduction strategy to accelerate evolutionary optimization. Appl. Soft Comput. J. **61**, 283–293 (2017)
24. Wu, G., Pedrycz, W., Suganthan, P., Mallipeddi, R.: A variable reduction strategy for evolutionary algorithms handling equality constraints. Appl. Soft Comput. J. **37**, 774–786 (2015)

Many-Objective EMO

Generating Uniformly Distributed Points on a Unit Simplex for Evolutionary Many-Objective Optimization

Kalyanmoy Deb[1] ⓘ, Sunith Bandaru[2]([✉]), and Haitham Seada[3]

[1] Michigan State University, East Lansing, MI, USA
kdeb@egr.msu.edu
[2] University of Skövde, Skovde, Sweden
sunith.bandaru@his.se
[3] Ford Motor Company, Dearborn, USA
seadahai@msu.edu

Abstract. Most of the recently proposed evolutionary many-objective optimization (EMO) algorithms start with a number of predefined reference points on a unit simplex. These algorithms use reference points to create reference directions in the original objective space and attempt to find a single representative near Pareto-optimal point around each direction. So far, most studies have used Das and Dennis's structured approach for generating a uniformly distributed set of reference points on the unit simplex. Due to the highly structured nature of the procedure, this method does not scale well with an increasing number of objectives. In higher dimensions, most created points lie on the boundary of the unit simplex except for a few interior exceptions. Although a level-wise implementation of Das and Dennis's approach has been suggested, EMO researchers always felt the need for a more generic approach in which any arbitrary number of uniformly distributed reference points can be created easily at the start of an EMO run. In this paper, we discuss a number of methods for generating such points and demonstrate their ability to distribute points uniformly in 3 to 15-dimensional objective spaces.

Keywords: Many-objective optimization · Reference points ·
Das and Dennis points · Diversity preservation

1   Introduction

Recent evolutionary many-objective optimization algorithms (EMO) use a set of reference directions as guides to parallelly direct their search to find a single Pareto-optimal solution along each direction. These so-called decomposition-based EMO methods, such as MOEA/D [18], NSGA-III [5], DBEA [1] are gaining popularity due to their success in handling three to 15-objective problems

© Springer Nature Switzerland AG 2019
K. Deb et al. (Eds.): EMO 2019, LNCS 11411, pp. 179–190, 2019.
https://doi.org/10.1007/978-3-030-12598-1_15

involving convex, non-convex, multi-modal, disjointed, biased density based, and non-uniformly scaled problems.

One of the requirements of these algorithms is the initial supply of a set of reference directions, a matter which has not been pursued much in the literature. Most studies use Das and Dennis's [4] structured method in which first a set of points are initialized on a M-dimensional unit simplex (where M is the number of objectives): $\mathbf{z} \in [0,1]^M$ satisfying $\sum_{i=1}^{M} z_i = 1$. Thereafter, a reference direction is constructed by a vector originating from the origin and connected to each of these points. The number of points on the unit simplex is determined by a parameter p, which indicates the number of divisions along each objective axis. It turns out that the total number of points on the unit simplex is $\binom{M+p-1}{p}$. For example, if $p = 10$ is chosen for an $M = 3$-objective problem, then the total number of points on the unit simplex is $\binom{12}{10}$ or 66. The 66 points are well distributed on the unit simplex. If an EMO algorithm works well to find a single Pareto-optimal solution for each of these 66 reference lines (obtained by a vector originating from the origin and passing through each point), a well-distributed set of Pareto-optimal solutions will be expected at the end. If more points are desired, p can be increased by one (or, $p = 11$), and the total number of points must jump to 78. In other words, if exactly 70 points are desired on the unit simplex, there is no way we can use Das and Dennis's method to achieve them.

Besides the inability to construct an arbitrary number of points, there is another issue with Das and Dennis's method, which has been problematic. As p increases, the total number of points on the unit simplex increases rapidly, as shown for $M = 10$ in Fig. 1 – sublinear plot in the semilog scale indicates less than exponential behavior. This requires a large population size to find a single Pareto-optimal solution for each direction. Moreover, most of the structured points lie on the boundary of the unit simplex and very few points lie in the interior of the simplex. Calculations reveal that, when $p < M$, there is no interior point, and when $p = M$, there is exactly one interior point. With $p > M$ more points are in the interior, but the number of such points is only $\binom{p-1}{p-M}$, which is only a tiny fraction of all Das and Dennis's points. Figure 1 shows that for $M = 10$-objective problem, the proportion of interior Das and Dennis's points grow with p. For $p < M$, the proportion is zero and then it starts to grow, but the proportion is still very low compared to the total number of points created. For example, for $p = 15$, there are a total of 1,307,504 points, of which only 0.15% (or only 2,002) points are in the interior. The rest of the points lie on the boundary of the unit simplex. A fix-up to the above problem has been suggested in the literature [5] by applying the Das and Dennis's method *layer-wise*. In every layer, a small p is chosen to create a few points, but layers are shrunk consecutively so that more interior points are created in the process. Even with this layer-wise procedure, any arbitrary number of points cannot be created.

In this paper, we discuss a number of methods by which an arbitrary number of well-distributed points can be created on the unit simplex. Methods based on filling, construction and elimination are first described. The methods are then applied in 3 to 15-dimensional objective spaces to show their effectiveness.

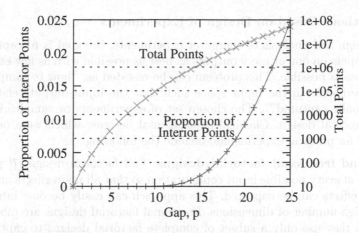

Fig. 1. Proportion of interior points compared to total Das and Dennis's points for $M = 10$-objective problem.

Section 2 describes a few standard space filling methods which can be used for the purpose. Section 3 then describes bottom-up approaches that build a set of reference points from a few seed points. Section 4 proposes an opposite scenario, in which a well-distributed set of reference points are chosen from an initial large set of points on the unit simplex. Results using some of these methods are shown in terms of the hypervolume measure in Sect. 5. Finally, conclusions are drawn in Sect. 6.

1.1 Motivation

The choice of reference directions in decomposition-based EMO methods is important, since the distribution of resultant Pareto-optimal solutions largely depends on them. If a near-uniform distribution of reference points can be supplied on a unit simplex, the corresponding reference directions are expected to produce a well-distributed set of Pareto-optimal points. In this paper, we propose a number of philosophies of finding n reference points $\mathbf{z}^{(i)}, i = 1, \ldots, n$ on a standard $(M-1)$-simplex[1], such that $\sum_{k=1}^{M} z_k^{(i)} = 1$ holds for each i. Being a preliminary study, we do not consider generating reference points on an arbitrary simplex [8] or on convex and concave manifolds [12].

2   Filling Methods

In these methods, the set of n points will be created on the unit simplex by using a standard filling technique. We describe a few techniques for this purpose.

---

[1] A standard $(M-1)$-simplex has M vertices in $\mathbb{R}^M$, each of which is one unit from the origin along each axis.

2.1   Methods Based on Design of Experiments

In the design and analysis of computer experiments, the goal is to capture the effect of inputs on one or more outputs as best as possible with as few expensive experiments as possible. This problem can be restated as: "how to sample computer experiments in the input space such that the input-output relationship are maximally captured?". The chosen set of experiments or samples is called an experimental design. Classical experimental designs, which were originally developed for physical experiments, include the following:

1. **Full and fractional factorial designs:** Full factorial design [7] places a sample at every possible input configuration so that all main effects and interaction effects can be captured. This approach can easily become intractable for a high number of dimensions. Fractional factorial designs are more practical as they use only a subset of complete factorial designs to capture just the main effects and low-order interactions.
2. **RSM designs:** Response surface methodology (RSM) [2] uses polynomials of various degrees to model the input-output relationship. RSM employs a variety of designs, such as full and fractional factorial, central composite design, Box-Behnken design and sequential method, which are aimed at minimizing errors at the design points and keeping biases in the estimated coefficients small.
3. **Optimal designs:** Optimal designs are aimed at optimizing various statistical criteria related to estimation and prediction. For example, D-optimal and A-optimal designs, respectively, minimize the determinant and trace of $X^T X$ (where X is the design matrix), which reduce estimation variance. On the other hand, G-optimal and Q-optimal designs respectively minimize the maximum and average prediction variance over the design points [3]. Here, we consider D-optimal design as a representative method of this class.

2.2   Space Filling Methods

Classical experimental designs described above are also called *model-dependent* designs because they require the knowledge of an underlying model. When no prior information about the input-output relationship is known, the general strategy is to assume that important features of the input-output relationship are equally likely to be present in all parts of the input space. In order to capture these features through an experimental design, the samples are spread evenly throughout the input space. Such designs are called *model-independent* or *space-filling* designs. The three important categories of space-filling designs are:

1. **Orthogonal arrays:** An orthogonal array [14] of *strength d* and *index* λ for κ factors ($\kappa > d$), each with s levels, is an experimental design that, upon projection onto any subset of d dimensions, resembles a full factorial, with each design point replicated λ times. The total number of designs required is therefore, $n = \lambda s^d$. The corresponding orthogonal array is denoted as (n, κ, s, d). $(s^3, 4, s, 2)$. Since, the number of points come as in a structured manner and cannot be set arbitrarily, we do not consider this method here.

2. **Latin hypercube design and sampling (LHS):** A Latin hypercube design is any orthogonal array of strength $d = 1$ and index $\lambda = 1$. Since $N = s$, this design gives the flexibility to generate an arbitrary number of samples when the factors are continuous. Latin hypercube sampling [13] involves dividing each factor into N equal intervals. Each of the N required designs is obtained by randomly selecting a previously unselected interval in each factor and sampling a single value from it.

3. **Number theoretic methods:** Number theoretic methods, originally developed for quasi-Monte Carlo integration, are aimed at creating points uniformly in the input space by minimizing different discrepancy metrics, which are measures of uniformity. Various low-discrepancy (also called quasi-random) sequences are available in literature. Popular among them are Halton set [9], Hammersley set [10], Sobol set [17] and Faure [6] sequences. When using low-discrepancy sequences, randomization of samples can be achieved by skipping over used sequences completely or partially. Here, we use three methods as representatives of this class.

Mapping onto Unit Simplex. All filling methods described above generate points in an M-dimensional hypercube ($z_k \in [0, 1]$). In order to map these points onto the $(M - 1)$-dimensional unit simplex, we adopt the following approach, suggested in [16]:

1. Generate points $\mathbf{z}^{(i)}, i = 1, \ldots, n$ in an $M - 1$ dimensional unit hypercube.
2. Set $i \leftarrow 1$
3. Sort ordinates $\{z_1^{(i)}, z_2^{(i)}, \ldots, z_{M-1}^{(i)}\}$ of $\mathbf{z}^{(i)}$ in ascending order.
4. Let $\{y_1^{(i)}, y_2^{(i)}, \ldots, y_{M-1}^{(i)}\}$ be the sorted ordinates.
5. Define $y_0^{(i)} = 0$ and $y_M^{(i)} = 1$, so that $y_0^{(i)} < y_1^{(i)} < y_2^{(i)} < \ldots < y_{M-1}^{(i)} < y_M^{(i)}$.
6. Then, set $z_k^{(i)} \leftarrow y_k^{(i)} - y_{k-1}^{(i)}$ for $k = 1, \ldots, M$ to form the mapped point $\mathbf{z}_k^{(i)}$.
7. If $i < n$, then set $i \leftarrow (i + 1)$ and go to Step 3, else Stop.

Note that $\sum_{k=1}^{M} z_k^{(i)} = 1$ is satisfied for each i.

In this paper, we will use the following filling methods because they allow us to choose an arbitrary number of points: D-optimal designs (DOD), Latin hypercube sampling (LHS), Halton set (HAL), Hammersley set (HAM) and Sobol set (SOB).

2.3 Structured Filling Methods

Das and Dennis's Method (DAS). As mentioned in Sect. 1, Das and Dennis's method is a structured approach which requires an integer gap parameter $p(\geq 1)$ and then creates $\binom{M+p-1}{p}$ points. The method is scalable to any number of objectives (M), but has several drawbacks:

– Number of points cannot be set arbitrarily,
– Most points lie on the boundary of the unit simplex, which may not be of interest to decision-makers, and

– The approach is not easily moldable to incorporate preference information.

Due to popular use of this method, we consider it in our comparison base and call this method DAS.

Layer-Wise Das and Dennis's Method. Layer-wise construction process is illustrated in Fig. 2. A relatively small value of p ($< M$) is used for each layer. The first layer covers the entire unit simplex, but the subsequent layers use a shrunk unit simplex, as shown in the figure. For example, instead of using $p = 12$ for a three-objective ($M = 3$) problem totalling 91 points, the use of three layers, as shown in the figure, with layer-wise $p = (3, 2, 1)$ requires $(10, 6, 3)$ or 19 points in total of which only 9 points are on the boundary. Points from Layer 2 and above are guaranteed to lie in the interior in such a construction, but the uniformity of the points gets lost in the process. In

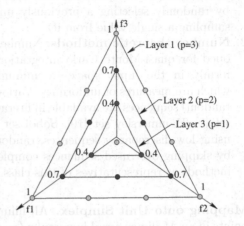

Fig. 2. Layer-wise construction of Das and Dennis's points for $M = 3$-objective problem.

addition, the layered approach can end up with different layouts for the same number of points. For example, with 5 objectives, generating 50 reference points can be done either using $p = (3, 2)$ or $p = (2, 2, 2, 1)$. Moreover, although using layers allows more flexibility in number of points than the original method, but still any arbitrary number of points is not possible to be created.

2.4   Probabilistic Filling Methods

A structured probability distribution for each objective can be chosen such that the sum of the objective values is exactly one. One such method was proposed in [11], in which the first objective z_1 is chosen in $[0, 1]$ with a probability distribution proportional to $(1 - z_1)^{M-2}$. This can be achieved by first choosing a random number $u_1 \in [0, 1]$ and then computing $z_1 = 1 - \sqrt[M-1]{u_1}$. Thereafter, z_k is computed with another random number $u_k \in [0, 1]$, as $z_k = \left(1 - \sum_{j=1}^{k-1} z_j\right)\left(1 - \sqrt[M-k]{u_k}\right)$. The process is continued until z_{M-1} and the final objective value is computed as $z_M = 1 - \sum_{j=1}^{M-1} z_j$. We call this method JAS in this paper.

Other Structured Methods. There exists a number of conformal mapping methods [15] in which uniformly distributed points on a hyperbox can be mapped into a unit simplex. Since we are interested in a near uniform distribution of points in the unit simplex, it may not be easy to use such methods efficiently.

3   Construction Methods

Construction methods uses a bottom-up approach in which the procedure starts with a single point $\mathbf{z}^{(0)}$ on the unit simplex. Thereafter, points are added one by one in stages until a set of n points are obtained with a near uniform distribution. The addition of points can be achieved by using a pre-defined procedure or by a sophisticated optimization procedure of maximizing the uniformity of points at every stage. At a stage, when k points are already found, the following optimization procedure can be applied to obtain the $(k+1)$-th point $\mathbf{z}^{(k+1)}$:

$$\begin{aligned}
&\text{Maximize } \min_{i=1}^{k} \text{Dist}(\mathbf{z}^{(k+1)}, \mathbf{z}^{(i)}),\\
&\text{Subject to } \sum_{j=1}^{M} z_j^{(k+1)} = 1,\\
&\qquad\qquad 0 \le z_j^{(k+1)} \le 1, \quad \text{for } j = 1, \ldots, M.
\end{aligned} \tag{1}$$

In this problem, there are only M variables $z_i^{(k+1)}$ and one constraint. Other diversity metrics can also be used instead of the minimum Euclidean distance to all existing k points. The only drawback of this approach is that the above optimization needs to be applied $(n-1)$ times and the computation gets expensive with increasing k. The final outcome of n points will depend on the initial point chosen, which can be a random point on the unit simplex, or its centroid.

3.1   Maximally Sparse Creation Method (MSC)

Instead of starting with a single initial point, the above procedure can be seeded with more than one well-distributed points on the unit simplex. For example, the process can be started with M vertices as initial points or with m points from Layer 1 specification (with a small p ($< M$)), as described in Sect. 2.3. The remaining $(n-M)$ or $(n-m)$ points, as the case may be, can be created by using the above optimization procedure in stages. In this study, we use the vertices as initial points and call this method as MSC. The optimization problem in (1) is solved using MATLAB's `fmincon()` function. The constraint is relaxed to an inequality $\sum_{j=1}^{M-1} z_j^{(k+1)} \le 1$ by solving the problem for first $M-1$ variables. The last variable is set to $z_M^{(k+1)} = 1 - \sum_{j=1}^{M-1} z_j^{(k+1)}$ once `fmincon()` terminates.

4   Elimination-based Methods

Contrary to construction methods, a completely opposite process can be devised. Starting with a large set $(\mathcal{S})$ of structured or random points on the unit simplex, a procedure can be devised to eliminate neighboring points. This can be done either by eliminating one point at a time similar to a pruning method or by eliminating multiple points at a time.

4.1   Maximally Sparse Selection Methods (MSS)

This approach starts by filling $\mathcal{S}$ with a large number of uniformly randomly generated points. Then $\mathcal{W}$ is initialized as the set of all extreme points $(1, 0, \ldots, 0)^T$, $(0, 1, \ldots, 0)^T$, $\ldots$, $(0, 0, \ldots, 1)^T$. Finally, the procedure selects the rest $(n - M)$ points one at a time. The point which is maximally away from the already selected set of points points $\mathcal{W}$ is picked, added to $\mathcal{W}$ and removed from $\mathcal{S}$. The procedure continues until a total of n reference points is reached. Thus, at a stage in which k points are already obtained, we choose the next point $\mathbf{z}^{(k+1)}$, as follows:

$$(k + 1) = \text{argmax}_{j \in \mathcal{S}, i \in \mathcal{W}} \sum_{i=1}^{k} \text{Dist}(\mathbf{z}^{(i)}, \mathbf{z}^{(j)}). \tag{2}$$

If the original large set $\mathcal{S}$ is a random set of points on the unit simplex [19], we call this method MSS-R and when the original set $\mathcal{S}$ is a large set of points created by Das and Dennis's method with a large p, we call it MSS-D.

Instead of starting with an extreme point, the point closest to the centroid of the entire set $\mathcal{S}$ can be used to start the procedure, as an alternative method.

4.2   Reductive Methods (RED)

In this method, we cluster the large set of points $\mathcal{S}$ into n separate clusters based on Euclidean distance. Then, we choose one representative point from each cluster to select exactly n points. If $\mathcal{S}$ is a random set of points on the unit simplex, we call it RED-R and if $\mathcal{S}$ is chosen using Das and Dennis's method with a large p, we call it RED-D.

These methods may lose boundary points. One way to overcome this issue, is to ensure that for boundary clusters, we choose a boundary point in order to have maximum coverage over the entire unit simplex. Alternatively, the point closest to the centroid of each cluster can be chosen first and then the set of n points can be stretched to extend to the unit simplex boundary. In RED-R and RED-D, once the un-stretched-yet points are generated, $\mathcal{W}'$, the smallest value for each objective i is subtracted from all i components of all the points in $\mathcal{W}'$. This step keeps pushing the points – that used to be on the unit simplex – towards the origin, until for each objective i at least one point exists whose i-th component is Zero. Then all the points are normalized again to fall back on the unit simplex, and these form the targeted $\mathcal{W}$. Since stretching does not guarantee keeping extreme points, we added an additional step to insert them (RED-DS) in place of their closest neighbors.

5   Results

In this section, we present results obtained from 13 different methods described in previous sections on $M = \{3, 5, 8, 10, 15\}$-objective problems for finding $n = \{50, 100, 150, 200, 250, 300\}$ points. The methods used in our study are (i) D-optimal designs (DOD), (ii) Latin hypercube sampling (LHS), (iii) Halton

(HAL), (iv) Hammersley (HAM), (v) Sobol (SOB), (vi) Jaszkiewicz (JAS), (vii) Das and Dennis (DAS), (viii) Maximally Sparse Creation (MSC), and (ix) Maximally Sparse Selection with random initial set (MSS-R), (x) Maximally Sparse Selection with Das and Dennis's initial set (MSS-D), (xi) Reduction method with random initial set (RED-R), (xii) Reduction method with Das and Dennis's initial set (RED-D), and (xiii) Reduction method with Das and Dennis's initial set and guaranteed extreme points (RED-DS).

To compare the performance of the methods, we have used the hypervolume measure using the vector $(1.01, \ldots, 1.01)^T$ as the reference vector, as we know that the unit simplex has a nadir point $(1, \ldots, 1)^T$. In all plots, we show a normalized hypervolume metric obtained by dividing the obtained hypervolume value for a method with the maximum hypervolume, computed as follows: N-HV $= \mathrm{HV}/\mathrm{HV}_{\max}$, where $\mathrm{HV}_{\max} = 1.01^M - 1/M!$.

Figure 3 compares all 13 methods in terms of box-plots for $M = 3$ objectives and $n = 20$ points. When $n = 50$ points are required, a different distribution will occur and resulting hypervolume values are plotted in Fig. 4 for all 13 methods. It is clear that the hypervolume is the best for RED methods.

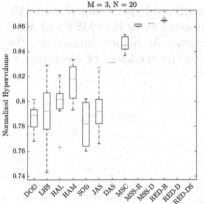 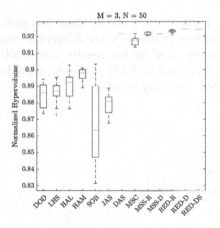

Fig. 3. Normalized hypervolume for $M = 3$ and $n = 20$.

Fig. 4. Normalized hypervolume for $M = 3$ and $n = 50$.

In order to show the sensitivity of the obtained distribution on the desired number of points (n), we use two elimination methods – RED-DS and MSS-D – $n = 50$ and 51 points for a three-objective problem, and obtain two distributions of points. Figures 5 and 6 show the difference in their distributions using RED-DS and MSS-D, respectively. The first plot indicates that an addition of an extra point in the set changes the arrangement in the intermediate part. Since MSS-D uses a sequential and deterministic selection method, the first 50 points for the $n = 51$ case will exactly be identical to the $n = 50$ case. Although both produce a well-distributed set of points, the clustering approach (in RED) has the dependence on n, and may be a better approach.

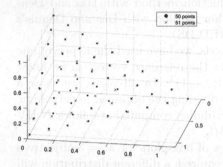

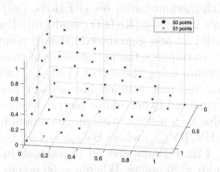

Fig. 5. Distribution of 50 and 51 points using RED-DS method for $M = 3$.

Fig. 6. Distribution of 50 and 51 points using MSS-D method for $M = 3$.

Figures 7, 8, 9 and 10 show box plots of normalized hypervolume for 5 to 15-objective problems with different n values. All plots show how MSS and MSC has the most robust results across all dimensions. At higher dimensions, the performance of RED methods degrades due to the tendency of clustering to avoid boundary points.

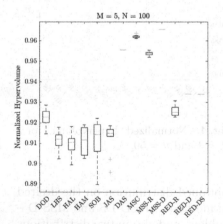

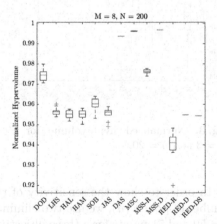

Fig. 7. Normalized hypervolume for $M = 5$ and $n = 100$.

Fig. 8. Normalized hypervolume for $M = 8$ and $n = 200$.

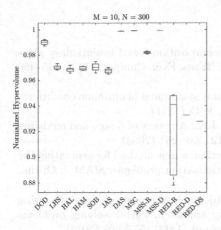

Fig. 9. Normalized hypervolume for $M = 10$ and $n = 300$.

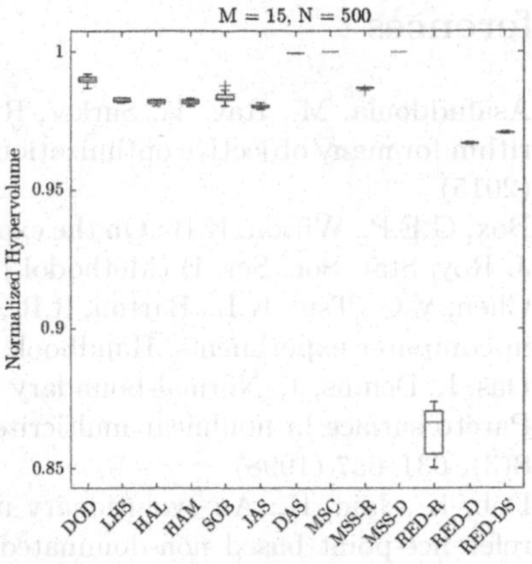

Fig. 10. Normalized hypervolume for $M = 15$ and $n = 500$.

6   Conclusions

In this paper, we have attempted to address an important issue related to evolutionary many-objective optimization (EMO) algorithms. While most EMO methods use a structured approach for allocating a set of reference points for leading the search, EMO researchers have always felt the need to make the process more flexible in specifying an arbitrary number of points on the unit simplex. In this paper, we have discussed a number of filling, construction and elimination methods for this purpose. Through extensive experiments in 3 to 15-dimensional objective spaces, we have compared the performance of a few of the proposed methods. Results based on the hypervolume metric have indicated the following:

1. Elimination methods (MSS-R, MSS-D, RED-R, and RED-D) are, in general, better than other methods considered in this paper at lower dimensions.
2. MSS-D procedure performs the best overall.
3. Construction method MSC and structured method DAS also perform well, particularly for higher dimensions.

In the future, we plan to make a comparison based on the computational complexity. Other methods, such as a simultaneous optimization of all n points, can also be included as global methods. The study can be extended to find a biased distribution of reference points, particularly if preference information is available. In some EMO applications, users may be interested in focusing on a particular part of the Pareto-optimal front. Methods that are able to create reference points on a specific region of the unit simplex will be another useful extension of this study.

References

1. Asafuddoula, M., Ray, T., Sarker, R.: A decomposition-based evolutionary algorithm for many objective optimization. IEEE Trans. Evol. Comput. **19**(3), 445–460 (2015)
2. Box, G.E.P., Wilson, K.B.: On the experimental attainment of optimum conditions. J. Roy. Stat. Soc. Ser. B (Methodol.) **13**(1), 1–45 (1951)
3. Chen, V.C., Tsui, K.L., Barton, R.R., Allen, J.K.: A review of design and modeling in computer experiments. Handbook Stat. **22**, 231–261 (2003)
4. Das, I., Dennis, J.: Normal-boundary intersection: a new method for generating the Pareto surface in nonlinear multicriteria optimization problems. SIAM J. Optim. **8**(3), 631–657 (1998)
5. Deb, K., Jain, H.: An evolutionary many-objective optimization algorithm using reference-point based non-dominated sorting approach, Part I: solving problems with box constraints. IEEE Trans. Evol. Comput. **18**(4), 577–601 (2014)
6. Faure, H.: Discrépance de suites associées à un système de numération (en dimension s). Acta Arith. **41**(4), 337–351 (1982)
7. Fisher, R.A.: The arrangement of field experiments. J. Minist. Agric. Great Br. **33**, 503–513 (1926)
8. Grimme, C.: Picking a uniformly random point from an arbitrary simplex. Technical report, University of Münster (2015)
9. Halton, J.H.: On the efficiency of certain quasi-random sequences of points in evaluating multi-dimensional integrals. Numer. Math. **2**(1), 84–90 (1960)
10. Hammersley, J.M.: Monte carlo methods for solving multivariable problems. Ann. New York Acad. Sci. **86**(3), 844–874 (1960)
11. Jaszkiewicz, A.: On the performance of multiple-objective genetic local search on the 0/1 knapsack problem - a comparative experiment. IEEE Trans. Evol. Comput. **6**(4), 402–412 (2002)
12. Zapotecas Martínez, S., Sosa Hernández, V.A., Aguirre, H., Tanaka, K., Coello Coello, C.A.: Using a family of curves to approximate the pareto front of a multi-objective optimization problem. In: Bartz-Beielstein, T., Branke, J., Filipič, B., Smith, J. (eds.) PPSN 2014. LNCS, vol. 8672, pp. 682–691. Springer, Cham (2014). https://doi.org/10.1007/978-3-319-10762-2_67
13. McKay, M.D., Beckman, R.J., Conover, W.J.: Comparison of three methods for selecting values of input variables in the analysis of output from a computer code. Technometrics **21**(2), 239–245 (1979)
14. Rao, C.R.: Factorial experiments derivable from combinatorial arrangements of arrays. Suppl. J. Roy. Stat. Soc. **9**(1), 128–139 (1947)
15. Schinzinger, R., Laura, P.A.A.: Conformal Mapping: Methods and Applications. Dover Books on Mathematics (2003)
16. Smith, N.A., Tromble, R.W.: Sampling uniformly from the unit simplex. Technical report, Johns Hopkins University (2004)
17. Sobol, I.M.: On the distribution of points in a cube and the approximate evaluation of integrals. USSR Comput. Math. Math. Phys. **7**(4), 86–112 (1967)
18. Zhang, Q., Li, H.: MOEA/D: a multiobjective evolutionary algorithm based on decomposition. IEEE Trans. Evol. Comput. **11**(6), 712–731 (2007)
19. Zhang, Q., Liu, W., Li, H.: The performance of a new version of MOEA/D on CEC09 unconstrained MOP test instances. In: IEEE Congress on Evolutionary Computation - CEC 2009, pp. 203–208. IEEE (2009)

On Timing the Nadir-Point Estimation and/or Termination of Reference-Based Multi- and Many-objective Evolutionary Algorithms

Dhish Kumar Saxena$^{(\boxtimes)}$ and Sarang Kapoor

Indian Institute of Technology Roorkee, Roorkee 247667, Uttarakhand, India
dhishfme@iitr.ac.in, skapoor@cs.iitr.ac.in

Abstract. There is considerable evidence that the Multi- and Many-objective Evolutionary Algorithms (jointly referred as MâOEAs, here) are mostly run for arbitrarily fixed number of generations. The absence of any justification for the same raises more questions than answers, and it is plausible to infer that the choices made for different problems coincide with the best-observed results. *Reference-based* MâOEAs (RMâOEAs) are a prominently emerging class of MâOEAs, where the diversity maintenance is assisted by externally provided reference vectors or points. However, the performance of most existing RMâOEAs is impacted by the efficacy with which the population is normalized along the *search*. This paper presents a novel and computationally efficient Termination Algorithm which under different parameter settings (*strong* and *mild*) not only determines the appropriate timing for RMâOEAs' termination but also the intermittent timings at which the population ought to be normalized. The proposed Algorithm can be tuned to integrate with different RMâOEAs. An instance of it is demonstrated here, with respect to NSGA-III. Experimental Results on the call for final termination of NSGA-III have been validated through Hypervolume measures. The results also establish that the performance of NSGA-III could be improved just by changing the frequency of Nadir-point estimates (used for population normalization). While several efforts have been made on *how* to estimate the Nadir-point, this to the best of the authors' knowledge is one of the rarest studies that explores *when* to estimate the Nadir-point.

Keywords: Many-objective Optimization · Termination criterion · Nadir-point

1 Introduction

The goal of a Multi-objective Evolutionary Algorithms (MOEAs) is to evolve a finite set of random solutions over several generations, to a set of solutions that approximates well the true Pareto-Front (PF) for a given problem in terms

© Springer Nature Switzerland AG 2019
K. Deb et al. (Eds.): EMO 2019, LNCS 11411, pp. 191–202, 2019.
https://doi.org/10.1007/978-3-030-12598-1_16

of *convergence* and *diversity* [6]. During the last two to three decades, several MOEAs were developed and their efficacy was demonstrated on test and real-world problems, dominantly with two or three objectives ($M = 2$ or 3). Though problems with $M \geq 4$ were reported earlier also, it was largely around the Year-2000 and onward, that the specific challenges posed by such problem to most existing MOEAs, were and are being recognized [4,9,11,14]. This perhaps explains why problems with $M \geq 4$ began to be referred as Many-objective Optimization problems (MaOPs), and Many-objective Evolutionary Algorithms (MaOEAs) gained importance. Since most MaOEAs also report their performance for $M \leq 3$, the Multi- and Many-objective Evolutionary Algorithms are collectively referred to as MâOEAs, here.

Reference-based RMâOEAs are a promising class of MâOEAs, where the diversity maintenance is assisted by externally provided reference vectors or points. As articulated in [2], these reference vectors/points can serve as the search targets, corresponding to each of which optimal solutions can be found. Prominent examples of RMâOEAs include NSGA-III [2] and its variants in [10, 19]; and MOEA/D [20] and its variants [1,12,18]. One major challenge associated with RMâOEAs relates to the efficacy of its normalization procedure which in turn impacts both the convergence and diversity in the PF-approximation.

In this context, this paper presents a novel and computationally efficient Termination Algorithm which under different parameter settings, not only determines the appropriate timing for RMâOEAs' termination but also the intermittent timings at which the population ought to be normalized. The remaining paper is structured as follows. Section 2 summarizes the existing Termination Algorithms for MâOEA. The proposed Termination Algorithm is presented in Sect. 3 and its integration with NSGA-III is discussed in Sect. 4. This is followed by experimental results on the *timing* of Nadir-point estimation and final termination of NSGA-III in Sect. 5. The paper concludes with Sect. 6.

2   Related Work on MâOEA Termination Algorithms

Considering that the question as to when to terminate an MâOEA, is a fundamental question faced by researchers and practitioners, it is ironical that the associated literature is rather sparse, and that too has failed to permeate into practice. Evidently, the number of generations for an MâOEA run are arbitrarily fixed *apriori*, given which an imbalance between quality of PF-approximation and computational cost is inevitable. While this imbalance could be ignored in test problems (known PF), it may have punitive implications in real-world problems where the quality of PF-approximation can not be gauged on-the-fly.

The available termination Criteria/Algorithms operate in two phases, namely, *evidence-gathering* phase (where information on some performance indicators is gathered) and *decision-making* phase (relying on statistical thresholds on gathered evidence) [16,17]. Notably, their applications in literature are dominantly restricted to $M \leq 3$; and their potential limitations in the context of MaOPs are highlighted in [15]. The rare studies considering their applications for MaOPs, include: (i) use of a global criterion based on hypervolume, ϵ, and

Mutual domination rate; Kalman filter based termination; and performance evaluation with $M = 3$ and 10, against DTLZ3, DTLZ6–7, and WFG1–9 [13], and (ii) an entropy based on-the-fly termination algorithm, demonstrated on 23 instances of two-objective problems, 12 instances of non-redundant MaOPs, 13 instances of redundant MaOPs [15], and real-world MaOPs in [7].

3   Proposed Termination Algorithm

The proposed Termination Algorithm, in principle, relies on capturing the *stability* of RMâOEA populations over successive generations. The stability is quantitatively assessed through a distance measure proposed in this paper. When the variation in the mean and standard deviation of this distance measure (to be computed over generations along an RMâOEA run, starting with the initial generation) falls below a threshold controlled by prescribed parameters, the RMâOEA's termination is called for.

With reference to Fig. 1, consider population sets P and Q from two successive generations of an RMâOEA operating with five reference vectors (set V). Let each solution $s \in P$ and Q associate itself to the *closest* vector $v \in V$, implying that $d^{\perp}(s, v) = s - v^T s / ||v||$ is minimum. This association would lead to the following scenarios, where there are vectors associated with:

1. exactly one member from each population (e.g., V_5): here, referring to members of P and Q associated with vector v, as p and q, respectively, compute the normalized distance between p and q along v, as given by Eq. 1. In that, $\mathcal{D}(v) = 0$ implies that members from two successive generations coincide (indicating stability) w.r.t. v.
2. one or more member(s) from each population (e.g., V_1): here, Eq. 1 holds again with the qualifier that p and/or q are to be treated as a representative solution of associated members from P and/or Q, respectively. In that, the objective vector for p is the average of the objective vectors of all members of P associated with v. The same holds for q.
3. one or more members from one population, but none from the other (e.g., V_3, $V4$): here, $\mathcal{D}(v)$ is set to 1 (maximum possible value), since absence of any member from either P or Q implies that the RMâOEA has not stabilized.
4. no member from any population (e.g., V_2): such a vector does not contribute to the *current* $\mathcal{D}(v)$ computation (v is not permanently dropped).

Following the $\mathcal{D}(v)$ computations, $\mu\mathcal{D}$ of the population P relative to Q is given by Eq. 2, where V^A is the set of those vectors which had at least one solution associated with them. This concludes the *evidence-gathering* phase of the proposed Termination Algorithm (Algorithm 1).

Notably, an RMâOEA starts from a randomly initialized population and evolves it over generations in pursuit of a good PF-approximation. In that, the disparity among successive populations is quite significant during early generations, and that subsides gradually. Considering this, towards the *decision-making* phase of Algorithm 1, it is proposed that: (i) the mean and standard deviation of $\mu\mathcal{D}$ be computed from the first to the tth (current) generation, as per Eqs. 3 and

4, respectively, and (ii) if the μ_t and σ_t measures in a pre-specified number of successive generations (n_s) of the RMâOEA may coincide up to a pre-specified number of decimal places (n_p), then the underlying RMâOEA be terminated, and the last generation be reported as N_{gt}. The above propositions are summarized in Algorithm 1, where vector association (L3–4) is followed by computation of $\mu\mathcal{D}$ (L5–13), leading to final termination criterion check (L15–19). Notably, the user-input parameters n_s and n_p would be collectively denoted by $T_{par} \equiv (n_p, n_s)$ in Sects. 4 and 5.

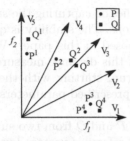

$$\mathcal{D}(v) = \frac{|\hat{v}^T \boldsymbol{p} - \hat{v}^T \boldsymbol{q}|}{\max(\hat{v}^T \boldsymbol{p}, \ \hat{v}^T \boldsymbol{q})} \quad (1)$$

$$\mu\mathcal{D} = \frac{1}{|V^A|} \sum_{v \in V^A} \mathcal{D}(v) \quad (2)$$

$$\mu_t = \frac{1}{t} \sum_{i=1}^{t} \mu\mathcal{D}_i, \ t \geq 1 \quad (3)$$

Fig. 1. Symbolic Vector association with successive Populations P and Q

$$\sigma_t = \frac{1}{t} \sum_{i=1}^{t} (\mu\mathcal{D}_i - \mu_i)^2; \ t \geq 1 \quad (4)$$

Algorithm 1. Proposed Termination Algorithm

---

1 **Procedure** $CalculateTerminationCriteria(P, Q, V, t, n_p, n_s)$
2 **begin**
3      $c_1 = false, \ c_2 = false$
4      $P_V \leftarrow Associate(P, V), \ Q_V \leftarrow Associate(Q, V)$
5      $\mu\mathcal{D}_t \leftarrow 0.0$
6   **foreach** $v \in V$ **do**
7   **if** $P_V[v] \neq \phi$ **and** $Q_V[v] \neq \phi$ **then**
8                $p = mean(P_V[v]), \ q = mean(Q_V[v])$
9                $\mu\mathcal{D}_t \leftarrow \mu\mathcal{D}_t + \mathcal{D}(v)$ (Equation 1)
10   **else if** $P_V[v] \neq \phi$ **or** $Q_V[v] \neq \phi$ **then** $\mu\mathcal{D}_t \leftarrow \mu\mathcal{D}_t + 1.0$;
11     $\mu\mathcal{D}_t \leftarrow \mu\mathcal{D}_t / |V^A|$ (Equation 2)
12   Compute μ_t and σ_t using Equation 3
13     $D_t = round(\mu_t, n_p), \ S_t = round(\sigma_t, n_p)$
14   **if** $[D_t = D_{t-1} = \cdots = D_{t-n_s}]$ **then** $c1 = true$;
15   **if** $[S_t = S_{t-1} = \cdots = S_{t-n_s}]$ **then** $c2 = true$;
16   **if** $c1 == true$ **and** $c2 == true$ **then return** $true$;
17   **else return** $false$;

---

4   Integration with NSGA-III

The proposed Termination Algorithm (Algorithm 1) integrated within NSGA-III is presented in Algorithm 2. For brevity, the discussion below focuses mainly on the interfacing of the Algorithm 1 with NSGA-III, in view of the dual goals of determining the *appropriate* timing for (a) Nadir-point estimation (Flag T_{nad} in Algorithm 2), and (b) final termination (Flag T_{stop} in Algorithm 2). Notably, NSGA-III uses the *Extreme-point-to-Nadir* approach [3], and the Nadir-point estimates are used to normalize the population in conjunction with *translation* (origin on to the ideal point). However, the authors' submissions here, include:

- normalization impacts the distribution of reference vectors, the association of solutions with those vectors, and eventually the diversity among solutions
- the rapidly evolving population specially in the early generations of NSGA-III may enforce too unstable Nadir-point estimates. On the contrary, the ideal point would remain relatively stable given that it is considered to be the best point found so far (not just in the current generation)
- that instead of estimating the Nadir-point at every generation of NSGA-III, its estimate be limited only to those generations where Algorithm 1 points to termination under *mild stabilization* within the current search hypercube (hypercube - post normalization, after the first Nadir-point estimation)
- *mild stabilization*: this is recommended, as the aim is just to negate the destabilizing effect of extreme-point fluctuations. If Nadir-point estimation were to be under *strict stabilization*, it may lead to poor diversity, owing to the *search* with non-uniformly distributed reference vectors for far too long
- hence, it is prudent to evaluate Flag T_{nad} under mild T_{par} settings, and the Flag T_{stop} ought to be evaluated under stricter T_{par} settings.

In the wake of the above, it may be noted that the Algorithm 2 retains the overall architecture of NSGA-III, except the changes around Flag T_{nad}. If the Flag T_{nad} is enabled through *isTNenabled* (L23), then the Nadir-point estimation will be *timed* by Algorithm 1, and NSGA-III shall be distinguished through a different abbreviation, as NSGA-III$_{TN}$. In that, the normalization (L19) is accomplished through translation of population (L11, 12) and updating of the Nadir-point (L13–15). The Flag T_{stop} is evaluated at the end of each NSGA-III generation (L25), regardless of whether the Flag T_{nad} is enabled or not. This illustrates the modular architecture of Algorithm 2. In that, if Flag T_{nad} is not enabled, then Algorithm 2 becomes identical to NSGA-III.

Overall Complexity: Given the number of vectors, H and number of objectives, M, the computation of μD takes the complexity of $O(MH)$. But the bottleneck lies in the association of solutions to vectors, which requires $O(MNH)$ complexity as the perpendicular distance of each solution in the population of size N from every vector, v needs to be computed. Thus, the overall complexity is $O(MNH + MH) = O(MH(N + 1))$. Since, $H \approx N$, therefore the final complexity is $O(MN^2)$. The information of associations can be re-utilized from the vector based MâOEA, of which it is a necessity. Thus, upon integration with MâOEA (NSGA-III here), the overall complexity can be reduced to $O(MN)$.

5   Experimental Settings and Results

In this paper, the non-redundant versions of the DTLZ [5] and WFG [8] problems are used. Experiments are done for $M = 5, 10$ and 15, with corresponding population size of $212, 276$ and 136, respectively, guided by the reference vector generation scheme used in [2].

Algorithm 2. Proposed Termination Algorithm integrated with NSGA-III

Input: H supplied reference Points, V; randomly generated population, P_1 with size, N; termination evaluation parameters, n_p^1, n_s^1, n_p^2, n_s^2; and timed Nadir-point estimation enabler, $isTNenabled()$

1 **begin**
2    $T_{nad} = false$, $T_{stop} = false$, $t_1 = 1$, $t_2 = 1$, $z^{min} = \phi$, $a = \phi$
3   **do**
4      $S_t = \phi$, $i = 1$
5      $Q_t = $ Recombination+Mutation(P_t), $R_t = P_t \bigcup Q_t$
6      $(F_1, F_2, \dots) = $ Non-dominated-sort(R_t)
7   **repeat**
8        $\mid$ $S_t = S_t \bigcup F_i$ and $i \leftarrow i + 1$
9   **until** $|S_t| \geq N$;
10   Last front to be included: $F_l = F_i$
11   Update z^{min} : $z_i^{min} = Min_{x \in R_t} f_i(x), i \in M$
12   Translate Pop: $s_i = s_i - z_i^{min}$, where $i \in M$, $s \in S_t$
13   **if** $T_{nad} == true$ **or** $isTNenabled() == false$ **then**
14        $a = UpdateIntercepts(S_t, z^{min})$
15        $T_{nad} = false$ and reinitialize T_{nad} with $t_1 = 1$
16   **if** $|S_t| = N$ **then** $P_{t+1} = S_t$, break;
17   **else**
18        $P_{t+1} = \bigcup_{j=1}^{l-1} F_j \bigcup \{F_l : K = N - |P_t + 1|\}$
19   **if** $a \neq \phi$ **then** Normalize Population: $s = s/a$, $s \in S_t$;
20   Associate each member s of S_t with a reference point
21   Compute niche count of reference point
22   Choose K members one at a time from F_l to construct P_{t+1}
23   **if** $isTNenabled() == true$ **then**
24        $\mid$ $T_{nad} = CalculateTerminationCriteria(P_t, P_{t+1}, V, t_1, n_p^1, n_s^1)$
25      $T_{stop} = CalculateTerminationCriteria(P_t, P_{t+1}, V, t_2, n_p^2, n_s^2)$
26      $t_1 \leftarrow t_1 + 1, t_2 \leftarrow t_2 + 1$
27   **while** $T_{stop} \neq true$;
28    $N_{gt} = t_2$
1 **Procedure** $UpdateIntercepts(P, z^{min})$
2 **begin**
3   Translate Population: $s_i = s_i - z_i^{min}$, where $i \in M$, $s \in P$
4   Find extreme points: $z_j^{max} = s : argmin_{s \in P} ASF(s, w_j)$, where
     $ASF(x, w) = \max_{i=1}^M f_i'(x)/w_i$, $w_j = (\epsilon, \dots, \epsilon)^T$; $\epsilon = 10^{-6}$; $w_j^j = 1$
5   Compute intercepts a_j for $j = 1, \dots, M$
6   **return** a

For DTLZ problems: the number of variables $n = M + k - 1$, where $k = 5$ for DTLZ1; and $k = 10$ for DTLZ2–4. For WFG problems: the number of variables are fixed at $n = 24$, of which the position variables $\rho = M - 1$ and distance variables, $k = 24 - \rho$. The performance indicators used include: (i) IGD (Inverted Generational Distance) [2], for DTLZs, and (ii) Hypervolume[1] for the normalized population with reference point as $(1.1, \ldots, 1.1)^T$, for WFGs.

Furthermore, the parameter settings used for NSGA-III include: (a) crossover probability and distribution index of 1.0 and 30, respectively, and (b) mutation probability and distribution index of $1/n$ and 20, respectively. $T_{par} \equiv (1, 20)$ is used to point to the need for Nadir-point estimation, while experiments for final termination of the NSGA-III are done with $T_{par} \equiv (2, 20)$ and $(3, 20)$.

5.1   Results on Final Termination of NSGA-III

The results presented in the Table 1 (T_{nad}: disabled) for DTLZ problems are self explanatory, in that, the IGD values near-zero suggest that when called for final termination, NSGA-III had offered a good PF-approximation. The results also reveal that for a fixed T_{par} setting and a particular problem, the N_{gt} (number of generations till termination) may *not necessarily* increase with an increase in the number of objectives. While this argument may seem flawed in the case of dominance-based MâOEAs, it is plausible in the case of RMâOEAs given that N_{gt} would also depend on (i) the population size, (ii) the number of reference vectors used, and (ii) the chance-success of vector-solution association.

For WFG problems, while the sample results are presented in Table 1, some of these (randomly chosen) have been validated through Hypervolume measures, as depicted in Fig. 2. It can be seen that for both the settings of T_{par} used, the proposed Algorithm 1 happened to call for the final termination only after the Hypervolume measures have reasonably stabilized. It is only fair that the termination results with $T_{par} \equiv (3, 20)$ are more reliable than those with $T_{par} \equiv (2, 20)$, given that the former seeks higher *degree of stability* than the latter, before termination can be called for.

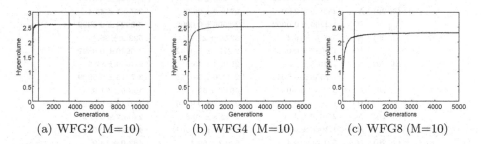

(a) WFG2 (M=10)   (b) WFG4 (M=10)   (c) WFG8 (M=10)

Fig. 2. Recommended terminations (given by left- and right-vertical lines for $T_{par} \equiv (2, 20)$ and $(3, 20)$, respectively) w.r.t. Hypervolume profile over NSGA-III generations

---

[1] The source code used can be found at: https://esa.github.io/pagmo2/.

Table 1. DTLZ and WFG problems: NSGA-III's termination as prescribed by the Algorithm 1 (at N_{gt} generations), and corresponding performance indicators

DTLZ	T_{par}		M		
			5	10	15
1	(2, 20)	N_{gt}	881.2 ± 63.6	951.2 ± 68.0	997.9 ± 83.8
		IGD	$9.03e{-}04 \pm 6.42e{-}04$	$1.45e{-}02 \pm 3.80e{-}04$	$7.44e{-}03 \pm 4.62e{-}03$
	(3, 20)	N_{gt}	4073.5 ± 58.0	4396.4 ± 299.3	4580.0 ± 254.5
		IGD	$8.28e{-}05 \pm 1.02e{-}04$	$1.29e{-}02 \pm 1.87e{-}03$	$1.10e{-}03 \pm 1.19e{-}04$
2	(2, 20)	N_{gt}	569.1 ± 14.4	578.8 ± 18.4	571.8 ± 39.7
		IGD	$4.38e{-}03 \pm 1.34e{-}03$	$4.14e{-}02 \pm 1.44e{-}03$	$2.98e{-}02 \pm 2.87e{-}03$
	(3, 20)	N_{gt}	2640.6 ± 36.3	2703.7 ± 77.4	2653.8 ± 107.2
		IGD	$5.49e{-}04 \pm 2.93e{-}04$	$2.89e{-}02 \pm 1.85e{-}04$	$5.23e{-}03 \pm 4.39e{-}04$
3	(2, 20)	N_{gt}	1033.9 ± 61.1	1050.6 ± 70.3	1094.1 ± 228.7
		IGD	$3.65e{-}03 \pm 4.58e{-}03$	$5.97e{-}02 \pm 1.84e{-}02$	$3.06e{-}01 \pm 5.65e{-}01$
	(3, 20)	N_{gt}	4146.5 ± 98.0	4285.7 ± 423.0	4491.5 ± 713.6
		IGD	$1.08e{-}03 \pm 8.13e{-}04$	$5.64e{-}02 \pm 2.78e{-}02$	$2.62e{-}02 \pm 3.28e{-}02$
4	(2, 20)	N_{gt}	583.6 ± 21.5	558.1 ± 14.9	533.7 ± 16.1
		IGD	$2.68e{-}03 \pm 1.06e{-}03$	$3.78e{-}02 \pm 8.80e{-}04$	$1.99e{-}02 \pm 9.62e{-}04$
	(3, 20)	N_{gt}	2726.9 ± 83.7	2580.3 ± 38.7	2445.4 ± 42.3
		IGD	$3.75e{-}04 \pm 3.24e{-}04$	$2.94e{-}02 \pm 3.77e{-}04$	$8.33e{-}03 \pm 1.15e{-}03$

WFG	T_{par}		M		
			5	10	15
1	(2, 20)	N_{gt}	618.6 ± 60.6	481.4 ± 63.3	630.5 ± 45.8
		HV	0.7391 ± 0.0622	1.3514 ± 0.1744	3.8516 ± 0.1697
	(3, 20)	N_{gt}	3699.3 ± 342.5	3183.0 ± 425.6	3194.4 ± 551.1
		HV	1.4720 ± 0.0467	2.3940 ± 0.0803	3.9783 ± 0.0382
2	(2, 20)	N_{gt}	688.3 ± 58.9	711.6 ± 273.8	497.8 ± 88.0
		HV	1.5962 ± 0.0019	2.5756 ± 0.0077	4.1210 ± 0.0189
	(3, 20)	N_{gt}	3071.1 ± 88.2	4977.8 ± 2773.9	3657.0 ± 1145.9
		HV	1.6065 ± 0.0006	2.5901 ± 0.0026	4.1701 ± 0.0071
4	(2, 20)	N_{gt}	539.5 ± 14.4	573.1 ± 27.7	675.8 ± 88.6
		HV	1.2730 ± 0.0048	2.4269 ± 0.0097	4.0522 ± 0.0568
	(3, 20)	N_{gt}	2532.5 ± 32.2	2712.8 ± 86.4	2836.1 ± 150.6
		HV	1.3071 ± 0.0004	2.5133 ± 0.0008	4.1367 ± 0.0003
5	(2, 20)	N_{gt}	539.3 ± 10.5	619.1 ± 78.0	557.0 ± 62.2
		HV	1.2168 ± 0.0022	2.3235 ± 0.0044	3.7990 ± 0.0121
	(3, 20)	N_{gt}	2539.0 ± 23.0	2718.9 ± 96.9	2552.1 ± 45.4
		HV	1.2269 ± 0.0002	2.3506 ± 0.0002	3.8324 ± 0.0002
6	(2, 20)	N_{gt}	538.5 ± 16.4	537.7 ± 27.4	522.2 ± 35.3
		HV	1.2087 ± 0.0055	2.2778 ± 0.0162	3.6940 ± 0.0897
	(3, 20)	N_{gt}	2513.3 ± 34.1	2546.4 ± 45.5	2356.9 ± 87.5
		HV	1.2358 ± 0.0042	2.3276 ± 0.0178	3.7315 ± 0.0929
7	(2, 20)	N_{gt}	534.3 ± 16.0	604.6 ± 67.1	493.5 ± 71.0
		HV	1.2845 ± 0.0022	2.4677 ± 0.0106	4.0178 ± 0.0398
	(3, 20)	N_{gt}	2507.9 ± 39.0	2612.8 ± 60.8	2403.7 ± 383.2
		HV	1.3060 ± 0.0006	2.5145 ± 0.0004	4.1279 ± 0.0083
8	(2, 20)	N_{gt}	531.0 ± 19.4	520.7 ± 59.4	483.0 ± 14.2
		HV	1.1711 ± 0.0038	2.1933 ± 0.0152	3.6841 ± 0.0331
	(3, 20)	N_{gt}	2488.2 ± 39.8	2387.9 ± 196.7	2162.2 ± 188.3
		HV	1.2074 ± 0.0016	2.3212 ± 0.0317	3.8180 ± 0.1013
9	(2, 20)	N_{gt}	589.0 ± 44.4	753.8 ± 44.4	497.8 ± 88.0
		HV	1.2050 ± 0.0066	2.2866 ± 0.0343	4.1210 ± 0.0189
	(3, 20)	N_{gt}	2707.4 ± 53.3	3122.4 ± 159.9	3657.0 ± 1145.9
		HV	1.2502 ± 0.0036	2.3788 ± 0.0183	4.1701 ± 0.0071

5.2   Results on Timed Nadir-Point Determination by NSGA-III

The results presented in Table 2 (T_{nad}: enabled) reveal that the intermittently *timed* Nadir-point estimation in NSGA-III$_{TN}$, in general facilitates better IGD and Hypervolume measures in the case of DTLZ and WFG problems, respectively. This assumes higher significance, given that intermittently timed Nadir-point estimation consuming $O(MN)$ computations, helps to do away with computational complexity of $O(M^2N)$ required for determination of extreme points through minimization of Achievement Scalarization Function (ASF, as in NSGA-III) in every generation. Some interesting observations can be made by comparing the respective results in Tables 2 and 1. For instance:

- in the case of DTLZ1(M=5): the $N_{gt}^{TN} \equiv 3517 \pm 65.5$ and $N_{gt} \equiv 4073.5 \pm 58$. Even through $N_{gt}^{TN} < N_{gt}$, the IGD obtained from NSGA-III$_{TN}$ at N_{gt}^{TN} is better than that obtained with NSGA-III at N_{gt}.
- the same trend holds for most problems experimented with and reported in Tables 2 and 1.

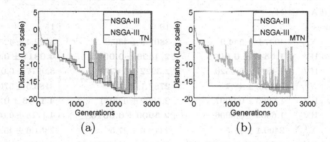

Fig. 3. DTLZ4 (M = 5): Effect of different Nadir-point update schemes

In terms of further improvement in quality of PF-approximation, the authors' envisage the following plausible interventions in the proposed Algorithm 1:

- *evidence-gathering*: Figure 3(a), corresponding to a particular seed, shows the departure (in Euclidean distance) of the estimated Nadir-point from the true Nadir-point (corresponding to the true PF), across generations. Evidently, this departure at a subsequent generation could be worse than that at a previous generation (negative fluctuation). This occurs in the current scheme, since the Nadir-point is updated by considering the extreme points corresponding to *only* that generation where T_{nad} (mild termination) is suggested. As an alternative, while the nadir point may still be updated after T_{nad} is suggested, but for the determination of extreme points, all the populations from previous to the current T_{nad} generations may be considered (to be referred as NSGA-III$_{MTN}$). This shall in principle help avoid the negative fluctuations. As a proof-of-concept, this scheme when implemented for the same

Table 2. DTLZ and WFG problems: performance of NSGA-III$_{TN}$ versus Original NSGA-III (abbreviated below as NS$_{TN}$ and NS, respectively) at N_{gt}^{TN} generations where NSGA-III$_{TN}$ is prompted for termination by the proposed Algorithm, with $T_{par} = (3, 20)$. Results are averaged over 10 independent runs of the underlying RMâOEA. IGD and HV denote Inverted Generational Distance and Hypervolume.

DTLZ	RMâOEA/		M		
	Measures		5	10	15
1		N_{gt}^{TN}	3517.0 ± 65.5	3163.7 ± 86.9	3722.4 ± 427.6
	NS$_{TN}$	IGD	**4.22e−05 ± 2.52e−05**	**1.20e−02 ± 2.34e−05**	**1.36e−03 ± 1.99e−04**
	NS	IGD	$6.42e{-}03 \pm 1.91e{-}02$	$1.40e{-}02 \pm 5.68e{-}03$	$5.78e{-}03 \pm 7.46e{-}03$
2		N_{gt}^{TN}	2485.2 ± 39.0	2383.0 ± 30.8	2324.5 ± 24.7
	NS$_{TN}$	IGD	**4.11e−04 ± 1.74e−04**	**2.93e−02 ± 4.31e−04**	**6.02e−03 ± 5.03e−04**
	NS	IGD	$7.23e{-}04 \pm 3.37e{-}04$	$3.23e{-}02 \pm 1.43e{-}03$	$8.90e{-}03 \pm 1.35e{-}03$
3		N_{gt}^{TN}	3740.2 ± 101.1	3282.6 ± 113.5	2819.3 ± 91.3
	NS$_{TN}$	IGD	**1.28e−02 ± 3.41e−02**	**2.86e−02 ± 2.28e−04**	**1.75e−02 ± 2.21e−02**
	NS	IGD	$3.86e{-}02 \pm 6.07e{-}02$	$4.29e{-}02 \pm 3.56e{-}02$	$4.68e{-}02 \pm 4.94e{-}02$
4		N_{gt}^{TN}	2739.9 ± 203.7	2428.7 ± 10.1	2363.8 ± 24.3
	NS$_{TN}$	IGD	**4.42e−04 ± 4.84e−04**	**2.96e−02 ± 3.65e−04**	**8.67e−03 ± 1.18e−03**
	NS	IGD	$4.97e{-}04 \pm 3.62e{-}04$	$2.98e{-}02 \pm 6.40e{-}04$	$1.25e{-}02 \pm 2.28e{-}03$

WFG	RMâOEA/		M		
	Measures		5	10	15
1		N_{gt}^{TN}	4037.0 ± 544.9	3807.9 ± 439.9	3722.4 ± 427.6
	NS$_{TN}$	HV	**1.5547 ± 0.0005**	**2.4496 ± 0.0070**	**4.0817 ± 0.0481**
	NS	HV	1.4871 ± 0.0520	2.3922 ± 0.0774	3.9977 ± 0.0529
2		N_{gt}^{TN}	2813.9 ± 48.7	2787.1 ± 47.2	3183.5 ± 570.9
	NS$_{TN}$	HV	1.6063 ± 0.0007	2.5902 ± 0.0011	**4.1743 ± 0.0015**
	NS	HV	1.6063 ± 0.0006	**2.5909 ± 0.0010**	4.1718 ± 0.0029
4		N_{gt}^{TN}	2400.1 ± 18.5	2418.0 ± 27.5	2351.6 ± 45.6
	NS$_{TN}$	HV	**1.3068 ± 0.0005**	**2.5122 ± 0.0006**	**4.1351 ± 0.0008**
	NS	HV	1.3067 ± 0.0003	2.5116 ± 0.0010	4.1342 ± 0.0008
5		N_{gt}^{TN}	2445.1 ± 26.7	2435.5 ± 25.3	2347.3 ± 25.7
	NS$_{TN}$	HV	1.2270 ± 0.0002	2.3507 ± 0.0001	**3.8330 ± 0.0001**
	NS	HV	1.2270 ± 0.0002	2.3507 ± 0.0003	3.8325 ± 0.0002
6		N_{gt}^{TN}	2451.4 ± 28.2	2423.7 ± 38.7	2223.6 ± 118.3
	NS$_{TN}$	HV	**1.2351 ± 0.0028**	**2.3342 ± 0.0204**	**3.7385 ± 0.1285**
	NS	HV	1.2336 ± 0.0039	2.3271 ± 0.0176	3.7313 ± 0.0931
7		N_{gt}^{TN}	2458.2 ± 27.0	2396.1 ± 39.6	2309.8 ± 20.0
	NS$_{TN}$	HV	**1.3062 ± 0.0005**	**2.5138 ± 0.0007**	**4.1344 ± 0.0004**
	NS	HV	1.3059 ± 0.0006	2.5137 ± 0.0006	4.1253 ± 0.0075
8		N_{gt}^{TN}	2389.8 ± 21.7	2276.7 ± 38.7	2253.2 ± 159.3
	NS$_{TN}$	HV	**1.2078 ± 0.0018**	**2.3520 ± 0.0400**	**3.8831 ± 0.0535**
	NS	HV	1.2069 ± 0.0017	2.3191 ± 0.0346	3.8133 ± 0.0953
9		N_{gt}^{TN}	2417.0 ± 32.0	2409.8 ± 33.7	2345.7 ± 130.0
	NS$_{TN}$	HV	**1.2474 ± 0.0054**	**2.3675 ± 0.0084**	**3.7494 ± 0.2583**
	NS	HV	1.2467 ± 0.0038	2.3604 ± 0.0243	3.7420 ± 0.1036

seed, results in Fig. 3(b). The new IGD value for this run, comes out to be $2.00e-04$, marking an improvement in the PF-approximation, though at $O(M^2N)$ computational cost as in NSGA-III.

- *decision-making*: instead of the simplistic measures of mean and standard deviation, better statistical indicators may be used.

6   Conclusion

In this paper, a generic, computationally efficient, and implementable *on-the-fly* Termination Algorithm has been proposed dedicatedly for Reference-based MâOEAs. Post integration with NSGA-III, it is established that under different parameter settings, the same Algorithm successfully accomplishes its dual goals of determining the appropriate *timing* for Nadir-point estimation and final termination. While the former is critically important for effective convergence and diversity of the PF-approximation, the latter helps balance the trade-off between the quality of PF-approximation and computational cost. Experimental results have revealed that NSGA-III implemented with intermittently *timed* Nadir-point estimation as prescribed by the proposed Algorithm not only terminated faster than the original NSGA-III but simultaneously offered better convergence and diversity characteristics. The authors hope, the revelation in this paper that NSGA-III's performance could be improved merely by changing the frequency of Nadir-point estimation, will open new doors for research towards more effective normalization of population in other RMâOEAs.

References

1. Asafuddoula, M., Ray, T., Sarker, R.: A decomposition-based evolutionary algorithm for many objective optimization. IEEE Trans. Evol. Comput. **19**(3), 445–460 (2015)
2. Deb, K., Jain, H.: An evolutionary many-objective optimization algorithm using reference-point-based nondominated sorting approach, part I: solving problems with box constraints. IEEE Trans. Evol. Comput. **18**(4), 577–601 (2014). https://doi.org/10.1109/TEVC.2013.2281535
3. Deb, K., Miettinen, K.: A review of nadir point estimation procedures using evolutionary approaches: a tale of dimensionality reduction. In: Multiple Criterion Decision Making (MCDM) Conference, pp. 1–14. Springer, Berlin (2008)
4. Deb, K., Saxena, D.K.: Searching for Pareto-optimal solutions through dimensionality reduction for certain large-dimensional multi-objective optimization problems. In: IEEE Congress on Evolutionary Computation, pp. 3353–3360 (2006)
5. Deb, K., Thiele, L., Laumanns, M., Zitzler, E.: Scalable test problems for evolutionary multi-objective optimization. In: Abraham, A., Jain, R., Goldberg, R. (eds.) Evolutionary Multiobjective Optimization: Theoretical Advances and Applications. AI&KP, pp. 105–145. Springer, London (2005). https://doi.org/10.1007/1-84628-137-7_6
6. Deb, K.: Multi-Objective Optimization Using Evolutionary Algorithms. Wiley, New York (2001)

7. Duro, J.A., Saxena, D.K.: Timing the decision support for real-world many-objective optimization problems. In: Trautmann, H., et al. (eds.) EMO 2017. LNCS, vol. 10173, pp. 191–205. Springer, Cham (2017). https://doi.org/10.1007/978-3-319-54157-0_14

8. Huband, S., Hingston, P., Barone, L., While, L.: A review of multiobjective test problems and a scalable test problem toolkit. IEEE Trans. Evol. Comput. **10**(5), 477–506 (2006)

9. Hughes, E.J.: Evolutionary many-objective optimisation: many once or one many? In: IEEE Congress on Evolutionary Computation, pp. 222–227 (2005)

10. Jain, H., Deb, K.: An evolutionary many-objective optimization algorithm using reference-point based nondominated sorting approach, part II: handling constraints and extending to an adaptive approach. IEEE Trans. Evol. Comput. **18**(4), 602–622 (2014). https://doi.org/10.1109/TEVC.2013.2281534

11. Khare, V., Yao, X., Deb, K.: Performance scaling of multi-objective evolutionary algorithms. In: Fonseca, C.M., Fleming, P.J., Zitzler, E., Thiele, L., Deb, K. (eds.) EMO 2003. LNCS, vol. 2632, pp. 376–390. Springer, Heidelberg (2003). https://doi.org/10.1007/3-540-36970-8_27

12. Li, H., Zhang, Q.: Multiobjective optimization problems with complicated Pareto sets, MOEA/D and NSGA-II. IEEE Trans. Evol. Comput. **13**(2), 284–302 (2009). https://doi.org/10.1109/TEVC.2008.925798

13. Marti, L., Garcia, J., Berlanga, A., Molina, J.M.: A stopping criterion for multi-objective optimization evolutionary algorithms. Inf. Sci. **367–368**, 700–718 (2016)

14. Saxena, D., Duro, J., Tiwari, A., Deb, K., Zhang, Q.: Objective reduction in many-objective optimization: linear and nonlinear algorithms. IEEE Trans. Evol. Comput. **17**(1), 77–99 (2013)

15. Saxena, D.K., Sinha, A., Duro, J.A., Zhang, Q.: Entropy-based termination criterion for multiobjective evolutionary algorithms. IEEE Trans. Evol. Comput. **20**(4), 485–498 (2016)

16. Trautmann, H., Wagner, T., Naujoks, B., Preuss, M., Mehnen, J.: Statistical methods for convergence detection of multi-objective evolutionary algorithms. Evol. Comput. **17**(4), 493–509 (2009). https://doi.org/10.1162/evco.2009.17.4.17403. pMID: 19916777

17. Wagner, T., Trautmann, H., Martí, L.: A taxonomy of online stopping criteria for multi-objective evolutionary algorithms. In: Takahashi, R.H.C., Deb, K., Wanner, E.F., Greco, S. (eds.) EMO 2011. LNCS, vol. 6576, pp. 16–30. Springer, Heidelberg (2011). https://doi.org/10.1007/978-3-642-19893-9_2

18. Wang, R., Zhang, Q., Zhang, T.: Decomposition-based algorithms using Pareto adaptive scalarizing methods. IEEE Trans. Evol. Comput. **20**(6), 821–837 (2016)

19. Yuan, Y., Xu, H., Wang, B.: An improved NSGA-III procedure for evolutionary many-objective optimization. In: Proceedings of the 2014 Annual Conference on Genetic and Evolutionary Computation, GECCO 2014, pp. 661–668. ACM, New York (2014)

20. Zhang, Q., Li, H.: MOEA/D: a multiobjective evolutionary algorithm based on decomposition. IEEE Trans. Evol. Comput. **11**(6), 712–731 (2007)

Variation Rate: An Alternative to Maintain Diversity in Decision Space for Multi-objective Evolutionary Algorithms

Oliver Cuate[1(✉)] and Oliver Schütze[2]

[1] Computer Science Department, Cinvestav-IPN, Av. Instituto Politécnico Nacional
No. 2508 Col. San Pedro Zacatenco, Mexico City, Mexico
ocuate@computacion.cs.cinvestav.mx
[2] Department of Applied Mathematics and Systems,
UAM Cuajimalpa Dr. Rodolfo Quintero Chair., Mexico City, Mexico
schuetze@cs.cinvestav.mx

Abstract. In almost all cases the performance of a multi-objective evolutionary algorithm (MOEA) is measured in terms of its approximation quality in objective space. As a consequence, most MOEAs focus on such approximations while neglecting the distribution of the individuals in decision space. This, however, represents a potential shortcoming in certain applications as in many cases one can obtain the same or a very similar qualities (measured in objective space) in several ways (measured in decision space) which may be very valuable information for the decision maker for the realization of a project.

In this work, we propose the variable-NSGA-III (vNSGA-III) an algorithm that performs an exploration both in objective and decision space. The idea behind this algorithm is the so-called variation rate, a heuristic that can easily be integrated into other MOEAs as it is free of additional design parameters. We demonstrate the effectiveness of our approach on several benchmark problems, where we show that, compared to other methods, we significantly improve the approximation quality in decision space without any loss in the quality in objective space.

Keywords: Evolutionary computation ·
Multi-objective optimization · Decision space diversity

1 Introduction

In many areas such as Economy, Finance, or Industry the problem arises naturally that several conflicting objectives have to be optimized concurrently. This leads to multi-objective optimization problems (MOPs). The solution of this kind of problems is a set of vectors that are incomparable to each other in terms

The authors acknowledge support for CONACyT project No. 285599.

© Springer Nature Switzerland AG 2019
K. Deb et al. (Eds.): EMO 2019, LNCS 11411, pp. 203–215, 2019.
https://doi.org/10.1007/978-3-030-12598-1_17

of their objective values. For some of these problems obtaining the greatest benefit from limited resources is essential. Such resources are typically represented as the variables of the problem, as the objective functions depends on them. Although by using constraints it is possible to control the value of decision variables, this would entail the loss of optimal solutions and that is not desirable for the decision-making process. For instance, in real-world problems where the value of some variables is crucial, the decision maker may prefer, among the set of optimal solutions, those that are easiest to implement as this can mean a saving in resources.

However, in almost all cases the performance of a MOEA is only measured in terms of its approximation quality in objective space. As a consequence, most MOEAs focus on such approximations while neglecting the distribution of the individuals in decision space. This represents a potential shortcoming in certain applications as in many cases one can obtain the same or a very similar quality (measured in objective space) in several ways (measured in decision space) which may be very valuable information for the decision maker for the realization of a project. In this context, there exists an additional challenge in solving a MOP, since we must find an approximation to the optimal set both in objective and decision space, in order to provide all these possible regions to the decision maker.

In this work, we propose the variable-NSGA-III (vNSGA-III) an algorithm that performs an exploration both in objective and decision space. The idea behind this algorithm is the so-called variation rate, a heuristic that can easily be integrated into other MOEAs as it is free of additional design parameters. We demonstrate the effectiveness of our approach on several benchmark problems, where we show that, compared to other methods, we significantly improve the approximation quality in decision space without any loss in the quality in objective space.

The rest of the paper is organized as follows, in Sect. 2, we present the background and the related work. In Sect. 3, a detailed description of the proposed algorithm (along with pseudo-codes) is presented. In Sect. 4, numerical results are provided. Finally, in Sect. 5, we discuss the advantages of the proposed algorithm and we discuss the possible future improvements to the algorithm.

2   Background and Related Work

Optimization refers to finding the best possible solution to a problem given a set of constraints [2]. MOP refers to the simultaneous optimization of multiple and usually conflicting objectives; as a result, a set of optimal solutions are obtained instead of having a single optimal solution. The MOP with k objectives is mathematical defined as:

$$\min_{x \in D} F(x),\tag{1}$$

where $D \subset \mathbb{R}^n$ is the domain and $F : D \subset \mathbb{R}^n \to \mathbb{R}^k$ is the objective function.

The optimality of a MOP is defined by the concept of *dominance*. Let $v, w \in \mathbb{R}^k$, the vector v is *less than* w ($v <_p w$), if $v_i < w_i$ for all $i \in \{1, \ldots, k\}$; the relation $\leq_p$ is defined analogously. A vector $y \in D$ is *dominated* by a vector $x \in D$ ($x \prec y$) with respect to (1) if $F(x) \leq_p F(y)$ and $F(x) \neq F(y)$, else y is called non-dominated by x. A point $x^* \in \mathbb{R}^n$ is Pareto optimal to (1) if there is no $y \in D$ which dominates x. The set of all the Pareto optimal points is called the Pareto set and its image is the Pareto front.

Unlike evolutionary algorithms for single objective optimization problems (SOP), maintaining diversity in decision space is not a priority for most MOEAs; even the performance indicators are developed in order to measure the accuracy based only on the objective function (e.g., the hypervolume [15] and the DOA [7]). As an exception, we have the Δ_p indicator [12], which is the averaged Hausdorff distance, and it actually measures the distance between two general sets. For this reason, we can use it both in objective space as well as in decision space.

Although works that explicitly consider at the same time variables and objectives are scarce, one can find some related work on this topic. For instance, the NSGA (the algorithm that precedes the well-known NSGA-II [3]) uses fitness sharing in decision space. In [9], some possible techniques are proposed to spread out solutions both in objective and decision space: *pointwise expansion, threshold sharing, sequential sharing, simultaneous sharing multiplicative,* and *simultaneous sharing additive*. It is important to point out that the above approaches are only part of the discussion of the paper and they were not implemented; the implemented algorithm was the Niched Pareto GA, a method with phenotypic sharing. Besides, all of the described techniques depend on the normal fitness sharing method, that is, two additional parameters must be provided or approximated (the niche radius σ_{share} in each space).

The omni-optimizer algorithm [6] is proposed as a procedure that aims at solving a wide variety of optimization problems (single or multi-objective and uni- or multi-modal problems). The authors argue that to solve different kinds of problems it is necessary to know different specialized algorithms. Thus, it is desirable to have an algorithm which adapts itself for handling any number of conflicting objectives, constraints, and variables. The omni-optimizer is important in the context of this work as it uses a two-tier fitness assignment scheme based on the crowding distance of the NSGA-II. The primary fitness is computed using the phenotypes (objectives and constraint values) and the secondary fitness is computed using both phenotypes and genotypes (decision variables). The modified crowding distance computes the average crowding distance of the population both in objectives and variables. If the crowding value for some individual above average (at any space), it is assigned the larger of the two distances; else the smaller of the two distances is assigned. However, omni-optimizer has a more general purpose.

Recently, the *MOEA/D with Enhanced Variable-Space Diversity (MOEA/D-EVSD)* has been proposed in [1]. This method is an extension of the MOEA/D [14] that explicitly promotes the diversity of the decision space via an enhanced variable-space diversity control. First generations of MOEA/D-EVSD try to induce a larger diversity via promoting the mating of dissimilar individuals. Similarly to MOEA/D, a new individual is created for each subproblem. Then, instead of randomly selecting two individuals of the neighborhood, a pool of α candidate parents is randomly filled from the neighborhood with probability δ, whereas it is randomly selected from the whole population with probability $1 - \delta$. Thus, the two selected parents are the ones that had the largest distance. As the δ parameter is dynamically set a gradual change between exploration and exploitation can be induced. Additionally, a final phase to further promote intensification is included, which is just a traditional MOEA/D with DE operators. For last generations of MOEA/D-EVSD the traditional mating selection of MOEA/D is conserved together with the Rand/1/bin scheme for the DE operators. Authors show that by inducing a gradual loss of diversity in the decision space, the state-of-the-art of MOEAs can be improved.

Finally, in [10], authors identify four different Pareto set and Pareto front type combinations: *Type I*, one Pareto set and one Pareto front; *Type II*, one Pareto set and multiple Pareto front parts; *Type III*, multiple Pareto subsets and one Pareto front; and *Type IV*, multiple Pareto subsets and Pareto front parts. In this work, a multi-start approach is proposed to solve problems of Type III, as authors argue that this kind of problems are rarely investigated and that standard MOEAs are not effective to preserve all Pareto subsets of equivalent quality. On the other hand, in [11], as a result of the study of multimodal problems, the authors conclude that a search in decision space is necessary to correctly solve them. In [13], a recovery technique in decision space is used to solve a problem of Type IV.

3   Proposed Algorithm

From the analysis of the combinations of Pareto set and front previously stated, it follows that, although most of MOEAs operate only in objective space, this is not a problem in cases like Types I, II, and IV. However, we must not lose sight of the fact that the value of the objectives depends on the variables. Thus, it is vital to maintaining the diversity of solutions in such space. In this way, when some method evaluates distribution considering only the objectives, potential search regions could be ruled out (Type III). On the other hand, if a method only takes into account the values of the variables without the objectives, then it could easily lose solutions along the Pareto front, what is highly penalized by the performance indicators. Our proposal seeks to perform an adequate grouping via a density estimator that allows obtaining a good distribution both in objective and decision space. The idea is to improve a classical density estimator, which groups the population in neighborhoods based only in the objectives. Such neighborhood structure is used to define a *variation rate* for each element in the neighborhood, according to some reference value in objective space, and certain measurement in decision space. In this way, the first grouping phase identifies promising solutions in objective space, meanwhile, the second phase favors solutions with the most different values in decision space. Thus, this variation rate represents the trade-off between these two aspects.

We consider the *variation rate* as the relation between the objective and the decision spaces. In order to properly define this rate, we need to group elements based on the value of their objectives (e.g., with the association method of the NSGA-III, the neighborhood structure of MOEA/D, the distance to a reference point, etc.). That is, we numerically assigned a reference value to the elements of each group for their ordering in objective space. Subsequently, the average distance in **decision space** of each element is computed against the rest elements in the group. Thus, the *variation rate* of each element is the quotient between its *reference value* in objective space and its *average distance* in decision space.

Let I denote a neighborhood (grouped in objective space) with v_i as the reference value for each $i \in I$, then the variation rate r_i is stated as follows:

$$r_i = \frac{v_i}{\texttt{distP}(i, I)} \tag{2}$$

where $\texttt{distP}(i, I)$ represents the average distance between each element $i \in I$ and the rest of elements in I, that is,

$$\texttt{distP}(i, I) = \frac{1}{|I| - 1} \sum_{j \in I \setminus \{i\}} d(i, j). \tag{3}$$

Equation (3) depends of a function $d(i, j)$, which is the used method to measure the distance between $i, j \in I$ in the decision space. Thus, $\texttt{distP}$ can vary according to the codification or the used norm.

As an example of how variation rate preserves the diversity, consider the following case. In Fig. 1, we can see a Type III Pareto Set/Front, that is, two lines in decision space –the Pareto set– map to the same Pareto front. Here, we have a neighborhood with three points ($\blacktriangledown$, $\blacklozenge$, and $\bigstar$) as their respective images are associated to the reference point (Z). Suppose that distance of $F(x_{\blacktriangledown})$, $F(x_{\blacklozenge})$ and $F(x_{\bigstar})$ to the reference Z is equal to one. On the other hand, $d(x_{\blacklozenge}, x_{\bigstar}) = 1$, $d(x_{\blacklozenge}, x_{\blacktriangledown}) = 2$, and $d(x_{\blacktriangledown}, x_{\bigstar}) = 2$, then,

- Variation rate of $\bigstar = \frac{1}{3/2} = \frac{2}{3}$
- Variation rate of $\blacklozenge = \frac{1}{3/2} = \frac{2}{3}$
- Variation rate of $\blacktriangledown = \frac{1}{4/2} = \frac{1}{2}$

In this way, if we select the point with the *less* variation rate, then we will conserve the most different individual in decision space with a good quality in objective space. In other words, elements with the minimum variation rate have the desired behavior.

Notice that, for problems of type I, II, and IV, it is expected that solutions in the same neighborhood have similar reference values in objective space and average distance in decision space, then the solution with the best reference value in objective space will be preferred. On the other hand, in type III problems the elements of the neighborhood will have a similar reference value in objectives but the average distance will be bigger for the most different solution in decision space, then its quotient (variation rate) will tend to be smaller than the rest of

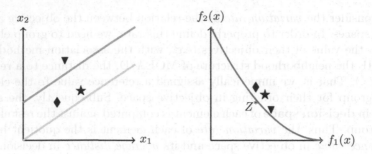

Fig. 1. Example of the computation of variation rate.

quotients. Thus, through the variation rate, we have a way to relate the objective and the decision spaces in order to choose the best element of each group.

The proposed method, called *variable-NSGA-III (vNSGA-III)*, is a modification of the NSGA-III algorithm [4] that seeks to solve the problem of diversity in the variables previously raised. Although this method is based on the NSGA-III, the *variation rate* can easily be adapted to other methods. The idea is to take advantage of the **association** method of the NSGA-III, which defines a "neighborhood". The association method assigns each element of F_j (the last front by classifying after the non-dominated sorting) to the nearest induced line by some weight $w_i \in Z$, where Z is a set of reference points. A weight can have more than one associated element, forming a neighborhood.

In the original NSGA-III, the **niching** is made by sort in ascending order the obtained groups in the association stage according to its cardinality. The element with less distance to the induced line in each group is selected, and it continues with the next group until filling the population. The proposed **niching** method does not prefer the element with the less distance value, instead it prefers the one with the smallest *variation rate*. The complete pseudocode of the vNSGA-III algorithm is shown in Algorithm 1.

4   Numerical Results

For the numerical results, we employ the following methodology. First, we compare vNSGA-III against two of the most widely used methods in the literature, NSGA-II and MOEA/D, in order to demonstrate that vNSGA-III improves the distribution in decision space without losing quality in objective space. Later, we compare our method with the MOEA/D-EVSD, that is a method with a similar purpose; (available code of MOEA/D-EVSD[1] is only for bi-objective problems, then this comparison is restricted to that kind of problems). Finally, we make a brief scalability test with the vNSGA-III and the original NSGA-III, to show how the variation rate can also improve the performance of this algorithm for many objective optimization problems (MaOPs), that is, problems with more than

---

[1] https://github.com/joelchaconcastillo/GECCO17_MOEA_D_MATING.

Algorithm 1. Iteration of the vNSGA-III

Require: Reference points Z, current population P_t
Ensure: Next population P_{t+1}
1: $S_t = \emptyset$, $i = 1$
2: $Q_t =$ apply variation operators to P_t
3: $M_t = P_t \cup Q_t$
4: $(F_1, F_2 \ldots,) =$ non-dominated-sort(M_t)
5: **while** $|S_t| \leq N$ **do**
6:     $t = S_t \cup F_i$
7:     $i = i + 1$
8: **end while**
9: Add first fronts to P_{t+1}
10: $F_i :=$ last added front
11: Normalize F_i
12: Associate elements of F_i with each Z
13: Niching of F_i (according with the variation rate)
14: $V_t :=$ best niching elements
15: $P_{t+1} : S_t \cup V_t$

---

three objectives. As we test stochastic algorithms, each execution was repeated 30 times with different seeds to obtain statistical significance. The parameter settings of all the used algorithms are in Table 1.

Table 1. Parameter configuration for each algorithm. Mutation probability m_p, crossover probability c_p, neighborhood size α, first phase percent P_f, additional parameters for MOEA/D-EVSD T_{r1} and T_{r2}, and number of reference points $\#Z$.

Parameter	vNSGA-III	MOEA/D-EVSD	NSGA-II	MOEA/D
m_p	$1/n$	0.3	$1/n$	0.1
c_p	1.0	0.9	0.8	1.0
α	-	20	-	10
P_f	-	80%	-	-
T_{r1}	-	2	-	-
T_{r2}	-	25	-	-
$\#Z$	50	-	-	-

For the first comparison, we consider the problems DTLZ 1–3. We use the hypervolume indicator and Δ_p [12] to know how different is the approximation in variable and objective space. Results are shown in Table 2.

Next, we compare vNSGA-III against MOEA/D-EVSD and omni-optimizer. Test considered problems are the first four WFG tests proposed in [8] and the following bi-objective problem [6]

Table 2. Hypervolume (HV) and Δ_p (objective and decision space); best, average with standard deviation, and worst values are showed. The best value is put in bold.

Problem	Indicator	vNSGA-III	NSGA-II	MOEA/D
DTLZ1	HV	0.040700	0.076781	**0.079128**
		0.073839 (0.011847)	0.078690 (0.000675)	**0.079425** (0.000170)
		0.080123	0.079552	0.079654
	O-Δ_p	**0.015422**	0.017439	0.025133
		0.026077 (0.0683503)	0.041077 (0.016282)	**0.025495** (0.000304)
		0.056208	0.070908	**0.026016**
	D-Δ_p	**0.038723**	0.042297	0.048348
		0.075532 (0.038904)	0.080363 (0.025395)	0.054312 (0.002910)
		0.147971	0.126349	**0.058551**
DTLZ2	HV	0.379642	0.413363	**0.416942**
		0.414960 (0.008706)	0.417088 (0.001976)	**0.417757** (0.000559)
		0.420359	0.419725	0.418694
	O-Δ_p	**0.044592**	0.041514	0.044972
		0.045075 (0.012895)	0.045373 (0.001998)	0.045086 (0.000094)
		0.505410	0.049205	**0.045257**
	D-Δ_p	**0.029387**	0.074172	0.034619
		0.036727 (0.006681)	0.078942 (0.002572)	0.037074 (0.001521)
		0.051319	0.082206	**0.040618**
DTLZ3	HV	0.287586	0.365446	**0.410906**
		0.399156 (0.034941)	0.404309 (0.013330)	**0.418067** (0.002929)
		0.418205	**0.422514**	0.422293
	O-Δ_p	0.043525	0.040273	**0.040057**
		0.047231 (0.0867439)	0.071554 (0.098862)	**0.041026** (0.000499)
		0.485502	0.488509	**0.041650**
	D-Δ_p	0.041069	0.046354	**0.035459**
		0.053028 (0.023238)	0.051261 (0.004364)	**0.037630** (0.001252)
		0.083221	0.062031	**0.039935**

$$f_1(x) = \sum_{i=1}^{n} \sin(\pi x_i), \quad f_2(x) = \sum_{i=1}^{n} \cos(\pi x_i), \tag{4}$$

where $0 \leq x_i \leq 6$, and $i = 1, 2, \ldots, n$. This problem, denoted as OMNI1 in this work, is a type III combination of Pareto set/front. The used configuration for the WFG problems was the following: the stopping criterion was set to 250 generations, the population size was fixed to 200, and they were configured with two objectives and 24 parameters (20 distance parameters and 4 position parameters). On the other hand, for the OMNI1 problem the stopping criterion was set to 200 generations, the population size was fixed to 100, and $n = 5$ (number of decision variables). Numerical results are shown in Table 3.

Table 3. Hypervolume values, best, average with standard deviation, and worst values are showed. The best value is put in bold and the statistical significance is indicated, according to the Wilcoxon test with $p = 0.05$, when appropriate.

Problem	vNSGA-III	MOEA/D-EVSD	Omni-Optimizer
OMNI1$^\downarrow$	22.485004	22.458890	**22.530915**
	22.525381 (0.019402)	22.531425 (0.029448)	**22.567425** (0.017941)
	22.564005	22.564666	**22.592098**
WFG1†	**4.050763**	2.816998	3.665387
	4.252966 (0.215981)	3.723304 (0.410585)	4.039288 (0.203742)
	5.042717	4.175823	4.211367
WFG2†	**5.293999**	4.274958	4.681840
	5.428045 (0.092903)	4.810172 (0.230530)	4.763940 (0.043061)
	5.511280	5.043839	4.876360
WFG3†	**4.877052**	3.915621	4.214724
	4.891410 (0.009166)	4.412886 (0.151054)	4.276409 (0.028887)
	4.906714	4.567030	4.331920
WFG4†	**2.403976**	2.129897	1.959788
	2.411151 (0.009000)	2.181784 (0.029585)	1.974326 (0.011972)
	2.417486	2.228117	2.003731

It is clear that our approach requires less additional parameters than the MOEA/D-EVSD and omni-optimizer. Actually, no additional parameters than the original NSGA-III are needed.

According with the values of Table 3, it is clear that our approach converges faster on the WFG problems. Although, the hypervolume value is worse for the for the OMNI1 problem, for such problem there are not significance according the Wilcoxon test. Moreover, in Fig. 2, we can see that the distribution in decision space, the main goal of this work, is better distributed with our method.

Finally, a brief scalability test is performed. We decide to compare our approach against the original NSGA-III, as this method is made to deal with MaOPs. We test on the DTLZ2 [5] problem with different number of objectives. The used parameters was the same that the used on the original paper of NSGA-III [4]. Numerical results are shown in Table 4.

From Table 4 we can see that the large the number of objectives, the better is our approach, at least for the DTLZ2 problem. With this results, we expect that this approach could be successfully applied for MaOPs. However, an extensive analysis must be performed in order to conclude something about scalability.

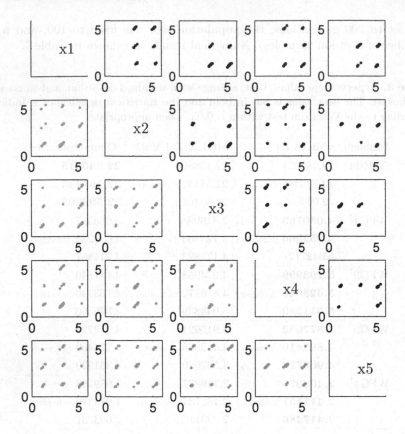

(a) Decision space. Left-down vNSGA-III, right-up Omni-optimizer

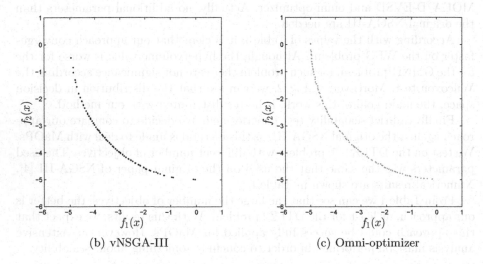

(b) vNSGA-III

(c) Omni-optimizer

Fig. 2. Graphical results of the run with the median a values for the OMNI1 function.

Table 4. Δ_p values in objective space for the DTLZ2 problem with different number of objectives. The best, average with standard deviation, and worst values are showed; the best value for each case is put in bold.

k	vNSGA-III	NSGA-III
3	0.088196	**0.081690**
	0.107123 (0.013942)	**0.095862** (0.012272)
	0.148009	**0.133477**
5	0.229096	**0.210055**
	0.253322 (0.023562)	**0.233658** (0.018896)
	0.308298	**0.284361**
8	**0.293232**	0.366762
	0.379999 (0.046295)	0.469987 (0.036129)
	0.456742	0.538956
10	**0.305149**	0.534823
	0.463134 (0.069997)	0.600828 (0.035696)
	0.578061	0.686215

5   Conclusions and Future Work

In this work, the *variation rate*, a heuristic to explicitly handle the diversity in decision space, is proposed. This is an original proposal, In general, there are a few related work and the algorithms in the state of the art preserve diversity in decision space in distinct ways. Although two proposed techniques in [9] are kind of similar, the simultaneous sharing multiplicative and additive, they use a sum and a multiplication, respectively, instead of a quotient. Moreover, such methods depend on the σ-shared, while the variation rate if more flexible about the grouping method. We test this method via vNSGA-III, an extension of NSGA-III which uses the variation rate. The results are very promising in this field and our approach presents some advantages over others proposals.

In contrast with omi-optimizer and MOEA/D-EVSD, the implementation of the variation rate is free of additional parameters. This fact allows that this proposal can be easy include more algorithms, for instance, MOEA/D. It is only necessary to conserve certain neighborhood structure and explore the elements in each group in decision space. Moreover, the presented numerical results in the original paper, both in omi-optimizer and MOEA/D-EVSD, report a huge number of function evaluations; 500 generations for 1,000 individuals and 50,000 generations for 250 individuals, respectively.

Numerical results also show that variation rate improves the performance of the NSGA-III when the number of objectives becomes to increase. In principle, the MOEA/D-EVSD can also solve problems with any number of objectives, but it is restricted by the capabilities of the MOEA/D algorithm. On the other hand, the omni-optimizer depends on the crowding distance, and such a method is not scalable for high dimensions. Of course, variation rate by itself is not enough for the treatment of MaOPs, this depends on the operators of the NSGA-III. That is, variation rate can enhance the overall performance of a certain algorithm, but if such algorithm is not conceived to deal with MaOPs, then the addition of variation rate would be not enough to solved MaOPs.

As future work, it is necessary to develop an indicator for problems of Type III. In general, performance indicators evaluate an approximation based on the value of the objectives, but for problems as OMNI1 this is not provided enough information. Once we have such indicator, we can validate the better performance of our methods in decision space. However, it is not clear what property has to be satisfied with this approximation. We also need to test this approach in problems with different properties in decision space, in particular problems with disconnected Pareto set. Finally, the adaptation of the variation rate into a different MaOPs will allow studying the effect of this heuristic.

References

1. Castillo, J.C., Segura, C., Aguirre, A.H., Miranda, G., León, C.: A multi-objective decomposition-based evolutionary algorithm with enhanced variable space diversity control. In: Proceedings of GECCO 2017, pp. 1565–1571. ACM, New York (2017)
2. Coello, C.A.C., Van Veldhuizen, D.A., Lamont, G.B.: Evolutionary Algorithms for Solving Multi-objective Problems, vol. 242. Springer, Berlin (2002). https://doi.org/10.1007/978-0-387-36797-2
3. Deb, K., Pratap, A., Agarwal, S., Meyarivan, T.: A fast and elitist multiobjective genetic algorithm: NSGA-II. IEEE Trans. Evol. Comput. **6**(2), 182–197 (2002)
4. Deb, K., Jain, H.: An evolutionary many-objective optimization algorithm using reference-point-based nondominated sorting approach, part I: solving problems with box constraints. IEEE Trans. Evol. Comput. **18**(4), 577–601 (2014)
5. Deb, K., Thiele, L., Laumanns, M., Zitzler, E.: Scalable multi-objective optimization test problems. In: 2002 Proceedings of the 2002 Congress on Evolutionary Computation. CEC 2002, vol. 1, pp. 825–830. IEEE (2002)
6. Deb, K., Tiwari, S.: Omni-optimizer: a generic evolutionary algorithm for single and multi-objective optimization. Eur. J. Oper. Res. **185**(3), 1062–1087 (2008)
7. Dilettoso, E., Rizzo, S.A., Salerno, N.: A weakly pareto compliant quality indicator. Math. Comput. Appl. **22**(1), 25 (2017)
8. Huband, S., Barone, L., While, L., Hingston, P.: A scalable multi-objective test problem toolkit. In: Coello, C.A.C., Hernández Aguirre, A., Zitzler, E. (eds.) EMO 2005. LNCS, vol. 3410, pp. 280–295. Springer, Heidelberg (2005). https://doi.org/10.1007/978-3-540-31880-4_20
9. Horn, J., Nafpliotis, N., Goldberg, D.E.: Multiobjective optimization using the niched Pareto genetic algorithm. IlliGAL report (93005), 61801–2296 (1993)
10. Preuss, M., Naujoks, B., Rudolph, G.: Pareto set and EMOA behavior for simple multimodal multiobjective functions. In: Runarsson, T.P., Beyer, H.-G., Burke, E., Merelo-Guervós, J.J., Whitley, L.D., Yao, X. (eds.) PPSN 2006. LNCS, vol. 4193, pp. 513–522. Springer, Heidelberg (2006). https://doi.org/10.1007/11844297_52

11. Rudolph, G., Naujoks, B., Preuss, M.: Capabilities of EMOA to detect and preserve equivalent pareto subsets. In: Obayashi, S., Deb, K., Poloni, C., Hiroyasu, T., Murata, T. (eds.) EMO 2007. LNCS, vol. 4403, pp. 36–50. Springer, Heidelberg (2007). https://doi.org/10.1007/978-3-540-70928-2_7

12. Schütze, O., Esquivel, X., Lara, A., Coello, C.A.C.: Using the averaged Hausdorff distance as a performance measure in evolutionary multiobjective optimization. IEEE Trans. Evol. Comput. **16**(4), 504–522 (2012)

13. Schütze, O., Witting, K., Ober-Blöbaum, S., Dellnitz, M.: Set oriented methods for the numerical treatment of multiobjective optimization problems. In: Tantar, E., et al. (eds.) EVOLVE - A Bridge between Probability, Set Oriented Numerics and Evolutionary Computation. SCI, vol. 447, pp. 187–219. Springer, Heidelberg (2013). https://doi.org/10.1007/978-3-642-32726-1_5

14. Zhang, Q., Li, H.: MOEA/D: a multiobjective evolutionary algorithm based on decomposition. IEEE Trans. Evol. Comput. **11**(6), 712–731 (2007)

15. Zitzler, E., Thiele, L.: Multiobjective evolutionary algorithms: a comparative case study and the strength Pareto approach. IEEE Trans. Evol. Comput. **3**(4), 257–271 (1999)

Indicator-Based Weight Adaptation for Solving Many-Objective Optimization Problems

Auraham Camacho[1], Gregorio Toscano[1], Ricardo Landa[1], and Hisao Ishibuchi[2(✉)]

[1] CINVESTAV-Tamaulipas, 87130 Ciudad Victoria, Tamaulipas, Mexico
{acamacho,gtoscano,rlanda}@tamps.cinvestav.mx
[2] Southern University of Science and Technology,
Shenzhen 518055, Guangdong, China
hisao@sustc.edu.cn

Abstract. Weight adaptation methods can enhance the diversity of solutions obtained by decomposition-based approaches when addressing irregular Pareto front shapes. Generally, these methods adapt the location of each weight vector during the search process. However, early adaptation could be unnecessary and ineffective because the population does not provide a good Pareto front approximation at early generations. In order to improve its performance, a better approach would be to trigger such adaptation only when the population has reached the Pareto front. In this paper, we introduce a performance indicator to assist weight adaptation methods, called the median of dispersion of the population (MDP). The proposed indicator provides a general snapshot of the progress of the population toward the Pareto front by analyzing the local progress of each subproblem. When the population becomes steady according to the proposed indicator, the adaptation of weight vectors starts. We evaluate the performance of the proposed approach in both regular and irregular test problems. Our experimental results show that the proposed approach triggers the weight adaptation when it is needed.

Keywords: Weight adaptation · Many-objective optimization · Decomposition

1 Introduction

Multiobjective optimization problems (MOPs) involve several objective functions to be optimized simultaneously. An MOP can be defined as follows [18]:

$$\text{Minimize } \mathbf{f}(\mathbf{x}) = [f_1(\mathbf{x}), \ldots, f_m(\mathbf{x})]^T \tag{1}$$
$$\text{subject to } \mathbf{x} \in X$$

where $\mathbf{f}$ comprises m objective functions, $f_i : \mathbb{R}^n \to \mathbb{R}$, for $i = 1, \ldots, m$. The decision variable vector $\mathbf{x} = [x_1, \ldots, x_n]^T$ belongs to the feasible set X, which

© Springer Nature Switzerland AG 2019
K. Deb et al. (Eds.): EMO 2019, LNCS 11411, pp. 216–228, 2019.
https://doi.org/10.1007/978-3-030-12598-1_18

is a subset of the decision variable space $\mathbb{R}^n$. For each $\mathbf{x} \in X$, there is an objective vector $\mathbf{z} = \mathbf{f}(\mathbf{x})$, which belongs to the objective space $Z \subseteq \mathbb{R}^m$. Problems with more than three objectives are called many-objective optimization problems (MaOPs) [11,15]. Often, the objectives are in conflict. That is, any improvement in a given objective f_i must lead to a deterioration of at least one other objective f_j [16]. Hence, this class of problems does not have a unique solution that optimizes the m objectives simultaneously. Instead, an optimizer is aimed to find a set of solutions with the best trade-offs among objectives. Such a set is called Pareto set in the decision space and Pareto front in the objective space [2]. MOPs with degenerated, disconnected, or inverted Pareto front shapes are considered irregular. Some optimizers could face difficulties when addressing these problems.

Evolutionary algorithms (EAs) are search methods based on natural evolution to address optimization problems [8]. Given the population-based nature, EAs are suitable to address MOPs, because they can produce a Pareto front approximation in a single run [3,22]. Several multiobjective evolutionary algorithms (MOEAs) for solving MaOPs have been proposed in recent years [5,16,23,24]. Among them, decomposition-based approaches have gained attention in the evolutionary multiobjective optimization (EMO) community. A decomposition-based approach transforms an MOP into a set of subproblems by means of a scalarizing function and a set of weight vectors. The Multiobjective Evolutionary Algorithm based on Decomposition (MOEA/D) [25] is considered a representative approach of this class of methods. Recent studies have been focused on determining which properties of MOPs are difficult to address by MOEAs [2,10,17,20]. Among such properties, the shape of the Pareto front can influence on the performance of decomposition [10]. When the shape of the Pareto front is regular (e.g., linear), decomposition-based methods have been found to be effective to find a well-distributed set of solutions on the Pareto front. Nevertheless, when such a shape is irregular, it is difficult to obtain a good Pareto front approximation via decomposition. This issue is related to the weight vectors. When the locations of the weight vectors do not reflect the shape of the Pareto front, the diversity of solutions obtained by decomposition-based approaches is deteriorated [10].

To alleviate such a diversity issue, a few weight adaptation methods have been proposed in the literature [2,13,17,20]. An early approach was introduced by Deb and Jain in the NSGA-III, which allows NSGA-III to handle irregular problems [12,13]. Their approach is based on two main operations: addition and deletion of weight vectors. It first identifies overcrowded weight vectors. In this context, a weight vector is considered overcrowded if there are multiple solutions associated with it. On the other hand, a weight vector is marked as unhelpful if it is not associated with any solution after the approach has converged. Ideally, every weight vector should be associated with a single solution to achieve good diversity. Additional vectors are created uniformly around overcrowded vectors to reallocate solutions. In order to keep a pre-specified number of weight vectors, unhelpful vectors are removed. Both operations, addition and deletion, are performed continuously during the search process. This way, the weight vectors are adapted according to the shape of the Pareto front. Recent weight adaptation methods consider additional procedures, including the use of an archive of non-dominated solutions to reflect the Pareto front shape [17,20], and the analysis of the range of the objective values for weight adaptation [2]. Most of those weight adaptation methods share two aspects in common:

- The population is employed for weight adaptation.
- The adaptation of weight vectors is performed continuously during the search process.

The population is employed as reference of the Pareto front shape before weight adaptation. However, it is unlikely that the population provides a good approximation of such a shape at early generations. Thus, early weight adaptation could be ineffective. A better approach would be to perform the weight adaptation only when the population has reached the Pareto front, in order to start the redistribution of weight vectors and, therefore, provide a better approximation to the Pareto front. Such an approach would reduce the computational overhead associated to weight adaptation as it is not required before finding the Pareto front.

In this paper, we propose an indicator to assist weight adaptation methods, called the median of dispersion of the population (MDP). This indicator relies on the local progress made for each subproblem after a given interval, also known as relative improvement [26]. By analyzing the local progress of every subproblem, the proposed indicator provides a general snapshot of the global progress of the population toward the Pareto front. When the subproblems cannot be improved significantly according to the proposed indicator, we assume that the population is steady, and can provide a good Pareto front approximation for weight adaptation. This way, early weight adaptation is avoided, reducing the computational burden. The main contributions of this paper are listed below:

- We introduce an indicator to assist weight adaptation methods. This indicator triggers the weight adaptation when the population becomes steady.
- As a proof of concept, we combine the proposed indicator into a weight adaptation strategy similar to the approach proposed by Deb and Jain [12,13]. Of course, other weight adaptation methods can be employed.
- We demonstrate the effectiveness of our idea by using MOEA/D. The proposed indicator along with the weight adaptation strategy has been integrated into MOEA/D. The resulting approach, called AMOEA/D, is evaluated through computational experiments on regular and irregular problems. We show that the proposed approach is able to improve the performance of MOEA/D on irregular problems, without deteriorating its performance on regular problems.

The remainder of this paper is organized as follows. Section 2 introduces background knowledge. The details of the proposed approach are given in Sect. 3. The experimental design and results are detailed in Sects. 4 and 5, respectively. Section 6 concludes this paper.

2   Background

In this section, we introduce preliminary concepts related to decomposition, as well as a description of the weight adaptation method proposed by Deb and Jain [12,13].

2.1   Decomposition

Definition 1 (Scalarizing function). *A scalarizing function is a parameterized function* $g : \mathbb{R}^m \to \mathbb{R}$. *Thus, an MOP is transformed into the following scalar problem:*

$$\begin{aligned} Optimize \; g(\mathbf{f}(\mathbf{x}); \mathbf{w}) \\ subject \; to \; \mathbf{x} \in X \end{aligned} \tag{2}$$

where $\mathbf{w} = [w_1, \ldots, w_m]^T$ is a weight vector. Given a scalarizing function g and a set of N weight vectors, denoted as $W = \{\mathbf{w}_1, \ldots, \mathbf{w}_N\}$, an MOP is decomposed into N scalar subproblems. The subproblem j is denoted by $g_j(\mathbf{f}(\mathbf{x}); \mathbf{w}_j)$, or g_j for simplicity. Each vector $\mathbf{w} \in W$ is defined as follows [10]:

$$\sum_{i=1}^{m} w_i = 1, \text{ and } w_i \geq 0, \text{ for } i = 1, \ldots, m \tag{3}$$

$$w_i \in \left\{ 0, \frac{1}{H}, \frac{2}{H}, \ldots, \frac{H}{H} \right\}, \text{ for } i = 1, \ldots, m \tag{4}$$

Several scalarizing functions have been proposed in the literature [18]. In this paper, we employ the Penalty-based Boundary Intersection [25] for decomposing a problem. A common approach to generate an evenly distributed set of vectors was proposed by Das and Dennis [4], where the total number of vectors is given by

$$\binom{H + m - 1}{m - 1} \tag{5}$$

where H is a positive integer. This parameter controls the number of divisions along each objective [5]. In [10], the authors suggested to use a small weight value, such as 10^{-6}, when $w_i = 0$. We follow the suggestion in this paper.

2.2   Addition of Weight Vectors

Deb and Jain [5,12] proposed a method to increase uniformly the number of weight vectors to enhance the coverage of NSGA-III over irregular Pareto front shapes. This procedure involves translating an m-dimensional simplex. Additional vectors are created by moving the simplex around a central vector. Notice that the locations of the new vectors depend on the central vector, as shown in Fig. 1(a)–(c). This procedure can be employed to increase the number of weight vectors uniformly, as shown in Fig. 1(d).

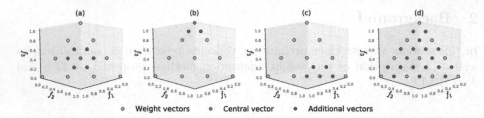

Fig. 1. Addition of weight vectors. (a)–(c) The locations of the additional vectors depend on the central vector. (d) This procedure can be employed to create a new set of uniformly distributed vectors.

2.3   Density Estimation

Deb and Jain [5] proposed a procedure to associate each individual of the population with its nearest weight vector for density estimation. The individual $\mathbf{x}_i$ is associated with the weight vector $\mathbf{w}_j$, denoted as $\pi(i) = j$, as follows:

$$\pi(i) = \arg\min_j \left(d^\perp(\mathbf{f}(\mathbf{x}_i), \mathbf{w}_j)\right) \tag{6}$$

where $d^\perp(\mathbf{u}, \mathbf{v})$ denotes the perpendicular distance between $\mathbf{u}$ and $\mathbf{v}$. Thus, $\mathbf{x}_i$ is associated with $\mathbf{w}_j$ if the distance between its image $\mathbf{f}(\mathbf{x}_i)$ and $\mathbf{w}_j$ is the smallest among the set of weight vectors. We employ this method to estimate the density of each weight vector. A counter ρ_j is defined, where ρ_j is the number of individuals associated with $\mathbf{w}_j$:

$$\rho_j = |\{i : \pi(i) = j, i = 1, 2, \ldots, N\}|, \text{ for } j = 1, 2, \ldots, N \tag{7}$$

This way, we can estimate the density of the niche (defined by $\mathbf{w}_j$) by counting the individuals in their vicinity. We can also determine whether $\mathbf{w}_j$ is needed to cover the Pareto front. Figure 2(a) shows the approximation achieved by MOEA/D on an irregular problem called inv-DTLZ1 [12,13]. After density estimation, we can identify the weight vectors that better approximate the Pareto front, as illustrated in Fig. 2(b). Every vector $\mathbf{w}_j$ highlighted as an associated vector in this figure is associated with at least one individual, that is, $\rho_j > 0$.

3   Proposed Indicator

The weight adaptation method proposed by Deb and Jain improves the diversity of solutions in NSGA-III [12,13]. Although this scheme could be embedded into MOEA/D as well, the additional overhead should be considered. In this paper, we follow a different approach by performing such a weight adaptation method only once in the search process in order to reduce the computational overhead. We propose a performance indicator, called MDP, to trigger the addition of weight vectors. This indicator provides a snapshot of the progress of the population toward the Pareto front by analyzing the local progress of each subproblem. The pseudocode of the proposed indicator is shown in Algorithm 1.

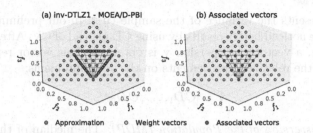

Fig. 2. Density estimation. (a) Approximation obtained by MOEA/D in inv-DTLZ1. (b) The association procedure identifies the weight vectors associated with at least one solution.

Relative Improvement: Let $\mathbf{x}_{j,t}$ and $\mathbf{x}_{j,t-\Delta T}$ denote the solutions of subproblem g_j in the current generation t and the $(t - \Delta T)$-th generation, respectively. The relative improvement of subproblem g_j in the last ΔT generations is defined as follows [26]:

$$u_j = \frac{g_j(\mathbf{x}_{j,t-\Delta T}; \mathbf{w}_j) - g_j(\mathbf{x}_{j,t}; \mathbf{w}_j)}{g_j(\mathbf{x}_{j,t-\Delta T}; \mathbf{w}_j)} \tag{8}$$

This indicator is employed to infer the local progress made for each subproblem in the last ΔT generations. If u_j is close to zero, then the solution $\mathbf{x}_j$ of subproblem g_j may be stagnant. The relative improvement u_j of subproblem g_j is computed every $\Delta T = 10$ generations in this paper.

Time Frame of Relative Improvement: In order to gain a better insight into the progress of the population, we collect the last l relative improvements for each subproblem to create a time frame $\mathbf{u}_j = [u_{j,1}, \dots, u_{j,l}]^T$. In this paper, the size of the time frame is defined as $l = 5$.

Description of Time Frames of Relative Improvement: The time frame $\mathbf{u}_j$ acts as a buffer to track the progress of subproblem g_j. This frame is reduced to a single scalar value by using a measure of spread. If the spread of $\mathbf{u}_j$ is low, then the progress of subproblem g_j could be stagnant. We propose a spread measure called the coefficient of dispersion (CD), based on the coefficient of variation (CV). This measure is defined as follows [9]:

$$CV = \frac{s}{\bar{x}} \tag{9}$$

where s and $\bar{x}$ denote the standard deviation and mean of the sample ($\mathbf{u}_j$), respectively. The CV expresses the variability of a sample in the units of its mean. This is a useful descriptive statistic to compare the dispersion among data sets (when the means are different across the data sets) [9,14]. Given the definition of the CV, we propose the following spread measure:

$$CD = \frac{v}{\bar{x}} \tag{10}$$

where v represents the variance of the sample ($\mathbf{u}_j$). In our preliminary experimentation, we noticed better results by using CD instead of CV. After computing the CD of $\mathbf{u}_j$, a vector of dispersion $\mathbf{v}$ is obtained. This vector represents the dispersion of the relative improvements on the N subproblems:

$$\mathbf{v} = [CD_1, \ldots, CD_N]^T \tag{11}$$

Median of Dispersion of the Population (MDP): The median of the dispersion vector $\mathbf{v}$ is employed to estimate the overall improvement of the population. A value close to zero means that most of the subproblems have no significant progress toward the ideal vector, which suggests stagnation.

Algorithm 1. Median of Dispersion of the Population

Require:
    Population P.
    Set of weight vectors W.
Ensure:
    MDP value of P.
1: **for** each subproblem $j = 1, \ldots, N$ **do**
2:      Create the time frame $\mathbf{u}$ with the last l relative
       improvements using Eq. (8):
       $\mathbf{u}_j = [u_{j,1}, \ldots, u_{j,l}]^T$
3:      Compute the CD of $\mathbf{u}_j$ to obtain its dispersion:
       $CD_j = \frac{Var(\mathbf{u}_j)}{Mean(\mathbf{u}_j)}$
4: **end for**
5: Create the vector of dispersion $\mathbf{v}$:
      $\mathbf{v} = [CD_1, \ldots, CD_N]^T$
6: Compute the median of dispersion of the population:
      $MDP = Median(\mathbf{v})$

Figure 3 illustrates the proposed indicator. This figure shows the mean and standard deviation of MDP for MOEA/D. The shaded area in this figure refers to the standard deviation. MOEA/D was evaluated 30 independent times using DTLZ1 [7] as a test problem. This figure shows that the MDP is able to capture the dynamics of the population during the search process. After 250 generations, the population becomes steady, as suggested by the low value of the indicator. This implies that the subproblems cannot be improved considerably in subsequent generations.

4   Experimental Design

This section describes the experimental design employed to evaluate the performance of the proposed approach. We have embedded the proposed indicator along with the addition operation of weight vectors into MOEA/D. The addition

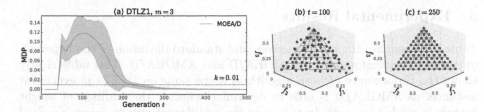

Fig. 3. Indicator MDP. (a) Mean and standard deviation of the proposed indicator. As the indicator approaches zero, the population becomes steady. Population (b) Before and (c) After reaching the threshold $k = 0.01$.

of weight vectors is performed if (a) the MDP value is below a given threshold k, and (b) there are overcrowded weight vectors. The first condition is needed to get a better approximation of the Pareto front shape. If it is achieved, the second condition is evaluated. This condition is needed to determine whether the addition of vectors is required: If every weight vector is associated with one solution, then the addition is not required; otherwise, there are overcrowded vectors and the addition of weight vectors can be performed. The resulting approach is called AMOEA/D. To evaluate its performance, we have adopted regular (DTLZ1 [7]) and irregular (inv-DTLZ1 [12,13]) problems from the literature. Our goal is to determine whether the proposed approach improves the diversity of solutions obtained by MOEA/D for inv-DTLZ1, without deteriorating its performance on DTLZ1. In our preliminary experiments, we examined five settings of k: 0.02, 0.01, 0.005, 0.0025, 0.00125. In this paper, we report the results from the best setting: $k = 0.01$ for 3–8 objectives, and $k = 0.005$ for 10 objectives. Each algorithm was evaluated 30 independent times on every test instance. Table 1 shows the parameters used for each problem. Simulated binary crossover (SBX) and polynomial mutation [6] were employed as variation operators with the following configuration: $\eta_c = 20$, and $p_c = 1.0$, for SBX, and $\eta_m = 15$, and $p_m = 1/n$ for polynomial mutation. To assess the performance of each algorithm, the hypervolume [27] and Solow-Polasky (SP) [19,21] indicators were adopted in this paper. A recent method for computing the hypervolume indicator was employed [1]. All hypervolume values presented in this paper were normalized to the range $[0, 1]$ as recommended by Cheng et al. [2].

Table 1. Population size and maximum number of generations for each test instance.

Objectives	Divisions	Population size	Generations	
(m)	(H)	(N)	DTLZ1	inv-DTLZ1
3	12	91	300	600
5	6	210	600	1200
8	3	120	800	1600
10	3	220	1000	2000

5   Experimental Results

Table 2 summarizes the mean, median, and standard deviation of the hypervolume and SP indicators for both MOEA/D and AMOEA/D. This table clearly shows the performance of both algorithms is the same on DTLZ1 in every test instance, as AMOEA/D is able to determine whether the addition of weight vectors is needed (or not). In this case, the addition of weights is not required because in this test problem, every weight vector is associated with a single solution, as shown in Fig. 4. This figure graphically confirms that both algorithms perform the same on DTLZ1 for $m = 3$. On the other hand, AMOEA/D achieved better results than MOEA/D according to the hypervolume and Solow-Polasky performance measures on the irregular problem as Table 2 clearly states for 5, 8, and 10 objectives. Figure 5 depicts the approximations obtained by both algorithms for $m = 3$. As it can be seen, a better coverage of the Pareto front was obtained by AMOEA/D than that achieved by MOEA/D (which can be confirmed by the Solow-Polasky performance measure). In this case, the proposed approach was able to determine that the addition of weight vectors was required. Thus, poorly-crowded niches were populated successfully.

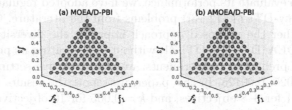

Fig. 4. Approximations obtained by (a) MOEA/D and (b) AMOEA/D on DTLZ1.

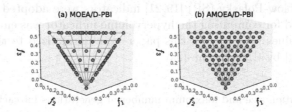

Fig. 5. Approximations obtained by (a) MOEA/D and (b) AMOEA/D on inv-DTLZ1.

Table 2. Mean, median, and standard deviation for hypervolume and Solow-Polasky indicators in test instances. Best performance for each indicator is shown in bold.

Problem	m	Hypervolume		Solow-Polasky	
		MOEA/D	AMOEA/D	MOEA/D	AMOEA/D
DTLZ1	3	0.9794	0.9794	1.5337	1.5337
		0.9795	0.9795	1.5330	1.5330
		±0.0005	±0.0005	±0.0043	±0.0043
	5	0.9994	0.9994	1.7333	1.7333
		0.9994	0.9994	1.7334	1.7334
		±0.0000	±0.0000	±0.0009	±0.0009
	8	1.0000	1.0000	1.8806	1.8806
		1.0000	1.0000	1.8830	1.8830
		±0.0000	±0.0000	±0.0067	±0.0067
	10	1.0000	1.0000	1.9596	1.9596
		1.0000	1.0000	1.9631	1.9631
		±0.0000	±0.0000	±0.0078	±0.0078
inv-DTLZ1	3	**0.7345**	0.7339	1.5252	**1.5300**
		0.7346	0.7341	1.5252	**1.5298**
		±0.0002	±0.0007	±0.0003	±0.0008
	5	0.3150	**0.3614**	1.5607	**1.6355**
		0.3147	**0.3611**	1.5607	**1.6355**
		±0.0009	±0.0015	±0.0002	±0.0011
	8	0.0338	**0.0412**	1.3950	**1.4143**
		0.0336	**0.0416**	1.3918	**1.4291**
		±0.0011	±0.0025	±0.0158	±0.0519
	10	0.0101	**0.0130**	1.3945	**1.4760**
		0.0100	**0.0126**	1.3918	**1.4625**
		±0.0003	±0.0009	±0.0162	±0.0272

6 Conclusion

In this paper, we proposed an indicator, called MDP, to assist the adaptation of weight vectors in MOEA/D to handle irregular Pareto fronts. This indicator analyzes the relative improvement of each subproblem to estimate the overall improvement of the population. When the subproblems cannot be improved significantly, the population becomes steady and can provide a better Pareto front approximation for weight adaptation. This way, early adaptation is avoided and its effectiveness can be improved. We have evaluated the performance of the proposed approach by using MOEA/D in both regular and irregular test problems. From our experimental results, we have found that the proposed approach was able to trigger weight adaptation when it was required for addressing irregular problems. Hence, the diversity of MOEA/D was improved. Although our evaluation was limited to a few test problems, the proposed approach seems to be a promising alternative to enhance weight adaptation methods. As future work, we consider the incorporation of the proposed approach into other weight adaptation methods [2,13,17,20]. Also, further experimentation is needed to assess the performance of the proposed approach in a variety of test problems with respect to both its search ability and its efficiency.

Acknowledgment. Auraham Camacho acknowledges support from CONACyT through a scholarship to pursue his studies. Gregorio Toscano and Ricardo Landa gratefully acknowledge support from SEP-Cinvestav project No. 262. Hisao Ishibuchi would like to thank the Shenzhen Peacock Plan (Grant No. KQTD2016112514355531), the Program for Guangdong Introducing Innovative and Entrepreneurial Teams (Grant No. 2017ZT07X386), the Science and Technology Innovation Committee Foundation of Shenzhen (Grant No. ZDSYS201703031748284), and the Program for University Key Laboratory of Guangdong Province (Grant No. 2017KSYS008).

References

1. Bradstreet, L., While, L., Barone, L.: A fast many-objective hypervolume algorithm using iterated incremental calculations. In: IEEE Congress on Evolutionary Computation, pp. 1–8, July 2010. https://doi.org/10.1109/CEC.2010.5586344
2. Cheng, R., Jin, Y., Olhofer, M., Sendhoff, B.: A reference vector guided evolutionary algorithm for many-objective optimization. IEEE Trans. Evol. Comput. **20**(5), 773–791 (2016). https://doi.org/10.1109/TEVC.2016.2519378
3. Coello, C., Lamont, G., Veldhuizen, D.V.: Evolutionary Algorithms for Solving Multi-Objective Problems, 2nd edn. Springer, Heidelberg (2007). https://doi.org/10.1007/978-0-387-36797-2
4. Das, I., Dennis, J.: Normal-boundary intersection: a new method for generating the Pareto surface in nonlinear multicriteria optimization problems. SIAM J. Optim. **8**(3), 631–657 (1998). https://doi.org/10.1137/S1052623496307510
5. Deb, K., Jain, H.: An evolutionary many-objective optimization algorithm using reference-point-based nondominated sorting approach, Part I: Solving problems with box constraints. IEEE Trans. Evol. Comput. **18**(4), 577–601 (2014). https://doi.org/10.1109/TEVC.2013.2281535
6. Deb, K., Pratap, A., Agarwal, S., Meyarivan, T.: A fast and elitist multiobjective genetic algorithm: NSGA-II. IEEE Trans. Evol. Comput. **6**(2), 182–197 (2002). https://doi.org/10.1109/4235.996017
7. Deb, K., Thiele, L., Laumanns, M., Zitzler, E.: Scalable test problems for evolutionary multiobjective optimization. In: Abraham, A., Jain, L., Goldberg, R. (eds.) Evolutionary Multiobjective Optimization: Theoretical Advances and Applications. AI&KP, pp. 105–145. Springer, London (2005). https://doi.org/10.1007/1-84628-137-7_6
8. Eiben, A., Smith, J.: Introduction to Evolutionary Computing. Springer, Heidelberg (2015). https://doi.org/10.1007/978-3-662-44874-8
9. Havbro, M.: Statistics and Probability Theory. Springer, Heidelberg (2012). https://doi.org/10.1007/978-94-007-4056-3

10. Ishibuchi, H., Setoguchi, Y., Masuda, H., Nojima, Y.: Performance of decomposition-based many-objective algorithms strongly depends on Pareto front shapes. IEEE Trans. Evol. Comput. **PP**(99), 1 (2016). https://doi.org/10.1109/TEVC.2016.2587749
11. Ishibuchi, H., Tsukamoto, N., Nojima, Y.: Evolutionary many-objective optimization: a short review. In: 2008 IEEE Congress on Evolutionary Computation (IEEE World Congress on Computational Intelligence), pp. 2419–2426, June 2008. https://doi.org/10.1109/CEC.2008.4631121
12. Jain, H., Deb, K.: An improved adaptive approach for elitist nondominated sorting genetic algorithm for many-objective optimization. Technical report, Indian Institute of Technology, Kanpur, India. Department of Mechanical Engineering (2013)
13. Jain, H., Deb, K.: An evolutionary many-objective optimization algorithm using reference-point based nondominated sorting approach, Part II: Handling constraints and extending to an adaptive approach. IEEE Trans. Evol. Comput. **18**(4), 602–622 (2014). https://doi.org/10.1109/TEVC.2013.2281534
14. Lee, H.: Foundations of Applied Statistical Methods. Springer, Heidelberg (2014). https://doi.org/10.1007/978-3-319-02402-8
15. Li, B., Li, J., Tang, K., Yao, X.: Many-objective evolutionary algorithms: a survey. ACM Comput. Surv. **48**(1), 13:1–13:35 (2015). https://doi.org/10.1145/2792984
16. Li, K., Deb, K., Zhang, Q., Kwong, S.: An evolutionary many-objective optimization algorithm based on dominance and decomposition. IEEE Trans. Evol. Comput. **19**(5), 694–716 (2015). https://doi.org/10.1109/TEVC.2014.2373386
17. Li, M., Yao, X.: What weights work for you? Adapting weights for any Pareto front shape in decomposition-based evolutionary multi-objective optimisation. CoRR abs/1709.02679 (2017). http://arxiv.org/abs/1709.02679
18. Miettinen, K.: On the methodology of multiobjective optimization with applications. Ph.D. thesis, University of Jyväskylä, Department of Mathematics (1994)
19. Solow, A., Polasky, S.: Measuring biological diversity. Environ. Ecol. Stat. **1**(2), 95–103 (1994). https://doi.org/10.1007/BF02426650
20. Tian, Y., Cheng, R., Zhang, X., Cheng, F., Jin, Y.: An indicator based multi-objective evolutionary algorithm with reference point adaptation for better versatility. IEEE Trans. Evol. Comput. **PP**(99), 1 (2017). https://doi.org/10.1109/TEVC.2017.2749619
21. Ulrich, T., Thiele, L.: Maximizing population diversity in single-objective optimization. In: Proceedings of the 13th Annual Conference on Genetic and Evolutionary Computation, GECCO 2011, pp. 641–648. ACM, New York (2011). https://doi.org/10.1145/2001576.2001665
22. Wang, R., Zhang, Q., Zhang, T.: Decomposition-based algorithms using Pareto adaptive scalarizing methods. IEEE Trans. Evol. Comput. **20**(6), 821–837 (2016). https://doi.org/10.1109/TEVC.2016.2521175
23. Xiang, Y., Zhou, Y., Li, M., Chen, Z.: A vector angle-based evolutionary algorithm for unconstrained many-objective optimization. IEEE Trans. Evol. Comput. **21**(1), 131–152 (2017). https://doi.org/10.1109/TEVC.2016.2587808
24. Yuan, Y., Xu, H., Wang, B., Yao, X.: A new dominance relation-based evolutionary algorithm for many-objective optimization. IEEE Trans. Evol. Comput. **20**(1), 16–37 (2016). https://doi.org/10.1109/TEVC.2015.2420112
25. Zhang, Q., Li, H.: MOEA/D: a multiobjective evolutionary algorithm based on decomposition. IEEE Trans. Evol. Comput. **11**(6), 712–731 (2007). https://doi.org/10.1109/TEVC.2007.892759

26. Zhou, A., Zhang, Q.: Are all the subproblems equally important? Resource allocation in decomposition-based multiobjective evolutionary algorithms. IEEE Trans. Evol. Comput. **20**(1), 52–64 (2016). https://doi.org/10.1109/TEVC.2015.2424251

27. Zitzler, E., Thiele, L.: Multiobjective evolutionary algorithms: a comparative case study and the strength Pareto approach. IEEE Trans. Evol. Comput. **3**(4), 257–271 (1999). https://doi.org/10.1109/4235.797969

Investigating the Normalization
Procedure of NSGA-III

Julian Blank(✉) ⓘ, Kalyanmoy Debⓘ, and Proteek Chandan Royⓘ

Michigan State University, East Lansing, MI, USA
{blankjul,kdeb,royprote}@msu.edu
http://www.coin-laboratory.com

Abstract. Most practical optimization problems are multi-objective in
nature. Moreover, the objective values are, in general, differently scaled.
In order to obtain uniformly distributed set of Pareto-optimal points, the
objectives must be normalized so that any distance metric computation
in the objective space is meaningful. Thus, normalization becomes a crucial component of an evolutionary multi-objective optimization (EMO)
algorithm. In this paper, we investigate and discuss the normalization
procedure for NSGA-III, a state-of-the-art multi- and many-objective
evolutionary algorithm. First, we show the importance of normalization
in higher-dimensional objective spaces. Second, we provide pseudo-codes
which presents a clear description of normalization methods proposed in
this study. Third, we compare the proposed normalization methods on
a variety of test problems up to ten objectives. The results indicate the
importance of normalization for the overall algorithm performance and
show the effectiveness of the originally proposed NSGA-III's hyperplane
concept in higher-dimensional objective spaces.

Keywords: Many-objective optimization · NSGA-III · Normalization

1 Introduction

The need to optimize several objectives at a time has been investigated for years,
and various algorithms have been proposed [21]. The desired result is a non-
dominated set of solutions close to the true Pareto-optimal front [13], instead of a
single optimal solution. The non-dominated set of solutions gives us the possibility to make a suitable decision for choosing a single preferred solution following
the algorithm's execution and provides useful information about optimal solutions
with respect to different preferences. Also, the decision maker can compare the
trade-offs between different solutions and therefore justify his/her choice.

However, the fact that the target space has more than one dimension brings
new challenges which must be addressed in designing the optimization algorithm.
To deal with multiple dimensions in the objective space, reference directions
express the trade-off between solutions with respect to each objective. Usually,
either the user provides them directly or they are sampled uniformly in the unit
space. If a uniformly distributed set of reference directions can be supplied and

ⓒ Springer Nature Switzerland AG 2019
K. Deb et al. (Eds.): EMO 2019, LNCS 11411, pp. 229–240, 2019.
https://doi.org/10.1007/978-3-030-12598-1_19

an EMO algorithm can find one or more Pareto-optimal solutions close to each reference direction, a widely distributed set will be achieved at the end.

Clearly, such a process will involve distance computations in the objective space, thereby necessitating a normalization procedure within the algorithm, which will consider the range of each objective on a same scale. In contrast to test problems where variables and objectives are already nicely scaled, in practical problems, the objective space range for each objective may differ by several magnitudes. Therefore, for any distance or trade-off calculation, normalization of objectives becomes an inevitable task.

In this paper, we investigate and discuss the normalization procedure of NSGA-III, a state-of-the-art evolutionary multi- and many-objective algorithm. In addition to the originally proposed normalization procedure of NSGA-III, we suggest a few other normalization methods. Our purpose in this paper is to: (i) show the importance of normalization in the objective space for high-dimensional multi-objective problems, (ii) compare different normalization procedures, and (iii) provide pseudo-codes for different normalization procedure.

In the remainder of this paper, we will first present a review of some past studies expanding upon or applying NSGA-III. Thereafter, in Sect. 3 we provide a brief description of the algorithm including the role of normalization. Then, different methodologies for normalization are discussed in depth, and a hands-on example is provided in Sect. 4. Afterwards, in Sect. 5, we present our results evaluated on a variety of test problems with up to ten objectives. Finally, conclusions of the study are presented in Sect. 6.

2 Related Studies

The need of optimizing more than one objective at a time brought attention of the multi-objective optimization research area. Also, normalization is often assumed implicitly and not discussed in detail, the importance to solve practical problems is indisputable.

The normalization procedure for MOEA/D [20] was investigated in [11]. The normalization was based on the PBI (penalty-based boundary intersection) measure by considering lower and upper bound estimations. The study showed that the normalization has both positive and negative effects on the performance on test problems. Interestingly, the normalization showed positive effects for test problems that do not need any normalization. Furthermore, three representative strategies for estimating the ideal point in MOEA/D were studied [17]. The ϵ value, which is subtracted from the minimum of each objective in the current population, is varied: small (pessimistic), large (optimistic), or decreasing over time (dynamic). The authors found out that the strategy has an effect on the exploration and exploitation of the algorithm and suggest to use the dynamic strategy for unknown problems. Also, the effect of local optimization to improve solutions contributing to the ideal point has been investigated [15]. The study showed that the local search helps to improve the diversity of the final non-dominated population for certain problems by converging close to the true ideal point in an early phase.

Moreover, NSGA-III [5,12], designed to solve problems with more than three objectives, was investigated since its publication in 2014, and some extensions and improved versions were proposed. For instance, a unified approach for mono-, multi- and many-objective problems, U-NSGA-III [14], introduces more selection pressure during the mating selection. Moreover, NSGA-III-OSD [3] decomposes the objective space into several subspaces by clustering the reference directions uniformly. Each subspace has its own population and PBI as decomposition method. Additionally, EliteNSGA-III [10] improves the diversity and accuracy of the resulting Pareto-front. An elite population archive is maintained to preserve previously generated elite solutions that would probably be eliminated by the reference survival selection procedure.

Moreover, NSGA-III has been applied to industry problems, for instance environmental dispatch problem [2], hydro thermal wind scheduling problem [19], and car engine design problem [9]. Also, NSGA-III has been implemented in different programming languages and popular optimization frameworks, such as jMetal [8], moeaframework [1], and PlatEMO [16].

3 NSGA-III

In the following, NSGA-III is explained and the role of normalization during the survival selection is illustrated. The basic framework remains similar to NSGA-II [6] with significant modifications to the mating and survival selection. In NSGA-III, parents to be used for recombination are selected randomly. The survival selection considers the M-dimensional objective space by using the reference direction concept. Reference directions Z represent trade-offs between solutions regarding their objective values. They are either provided a priori by the user or created uniformly, commonly executed using the Das and Dennis's technique [4].

An outline of the survival selection is shown in Algorithm 1. Considering an optimization problem with M objectives and an evolutionary algorithm with a population size of N, generation t begins with the current population $P^{(t)}$ known as the parent population, creates an offspring population $Q^{(t)}$ through recombination and mutation, and merges two populations together to create $R^{(t)} = P^{(t)} \cup Q^{(t)}$. The survival selection has to return $P^{(t+1)}$ – the next generation population of size N. The creation of $P^{(t+1)}$ is as follows. First, the individuals of the merged population $R^{(t)}$ are sorted by non-dominated rank which results in a list of fronts $(F_1, F_2, \ldots)$. To do this, the set of surviving solutions S is initialized as an empty set. Thereafter, it is iterated through the list of fronts and the current front F_i is appended to S, if the resulting number of individuals does not exceed N. The front where $|S \cup F_i| \geq N$ is the potential splitting front F_L. In case, $|S| + |F_L| = N$ no splitting is necessary and all surviving individuals are already determined. Otherwise, a niching method is employed to choose those F_L members that are associated with the least represented reference directions already associated by individuals in S. To assign individuals to the reference directions Z, S is normalized by using $\hat{z}^*$ as a lower and the nadir point estimation $\hat{z}^{nad}$ as an upper bound. Therefore, each already selected individual k in S is assigned to the closest reference direction π_k having a perpendicular distance of d_k. The niche count ρ is kept track of and incremented

---

Algorithm 1. NSGA-III Survival Selection

Input: Merged Population $R^{(t)}$, Number of surviving individuals N, Reference
       Directions Z, Ideal Point Estimation $\hat{z}^*$, Nadir Point Estimation $\hat{z}^{nad}$
Output: Surviving Individuals $P^{(t+1)}$
1 $(F_1, F_2, \ldots) \leftarrow \texttt{non\_dominated\_sort}(R^{(t)})$
2 $S = \emptyset, \quad i = 1$
3 **while** $|S| + |F_i| < N$ **do** $S \leftarrow S \cup F_i; \quad i = i + 1$
4 $F_L \leftarrow F_i$
5 **if** $|S| + |F_L| = N$ **then** $S \leftarrow S \cup F_L$
6 **else**

    /* Normalize objectives space and update boundary estimation   */
7     $\bar{S}, \bar{F}_L, \hat{z}^*, \hat{z}^{nad} \leftarrow \texttt{normalize}(S, F_L, \hat{z}^*, \hat{z}^{nad})$

    /* niche count, assigned Z_i, perpendicular dist to Z_i */
8     $\rho, \pi, d \leftarrow 0$
9   **for** $k \leftarrow 1$ *to* $|S|$ **do**
10 | $\pi_k, d_k \leftarrow \texttt{associate}(\bar{S}_k, Z); \quad \rho_{\pi_k} \leftarrow \rho_{\pi_k} + 1$
11   **end**

    // Remaining individuals from F_L to fill up S
12     $S \leftarrow S \cup \texttt{niching}(\bar{F}_L, n - |S|, \rho, \pi, d)$
13 **end**
14 $P^{(t+1)} \leftarrow S$
15 **return** $P^{(t+1)}$

---

by one for each assignment. Finally, the `niching` method selects from $\bar{F}_L$ the remaining $N - |S|$ individuals using ρ, π, d. A population member associated with an under-represented or un-represented reference direction is immediately preferred. With a continuous stress for emphasizing non-dominated individuals, the whole process is then expected to find one population member corresponding to each supplied reference direction close to the Pareto-optimal front.

4 Normalization Procedure

In this section, we investigate different normalization procedures for NSGA-III. The normalization relies on lower and upper boundaries in the objective space that correspond to the estimated ideal point $\hat{z}^*$ and the estimated nadir point $\hat{z}^{nad}$. Therefore, it is sufficient to provide the estimation for both points in order to normalize. The normalized value $\bar{a}_i$ in the i-th objective is then calculated by

$$\bar{a}_i = \frac{a_i - \hat{z}_i^*}{\hat{z}_i^{nad} - \hat{z}_i^*}. \tag{1}$$

The open question is how to find estimation the boundary points $\hat{z}^*$ and $\hat{z}^{nad}$, so that non-dominated solutions are properly emphasized. The ideal point estimation $\hat{z}^*$ is rather simple and the calculation is based on the smallest value in each objective we have observed since the start of the optimization run:

$$\hat{z}_j^* = \min(\hat{z}_j^* \cup R_j), \tag{2}$$

where R_j denotes the j-th objective of the merged population. Please note that the ideal point should not be calculated from R at each generation, but being updated. The survival selection of NSGA-III does not guarantee each individual contributing to an ideal point to survive in higher dimensions. For this reason, an update is necessary for a correct estimation of the ideal point.

The nadir point estimation is more tricky and is one of the main cruxes of this study. Let us first discuss the requirements and goals for estimating the nadir point in the context of many-objective evolutionary algorithms.

(i) **Estimated ideal point must dominate estimated nadir point:** Since it is normalized between the ideal and nadir point estimation, we need to make sure that $\forall i \in [1,..,M] : \hat{z}_i^{nad} > \hat{z}_i^*$. In practice, the formulation should be more strict where $\forall i \in [1,..,M] : \hat{z}_i^{nad} - \hat{z}_i^* > \epsilon_{nad}$ with ϵ_{nad} being our assumption about the minimum range of the Pareto-front for all objectives. A minimum difference ϵ_{nad} prevents having floating point issues and loosing the diversity during the survival selection.

(ii) **Estimated nadir point must converge to the true nadir point with generations:** Finally, when the population converges to the true optimum, the estimated nadir point should converge to the true nadir point. When both ideal and nadir points are estimated close to their true values, the EMO algorithm gets stabilized and works efficiently to find a well-distributed set of near Pareto-optimal points.

(iii) **Estimated nadir point must gradually change from one generation to the next:** This requirement is especially important in an evolutionary context, because the normalization is applied before assigning to reference directions and directly influences the survival method. An abrupt change of the normalization process will make previous generation's non-dominated solutions meaningless, thereby creating a restart situation.

In the following, we will suggest a number of possible normalization procedures. Towards this goal, we shall revise the hyperplane concept in Sect. 4.3 which was proposed in the original publication and present corner cases that must be handled on an implementation level.

4.1   Maximum of Non-dominated Front (MNDF)

Straightforwardly, we can concatenate the maximum of each objective of the non-dominated front of each generation and construct the nadir point. Assuming the algorithm converges eventually to the entire Pareto-optimal front, the estimated nadir point will be equal to the true nadir point. Since the method is based on the current non-dominated front in each generation, some special degenerate cases must be addressed. If the population has only one non-dominated solution, this solution might also be equal to the ideal point. This will cause a division by zero problem to Eq. 1. In this case, we propose to consider the next non-dominated front for the estimation of the nadir point. This process can continue until the difference between estimated nadir and ideal points becomes larger than a pre-specified threshold ϵ_{nad}.

4.2 Maximum of Extreme Points (ME)

We can use the achievement scalarization function (ASF) [18] along the axes to find M extreme points. The ASF function is defined by:

$$ASF(f(x), w, \hat{z}^*) = \max_{i=1}^{M} \frac{f_i(x) - \hat{z}_i^*}{w_i}, \tag{3}$$

where the weight vector w of the k-th objective is $w_k = 1$ and $w_i = \epsilon_{asf}$, if $i \neq k$. For our experiments we set $\epsilon_{asf} = 10^{-6}$. The procedure to update the extreme points each generation is presented in Algorithm 2. We find the extreme points by combining the merged population with the current extreme points. This ensures that the extreme points get an update, instead of a straightforward replacement all the time. Then, the ASF function scalarizes multiple objectives into a single value. We simply choose the solution with the minimum ASF value having at least ϵ_{nad} different in the i-th objective. Finally, we set the extreme point $e^{(i)}$ to the objective vector of the index found.

Algorithm 2. Maximum of Extreme Points

Input: R, $\hat{z}^*$, Current Extreme Points e
Output: Updated Extreme Points e

1 $A \leftarrow R \cup e$
2 **for** $i \leftarrow 1$ *to* M **do**
3 $\quad$ $w \leftarrow (\epsilon_{asf}^1, \ldots, \epsilon_{asf}^M)$
4 $\quad$ $w_i \leftarrow 1$
5 $\quad$ $k \leftarrow \mathrm{argmin}(\mathrm{ASF}(A, w, \hat{z}^*), \epsilon_{nad})$,
6 $\quad$ $e^{(i)} \leftarrow A^{(k)}$
7 **end**

4.3 Revised Hyperplane Through Extreme Points (HYP)

In the following, we revise the idea implemented in the original NSGA-III. We analyze the hyperplane concept on implementation level, where a number of exceptions must be handled to ensure the algorithm will not fail – a number of which we describe below.

Negative Intercepts. The hyperplane is found in the translated space $e' \leftarrow e - \hat{z}^*$ and the intercepts I are the intersections with the coordinate axes. The intercepts with the axes in the translated space represent the estimation of the range of the Pareto-front. For this reason, the intercepts are required to be positive. However, the hyperplane through the extreme points might intersect the axis not in the positive orthant. Let us consider the following scenario, illustrated in Fig. 1a, with solutions in the objective space $f^{(1)} = (1.0, 0.2, 0.0)$, $f^{(2)} = (0.4, 0.1, 0.4)$, and $f^{(3)} = (0.1, 0.0, 1.0)$. Each point is an extreme point with respect to the achievement scalarization function along an axis: $f^{(1)}$ along f_1, $f^{(2)}$ along f_2, and $f^{(3)}$ along f_3. The intercepts of the hyperplane will result in

$I = (-1.4, 0.1167, 0.933)$. Obviously, the intercept with the first axis is negative and cannot be used for normalization.

No Unique Hyperplane Exists. A unique M-dimensional hyperplane through the extreme points can only be found if all points are linearly independent from each other. On implementational level, a matrix with the extreme points as row vectors has to be inverted to obtain the axis intercepts. Linearly dependent rows form a singular matrix, where an exception will be thrown during the inversion. Moreover, the extreme point selection does not guarantee to select different points for different axes. For instance, let us consider three non-dominated points $f^{(1)} = (0.8, 0.5, 0.5)$, $f^{(2)} = (0.1, 0.3, 0.9)$, and $f^{(3)} = (0.4, 0.1, 0.9)$, as shown in Fig. 1b. The extreme points are selected using the achievement scalarization function for each axis. Here, $f^{(1)}$ will be chosen for both f_1 and f_2 axes, and $f^{(2)}$ for f_3. Because $f^{(1)}$ is chosen for two axes, the matrix to be inverted is singular and a unique hyperplane does not exist. Additionally, numerical instability through floating point calculations depending on the library used for the inversion must be addressed as well.

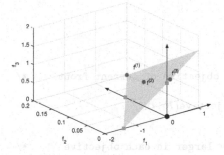 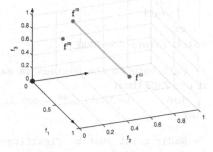

(a) Negative Intercepts in the objective space: $f^{(i)}$ are illustrated by blue dots and axes intercepts by red squares.

(b) No unique hyperplane exists: Any hyperplane through the line is possible.

Fig. 1. Two degenerate cases of HYP normalization method. (Color figure online)

Pseudo-code of HYP. For any implementation using the hyperplane idea, the above presented scenarios must be addressed. In these cases, we propose the algorithm to fall back to the worst point of the current non-dominated front or of the population. Algorithm 3 illustrates the procedure using the hyperplane idea and handling the degenerate cases. Note, that our procedure requires the worst point estimation $\hat{z}^w$. In contrast to the ideal point estimation $\hat{z}^*$, we define $\hat{z}^w$ having the largest value observed so far, for each objective. We use $\hat{z}^w$ as an upper bound of $\hat{z}^{nad}$. With the check, if any intercept is smaller than ϵ_{nad}, we ensure that the hyperplane has no negative intercepts and the resulting nadir point estimation is significantly larger than $\hat{z}^*$. Also, we make sure that the resulting nadir point estimation $I_k + \hat{z}_k^*$ is not larger than our upper bound $\hat{z}_k^w$. If one of these requirements is not met, we declare the hyperplane as not useful for the normalization purpose and, therefore, the nadir point estimation is set as the maximum of each objective of current non-dominated population members.

Finally, we make sure the nadir point estimation satisfies the first requirement, being dominated by the estimated ideal point. If it is not, then we use the maximum of corresponding violating objective in the population.

Algorithm 3. Hyperplane through extreme points.

Input: Merged Population R, Non-dominated Fronts $(F_1, .., F_K)$, $\hat{z}^*$, Worst
  Point Estimation $\hat{z}^w$, Extreme Points e
Output: Nadir Point Estimation $\hat{z}^{nad}$

1 $e \leftarrow$ update_extreme_points$(R, \hat{z}^*, e)$
2 $b \leftarrow$ FALSE
3 **try:**
4 $\quad A' \leftarrow$ find_hyperplane$(e, \hat{z}^*)$
5 $\quad I \leftarrow$ find_intercepts(A')
6 $\quad$ **for** $k \leftarrow 1$ to M **do**
7 $\quad\quad \hat{z}_k^{nad} \leftarrow I_k + \hat{z}_k^*$
8 $\quad\quad$ **if** $I_k < \epsilon_{nad}$ *or* $\hat{z}_k^{nad} > \hat{z}_k^w$ **then**
9 $\quad\quad\quad b \leftarrow$ TRUE
10 $\quad\quad\quad$ **break**
11 $\quad\quad$ **end**
12 $\quad$ **end**
13 **catch** *Error*: $b \leftarrow$ TRUE

/* Fall back to the maximum in each objective of current front   */
14 **if** $b =$ *TRUE* **then**
15 $\quad$ **for** $i \leftarrow 1$ to M **do** $\hat{z}_i^{nad} \leftarrow$ max_in_objective(F_1, i)
16 **end**

/* Nadir point must be significantly larger in each objective   */
17 **for** $i \leftarrow 1$ to M **do**
18 $\quad$ **if** $\hat{z}_i^{nad} - \hat{z}_i^* < \epsilon_{nad}$ **then** $\hat{z}_i^{nad} \leftarrow$ max_in_objective$((F_1 \cup ... \cup F_k), i)$
19 **end**

---

5 Results

In this section, we present the simulation results of NSGA-III using the proposed normalization procedures[1]. First, we analyze the ideal and nadir point estimation error over generations. Second, we present the performance of the proposed normalization procedures on test problems. We use the scalable multi-objective optimization test problems suite, DTLZ [7], for our evaluation. Also, we investigate scaled versions of these problems, where objectives are multiplied with increasing factors. We conducted the experiment analogous to the original

---

[1] The source code is freely available at https://github.com/msu-coinlab/pymoo.

publication of NSGA-III. Therefore, we refer to [5] for details about the experimental setup and algorithm parameters, such as references lines, population size, number of generation and recombination operators. We run each algorithm 50 times on each test problem.

We compute the following squared ideal point and nadir point estimation errors to track the progress of them through generations:

$$\hat{e}^* = \sum_{i=1}^{m} \left(\frac{\hat{z}_i^* - z_i^*}{z_i^{nad} - z_i^*} \right)^2 \qquad \hat{e}^{nad} = \sum_{i=1}^{m} \left(\frac{\hat{z}_i^{nad} - z_i^{nad}}{z_i^{nad} - z_i^*} \right)^2 \qquad (4)$$

The squared ideal point estimation error $\hat{e}^*$ is calculated by summing up the normalized squared difference in each objective. The squared nadir point estimation error $\hat{e}^{nad}$ is also defined accordingly. Figure 2a shows the median $\hat{e}^*$ during the first 50 generations. Note that the error decreases below one percent after at most 20 generations for all considered test problems. This confirms the hypothesis, that the ideal point estimation is quick, less problematic and the assumption to use the smallest values for each objective reduces the estimation error effectively. Analogous, we illustrate $\hat{e}^{nad}$ in Fig. 2b. Clearly, the overall estimation error is higher than the ideal point estimation error and the convergence is slower. Furthermore, we can cluster the error into two groups, where DTLZ1 and DTLZ3 start with a smaller estimation error compared to DTLZ2 and DTLZ4. This is caused by the multimodality introduced through the convergence function for DTLZ1 and DTLZ3.

Next, let us discuss the performance of the different normalization procedures. We use the inverse generational distance (IGD) as a performance metric for our study. For scaled problems (SDTLZ), we used the weighted Euclidean distance for the IGD computation, where the distance in objective k is divided by $z_k^{nad} - z_k^*$. Figure 3 shows the box plots of the IGD values on the DTLZ test problem suite for three, five, and ten objectives. In addition to the proposed methods, we evaluate a normalization procedure (TRUE) where the true boundary points $\hat{z}^* = z^*$ and $\hat{z}^{nad} = z^{nad}$ are used all along. The following observations are made:

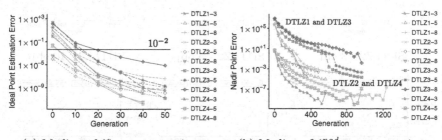

(a) Median of $\hat{e}^*$ over generations.   (b) Median of $\hat{e}^{nad}$ over generations.

Fig. 2. Variation of ideal and nadir point estimation errors using HYP.

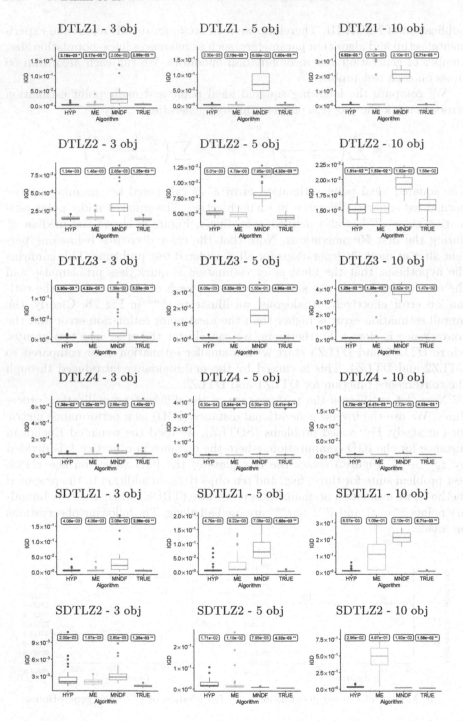

Fig. 3. Box plots showing IGD values for DTLZ (normalized IGD for SDTLZ) problems. The median values are presented for each algorithm annotated by ** if best and * if not significantly worse than best (according to Wilcoxon signed-rank test with $p = 0.05$)

- It can be concluded from the experiments, that MNDF is too naive and performed mostly worse than the other approaches. Due to the fact that the current non-dominated front may not be close to the true Pareto-optimal front, the search is easily biased to specific regions of the objective space.
- For scaled DTLZ problems, HYP has more outliers compared to the other problems. For SDTLZ2, MNDF shows surprisingly good results. Because the convergence function is rather simple, the non-dominated front seems to be a good representative of the true Pareto-optimal front.
- By comparing the median IGD values, we can observe that TRUE performs 17 out of 18 times the best. However, in practice the boundary points of the Pareto-front are unknown and this information can not be utilized.
- ME and MNDF showed more outliers and significantly higher median performances compared to HYP. Moreover, seven times HYP did not perform significantly worse than TRUE. We recommend using HYP whenever the true boundary points of the Pareto front are unknown.

6   Conclusions

In this paper, different normalization procedures for NSGA-III for solving many-objective optimization problems have been investigated. It has been shown that normalization is a crucial component for multi-objective algorithms and necessary to solve problems where objectives are scaled differently. The original proposed normalization method has been analyzed and degenerate cases, such as negative intercepts and no unique hyperplanes, have been discussed. The proposed methods have been applied to test problems up to ten objectives, where the ideal as well as the nadir point estimation errors over generations has been analyzed. The results confirm that the ideal point estimation is less problematic and gets settled quickly, whereas the nadir point estimation is tricky and requires a large number of generations to get settled. Moreover, the overall performance of NSGA-III with different normalization procedures has been evaluated. Although the original proposed hyperplane concept HYP must handle degenerate cases carefully (see Algorithm 3), it shows the best performance besides TRUE. The hyperplane concept is not only applicable for NSGA-III and can now be tested with other multi- and many-objective algorithms where normalization is not addressed properly or naively implemented. Moreover, the effect of normalization on different shapes of the Pareto-optimal front must be studied next.

References

1. Moeaframework. http://moeaframework.org. Accessed 26 Sept 2018
2. Bhesdadiya, R.H., Trivedi, I.N., Jangir, P., Jangir, N., Kumar, A.: An NSGA-III algorithm for solving multi-objective economic/environmental dispatch problem. Cogent Eng. **3**(1), 1269383 (2016)
3. Bi, X., Wang, C.: An improved NSGA-III algorithm based on objective space decomposition for many-objective optimization. Soft Comput. **21**(15), 4269–4296 (2017)

4. Das, I., Dennis, J.E.: Normal-boundary intersection: a new method for generating the pareto surface in nonlinear multicriteria optimization problems. SIAM J. Optim. **8**(3), 631–657 (1998)
5. Deb, K., Jain, H.: An evolutionary many-objective optimization algorithm using reference-point-based nondominated sorting approach, part i: solving problems with box constraints. IEEE Trans. Evol. Comput. **18**(4), 577–601 (2014)
6. Deb, K., Pratap, A., Agarwal, S., Meyarivan, T.: A fast and elitist multiobjective genetic algorithm: NSGA-II. Trans. Evol. Comput. **6**(2), 182–197 (2002)
7. Deb, K., Thiele, L., Laumanns, M., Zitzler, E.: Scalable test problems for evolutionary multiobjective optimization. In: Abraham, A., Jain, L., Goldberg, R. (eds.) Evolutionary Multiobjective Optimization. AI&KP, pp. 105–145. Springer, London (2005). https://doi.org/10.1007/1-84628-137-7_6
8. Durillo, J., Nebro, A., Alba, E.: The jmetal framework for multi-objective optimization: design and architecture. In: CEC 2010, Barcelona, Spain, pp. 4138–4325, July 2010
9. Gaur, A., Talukder, A.K.M.K., Deb, K., Tiwari, S., Xu, S., Jones, D.: Finding near-optimum and diverse solutions for a large-scale engineering design problem. In: 2017 IEEE Symposium Series on Computational Intelligence (SSCI), pp. 1–8, November 2017
10. Ibrahim, A., Rahnamayan, S., Martin, M.V., Deb, K.: EliteNSGA-III: an improved evolutionary many-objective optimization algorithm. In: 2016 IEEE Congress on Evolutionary Computation (CEC), pp. 973–982, July 2016
11. Ishibuchi, H., Doi, K., Nojima, Y.: On the effect of normalization in MOEA/D for multi-objective and many-objective optimization. Complex Intell. Syst. **3**(4), 279–294 (2017)
12. Jain, H., Deb, K.: An evolutionary many-objective optimization algorithm using reference-point based nondominated sorting approach, part ii: handling constraints and extending to an adaptive approach. IEEE Trans. Evol. Comput. **18**(4), 602–622 (2014)
13. Miettinen, K.: Nonlinear Multiobjective Optimization. Kluwer, Boston (1999)
14. Seada, H., Deb, K.: A unified evolutionary optimization procedure for single, multiple, and many objectives. IEEE Trans. Evol. Comput. **20**(3), 358–369 (2016)
15. Singh, H.K., Yao, X.: Improvement of reference points for decomposition based multi-objective evolutionary algorithms. In: Shi, Y., et al. (eds.) SEAL 2017. LNCS, vol. 10593, pp. 284–296. Springer, Cham (2017). https://doi.org/10.1007/978-3-319-68759-9_24
16. Tian, Y., Cheng, R., Zhang, X., Jin, Y.: Platemo: a matlab platform for evolutionary multi-objective optimization [educational forum]. IEEE Comput. Intell. Mag. **12**(4), 73–87 (2017)
17. Wang, R., Xiong, J., Ishibuchi, H., Wu, G., Zhang, T.: On the effect of reference point in MOEA/D for multi-objective optimization. Appl. Soft Comput. **58**, 25–34 (2017)
18. Wierzbicki, A.P.: The use of reference objectives in multiobjective optimization. In: Fandel, G., Gal, T. (eds.) Multiple Criteria Decision Making Theory and Applications. LNEMS, vol. 177, pp. 468–486. Springer, Berlin (1980)
19. Yuan, X., Tian, H., Yuan, Y., Huang, Y., Ikram, R.M.: An extended NSGA-III for solution multi-objective hydro-thermal-wind scheduling considering wind power cost. Energy Convers. Manag. **96**, 568–578 (2015)
20. Zhang, Q., Li, H.: MOEA/D: a multiobjective evolutionary algorithm based on decomposition. IEEE Trans. Evol. Comput. **11**(6), 712–731 (2007)
21. Zhou, A., Qu, B.Y., Li, H., Zhao, S.Z., Suganthan, P.N., Zhang, Q.: Multiobjective evolutionary algorithms: a survey of the state of the art. Swarm Evol. Comput. **1**(1), 32–49 (2011)

MAC: Many-objective Automatic Algorithm Configuration

Hojjat Rakhshani[✉], Lhassane Idoumghar, Julien Lepagnot,
and Mathieu Brévilliers

Université de Haute-Alsace, IRIMAS-UHA, 68093 Mulhouse, France
{hojjat.rakhshani,lhassane.idoumghar,julien.lepagnot,
mathieu.brevilliers}@uha.fr

Abstract. State-of-the-art optimization algorithms often expose many
parameters that should be configured to improve empirical performance.
Manually tuning of such parameters is synonymous with tedious experiments which tend to lead to unsatisfactory outcomes. Accordingly,
researchers developed several frameworks to tune the parameters of a
given algorithm over a class of problems. Until very recently, however,
these approaches are not testified and applied to many-objective algorithms. This study formulates a many-objective algorithm configuration
(MAC) method which is available for the Matlab and Python. In MAC,
we take into account the importance of a given configuration by building a conditional probability graph. In this light, the introduced algorithm aims to explore more important variables using an undirected fully-
connected graph. Experimental results reveal that MAC performs better
in comparison with state-of-the-art F-Race and SMAC frameworks.

Keywords: Algorithm configuration · Many-objective optimization ·
Machine learning

1 Introduction

There is no doubt that optimization algorithms have gained immense popularity
in recent years. The adoption of these algorithms to unseen $\mathcal{NP}$-hard problems,
however, is severely hampered by choosing a set of optimal parameters associated
with them. The learning rate in stochastic gradient descent or the mutation rate
in the genetic algorithm are examples of these parameters. In particular, we can
point out parameters of meta-heuristics whose configurations have a high impact
on their overall performance on a given class of instances. These configurations
are correlated in non-intuitive ways which makes it difficult and tedious to tune
them manually.

Automatic algorithm configuration deals with optimizing parameters of an
algorithm so as to perform well across a broad range of instance types. In this
regard, standard optimization algorithms like meta-heuristics may need hundreds of evaluations to locate a near-optimal solution which is a major challenge

© Springer Nature Switzerland AG 2019
K. Deb et al. (Eds.): EMO 2019, LNCS 11411, pp. 241–253, 2019.
https://doi.org/10.1007/978-3-030-12598-1_20

to their successful application. This is primarily due to the expensive computational cost associated with them which often consume many minutes to even days of CPU time. In this context, the advantages of so-called model-based algorithms become clear [12]. They construct computationally cheap-to-evaluate surrogate models in order to provide a fast approximation of the expensive fitness evaluations during the search process. By leveraging surrogate models, the computational cost can be greatly reduced since the time overhead of training and building surrogate models is insignificant compared to evaluating the exact fitness function. To this fact, state-of-the-art frameworks such as SMAC [12] and F-RACE [2] have focused on model-based optimization.

To the best of our knowledge, automatic configuration methods have not been applied to many-objective optimization and researchers were more interested on single-objective and multi-objective cases. In contrast to conventional multi-objective approaches, many-objective optimization poses a great challenge due to the ineffectiveness of Pareto dominance, inefficiency of recombination operation, rapid increase of computational time and parameter sensitivity. In this study, we apply a new algorithm based on *optimal contraction theorem* [4] for adjusting parameters of many-objective optimization algorithms. A series of experiments using CEC2018 benchmark problems [6] are conducted to demonstrate the effectiveness of the MAC. The proposed algorithm is validated against state-of-the-art algorithm configuration approaches from the literature. The statistical comparisons of experimental results show that MAC has a superior performance in terms of solution accuracy over the considered problems and algorithms.

The rest of the paper is organized as follows. Section 2 provides us with a brief review on the related works. Section 3 gives a brief description of the algorithm configuration problem. Section 4 elaborates technical details of our proposed approach. In Sect. 5, the performance of the introduced MAC is investigated by conducting a set of experiments. The last section summarizes the paper and draws conclusions.

2   Related Works

Sequential Model-based Algorithm Configuration (SMAC) [12], Spearmint [21], F-RACE [2] and Tree-structure Parzen Estimator (TPE) [1] are examples of well known methods for automatic configuration task. A large class of such methods is characterized by modeling a conditional probability $p(y|\varphi)$ of a m-dimensional configuration φ, given n observations s with the corresponding evaluation metrics y. SMAC adopted a random forests model and Expected Improvement (EI) to compute $p(y|\varphi)$. It applies a multi-start local search and selects resulting configurations with locally maximal EI. The exploration property of SMAC is enhanced by the fact that EI conditioned on points with large uncertainty and low values of predictive mean. Similarly, TPE et al. [1] defined a configuration algorithm based on tree-structure Parzen estimator and EI. To tackle the *curse of dimensionality*, TPE assigns particular values of other elements to the configurations which are known to be irrelevant. Ilievski et al. [13] proposed a deterministic method which employs dynamic coordinate search and radial basis functions (RBFs) to find most promising configurations. By using the RBFs [16] as surrogate model, they mitigated some of the requirements for inner acquisition function optimization. In another work [22], the authors put forward neural networks as an alternative to Gaussian process for modeling distributions over functions. They show that their introduced method is competitive with state-of-the-art GP-based approaches while it scales linearly with the data size rather than cubically. Blot et al. [3] introduced a multi-objective extension of the well-known ParamILS configuration framework and they demonstrate that it gives promising results on several challenging bi-objective scenarios. Interestingly, *Google* introduced *Google Vizier* [11], an internal service which incorporates Batched Gaussian Process Bandits along with the EI acquisition function.

3   Automatic Algorithm Configuration

A general definition of the algorithm configuration problem can be presented by a tuple $<I, \Theta, \Lambda, \zeta>$ as follows:

$$\theta^* = \arg \max_{\theta \in \Theta} u(\theta), \text{where } u(\theta) = f(\theta | I, P_I, P_\zeta, t) \tag{1}$$

where I is a set of problem instances which is given by a distribution P_I over admissible instances; Λ is an algorithm which should solve the problem class I, with input configurations $\theta = (p_1, \cdots, p_k) \in \Theta$. Here, $\Lambda(\theta)$ is the instance of algorithm Λ configured with θ; $\zeta(\theta, i, t) = \zeta(\Lambda(\theta), i, t)$ assigns a cost value to each configuration θ when running $\Lambda(\theta)$ on instance $i \in I$ for time t. It could be modeled as $\zeta \sim P_\zeta(\zeta | \theta, i, t)$; and finally Θ is a set of all possible combinations of values of p_i. A hyperparameter approach then should try to find configuration $\theta^* \in \Theta$ such that $\Lambda(\theta)$ yields the best utility u.

4   The Proposed Method

This section discusses in detail the main components of the proposed MAC method. In brief, MAC consist of two main phases:

- **Exploration:** The algorithm tries to learn probabilistically about the relevance of configurations and the model's performance during the optimization process. In another word, it expects to find reasons why a collection of past solutions is superior to others. To do so, MAC is equipped by a linkage learning component which periodically acquires information about the problem at hand to find most informative configurations. The aforementioned schema encodes the underlying dependencies between variables using an undirected graph, where nodes denote configurations and edges show the probability that two nodes are relevant. We adopted the idea of Eigenvector centrality feature selection [19] to learn the factor graph.

- **Exploitation:** The collected information from the previous step are then processed and used to generate new solutions. The introduced informed component guides the algorithm toward the search space that are likely to contain the promising solutions. This orthogonal technique prevents MAC to uniformly consider all configurations and bias the search process toward the good configurations.

We extend the idea of stochastic RBF [17] to be suitable for the algorithm configuration task. It is a model-based algorithm that cycles from emphasis on the objective to emphasis on the distance using a weighting strategy. Compared to the evolutionary algorithms like genetic algorithm, stochastic RBF need less computational time by virtue of surrogate modeling techniques. On the other hand, it mitigated some of the requirements for inner acquisition function optimization in comparison with well-know efficient global optimization (EGO) algorithm [14]. Hence, we focused on proposing a new algorithm configuration approach based on stochastic RBF. A generic framework for MAC includes some basic steps which can be stated follows: (1) generating a set of initial configurations θ^i ($i = 1, 2, ..., n$) using design of experiments (DoE) and compute the cost value for each configuration; (2) Building an initial surrogate model based on the sampled configurations θ^i in the first step; (3) Finding the current best configuration $conf_{best}$; (4) Generating a set of random perturbations ρ based on exploration/exploitation modes; (5) Generating a set of new configurations $confs_{new}$ around $conf_{best}$ using ρ; (6) Use the surrogate model to select the best configuration $conf_{new}$; (7) Evaluating $conf_{new}$ using exact cost function; (8) Updating the surrogate model based on $conf_{new}$; (9) Checking the stopping criteria: if some stopping criteria are satisfied go to Step 10; otherwise go to Step 3; (10) Post-processing the results

4.1   Initial Design

In MAC, the first step involves generating a set of random configurations θ^i ($i = 1, 2, ..., n$) (i.e., initial population). Here, the algorithm might possibly miss a considerable portion of the promising area due to the high dimensionality of the configuration space (it should be noticed that we have a small and a fix computational budget and increasing size of the initial population cannot remedy the issue). Furthermore, it is crucial for a model-based algorithm to efficiently explore the search space so as to approximate the nonlinear behavior of the objective function. For these reasons, as with many model-based algorithms, MAC adopts DoCE methods to partially mitigate high dimensionality of the search space. Among them, MAC uses the Latin Hypercube Sampling (LHS) [15] to provide a uniform cover in the search space using a minimum number of population. The main advantage of LHS is that it does not require an increased initial population size for more dimensions.

4.2   Approximation Model

As the next step, we evaluate all the generated configurations θ^i $(i = 1, 2, ..., n)$ to build an approximate model of the cost function. This computationally cheap-to-evaluate model can provide a fast approximation of the expensive fitness evaluations during the search process. MAC tries to model conditional probability $p(y|\varphi)$ of a d-dimensional configuration φ given n observations $\mathbf{S}$ with the corresponding cost metrics $\mathbf{y}$:

$$\mathbf{S} = \left[\theta^{(1)}, ..., \theta^{(n)} \right]^{\mathrm{T}} \in \mathbb{R}^{n \times d}, \theta = \{\theta_1, ..., \theta_d\} \in \mathbb{R}^d \tag{2}$$

To do so, it offers surrogate models which are a set of mathematical tools for predicting the output of an expensive objective function. Particularly, they are designed to predict the fitness function value for any unseen configuration $\hat{\theta}$ according to computed data points (θ^i, y^i). Given a set of distinct configurations $\theta^1, ..., \theta^n \in \mathbb{R}^d$ with known values y^i, the RBF interpolant form then is computed as belows [17]:

$$\tilde{f}(\hat{\theta}) = \sum_{i=1}^{n} \lambda_i \phi(\|\hat{\theta} - \theta^i\|) + p(\hat{\theta}), \ \hat{\theta} \in \mathbb{R}^d \tag{3}$$

In (3), $\|.\|$ is the Euclidean norm, $\lambda_i \in \mathbb{R}$ for $i = 1, ..., n$, $p \in \prod_m^d$ denotes the linear space of polynomials in d variables of degree which is less than or equal to m, and ϕ is a RBF with one of the *surface splines* $(\phi(r) = r^k$ where $k \in \mathbb{N}$ is an odd number, or $\phi(r) = r^k log(r)$ where k is an even number), *multiquadrics* $(\phi(r) = (r^2 + \gamma^2)^k$ where $k > 0$ and $k \notin \mathbb{N})$, *inverse multiquadrics* $(\phi(r) = (r^2 + \gamma^2)^k$ where $k < 0$ and $k \notin \mathbb{N})$ and Gaussians $(\phi(r) = e^{-\gamma r^2})$ forms. Here, $r \geq 0$ and $\gamma > 0$.

Following [17], MAC selected the *surface splines* form with $k = 3$ as the RBF. Having this in mind, we can compute a matrix $\Im \in \mathbb{R}^{n \times n}$ by $\Im_{i,j} = \phi(\|\theta^i - \theta^j\|)$; $i, j = 1...n$. Assume that $\hat{m}$ be the dimension of the linear space $\prod_m^d$ such that $m \geq= \lfloor k/2 \rfloor$. Accordingly, we have another matrix $\mathbf{P} \in \mathbb{R}^{n \times \hat{m}}$ such that: $P_{ij} = p^{(i)}(\theta^{(i)})$, $i = 1..n; j = 1..\hat{m}$. The approximated model then can be obtained by solving the system as presented in (4), where $\mathbf{c} = (c_1, ..., c_{\hat{m}})^T \in \mathbb{R}^{\hat{m}}$.

$$\begin{pmatrix} \Im & \mathbf{P} \\ \mathbf{P}^\top & 0 \end{pmatrix} \begin{pmatrix} \gamma \\ \mathbf{c} \end{pmatrix} = \begin{pmatrix} \mathbf{y} \\ 0_{\hat{m}} \end{pmatrix} \tag{4}$$

4.3   Exploration

The original stochastic RBF method generates a set of candidate points by adding random perturbations ρ to the best obtained solution (i.e., configuration) to guide the search process. This trial-and-error procedure does not take into account the interactions between the generated configurations and the obtained objective values. We note that the performance of stochastic RBF depends on this *random* points and a more informed scheme can be beneficial to enhance the robustness of the algorithm. Indeed, this is the same desired property in *optimal contraction theorem* [4] which states an optimal optimizer should dynamically considers useful information about the problem at hand. Motivated by this finding, MAC incorporates an adaptive control strategy which keeps a historical memory of the ρ perturbations to guide the generation of future configurations. In the exploration phase, MAC generates a diverse set of random perturbations ρ and tends to increase global search to prevent algorithm from being trapped in a local minimum. We adopted Student's t-distribution to do so, which is symmetric and bell-shaped family of distributions like the normal distribution. In contrast, however, it has heavier tails which let MAC to explore the points that fall far from the distribution's mean. At each iteration t, MAC archives the generated perturbations for the exploitation phase.

4.4   Exploitation

After half of the iterations, MAC employs a feature selection algorithm method to acquire information about the performance of each of those randomly generated perturbations in the previous phase. It uses this information to dynamically make a balance between exploration and exploitation. In other words, MAC transforms the task of learning the optimal feature in feature selection algorithms into the search for an efficient and adaptive optimization behavior. This enables MAC to take into account the underlying correlations between the generated perturbations and domain-specific search knowledge of the problem.

Following the [18], MAC creates an undirected graph $G = <V, E>$ according to which nodes represent the generated random perturbation $\rho^{(t)}$ and edges denote relationships among pairs of nodes. All the archived perturbation $\rho^{(t)}$ are ranked in descending order according to their associated cost values: the first best 50% solutions are labeled as *promising* and the other solutions are labeled as *non-promising*. This consideration address the imbalanced training set and prevent of biasing against the minority class.

Given the above-mentioned training set, an adjacency matrix A is associated with G in order to define relationships between the nodes. The G is represented through an adjacent matrix A, where each element $a_{i,j}$ shows pairwise relations among feature distributions. The $a_{i,j}$ elements are defined as follows:

$$a_{i,j} = \alpha\sigma_{i,j} + (1 - \alpha)c_{i,j}; 1 \leqslant i, j \leqslant t \tag{5}$$

In (5), α is a scaling factor $\in [0, 1]$, $\sigma_{i,j} = max(\sigma_i, \sigma_j)$ where σ_i denotes the standard deviation over the ρ and $c_{i,j}$ is a kernel. To compute the $c_{i,j}$, first the Fisher criterion should be applied [18]:

$$f_i = \frac{|\mu_{i,1} - \mu_{i,2}|^2}{\sigma_{i,1}^2 + \sigma_{i,2}^2} \tag{6}$$

In (6), discriminate classes *promising* and *non-promising* are labeled as 1 and 2, respectively. Also, $\sigma_{i,c}^2$ and $\mu_{i,c}$ are the mean and standard deviation of the i-th feature for class c. The k is then can be obtained as $k = (f.m^{\top})$ where the mutual information m is [18]:

$$m_i = \sum_{y \in Y} \sum_{z \in \rho^{(i)}} p(z, y) log(\frac{p(z, y)}{p(z)p(y)}) \tag{7}$$

In (7), Y shows class labels and p denotes the joint probability distribution. Now, MAC computes the eigenvalues v and eigenvectors v of A. The obtained weigh for generating the new configurations is equal to the eigenvector associated to $\eta_0 = \max_{\eta \in v} abs(\eta)$.

5   Experimental Results

In this section, we investigate the performance of the introduced MAC method using two different many-objective algorithm configuration scenarios. The first scenario is designed to optimize the configuration space of NSGA-II [9] defined by three control parameters, while the second one adopts MOPSO [7] which is defined by six control parameters. The adopted algorithms are introduced to measure search performance of the MAC under different dimensions. The NSGA-II is a low dimensional configuration problem, while MOPSO is a medium dimension problem. The considered configurations are presented in Tables 1 and 2.

Table 1. The considered configurations of NSGA-II for a D-dimensional problem

Name	Type	Range	Default values [10]
Population size	Integer	[100, 500]	100
Crossover rate	Continuous	[0.1, 1]	1
Mutation rate	Continuous	[0.1, 1]	$1/D$

Table 2. The considered configurations of MOPSO for a D-dimensional problem

Name	Type	Range	Default values [20]
Population size	Integer	[100, 500]	100
Inertia weight	Continuous	[0.1, 1]	0.9
C1	Continuous	[0.1, 2]	1.8
C2	Continuous	[0.1, 2]	1.8
V	Continuous	[0.1, 1]	0.6
Mutation rate	Continuous	[0.1, 1]	$1/D$

It is curious that the authors choose NSGA-II as one of the algorithms, since it is typically thought not to perform well on problems with more than 4 objectives - at least when given its default parameterisation from the literature. Indeed, the main goal is to show how manual parameter configuration of many-objective methods may lead to inferior results. We will show that a well-tuned NSGA-II can perform better than a standard NSGA-III many-objective optimization method. Meanwhile, hyperparameter optimization results of the NSGA-III and RVEA [5] as two many-objective methods will be made available at https://github.com/ML-MHs/MAC.

MAC is implemented in Python and Maltab and can be easily installed. Although it supports to directly configure the algorithms in PlatEMO [23], Platypus and PyGMO packages, but like the SMAC and F-RACE it can be linked with other frameworks such as jMetal by means of a Wrapper function. MAC supports automatic generation of LaTeX tables, applying statistical pairwise comparison, and graphical visualization. The MAC can be executed in parallel on multiple cores. The MAC will be made available on request to the academic community.

We considered CEC'2018 benchmark suite to compare the competitive configuration algorithms. It should be mentioned that the number of objectives is set to $M = 5$. Our experimental procedure follows two steps, namely *training* and *test*. In the *training* step, each evaluation involves running the NSGA-II/MOPSO on the training problem instances MaF1, MaF2, MaF5 and MaF6-10 (the size of training and test instances is small due to small number of CEC'2018 benchmark problems.) for 10 runs. After finishing the automatic configuration step, the configurations obtained from the *training* step are applied to rest of the problems in order to validate the performance of the optimized configurations on the unseen test instances. In the case of NSGA-II and MOPSO, number of fitness evaluations is set to be $max(1e5, D \times 1e4)$, where D is the default dimensionality of the problem.

Experiments are conducted based on random search, SMAC and F-RACE methods. Empirical evidence reveals that random search can outperform Grid search within a small fraction of the computation time. Furthermore, SMAC and F-RACE are two state-of-the-art automatic configuration frameworks and to our best knowledge this is the first study which reports their performance for many-objective problems. For each algorithm, stopping criteria is when algorithm exceeds 5×200 evaluations, or when computational time reaches to 5×24 h. We used Hyper Volume (HV) in order to compare the proximity and diversity of the obtained results. The HV indicator should be maximized during the configuration process.

The obtained results of 5 independent runs are summarized in Tables 3 and 4. In these tables, NSGA-II, NSGA-III [8] and MOPSO denotes the obtained results by using the default configurations. The best results are indicated in boldface.

Arguably, NSGA-II is one of the well-known methods which mimics the same developmental process in the standard GA: Selection, reproduction and evaluation. It is worth tuning the population size, mutation and crossover probabilities of the NSGA-II to find reasonable settings for the problem at hand. A small population size will lead to the early convergence problem, while using a large population increases the computational cost. The configurations tuning of NSGA-II

Table 3. Average Hypervolume values for final test fronts of NSGA-II algorithm and NSGA-III

Problem	Random search	SMAC	F-RACE	NSGA-II	MAC	NSGA-III
MaF1	0.0132	**0.0141**	0.0109	0.0072	0.0135	0.00320
MaF2	0.0440	0.0448	0.0407	0.0342	**0.0449**	0.03990
MaF5	32400.0000	32759.7820	29971.3091	20900.0000	27200.0	**37963.0**
MaF6	**0.0093**	**0.0093**	**0.0093**	0.0092	**0.0093**	0.008
MaF7	1.8471	1.8914	1.7851	1.7035	**1.9127**	1.835
MaF8	**4.7763**	4.7708	1.5404	3.9464	4.7749	2.6919
MaF9	7.5106	7.9908	7.4298	3.7968	**8.0242**	5.5338
MaF10	2500.0000	2810.4974	2470.0709	2464.2300	2470.0	**6029.70**
MaF11	6080.0000	6102.3614	6070.2643	5970.0000	6100.0	**6129.60**
MaF12	3710.0000	3869.2988	3657.2979	2825.0000	3880.0	**4557.0**
MaF13	0.4376	0.4446	0.4470	0.2080	**0.4547**	0.26600
MaF14	0.1464	0.1464	0.0732	0.1464	**0.1513**	0.09770

Table 4. Average Hypervolume values for final test fronts of MOPSO algorithm

Problem	Random search	SMAC	F-RACE	MOPSO	MAC
MaF1	0.0106	**0.0116**	0.0154	0.0030	0.0107
MaF2	**0.0417**	0.0396	0.0389	0.0315	0.0407
MaF5	29100.0000	**33163.6115**	32753.9500	16600.0000	32900.0000
MaF6	**0.0093**	**0.0093**	**0.0093**	0.0089	**0.0093**
MaF7	1.5039	**1.7304**	0.6107	1.0984	1.6691
MaF8	4.3301	4.5889	4.6987	4.3521	**4.7207**
MaF9	7.9267	8.7605	8.7001	2.5895	**9.1728**
MaF10	1440.0000	1513.8214	1485.7750	1465.0000	**1570.0000**
MaF11	5540.0000	5471.3664	5570.2050	4915.0000	**5600.0000**
MaF12	3190.0000	3143.7161	3181.9100	1440.0000	**3450.0000**
MaF13	0.0475	0.3558	0.3705	0.0219	**0.3710**
MaF14	**0.1464**	**0.1464**	**0.1464**	0.1369	**0.1464**

becomes even more challenging due to the fact that there is a correlation between its control parameters. For example, mutation is more effective on smaller population sizes while crossover is likely to benefit from large populations. All the mentioned reasons make the NSGA-II a challenging algorithm for benchmarking the performance of the MAC. However, Table 3 shows the automatic algorithm configuration methods enhanced the performance of the NSGA-II over the problems. Meanwhile, it is worth mentioning that MAC exhibits more promising performance other competitive methods. It is quite interesting to note that how the tuned NSGA-II algorithm by MAC can outperform the NSGA-III on this

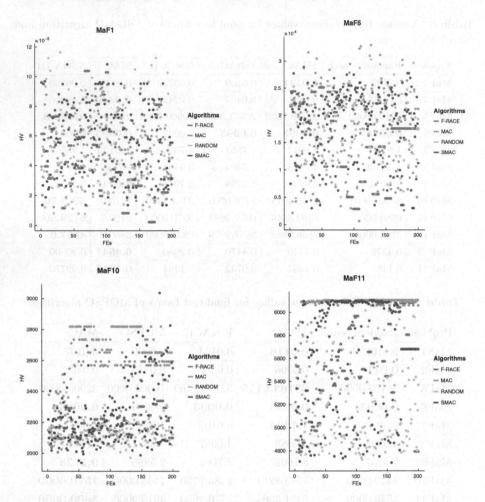

Fig. 1. HV values v.s. number of function evaluations (FEs) of different methods for optimizing 6 configurations of MOPSO (first row) and 3 configurations of NSGA-II (second row). One dot represents HV value of an algorithm at the corresponding evaluation number

set of benchmarks (the configurations of NSGA-III are taken from [23]). The same situation could happen for the considered MOPSO algorithm. As can also be seen from Table 4, MAC finds considerably better configurations in terms of HV indicator. Figure 1 provides additional details by showing the behavior of the considered methods over 4 different problems. This figure shows how the introduced methods adopt their tuning behavior for NSGA-II and MOPSO algorithms. Altogether, with respect to the performed experiments, we can say that MAC and other frameworks has achieved promising results. From the illustrated correlation matrix (results are recorded by MAC) in Fig. 2, it can also be con-

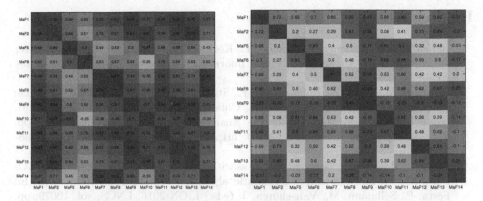

Fig. 2. The obtained correlations between the configurations of different problems for the NSGA-II (left) and MOPSO (right) algorithms

cluded that well-tuning the algorithms even with very small training instances can enhance the search performance of the many-objective approaches.

6   Conclusion

In this study, we present a framework for automatic algorithm configuration of many-objective optimization methods. The introduced MAC incorporated the idea of feature selection into the stochastic RBF method using an undirected graph. The MAC is proposed in the interest of integrating the optimization methods and machine learning techniques. The application of MAC to very recent CEC 2018 benchmarks against 3 state-of-the-art competitors, at most contributes to top performances. The results show how manual parameter configuration of NSGA-III many-objective algorithms can decrease its performance. Interestingly, the tuned NSGA-II algorithm by MAC, outperforms the state-of-the-art NSGA-III method. We believe that MAC could be very useful in real-world applications due to the fact that there is no tedious way to tune the parameter of many-objective algorithms by considering multiple performance criteria. As the future work, we are intended in using many-objective versions of model-based algorithms for parameter tuning tasks. It would be interesting to investigate how many-objective methods can cover a range of trade-offs in comparison with single-objective ones.

Acknowledgments. This research was supported in part through computational resources provided by Mésocentre of Strasbourg.

References

1. Bergstra, J.S., Bardenet, R., Bengio, Y., Kégl, B.: Algorithms for hyper-parameter optimization. In: Advances in Neural Information Processing Systems, pp. 2546–2554 (2011)
2. Birattari, M., Yuan, Z., Balaprakash, P., Stützle, T.: F-Race and iterated F-Race: an overview. In: Bartz-Beielstein, T., Chiarandini, M., Paquete, L., Preuss, M. (eds.) Experimental Methods for the Analysis of Optimization Algorithms, pp. 311–336. Springer, Heidelberg (2010). https://doi.org/10.1007/978-3-642-02538-9_13
3. Blot, A., Hoos, H.H., Jourdan, L., Kessaci-Marmion, M.É., Trautmann, H.: MO-ParamILS: a multi-objective automatic algorithm configuration framework. In: Festa, P., Sellmann, M., Vanschoren, J. (eds.) LION 2016. LNCS, vol. 10079, pp. 32–47. Springer, Cham (2016). https://doi.org/10.1007/978-3-319-50349-3_3
4. Chen, J., Xin, B., Peng, Z., Dou, L., Zhang, J.: Optimal contraction theorem for exploration-exploitation tradeoff in search and optimization. IEEE Trans. Syst. Man Cybern.-Part A: Syst. Hum. 39(3), 680–691 (2009)
5. Cheng, R., Jin, Y., Olhofer, M., Sendhoff, B.: A reference vector guided evolutionary algorithm for many-objective optimization. IEEE Trans. Evol. Comput. 20(5), 773–791 (2016)
6. Cheng, R., et al.: Benchmark functions for the CEC'2018 competition on many-objective optimization. Technical report, University of Birmingham, United Kingdom (2018)
7. Coello, C.A.C., Pulido, G.T., Lechuga, M.S.: Handling multiple objectives with particle swarm optimization. IEEE Trans. Evol. Comput. 8(3), 256–279 (2004)
8. Deb, K., Jain, H.: An evolutionary many-objective optimization algorithm using reference-point-based nondominated sorting approach, part i: solving problems with box constraints. IEEE Trans. Evol. Comput. 18(4), 577–601 (2014)
9. Deb, K., Pratap, A., Agarwal, S., Meyarivan, T.: A fast and elitist multiobjective genetic algorithm: NSGA-II. IEEE Trans. Evol. Comput. 6(2), 182–197 (2002)
10. Durillo, J.J., Nebro, A.J.: jmetal: a Java framework for multi-objective optimization. Adv. Eng. Softw. 42(10), 760–771 (2011)
11. Golovin, D., Solnik, B., Moitra, S., Kochanski, G., Karro, J., Sculley, D.: Google vizier: a service for black-box optimization. In: Proceedings of the 23rd ACM SIGKDD International Conference on Knowledge Discovery and Data Mining, pp. 1487–1495. ACM (2017)
12. Hutter, F., Hoos, H.H., Leyton-Brown, K.: Sequential model-based optimization for general algorithm configuration. In: Coello, C.A.C. (ed.) LION 2011. LNCS, vol. 6683, pp. 507–523. Springer, Heidelberg (2011). https://doi.org/10.1007/978-3-642-25566-3_40
13. Ilievski, I., Akhtar, T., Feng, J., Shoemaker, C.A.: Efficient hyperparameter optimization for deep learning algorithms using deterministic RBF surrogates. In: AAAI, pp. 822–829 (2017)
14. Jones, D.R., Schonlau, M., Welch, W.J.: Efficient global optimization of expensive black-box functions. J. Global Optim. 13(4), 455–492 (1998)
15. Olsson, A., Sandberg, G., Dahlblom, O.: On latin hypercube sampling for structural reliability analysis. Struct. Saf. 25(1), 47–68 (2003)
16. Park, J., Sandberg, I.W.: Universal approximation using radial-basis-function networks. Neural Comput. 3(2), 246–257 (1991)

17. Regis, R.G., Shoemaker, C.A.: A stochastic radial basis function method for the global optimization of expensive functions. INFORMS J. Comput. **19**(4), 497–509 (2007)
18. Roffo, G., Melzi, S.: Features selection via eigenvector centrality. In: Proceedings of New Frontiers in Mining Complex Patterns (NFMCP 2016), October 2016 (2016)
19. Roffo, G., Melzi, S., Castellani, U., Vinciarelli, A.: Infinite latent feature selection: a probabilistic latent graph-based ranking approach. In: Computer Vision and Pattern Recognition (2017)
20. Sierra, M.R., Coello Coello, C.A.: Improving PSO-based multi-objective optimization using crowding, mutation and ∈-Dominance. In: Coello, C.A.C., Hernández Aguirre, A., Zitzler, E. (eds.) EMO 2005. LNCS, vol. 3410, pp. 505–519. Springer, Heidelberg (2005). https://doi.org/10.1007/978-3-540-31880-4_35
21. Snoek, J., Larochelle, H., Adams, R.P.: Practical Bayesian optimization of machine learning algorithms. In: Advances in Neural Information Processing Systems, pp. 2951–2959 (2012)
22. Snoek, J., et al.: Scalable Bayesian optimization using deep neural networks. In: International Conference on Machine Learning, pp. 2171–2180 (2015)
23. Tian, Y., Cheng, R., Zhang, X., Jin, Y.: PlatEMO: a MATLAB platform for evolutionary multi-objective optimization [educational forum]. IEEE Comput. Intell. Mag. **12**(4), 73–87 (2017)

A Parallel Tabu Search Heuristic to Approximate Uniform Designs for Reference Set Based MOEAs

Alberto Rodríguez Sánchez[1]([✉]), Antonin Ponsich[1], Antonio López Jaimes[2], and Saúl Zapotecas Martínez[2]

[1] Dpto. de Sistemas, Universidad Autónoma Metropolitana Azcapotzalco,
Mexico City, Mexico
{ars,aspo}@azc.uam.mx
[2] Dpto. de Matemáticas Aplicadas y Sistemas,
Universidad Autónoma Metropolitana Cuajimalpa, Mexico City, Mexico
alopez@cua.uam.mx, saul.zapotecas@gmail.com

Abstract. Recent Multi-objective Optimization (MO) algorithms such as MOEA/D or NSGA-III make use of an uniformly scattered set of reference points indicating search directions in the objective space in order to achieve diversity. Apart from the mixture-design based techniques such as the simplex lattice, the mixture-design based techniques, there exists the Uniform Design (DU) approach, which is based on based on the minimization of a discrepancy metric, which measures how well equidistributed the points are in a sample space. In this work, this minimization problem is tackled through the L_2 discrepancy function and solved with a parallel heuristic based on several Tabu Searches, distributed over multiple processors. The computational burden does not allow us to perform many executions but the solution technique is able to produce nearly Uniform Designs. These point sets were used to solve some classical MO test problems with two different algorithms, MOEA/D and NSGA-III. The computational experiments proves that, when the dimension increases, the algorithms working with a set generated by Uniform Design significantly outperform their counterpart working with other state-of-the-art strategies, such as the simplex lattice or two layer designs.

Keywords: NSGA-III · MOEA/D · Uniform design · Weight vector · Multi-objective evolutionary algorithm (MOEA)

1 Introduction

The solution of Multi-objective Optimization Problems (MOPs) has generated a great interest and many classes of strategies have been developed to solve it. First, dominance-based algorithms, such as NSGA-III [3], use Pareto ranking mechanisms to compute the fitness of a solution. In turn, other techniques evaluate the current approximation of the real front through indicators and try to

© Springer Nature Switzerland AG 2019
K. Deb et al. (Eds.): EMO 2019, LNCS 11411, pp. 254–265, 2019.
https://doi.org/10.1007/978-3-030-12598-1_21

maximize their quality [17]. A third class is constituted by decomposition-based techniques, which perform a number of single-objective searches along different search directions evenly distributed over the objective space [19].

The use of a set of search directions, or weight vectors, or reference set, attracted much attention after the publication of MOEA/D [18], and this methodology was integrated within other algorithms, for instance NSGA-III [3], which is still based on dominance but employs a set of reference vectors to achieve diversity during the search process.

However, as some studies have shown, the distribution of the solutions highly depends on several factors such as: the scalarization function, the geometry of the Pareto front, and, particularly, on the generation of the reference points. This latter might be the most important factor that determines the distribution of the solutions found. Das and Dennis [2] proposed the simplex lattice technique, which was used in many of the first decomposition MOEAs proposed.

One method that has shown promising results is the uniform design (UD). Therefore, in this paper we present an algorithm to generate reference points for this technique. In order to generate the points, the algorithm includes, as an embedded mechanism, the good lattice point technique (GLP), that provides a good distribution, although showing a high computing time and memory complexity. For solving this problem in reasonable CPU time, a parallel tabu search (TS) heuristic was adopted. To validate the performance of our proposal, the results are compared with those obtained by the simplex lattice design and two layer design. The results show that, independently of the MOEA employed, UD provides a better distribution, specially when the number of objectives is high.

The rest of the paper is structured as follows. Section 2 presents a short overview regarding weight vector construction for MOEAs and the statement of the corresponding optimization problem is introduced in Sect. 3. The solution technique, a parallel algorithm based on Tabu Search, is described in Sect. 4 while Sect. 5 discusses the computational results obtained. Finally, some conclusions are drawn in Sect. 6.

2 Weight Vector Designs for Reference-Based MOEAs

The interest on sets of uniformly distributed reference points within MOEAs is quite recent. Nonetheless, this topic has been previously studied in the framework of mixture designs, for which a set of m ingredients should be combined to meet specific criteria and, therefore, the sum of their respective composition must equal 1. Among the many generation techiques proposed, two classes emerge, first those methods based on geometric concepts and, on the other hand, techniques based on the minimization of discrepancy functions.

2.1 Mixture Design

Experiments with mixtures are experiments in which the variables are the ingredient proportions in a mixture. An example is an experiment for determining the proportion of ingredients in a polymer mixture that will produce plastics products with the highest tensile strength. Similar experiments are commonly encountered in the industry. The problem of deciding how to mix ingredients to optimize some criterion is called experimental design with mixtures. A design of n runs for mixtures of m ingredients is a set of n points in the domain:

$$T_m = \{(\lambda_1, \ldots, \lambda_m) : \forall j \in \{1, \ldots, m\}, \lambda_j \geq 0, \lambda_1 + \cdots + \lambda_m = 1\}. \quad (1)$$

A great variety of works from the statistics literature have proposed many kinds of designs. Scheffé introduced simplex lattice designs and the corresponding polynomial models [11]. Subsequently, he developed an alternative design to the general simplex lattice, the simplex centroid design. Later, Cornell [1] suggested the axial design and proposed a comprehensive review of nearly all the statistics works dealing with experimental designs with mixtures and data analysis. The two following strategies are of particular relevance for this work.

Simplex Lattice Design. The simplex lattice design is a space filling design that creates a triangular grid of runs, which does not necessarily include the centroid. Let the mixture involve m components, H be a positive integer and furtherly suppose that each component can take $(H+1)$ equally spaced positions between 0 to 1 (included). Then:

$$\forall i \in \{1, \ldots, m\}, \lambda_i \in \{0, \frac{1}{H}, \frac{2}{H}, \ldots, \frac{H-1}{H}, 1\}. \quad (2)$$

This design can be efficiently computed using Das and Dennis systematic approach [2], and it is widely used within a majority of MOEAs based on reference points.

Two Layer Simplex Lattice Design. A recognized drawback of the simplex lattice design is that it generates many points on the boundary of the simplex, when the number of points H or the number of dimensions m increases. To avoid this problem, Deb et al. proposed in [3] the union of two simplex lattices: the first one maintains the original shape and size of the classical simplex lattice (external layer) while the other is scaled according to a factor $f \in [0,1]$, in order to further cover the central region of the simplex. This new design was originally introduced within the NSGA-III framework to generate the reference points used to deal with many-objective problems.

2.2   Weight Vector Designs Based on Discrepancy Functions

The discrepancy theory (also called theory of distribution irregularities) is a branch of mathematics addressing the problem of distributing points uniformly over some geometric object and evaluating the inevitably arising errors. This theory was ignited by theoretical contributions such as Weyl's equidistribution theorem and Roth's theorem [10]. The discrepancy of a point set measures the non-uniformity of such points placed (without loss of generality) in a unit cube $[0, 1]^m$, where $m > 0$ denotes the dimension of the unit cube.

The low-discrepancy sequences are also called quasi-random or sub-random sequences, due to their common use as a replacement of uniformly distributed random numbers. In [15], the use of low-discrepancy sequences as weight vectors is introduced in the MOEA/D framework.

From this point of view, other generation techniques such as Good Lattice Point and Uniform Design are able to produce sets with generally low discrepancies. This is demonstrated in [8], particularly in the case of the Good Lattice Point method. Although both are space filling methods, the uniform design can be formulated as an optimization problem. Tan et al. [12] introduce the use of uniform design as weight vectors in MOEA/D with promising results, but restricting the number of objectives to $2 \leq m \leq 4$, mostly due to the complexity associated with the computation of uniform designs for higher numbers of points and dimensions.

3   Uniform Design: Problem Statement

Fang et al. [4] proposed the Uniform Design (UD) as a kind of statistical experimental design. UD builds the experimental points in such a way that they are uniformly scattered over an experimental domain in the sense of low discrepancy [4]. UD does not generate points in the border of the simplex. In [14] it was demonstrated that UD is robust to changes in the model.

Let be m the number of factors (or objectives in the context of MOP) over the standard domain T_m. The aim is to choose a set of n points $P_n = \{\lambda^1, \ldots, \lambda^n\} \subset T_m$ such that those points are uniformly scattered on T_m. Let $D(P_n)$ be a measure of the non-uniformity (discrepancy) of P_n. The objective is to determine a set P_n^* minimizing D or, equivalently, maximizing the uniformity of n points over T_m. The associated optimization problem can be described as follows:

$$\min D(P_n) \tag{3}$$
$$\text{s.t. } P_n \in T_m$$

The base element of UD is the U-type design. A U-type design U is a $n \times m$ matrix $U = (u_{ij})$ in which each column has q entries $1, \ldots, q$ appearing equally often. The induced matrix of U, denoted by $X = (x_{ij})$, is defined by $x_{ij} = (u_{ij} - 0.5)/q$. A U-type design U becomes a UD when its induced matrix X has the smallest discrepancy in the set of all the possible induced matrices in domain T_m.

It was demonstrated that the problem of finding a uniform design under a given discrepancy metric D is $NP - hard$ [5] when the number of runs $n \to \infty$ and the number of factors $m > 1$. Many construction methods for uniform designs or nearly uniform designs, such as the Good Lattice Point (GLP) method, optimization methods have been proposed. The GLP approach is adopted in this paper.

3.1 Good Lattice Point

The Good Lattice Point (GLP) is a method to generate a set of uniformly distributed reference points. It was proposed by Korobov for the numerical evaluation of multivariate integrals [13]. The point sets generated using this method are widely used in quasi-Monte Carlo methods, uniform designs and computer experiments. To generate a U-type matrix using GLP, the first step consists in computing a set $H_n \subset \mathbb{N}$ such that:

$$H_n = \{h : h < n | gcd(h, n) = 1\}. \tag{4}$$

The greatest common divisor (gcd) equals to one ensures that coprime condition. If $k = |H_n|$, for any m distinct elements of $H_n = \{h_1, h_2, \ldots, h_k\}$, an $n \times m$ matrix $U = (u_{ij})$ is generated where $u_{ij} = i \cdot ((h_j \mod n) + 1)$. Finally, the last row of the GLP generated matrix should be deleted to obtain the U-type matrix.

3.2 CD_2 Discrepancy Function and Optimization Problem for UD

Several discrepancy measures, which determine the non-uniformity of a point set, are proposed in the specialized literature. There is more than one definition of discrepancy for D to measure the non-uniformity of P_n, for instance: the star discrepancy, the L_2-discrepancy (CD, the most commonly used) or the wrap around L_2-discrepancy (WD). The centered L_2-discrepancy, denoted by CD_2, is used in the implementation presented here, because its computing process is not complex and because this technique is invariant under relabeling of the coordinate axes [6].

As a consequence of the above explanations, the problem of determining uniform designs can be defined using the induced matrix X of U as follows:

$$\operatorname{argmin} CD_2(X) = \left(\frac{13}{12}\right)^m - \left(\frac{2}{n}\right) \sum_{k=1}^{n} \prod_{i=1}^{m} \left(1 + \frac{1}{2}|X_{ki} - \frac{1}{2}| - \frac{1}{2}|X_{ki} - \frac{1}{2}|^2\right) \tag{5}$$

$$+ \left(\frac{1}{n^2}\right) \sum_{k=1}^{n} \sum_{j=1}^{n} \prod_{i=1}^{m} \left(1 + \frac{1}{2}|X_{ki} - \frac{1}{2}| + \frac{1}{2}|X_{ji} - \frac{1}{2}| - \frac{1}{2}|X_{ki} - X_{ji}|\right)$$

The U-type design U^* whose induced matrix X^* has a minimum discrepancy under measure CD_2 is called a uniform design. Also, a U-type design with a low discrepancy measure can be denoted as a nearly uniform design [9].

4 Parallel Tabu Search Based Heuristic for UD Generation

Since solution evaluation in the optimization problem described above is time-consuming, a trajectory-based method was selected to solve the problem and to provide nearly Uniform Designs. However, the exploration of a solution space whose size dramatically increases with both the number of dimensions and generated points (number of m-combinations of set H_n) is an issue. As a simple solution, a parallel implementation allowed to distribute several search processes over different processors, in order to enhance exploration with reasonable CPU time. Another populational methods like GA requires more computing time becoming non-viable for our purposes.

Tabu Search is a metaheuristic technique introduced in the 1980s by Fred Glover [7]. It is an extension of a simple hill-climber, with additional features to avoid getting trapped in local optima:

- An exhaustive evaluation of the neighborhood of the current solution is carried out in order to identify the best neighbor, which is always accepted as the new solution.
- Since the above mentioned operating mode can lead to cycles in the search path, a tabu list registers either the last solutions visited or the last moves performed. Hence, the next candidate is chosen among those neighbors of the current solution that are not tabu. When the tabu list reports forbidden moves, an aspiration criterion is generally included to break the tabu status of certain moves that could lead to unvisited solutions.

4.1 Specific Features

The implementation developed in this work includes the Good Lattice Point as an embedded mechanism for generating the U-type design with a specified input. As mentioned in Sect. 3.1, this input is a set of m elements selected from H_n (see Eq. 4), which is subsequently used to compute a U-type matrix, its induced matrix X and the corresponding discrepancy function $CD_2(X)$ (see Eq. 5). It is worth mentioning that this process is expensive because of its quadratic growth in terms of the number of points generated (800 for 8 dimensions).

Thus, the solution encoding is simply the input of the GLP technique, i.e. a set of m indexes indicating which elements of H_n are selected to compute the U-type design. A solution S can be formulated as $S = \{s_1, \ldots, s_m\}$ with $\forall i, j \in \{1, \ldots, m\}$, $s_i \in H_n$ and $s_i \neq s_j$. For initialization, solutions are randomly generated. When a move is to be performed, a random index $j \in \{1, \ldots, m\}$ is chosen and s_j is replaced by an element randomly drawn from H_n, different from all the elements s_i in the current solution.

First, in order to promote the exploration of the search space, the process is distributed over W workers, or agents. All the agents work independently from each other and are assigned to their own processor to perform asynchronous parallel searches. The only information they share is through the tabu list, which is common for all of them in order to avoid multiple evaluations of the same solution. Therefore, some procedures, such as the production of new solutions (which makes use of the tabu list) or the tabu list truncation, cannot be carried out simultaneously for several agents (these processes are identified with the comment "Critical section" in the following pseudocode).

Besides, the reason for sharing the tabu list among all the agents is that each item reports the complete solution $S = \{s_1, \ldots, s_m\}$, instead of recording those elements involved in a move as commonly done in many TS implementations. This is possible because of the reasonable memory use to store this vector (m equals at most 8 in this study). Thus, agents share the same tabu list in order to prevent evaluating again a solution already visited by another agent. Finally, no aspiration criterion is needed since the information contained in the tabu list (solutions instead of moves) does not prevent considering unvisited solutions.

The last specific feature of the present implementation is that, due to the above-mentioned size of the search space, the thousands of neighbors of a solution cannot be generated since their evaluation would involve an unreasonable computational time. In order to reduce as much as possible this computational burden, only two neighbors are generated (and the best one replaces the current solution, even though this deteriorates the objective function).

Algorithm 4.1. Parallel Tabu Search

Data: *IterMax*: Stop Condition, W: workers, T_{max}: Tabu List size, n: Number of points wanted, *dim*: number of dimensions

Result: *NUD*:Nearly Uniform Design

1 $H_n \leftarrow$ SearchCoprime(N);
2 Solutions $\leftarrow$ RandomDisjoint(H_n,*dim*,W);
3 *bestdis* $\leftarrow \infty$, *bestsol* $\leftarrow \emptyset$, *TabuList* $\leftarrow \emptyset$;
4 **while** $\neg Stopping(IterMax)$ **do**
       /* Parallel For   */
5   **for** $i := 1$ *to* W **do**
6      $U_i \leftarrow$ GLP($Solutions_i$) ;
7      $X_i \leftarrow$ InducedMatrix(U_i) ;
8      $discrepancy_i \leftarrow CD_2(X_i)$;
9   **if** $discrepancy_i < bestdis_i$ and $Solutions_i \notin TabuList$ **then**
10 | $bestsol_i \leftarrow Solutions_i$;
11 | $bestdis_i \leftarrow discrepancy_i$;
12     $TabuList \leftarrow TabuList \cup Solutions_i$;
       /* Critical section   */
13     $Solutions_i \leftarrow$ UpdateSol($Solutions_i$,H_n);
14   **if** $|TabuList| > T_{max}$ **then**
15 | $TabuList \leftarrow$ MaintainTabuList($TabuList$) ;
16   $NUD \leftarrow \min(Solutions, bestdis)$;
17 **return** (NUD)

4.2 General Algorithm

The global process is described in Algorithm 4.1. First, set H_n is generated only once, at the beginning of the procedure (line 1). Then, the solution of each agent is built selecting m different elements from H_n, either randomly at the beginning of the search (line 2: W different solutions are produced) or when a neighbor of the current solution is produced (line 13). Subsequently, for each agent, the resulting U-type design is generated with the GLP strategy using as an input the m elements selected from H_n (line 6). The corresponding induced matrix X_i and the $CD_2(X_i)$ value are computed in lines 7–8 to evaluate the discrepancy of the generated design. The best solution found by each worker and the tabu list are updated in lines 9–12.

Then, within the "critical section", the following operations cannot be performed simultaneously since they involve access to the shared tabu list. Each worker produces two new candidates in the neighorhood of its current solution (line 13) and moves towards the best one (in terms of CD_2). This might cause a deterioration of the current objective function. Finally, the tabu list can be truncated to respect the allowed size, removing the oldest element.

5   Computational Experiments and Results

This Sections presents two sets of computational experiments. The first one describes the nearly Uniform Designs obtained when solving the optimization problem with the parallel TS based algorithm. Then, two state-of-the-art MOEAs, MOEA/D [18] (based on decomposition) and NSGA-III [3] (based on dominance), are applied to some classical MOPs, in order to evaluate their efficacy using UDs as a reference sets. For both experiments, the results obtained with Uniform Design are compared with those of two designs commonly used within MOEAs: simplex lattice (SLD) and two layer simplex lattice (2LD) designs.

5.1   Uniform Designs with the Tabu Search Based Heuristic

The Tabu Search based algorithm was executed for 3, 5 and 8 dimensions (objectives), generating respectively 300, 500 and 792 points. This number of points is the same as that used for the SLD and 2LD configurations (tuning the H factor in Eq. 2).

First, 10 executions were carried out for the 3-objective case, using $W = 4$ workers, $IterMax = 5,000$ iterations (implying 40,000 function evaluations) and the tabu list size is set to 1,000 solutions. Note that the minimum discrepancy can be known through an exhaustive evaluation of the search space of this small case: $CD_2^* = 2.835514900301206e^{-05}$. Within 10 executions, the parallel heuristic identifies 6 times the optimal solution, the mean value of CD_2 equals $2.8355149004366532e^{-05}$ and the standard deviation is 6.53^{-16}, which proves that the proposed algorithm can robustly find nearly UDs. The mean time is 20 min, when performed on HP ProLiant BL465c G7 with 24 AMD Opteron(tm) 6174 Processors and 128 GB in RAM.

For 5 and 8 dimensions, the parameters are: $W = 25$, $IterMax = 50,000$ (meaning 2.5×10^6 function evaluations) and a tabu list size equal to 10,000. This involves very long runs: approximately 7 h and 6 days for 5 and 8 objectives, respectively. Thus, it was only possible, due to time limitations, to perform one run and no statistical results are presented here. However, the final designs obtained can be compared with those computed by SLD and 2LD. Figure 1 shows the distribution of the resulting point sets in parallel-coordinate plots, as well as the corresponding hypervolume HV (using 1.18^m as a reference point).

The parallel-coordinate plots illustrate an expected trend: the SLD produces many points in the simplex boundary. On the other hand, 2LD has also many points on the simplex boundary (external layer), although the number of points inside the simplex is higher than in the SLD case (due to the internal layer). Finally, the UD misses some boundary points, but has a much more dense

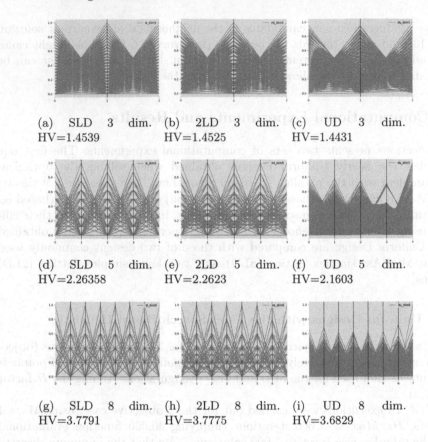

(a) SLD 3 dim.   (b) 2LD 3 dim.   (c) UD 3 dim.
HV=1.4539   HV=1.4525   HV=1.4431

(d) SLD 5 dim.   (e) 2LD 5 dim.   (f) UD 5 dim.
HV=2.26358   HV=2.2623   HV=2.1603

(g) SLD 8 dim.   (h) 2LD 8 dim.   (i) UD 8 dim.
HV=3.7791   HV=3.7775   HV=3.6829

Fig. 1. Parallel-coordinate plots and HV for different dimensions and design generation techniques

distribution than the simplex lattice design. As a consequence, it might be expected that the nearly uniform designs computed here with the parallel heuristic should provide better estimations for many objectives. Note that the same trend is captured by the HV indicator, which is always better for the SLD.

5.2 Experiments on Classical MOPs

In this second set of experiments, the nearly Uniform Designs obtained in the previous step were used as reference sets within two state-of-the-art MOEAs, namely MOEA/D and NSGA-III. The numerical results are compared with those obtained with reference sets built with SLD and 2LD. The test functions used present different Pareto front features: DTLZ2 has a concave front, DTLZ7 has several disconnected components, while the DTLZ1^{-1} and Kite functions (more recently proposed in [16]) are characterized by entire regions of the objective space without Pareto solutions.

For both MOEAs, the population size and generation number are set to 300 and 500, 500 and 1,000, 792 and 1,500 for 3, 5 and 8 dimensions respectively.

Table 1. Results with two MOEAs and SLD, 2LD and UD

	Ind.	Dim.	MOEA/D (Tchebycheff)			NSGA-III		
			SLD	2LD	UD	SLD	2LD	UD
DTLZ2	HV	3	**1** **(0.00068)**	0.9930 (0.00038)	0.9996 (0.00087)	**1** **(0.00045)**	0.9979 (0.00033)	0.9876 (0.00132)
		5	0.9104 (0.00013)	**1** **(0.00020)**	0.8817 (0.00167)	0.8499 (0.1329)	**1** **(0.19154)**	0.7098 (0.05080)
		8	0.9658 (0.0786)	**1** **(0.0451)**	0.6935 (0.2964)	0.9567 (0.0653)	**1** **(0.0619)**	0.6943 (0.3445)
	Δ	3	0.68341 (0.00412)	0.61814 (0.00757)	**0.42857** **(0.00324)**	**0.21691** **(0.00257)**	0.36447 (0.00455)	0.23627 (0.00856)
		5	1.83576 (0.00242)	0.94103 (0.43419)	**0.59685** **(0.00261)**	1.26537 (0.45764)	0.94103 (0.43419)	**0.86577** **(0.11821)**
		8	1.69124 (0.08990)	1.13190 (0.01297)	**0.43609** **(0.00232)**	0.65595 (0.47230)	**0.60710** **(0.48134)**	1.24396 (0.43613)
DTLZ7	HV	3	**1** **(0.0431)**	0.9999 (0.0467)	0.9969 (0.0279)	0.9997 (0.0005)	**1** **(0.0005)**	0.9955 (0.0022)
		5	**1** **(0.0929)**	0.9898 (0.0004)	0.9741 (0.0007)	**1** **(0.2511)**	0.9812 (0.2939)	0.8944 (0.2680)
		8	0.9648 (0.2206)	0.961 (0.1082)	**1** **(0.0454)**	0.9974 (0.4161)	**1** **(0.1226)**	0.9815 (0.0124)
	Δ	3	1.30312 (0.00653)	1.31041 (0.01789)	**1.16453** **(0.01207)**	1.67591 (0.24637)	**1.34075** **(0.24603)**	1.64339 (0.23005)
		5	1.52916 (0.01645)	1.54194 (0.13767)	**1.00264** **(0.00731)**	1.57990 (0.15099)	1.54194 (0.13767)	**1.43518** **(0.14890)**
		8	1.50361 (0.04593)	1.13237 (0.08790)	**0.92790** **(0.00337)**	1.12132 (0.02374)	1.078746 (0.02818)	**1.07546** **(0.02407)**
Kite	HV	3	**1** **(0.00002)**	0.9899 (0.01447)	0.9968 (0.01302)	0.9990 (0.00561)	**1** **(0.00540)**	0.9963 (0.00670)
		5	0.9866 (0.00017)	**1** **(0.00001)**	0.9484 (0.00002)	**1** **(0.00577)**	0.9964 (0.00614)	0.9251 (0.03326)
		8	0.9999 (0.0001)	**1** **(0.00011)**	0.9997 (0.00026)	0.9998 (0.00013)	**1** **(0.00001)**	0.9975 (0.11235)
	Δ	3	0.95595 (0.00017)	0.45136 (0.02933)	**0.45036** **(0.01963)**	1.42502 (0.19702)	1.39200 (0.18580)	**1.39040** **(0.20709)**
		5	1.51380 (0.11604)	1.52806 (0.13389)	**1.25863** **(0.07760)**	1.98942 (0.04557)	2.00077 (0.06196)	**1.92650** **(0.19005)**
		8	1.16733 (0.00004)	**0.77055** **(0.00083)**	1.07569 (0.24373)	1.07546 (0.02407)	1.07546 (0.02818)	**1.0758** **(0.02374)**
DTLZ1^{-1}	HV	3	**1** **(0.0075)**	0.9905 (0.0069)	0.9801 (0.0089)	0.9883 (0.0195)	**1** **(0.0222)**	0.9870 (0.0316)
		5	0.8384 (0.0015)	**1** **(0.2099)**	0.8831 (0.0267)	0.9912 (0.0109)	0.8347 (0.3843)	**1** **(0.4464)**
		8	**1** **(0.4502)**	0.9852 (0.4366)	0.9999 (0.4493)	**1** **(0.3035)**	0.9386 (0.3109)	0.9858 (0.4125)
	Δ	3	1.05003 (0.00114)	0.99743 (0.00210)	**0.42752** **(0.02296)**	1.42844 (0.18526)	**1.09888** **(0.19316)**	1.30346 (0.22475)
		5	1.98942 (0.04557)	2.00077 (0.06196)	**1.92650** **(0.19005)**	1.51380 (0.11604)	1.52806 (0.13389)	**1.25863** **(0.07760)**
		8	0.44218 (0.00005)	0.49086 (0.00031)	**0.37120** **(0.00062)**	1.99657 (0.04208)	1.99691 (0.06690)	**1.98622** **(0.05259)**

Furthermore SBX and polynomial mutation (with $\eta_c = \eta_m = 20$) are employed, with a mutation rate $1/NVars$. Finally, the number of neighbors for MOEA/D is 20% of the population size. The obtained results are compared in terms of

HV (with reference point 1.18^m for functions DTLZ2 and Kite, 0^m DTLZ1^{-1} and 9^m for DTLZ7) and of the Δ-metric for diversity assessment of the final population. Each configuration {algorithm–reference set} is executed 50 times for each MOP.

The results are provided in Table 1. For each function and each algorithm, the HV values are normalized with respect to the best HV found by a reference set generation technique (i.e., the technique having the best HV obtains 1). On the other hand, the number in parenthesis represent the variation coefficient for HV (standard deviation divided by the corresponding mean value) and the standard deviation for the Δ-diversity indicator. This table first highlights that the SLD and, in a marginally, the 2LD, obtain the best HV results. Even for the highest dimensions, UD is not able to provide such a good HV. However, when considering the diversity indicator, it is clear that UD outperforms SLD and 2LD in many cases. This observation is particularly true when the objective number increases. Finally, it is worth mentioning that the above observations are independent from the MOEA used: no matter the search strategy (dominance or decomposition), the benefits of UD for diversity, particularly for high m values, is demonstrated with this experimental study.

6   Conclusions

This work explores an alternative technique for generating a precalculated set of reference-points used by many recent MOEAs to guide the search and promote diversity within the approximated Pareto front. In particular, the generation of Uniform Designs can be formulated as an optimization problem that minimizes the corresponding point set discrepancy, measured through the classical CD_2 function. Due to the size of the associated search space, a parallel heuristic based on Tabu Search is developed in this framework in order to produce nearly UDs.

The obtained uniform sets show a dense distribution of the vectors inside the simplex, while somehow disregarding the points on the simplex boundary. This feature allows to obtain a better diversity, when the UD is used within classical MOEAs, than that obtained by the Simplex Lattice or Two Layer designs. This observation is true for both dominance-based and decomposition-based MOEAs and is strengthened when the dimension increases. As future work, the efficiency of the TS based heuristic should be improved to tackle higher objective numbers. Besides, the benefits of UD as a reference set for solving MOPS should be evaluated on more test functions, involving for example degenerated fronts.

References

1. Cornell, J.A.: Experiments with Mixtures: Designs, Models, and the Analysis of Mixture Data, vol. 403. Wiley, Hoboken (2011)
2. Das, I., Dennis, J.E.: Normal-boundary intersection: a new method for generating the Pareto surface in nonlinear multicriteria optimization problems. SIAM J. Optim. **8**(3), 631–657 (1998). https://doi.org/10.1137/S1052623496307510

3. Deb, K., Jain, H.: An evolutionary many-objective optimization algorithm using reference-point-based nondominated sorting approach, part i: solving problems with box constraints. IEEE Trans. Evol. Comput. **18**(4), 577–601 (2014). https://doi.org/10.1109/TEVC.2013.2281535

4. Fang, K.T.: Uniform design: application of number-theoretic methods in experimental design. Acta Math. Appl. Sin. **3**, 363–372 (1980)

5. Fang, K.T., Qin, H.: A note on construction of nearly uniform designs with large number of runs. Stat. Prob. Lett. **61**(2), 215–224 (2003)

6. Fang, K., Lin, D.: J. Uniform designs and their application in industry. In: Handbook on Statistics in Industry, pp. 131–170. Elsevier, Amsterdam (2003)

7. Glover, F.: Future paths for integer programming and links to artificial intelligence. Comput. Oper. Res. **13**(5), 533–549 (1986). https://doi.org/10.1016/0305-0548(86)90048-1

8. Hua, L.K., Wang, Y.: Applications of Number Theory to Numerical Analysis. Springer, Heidelberg (2012)

9. Ma, C., Fang, K.T.: A new approach to construction of nearly uniform designs. Int. J. Mater. Product Technol. **20**(1–3), 115–126 (2004)

10. Roth, K.F.: Rational approximations to algebraic numbers. Mathematika **2**(1), 1–20 (1955)

11. Scheffé, H.: Experiments with mixtures. J. R. Stat. Soc. Ser. B (Methodol.) 344–360 (1958)

12. Tan, Y.Y., et al.: MOEA/D+ uniform design: a new version of MOEA/D for optimization problems with many objectives. Comput. Oper. Res. **40**(6), 1648–1660 (2013)

13. Wang, Y., Hickernell, F.J.: An historical overview of lattice point sets. In: Fang, K.T., Niederreiter, H., Hickernell, F.J. (eds.) Monte Carlo and Quasi-Monte Carlo Methods, pp. 158–167. Springer, Heidelberg (2002). https://doi.org/10.1007/978-3-642-56046-0_10

14. Xie, M.Y., Fang, K.T.: Admissibility and minimaxity of the uniform design measure in nonparametric regression model. J. Stat. Plan. Inference **83**(1), 101–111 (2000)

15. Zapotecas-Martínez, S., Aguirre, H.E., Tanaka, K., Coello Coello, C.A.: On the low-discrepancy sequences and their use in MOEA/D for high-dimensional objective spaces. In: 2015 IEEE Congress on Evolutionary Computation (CEC), pp. 2835–2842. IEEE (2015)

16. Zapotecas-Martínez, S., Coello, C.A.C., Aguirre, H.E., Tanaka, K.: A review of features and limitations of existing scalable multi-objective test suites. IEEE Trans. Evol. Comput. 1 (2018). https://doi.org/10.1109/TEVC.2018.2836912

17. Zapotecas-Martínez, S., López-Jaimes, A., García-Nájera, A.: LIBEA: a Lebesgue indicator-based evolutionary algorithm for multi-objective optimization. Swarm Evol. Comput. (2018). https://doi.org/10.1016/j.swevo.2018.05.004, http://www.sciencedirect.com/science/article/pii/S2210650217307216

18. Zhang, Q., Li, H.: MOEA/D: a multiobjective evolutionary algorithm based on decomposition. IEEE Trans. Evol. Comput. **11**(6), 712–731 (2007). https://doi.org/10.1109/TEVC.2007.892759

19. Zhang, Q., et al.: Multiobjective optimization test instances for the CEC 2009 special session and competition. University of Essex, Colchester, UK and Nanyang technological University, Singapore, special session on performance assessment of multi-objective optimization algorithms, Technical report 264 (2008)

Comparison of Reference- and Hypervolume-Based MOEA on Solving Many-Objective Optimization Problems

Dani Irawan[✉][iD] and Boris Naujoks

Institute for Data Science, Engineering and Analytics,
TH-Köln - University of Applied Sciences, Cologne, Germany
{dani.irawan,boris.naujoks}@th-koeln.de

Abstract. Hypervolume-based algorithms are not widely used for solving many-objective optimization problems due to the bottleneck of hypervolume computation. Approximation methods can alleviate the problem and are discussed and tested in this work. Several MOEAs are considered, but after pre-experimental tests, only two variants of SMS-EMOA are considered further. These algorithms are compared to NSGA-III, a reference-based algorithm. The results show that SMS-EMOA with hypervolume approximation is viable for many-objective optimization problems and is faster in convergence towards the Pareto-front.

Keywords: Hypervolume approximation · MOEA · Reference vector · Many-objective optimization

1 Introduction

Many-objective Optimization Problems (MaOPs) bring challenges to evolutionary algorithms. The first challenge is the loss of pressure to find the Pareto front. With increasing dimensionality, non-dominatedness is easier to achieve, leading to smaller selection pressure to converge to the Pareto front [16]. Another big challenge in the field is how to handle the "curse of dimensionality": as the problem dimension increases, the computational effort required also increases.

Recent research in the field tried to reduce the computational cost by developing algorithms which reduce the effect of dimensionality. One such example is the NSGA-III [7] algorithm which attempts to reduce the curse of dimensionality effect by using reference points. NSGA-III is an improvement to a method which is regularly used to solve multi-objective optimization problems (MOPs): NSGA-II [8]. Compared to NSGA-II, the number of reference points and the population size in NSGA-III have a relatively more significant influence on the computation time rather than the dimensionality. The computational cost would still grow as the dimension increases, but the growth will be slower.

Other research attempts to reduce computation time by using approximations. Algorithms relying on hypervolume (HV) computations have this bottleneck where it is expected that no algorithm can compute the HV in polynomial

© Springer Nature Switzerland AG 2019
K. Deb et al. (Eds.): EMO 2019, LNCS 11411, pp. 266–277, 2019.
https://doi.org/10.1007/978-3-030-12598-1_22

time [5]. The HV approximation methods can decrease the computation time by one or two orders of magnitudes [5, 6].

This work is intended to compare the performance of the above two approaches to handle the curse of dimensionality in terms of number of evaluations required. This is important especially when expensive objective functions are considered. Li, et al. has done similar work [20]. However, they focused on the algorithms' performance on a limited budget (100 000 evaluations) while here the budget was not limited, instead, a performance-target is set. The algorithms are tested on box-constrained test problems from the DTLZ [9] and the WFG [12] test suites.

Hypervolume-Based Algorithms. The HV indicator is the most commonly used performance indicator in multi-objective optimization. HV-based algorithms are quite straightforward: to approach the real Pareto front, maximize the HV indicator because it is Pareto compliant [2, 24]. Several HV-based algorithms were proposed in the past, such as SMS-EMOA [3, 10], HypE [2], and MO-CMA-ES [15].

HV-based algorithms, however, are not as widely used for many-objective optimization problem because the computation cost grows exponentially with dimension. It is expected that no algorithm can compute the HV in polynomial time [5]. To circumvent the problem, approximation methods were proposed to speed-up the computation. Algorithms already using approximations are HypE and an SMS-EMOA implementation by Ishibuchi [17].

Approximation methods are usually limited to only approximate the least contributor to the total HV, not the total HV itself. Examples of these methods are the methods by Ishibuchi [17] and Bringmann-Friedrich [6]. However, there are indeed some methods to approximate the total HV such as FPRAS [5] and the method proposed by Tang, et al. [22].

Reference-Based Algorithms. Recent algorithms use reference or target vectors to guide solutions to the Pareto front. Examples of algorithms using reference-vectors are NSGA-III [7], MOEA/D [25], MSOPS [13], and MSOPS-II [14]. This approach is favored because it is inherently diversity-preserving. This approach can even be done by decomposing the many-objective problem into single-objective problems such as the one done in MOEA/D.

2   Pre-experimental Phase

Before the experiments were executed, the algorithms to be used are tested to see whether to continue testing them is viable or not. For this pre-experimental phase, the algorithms were tested on the DTLZ2 test problem because it can be considered as the easiest among the test problems. From Table 1, it can be seen that the DTLZ2 problem is unimodal which makes it easier than the multimodal DTLZ1 and DTLZ3. It is very similar to DTLZ4 except that it is unbiased. It also has the same magnitude in both objective and search space so scaling should not be an issue for any algorithm.

For the tests, reference lines are spread over the objective space following Das and Dennis systematic distribution as explained in [7]. The intersections of the lines with the Pareto front are used as reference points to create a target HV. The termination criterion for the tests is to achieve 99.9% HV formed by a reference point at $(2, \ldots, 2)$ and the intersection of the reference lines with the Pareto front. The algorithms must achieve the target HV within 24 h.

The following algorithms were considered in the pre-experimental phase.

2.1 MOEA/D

MOEA/D is an algorithm designed to solve multi-objective optimization problems (MOPs). MOEA/D works by decomposing the MOP into several single objective problems (SOPs) using scalarizing methods. Each SOP then has its own search direction in objective space depending on its corresponding weight vector. The Chebyshev decomposition were used and no external population is maintained to make MOEA/D comparable to other algorithms used in the study.

To create variations, MOEA/D limits the interaction between individuals by setting neighborhoods. A neighborhood is a set of subproblems with similar weight vectors. Each individual has its own neighbors which are determined *a priori*. In this work, the algorithm makes use of the SBX recombination operator and polynomial mutation (see [1]).

The pre-experimental tests showed that the algorithm can easily achieve 95% of the target HV in minutes. However, it fails to achieve the target HV before the time limit. Further check reveals that the solutions are actually located near the Pareto set, but in the objective space, their mapping were not well distributed hence the HV is small. It is concluded that MOEA/D should not be used further in the experiment because the behavior is expected to emerge in all experiments as discussed in [21] and it will always struggle to achieve the target HV.

2.2 NSGA-III

NSGA-III has gained popularity in the recent period. NSGA-III uses non-dominated sorting and reference lines as guidance. The reference lines are formed by connecting the ideal point with some reference points defined by the user.

In this work, the original, standard algorithm is implemented. SBX recombination operator and polynomial mutation are used to create offspring. The pre-experimental test shows that a single optimization process can finish in less than five minutes on our machine. We consider this very fast; therefore, the algorithm will be tested further in the experiment.

2.3 SMS-EMOA and GSMS-EMOA

SMS-EMOA is one of the earliest methods employing HV as a selection criteria instead of just a performance metric. The goal of the algorithm is to maximize the HV value of the population. The procedure involved is non-dominated sorting and eliminating individuals with the smallest HV contribution.

Two implementations of SMS-EMOA are tested. The first variant is the basic SMS-EMOA. The second variant is the "start-small grow-big" SMS-EMOA [11] with an adaptive population size scheme which further will be addressed as growing SMS-EMOA (GSMS-EMOA). For both implementations, the SBX and polynomial mutation are used.

In adaptive GSMS-EMOA, the population will grow when the R-indicator is below a threshold R_{reset}. This means that the population has not been able to find better solutions for a while. The parameters involved in determining the R-indicator is the initial value R_{init}, the threshold R_{reset} in which the population will grow and R will reset to its initial value, and the learning rate α. The values used are shown in Table 2.

The pre-experimental tests concluded that both algorithms can finish within approximately 5 h on our machine using exact HV computation. It is expected to take considerably more time than NSGA-III due to HV computations.

2.4 MO-CMA-ES

The Covariance Matrix Adaptation Evolution Strategies (CMA-ES) is one of the most successful evolutionary algorithms. In CMA-ES the offspring are sampled from a multivariate normal distribution based on the parents and an adapting covariance matrix. This procedure can be considered as a mutation operation. The covariance matrix is updated in every generation based on an evolution path which represents a sequence of selected mutations.

Multi-objective implementation of CMA-ES is known as MO-CMA-ES, using several $(1+\lambda)$-CMA-ES to create a set of solutions [15]. The solutions are then ranked using multi-objective optimization performance metrics (such as HV, IGD, crowding distance, etc.) and the best ones are kept. In the pre-experimental phase, MO-CMA-ES variant introduced in [23] is used. This variant allows us to use approximation of the least HV contributor and reference-line based selection enabling the comparison of both approaches.

MO-CMA-ES parents have more information to convey: not only the solution/design point, but also the covariance matrix, step size, average success rate, and evolution path. For recombination, the SBX is used with all objects (aside from the design point) taken as the element-wise average from the two parents. This ensures all objects remain valid (e.g. the covariance matrices are non-negative definite, step sizes are above their minimum).

One last important aspect to consider in using MO-CMA-ES is the constraint handling. Unlike the constrained polynomial mutation operator, offspring from CMA mutation may be infeasible. The death-penalty method is used. In death-penalty method, non-feasible solutions are not evaluated and count as failing offspring, reducing the step size and a feasible solution is more likely to emerge at the next generation. Kramer [19] mentioned death-penalty is inefficient, but CMA-ES could cope quite well with it. Kramer noted that a minimum step-size is required to prevent premature step-size reduction.

The pre-experimental tests show that using CMA, most suggested-solutions are found infeasible thus the step sizes drop very quickly to the minimum value.

Furthermore, if the step sizes are too low, then the mutations will be negligible, undermining the main feature of MO-CMA-ES. If the step sizes are too large the offspring success rates are small. Tuning of the step sizes is then required, including tuning the initial values. It is concluded that more research on determining the proper operators and their parameters are required and MO-CMA-ES would not be used in the experiment.

3   Numerical Experiments

To compare the algorithms performances, the DTLZ [9] and WFG [12] test suites are used. The characteristics of each test function used are shown in Table 1. DTLZ and WFG are chosen because both are easily scalable. Both are also popular test functions widely used in benchmarking many-objective optimization algorithms. Additionally, the convex DTLZ2 is also used and abbreviated as the cDTLZ2 test problem.

Table 1. Characteristics of the test functions [12].

Test problem	Shape	Separability	Modality	Contain bias
DTLZ1	Linear	Separable	Multimodal	×
DTLZ2	Concave	Separable	Unimodal	×
cDTLZ2	Convex	Separable	Unimodal	×
DTLZ3	Concave	Separable	Multimodal	×
DTLZ4	Concave	Separable	Unimodal	✓
WFG4	Concave	Separable	Multimodal	×
WFG5	Concave	Separable	Deceptive	×
WFG6	Concave	Non-separable	Unimodal	×
WFG7	Concave	Separable	Unimodal	✓
WFG8	Concave	Non-separable	Unimodal	✓
WFG9	Concave	Non-separable	Multimodal, deceptive	✓

The target HV for the tests are created in the same way as in Sect. 2. However, due to limited computing resources, the tests are also time limited. If an experiment fails to achieve the target HV before a set time limit expires, no result is recorded for the experiment and it is marked as "fail". The time limit is 24 h for 5-objective problems and 240 h for 8-objective problems. Other parameters for the algorithms are shown in Table 2.

Usage of Approximation Methods

As mentioned in Sect. 1, there are two kinds of approximations that can be conducted: approximation of the least HV-contributor and approximation of the

Table 2. Values of parameters used, n is the number of objectives.

Parameter	Value
Distribution Index, SBX η_c	30
Distribution Index, Poly-mutation η_m	20
Crossover Probability p_c	1.0
Mutation Probability p_m	$\frac{1}{n}$
GSMS-EMOA Initial adaptation factor R_{init}	0.5
GSMS-EMOA adaptation factor threshold R_{reset}	0.2
GSMS-EMOA learning rate α	0.025

total HV. In this work, approximation of the least HV-contributor is used for selection purpose and conducted using the method proposed by Bringmann and Friedrich [6]. Approximation of the total HV is used for the stopping criterion and conducted using the FPRAS method. The PyGMO package [4] contains implementations of both methods and they are used in this work.

NSGA-III does not use HV for its selection; therefore, approximations are used only for checking whether or not it has achieved the target HV. In SMS-EMOA, the selection procedure removes the least HV-contributor from the population; therefore, SMS-EMOA uses both approximation methods.

By using HV approximation on SMS-EMOA, the algorithm became similar to a steady-state implementation of SIBEA [26], albeit in this work different approximation methods are used.

4 Results

4.1 Success Rate

5-Objective Problems. Some experiments indeed cannot achieve the target HV before the time limit expires and hence no data is available. This is most prominent for the DTLZ4, WFG5, WFG6 and WFG9 test functions. Table 3 lists the percentage of optimization runs that achieved the target HV within the allocated time. Due to the poor performance of all algorithms in DTLZ4, WFG5, WFG6, and WFG9, the 4 test functions are not considered for benchmarking.

The DTLZ4 problem is biased and a dense set of solution exist in one of the hyperplane [9]. All algorithms experience difficulty to spread the solutions and experience premature convergence.

The WFG5 and WFG9 test functions are deceptive problems [12] and all methods seems to always fall to the deceptive front instead of the Pareto front. Hence, all experiments never reached the target HV.

In the case of WFG6, its distinct difference with other problems is that it is highly nonseparable and this feature causes the optimizers to struggle. The same should happen on WFG8 as the bias should increase the difficulty, but the opposite is observed.

Table 3. Percentage of successful runs out of 25 samples in 5-objective problems. Cells with 'NT' indicates it was not tested.

Test problem	NSGA-III		SMS-EMOA		GSMS-EMOA	
	Exact HV (I)	Approx HV (II)	Exact HV (III)	Approx HV (IV)	Exact HV (V)	Approx HV (VI)
DTLZ1	100%	100%	100%	100%	100%	100%
DTLZ2	100%	100%	100%	100%	100%	100%
cDTLZ2	100%	100%	100%	100%	100%	100%
DTLZ3	100%	100%	100%	100%	100%	100%
DTLZ4	0%	NT%	0%	NT%	0%	NT%
WFG4	0%	0%	100%	100%	12%	8%
WFG5	0%	NT	0%	NT	0%	NT
WFG6	0%	NT	0%	NT	0%	NT
WFG7	100%	100%	100%	100%	96%	96%
WFG8	100%	100%	100%	100%	100%	92%
WFG9	0%	NT	0%	NT	0%	NT

8-Objective Problems. Table 4 shows the success rate on 8-objective problems. Again, it can be seen that not all optimization runs are able to achieve the target HV within the allocated time.

In Table 4 NSGA-III is able to find the Pareto front of 8-objective WFG4 where it completely fails in the 5-objective WFG4. No further investigation was conducted on why this occurs. SMS-EMOA is successful in all test problems. GSMS-EMOA, on the other hand, performs poorly in 3 out of 7 test problems indicating a problem in a growing scheme.

Table 4. Percentage of successful runs out of 25 samples in 8-objective problems. For all algorithms, only the implementations with approximated HV are tested.

Test problem	NSGA-III (II)	SMS-EMOA (IV)	GSMS-EMOA (VI)
DTLZ1	100%	100%	100%
cDTLZ2	100%	100%	100%
DTLZ2	100%	100%	100%
DTLZ3	100%	100%	100%
WFG4	96%	100%	8%
WFG7	100%	100%	8%
WFG8	100%	100%	4%

4.2   Evaluation to Convergence

Figures 1 and 2 show the boxplots of the number of evaluations required to achieve the target HV for 5- and 8-objective problems respectively. Failed tests are omitted, hence the boxplots are produced from different data sizes. Some boxes are missing because data are not available or completely omitted. Figure 1 shows that using approximations to compute HV always yield faster convergence.

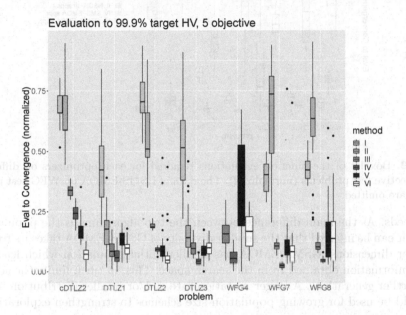

Fig. 1. Boxplot of the evaluations required for each optimizers on different 5-objective test problems (normalized). The methods are NSGA-III exact and with approximation, SMS-EMOA exact and with approximation, and GSMS-EMOA exact and with approximation respectively from I to VI.

5   Analysis

5.1   Success Rate

We start by looking at the success rates in 5-objective problems in Table 3. Overall, all algorithms have similar success rate. The only noteworthy performance is SMS-EMOA being able to solve WFG4 while the other two cannot. GSMS-EMOA started with smaller population, it is lacking diversity and is easily trapped in local optima.

Table 4 provides the result for 8-objective problems. First it can be seen that HV-based algorithms can solve the test problems within the allocated time so they are quite viable for solving many-objective problems. However, it can be seen that GSMS-EMOA fails in WFG4, WFG7, and WFG8 while SMS-EMOA

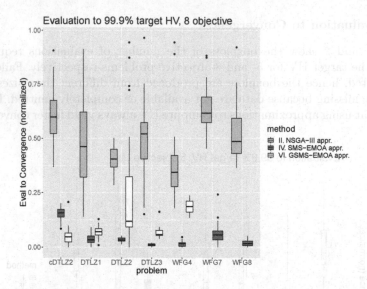

Fig. 2. Boxplot of the function evaluations required for each optimizers on different 8-objective test problems (normalized). The data of GSMS-EMOA on WFG test problems are omitted.

succeeds. As the main difference between the two algorithms is the population size, it can be inferred that the smaller diversity of GSMS-EMOA takes its toll on higher dimensions. GSMS-EMOA has smaller initial population which leads to less information obtained from the search space. This, in turn, limits the search in further generations. A larger mutation rate and/or smaller distribution index should be used for growing population-size schemes to strengthen exploration.

5.2   Convergence Speed

The first interesting result from Fig. 1 is in comparing NSGA-III (I & II) and SMS-EMOA (III & IV). In all tests, SMS-EMOA is able to achieve similar or better performance constantly. To check whether the performance difference is significant, one-tailed pairwise-Wilcoxon tests are performed. The results are first grouped into 3 categories: 5-objective problems solved with exact method, 5-objective problems solved with approximations, and 8-objective problems solved with approximations.

Table 5 summarize the statistical tests results. The first column in each category (Ref < HV) indicate whether NSGA-III is outperformed by the other two algorithms (checkmark) or only by SMS-EMOA ('1') in statistically significant way. The second column indicates whether SMS-EMOA is outperformed by GSMS-EMOA in statistically significant way.

SMS-EMOA is rarely outperformed by GSMS-EMOA which agrees to the results from [11] that using GSMS-EMOA algorithm has little effect on the number of function evaluations. The benefit of using GSMS-EMOA is due to fewer points being considered in the early phase, reducing the total time consumed.

Table 5. One-tailed pairwise-Wilcoxon test results.

Test Problem	5-obj. exact		5-obj. appr.		8-obj. appr.	
	Ref < HV	SMS < GSMS	Ref < HV	SMS < GSMS	Ref < HV	SMS < GSMS
DTLZ1	✓	×	✓	×	✓	×
DTLZ2	✓	✓	✓	✓	✓	×
cDTLZ2	✓	✓	✓	✓	✓	✓
DTLZ3	✓	×	✓	×	✓	×
WFG4	1	×	✓	×	1	×
WFG7	✓	×	✓	×	1	×
WFG8	✓	×	✓	×	1	×

SMS-EMOA always outperform NSGA-III in all problems. GSMS-EMOA performs poorly in WFG4, WFG7 and WFG8, but in other test problems it also outperforms NSGA-III. The results imply that in terms of number of evaluations to achieve large HV, HV-based algorithms can outperform reference-based algorithms. One could argue that NSGA-III is not designed to optimize the HV measure, but we try to analize what deters NSGA-III from achieving large HV. This could be attributed to the reference point distribution.

Das and Dennis's systematic method may have created sub-optimally distributed reference points such that it is easy to achieve higher HV by using different point-distribution. While NSGA-III and reference-based algorithms will always follow their references, HV-algorithms are free to distribute their solutions mapping in the objective space. Looking back to the pre-experimental test, MOEA/D also struggles in terms of distribution and the proposed remedy in [21] is by adjusting the reference distribution. These findings signify that while reference-based algorithms can utilize specific preference information easily, they are also restricted by these preferences, limiting their search power.

Although HV-based algorithms are still having the bottleneck on HV computations, the advances in HV approximation methods are making the algorithms viable for many-objective optimization. They are faster in converging and distributing the solutions thus if the costs for evaluating the objectives are very high compared to HV approximations, the algorithms will gain the upper hand.

6 Conclusion and Future Works

From the pre-experimental tests, two main results are found. First, MOEA/D cannot be expected to achieve high HV-value. Second, extending MO-CMA-ES to solve many-objective problems is non-trivial. MO-CMA-ES requires careful selection of operators and their parameters.

We also found that using an growing population size has a disadvantage in problems with higher dimensions. The scheme performs not as good as expected due to lack of diversity in early population. We suggest that growing schemes should be given stronger exploration power from the genetic operators.

We have also shown that HV-based algorithms are viable options to solve MaOPs with the aid of approximation methods. The current state of the work only compare NSGA-III with SMS-EMOA as representatives of reference-based algorithms and HV-based algorithms respectively, but in this paper, we have shown that HV-based algorithms can solve MaOPs in mere hours. Normally it would take hours for a single HV computation, but here, we were able to perform thousands of HV computations. HV-based algorithms have the advantage of requiring fewer function evaluations to achieve large HV.

As a future work, it would be interesting to see how adaptive-NSGA-III [18] scheme would behave if a target HV is set as in this experiment. The scheme will adaptively relocate reference points and enables NSGA-III to explore areas not covered by initial references. It is interesting to also see the effect of these adaptively changing reference points on convergence speed. It is worth to test the algorithms' performances if lower target HV are subscribed which will give insight on the strength of each algorithms and possibilities for hybrid methods.

Acknowledgments. This work is funded by the European Commission's H2020 programme through the UTOPIAE Marie Curie Innovative Training Network, H2020-MSCA-ITN-2016, under Grant Agreement No. 722734 as well as through the Twinning project SYNERGY under Grant Agreement No. 692286.

References

1. Agrawal, R.B., Deb, K.: Simulated binary crossover for continuous search space. Complex Syst. **9**(2), 115–148 (1995)
2. Bader, J., Zitzler, E.: Hype: an algorithm for fast hypervolume-based many-objective optimization. Evol. Comput. **19**(1), 45–76 (2011)
3. Beume, N., Naujoks, B., Emmerich, M.: SMS-EMOA: multiobjective selection based on dominated hypervolume. Eur. J. Oper. Res. **181**(3), 1653–1669 (2007)
4. Biscani, F., Izzo, D.: esa/pagmo2: pagmo 2.8 (2018). https://doi.org/10.5281/zenodo.1311209
5. Bringmann, K., Friedrich, T.: Approximating the volume of unions and intersections of high-dimensional geometric objects. Comput. Geom. **43**(6), 601–610 (2010)
6. Bringmann, K., Friedrich, T.: Approximating the least hypervolume contributor: NP-hard in general, but fast in practice. Theor. Comput. Sci. **425**, 104–116 (2012)
7. Deb, K., Jain, H.: An evolutionary many-objective optimization algorithm using reference-point-based nondominated sorting approach, part I: solving problems with box constraints. Trans. Evol. Comput. **18**(4), 577–601 (2014)
8. Deb, K., Pratap, A., Agarwal, S., Meyarivan, T.: A fast and elitist multiobjective genetic algorithm: NSGA-II. Trans. Evol. Comput. **6**(2), 182–197 (2002)
9. Deb, K., Thiele, L., Laumanns, M., Zitzler, E.: Scalable multi-objective optimization test problems. In: Congress on Evolutionary Computation (CEC), pp. 825–830. IEEE Press, Piscataway (2002)
10. Emmerich, M., Beume, N., Naujoks, B.: An EMO algorithm using the hypervolume measure as selection criterion. In: Coello Coello, C.A., Hernández Aguirre, A., Zitzler, E. (eds.) EMO 2005. LNCS, vol. 3410, pp. 62–76. Springer, Heidelberg (2005). https://doi.org/10.1007/978-3-540-31880-4_5
11. Glasmachers, T., Naujoks, B., Rudolph, G.: Start small, grow big? Saving multi-objective function evaluations. In: Bartz-Beielstein, T., Branke, J., Filipič, B., Smith, J. (eds.) PPSN 2014. LNCS, vol. 8672, pp. 579–588. Springer, Cham (2014). https://doi.org/10.1007/978-3-319-10762-2_57

12. Huband, S., Hingston, P., Barone, L., While, L.: A review of multiobjective test problems and a scalable test problem toolkit. Trans. Evol. Comput. **10**(5), 477–506 (2006)
13. Hughes, E.J.: Multiple single objective Pareto sampling. In: Congress on Evolutionary Computation (CEC), vol. 4, pp. 2678–2684. IEEE Press, Piscataway (2003)
14. Hughes, E.J.: MSOPS-II: a general-purpose many-objective optimiser. In: Congress on Evolutionary Computation (CEC), pp. 3944–3951. IEEE Press, Piscataway (2007)
15. Igel, C., Hansen, N., Roth, S.: Covariance matrix adaptation for multi-objective optimization. Evol. Comput. **15**(1), 1–28 (2007)
16. Ishibuchi, H., Tsukamoto, N., Nojima, Y.: Evolutionary many-objective optimization: a short review. In: Congress on Evolutionary Computation (CEC), pp. 2419–2426. IEEE Press, Piscataway (2008)
17. Ishibuchi, H., Tsukamoto, N., Sakane, Y., Nojima, Y.: Indicator-based evolutionary algorithm with hypervolume approximation by achievement scalarizing functions. In: Genetic and Evolutionary Computation (GECCO), pp. 527–534. ACM, New York (2010)
18. Jain, H., Deb, K.: An improved adaptive approach for elitist nondominated sorting genetic algorithm for many-objective optimization. In: Purshouse, R.C., Fleming, P.J., Fonseca, C.M., Greco, S., Shaw, J. (eds.) EMO 2013. LNCS, vol. 7811, pp. 307–321. Springer, Heidelberg (2013). https://doi.org/10.1007/978-3-642-37140-0_25
19. Kramer, O.: A review of constraint-handling techniques for evolution strategies. Appl. Comput. Intell. Soft Comput. 3:1–3:19 (2010)
20. Li, M., Yang, S., Liu, X., Shen, R.: A comparative study on evolutionary algorithms for many-objective optimization. In: Purshouse, R.C., Fleming, P.J., Fonseca, C.M., Greco, S., Shaw, J. (eds.) EMO 2013. LNCS, vol. 7811, pp. 261–275. Springer, Heidelberg (2013). https://doi.org/10.1007/978-3-642-37140-0_22
21. Qi, Y., Ma, X., Liu, F., Jiao, L., Sun, J., Wu, J.: MOEA/D with adaptive weight adjustment. Evol. Comput. **22**(2), 231–264 (2014)
22. Tang, W., Liu, H., Chen, L.: A fast approximate hypervolume calculation method by a novel decomposition strategy. In: Huang, D.-S., Bevilacqua, V., Premaratne, P., Gupta, P. (eds.) ICIC 2017. LNCS, vol. 10361, pp. 14–25. Springer, Cham (2017). https://doi.org/10.1007/978-3-319-63309-1_2
23. Voß, T., Hansen, N., Igel, C.: Improved step size adaptation for the MO-CMA-ES. In: Genetic and Evolutionary Computation (GECCO), pp. 487–494. ACM, New York (2010)
24. Wagner, T., Beume, N., Naujoks, B.: Pareto-, aggregation-, and indicator-based methods in many-objective optimization. In: Obayashi, S., Deb, K., Poloni, C., Hiroyasu, T., Murata, T. (eds.) EMO 2007. LNCS, vol. 4403, pp. 742–756. Springer, Heidelberg (2007). https://doi.org/10.1007/978-3-540-70928-2_56
25. Zhang, Q., Li, H.: MOEA/D: a multiobjective evolutionary algorithm based on decomposition. Trans. Evol. Comput. **11**(6), 712–731 (2007)
26. Zitzler, E., Brockhoff, D., Thiele, L.: The hypervolume indicator revisited: on the design of pareto-compliant indicators via weighted integration. In: Obayashi, S., Deb, K., Poloni, C., Hiroyasu, T., Murata, T. (eds.) EMO 2007. LNCS, vol. 4403, pp. 862–876. Springer, Heidelberg (2007). https://doi.org/10.1007/978-3-540-70928-2_64

Diversity over Dominance Approach for Many-Objective Optimization on Reference-Points-Based Framework

Deepak Sharma$^{(\boxtimes)}$ (iD), Syed Zaheer Basha, and Sandula Ajay Kumar

Department of Mechanical Engineering, Indian Institute of Technology Guwahati,
Guwahati 781039, Assam, India
{dsharma,b.sayed,sandula}@iitg.ernet.in

Abstract. Evolutionary multi-objective optimization (EMO) algorithms are designed to achieve a balance between convergence and diversity. However, these algorithms confront major challenges when all of their individuals become non-dominated while solving many-objective optimization problems. Although the appreciable efforts have been made by using the reference-points-based framework coupled with the Pareto-dominance ranking in the literature, selection of a diverse set of individuals, sometimes preferring isolated and dominated individuals over non-dominated individuals, needs to be addressed. In this paper, we propose the diversity over dominance (DoD) approach in which the diversity is preserved first by making clusters of individuals that are made by associating individuals to their nearest line using the reference-points-based framework. The Pareto-dominance ranking is then used to rank the individuals separately for each cluster. The environmental selection is then developed that selects individuals from each cluster. The DoD approach is tested on DTLZ and WFG problem instances and the results demonstrate its competitive performance over the existing EMO algorithms.

Keywords: Diversity · Dominance · Many objective optimization ·
Evolutionary algorithm · Evolutionary multiobjective optimization

1 Introduction

Many real-world problems often have multiple objectives that are conflicting in nature such as in crashworthiness of vehicle [11], bulldozer blade-design [1] to name a few. For such problems, a set of solutions is optimal, which are referred as Pareto-optimal (PO) solutions. Evolutionary multi-objective optimization (EMO) algorithms are the ideal choice for solving these problems because a set of PO solutions can be generated in one run.

From last few years, EMO algorithms for solving many-objective optimization problems (generally more than three-objective problems) are getting attention worldwide. The most successful ones like NSGA-III [3], θ−DEA [15], MOEA/DD [10] to name a few, are developed using the reference-points-based framework in

ⓒ Springer Nature Switzerland AG 2019
K. Deb et al. (Eds.): EMO 2019, LNCS 11411, pp. 278–290, 2019.
https://doi.org/10.1007/978-3-030-12598-1_23

which the convergence is achieved by performing the Pareto-dominance ranking and the diversity is preserved by selecting individuals based on the reference points generated on a unit hyperplane [2]. Since for many-objective optimization problems, almost all individuals of a population become non-dominated, the Pareto-dominance ranking fails to provide enough selection pressure for convergence [12]. At this stage, the environmental selection based on the reference-points framework plays a major role in selecting individuals. Therefore, NSGA-III introduced the niching procedure for selecting individuals representing the lines that are drawn using the reference points. To give preference to isolated individuals, MOEA/DD [10] introduced a uniform paradigm of dominance and decomposition approaches in which only one offspring individual at a time gets a chance for its survival. On the similar line, SPEA/R [8] proposed a composite fitness function so that individuals from each subregion (defined using the reference points) can be selected. In [7], an external archive was maintained for those individuals which may get eliminated using NSGA-III's environmental selection. An individual with minimum distance between the ideal point and the line drawn from the reference point was selected to update the archive. In all these studies, the environmental selection gave first emphasis on dominance-based selection followed by diversity for each subregion constructed from the directions through the reference points and the origin.

On the contrary to the above EMO algorithms, Jiang and Yang [9] suggested performing diversity-first sorting approach. The environmental selection for diversity was performed first and then Pareto-ranking was used to select individuals. The diversity-first sorting based evolutionary algorithm (DBEA) outperformed NSGA-III on many-objective optimization instances of WFG [6] problems. On the similar line, $\theta-$DEA [15] performs $\theta-$dominance sorting on the clusters of individuals which are made using the reference-points framework for diversity. $\theta-$DEA showed better results than NSGA-III over DTLZ [4] and WFG problem instances. Motivated by these approaches, we propose a dominance over decomposition approach, refer as DoD, in which individuals in a population are first clustered based on the directions from the reference points and the origin. Thereafter, the non-dominated sorting is performed to each clustered independently. The main contribution of DoD approach is the environmental selection that selects a diverse set of individuals by preferring isolated individuals from each cluster and sometimes selecting dominated individuals over the crowded non-dominated individuals. In the remaining paper, the challenges with dominance-based EMO algorithms are discussed in Sect. 2. The DoD approach is described followed by its implementation in Sect. 3. The results are discussed and compared with the existing EMO algorithms in Sect. 4. The paper is concluded in Sect. 5 with the future work.

2 Challenges with Dominance-Based Environmental Selection

The challenges with the environmental selection of dominance-based EMO algorithms are shown using three cases in Fig. 1. In all cases, the objective space

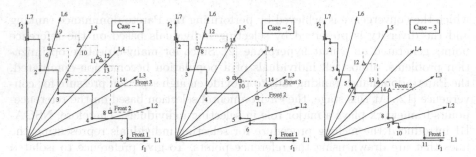

Fig. 1. Case-1: when the number of non-dominated individuals is less than N. Case-2: when the number of non-dominated individuals is equal to N. Case-3: when the number of non-dominated individuals is more than N.

has $N = 7$ reference lines ($L1, \ldots, L7$), which are drawn using the structured reference points [2]. Among $2N$ individuals, N individuals need to be selected. In Case-1, the number of non-dominated individuals (currently five) is less than N. The dominance-based approach, which prefers dominance followed by diversity, selects all non-dominated individuals from the front-1 and the rest of two individuals will be selected from the front-2. In this case, no individual representing lines $L3$ and $L4$ is selected. In case-2, the number of non-dominated individuals is equal to N. The dominance-based approach selects all individuals from the front-1. Again, there is no individual representing lines $L3$ and $L4$. In Case-3, the number of non-dominated individuals are more than N. In this case, the dominance-based approach will select individuals based on the diversity preserving mechanism since the ranking through non-dominated sorting cannot differentiate individuals. For example, the dominance approach can select individuals nearest to their respective lines. For example, individuals marked as 1, 4, 6, 9, and 11 are selected. The remaining two individuals are selected only from the front-1. In this case also, there is no individual representing lines $L3$ and $L4$.

The challenges described above leads to the motivation of the present work in which the DoD approach is proposed to select a diverse set of individuals, especially when a large number of individuals is non-dominated. Moreover, an additional emphasis is given to select isolated and sometimes dominated individuals over crowded non-dominated individuals for better convergence and diversity.

3   DoD Approach and Its Implementation

The DoD approach is described using Fig. 2 in which the DoD approach selects diverse individuals for the same three cases as presented in Fig. 1. For Case-1, the clusters are made for every line as shown in Fig. 2. Then, the Pareto-dominance ranking is applied to rank the individuals for every cluster separately. The non-dominated individual within each cluster is then chosen. For example, all non-dominated individuals from the front-1 are selected along with individuals (isolated, and dominated as per the dominance-based approach) marked

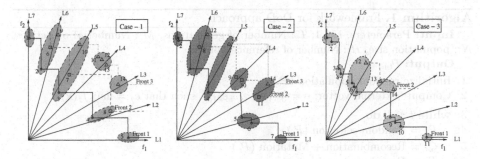

Fig. 2. DoD approach for selecting individuals for three cases presented in Fig. 1.

as 13 and 14 in the figure. For Case-2, the DoD approach selects individuals marked as 1, 3, 4, 6, 7 from the front-1. It is noted that individual marked as 1 is preferred over 2 in the same cluster. Basically, when a cluster has more than one non-dominated individual, the individual nearest to the line gets selected. The other individuals marked as 10 and 11 are selected from different clusters. For Case-3, the DoD approach selects individuals marked as 1, 4, 6, 9 and 11 from the front-1 and 13 from the front-2 from their respective clusters. In this case, there is no associated individual with line $L3$. The DoD approach first associates the nearest individual to line $L3$, that is, individual marked as 14 and then selects the same individual. From the above discussion, it can be observed that the DoD approach can select a diverse set of individuals representing every line. This environmental selection can keep enough selection pressure for better convergence and diversity of EMO algorithms. The major limitation of this approach is preferring dominated individuals over non-dominated individuals, which may cause convergence issue.

The DoD approach is implemented using the reference-points-based framework, which is shown in Algorithm 1[1]. At any generation t, the individuals are randomly selected from the parent population (P_t) on which simulated binary crossover and polynomial mutation operators are applied to create a new population, which is referred as offspring population (Q_t). Both P_t and Q_t are merged into R_t. The next generation parent population (P_{t+1}) is then chosen using the DoD environmental selection. Algorithm 2 shows steps of the DoD environmental selection in which R_t is normalized and then the individuals are associated with the reference lines. In this paper, an external vector **e** is maintained for storing the extreme objective function values for normalizing the population. As it can be seen in step 2 of Algorithm 1 that **e** is initialized by maximum objective function values. In normalization, **e** is updated or kept same based on the intercept and extreme points found. Algorithm 3 shows steps required for normalizing R_t, which include computing ideal point, extreme points and intercepts on each objective axis. Since some degenerate cases like unavailability of distinct extreme points from R_t or negative intercept can occur, the Nadir point is found from

---

[1] Source code at http://www.iitg.ac.in/dsharma/pub.html.

Algorithm 1. Framework for DoD approach

Input: Parameters, $t = 1$, T : Number of generations, M : Number of objectives, N : population size, H : Number of reference points

Output: P_{t+1}

1: Initialize random population (P_t)
2: Compute external vector: $\mathbf{e} = (e_1, e_2, \ldots, e_M)^T$ such that $e_j = \max\limits_{\mathbf{x} \in P_t} f_j(\mathbf{x})$
3: **while** $t \leq T$ **do**
4:　　$P_t' = $ Random selection (P_t)
5:　　$Q_t = $ Recombination + Mutation (P_t')
6:　　$R_t = P_t \cup Q_t$
7:　　$P_{t+1} = $ DoD Environmental selection (R_t)
8:　　$t = t + 1$
9: **end while**

R_t. The external vector $\mathbf{e}$ is updated when any component of the Nadir point is better than the corresponding component of $\mathbf{e}$ as shown in step 8 of Algorithm 3. Otherwise, the external vector $\mathbf{e}$ is updated completely by the intercepts found in step 11. Finally, each objective of all individuals of R_t is normalized using $\mathbf{e}$ at step 13.

Once R_t is normalized, the individuals are then associated with their nearest reference lines. The association procedure is shown in Algorithm 4 for which the normalized $\bar{R}_t$ and H are required. The structured reference points are created on a unit hyperplane using Das and Dennis approach [2]. These reference points are then used to compute reference lines (step 2) which pass from the origin and the reference point. These reference lines are stored in Z^r for associating individuals of R_t in the normalized objective space. As can be seen from step 4, a set $C_j \in C$ is initialized empty, which will store the individuals associated with a reference line j. Also, the niche count ρ_j for all reference lines is set zero that signifies a number of individuals associated with a reference line j. Inside the loop at step 5, each individual $\mathbf{r}$ is associated with its nearest reference line $(\pi(\mathbf{r}))$ based in its distance as shown in step 9. Thereafter, an individual $\mathbf{r}$ is stored in the cluster of $\pi(\mathbf{r})$ reference line. Also, the niche count of reference line $\pi(\mathbf{r})$ is incremented by one.

After normalization and association, the non-dominated sorting is performed for the individuals stored in a cluster C_j for a reference line j, which has at least one associated individual (refer step 3 of Algorithm 2). If a number of non-dominated individuals in C_j is more than one, then an individual $\mathbf{x}$ is selected, which is nearest to a reference line j. Otherwise, the only non-dominated individual is selected. The selected individual is then copied to P_{t+1} and it is removed from R_t. These steps are followed for all the lines, which has a niche count $\rho_j > 0$.

In addition to the above steps, if any reference line has no individual associated (meaning $\rho_j = 0$ and $C_j = \phi$, refer step 12 of Algorithm 2), the individual $\mathbf{x}$ nearest to a reference line j is chosen from the remaining individuals of R_t. This individual $\mathbf{x}$ is then stored in C_j and the niche count is increased by one. The same individual $\mathbf{x}$ is then copied to P_{t+1} and it is removed from R_t. These steps are then followed for those reference lines which have their ρ_j's zero.

Algorithm 2. DoD Environmental Selection (R_t)

Input: R_t, H, $\mathbf{e}$
Output: P_{t+1}

1: $\bar{R}_t := \text{Normalize}(R_t, \mathbf{e})$
2: $(C, \rho, Z^r) := \text{Associate}(\bar{R}_t, H)$ $\%C = \{C_1, \ldots, C_H\}$, $\rho = (\rho_1, \ldots, \rho_H)^T$, Z^r : set of reference lines.
3: **for** each $j \in Z^r$ and $\rho_j > 0$ **do**
4:     Non-dominated sorting of individuals $\in C_j$
5:     **if** Number of the best ranked individuals > 1 **then**
6:         Select the individual $\mathbf{x} \in C_j$ which is nearest to the reference line j
7:     **else**
8:         Select the best ranked individual $\mathbf{x} \in C_j$
9:     **end if**
10:     Include $P_{t+1} = P_{t+1} \cup \mathbf{x}$ and update $R_t = R_t \setminus \mathbf{x}$
11: **end for**
12: **for** each $j \in Z^r$ and $\rho_j == 0$ **do**
13:     Associate the closest individual $\mathbf{x}$ from the remaining R_t to the reference line j and update $I_j = I_j \cup \mathbf{x}$, and $\rho_j = 1$
14:     Include $P_{t+1} = P_{t+1} \cup \mathbf{x}$ and update $R_t = R_t \setminus \mathbf{x}$
15: **end for**
16: **while** $|P_{t+1}| < N$ **do**
17:     Select a random reference line $j \in Z^r : \rho_j > 1$
18:     Select the best individual $\mathbf{x}$ associated to the reference line j such that $\mathbf{x} \notin P_{t+1}$
19:     Include $P_{t+1} = P_{t+1} \cup \mathbf{x}$ and update $R_t = R_t \setminus \mathbf{x}$
20: **end while**

Algorithm 3. Normalize (R_t)

Input: R_t, $\mathbf{e}$
Output: $\bar{R}_t$: Normalized population

1: Determine ideal point, $\mathbf{z^I} = (z_1^I, z_2^I, \ldots, z_M^I)^T$ such that $z_j^I = \min\limits_{\mathbf{r} \in R_t} f_j(\mathbf{r})$
2: Translate objectives, $\mathbf{f}'(\mathbf{r}) = (f_1'(\mathbf{r}), f_2'(\mathbf{r}), \ldots, f_M'(\mathbf{r}))^T$ such that $f_j'(\mathbf{r}) = f_j(\mathbf{r}) - z_j^I$, $\forall \mathbf{r} \in R_t$
3: Compute extreme solutions, $Z = (\mathbf{z_1^e}, \mathbf{z_2^e}, \ldots, \mathbf{z_M^e})$ such that $\mathbf{z_j^e} = \mathbf{f}'(\mathbf{r})$, $\mathbf{r}$: $\min\limits_{\mathbf{r} \in R_t} \left(\max\limits_{i=1}^{M} f_i'(\mathbf{r})/w_i \right)$
4: Compute intercept a_j for $j = 1, \ldots, M$.
5: **if** Degenerate case or negative intercept found **then**
6:     Compute Nadir point, $\mathbf{z^N} = (z_1^N, z_2^N, \ldots, z_M^N)^T$ such that $z_j^N = \max\limits_{\mathbf{r} \in R_t^*} f_j(\mathbf{r})$ and $R_t^* \in R_t$ is the set of the non-dominated individuals.
7:     **if** $z_j^N < e_j$, where $j \in \{i, \ldots, M\}$ **then**
8:         $e_j = z_j^N$
9:     **end if**
10: **else**
11:     Update $e_j = a_j$, $\forall j \in \{i, \ldots, M\}$
12: **end if**
13: Normalize objective $\bar{f}_j(\mathbf{r}) = f_j'(\mathbf{r})/e_j, \forall \mathbf{r} \in R_t$, $\forall j \in \{i, \ldots, M\}$ and return $\bar{R}_t$

Algorithm 4. Associate (R_t) with the reference lines

Input: $\bar{R}_t$, H
   Output: C, ρ, Z^r
1: **for all** $\mathbf{r} \in H$ **do**
2:    Compute reference line $\mathbf{w}$ and $Z^r = Z^r \cup \mathbf{w}$
3: **end for**
4: Initialize $C_j = \emptyset$ $\forall j \in H$ and $\rho = (0, 0, \ldots, 0)^T$
5: **for all** $\mathbf{r} \in R_t$ **do**
6:    **for all** $\mathbf{w} \in Z^r$ **do**
7:       Compute $dist(\mathbf{r}, \mathbf{w}) = \|(\mathbf{r} - \mathbf{w}^T \mathbf{r} \mathbf{w} / \|\mathbf{w}\|^2)\|$
8:    **end for**
9:    $\pi(\mathbf{r}) = \mathbf{w} : \operatorname{argmin} dist(\mathbf{r}, \mathbf{w})$
10:    $d(\mathbf{r}) = dist(\mathbf{r}, \pi(\mathbf{r}))$
11:    $C_{\pi(\mathbf{r})} = C_{\pi(\mathbf{r})} \cup \mathbf{r}$
12:    $\rho_{\pi(\mathbf{r})} = \rho_{\pi(\mathbf{r})} + 1$
13: **end for**

Since the structure reference points are generated for reference lines, sometimes H (number of reference points) is less than N (population size). In this scenario, a few of individuals are selected from those lines which has $\rho_j > 1$. It is because the best individual from each cluster is already selected earlier, which cannot be copied again to P_{t+1}. Satisfying the conditions given in step 17 of Algorithm 2, a random reference line j is chosen and then select the best individual $\mathbf{x}$ associated to the reference line j such that $\mathbf{x} \notin P_{t+1}$. The selected individual is then copied to P_{t+1} and it is removed from R_t. The loop at step 16 of Algorithm 2 is active till $|P_{t+1}|$ become N.

It can be observed that the condition at step 3 of Algorithm 2 is imposed to select the nearest non-dominated individual from each cluster of all reference lines. It means that the diversity driven by the reference-points approach is maintained. Since every reference line is important for maintaining diversity among the individuals of a given population, the condition at step 12 of Algorithm 2 is imposed to select individuals for empty reference lines, which can be isolated and sometimes, dominated individual.

The computational complexity of the DoD approach remains same as NSGA-III that is $\max(O(N^2 \log^{M-2} N), O(N^2 M))$ when almost all individuals are associated with a single line. First association requires $O(N^2 M)$ operations and then, the non-dominated sorting for this cluster requires $O(N^2 \log^{M-2} N)$ operations.

4   Results and Discussion

The proposed DoD approach is compared with the existing EMO algorithms, such as NSGA-III [3] and MOEA/D [16] on DTLZ problems [4] with $M \in \{3, 5, 8, 10, 15\}$ objective test instances and WFG problems [6] with $M \in \{3, 5, 8, 10\}$ objective instances. For DTLZ problems, the number of decision variables is given as $n = M + k - 1$, where $k = 5$ for DTLZ1, and $k = 10$ for DTLZ2-4 problems. For WFG6-7 problems, the number of decision variables is set to $n = k + l$ in which the position-related variable is $k = 2 \times (M - 1)$, and the distance-related variable is $l = 20$. The inverse generalized distance (IGD) indicator [16] and hypervolume (HV) indicator [13] are used for the performance evaluation of EMO algorithms. All EMO algorithms are run for 20 times with different initial population. Moreover, a difference for statistical significance is tested using the Wilcoxon signed-rank test [14] at 5% significance level for the assessment of obtained results from competing EMO algorithms.

Table 1 presents the population sizes, divisions and a number of reference points for EMO algorithms. For more than 5-objective instances, the two-layered reference points are generated similar to [3]. The table also summarizes termination conditions for all problems, which is kept similar to [3].

Table 1. Input parameters for EMO algorithms.

Population				Termination				
M	Divisions	H	N	DTLZ1	DTLZ2	DTLZ3	DTLZ4	WFG6-7
3	12	91	92	400	250	1000	600	1000
5	6	210	210	600	350	1000	1000	1250
8	(3, 2)	156	156	750	500	1000	1250	1500
10	(3, 2)	275	276	1000	750	1500	2000	2000
15	(2, 1)	135	136	1500	1000	2000	3000	3000

Table 2 presents the IGD values obtained from three EMO algorithms. A smaller IGD value refers better performance. It can be seen that the DOD approach is superior to both EMO algorithms in DTLZ2, DTLZ4, and WFG7 instances. For DTLZ3, the DoD approach is found to be better in lower objective instances. NSGA-III is better than both EMO algorithms in WFG6 instances. Table 3 presents HV values in which it can be seen that the DoD approach is better than both EMO algorithms in almost all instances of DTLZ and WFG problems. Since HV values are close to one, the DoD approach showed its efficacy in selecting a diverse set of solutions.

Table 2. Best, median and worst IGD values obtained by DoD approach and other algorithms on DTLZ and WFG instances with different number of objectives. Best performances are highlighted in bold face with gray background. NSGA-III results are obtained from [3], and MOEA/D results are obtained using [5].

	M	NSGA-III	MOEA/D	DoD
DTLZ1	3	4.880E-04	2.607E-02	**3.333E-04**
		1.308E-03	4.713E-02+	**1.106E-03**
		4.880E-03	3.954E-01	5.770E-03
	5	5.116E-04	1.582E-02	5.567E-04
		9.799E-04	3.071E-02+	1.354E-03
		1.979E-03	6.377E-02	1.167E-02
	8	2.044E-03	1.798E-02	2.176E-03
		3.979E-03	2.721E-02+	**3.546E-03**
		8.721E-03	5.509E-02	9.393E-03
	10	2.215E-03	2.168E-02	2.219E-03
		3.462E-03	3.007E-02+	**3.034E-03**
		6.869E-03	4.202E-02	**6.482E-03**
	15	**2.649E-03**	4.782E-02	3.740E-03
		5.063E-03	5.338E-02-	2.778E-01
		1.123E-02	6.177E-02	3.878E-01
DTLZ2	3	1.262E-03	1.056E-02	**1.162E-03**
		1.357E-03	1.469E-02+	1.509E-03
		2.114E-03	2.243E-02	5.328E-03
	5	4.254E-03	1.321E-02	**3.797E-03**
		4.982E-03	1.675E-02+	**4.630E-03**
		5.862E-03	2.295E-02	**5.562E-03**
	8	1.371E-02	3.001E-02	**1.141E-02**
		1.571E-02	3.453E-02+	**1.410E-02**
		1.811E-02	4.046E-02	1.848E-02
	10	1.350E-02	2.509E-02	**1.116E-02**
		1.528E-02	3.974E-02+	**1.265E-02**
		1.697E-02	4.348E-02	**1.532E-02**
	15	1.360E-02	2.248E-02	**1.063E-02**
		1.726E-02	6.526E-02+	**1.304E-02**
		2.114E-02	1.917E-01	**1.686E-02**

	M	NSGA-III	MOEA/D	DoD
DTLZ3	3	9.751E-04	2.491E-02	**6.966E-04**
		4.007E-03	4.477E-01+	**2.144E-03**
		6.665E-03	1.691E+01	**5.898E-03**
	5	3.086E-03	2.311E-02	**1.698E-03**
		5.960E-03	2.303E-01+	**5.181E-03**
		1.196E-02	4.304E-01	7.762E-02
	8	**1.244E-02**	7.445E-02	1.828E-02
		2.375E-02	6.251E-01+	3.478E-02
		9.649E-02	1.151E+00	2.033E+00
	10	**8.849E-03**	4.514E-02	9.529E-03
		1.188E-02	2.744E-01+	1.579E-02
		2.083E-02	1.161E+00	2.598E-02
	15	1.401E-02	1.864E-01	**1.004E-02**
		2.145E-02	1.281E+00+	**1.672E-02**
		4.195E-02	1.300E+00	6.676E-01
DTLZ4	3	**2.915E-04**	7.446E-03	3.535E-04
		5.970E-04	5.307E-01+	**4.469E-04**
		4.286E-01	9.503E-01	**6.577E-04**
	5	9.849E-04	1.475E-02	**3.741E-04**
		1.255E-03	3.095E-02+	**4.632E-04**
		1.721E-03	6.050E-01	**5.623E-04**
	8	5.079E-03	3.161E-02	**3.123E-03**
		7.054E-03	2.936E-01+	**3.546E-03**
		6.051E-01	6.410E-01	**4.695E-03**
	10	5.694E-03	4.741E-02	**3.448E-03**
		6.337E-03	1.856E-01+	**4.252E-03**
		1.076E-01	3.959E-01	**5.031E-03**
	15	7.110E-03	5.447E-02	**5.404E-03**
		3.431E-01	2.656E-01+	**7.290E-03**
		1.073E+00	6.714E-01	**9.265E-03**

	M	NSGA-III	MOEA/D	DoD
WFG6	3	**4.828E-03**	7.550E-02	1.962E-02
		1.224E-02	8.163E-02	2.847E-02
		5.486E-02	1.242E-01	**3.633E-02**
	5	**5.065E-02**	3.159E-01	2.604E-02
		1.965E-02	4.418E-01	3.381E-02
		4.475E-02	5.407E-01	**4.237E-02**
	8	**1.009E-02**	9.031E-01	3.465E-02
		2.922E-02	9.362E-01	4.114E-02
		7.098E-02	9.716E-01	**5.024E-02**
	10	**1.060E-02**	9.487E-01	2.781E-02
		2.491E-02	1.008E+00	3.562E-02
		6.129E-02	1.031E+00	**4.480E-02**
	15	**1.368E-02**	1.120E+00	2.486E-02
		2.877E-02	1.240E+00	3.522E-02
		6.970E-02	1.250E+00	2.028E-01

	M	NSGA-III	MOEA/D	DoD
WFG7	3	2.789E-03	9.344E-02	**2.309E-03**
		3.692E-03	1.049E-01	**2.891E-03**
		4.787E-03	1.182E-01	**3.696E-03**
	5	8.249E-03	3.613E-01	**6.549E-03**
		9.111E-03	3.950E-01	**8.103E-03**
		1.050E-02	4.315E-01	2.224E-02
	8	2.452E-02	8.977E-01	**1.665E-02**
		2.911E-02	9.303E-01	**2.089E-02**
		6.198E-02	9.595E-01	**2.399E-02**
	10	3.228E-02	9.368E-01	**2.091E-02**
		4.292E-02	9.533E-01	**2.309E-02**
		9.071E-02	1.006E+00	**2.533E-02**
	15	**3.457E-02**	1.212E+00	8.945E-02
		5.450E-02	1.216E+00	5.559E-01
		8.826E-02	1.222E+00	6.990E-01

+,− and = indicate that DoD approach performs significantly better, significantly bad, and equivalent to the corresponding EMO algorithm.

Table 3. Best, median and worst HV values obtained by DoD approach and other algorithms on DTLZ and WFG instances with different number of objectives. Best performances are highlighted in bold face with gray background. NSGA-III results are obtained from [10], and MOEA/D results are obtained using [5].

	M	NSGA-III	MOEA/D	DoD		M	NSGA-III	MOEA/D	DoD
DTLZ1	3	9.73519E-01	9.66870E-01	**9.73627E-01**	DTLZ3	3	9.26480E-01	3.41019E-03	**9.26669E-01**
		9.73217E-01	9.57697E-01=	**9.73509E-01**			9.25805E-01	6.94952E-03=	**9.26328E-01**
		9.71931E-01	6.14190E-01	**9.73198E-01**			9.24234E-01	1.89791E-01	**9.25428E-01**
	5	9.98971E-01	9.98629E-01	**9.98981E-01**		5	9.90453E-01	9.90009E-01	**9.90565E-01**
		9.98963E-01	9.98330E-01=	**9.98971E-01**			9.90344E-01	9.76349E-01=	**9.90446E-01**
		9.98673E-01	9.97441E-01	**9.98942E-01**			9.89510E-01	9.43850E-01	**9.90256E-01**
	8	**9.99975E-01**	9.99645E-01	9.99974E-01		8	9.99300E-01	9.99122E-01	**9.99308E-01**
		9.93549E-01	9.99370E-01=	**9.99970E-01**			9.24059E-01	7.76470E-01=	**9.99253E-01**
		9.66432E-01	9.98375E-01	**9.99962E-01**			**9.04182E-01**	5.03871E-01	6.43785E-02
	10	9.99991E-01	9.99934E-01	**9.99998E-01**		10	**9.99921E-01**	9.99865E-01	9.99920E-01
		9.99985E-01	9.99875E-01=	**9.99997E-01**			**9.99918E-01**	9.99144E-01=	9.99916E-01
		9.99969E-01	9.99672E-01	**9.99994E-01**			**9.99910E-01**	5.10243E-01	9.99908E-01
DTLZ2	3	9.26660E-01	9.25292E-01	**9.26666E-01**	DTLZ4	3	9.26659E-01	9.26587E-01	**9.26774E-01**
		9.26536E-01	9.24412E-01=	**9.26632E-01**			9.26705E-01	8.00983E-01=	**9.26728E-01**
		9.26395E-01	9.22765E-01	**9.26497E-01**			7.99572E-01	5.00000E-01	**9.26716E-01**
	5	9.90459E-01	9.90426E-01	**9.90493E-01**		5	**9.91102E-01**	9.90611E-01	9.90586E-01
		9.90400E-01	9.90271E-01=	**9.90460E-01**			9.90413E-01	9.90564E-01=	**9.90575E-01**
		9.90328E-01	9.90013E-01	**9.90431E-01**			9.90156E-01	9.12068E-01	**9.90570E-01**
	8	9.99320E-01	9.99323E-01	**9.99335E-01**		8	9.99363E-01	**9.99383E-01**	9.99364E-01
		9.78936E-01	9.99315E-01=	**9.99327E-01**			9.99361E-01	9.99131E-01=	**9.99364E-01**
		9.19680E-01	9.99298E-01	**9.99319E-01**			9.94784E-01	9.86416E-01	**9.99363E-01**
	10	9.99918E-01	**9.99919E-01**	9.99919E-01		10	9.99915E-01	**9.99926E-01**	9.99924E-01
		9.99916E-01	9.99876E-01=	**9.99918E-01**			9.99910E-01	9.99917E-01=	**9.99923E-01**
		9.99915E-01	9.99868E-01	**9.99916E-01**			9.99827E-01	9.99430E-01	**9.99923E-01**

	M	MOEA/D	DoD		M	MOEA/D	DoD
WFG6	3	8.90380E-01	**9.11580E-01**	WFG7	3	9.08050E-01	**9.25330E-01**
		8.83410E-01=	**9.04850E-01**			8.96660E-01=	**9.24910E-01**
		8.10510E-01	**8.98480E-01**			8.67760E-01	**9.24030E-01**
	5	9.13140E-01	**9.68380E-01**		5	9.33350E-01	**9.87230E-01**
		8.16580E-01=	**9.61910E-01**			9.03530E-01=	**9.86740E-01**
		7.14410E-01	**9.54090E-01**			8.60340E-01	**9.85380E-01**
	8	7.08280E-01	**9.72600E-01**		8	7.38580E-01	**9.95360E-01**
		6.55160E-01=	**9.63970E-01**			6.73050E-01=	**9.93610E-01**
		6.08160E-01	**9.52200E-01**			6.22420E-01	**9.91490E-01**
	10	7.69000E-01	**9.75300E-01**		10	8.03100E-01	**9.96890E-01**
		6.71860E-01=	**9.66450E-01**			7.81430E-01=	**9.96310E-01**
		6.18670E-01	**9.58170E-01**			6.85380E-01	**9.95490E-01**

+,− and = indicate that DoD approach performs significantly better, significantly bad, and equivalent to the corresponding EMO algorithm.

The non-dominated solutions obtained corresponding to the median IGD value run are shown in Fig. 3 for DTLZ problems. A well-distributed front can be seen from the DoD approach, whereas MOEA/D is unable to generate similar fonts for DTLZ1 and DTLZ4 problems. Figure 4 presents parallel coordinates for 10-objective DTLZ problems. It can be seen that a well-distributed set of solutions is generated by the DoD approach against MOEA/D.

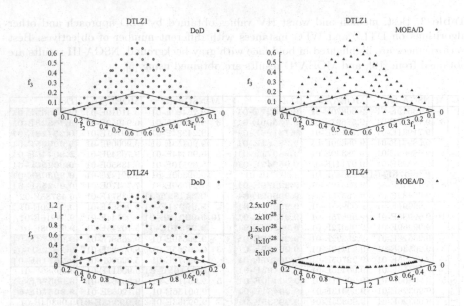

Fig. 3. Non-dominated solutions obtained using the DoD approach and MOEA/D for DTLZ1 and DTLZ4 problems.

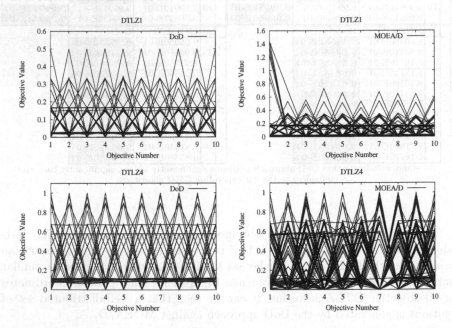

Fig. 4. Parallel coordinates of non-dominated front obtained from the DoD approach and MOEA/D for DTLZ1 and DTLZ4 problems.

5   Conclusions

The purpose of DoD approach was to select a diverse set of individuals in the environmental selection using the reference-points-based framework. Since almost all individuals became non-dominated for many-objective optimization, the DoD approach showed its superiority over the environmental selection of NSGA-III by solving many test instances of DTLZ and WFG problems. The IGD values obtained using the DoD approach were found to be better than NSGA-III in many instances and better in all instances against MOEA/D. The HV values indicated that the DoD approach served its purpose of selecting diverse individuals and showed its efficacy against NSGA-III and MOEA/D in almost all test instances. In future, the DoD approach can be improved further to design selection rules that can emphasis non-dominated individuals over dominated individuals without losing its core idea.

References

1. Barakat, N., Sharma, D.: Evolutionary multi-objective optimization for bulldozer and its blade in soil cutting. Int. J. Manag. Sci. Eng. Manag. 1–11 (2018, to appear). https://doi.org/10.1080/17509653.2018.1500953

2. Das, I., Dennis, J.E.: Normal-boundary intersection: a new method for generating the Pareto surface in nonlinear multicriteria optimization problems. SIAM J. Optim. 8(3), 631–657 (1998). https://doi.org/10.1137/S1052623496307510

3. Deb, K., Jain, H.: An evolutionary many-objective optimization algorithm using reference-point-based nondominated sorting approach, part i: solving problems with box constraints. IEEE Trans. Evol. Comput. 18(4), 577–601 (2014). https://doi.org/10.1109/TEVC.2013.2281535

4. Deb, K., Thiele, L., Laumanns, M., Zitzler, E.: Scalable test problems for evolutionary multiobjective optimization. In: Abraham, A., Jain, L., Goldberg, R. (eds.) Evolutionary Multiobjective Optimization. AI&KP, pp. 105–145. Springer, London (2005). https://doi.org/10.1007/1-84628-137-7_6

5. Durillo, J.J., Nebro, A.J.: jMetal: a Java framework for multi-objective optimization. Adv. Eng. Softw. 42, 760–771 (2011). http://www.sciencedirect.com/science/article/pii/S0965997811001219, https://doi.org/10.1016/j.advengsoft.2011.05.014

6. Huband, S., Hingston, P., Barone, L., While, L.: A review of multiobjective test problems and a scalable test problem toolkit. IEEE Trans. Evol. Comput. 10(5), 477–506 (2006). https://doi.org/10.1109/TEVC.2005.861417

7. Ibrahim, A., Rahnamayan, S., Martin, M.V., Deb, K.: EliteNSGA-III: an improved evolutionary many-objective optimization algorithm. In: 2016 IEEE Congress on Evolutionary Computation (CEC), pp. 973–982, July 2016. https://doi.org/10.1109/CEC.2016.7743895

8. Jiang, S., Yang, S.: A strength Pareto evolutionary algorithm based on reference direction for multiobjective and many-objective optimization. IEEE Trans. Evol. Comput. 21(3), 329–346 (2017). https://doi.org/10.1109/TEVC.2016.2592479

9. Jiang, S., Yang, S.: Convergence versus diversity in multiobjective optimization. In: Handl, J., Hart, E., Lewis, P.R., López-Ibáñez, M., Ochoa, G., Paechter, B. (eds.) PPSN 2016. LNCS, vol. 9921, pp. 984–993. Springer, Cham (2016). https://doi.org/10.1007/978-3-319-45823-6_92

10. Li, K., Deb, K., Zhang, Q., Kwong, S.: An evolutionary many-objective optimization algorithm based on dominance and decomposition. IEEE Trans. Evol. Comput. **19**(5), 694–716 (2015). https://doi.org/10.1109/TEVC.2014.2373386

11. Liao, X., Li, Q., Zhang, W., Yang, X.: Multiobjective optimization for crash safety design of vehicle using stepwise regression model. Struct. Multi. Optim. **35**(6), 261–569 (2008)

12. Purshouse, R.C., Fleming, P.J.: On the evolutionary optimization of many conflicting objectives. IEEE Tran. Evol. Comput. **11**(6), 770–784 (2007). https://doi.org/10.1109/TEVC.2007.910138

13. While, L., Bradstreet, L., Barone, L.: A fast way of calculating exact hypervolumes. IEEE Trans. Evol. Comput. **16**(1), 86–95 (2012). https://doi.org/10.1109/TEVC.2010.2077298

14. Wilcoxon, F.: Individual comparisons by ranking methods. Biom. Bull. **1**(6), 80–83 (1945)

15. Yuan, Y., Xu, H., Wang, B., Yao, X.: A new dominance relation-based evolutionary algorithm for many-objective optimization. IEEE Trans. Evol. Comput. **20**(1), 16–37 (2016). https://doi.org/10.1109/TEVC.2015.2420112

16. Zhang, Q., Li, H.: MOEA/D: a multiobjective evolutionary algorithm based on decomposition. IEEE Trans. Evol. Comput. **11**(6), 712–731 (2007). https://doi.org/10.1109/TEVC.2007.892759

A Two-Stage Evolutionary Algorithm for Many-Objective Optimization

Yi Wu[1], Bin Li[1(✉)], Sanchao Ding[2], and Yinda Zhou[1]

[1] School of Information Science and Technology,
University of Science and Technology of China, Hefei, Anhui, China
986879902@qq.com, binli@ustc.edu.cn, zhouyd@mail.ustc.edu.cn
[2] School of Computer Science and Technology,
University of Science and Technology of China, Hefei, Anhui, China
dingsc@mail.ustc.edu.cn

Abstract. Many-objective optimization is of top challenge in multi-objective optimization research community. Various many-objective optimization problems exhibit difference characteristics and require the algorithm to be effective and robust to deal with as many as possible of them. In this paper, a two-stage evolutionary algorithm for many-objective optimization is presented with the purpose to be effective and robust to various problems. In the first stage of the algorithm, NSGA-III is adopted to explore the shape of the Pareto Front, a variant of NSGA-III is designed to select solutions for better diversity by maximizing the angle to their own neighbor solution set. When the improvement of solutions with good diversity slows down to a certain degree, the first stage is terminated and the weight vectors are adapted for the use in the second stage. In the second stage, MOEA/DD is adopted to converge solutions to the Pareto Front. The proposed algorithm is tested on DTLZ test suite and WFG test suite with up to 15 objectives, and compared with two related state-of-art algorithms, NSGA-III and MOEA/DD. Experiment results show that while NSGA-III and MOEA/DD work well on different types of problems, the proposed algorithm has competitive or better performance on all problems considered in this paper.

Keywords: Many-objective optimization · Two-stage algorithm ·
Convergence · Diversity · Weight vector adaptation

1 Introduction

In real applications, the optimization problems often have more than one objective. Multi-objective optimization problems (MOPs) can be formulated as

$$
\begin{aligned}
\min \quad & \mathbf{F}(\mathbf{x}) = (f_1(\mathbf{x}), \ldots, f_m(\mathbf{x})) \\
s.t. \quad & \mathbf{x} \in \Omega
\end{aligned}
\tag{1}
$$

© Springer Nature Switzerland AG 2019
K. Deb et al. (Eds.): EMO 2019, LNCS 11411, pp. 291–304, 2019.
https://doi.org/10.1007/978-3-030-12598-1_24

where $\Omega = \prod_{i=1}^{n} [a_i, b_i] \subseteq \mathbb{R}^n$ is the decision (variable) space, $\mathbf{x} = (x_1, \ldots, x_n)^T \in \Omega$ is a candidate solution. $\mathbf{F} : \Omega \to \mathbb{R}^m$ constitutes m conflicting objectives, and $\mathbb{R}^m$ denotes the objective space.

Over the last two decades, evolutionary algorithms (EAs) have shown excellent performance on MOPs [8]. These traditional evolutionary multi-objective algorithms (MOEAs), like NSGA-II [5] and MOEA/D [18], are effective and efficient for solving MOPs with two or three objectives, while when the number of objectives increases up to more than 5, their performance would deteriorate dramatically. Many-objective optimization problems (MaOPs) have been one of the popular topics in MOEAs research community in recent years. A variety of approaches can be roughly divided into three categories [3].

The first category covers various convergence-driven approaches. With the number of objectives increasing, most traditional MOEAs do not have enough ability to distinguish solutions for convergence. Dominance relationship is modified to steer solutions toward the Pareto Front (PF), like ε-dominance [6]. In GrEA [16], grid dominance is adopted to enhance the convergence while maintaining diversity by three grid criteria. Some add other convergence criteria in addition to the original Pareto dominance comparison. A knee point-driven evolutionary algorithm (KnEA) [19] is proposed in which preference over knee points can bring larger performance indicator value. A shift-based density estimation strategy (SDE) [10] is proposed to put solutions with poor convergence into crowded regions.

The second category covers various decomposition-based approaches. One idea is to divide a MOP into a number of single-objective problems and solve them in a collaborative manner. A number of collaboration strategies and aggregate functions have been introduced. A scalarization approach named angle-penalized distance is adopted to balance convergence and diversity in [3]. Another idea is to divide a MOP into a series of small-scale sub-MOPs. In MOEA/D-M2M [12], a sub-MOP, defined by a weight vector, has its own solutions. Some algorithms, such as NSGA-III [4], MOEA/DD [9] and RVEA [3], use both dominance comparison and decomposition.

The third category covers performance indicator-based approaches by using the value of the indicator to guide the search process. The S-metric selection-based evolutionary multi-objective algorithm (SMS-EMOA) [2] and an algorithm for fast hypervolume-based many-objective optimization (HypE) [1] are classic in this category. These approaches can balance both convergence and diversity for MaOPs since these indicators are originally designed to measure convergence and diversity. However, when the number of objectives is large, the expensive cost is unavoidable for calculating the value of the performance indicator.

There are still a few other categories of approaches showing competitive performance. Some algorithms have been proposed for generating solutions toward users' preferred regions like [15]. An improved two-archive algorithm (Two_Arch2) [7] has been proposed to obtain good convergence, good diversity and acceptable computational cost. Objective reduction in many-objective optimization [17] is to avoid great difficulties [4] of MaOPs.

Weight vectors are adopted in most algorithms for MaOPs. The uniform distribution of weight vectors does not guarantee that the final approximate PFs are uniformly distributed. To obtain the solutions with uniform distribution, weight vector adaptation strategies have been designed and adopted in algorithms, such as RVEA [3] and MOEA/D-AM2M [11].

In this paper, a two-stage evolutionary algorithm for many-objective optimization (TSMOEA) is proposed. In the first stage of TSMOEA, NSGA-III and one of its variants are adopted to explore the shape of the PF. In the second stage, MOEA/DD with adapted weight vectors is adopted to steer population convergence to PF while maintaining diversity.

The rest of this paper is organized as follows. In Sect. 2, related works are introduced. A two-stage evolutionary algorithm for many-objective optimization is proposed in Sect. 3. Experimental results are presented in Sect. 4 to compare the performance of the proposed algorithm with two related algorithms for solving different classes of MaOPs. The conclusion is given in Sect. 5.

2 Related Works

2.1 NSGA-III

In each generation of NSGA-III [4], N individuals are produced once then N individuals are selected from the whole population. With the Pareto dominance relationship as the first selection criteria, the non-dominated solutions are more likely to be selected as the parents of the next generation. When the first criterion fails to provide sufficient selection pressure, the diversity criterion as second selection criteria is activated to distinguish solutions, which is called the Active Diversity Promotion (ADP) phenomenon [13].

The approximate PF obtained by NSGA-III is well distributed since objective normalization can be carried out dynamically in each generation. However, some experimental observations [13] have shown that the APD phenomenon can maintain good diversity, but it may lead to poor convergence of the final solutions.

2.2 MOEA/DD

In each generation of MOEA/DD [9], an offspring is produced and used to update the population one by one, which is different from NSGA-III. Pareto dominance relationship and penalty-based boundary intersection (PBI) approach are used successively. To maintain diversity, the solution, associated with an isolated subregion, should be selected and kept even if it is in the last non-domination level.

The performance of MOEA/DD is excellent in terms of convergence and diversity when all the weight vectors are evenly distributed in PF, instead of the whole objective space. MOEA/DD is not suitable for solving degenerated, scalable MaOPs with disconnected PFs, like WFG3, since some solutions, away from the PF, are kept for diversity preservation, but they have no need.

3 Proposed Algorithm: TSMOEA

Previous experimental investigations have shown that different approaches have different search fortes, and none of them outperforms the others on all types of problems. For an unknown problem, a MOEA should be effective and robust, and the algorithm should be adaptable to keep efficiency at different stages. Based on the merits and demerits of NSGA-III and MOEA/DD shown on various types of problems, we can combine them into one framework to balance both convergence and diversity. In this paper, The two-stage evolutionary algorithm for MaOPS (TSMOEA) is presented following this idea.

In the first stage of TSMOEA, NSGA-III and one of its variants are adopted to explore the shape of the PF. A variant of NSGA-III is presented to maximize the angle between solutions during diversity promotion, which is good for weight vector adaptation. At the end of the first stage, weight vector adaptation is activated and weight vectors are updated for the second stage. In the second stage, MOEA/DD is adopted to converge solutions to PF while maintaining diversity. To obtain more accurate shape of PF, the first stage should be carried out as many generations as possible. However, since the computational resources are limited and the convergence must be considered, the condition to trigger the second stage is of great importance.

Algorithm 1. The framework of TSMOEA

Input: the maximal number of generations Gen_max; $Gen = 1$; a population size N
    ($Gen\_max \times N$ = the maximal number of fitness evaluations); $first\_stage = true$;
    the generation Gen_J to trigger the judgement operation.
Output: Population **P**
1: $[\mathbf{P}, \mathbf{W}, \mathbf{B}]$ = Initiallization(N) //**W**: weight vector set, **B**: neighborhood index set
2: **while** $Gen \leq Gen\_max$ **do**
3:    **if** $first\_stage$ **then**
4:      **if** $Gen < Gen\_J$ **then**
5:        Reproduction+Selection by NSGA-III
6:      **else**
7:        Reproduction+Selection by a variant of NSGA-III
8:        $first\_stage$ = Judgement(**P**) //if the judgement is activated
9:        **if** !$first\_stage$ **then**
10:            $[\mathbf{W}, \mathbf{B}]$= Weight_vector_adaptation(**W**, **P**)
11:        **end if**
12:      **end if**
13:    **else**
14:      Reproduction+Selection by MOEA/DD
15:    **end if**
16:    $Gen = Gen + 1$
17: **end while**
18: **return P**

3.1 Framework of Proposed Algorithm

The framework of TSMOEA is presented in Algorithm 1. In initialization, N solutions are initialized at random and N weight vectors are generated based on the two-layer weight vector generation method [9]. Two outstanding algorithms, NSGA-III and MOEA/DD, are adopted to implement the search in the two stages respectively. For better diversity, a variant of NSGA-III runs later in the first stage. The judgement of whether to end the first stage is activated per several generations when $Gen > Gen\_J$ in the first stage. Once the condition parameter $first\_stage = false$, weight vector adaptation is activated.

3.2 A Variant of NSGA-III

In the diversity promotion procedure of NSGA-III [4], the values of all objectives are normalized to $[0, 1]$ and solutions in last acceptable non-domination level ($\mathbf{F}_l$) are selected only by the niche count (ρ) of weight vectors. One weight vector with minimum niche count is chosen at random, then one of solutions associated with it is also chosen at random if niche count is non-zero. The selection operation may cause poor diversity in a subregion in objective space if a few very close solutions, associated with the same weight vector, are chosen when solving degenerated MaOPs especially. The procedure of the variant of NSGA-III remains same as the original NSGA-III, but the diversity promotion of $\mathbf{F}_l$ is different which is detailed in Algorithm 2.

Algorithm 2. A Variant of Diversity Promotion of $\mathbf{F}_l$

1: $[\mathbf{S}, \mathbf{F}_l]$=dominance_selection($\mathbf{P}$) // $\mathbf{S}$: selected set
2: **while** $|\mathbf{S}| < N$ **do**
3:     Compute niche count ρ for $\mathbf{S}$
4:     **if** $min(\rho) == 0$ **then**
5:         Choose a weight vector with minimum niche count at random
6:         Add the solution s with minimum d_2 to $\mathbf{S}$
7:     **else**
8:         **for** $i = 1$ to j ($No.$ of weight vectors with $min(\rho)$) **do**
9:             Find the chosen solution subset in its neighborhood subregions
10:            **for** $k = 1$ to h ($No.$ of solutions in i-th subregion and $\mathbf{F}_l$) **do**
11:                Calculate the smallest angle between a solution and the corresponding solution subset
12:            **end for**
13:            Find i-th solution with the largest angle from h solutions
14:        **end for**
15:        Add the solution s with the largest angle from j solutions to $\mathbf{S}$
16:    **end if**
17:    $\rho_{\bar{j}} = \rho_{\bar{j}} + 1$ //the chosen weight vector index is $\bar{j}$
18:    $\mathbf{F}_l = \mathbf{F}_l \setminus s$
19: **end while**

If the minimum niche count is zero, a solution with minimum vertical distance d_2 to its weight vector is chosen (lines 5–6). Otherwise, we choose the solution with the largest angle to the chosen solution subset in its neighborhood subregions with $min(\rho)$. Considering the scalable MaOPs, the acute angle between two solutions is calculated in original objective space, not in normalized objective space. For i-th weight vector with $min(\rho)$, we calculate the acute angle between a solution in it and each member of the corresponding chosen solution subset, and keep the smallest acute angle as the solution's angle. The solution with the largest angle is regarded as the best solution of the subregion to choose, called i-th solution (line 17). Then a solution with largest angle from the j subregions is added to the selected set $\mathbf{S}$. The above steps are repeated until N solutions are selected. The variant of NSGA-III runs later in the first stage. In this paper, it is activated when $Gen \geq Gen\_J$.

3.3   Switching from 1st Stage to 2nd Stage

In terms of convergence, it has been observed [19] that NSGA-III improves slowly in the later generations due to ADP phenomenon [13]. From the related research results, we found that the shape of PF can be approximately determined by NSGA-III generally after 70% of maximum generations, so the judgement operation can be triggered at this time, that is $Gen\_J = 70\% \times Gen\_max$. However, the performance of the same algorithm is usually different on different problems. To adaptively make judgment that if the first stage should be ended, the hypervolume (HV) performance metric is adopted to measure the performance, mainly convergence.

We adopt $\mathbf{z}^{nad}$ of the Gen_J generation as the reference point for HV metric of the judgement, in which $\mathbf{z}^{nad}$ is defined as a nadir point with maximum objective values. To reduce the computational cost of HV metric, the number of interval generations to trigger the judgement is set to $2 \times m$. If the slope of HV value is less than $slope\_HV$, we end the first stage. The parameter $slope\_HV$ is of great importance and related to the problems. If it is too small, the proposed algorithm is similar to NSGA-III. In contrast, it may end the first stage early. In this paper, $slope\_HV = 0.005 \times m$.

3.4   Weight Vector Adaptation Strategy

At the end of the first stage, niche count of every weight vector is calculated in original objective space. To deal with the general problems including scalable MaOPs, weight vectors need to be scaled by (2) like the inverse operation of objective normalization in NSGA-III.

$$\mathbf{w}^i_{scaled} = \mathbf{w}^i \circ (\mathbf{z}^{nad} - \mathbf{z}^*) \tag{2}$$

where $\mathbf{z}^*$ is an ideal point with minimum objective values, the $\circ$ operator denotes the Hadamard product. Assume that there are K legal weight vectors whose niche count is non-zero. If K and N are approximately equal, there is no need to update weight vectors and weight vector set $\mathbf{W} = \mathbf{W}_{scaled}$, $\mathbf{W}_{scaled} = \{\mathbf{w}_{scaled}^1, \ldots, \mathbf{w}_{scaled}^N\}^T$. Otherwise, weight vector adaptation strategy is to add and delete vectors. The procedure is detailed in Algorithm 3.

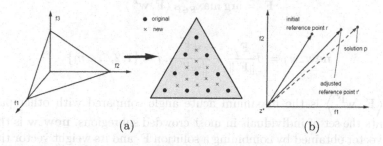

Fig. 1. Illustration of addition. (a) by reference points. (b) by a initial reference point and the solution.

Add new vectors evenly around the every legal vector. For instance, for $m = 3$ objectives, there are some new vectors evenly around a vector as shown in the Fig. 1(a). If the niche count of a weight vector is non-zero, we place m new vectors evenly around it in the hyper-plane, where the distance of the two new points equals the distance of the two original adjacent points generated in initialization procedure. We only need the unique vectors with all non-zero values. Assume that there are $K1$ vectors after deleting the new vectors whose niche counts are zero. We need to ensure that there are exactly N weight vectors for MOEA/DD. If $K1 > N$, some new vectors are randomly selected to delete. When $K1 < N$, we should add other vectors until the number of weight vectors is N.

Algorithm 3. Weight Vector Adaptation Strategy

1: Calculate K weight vectors whose niche counts are non-zero
2: Add new vectors evenly around the legal vectors
3: Delete the new vectors whose niche counts are 0 // $K1$ vectors after deletion
4: Delete $max(0, K1 - N)$ new vectors at random.
5: **for** $i = 1$ to $N - K1$ **do**
6:     $\mathbf{F}' = \arg\max_{\mathbf{F} \in \bar{\mathbf{P}}} \langle \mathbf{F}, \mathbf{w}^{\mathbf{F}} \rangle$
7:     **for** $j = 1$ to m **do**
8:         $new\_w_j^i = \beta \frac{\mathbf{F}_j'}{\|\mathbf{F}'\|} + \frac{\mathbf{w}_j^{\mathbf{F}'}}{\|\mathbf{w}^{\mathbf{F}'}\|}$
9:     **end for**
10: **end for**
11: **for** $i = 1$ to N **do**
12:     $\mathbf{B}^i = \{i_1, \ldots, i_T\}$, where $\mathbf{w}^{i_1}, \cdots, \mathbf{w}^{i_T}$ are the T closest weight vectors to $\mathbf{w}^i$
13: **end for**

The approximate PF represents partly the real PF, the crowded subregions need more attention. The solution in most crowded subregion, with maximum acute angle between it and its weight vector, is selected to combine with the weight vector to produce a new vector, which is shown in the Fig. 1(b). Combination method is define as

$$\mathbf{F}' = \arg\max_{\mathbf{F} \in \bar{P}} \left\langle \mathbf{F}, \mathbf{w}^{\mathbf{F}} \right\rangle \tag{3}$$

$$\mathbf{new\_w}_j = \beta \frac{\mathbf{F}'_j}{\|\mathbf{F}'\|} + \frac{\mathbf{w}_j^{\mathbf{F}'}}{\|\mathbf{w}^{\mathbf{F}'}\|}, j = \{1, 2, \ldots, m\} \tag{4}$$

where $\left\langle \mathbf{F}', \mathbf{w}^{\mathbf{F}'} \right\rangle$ is the maximum acute angle compared with other pairs. $\bar{P}$ represents the set of individuals in most crowded subregions. $\mathbf{new\_w}$ is the new weight vector obtained by combining a solution $\mathbf{F}'$ and its weight vector through (4). The weight coefficients of (4) are great different because the solutions represent the PF approximately, the weight vectors should be consistent with them greatly, but not equal to them. Based on PBI function in MOEA/DD, the solutions are hard to evolve if weight vectors are solutions. Finally, we need to update the neighborhood index set $\mathbf{B}$ through weight vectors for MOEA/DD. In this paper, $\beta = m - 1$.

4   Experimental Results

We compare the performance of the proposed algorithm[1] with those of NSGA-III and MOEA/DD. The hypervolume (HV) metric [20] is chosen as a performance metric. Let $\mathbf{z}^r = (\mathbf{z}_1^r, \ldots, \mathbf{z}_m^r)$ be a reference point for HV metric that is dominated by all Pareto optimal (objective) vectors. HV metric measures the hypervolume dominated by the obtained solutions in objective space and bounded by $\mathbf{z}^r$. The experiments are conducted on the recently developed software platform PlatEMO [14].

4.1   Setting

Several parameters are summarized as follows. Unspecified parameters use the default values of PlatEMO.

- The number of objectives $m \in \{3, 5, 8, 10, 15\}$ and the corresponding population size $N = \{91, 210, 156, 275, 135\}$. DTLZ1 to DTLZ4 from the DTLZ test suite are chosen for our experiment. The number of generations for different DTLZ test problems are shown in Table 1. In addition, WFG1 to WFG9 from WFG test suite are also chosen. The number of objective $m \in \{3, 5, 8, 10\}$, and the number of generations for different WFG test problems are all 1000.
- The simulated binary crossover (SBX) operator is used for crossover, and polynomial mutation is applied for mutation, with distribution indexes both set to 20.

---

[1] https://github.com/wuyiaishenghuo/TSMOEA.

Table 1. Number of generations for different DTLZ test problems

Problem	m = 3	m = 5	m = 8	m = 10	m = 15	Problem	m = 3	m = 5	m = 8	m = 10	m = 15
DTLZ1	400	600	750	1,000	1,500	DTLZ2	250	350	500	750	1,000
DTLZ3	1,000	1,000	1,000	1,500	2,000	DTLZ4	600	1,000	1,250	2,000	3,000

- In MOEA/DD, the penalty parameter θ of the PBI function is initially set to 5. If the maximum value of two conflicting objectives varies greatly, it may cause the objective with larger maximum value to lose larger value and poor diversity. In this case, we need to increase the value of θ ($\theta \leq 8$) for diversity. However, in real problems, we tend to prefer internal solutions, and we should not be overly interested in diversity or extreme solutions. The penalty parameter should be determined by the problem.
- The neighborhood size T is set to $N/10$, same in a variant of NSGA-III; the neighborhood selection probability δ is set to 0.9.
- The final results are obtained by executing 20 independent runs of each algorithm.

4.2 Performance Comparisons on DTLZ Test Suite

The statistical results of the HV values of DTLZ test suite obtained by the three algorithms over 20 independent runs are summarized in Table 2, where the best results are highlighted with gray background. It can be seen that TSMOEA, together with MOEA/DD, shows best performance on the DTLZ1, DTLZ2 and DTLZ4 test problems in general. The performance of MOEA/DD is so excellent since the features of the four test problems are non-scalable MaOPs with connected PFs. Compared with MOEA/DD, the number of the instances that the proposed algorithm shows significantly better/worse performance using the Wilcoxon rank sum test (0.05 significance level) are 1 and 4 among 20 instances respectively. NSGA-III is poor in convergence compared to the above two algorithms. In terms of standard variance, it is the fact that the performance of NSGA-III is not very stable on DTLZ2 to DTLZ4, especially on high dimensional instances. Compared with NSGA-III, the number of the instances that the proposed algorithm shows significantly better/worse performance are 13 and 0 among 20 instances respectively.

For DTLZ1 and DTLZ3, which are multi-modal problems, the three algorithms perform closely on three- to ten-objective DTLZ1 instances and three- to five-objective DTLZ3 instances. NSGA-III performs worst when the number of objective is large, especially on DTLZ3 instances. DTLZ2 and DTLZ4 problems are concave and uni-modal. NSGA-III performs worst and unsteadily on almost all instances because of poor convergence. TSMOEA performs worse than MOEA/DD on three-objective DTLZ4 instance due to NSGA-III.

Table 2. HV Values (Mean and Sd) Obtained by TSMOEA, NSGA-III and MOEA/DD on DTLZ1-DTLZ4 Instances with Different Number of Objectives. Best Performance is Highlighted in Bold Face with Gray Background.

	m	TSMOEA	NSGA-III	MOEA/DD		TSMOEA	NSGA-III	MOEA/DD
DTLZ1	3	1.3977e-1 ((2.88e-4)	1.3970e-1 (3.16e-4)≈	**1.3980e-1 (2.78e-4)≈**	DTLZ2	**7.4433e-1 (1.93e-4)**	7.4382e-1 (2.82e-4)−	7.4429e-1 (1.49e-4)≈
	5	4.9311e-2 (6.97e-6)	4.9309e-2 (9.24e-6)≈	**4.9314e-2 (8.35e-6)≈**		**1.3082e+0 (7.41e-4)**	1.3075e+0 (5.39e-4)−	1.3079e+0 (5.55e-4)≈
	8	8.3503e-3 (7.51e-6)	8.3511e-3 (6.16e-6)≈	**8.3525e-3 (5.51e-7)**		1.9795e+0 (5.50e-4)	1.9659e+0 (4.12e-3)−	**1.9800e+0 (6.71e-4)≈**
	10	2.5321e-3 (1.54e-7)	2.5320e-3 (5.45e-7)≈	**2.5321e-3 (3.71e-8)≈**		**2.5153e+0 (4.72e-4)**	2.5042e+0 (5.20e-4)−	2.5152e+0 (3.58e-4)≈
	15	**1.2747e-4 (1.36e-9)**	1.2648e-4 (1.56e-6)−	1.2743e-4 (5.49e-9)−		4.1381e+0 (3.88e-4)	4.1371e+0 (1.76e-3)−	**4.1382e+0 (4.10e-4)≈**
DTLZ3	3	7.3912e-1 (4.67e-3)	7.3833e-1 (2.83e-3)≈	**7.4017e-1 (2.31e-3)≈**	DTLZ4	7.3033e-1 (6.48e-2)	6.8654e-1 (1.30e-1)−	**7.4483e-1 (1.57e-5)+**
	5	1.3061e+0 (2.83e-3)	1.3058e+0 (2.10e-3)≈	**1.3072e+0 (2.58e-3)≈**		**1.3088e+0 (6.43e-4)**	1.3082e+0 (7.95e-4)≈	1.3087e+0 (3.39e-4)≈
	8	1.9697e+0 (5.04e-2)	1.9003e+0 (4.73e-1)−	**1.9756e+0 (5.91e-3)+**		**1.9808e+0 (5.99e-4)**	1.9701e+0 (6.26e-4)−	1.9804e+0 (5.89e-4)≈
	10	2.5123e+0 (6.20e-3)	2.5003e+0 (1.83e-3)−	**2.5153e+0 (4.93e-4)+**		**2.5156e+0 (5.34e-4)**	2.5087e+0 (9.78e-4)−	2.5155e+0 (5.17e-4)≈
	15	4.1306e+0 (3.62e-2)	4.0581e+0 (3.17e-2)−	**4.1382e+0 (5.34e-4)+**		4.1378e+0 (8.22e-4)	4.1277e+0 (7.28e-3)−	**4.1380e+0 (5.81e-4)≈**

$+/-$: TSMOEA shows significantly worse/better performance in the comparison.
$\approx$: There is no significant difference between the compared results.

4.3   Performance Comparisons on WFG Test Suite

The HV values of the three algorithms on WFG test suite are shown in Table 3. WFG problems are scalable to any objective. It shows that TSMOEA and NSGA-III perform best on WFG test suite. TSMOEA has shown the most competitive performance on WFG8 and low dimensional objective space instances ($m \leq 5$). NSGA-III has an overwhelming advantage over both of WFG1 and WFG2 problems. Compared with NSGA-III, the number of the instances that the proposed algorithm shows significantly better/worse performance are 5 and 7 among 36 instances respectively. However, the performance of MOEA/DD is poor on all WFG problems. Due to the inappropriate predefined weight vectors, MOEA/DD does not show its advantages.

WFG1 is a problem with flat bias and a mixed structure of the PF. WFG2 is a test problem with a disconnected PF. NSGA-III outperforms on all instances except one instance, and has best diversity. The proposed algorithm outperforms on three-objective WFG2 instances because of good convergence compared with NSGA-III. WFG3 is a difficult problem with a degenerated PF and the decision variables are non-separable. Hence, predefined weight vectors, with uniform distribution in $\mathbb{R}^m$, can greatly waste computational resources for diversity in MOEA/DD.

WFG4 to WFG9 share the same PF shape in the objective space, but have different difficulties in the decision space. TSMOEA and NSGA-III show generally competitive performance on WFG4 to WFG9. WFG8 and WFG9 are

Table 3. HV Values (Mean and Sd) Obtained by TSMOEA, NSGA-III and MOEA/DD on WFG1-WFG9 Instances with Different Number of Objectives.

	m	TSMOEA	NSGA-III	MOEA/DD			TSMOEA	NSGA-III	MOEA/DD
WFG1	3	**5.9732e+1** (2.20e-1)	**5.9975e+1** (4.45e-2)+	5.6854e+1 (3.78e+0)−	WFG2		**5.9675e+1** (4.07e-2)	5.9484e+1 (4.66e-2)−	5.9111e+1 (1.49e-1)−
	5	6.0104e+3 (6.77e+0)	**6.0405e+3** (9.63e-1)+	5.9357e+3 (7.33e+1)−			6.1371e+3 (4.26e+0)	**6.1656e+3** (2.87e+0)+	6.0635e+3 (2.52e+1)−
	8	2.0667e+7 (2.82e+4)	**2.0712e+7** (6.60e+3)+	2.0248e+7 (6.13e+5)−			2.2034e+7 (3.24e+4)	**2.2058e+7** (4.81e+4)+	2.1253e+7 (1.27e+5)−
	10	8.6313e+9 (9.36e+6)	**8.6612e+9** (2.02e+6)+	8.6199e+9 (8.91e+6)−			9.5988e+9 (1.29e+7)	**9.6136e+9** (1.27e+7)+	9.2355e+9 (4.74e+7)−
WFG3	3	**6.2348e+0** (5.91e-2)	6.2228e+0 (6.04e-2)≈	5.7582e+0 (2.62e-2)−	WFG4		3.5653e+1 (1.74e-1)	**3.5697e+1** (2.06e-2)≈	3.4960e+1 (1.14e-1)−
	5	**2.1288e+0** (1.42e-1)	2.0807e+0 (1.51e-1)≈	1.6996e+0 (2.58e-1)−			5.0002e+3 (4.77e+0)	**5.0016e+3** (5.02e+0)≈	4.8421e+3 (1.56e+1)−
	8	5.3631e-3 (4.01e-3)	7.2410e-3 (3.80e-3)≈	2.2878e-3 (3.15e-3)−			**2.0257e+7** (4.52e+4)	2.0149e+7 (4.56e+5)−	1.7283e+7 (3.89e+5)−
	10	3.9866e-6 (6.86e-6)	5.9016e-6 (8.11e-6)≈	0.0000e+0 (0.00e+0)−			9.2487e+9 (1.88e+7)	**9.2509e+9** (1.53e+7)≈	7.7873e+9 (1.19e+8)−
WFG5	3	3.3120e+1 (2.11e-3)	**3.3121e+1** (1.95e-3)≈	3.2514e+1 (1.11e-1)−	WFG6		3.1866e+1 (7.07e-1)	**3.2352e+1** (9.18e-1)≈	3.2164e+1 (1.02e+0)≈
	5	**4.7097e+3** (2.03e+0)	4.7091e+3 (2.22e+0)≈	4.5611e+3 (1.81e+1)−			**4.6061e+3** (6.86e+1)	4.5914e+3 (9.88e+1)≈	4.4389e+3 (1.36e+2)−
	8	1.9087e+7 (1.21e+4)	**1.9092e+7** (7.52e+3)≈	1.6968e+7 (3.39e+5)−			1.8444e+7 (3.92e+5)	**1.8700e+7** (3.00e+5)≈	1.5409e+7 (6.86e+5)−
	10	**8.7086e+9** (2.98e+6)	8.7073e+9 (2.68e+6)≈	7.3704e+9 (1.21e+8)−			**8.4569e+9** (1.73e+8)	8.4492e+9 (1.12e+8)≈	6.9705e+9 (2.71e+8)−
WFG7	3	3.5637e+1 (1.84e-1)	**3.5674e+1** (1.39e-2)≈	3.4808e+1 (1.29e-1)−	WFG8		**3.0454e+1** (1.40e-1)	3.0146e+1 (1.05e-1)−	3.0073e+1 (1.21e-1)−
	5	**5.0119e+3** (3.15e+0)	5.0074e+3 (2.66e+0)−	4.8405e+3 (1.51e+1)−			**4.3358e+3** (1.96e+1)	4.3141e+3 (2.02e+1)−	4.2171e+3 (5.07e+1)−
	8	2.0320e+7 (4.36e+4)	**2.0337e+7** (2.05e+4)+	1.8479e+7 (3.15e+5)−			1.7000e+7 (3.72e+5)	**1.7207e+7** (8.51e+5)+	1.0179e+7 (1.06e+6)−
	10	**9.2984e+9** (1.19e+7)	9.2620e+9 (1.31e+8)≈	8.2022e+9 (1.48e+8)−			8.1960e+9 (2.12e+8)	**8.2207e+9** (2.00e+8)≈	7.0545e+9 (6.63e+8)−
WFG9	3	3.4038e+1 (3.43e-1)	**3.4256e+1** (2.77e-1)≈	3.2841e+1 (2.07e+0)−		8	**1.8293e+7** (1.20e+6)	1.8055e+7 (1.42e+6)≈	1.6117e+7 (7.67e+5)−
	5	4.7412e+3 (2.15e+1)	**4.7434e+3** (2.58e+1)≈	4.5485e+3 (6.00e+1)−		10	**8.4675e+9** (4.64e+8)	8.3423e+9 (5.49e+8)≈	6.6760e+9 (4.42e+8)−

+/− : TSMOEA shows significantly worse/better performance in the comparison.
≈: There is no significant difference between the compared results.

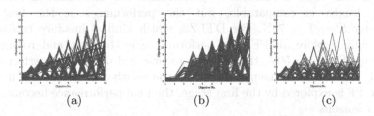

Fig. 2. Performance with median metric value obtained by three algorithms on the 10-objective WFG8 instances. (a)TSMOEA. (b)NSGA-III. (3)MOEA/DD.

designed with multi-modal and significant bias. The distance related parameters in WFG8 are dependent on position related parameters, while the position related parameters in WFG9 are dependent on distance related parameters. This type of dependency in WFG9 is not as difficult as WFG8. TSMOEA has best performance on three- and five-objective WFG8 instances. Performance with median metric value on 10-objective WFG8 instance obtained by TSMOEA, NSGA-III and MOEA/DD can be seen in Fig. 2. It shows that TSMOEA can better balance both diversity and convergence.

4.4   The Impact of the Length of the Two Stages

As discussed in Subsect. 3.3, it is the fact that when to trigger the second stage is of importance. Hence, we further study the impact of the length of the two stages. To study how TSMOEA is sensitive to the parameter, we have tried to cover a wide range of values to control the percentage of maximum generations for the run-time of the first stage, called *Percentage_Stage*1. Five values are considered: 50%, 60%, 70%, 80%, and 90%. We have taken DTLZ1 to DTLZ4 and WFG1 to WFG9 instances with five and ten objectives. The HV values of the proposed algorithm with different length on some instances with ten objectives are shown in Table 4.

Table 4. HV Values (Mean and Sd) Obtained by TSMOEA with different length on some instances with $m = 10$.

	DTLZ1	DTLZ2	DTLZ3	DTLZ4	WFG6	WFG7	WFG8	WFG9
50%	2.5314e-3 (1.64e-6)	**2.5153e+0** **(2.89e-4)**	2.5071e+0 (1.92e-2)	2.5154e+0 (5.16e-4)	8.4334e+9 (1.16e+8)	9.0870e+9 (4.19e+7)	7.7224e+9 (3.45e+8)	8.1611e+9 (3.90e+8)
60%	2.5321e-3 (1.19e-7)	2.5153e+0 (4.09e-4)	2.5095e+0 (1.14e-2)	2.5155e+0 (3.66e-4)	8.4482e+9 (1.96e+8)	9.1244e+9 (4.59e+7)	7.8526e+9 (3.14e+8)	8.2207e+9 (4.38e+8)
70%	**2.5322e-3** **(5.12e-8)**	2.5153e+0 (4.21e-4)	2.5107e+0 (2.25e-2)	**2.5157e+0** **(4.26e-4)**	8.4598e+9 (1.88e+8)	9.1653e+9 (4.17e+7)	7.9029e+9 (1.76e+8)	8.3432e+9 (6.26e+8)
80%	2.5321e-3 (3.72e-8)	2.5150e+0 (4.34e-4)	**2.5118e+0** **(1.59e-2)**	2.5156e+0 (4.06e-4)	8.4636e+9 (1.44e+8)	9.2436e+9 (2.82e+7)	8.1741e+9 (1.75e+8)	8.4185e+9 (3.89e+8)
90%	2.5321e-3 (5.28e-8)	2.5149e+0 (4.31e-4)	2.5094e+0 (1.12e-2)	2.5153e+0 (4.29e-4)	**8.4919e+9** **(1.19e+8)**	**9.2853e+9** **(1.26e+7)**	**8.2297e+9** **(1.69e+8)**	**8.5979e+9** **(7.43e+7)**

For DTLZ1 and DTLZ4, the performance of the same instance with different length is comparable, but the performance is the best when *Percentage_Stage*1 = 70%. For DTLZ2, which kind of instance can soon be able to fully explore all PF, the performance is the best and robust when *Percentage_Stage*1 = 50%, then becomes worse and worse. The performance of DTLZ3 is so sensitive to the parameter. As the length of the first stage increases, the more PF is explored by the first stage, the final performance becomes better and then worsens.

For WFG1 and WFG2, on condition that $m = 10$ and *Gen_max* = 1000, NSGA-III needs to run all the time to explore the whole PF. We either increase the number of generations or adopt the first stage all the time for this kind of problems. For WFG4 to WFG9, as the length of the first stage increases, the final performance becomes better. Moreover, the results of TSMOEA on WFG6 and WFG9 are significantly better than NSGA-III when *Percentage_Stage*1 = 90%. The HV values(Mean and Sd) of WFG8 with five objectives and different length are $4.3558e + 3(1.18e + 1)$, $4.3463e + 3(1.63e + 1)$, $4.3384e + 3(1.84e + 1)$, $4.3291e + 3(1.61e + 1)$ and $4.3208e + 3(1.53e + 1)$, respectively. It shows that the optimal length of the same instance with different numbers of objectives is still different.

In summary, the second stage does not trigger until the PF is fully explored, as a consequence, the proposed algorithm can obtain comparable and even the best performance. Once the whole PF is explored, the earlier the second stage is triggered, the better the performance of the proposed algorithm has. The optimal length of the two stages is different for each problem. Based on the characteristics of the problem, a more efficient algorithm can be selected to perform well.

5 Conclusion

In this paper, we have proposed a two-stage evolutionary algorithm for many-objective optimization, named TSMOEA. The motivation is to design a MOEA that is effective and robust for MaOPs of various characteristics by combining the merits of two state-of-art algorithms, NSGA-III and MOEA/DD. NSGA-III has objective normalization and the great ability of diversity promotion, while MOEA/DD can obtain the solutions with good convergence and diversity based on the appropriate weight vectors. In the first stage, NSGA-III and one of its variants are adopted to explore the shape of PF. At the end of the first stage, weight vectors are updated with not only the information of the original weight vectors, but also the final obtained solutions. In the second stage, MOEA/DD is adopt to enhance convergence while maintaining the diversity. The condition to trigger the second stage is of great importance and need to be more adaptively determined by the problem in the future work. Experimental results on DTLZ and WFG test suites show that TSMOEA has competitive performance and is suitable and robust for solving different types of MaOPs considered in this paper.

Acknowledgement. This work was supported in part by National Natural Science Foundation of China under Grant No. 61473271.

References

1. Bader, J., Zitzler, E.: Hype: an algorithm for fast hypervolume-based many-objective optimization. Evol. Comput. **19**(1), 45–76 (2011)
2. Beume, N., Naujoks, B., Emmerich, M.: SMS-EMOA: multiobjective selection based on dominated hypervolume. Eur. J. Oper. Res. **181**(3), 1653–1669 (2007)
3. Cheng, R., Jin, Y., Olhofer, M., Sendhoff, B.: A reference vector guided evolutionary algorithm for many-objective optimization. IEEE Trans. Evol. Comput. **20**(5), 773–791 (2016)
4. Deb, K., Jain, H.: An evolutionary many-objective optimization algorithm using reference-point-based nondominated sorting approach, part i: solving problems with box constraints. IEEE Trans. Evol. Comput. **18**(4), 577–601 (2014)
5. Deb, K., Pratap, A., Agarwal, S., Meyarivan, T.: A fast and elitist multiobjective genetic algorithm: NSGA-II. IEEE Trans. Evol. Comput. **6**(2), 182–197 (2002)

6. Laumanns, M., Thiele, L., Deb, K., Zitzler, E.: Combining convergence and diversity in evolutionary multiobjective optimization. Evol. Comput. **10**(3), 263–282 (2002)
7. Li, B., Li, J., Tang, K., Yao, X.: An improved two archive algorithm for many-objective optimization. In: 2014 IEEE Congress on Evolutionary Computation (CEC), pp. 2869–2876. IEEE (2014)
8. Li, K., Wang, R., Zhang, T., Ishibuchi, H.: Evolutionary many-objective optimization: a comparative study of the state-of-the-art. IEEE Access **6**, 26194–26214 (2018). https://doi.org/10.1109/ACCESS.2018.2832181
9. Li, K., Deb, K., Zhang, Q., Kwong, S.: An evolutionary many-objective optimization algorithm based on dominance and decomposition. IEEE Trans. Evol. Comput. **19**(5), 694–716 (2015)
10. Li, M., Yang, S., Liu, X.: Shift-based density estimation for Pareto-based algorithms in many-objective optimization. IEEE Trans. Evol. Comput. **18**(3), 348–365 (2014)
11. Liu, H.L., Chen, L., Zhang, Q., Deb, K.: Adaptively allocating search effort in challenging many-objective optimization problems. IEEE Trans. Evol. Comput. **22**(3), 433–448 (2018)
12. Liu, H.L., Gu, F., Zhang, Q.: Decomposition of a multiobjective optimization problem into a number of simple multiobjective subproblems. IEEE Trans. Evol. Comput. **18**(3), 450–455 (2014)
13. Purshouse, R.C., Fleming, P.J.: On the evolutionary optimization of many conflicting objectives. IEEE Trans. Evol. Comput. **11**(6), 770–784 (2007)
14. Tian, Y., Cheng, R., Zhang, X., Jin, Y.: PlatEMO: a matlab platform for evolutionary multi-objective optimization [educational forum]. IEEE Comput. Intell. Mag. **12**(4), 73–87 (2017)
15. Wang, R., Purshouse, R.C., Fleming, P.J.: Preference-inspired coevolutionary algorithms for many-objective optimization. IEEE Trans. Evol. Comput. **17**(4), 474–494 (2013)
16. Yang, S., Li, M., Liu, X., Zheng, J.: A grid-based evolutionary algorithm for many-objective optimization. IEEE Trans. Evol. Comput. **17**(5), 721–736 (2013)
17. Yuan, Y., Ong, Y.S., Gupta, A., Xu, H.: Objective reduction in many-objective optimization: evolutionary multiobjective approaches and comprehensive analysis. IEEE Trans. Evol. Comput. **22**(2), 189–210 (2018)
18. Zhang, Q., Li, H.: MOEA/D: a multiobjective evolutionary algorithm based on decomposition. IEEE Trans. Evol. Comput. **11**(6), 712–731 (2007)
19. Zhang, X., Tian, Y., Jin, Y.: A knee point-driven evolutionary algorithm for many-objective optimization. IEEE Trans. Evol. Comput. **19**(6), 761–776 (2015)
20. Zitzler, E., Thiele, L.: Multiobjective evolutionary algorithms: a comparative case study and the strength Pareto approach. IEEE Trans. Evol. Comput. **3**(4), 257–271 (1999)

Performance Metrics and Indicators

CRI-EMOA: A Pareto-Front Shape Invariant Evolutionary Multi-objective Algorithm

Jesús Guillermo Falcón-Cardona[1]([⊠]), Carlos A. Coello Coello[1], and Michael Emmerich[2]

[1] Computer Science Department,
CINVESTAV-IPN (Evolutionary Computation Group),
Av. IPN No. 2508, Col. San Pedro Zacatenco, 07300 México D.F., Mexico
jfalcon@computacion.cs.cinvestav.mx, ccoello@cs.cinvestav.mx
[2] Leiden Institute of Advanced Computer Science, Leiden University,
Niels Bohrweg 1, 2333 CA Leiden, The Netherlands
m.t.m.emmerich@liacs.leidenuniv.nl

Abstract. The use of multi-objective evolutionary algorithms (MOEAs) that employ a set of convex weight vectors as search directions, as a reference set or as part of a quality indicator has been widely extended. However, a recent study indicates that these MOEAs do not perform very well when tackling multi-objective optimization problem (MOPs), having different Pareto front geometries. Hence, it is necessary to propose MOEAs whose good performance is not strongly depending on certain Pareto front shapes. In this paper, we propose a Pareto-front shape invariant MOEA that combines the individual effect of two indicator-based density estimators. We selected the weakly Pareto-compliant IGD$^+$ indicator to promote convergence and the Riesz s-energy indicator that leads to uniformly distributed point sets for the large class of rectifiable d-dimensional manifolds. Our proposed approach, called CRI-EMOA, is compared with respect to MOEAs that adopt convex weight vectors (NSGA-III, MOEA/D and MOMBI2) as well as to MOEAs not using this set of vectors (Δ_p-MOEA and GDE-MOEA) on MOPs belonging to the test suites DTLZ, DTLZ^{-1}, WFG and WFG^{-1}. Our experimental results show that CRI-EMOA outperforms the considered MOEAs, regarding the hypervolume indicator and the Solow-Polasky indicator, on most of the test problems and that its performance does not depend on the Pareto front shape of the problems.

Keywords: Multi-objective optimization · Quality indicators · Multi-indicator density estimation

The first author acknowledges support from CONACyT and CINVESTAV-IPN to pursue graduate studies in Computer Science. The second author gratefully acknowledges support from CONACyT grant no. 2016-01-1920 (*Investigación en Fronteras de la Ciencia 2016*).

© Springer Nature Switzerland AG 2019
K. Deb et al. (Eds.): EMO 2019, LNCS 11411, pp. 307–318, 2019.
https://doi.org/10.1007/978-3-030-12598-1_25

1   Introduction

In the last 30 years, Multi-Objective Evolutionary Algorithms (MOEAs), which are population-based and gradient-free metaheuristics, have arisen as a popular approach to solve problems that involve the simultaneous optimization of several, often conflicting, objective functions [1]. These are the so-called multi-objective optimization problems (MOPs). MOEAs employ the principles of natural evolution to drive a set of objective vectors towards the Pareto optimal front that represents the solution to a MOP. In this regard, solving a MOP involves finding the best possible trade-offs among its objectives. The particular set that yields the best possible trade-offs among the objectives is known as the Pareto Optimal Set $(\mathcal{P}^*)$ and its image is known as the Pareto Optimal Front $(\mathcal{PF}^*)$.

Currently, there are different strategies for designing MOEAs, such as the decomposition of a MOP into several single-objective optimization problems [2], the use of reference sets to guide the population towards the Pareto front [3], and the generation of selection mechanisms based on (unary) quality indicators[1] [4]. A wide variety of state-of-the-art MOEAs based on the previously indicated strategies employ a set of convex weight vectors as search directions for the decomposition, in a method to construct reference sets, or as part of the definition of a quality indicator. A vector $\boldsymbol{w} \in \mathbb{R}^m$ is a convex weight vector if $\sum_{i=1}^m w_i = 1$ and $w_i \geq 0$ for all $i = 1, \ldots, m$. These weight vectors lie on an $(m-1)$-simplex. However, Ishibuchi et al. [5] empirically showed that the use of convex weight vectors overspecializes MOEAs on MOPs whose Pareto fronts are strongly correlated to the simplex formed by such weight vectors. In other words, such MOEAs are unable to produce good results when tackling MOPs whose Pareto fronts are not highly coupled with the $(m-1)$-simplex. In consequence, more general MOEAs need to be designed to avoid this overspecialization on specific benchmark problems such as the DTLZ and the WFG test suites.

There are MOEAs that do not use in any of their mechanisms a set of convex weight vectors. An example is the Nondominated Sorting Genetic Algorithm II (NSGA-II) [6] which uses Pareto dominance[2] in its main selection mechanism and crowding distance as its second selection mechanism. However, the selection pressure of NSGA-II dilutes when tackling MOPs having four or more objective functions. Additionally, the crowding distance density estimator cannot produce evenly distributed Pareto fronts in high dimensionality. Another example is the $\mathcal{S}$ Metric Selection Evolutionary Multi-Objective Algorithm (SMS-EMOA) [7] which is a steady-state MOEA that replaces the crowding distance of NSGA-II by the contribution of points to the hypervolume (HV) indicator. The HV is a performance indicator that measures convergence and maximum spread simultaneously. HV is the only unary indicator which is known to be

---

[1] A unary indicator I is a function that assigns a real value to set of points $\mathcal{A} = \{\boldsymbol{a}^1, \ldots, \boldsymbol{a}^N\}$, where $\boldsymbol{a}^i \in \mathbb{R}^m$.

[2] Given $\boldsymbol{u}, \boldsymbol{v} \in \mathbb{R}^m$, $\boldsymbol{u}$ Pareto dominates $\boldsymbol{v}$ (denoted as $\boldsymbol{u} \prec \boldsymbol{v}$) if and only if $\forall i = 1, \ldots, m, u_i \leq v_i$ and there exists at least an index $j \in \{1, \ldots, m\} : u_j < v_j$.

Pareto-compliant[3], but its use in MOEAs with many objectives is limited due to its high computational cost. In 2015, Menchaca-Méndez and Coello proposed an environmental selection mechanism based on the Generational Distance (GD) indicator [8] coupled with a diversity mechanism that adopts ϵ dominance to divide the objective space into hypercubes where the solutions are distributed. A clear disadvantage of GDE-MOEA is the determination of the ϵ value which is required to divide high-dimensional objective spaces and which has an impact on the generation of evenly distributed solutions. Finally, Δ_p-MOEA, proposed by Menchaca-Mendez et al. [9], is an improvement of GDE-MOEA in which instead of using GD in its selection mechanism, adopts the Δ_p indicator. Δ_p-MOEA improves the diversity of the solutions produced, but it still depends on the calculation of the ϵ value to construct a reference set.

In order to overcome the difficulties of MOEAs that do not use weight vectors, we propose here an MOEA that takes advantage of the combination/synergy of the individual effect of two density estimators: one based on the IGD$^+$ indicator [10] and another one based on the Riesz s-energy indicator [11]. The main idea of our Evolutionary Multi-Objective Algorithm based on the Combination of the Riesz s-energy and IGD$^+$ (CRI-EMOA) is to analyze the convergence behavior during the search process in a statistical manner. If convergence stagnates, the generation of evenly distributed solutions is promoted using Riesz s-energy; otherwise, the IGD$^+$-based density estimator will drive the population to $\mathcal{PF}^*$.

The remainder of this paper is organized as follows. Section 2 provides some basic definitions. Our proposed approach is described in Sect. 3. Our experimental results are discussed in Sect. 4. Finally, Sect. 5 outlines our main conclusions and some possible paths for future work.

2   Background

In this work, we focus, without loss of generality, on unconstrained MOPs that *minimize* all the objective functions. A MOP is formally defined as follows:

$$\min_{x \in \Omega} F(x) = (f_1(x), f_2(x), \ldots, f_m(x))^T, \tag{1}$$

where $x \in \Omega \subseteq \mathbb{R}^n$ is the vector of decision variables and Ω is the decision variable space. $f_i : \mathbb{R}^n \to \mathbb{R}, i = 1, 2, \ldots, m$ are the objective functions, where $m \geq 2$. MOPs having four or more objective functions are called many-objective optimization problems (MaOPs).

In the following, two unary quality indicators are described. For this purpose, let $\mathcal{A}$ represent an approximation to $\mathcal{PF}^*$ and $\mathcal{Z} \subset \mathbb{R}^m$ be a reference set. On the one hand, Ishibuchi et al. proposed the Inverted Generational Distance plus (IGD$^+$) indicator in 2015 [10]. This indicator measures the average distance between $\mathcal{Z}$ and $\mathcal{A}$, using a modified Euclidean distance that takes into account

---

[3] Let $\mathcal{A}$ and $\mathcal{B}$ be two non-empty sets of m-dimensional vectors and let I be a unary indicator. I is Pareto-compliant if and only if $\mathcal{A}$ dominates $\mathcal{B}$ implies $I(\mathcal{A}) > I(\mathcal{B})$ (assuming maximization of I).

Pareto dominance. Due to this modified distance, IGD$^+$ is a weakly Pareto compliant indicator. It is mathematically defined as follows:

$$\text{IGD}^+(\mathcal{A}, \mathcal{Z}) = \frac{1}{|\mathcal{Z}|} \sum_{z \in \mathcal{Z}} \min_{a \in \mathcal{A}} d^+(\boldsymbol{a}, \boldsymbol{z}), \qquad (2)$$

where $d^+(\boldsymbol{a}, \boldsymbol{z}) = \sqrt{\sum_{k=1}^m \max(a_i - z_i, 0)^2}$ is the proposed modified Euclidean distance. On the other hand, Hardin and Saff proposed the Riesz s-energy indicator [11] in order to measure the even distribution of a set of points in d-dimensional manifolds. Its mathematical definition is given by:

$$E_s(\mathcal{A}) = \sum_{i \neq j} \|\boldsymbol{a}_i - \boldsymbol{a}_j\|^{-s} \qquad (3)$$

where $s > 0$ is a fixed parameter that controls the degree of uniformity of the solutions in $\mathcal{A}$. Riesz s-energy has been found to lead to uniformly distributed point sets for the large class of rectifiable d-dimensional manifolds. Moreover, s is not a shape-dependent parameter [12].

3   Our Proposed Approach

Quality indicators can be integrated into MOEAs in three different ways: (1) in the environmental selection mechanism, (2) as an update rule for archives, and

---

Algorithm 1. CRI-EMOA general framework

Require: T_w, $\bar{\beta}$, $\bar{\theta}$
Ensure: Pareto front Approximation
1: Randomly initialize population P
2: $t \leftarrow 0$
3: **while** stopping criterion is not fulfilled **do**
4:     $q \leftarrow Variation(P)$
5:     $Q \leftarrow P \bigcup \{q\}$
6:     Normalize Q
7:     $\{L_1, L_2, \ldots, L_k\} \leftarrow$ nondominated-sorting(Q)
8:     $z_i^{\max} = \begin{cases} f_i^* = \max_{x \in L_1} f_i(\boldsymbol{x}), & f_i^* > z_i^{\max} \\ z_i^{\max}, & \text{otherwise} \end{cases}$
9:     $S_{\text{HV}}[t \bmod T_w] \leftarrow HV_{appr}(t)$
10:    Statistically analyze the last T_w samples in S_{HV} and generate β and θ
11:    **if** $k = 1$ and $\beta \leq \bar{\beta}$ and $\theta \in [-\bar{\theta}, \bar{\theta}]$ **then**
12:        $a_{\text{worst}} = \arg\max_{a \in L_1} C_{E_s}(\boldsymbol{a}, L_1)$
13:    **else**
14:        **if** $|L_k| > 1$ **then**
15:            $a_{\text{worst}} = \arg\min_{a \in L_k} C_{\text{IGD}+}(\boldsymbol{a}, L_k, L_1)|$
16:        **else**
17:            a_{worst} is equal to the sole individual in L_k
18:    $P \leftarrow Q \setminus \{a_{\text{worst}}\}$
19:    $t \leftarrow t + 1$
20: **return** P

---

(3) as density estimators (DEs). From these approaches, indicator-based DEs (IB-DEs) have been widely used. An IB-DE is the secondary selection mechanism of an MOEA. IB-DEs impose a total order among the solutions of an approximation set by calculating the individual contribution of each solution to the indicator value. Then, the worst-contributing solution is deleted from the population. In this work, we employed IGD$^+$ and Riesz s-energy as IB-DEs. Regarding IGD$^+$, the individual contribution C of a solution $a \in \mathcal{A}$ is defined as follows: $C_{\mathrm{IGD}^+}(a, \mathcal{A}, \mathcal{Z}) = |\mathrm{IGD}^+(\mathcal{A}, \mathcal{Z}) - \mathrm{IGD}^+(\mathcal{A}\backslash\{a\}, \mathcal{Z})|$. On the other hand, for Riesz s-energy, the individual contribution of $a \in \mathcal{A}$ is given by: $C_{E_s}(a, \mathcal{A}) = \frac{1}{2}[E_s(\mathcal{A}) - E_s(\mathcal{A}\backslash\{a\})]$. On the basis of the above equations, IGD$^+$-DEs and E_s-DE are respectively defined as follows: (1) $a_{\mathrm{worst}} = \arg\min_{a \in \mathcal{A}} C_{\mathrm{IGD}^+}(a, \mathcal{A}, \mathcal{Z})$, and (2) $a_{\mathrm{worst}} = \arg\max_{a \in \mathcal{A}} C_{E_s}(a, \mathcal{A})$, where a_{worst} denotes the solution having the wost-contributing value.

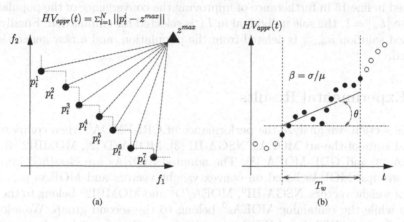

Fig. 1. (a) The hypervolume approximation adds up all the distances between the reference point and each nondominated solution, (b) linear model of the convergence behavior created using the last T_w measures of HV$_{\mathrm{appr}}$.

Algorithm 1 describes our proposed approach, called CRI-EMOA. It is a steady-state MOEA that adopts Pareto dominance in its environmental selection mechanism (using the nondominated sorting algorithm [6] in line 7) and an IB-DE as its secondary selection criterion. The main idea of CRI-EMOA is to exploit the properties of IGD$^+$ and Riesz s-energy by combining the individual effect of the corresponding IB-DEs. In other words, we want to drive the population towards the Pareto front using IGD$^+$-DE and, simultaneously, generating an evenly distributed approximation to the Pareto front through E_s-DE. To this end, CRI-EMOA switches between the two IB-DEs depending on a statistical analysis of the convergence behavior of the population, using an approximation to the hypervolume indicator (denoted as HV$_{\mathrm{appr}}$). HV$_{\mathrm{appr}}$ is a simplification of the proposal of Ishibuchi *et al.* [13] and it adds up all the distances between an anti-optimal reference point z^{max} and the set of current nondominated solutions in L_1 (see Fig. 1a). In line 8, each $z_i^{\max}, i = 1, \ldots, m$ is updated if and only a worse objective value in L_1 is found and, then, $\mathrm{HV}_{\mathrm{appr}}(t)$ is computed such that the obtained value is stored in a circular array S_{HV} of size T_w. After the first T_w generations, S_{HV} will be full, and we can statistically analyze at each iteration the last T_w samples of $\mathrm{HV}_{\mathrm{appr}}$ as shown in Fig. 1b. In line 10, the mean μ and the standard deviation σ of the samples are computed such that the coefficient[4] of variation $\beta = \sigma/\mu$ is calculated. Additionally, the angle θ of a linear regression model of the samples is computed. Based on β and θ, we can exploit the properties of a certain IB-DE. If the number k of ranks produced by the nondominated sorting algorithm is equal to one and it holds that $\beta \leq \bar{\beta}$ and $\theta \in [-\bar{\theta}, \bar{\theta}]$ (where $\bar{\beta}$ and $\bar{\theta}$ are user-supplied parameters), it means that the convergence behavior is stagnated since there is not too much variation of $\mathrm{HV}_{\mathrm{appr}}$ and the linear model cannot be considered as ascending or descending. In consequence, we have to promote diversity using E_s-DE in line 12. Otherwise, if $|L_k| > 1$, IGD$^+$-DE is selected in line 15 in furtherance of improving the convergence of the population. In case $|L_k| = 1$, the sole individual in L_k is selected for elimination. Finally, the selected solution a_{worst} is deleted from the population, and a new generation is created.

4   Experimental Results

In this section, we analyze the performance of CRI-EMOA[5] when compared to several state-of-the-art MOEAs: NSGA-III [3], MOEA/D [2], MOMBI2 [4], Δ_p-MOEA [9] and GDE-MOEA [8]. The adopted MOEAs are classified into two main groups: MOEAs based on convex weight vectors and MOEAs not using convex weight vectors. NSGA-III[6], MOEA/D[7] and MOMBI2[8] belong to the first group while the remaining MOEAs[9] belong to the second group. We adopted MOPs from the DTLZ and WFG test suites, as well as from the minus versions of them, denoted as DTLZ^{-1} and WFG^{-1} that were proposed by Ishibuchi *et al.* [5]. The use of the minus versions of the benchmarks is to determine the performance of the considered MOEAs on MOPs whose Pareto fronts are not correlated to the simplex formed by a set of convex weight vectors. Additionally, the Pareto fronts of these MOPs cover a wide range of geometries such as linear,

---

[4] β is a standardized measure of dispersion that shows the extent of variability to the mean of the population.

[5] The source code of CRI-EMOA is available at http://computacion.cs.cinvestav.mx/~jfalcon/CRI-EMOA.html.

[6] We used the implementation available at: http://web.ntnu.edu.tw/~tcchiang/publications/nsga3cpp/nsga3cpp.htm.

[7] We used the implementation available at: http://dces.essex.ac.uk/staff/zhang/webofmoead.htm.

[8] We used the implementation available at http://computacion.cs.cinvestav.mx/~rhernandez/.

[9] The source code of Δ_p-MOEA and GDE-MOEA was provided by its author, Adriana Menchaca Méndez.

concave, degenerated, disconnected and mixed. In each case, we employed 3, 5 and 10 objective functions. In order to assess the performance of our proposed CRI-EMOA and the other MOEAs adopted in our comparative study, we applied HV and the Solow-Polasky indicator [14] for assessing convergence and diversity, respectively. For each MOEA in each test instance, we performed 30 independent executions.

4.1 Parameters Settings

For a fair comparison, we set the population size N of all MOEAs, equals to the number of convex weight vectors that some of them employed, i.e., $N = C_{m-1}^{H+m-1}$, where m is the number of objective functions and H is a user-supplied parameter. Hence, in each case, the tuple (m, H, N) was set as follows: (3, 14, 120), (5, 5, 126), and (10, 3, 220). For the considered number of objective functions, we set 50×10^3, 70×10^3, and 120×10^3 function evaluations as our stopping criterion, respectively. Since our approach and all the considered MOEAs are genetic algorithms that use Simulated Binary Crossover and Polynomial-based Mutation as variation operators, we set the crossover probability (P_c), the crossover distribution index (N_c), the mutation probability (P_m), and the mutation distribution index (N_m) as follows. For MOPs having three objective functions $P_c = 0.9$ and $N_c = 20$, while for MaOPs $P_c = 1.0$ and $N_c = 30$. In all cases, $P_m = 1/n$, where n is the number of decision variables, and $N_m = 20$. Regarding both the WFG and the WFG^{-1} test problems with 3, 5 and 10 objectives, we set the number of variables as $n = 26, 30$ and 40, in each case using the following position-related parameters: 2, 4, and 9. Considering the DTLZ and DTLZ^{-1} instances, the number of variables is equal to $n = m + K - 1$, where $K = 5$ for DTLZ1 and DTLZ1^{-1}, $K = 10$ for DTLZ2, DTLZ5 and their minus versions, and $K = 20$ for DTLZ7 and DTLZ7^{-1}. For MOEA/D, the neighborhood size was set to 20 in all cases. Regarding CRI-EMOA, we employed $T_w = N$, $\bar{\beta} = 0.1$ and $\bar{\theta} = 0.25°$ for all instances.

4.2 Discussion of Results

Tables 1 and 2 show the mean and standard deviation (in parentheses) obtained by all the compared algorithms for the hypervolume and the Solow-Polasky[10] indicators, respectively. The two best values among the MOEAs are highlighted using gray scale, where the darker tone corresponds to the best value. Aiming to obtain the statistical confidence of our results, we performed a one-tailed Wilcoxon test using a significance level of 0.05. Based on the Wilcoxon test, the symbol # is placed when CRI-EMOA performs better than another MOEA in a statistically significant way.

---

[10] The Solow-Polasky indicator requires a parameter θ that was set to 10.

Regarding the hypervolume indicator, CRI-EMOA is the best algorithm since it obtained the first place in 50% of the test problems. The second place corresponds to NSGA-III because it was the best MOEA in 8 out of 42 problems. However, it is worth emphasizing that for the minus benchmarks, NSGA-III only obtained one first place, specifically for DTLZ7^{-1} with 3 objective functions. In this regard, MOEA/D and MOMBI2 have just one first place in these minus benchmarks, and the remaining of their first places belong to the original DTLZ and WFG test suites. In consequence, it is clear the overspecialization of MOEAs using convex weight vectors on these benchmarks. Considering Δ_p-MOEA and GDE-MOEA, their performance is not so high. In fact, GDE-MOEA never obtains the first place and Δ_p-MOEA is the best algorithm in four test instances.

The Solow-Polasky indicator supports the good results of CRI-EMOA. This indicator measures the number of species present in the population. Thus, a larger value of the indicator is better because it means a good diversity of solutions. Our proposed approach produces well-distributed Pareto fronts in 26 out of 42 test instances (see Fig. 2). As a matter of fact, in most cases, when CRI-EMOA obtains the best HV value, it also obtains the best Solow-Polasky value. Hence, this a first insight that the synergy between IGD$^+$ and Riesz s-energy is actually responsible of its good performance in both convergence and diversity. Regarding the other MOEAs, NSGA-III and Δ_p-MOEA tie in second place since they obtained the best indicator value in 5 problems. Once again, NSGA-III can only produce good results for the original DTLZ and WFG problems. The worst algorithm regarding this indicator is MOMBI2.

For DTLZ1 and DTLZ1^{-1}, which have a linear Pareto front, CRI-EMOA does not obtain the best HV value. However, the Solow-Polasky indicator reflects that our approach has a better diversity. The top part of Fig. 2 shows the DTLZ1^{-1} fronts produced by all the MOEAs, and it is evident that CRI-EMOA produces an evenly distributed front in comparison with the adopted MOEAs. MOEA/D and MOMBI2 generate numerous solutions in the boundary of the front, while Δ_p-MOEA, GDE-MOEA and NSGA-III do not produce well-distributed solutions. For convex problems, i.e., DTLZ2^{-1} and DTLZ5^{-1}, it is evident that CRI-EMOA has a good performance. This is because it entirely covers the Pareto front, unlike the other MOEAs which are unable to do the same. This effect is illustrated in the second row of Fig. 2. For more complicated problems such as DTLZ7 and WFG2^{-1} that assess the ability of a MOEA to manage subpopulations, it is evident that CRI-MOEA produces better results. In the light of these results, we can claim that CRI-EMOA is a more general optimizer because its performance is not strongly linked to certain types of benchmark problems.

Table 1. Mean and standard deviation (in parentheses) of the hypervolume indicator. A symbol # is placed when CRI-EMOA performed significantly better than the other approaches based on a one-tailed Wilcoxon test using a significance level of $\alpha = 0.05$. The two best values are shown in gray scale, where the darker tone corresponds to the best value.

MOP	Dim.	CRI-EMOA	NSGA-III	MOEA/D	MOMBI2	Δ_p-MOEA	GDE-MOEA
DTLZ1	3	9.739039e-01 (3.858675e-04)	9.741141e-01 (3.120293e-04)	9.740945e-01 (2.619649e-04)	9.663444e-01# (1.080932e-03)	9.413310e-01# (1.964370e-02)	9.676446e-01# (2.362618e-03)
	5	9.877798e-01 (3.117917e-03)	9.986867e-01 (3.379577e-05)	9.986355e-01 (3.735697e-05)	9.904662e-01 (1.120127e-03)	3.320501e-02# (8.565974e-02)	4.840903e-01# (4.857106e-01)
	10	9.963635e-01 (1.065991e-03)	9.999939e-01 (2.139857e-06)	9.996746e-01 (1.025281e-04)	9.961538e-01 (9.574496e-04)	3.040882e-02# (5.310077e-02)	0.000000e+00# (0.000000e+00)
DTLZ2	3	7.419537e+00 (3.056980e-03)	7.421572e+00 (6.064709e-04)	7.421715e+00 (1.372809e-04)	7.380040e+00# (7.076656e-03)	7.371981e+00# (3.875638e-02)	7.350569e+00# (2.220661e-02)
	5	3.157090e+01 (2.415933e-02)	3.166721e+01 (6.548007e-04)	3.166781e+01 (5.129480e-04)	3.149886e+01# (2.619865e-02)	3.145814e+01# (6.277721e-02)	3.139858e+01# (7.085084e-02)
	10	1.021699e+03 (4.906893e-01)	1.023905e+03 (1.423610e-03)	1.023902e+03 (4.192719e-03)	1.022163e+03 (4.299615e-01)	1.022172e+03 (3.206973e-01)	8.223136e+02# (4.847301e+01)
DTLZ5	3	6.103498e+00 (2.913259e-04)	6.086240e+00 (3.462620e-03)	6.046024e+00# (2.227008e-04)	6.018466e+00# (3.166178e-03)	6.083103e+00# (4.024434e-02)	6.070736e+00# (4.307412e-02)
	5	2.306362e+01 (2.295313e-01)	2.162912e+01 (9.476133e-01)	2.328373e+01 (1.640165e-02)	2.175597e+01# (2.378197e-01)	2.152316e+01# (1.422545e+00)	1.943602e+01# (1.234198e+00)
	10	6.453781e+02 (4.080592e+01)	6.172582e+02# (4.132326e+01)	7.043390e+02 (1.714256e+00)	6.054385e+02# (4.091687e+01)	5.909772e+02# (7.644220e+01)	9.641241e+01# (1.554238e+01)
DTLZ7	3	1.634605e+01 (5.285233e-02)	1.631926e+01 (1.253568e-02)	1.620770e+01# (1.240925e-01)	1.613885e+01# (3.101462e-02)	1.612577e+01# (1.553168e-01)	1.615480e+01# (1.492618e-01)
	5	1.281085e+01 (1.974810e-01)	1.284401e+01 (3.182259e-02)	6.515913e+00# (1.170945e+00)	1.269646e+01# (4.907749e-02)	1.255217e+01# (1.341411e-01)	1.234590e+01# (2.234605e-01)
	10	3.479852e+00 (2.403388e-01)	1.806637e+00# (4.781492e-01)	2.756082e-03# (7.839814e-03)	3.033892e+00# (5.070947e-02)	3.027342e+00# (9.110566e-02)	2.080502e+00# (4.312007e-01)
WFG1	3	5.056544e+01 (1.657420e+00)	4.917540e+01# (1.742752e+00)	4.994533e+01 (2.615320e+00)	5.250059e+01 (1.702362e+00)	3.624458e+01# (9.571499e-01)	3.857628e+01# (9.613983e-01)
	5	4.509188e+03 (1.444159e+02)	4.049661e+03# (1.445036e+02)	4.522924e+03 (1.145447e+02)	4.682300e+03 (7.687667e+01)	3.198417e+03# (8.802857e+01)	3.499936e+03# (7.077142e+01)
	10	5.037589e+09 (8.535179e+07)	4.333786e+09# (4.767509e+07)	4.626119e+09# (9.082857e+07)	5.028893e+09 (6.062765e+07)	3.422833e+09# (2.182108e+07)	3.554077e+09# (4.491835e+07)
WFG2	3	1.000262e+02 (2.196919e-01)	1.000303e+02 (2.020421e-01)	9.425491e+01# (1.887090e+00)	9.995196e+01# (2.218338e-01)	2.860787e+01# (1.562061e-01)	2.678405e+01# (3.147546e-02)
	5	1.008420e+04 (5.737764e+01)	1.022660e+04 (2.444328e+01)	9.147103e+03# (2.989196e+02)	1.021265e+04 (2.425440e+01)	2.356563e+03# (1.302041e+01)	2.352252e+03# (2.298487e+01)
	10	1.348499e+10 (4.708062e+07)	1.343510e+10# (5.838755e+07)	1.153362e+10# (4.307707e+08)	1.346239e+10 (6.456777e+07)	2.433110e+09# (1.405830e+07)	2.417620e+09# (3.423298e+07)
WFG3	3	7.306197e+01 (3.258533e-01)	7.359113e+01 (3.698540e-01)	6.949014e+01# (7.049197e+00)	7.476737e+01 (9.915001e-01)	2.974536e+01 (8.186136e-01)	3.026476e+01 (9.039859e-02)
	5	6.735962e+03 (9.568603e+01)	6.705622e+03 (6.623165e+01)	5.831355e+03# (1.740491e+02)	6.720322e+03 (8.790247e+01)	2.425136e+03# (2.737458e+01)	2.467475e+03# (5.330311e+00)
	10	8.262095e+09 (2.467236e+08)	7.851751e+09# (1.420734e+08)	3.407782e+09# (4.406816e+08)	7.150575e+09# (8.942471e+08)	2.435088e+09# (7.572200e+07)	2.460728e+09# (2.651078e+07)
DTLZ1^{-1}	3	2.237019e+07 (1.096230e+05)	2.044422e+07# (2.230718e+05)	1.708422e+07# (2.776295e+05)	1.754720e+07# (1.024912e+04)	2.249206e+07 (9.308520e+04)	2.178413e+07# (1.919526e+05)
	5	5.990400e+10 (5.969126e+09)	1.653440e+10# (7.395153e+09)	1.275157e+10# (5.929635e+09)	1.829497e+10# (1.178680e+08)	8.421535e+10 (5.019922e+09)	7.834908e+10 (5.592427e+09)
	10	2.331601e+15 (1.332180e+15)	1.690928e+16 (1.594681e+16)	2.068669e+10# (2.776909e+10)	3.254959e+17 (7.964585e+16)	4.163772e+17 (1.784438e+17)	1.959914e+17 (7.692566e+16)
DTLZ2^{-1}	3	1.255756e+02 (1.372903e-01)	1.226427e+02# (4.332124e-01)	1.241646e+02# (1.767939e-01)	1.246298e+02# (1.975120e-02)	1.202429e+02# (1.235826e+00)	1.232392e+02# (4.384877e-01)
	5	1.823404e+03 (5.652832e+00)	1.529187e+03# (3.829295e+01)	1.570781e+03# (5.466206e+00)	1.377041e+03# (2.801096e+00)	1.615070e+03# (3.622796e+01)	1.684100e+03# (2.422012e+01)
	10	3.952305e+05 (6.000728e+03)	2.480210e+05# (3.215706e+04)	1.837497e+05# (3.540744e+03)	1.941735e+05# (4.318334e+03)	4.467775e+05 (1.153133e+04)	4.295481e+05 (1.104582e+04)
DTLZ5^{-1}	3	1.240446e+02 (1.543643e-01)	1.212729e+02# (4.506920e-01)	1.230132e+02# (1.173182e-01)	1.233805e+02# (2.897257e-02)	1.191790e+02# (1.218659e+00)	1.217996e+02# (3.913095e-01)
	5	1.830136e+03 (8.376583e+00)	1.526551e+03# (4.186892e+01)	1.532378e+03# (6.612506e+00)	1.490703e+03# (3.599646e+00)	1.550531e+03# (3.545733e+01)	1.663295e+03# (2.143198e+01)
	10	5.043244e+05 (5.933536e+03)	2.353908e+05# (2.658733e+04)	1.618586e+05# (2.870596e+03)	1.786897e+05# (4.650613e+03)	3.841427e+05# (1.267929e+04)	3.788162e+05# (1.409232e+04)
DTLZ7^{-1}	3	2.139263e+02 (1.705184e+00)	2.144432e+02 (1.844494e-02)	2.144785e+02 (3.401603e-03)	2.144350e+02 (1.484695e-02)	2.141398e+02 (6.446048e-01)	2.117720e+02# (5.620357e+00)
	5	1.193104e+03 (7.463449e+00)	1.190442e+03# (4.159670e+00)	6.388549e+02# (5.254422e+01)	1.177244e+03 (5.760920e+00)	1.195714e+03 (1.560565e+00)	1.167397e+03# (3.067229e+01)
	10	6.493424e+04 (1.799575e+02)	6.282093e+04# (1.236603e+02)	7.555843e+03# (6.397426e+02)	6.278498e+04# (5.606912e+02)	6.374490e+04# (1.597907e+02)	6.336153e+04# (1.579373e+02)
WFG1^{-1}	3	4.721465e+02 (5.118363e+01)	5.214593e+02# (2.613138e+01)	3.653092e+02# (2.305800e+00)	4.717969e+02# (4.848793e+01)	4.289752e+02# (4.089696e+01)	4.226979e+02# (4.328855e+01)
	5	8.957760e+04 (1.295509e+04)	6.766707e+04# (3.634016e+03)	4.312409e+04# (1.486578e+03)	8.604789e+04# (1.028243e+04)	6.687040e+04# (8.125469e+03)	5.398842e+04# (6.022448e+03)
	10	1.920711e+11 (1.254828e+10)	1.167307e+11# (9.811876e+09)	7.403214e+10# (3.748511e+09)	5.753336e+10# (1.586430e+09)	1.037099e+11# (5.197695e+09)	8.712812e+10# (9.507560e+09)
WFG2^{-1}	3	7.318853e+02 (4.584376e-01)	7.256549e+02# (2.471515e+00)	7.318071e+02# (5.137348e-01)	7.277336e+02# (7.218694e-01)	3.548073e+02# (4.631427e-01)	3.549143e+02# (1.951948e-01)
	5	1.638383e+05 (1.165835e+03)	1.470928e+05# (8.586496e+03)	1.122933e+05# (1.197256e+04)	1.499384e+05# (4.291788e+02)	4.315723e+04# (1.487567e+02)	4.156049e+04# (6.526206e+02)
	10	7.365072e+11 (6.254171e+09)	3.658776e+11# (1.973606e+10)	2.462168e+11# (2.934157e+10)	8.919695e+10# (1.091716e+10)	7.359311e+10# (2.277690e+08)	7.165991e+10# (8.580112e+08)
WFG3^{-1}	3	6.701244e+02 (9.728569e-01)	6.581207e+02# (2.461272e+00)	6.559404e+02# (1.399701e-01)	6.678986e+02# (4.368737e-01)	3.901185e+02# (2.837691e+00)	3.929122e+02# (1.663457e+00)
	5	1.460039e+05 (2.618698e+03)	1.271888e+05# (5.065268e+03)	9.818104e+04# (4.519958e+03)	1.345863e+05# (2.667741e+02)	4.822825e+04# (1.017141e+03)	4.912237e+04# (6.555485e+02)
	10	6.613123e+11 (2.015972e+10)	3.003925e+11# (2.070638e+10)	1.932277e+11# (2.123809e+10)	1.572410e+11# (3.994016e+09)	8.120430e+10# (2.177319e+09)	8.405921e+10# (1.518237e+09)

Table 2. Mean and standard deviation (in parentheses) of the Solow-Polasky indicator. A symbol # is placed when CRI-EMOA performed significantly better than the other approaches based on a one-tailed Wilcoxon test using a significance level of $\alpha = 0.05$. The two best values are shown in gray scale, where the darker tone corresponds to the best value.

MOP	Dim.	CRI-EMOA	NSGA-III	MOEA/D	MOMBI2	Δ_p-MOEA	GDE-MOEA
DTLZ1	3	9.944808e+00 (7.332450e-01)	9.394548e+00# (2.930251e-01)	9.314418e+00# (3.914884e-02)	9.000566e+00# (2.446366e-02)	7.811889e+00# (9.608413e-01)	9.208526e+00# (7.142910e-01)
	5	1.338590e+01 (5.394744e-01)	1.927839e+01 (2.200570e-01)	1.910784e+01 (2.012103e-01)	1.784107e+01 (5.535436e-02)	1.258001e+02 (3.614573e-01)	7.251588e+01 (4.806114e+01)
	10	1.785253e+01 (8.881198e-01)	4.215677e+01 (2.267717e+00)	3.557264e+01 (6.064497e+01)	3.493408e+01 (2.073537e+00)	2.196627e+02 (4.229255e-01)	1.937667e+02 (5.463117e+00)
DTLZ2	3	3.395527e+01 (9.380927e-02)	3.394704e+01# (1.377030e-02)	3.393654e+01# (1.057577e-03)	3.320388e+01# (3.200128e-02)	3.071966e+01# (5.648283e-01)	3.130480e+01# (3.907121e-01)
	5	9.880242e+01 (3.075202e+00)	1.023559e+01 (2.316020e-01)	1.017397e+02 (4.330518e-03)	1.000214e+02 (9.376416e-02)	9.047203e+01# (1.071667e+00)	8.885177e+01# (1.407456e+00)
	10	2.144437e+02 (8.333968e-01)	2.144143e+02# (4.461039e-02)	2.140218e+02# (1.052798e-02)	2.134074e+02# (2.440550e-01)	2.073661e+02# (1.076644e+00)	2.149790e+02 (1.820800e+00)
DTLZ5	3	8.835302e+00 (8.683488e-03)	8.689954e+00# (4.814112e-02)	4.565503e+01 (6.372947e-01)	8.446415e+00# (1.275105e-02)	8.725615e+00# (1.118233e-01)	9.131640e+01 (8.893988e-01)
	5	5.453458e+01 (3.836635e+00)	7.846618e+01 (3.806546e+00)	2.193721e+01 (7.192604e+01)	1.733111e+01 (1.215347e+00)	6.458870e+01 (4.414063e+00)	9.229364e+01 (3.153601e+00)
	10	1.426916e+02 (1.105651e+01)	1.855864e+02 (4.441145e+00)	7.636613e+00 (7.127440e-02)	2.097795e+01 (1.899760e+00)	1.636387e+02 (1.190412e+01)	2.009986e+02 (2.726407e+00)
DTLZ7	3	4.693189e+01 (4.563587e-01)	4.248938e+01# (8.838503e-01)	3.411613e+01# (6.885687e+00)	3.750968e+01# (4.295088e-01)	3.356066e+01# (8.918332e+00)	3.791999e+01# (1.074318e+01)
	5	7.703740e+01 (2.640331e+00)	9.605921e+01 (4.006295e+00)	2.595428e+01# (3.104755e-01)	7.335971e+01 (1.892378e+00)	1.014229e+02 (7.384253e+00)	8.467007e+01 (2.946531e+01)
	10	2.083721e+02 (1.401193e+01)	3.401405e+01# (4.627073e+01)	6.635493e+00# (8.377910e-01)	1.539631e+02 (1.794040e+01)	2.161036e+02 (1.887145e+00)	1.635677e+02# (5.659825e+01)
WFG1	3	6.266729e+01 (4.306665e+00)	5.624993e+01# (4.311929e+00)	5.053063e+01# (2.764405e+00)	5.406056e+01# (2.296813e+00)	3.936107e+01# (2.712236e+00)	4.901870e+01# (2.752851e+00)
	5	7.766310e+01 (9.797998e+00)	9.244372e+01 (7.266040e+00)	7.480740e+01# (3.832994e+00)	7.292172e+01# (5.425443e+00)	5.404634e+01# (4.708150e+00)	9.197836e+01 (4.116442e+00)
	10	1.153389e+02 (1.285140e+01)	8.917693e+01# (8.545945e+00)	1.552376e+01# (3.169355e+00)	6.819405e+01# (8.992674e+00)	9.420152e+01# (6.434297e+00)	1.681839e+02 (7.642626e+00)
WFG2	3	1.031961e+02 (6.913412e-01)	9.475339e+01# (5.942618e+00)	7.243218e+01# (1.099197e+00)	8.113447e+01# (1.694539e+00)	1.566893e+01# (4.695226e-01)	1.597100e+01# (5.210876e+00)
	5	9.923778e+01 (3.753788e+00)	1.259866e+02 (5.442239e-01)	9.750359e+01# (2.449040e+00)	1.226234e+02 (1.081329e+00)	2.491924e+01# (1.910851e+00)	2.346945e+01# (2.689896e+00)
	10	1.981494e+02 (4.297874e+00)	2.034942e+02 (6.167357e+00)	2.746068e+01# (9.314055e+00)	1.826284e+02 (2.286544e+01)	5.897645e+01# (4.305811e+00)	5.040485e+01# (7.890364e+00)
WFG3	3	7.979549e+01 (8.271398e-01)	5.447458e+01# (3.954759e+00)	6.745390e+01# (1.429561e+00)	4.359786e+01# (9.246690e-01)	2.088260e+01# (5.548807e-01)	2.237972e+01# (2.670741e-01)
	5	1.207901e+02 (1.514908e+00)	9.114798e+01# (4.803291e+00)	1.203892e+02 (1.120195e+00)	3.884532e+01# (5.191645e+00)	3.640185e+01# (1.590306e+00)	3.986356e+01# (1.669298e+00)
	10	2.198151e+02 (1.511883e+00)	1.842494e+02# (6.381996e+00)	1.685512e+02# (8.180955e+01)	1.223302e+02 (2.606946e+01)	7.655449e+01# (7.601122e+00)	9.569073e+01# (5.324356e+00)
DTLZ1^{-1}	3	1.238722e+02 (6.478345e-01)	1.192301e+02 (9.769981e-01)	1.110656e+02 (2.067972e-01)	1.076858e+02 (1.612270e+00)	1.194494e+02 (7.076038e-01)	1.026058e+02 (2.385964e+00)
	5	1.261138e+02 (3.746123e-01)	1.276094e+02 (7.483806e-01)	1.160278e+02 (3.018754e+00)	6.760143e+01 (4.891955e+00)	1.256546e+02 (4.961652e-01)	1.091644e+02 (2.978643e+00)
	10	2.200000e+02 (6.478398e-01)	2.194982e+02# (5.550208e-01)	1.845834e+01# (3.362045e+01)	2.177230e+02 (1.598879e+00)	2.199297e+02# (1.852502e-01)	1.956693e+02# (3.843626e+00)
DTLZ2^{-1}	3	1.129425e+02 (2.079720e-01)	9.168006e+01# (2.394444e+00)	9.466441e+01# (8.426911e-02)	9.433643e+01# (1.896075e-01)	8.857439e+01# (2.604729e+00)	8.818635e+01# (2.087133e+00)
	5	1.259981e+02 (4.099458e-04)	1.134723e+02# (2.970256e+00)	1.247875e+02 (1.631879e+00)	4.888021e+01# (1.111632e+00)	1.185054e+02# (1.679308e+00)	1.078721e+02# (2.347639e+00)
	10	2.249876e+02 (1.368722e+00)	2.075064e+02# (3.824826e+00)	2.079851e+02# (1.899257e+00)	1.810577e+02# (3.309000e+00)	2.118042e+02# (2.291510e+00)	1.931317e+02# (4.552358e+00)
DTLZ5^{-1}	3	1.069905e+02 (2.945855e-01)	8.469885e+01# (1.989703e+00)	7.942908e+01# (2.711812e+00)	8.622124e+01# (1.733700e-01)	8.484735e+01# (2.558670e+00)	8.305258e+01# (1.583665e+00)
	5	1.250747e+02 (3.780629e-03)	1.041424e+02# (3.722763e+00)	1.229014e+02 (1.837627e+00)	4.957223e+01# (1.976910e+00)	1.180479e+02# (1.846989e+00)	1.076183e+02# (2.710724e+00)
	10	2.199997e+02 (6.289321e-05)	1.579241e+02# (1.854156e+01)	1.997485e+02# (1.822188e+00)	1.636386e+02# (7.545337e+00)	2.094613e+02# (2.217251e+00)	1.946215e+02# (3.565477e+00)
DTLZ7^{-1}	3	2.345500e+01 (6.661044e+00)	2.375280e+01 (1.117994e+00)	2.588876e+01 (3.461387e+00)	1.994117e+01# (4.519640e-01)	2.178525e+01# (8.341601e-01)	1.805087e+01# (8.596927e+00)
	5	5.660901e+01 (1.568211e+01)	7.238636e+01 (1.269825e+01)	1.211841e+01# (1.172186e+00)	4.067053e+01# (9.192226e+00)	8.003609e+01# (3.156330e+00)	3.632242e+01# (3.834467e+01)
	10	2.043557e+02 (1.375176e+01)	1.347251e+01# (4.083423e+00)	4.293619e+00# (1.198274e+01)	8.812385e+01# (2.123041e+01)	2.008713e+02# (2.214667e+00)	2.028099e+02# (5.987172e+00)
WFG1^{-1}	3	6.415681e+01 (4.459890e+00)	5.511082e+01# (2.648385e+00)	1.663876e+01# (1.535080e+00)	4.730483e+01# (1.092483e+00)	4.842279e+01# (5.020350e+00)	4.279700e+01# (1.538301e+01)
	5	1.210334e+02 (2.192520e+00)	5.596308e+01# (5.654765e+00)	7.815456e+00# (1.225035e+00)	3.289189e+01# (2.858823e+00)	1.098994e+02# (4.122587e+00)	5.939438e+01# (3.673750e+01)
	10	2.186105e+02 (3.132639e-01)	6.927510e+01# (2.258115e+01)	2.480353e+00# (2.063527e+00)	3.476224e+01# (4.296041e+00)	1.950445e+02# (6.121614e+00)	1.138247e+02# (5.884228e+01)
WFG2^{-1}	3	1.140860e+02 (4.325357e-01)	9.532850e+01# (1.992380e+00)	8.890140e+01# (1.794800e-01)	9.018353e+01# (4.679018e-01)	3.230763e+00# (4.485117e-02)	2.757698e+00# (2.020626e-01)
	5	1.234346e+02 (7.027142e-01)	9.827363e+01# (2.826007e+00)	4.956629e+01# (8.529213e+00)	3.689443e+01# (1.559225e+00)	5.922537e+00# (7.879377e-01)	2.841850e+00# (6.498607e-01)
	10	2.196197e+02 (9.739353e-02)	2.001692e+02# (4.030917e+00)	2.500717e+01# (2.754994e+00)	1.075588e+02# (1.284360e+01)	1.138578e+01# (8.309404e-01)	7.783812e+00# (1.059255e+00)
WFG3^{-1}	3	1.075596e+02 (2.777510e-01)	7.484580e+01# (2.494935e+00)	6.164392e+01# (7.387154e-02)	7.097018e+01# (1.699291e-01)	2.303097e+01# (7.406376e-01)	2.381595e+01# (2.661523e-01)
	5	1.259055e+02 (2.786019e-02)	8.633630e+01# (4.246132e+00)	6.654930e+01# (3.739241e+00)	3.937358e+01# (1.177063e+00)	3.626087e+01# (2.423541e+00)	4.063424e+01# (2.006301e+00)
	10	2.199995e+02 (8.710466e-04)	1.929382e+02# (8.857025e+00)	5.251797e+01# (3.065445e+00)	1.414077e+02# (1.791224e+01)	7.135125e+01# (7.053632e+00)	9.336338e+01 (4.573319e+00)

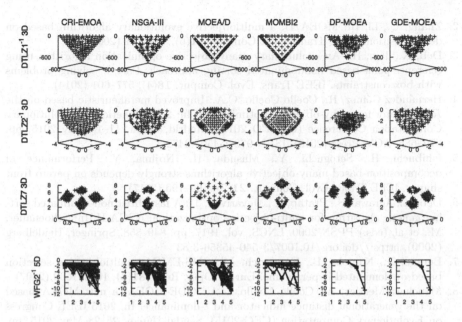

Fig. 2. Pareto fronts generated by CRI-EMOA and the adopted MOEAs. Each front corresponds to the median of the hypervolume value.

5 Conclusions and Future Work

In this paper, we propose an Evolutionary Multi-Objective Algorithm based on the combination of the Riesz s-energy and IGD^+ indicators. Our proposed approach, called CRI-EMOA, overcomes the overspecialization on certain benchmark problems of state-of-the-art MOEAs that employ a set of convex weight vectors as search directions, as a reference set or as part of a quality indicator. CRI-EMOA exploits the convergence property of IGD^+ and promotes evenly distributed solutions using Riesz s-energy. Our proposal was compared with MOEAs with and without the use of convex weight vectors. Our experimental results showed that our approach has a competitive performance on the DTLZ and WFG instances, while it outperforms the adopted MOEAs on the $DTLZ^{-1}$ and WFG^{-1} problems. These empirical results provide some evidence about CRI-EMOA being a more general multi-objective optimizer. As part of our future work, we are interested in improving the performance of CRI-EMOA on the original benchmark problems while preserving its good performance on the minus versions of the considered test suites.

References

1. Coello Coello, C.A., Lamont, G.B., Van Veldhuizen, D.A.: Evolutionary Algorithms for Solving Multi-Objective Problems, 2nd edn. Springer, New York (2007). https://doi.org/10.1007/978-0-387-36797-2. ISBN 978-0-387-33254-3

2. Zhang, Q., Li, H.: MOEA/D: a multiobjective evolutionary algorithm based on decomposition. IEEE Trans. Evol. Comput. **11**(6), 712–731 (2007)
3. Deb, K., Jain, H.: An evolutionary many-objective optimization algorithm using reference-point-based nondominated sorting approach, Part I: solving problems with box constraints. IEEE Trans. Evol. Comput. **18**(4), 577–601 (2014)
4. Hernández Gómez, R., Coello Coello, C.A.: Improved metaheuristic based on the R2 indicator for many-objective optimization. In: 2015 Genetic and Evolutionary Computation Conference (GECCO 2015), Madrid, Spain, 11–15 July 2015, pp. 679–686. ACM Press (2015). ISBN 978-1-4503-3472-3
5. Ishibuchi, H., Setoguchi, Y., Masuda, H., Nojima, Y.: Performance of decomposition-based many-objective algorithms strongly depends on pareto front shapes. IEEE Trans. Evol. Comput. **21**(2), 169–190 (2017)
6. Deb, K., Agrawal, S., Pratap, A., Meyarivan, T.: A fast elitist non-dominated sorting genetic algorithm for multi-objective optimization: NSGA-II. In: Schoenauer, M., et al. (eds.) PPSN 2000. LNCS, vol. 1917, pp. 849–858. Springer, Heidelberg (2000). https://doi.org/10.1007/3-540-45356-3_83
7. Beume, N., Naujoks, B., Emmerich, M.: SMS-EMOA: multiobjective selection based on dominated hypervolume. Eur. J. Oper. Res. **181**(3), 1653–1669 (2007)
8. Menchaca-Mendez, A., Coello Coello, C.A.: GDE-MOEA: a new MOEA based on the generational distance indicator and ε-dominance. In: 2015 IEEE Congress on Evolutionary Computation (CEC 2015), Sendai, Japan, 25–28 May 2015, pp. 947–955. IEEE Press (2015). ISBN 978-1-4799-7492-4
9. Menchaca-Mendez, A., Hernández, C., Coello Coello, C.A.: Δ_p-MOEA: a new multi-objective evolutionary algorithm based on the Δ_p indicator. In: 2016 IEEE Congress on Evolutionary Computation (CEC 2016), Vancouver, Canada, 24–29 July 2016, pp. 3753–3760. IEEE Press (2016). ISBN 978-1-5090-0623-9
10. Ishibuchi, H., Masuda, H., Tanigaki, Y., Nojima, Y.: Modified distance calculation in generational distance and inverted generational distance. In: Gaspar-Cunha, A., Henggeler Antunes, C., Coello, C.C. (eds.) EMO 2015. LNCS, vol. 9019, pp. 110–125. Springer, Cham (2015). https://doi.org/10.1007/978-3-319-15892-1_8
11. Hardin, D.P., Saff, E.B.: Discretizing manifolds via minimum energy points. Not. AMS **51**(10), 1186–1194 (2004)
12. Hardin, D.P., Saff, E.B.: Minimal riesz energy point configurations for rectifiable d-dimensional manifolds. Adv. Math. **193**(1), 174–204 (2005)
13. Ishibuchi, H., Tsukamoto, N., Sakane, Y., Nojima, Y.: Hypervolume approximation using achievement scalarizing functions for evolutionary many-objective optimization. In: 2009 IEEE Congress on Evolutionary Computation (CEC 2009), Trondheim, Norway, May 2009, pp. 530–537. IEEE Press (2009)
14. Emmerich, M.T.M., Deutz, A.H., Kruisselbrink, J.W.: On quality indicators for black-box level set approximation. In: Tantar, E., et al. (eds.) EVOLVE - A bridge between Probability, Set Oriented Numerics and Evolutionary Computation. Studies in Computational Intelligence, vol. 447, pp. 157–185. Springer, Heidelberg (2013). https://doi.org/10.1007/978-3-642-32726-1_4. Chap. 4. ISBN 978-3-642-32725-4

The Hypervolume Indicator as a Performance Measure in Dynamic Optimization

Sabrina Oliveira[1][(✉)], Elizabeth F. Wanner[1,2], Sérgio R. de Souza[1], Leonardo C. T. Bezerra[3], and Thomas Stützle[4]

[1] Centro Federal de Educação Tecnológica de Minas Gerais, Belo Horizonte, Brazil
{soliveira,efwanner,sergio}@decom.cefetmg.br
[2] Engineering and Applied Sciences, Aston University, Birmingham, England
e.wanner@aston.ac.uk
[3] IMD, Universidade Federal do Rio Grande do Norte, Natal, RN, Brazil
leobezerra@imd.ufrn.br
[4] IRIDIA, Université Libre de Bruxelles, Brussels, Belgium
stuetzle@ulb.ac.be

Abstract. In many real world problems the quality of solutions needs to be evaluated at least according to a bi-objective non-dominated front, where the goal is to optimize solution quality using as little computational resources as possible. This is even more important in the context of dynamic optimization, where quickly addressing problem changes is critical. In this work, we relate approaches for the performance assessment of dynamic optimization algorithms to the existing literature on bi-objective optimization. In particular, we introduce and investigate the use of the hypervolume indicator to compare the performance of algorithms applied to dynamic optimization problems. As a case study, we compare variants of a state-of-the-art dynamic ant colony algorithm on the traveling salesman problem with dynamic demands (DDTSP). Results demonstrate that our proposed approach accurately measures the desirable characteristics one expects from a dynamic optimizer and provides more insights than existing alternatives.

Keywords: Dynamic optimization · Multi-objective optimization · Performance assessment

1 Introduction

Real world optimization problems are often modeled as combinatorial optimization problems (COPs), which involve finding values for a set of discrete variables related to a given objective function. When the optimal solution cannot be efficiently obtained in practice, approximate algorithms such as heuristics and metaheuristics have been successfully applied to obtain near-optimal solutions.

© Springer Nature Switzerland AG 2019
K. Deb et al. (Eds.): EMO 2019, LNCS 11411, pp. 319–331, 2019.
https://doi.org/10.1007/978-3-030-12598-1_26

Dynamic environments, where instances are allowed to undergo some modifications over time, impose some additional challenges to COPs since such problems need to be re-optimized over time to ensure not only feasibility but also quality of the solutions.

Some well-established evolutionary computation and swarm intelligence techniques have been tailored to solve dynamic COPs (DCOPs) [6,8,11,15,21]. However, to guarantee the effectivenesses of the solution quality generated by an approximate algorithm, one has to execute it several times and solution quality of all executions should be compared in a way to prove whether their values are consistent. The literature is rich in examples of measures to assess the performance of algorithms applied to this context, but many measures (i) evaluate algorithms based solely on the quality of the final solution they produce [9,20], completely disregarding the performance of the algorithm during different re-optimization cycles (*environments*); (ii) require several measures to be combined in order to obtain knowledge about the behavior over dynamic changes and solution quality development [8,11], and/or; (iii) require the adaptation of measures that need a priori knowledge of the optimal solution for each problem change [10], since they were proposed in the context of artificially designed test problems where optimal solutions are known beforehand.

Aiming to overcome the issues discussed above, we propose a bi-objective formulation of DCOPs, where runtime and solution quality are considered objectives to be simultaneously optimized. This idea is largely inspired by [17], who adopt the hypervolume measure to assess the *anytime behavior* of heuristic algorithms. In this work, we extend that concept to comprehend problem changes, enabling the comparison of different algorithms between consecutive changes (*environment-wise analysis*) or during their entire execution (*scenario-wise analysis*). In addition, our approach is based on an unary version of the hypervolume indicator, which makes it scalable as to the number of algorithms considered in the analysis. Our formulation is also scalable as to the number of objectives considered, meaning one can use it to assess the performance of dynamic multi- objective optimizers; yet, in this work we focus on the assessment of traditional DCOP optimizers.

As a case study, we consider the traveling salesman problem with dynamic demands (TSPDD, [8,11]). Specifically, we assess the performance of different versions of the population-based ant colony algorithm (P-ACO, [5,8]), an effective approximate algorithm originally devised for DCOPs. Our proposed approach leads to interesting observations. In general, an algorithm that is tailored for DCOPs tends to be robust to problem changes, with similar performance patterns across different environments. Yet, the adoption of simple mechanisms such as parameter configuration and local search considerably affects its overall performance. More importantly, local search helps to very quickly re-optimize solutions when faced with problem changes, a counterintuitive result when one considers the computational overhead usually associated with local search components.

The remainder of the paper is outlined as follows. In Sect. 2, we briefly discuss the most important background concepts related to dynamic optimization, and present an overview of the measures most commonly adopted in this context. Next, we discuss the original anytime behavior formulation and its assessment through the hypervolume in Sect. 3, and extend it for the assessment of dynamic optimizers. We evaluate our proposal through a case study in Sect. 4, where we discuss both experimental setup and results. Finally, we conclude our work in Sect. 5, discussing future work possibilities.

2 Dynamic Optimization: Background and Measures

Without loss of generality, a *dynamic optimization problem* (DOP) can be defined as a problem in which the changes to its specifications are time-dependent. Different ways to address these changes can be found in literature. As an example, a DOP can be seen as a sequence of static optimization problems (SOP) over time. The goal for each SOP is to find a solution maximizing the fitness function of that SOP. Alternatively, a DOP can be also considered as the problem of adapting a solution to a changing fitness landscape. Whatever the interpretation, a dynamic *combinatorial* optimization problem (DCOP) can be defined as a straightforward variation of DOPs, where the problem has a discrete search space which consists of a finite set of solutions.

In this work, we refer to changes in problem specifications as *environment changes*, and to a time span between problem changes as an *environment*. At the beginning of each environment, the best solution from the previous environment most likely needs to be re-optimized, since it may be unfeasible and/or far from optimal. This likelihood is expected to increase as a function of the *degree of dynamism* of the problem, i.e., how strong environment changes are. The whole execution of an algorithm is dubbed a *scenario*, which may comprise different numbers of environments as a factor of the *frequency of change* the problem presents. Ideally, effective dynamic optimizers are those able to react to environment changes as fast as possible, despite how fast and/or strong these changes are. Many performance assessment measures have been proposed in the dynamic optimization literature and can be classified as either *final quality-based* or *behavior-based*. Below, we present a high-level overview of each category:

Final quality measures [2,3,7,9,15,19,20] are based on quality of the best solution found in each environment. For a scenario-wise analysis, performances across all environments are traditionally averaged. The major drawback with these approaches is the indifference to the search dynamics within environments. Specifically, a given algorithm may re-optimize solutions very quickly and still be considered equivalent to another that takes much longer to reach the same solution quality.

Behavioral measures [3,11,14,15,19] compare algorithms based on their search dynamics, providing more insights than final quality ones. Nonetheless, many of these measures may require the knowledge of the optimal solution to do so, or use an auxiliary measure that requires it. The only measure we identify without such restrictions is the *area between curves* (ABC, [1]), which considers that the performance of algorithms are time-quality fronts and measure the area between these fronts for each environment. Aggregation for a scenario-wise analysis is traditionally done through an algebraic sum of the areas identified in each environment.

Notice that the ABC measure is loosely related to the multi-objective optimization performance assessment literature. Specifically, when assessing a single environment this measure becomes close in spirit to the *binary hypervolume measure* [22]. Yet, differently from the hypervolume one cannot draw Pareto-compliant conclusions about the fronts being compared by the ABC, as we will later detail. In addition, this measure can only be applied to pairwise algorithm comparisons given its binary nature – the literature on binary measures for multi-objective optimization is clear that this is a non-scalable approach [22].

In the next section, we review another existing approach to a multi-objective formulation of algorithm performance, extend it to the context of dynamic optimization, and highlight the benefits of our approach over ABC.

3   Assessing the Anytime Behavior of a Dynamic Optimizer: The Hypervolume Approach

The *anytime behavior* of an algorithm is defined by [17] as the robustness of an algorithm to different stopping criteria. To compare different algorithms as to their anytime behavior, authors propose that the original optimization problem under investigation be reformulated as a bi-objective problem, through the addition of a resource-minimizing objective. In the most traditional scenario, one wants to optimize the *solution quality* of a target problem, and *runtime* is the resource whose consumption is to be minimized. Under this formulation, the performance of an algorithm $\mathcal{A}_i$ is a nondominated front comprising points $\langle t_i, q_i \rangle$, i.e., the solution quality q_i obtained at time t_i. Different algorithms can then be compared through the hypervolume they dominate, using a common reference point strictly dominated by all other points. Albeit simple, this approach is powerful in that (i) multiple algorithms can be simultaneously compared, and (ii) an algorithm that dominates a larger hypervolume cannot present a worse anytime behavior than one which dominates a smaller hypervolume (and vice-versa).

The application of the approach above to the context of dynamic optimization is straightforward when a single environment is considered. Yet, when problem changes are introduced, a few adjustments need to be made. To help illustrate these adjustments, Fig. 1 depicts the performance fronts of two dynamic optimizers (left-most plots) and the comparison of their hypervolumes (right-most plot). In all plots, runtime is given on the x-axis, while solution quality is given on the y-axis (w.l.o.g. we consider a solution quality minimization problem). The first issue for computing a scenario-wise hypervolume illustrated in this figure is that, if the whole run of an algorithm is considered as a single front, most of the front depicting a given environment will be considered dominated by the best solution of the previous environment.

An alternative is to consider environments separately and aggregate over environment-wise hypervolumes to draw scenario-wise conclusions. This approach requires a second adjustment in the methodology, namely that reference points for each environment be computed as x-axis translations of the scenario-wise reference point. In more detail, to ensure all environment- wise hypervolumes are comparable (and hence can be aggregated), all reference points considered must present the same solution-quality coordinate. In addition, the time coordinate of each reference point is computed such that it be strictly dominated by its environment front.[1] As we will later discuss, our case study presents fixed-duration environments and solution-quality ranges do not vary considerably across environments. Hence, we only apply a scenario-wise normalization to ensure both axes contribute equally to the hypervolumes. However, depending on the application considered axes normalization for each environment may also be necessary.

Concerning environment-wise analysis, our approach preserves the benefits of the original anytime behavior formulation, which greatly improve over the ABC metric. Specifically, the ABC metric is a particular case of the binary hypervolume metric where the reference point is only weakly dominated by the front assessed, and hence conclusions drawn from it cannot be guaranteed Pareto-compliant. More importantly, this poor choice of reference point (albeit implicit in the metric definition) means environment-final solutions are not properly valued. By contrast, as long as standard guidelines about the hypervolume are followed, these solutions are guaranteed to be properly valued.[2]

Regarding a scenario-wise analysis, the benefits of our approach vary as a function of the aggregation method considered. If one uses a rank sum analysis, one can understand how often one algorithm reacts more efficiently to problem changes than others. More importantly, if an algorithm A_1 presents larger hypervolumes than another algorithm A_2 on all environments, one can be sure that A_1 cannot present worse anytime behavior than A_2. Conversely, if one is more interested in average performance, it is straightforward to compare algorithms based on the average of the hypervolumes computed for each environment. This flexibility of aggregation approaches is another improvement over the ABC measure, specially given that an algebraic sum implicitly embedded in the ABC measure provides less information than the alternatives discussed here.

In the next section, we present a case study where we empirically evaluate our proposed approach.

---

[1] In practice, this requires isolating fronts from each environment before computing hypervolumes, since the reference point of a given environment may intersect with the next environment.

[2] One could argue that an application may require a custom importance distribution for the different stages of the run. This can be achieved through the *weighted hypervolume measure*, as proposed in [17].

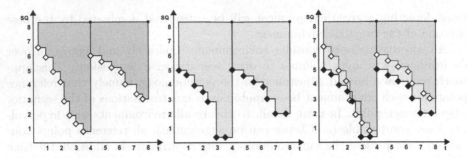

Fig. 1. The hypervolumes dominated by the performance curves of two different algorithms (left and center), computed as a function of reference points selected for each environment. The right-most plot shows a direct comparison between these algorithms of the hypervolume.

4   Experimental Study

The formulation and measure we propose to employ in this paper are general enough to assess the performance of any dynamic optimization algorithm on any given DOP. In this paper, we conduct an experimental evaluation using the traveling salesman problem with dynamic demands (TSPDD, [8,11]) as test benchmark and the population-based ant colony optimization algorithm (P-ACO, [5,8]) as target algorithm. We start our discussion by respectively reviewing the definitions of the TSPDD and P-ACO. Next, we detail the experimental setup we adopt, and later proceed to a discussion of the results observed.

4.1   Problem and Algorithm Definitions

The dynamic traveling salesman problem (TSP) is a variation of the static TSP in which the problem data changes over time [6,8,11,13]. In this paper, we consider the TSP with dynamic demands (TSPDD), which can be modeled by a sequence of graphs $G_s = (V_s, E_s)$, $s = 0, \ldots, S$, and two sequences of vertex sets A_s and D_s, $s = 1, \ldots, S - 1$, where A_s represents the set of new customers to be served and D_s the set of deleted customers. Each V_s is obtained by $(V_{s-1} \bigcup A_s)/D_s$ and $E_s = V_s \times V_s$. We follow [12] to produce problem changes, evenly splitting the set of customers into two sets called currentpool and sparepool; only currentpool customers must be visited in a given environment. An environment change consists of switching $\xi \cdot n$ vertices between currentpool and sparepool; parameter $\xi \in \{20\%, 40\%, 80\%\}$ is called *degree of dynamism*. A given run presents f environments, which we set in this paper to $f \in \{2, 10\}$. The maximum runtime allowed for a single run is evenly split between these environments, respectively meaning that changes occur after half and a tenth of the maximum allowed runtime.

P-ACO is a population-based ant colony optimization algorithm (P-ACO) that was originally developed for tackling the TSPDD [8] and has been used or solving DCOPs in general. Pheromone updates in P-ACO are only done based on solutions that enter or leave the solution archive. This scheme considerably reduces the computation time needed for the pheromone update when compared to classical ACO algorithms. In this work, we consider three P-ACO variants, and their parameter configurations are depicted in Table 1, where m is the number of ants, α and β respectively regulate the importance of pheromone and heuristic information, τ_{max} is the maximum pheromone deposit for a given edge, and K is the solution archive size. The first two variants, dubbed *static-* and *dynamic-default*, differ only as to their parameter configuration, using the values traditionally employed for solving static [16] and dynamic problems [11], respectively. The third variant, dubbed *static-default + ls*, differs from the previous two since it is the only variant that adopts a local search (LS) procedure. We remark that this configuration had previously been only applied to static problems, due to the expected computational overhead posed by local search procedures. Yet, we include this variant in our study given the important role that local search plays for ACO effectiveness on static COPs, providing algorithms a means to locally explore a neighborhood in the search space. Specifically, we adopt the 2-opt neighborhood operator with a first-improvement pivoting rule.

Table 1. Default parameters used in the literature for P-ACO. *Dynamic-default + ls* settings are not given because no study has yet investigated this setup.

Settings	m	α	β	τ_{max}	K
static-default	$n/4$	1	2	3	25
dynamic-default	50	1	5	3	3
static-default + ls	25	1	2	3	1

4.2   Experimental Setup

We adopt two sets of TSPDD instances, namely TSPLIB instances rl1323, u1817, rl1889, u2152, and pr2392 [18], and 3 subsets of 5 random uniform Euclidean (RUE) instances each, with sizes ranging from 3000 to 4000, created though the portgen generator from the 8th DIMACS implementation challenge [4]. Each instance is further parameterized by the degree of dynamism ($\xi \in \{20\%, 40\%, 80\%\}$) and frequency of change ($f \in \{2, 10\}$). Algorithms are allowed a maximum runtime of 1 000 s, which means environments last 500 s when $f = 2$ and 100 s when $f = 10$. To account for variance, each algorithm is run 20 times on each instance configuration, and results reported are averages of those runs.

The hypervolume computation requires a few pre-processing steps, as follows. First, a solution quality range is computed for each instance, and the runs of all algorithms are normalized so that the worst solution quality value ever found by an algorithm for a given instance corresponds to 0, and the best to 1. Time coordinates are normalized in the $[0, 1\,000]$ range, and hence both objectives contribute equally to the hypervolume. As previously discussed, the reference points for each environment are computed as x-axis translations of the global reference point $(1.1, 1.1)$. We ensure environment-specific reference points equally value extreme solutions by using the 1.1 ratio for both objectives. We assess scenario-wise conclusions for a given instance configuration using the most adequate aggregation approach, depending on the environment-wise results. For more general conclusions across all instance configurations, we conduct a rank sum analysis of the aggregation.

4.3   Results

We start our analysis with the help of solution quality over time (SQT) plots, given in Figs. 2 and 3, which respectively depict instances configured with $\xi = 20\%$ and $\xi = 80\%$. In both figures, a RUE instance is given on the top row, whereas a TSPLIB instance is given on the bottom row. On the left column, instances are configured with $f = 2$, whereas the right column depicts instances configured with $f = 10$. Note that the number of environments has a stronger influence on the final performance of the algorithm, since more environments mean less time for re-optimization. Conversely, the degree of dynamism influences the solution quality recovery after an environment, since the problem changes more or less drastically depending on this parameter.

We focus the remainder of the analysis on the most relevant insights from the direct comparison between (i) the two variants that do not use local search, and (ii) between the variants that do not use local search and the one that does.

Experiments without local search. The first, contrasting difference between the P-ACO variants that do not use local search, is that the one configured for dynamic optimization is much faster in (re-)optimizing solutions, yet reaches a much worse solution quality when compared to the variant configured for static optimization. This is a very interesting result, as it corroborates that dynamic algorithms need to be engineered with anytime behavior in mind, rather than just being fast or achieving a good final solution. Another interesting observation is that both variants react to the randomly produced changes is a very uniform pattern, and indeed the hypervolumes for all environments are very similar. In terms of anytime behavior, the variant configured for static optimization dominates a larger hypervolume, a fact confirmed by the environment-wise hypervolume computation, whatever instance configuration considered.

Experiments with local search. The overall improvement provided by local search is remarkable for most of the instance configurations considered. Indeed, the variant with local search is at least as fast as the variant configured for dynamic optimization, and reaches a final solution with at least the same quality as the variant configured for static optimization. A few factors affect this pattern to some extent. For instance, a larger number of environments further enhance the benefits of local search, a counterintuitive result

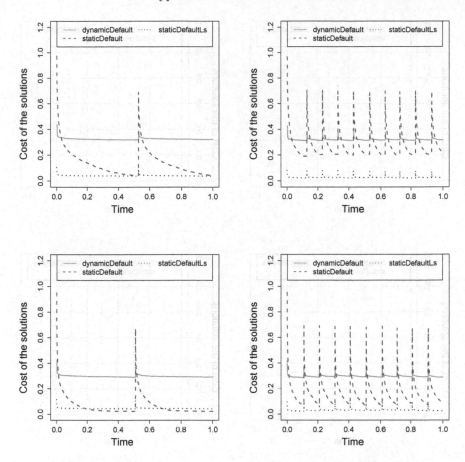

Fig. 2. SQT plot depicting the anytime performance of P-ACO, on RUE instance size 3500 (top row) and on TSPLIB instance pr2392 (bottom row). Left: scenario $\xi 20f2$. Right: scenario $\xi 20f10$.

given that less runtime is available and local search is known to be computationally costly. Conversely, the benefits for TSPLIB instances are less than for RUE instances, an understandable pattern given that the neighborhood operator we adopt is not particularly effective for TSPLIB. Indeed, when run on TSPLIB instances configured with $f = 2$, P-ACO retrieves better final solutions for each environment when not using local search. Yet, for all but two instance configurations considered, the environment-wise hypervolumes favor the variant that uses local search, indicating that it presents a better anytime behavior than the remaining variants.

The fact that the hypervolume indicates an algorithm as having better anytime behavior when it does not reach the best solution quality at the end of an environment (or the run) is a likely possibility. However, as already investigated elsewhere [17], it is possible to adopt the weighted variant of the hypervolume

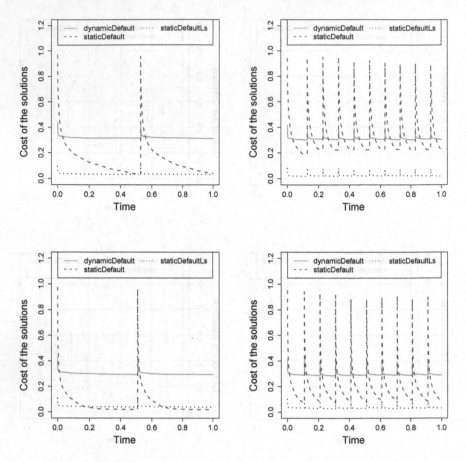

Fig. 3. SQT plot depicting the anytime performance of P-ACO, on RUE instance size 3500 (top row) and on TSPLIB instance pr2392 (bottom row). Left: scenario $\xi 80f2$. Right: scenario $\xi 80f10$.

indicator when one wants to change the importance distribution of the different regions of the SQT plot. A second important observation is that the clear performance patterns from each algorithm across environments mean that the algorithms that dominate larger hypervolumes can be guaranteed to not have worse anytime performance than algorithms with smaller hypervolumes. In this context, averaging hypervolumes from the different environments is only helpful for the rank sum analysis we conduct next.

Table 2 shows results that aggregate from all instance configurations considered. Each column depicts the rank sum achieved by each P-ACO variant assessed, and the last column (ΔR) gives the difference in ranks above which the lowest-ranked algorithm can be considered statistically significant better than the remaining algorithms, according to Friedman's non-parametrical test with 95% confidence. Results show a clear separation between variants, and as expected the one that uses local search achieves a lower average hypervolume much more often than the remaining ones.

Table 2. Statistical analysis of the different P-ACO variants on RUE and TSPLIB instances aggregate over all instance configurations considered. Each run of an algorithm is ranked according to the average of the environment-wise hypervolumes they dominated. ΔR gives the difference of the sum of ranks that is statistically significant according to Friedman's non-parametrical test with a confidence level of 95%. The best variant, that is significantly different from the others, is indicated in bold face.

Instances	$dynamicDefault$	$staticDefault$	$\textbf{staticDefaultLs}$	ΔR
RUE + TSPLIB	51	32	**19**	3.94

5 Conclusions

Dynamic optimization problems demand algorithms engineered to quickly produce high-quality solutions, particularly after problem data changes. This dichotomy can be more accurately addressed when formulated as a bi-objective optimization problem, where solution quality and runtime are the most prototypical objective examples. In this paper, we have extended this formulation and applied the concept of anytime behavior to dynamic optimizers. Through this formulation, one is able to use the hypervolume to compare any number of algorithms and, under certain circumstances, benefit from the Pareto-compliance property of this indicator. Together, these characteristics greatly improve over previous measures adopted in the dynamic optimization literature.

To empirically evaluate our proposed approach, we conducted an experimental study on the performance of different variants of the population-based ant colony optimization algorithm run on the traveling salesman problem with dynamic demands. Surprisingly, the variant configured for static optimization performed better than the one configured for dynamic optimization. More importantly, we have seen that local search is a critical component even in the context of dynamic optimization, leading to the best anytime behavior on most of the experiments conducted.

The formulation adopted in this work opens a number of possibilities for dynamic optimization performance assessment. One is to consider more objectives, which can be other resource consumption objectives, such as number of function evaluations, or even dynamic multi-objective optimization problems. Another important research direction is to adapt existing techniques employed in the assessment of static multi-objective optimizers to deal with dynamic problems, such as empirical attainment functions.

References

1. Alba, E., Sarasola, B.: ABC, a new performance tool for algorithms solving dynamic optimization problems. In: Proceedings of the 2010 IEEE Congress on Evolutionary Computation, pp. 1–7 (2010)
2. Ben-Romdhane, H., Alba, E., Krichen, S.: Best practices in measuring algorithm performance for dynamic optimization problems. Soft Comput. **17**(6), 1005–1017 (2013)
3. Cruz, C., Gonzalez, J.R., Pelta, D.A.: Optimization in dynamic environments: a survey on problems, methods and measures. Soft Comput. **15**(7), 1427–1448 (2011)
4. 8th DIMACS Implementation Challenge: The Traveling Salesman Problem (2018). http://dimacs.rutgers.edu/archive/Challenges/TSP/
5. Dorigo, M., Montes de Oca, M.A., Oliveira, S., Stützle, T.: Ant colony optimization. In: Cochran, J.J., Cox, L.A., Keskinocak, P., Kharoufeh, J.P., Smith, J.C. (eds.) Wiley Encyclopedia of Operations Research and Management Science. Wiley (2011)
6. Eyckelhof, C.J., Snoek, M.: Ant systems for a dynamic TSP. In: Dorigo, M., Di Caro, G., Sampels, M. (eds.) ANTS 2002. LNCS, vol. 2463, pp. 88–99. Springer, Heidelberg (2002). https://doi.org/10.1007/3-540-45724-0_8
7. Feng, W., Brune, T., Chan, L., Chowdhury, M., Kuek, C.K., Li, Y.: Benchmarks for testing evolutionary algorithms. In: Proceedings of the 3rd Asia-Pacific Conference on Control and Measurement, pp. 134–138 (1998)
8. Guntsch, M.: Ant algorithms in stochastic and multi-criteria environments. Ph.D. thesis, Universität Fridericiana zu Karlsruhe (2004)
9. Mavrovouniotis, M., Müller, F.M., Yang, S.: Ant colony optimization with local search for dynamic traveling salesman problems. IEEE Trans. Cybern. **47**(7), 1743–1756 (2017)
10. Mavrovouniotis, M., Yang, S.: Ant colony optimization with immigrants schemes in dynamic environments. In: Schaefer, R., Cotta, C., Kołodziej, J., Rudolph, G. (eds.) PPSN 2010. LNCS, vol. 6239, pp. 371–380. Springer, Heidelberg (2010). https://doi.org/10.1007/978-3-642-15871-1_38
11. Mavrovouniotis, M., Yang, S.: A memetic ant colony optimization algorithm for the dynamic travelling salesman problem. Soft Comput. **15**(7), 1405–1425 (2011)
12. Mavrovouniotis, M., Yang, S.: Ant colony optimization with memory-based immigrants for the dynamic vehicle routing problem. In: Proceedings of the 2012 IEEE Congress on Evolutionary Computation, pp. 2645–2652 (2012)
13. Melo, L., Pereira, F., Costa, E.: Multi-caste ant colony algorithm for the dynamic traveling salesperson problem. In: Tomassini, M., Antonioni, A., Daolio, F., Buesser, P. (eds.) ICANNGA 2013. LNCS, vol. 7824, pp. 179–188. Springer, Heidelberg (2013). https://doi.org/10.1007/978-3-642-37213-1_19
14. Mori, N., Kita, H., Nishikawa, Y.: Adaptation to changing environments by means of the memory based thermodynamical genetic algorithm. Trans. Inst. Syst. Control Inf. Eng. **14**(1), 33–41 (2001)
15. Nguyen, T.T., Yang, S., Branke, J.: Evolutionary dynamic optimization: a survey of the state of the art. Swarm Evol. Comput. **6**, 1–24 (2012)
16. Oliveira, S.M., Hussin, M.S., Stützle, T., Roli, A., Dorigo, M.: A detailed analysis of the population-based ant colony optimization algorithm for the TSP and the QAP. In: Proceedings of the 13th GECCO, pp. 13–14 (2011)

17. Radulescu, A., López-Ibáñez, M., Stützle, T.: Automatically improving the anytime behaviour of multiobjective evolutionary algorithms. In: Purshouse, R.C., Fleming, P.J., Fonseca, C.M., Greco, S., Shaw, J. (eds.) EMO 2013. LNCS, vol. 7811, pp. 825–840. Springer, Heidelberg (2013). https://doi.org/10.1007/978-3-642-37140-0_61

18. TSPLIB (2008). http://comopt.ifi.uni-heidelberg.de/software/TSPLIB95/

19. Weicker, K.: Performance measures for dynamic environments. In: Guervós, J.J.M., Adamidis, P., Beyer, H.-G., Schwefel, H.-P., Fernández-Villacañas, J.-L. (eds.) PPSN 2002. LNCS, vol. 2439, pp. 64–73. Springer, Heidelberg (2002). https://doi.org/10.1007/3-540-45712-7_7

20. Yang, S.: Memory-based immigrants for genetic algorithms in dynamic environments. In: Proceedings of the 7th GECCO, pp. 1115–1122 (2005)

21. Yang, S., Jiang, Y., Nguyen, T.T.: Metaheuristics for dynamic combinatorial optimization problems. IMA J. Manag. Math. 24(4), 451–480 (2012)

22. Zitzler, E., Thiele, L., Laumanns, M., Fonseca, C.M., da Fonseca, V.G.: Performance assessment of multiobjective optimizers: an analysis and review. IEEE Trans. Evol. Comput. 7, 117–132 (2003)

Comparison of Hypervolume, IGD and IGD+ from the Viewpoint of Optimal Distributions of Solutions

Hisao Ishibuchi[1]($\boxtimes$), Ryo Imada[2], Naoki Masuyama[2], and Yusuke Nojima[2]

[1] Shenzhen Key Laboratory of Computational Intelligence,
University Key Laboratory of Evolving Intelligent Systems of Guangdong
Province, Department of Computer Science and Engineering,
Southern University of Science and Technology, Shenzhen 518055, China
hisao@sustc.edu.cn
[2] Department of Computer Science and Intelligent Systems,
Graduate School of Engineering, Osaka Prefecture University,
1-1 Gakuen-cho, Naka-ku, Sakai, Osaka 599-8531, Japan
ryo.imada@ci.cs.osakafu-u.ac.jp,
{masuyama,nojima}@cs.osakafu-u.ac.jp

Abstract. Hypervolume (HV) and inverted generational distance (IGD) have been frequently used as performance indicators to evaluate the quality of solution sets obtained by evolutionary multiobjective optimization (EMO) algorithms. They have also been used in indicator-based EMO algorithms. In some studies on many-objective problems, only the IGD indicator was used due to a large computation load of HV calculation. However, the IGD indicator is not Pareto compliant. This means that a better solution set in terms of the Pareto dominance relation can be evaluated as being worse. Recently the IGD plus (IGD+) indicator has been proposed as a weakly Pareto compliant version of IGD. In this paper, we compare these three indicators from the viewpoint of optimal distributions of solutions. More specifically, we visually demonstrate similarities and differences among the three indicators by numerically calculating near-optimal distributions of solutions to optimize each indicator for some test problems. Our numerical analysis shows that IGD+ is more similar to HV than IGD whereas the formulations of IGD and IGD+ are almost the same.

Keywords: Indicator-based multiobjective algorithms · Hypervolume (HV) ·
Inverted generational distance (IGD) ·
Inverted generational distance plus (IGD+)

1 Introduction

The hypervolume (HV) indicator [28] has been frequently used to evaluate the quality of solution sets obtained by evolutionary multiobjective optimization (EMO) algorithms (i.e., to compare EMO algorithms). Whereas a number of performance indicators have been proposed in the literature [29], there is no other Pareto compliant unary

© Springer Nature Switzerland AG 2019
K. Deb et al. (Eds.): EMO 2019, LNCS 11411, pp. 332–345, 2019.
https://doi.org/10.1007/978-3-030-12598-1_27

indicator known so far [27]. This is the main reason why the HV indicator has been almost always used for performance comparison of EMO algorithms. The HV indicator has also been frequently used in indicator-based algorithms such as SMS-EMOA [3, 9], HypE [2] and FV-MOEA [19]. As reported in the literature [25], HV-based algorithms often have higher search ability for many-objective problems than Pareto dominance-based algorithms (e.g., NSGA-II [6]).

However, HV calculation needs a huge computation load for a large solution set of a many-objective problem (e.g., 500 non-dominated solutions of a 15-objective problem). As a result, usually the HV indicator is not used in performance comparison of EMO algorithms for many-objective problems with more than ten objectives. For such a many-objective problem, the inverted generational distance (IGD [5, 22]) indicator is usually used. The IGD indicator has also been used in indicator-based EMO algorithms [23, 24]. The main advantage of IGD over HV is its computational efficiency. However, the IGD indicator is not Pareto compliant. This means that a better solution set in terms of the Pareto dominance relation can be evaluated as being worse by the IGD indicator (see [29] for the Pareto dominance-based "better" relation between solution sets).

Recently the IGD plus (IGD$^+$) indicator was proposed in [15] as a weakly Pareto compliant version of IGD. Performance comparison results by the IGD$^+$ indicator are never inconsistent with the Pareto dominance relation. That is, when a solution set A is better than another solution set B in terms of the Pareto dominance relation, B is never evaluated as being better than A by the IGD$^+$ indicator. In this case, B can be evaluated as being better than A by the IGD indicator (as explained later in this paper). The weak Pareto compliance is the main advantage of IGD$^+$ over IGD. The IGD$^+$ indicator has also been used in indicator-based algorithms (e.g., [20]). A similar idea to IGD$^+$ was utilized to modify IGD in the indicator-based algorithm of Sun et al. [23].

Comparison of different indicators has been performed by examining the consistency among performance comparison results through computational experiments (e.g., [18, 21]). In such a computational experiment, multiple solution sets were compared and ranked by each indicator. Then the consistency in the ranking by each indicator was analyzed. Whereas we can observe similarities (and dissimilarities) among performance indicators from the consistency analysis of ranking results, characteristic features of each indicator are still unclear. In this paper, we visually compare the HV, IGD and IGD$^+$ indicators by showing near-optimal distributions of solutions for each indicator. This is to clearly explain what type of solution sets will be highly evaluated (i.e., favored) by each indicator. This is also to clearly explain what type of solution sets will be obtained by using each indicator in EMO algorithms.

This paper is organized as follows. In Sect. 2, the difference between the IGD and IGD$^+$ indicators is briefly explained. In Sect. 3, we discuss optimal distributions of solutions for two-objective problems with linear, convex and concave Pareto fronts. We show numerically obtained near-optimal distributions of solutions for each indicator. Our experimental results demonstrate a clear similarity in the obtained solution sets for optimizing the HV and IGD$^+$ indicators. In Sect. 4, we discuss optimal distributions of solutions for three-objective problems with six types of Pareto fronts, which are generated by combining three curvature properties (i.e., linear, convex and concave) and two shape properties (i.e., triangular and inverted triangular). As in the case of two objectives in Sect. 3, a clear similarity between the HV and IGD$^+$ indicators is observed in Sect. 4 for the case of three objectives. Our experimental results also demonstrate a strong dependency of the optimal distribution of solutions for HV maximization on the choice of a reference point for HV calculation when test problems have inverted triangular Pareto fronts. In Sect. 5, we conclude this paper. We also discuss future research directions in Sect. 5.

2   IGD and IGD$^+$ Indicators

Both the IGD and IGD$^+$ indicators need a set of reference points on the Pareto front. Let $Z = \{z_1, z_2, \ldots, z_{|Z|}\}$ be a set of reference points where $z_j = (z_{j1}, z_{j2}, \ldots, z_{jm})$ is a point on the Pareto front in an m-dimensional objective space. The IGD value of a non-dominated solution set $A = \{a_1, a_2, \ldots, a_{|A|}\}$ is calculated as follows (where $a_i = (a_{i1}, a_{i2}, \ldots, a_{im})$ is a solution in the m-dimensional objective space):

$$IGD(A) = \frac{1}{|Z|} \sum_{j=1}^{|Z|} \min_{a_i \in A} d(a_i, z_j), \tag{1}$$

where $d(a_i, z_j)$ is the Euclidean distance between a_i and z_j in the objective space. This definition shows that the IGD value is the average distance from each reference point to the nearest solution. The Euclidean distance between a solution $a = (a_1, a_2, \ldots, a_m)$ and a reference point $z = (z_1, z_2, \ldots, z_m)$ is calculated in the IGD indicator as

$$d(a, z) = \sqrt{\sum_{k=1}^{m} (a_k - z_k)^2}. \tag{2}$$

In the IGD$^+$ indicator, the distance calculation is slightly modified. For minimization problems, the distance between a and z is calculated as follows [15]:

$$d_{IGD^+}(a, z) = \sqrt{\sum_{k=1}^{m} (\max\{a_k - z_k, 0\})^2}. \quad \text{(Minimization Problems)} \tag{3}$$

When the solution a is dominated by the reference point z, this is exactly the same as the Euclidean distance since $a_k \geq z_k$ for all k. In (3), when the solution a is not inferior to the reference point z with respect to the kth objective (i.e., when $a_k \leq z_k$ for minimization problems), the kth objective has no effect on the distance calculation.

For maximization problems, the distance between the solution a and the reference point z is calculated in the IGD$^+$ indicator as follows [15]:

$$d_{IGD^+}(a, z) = \sqrt{\sum_{k=1}^{m} (\max\{z_k - a_k, 0\})^2}. \quad \text{(Maximization Problems)} \tag{4}$$

As in (3), when the solution a is dominated by the reference point z, this distance is exactly the same as the Euclidean distance since $a_k \leq z_k$ for all k. When the solution a is not inferior to the reference point z with respect to the kth objective (i.e., when $a_k \geq z_k$ for maximization problems), the kth objective has no effect in (4).

The distance calculation in the IGD and IGD$^+$ indicators is illustrated in Fig. 1 for a minimization problem where six reference points and four solutions are given. The IGD indicator is the average distance from each reference point to the nearest solution (i.e., the average length of the six arrows in Fig. 1(a)). The IGD$^+$ indicator is the average length of the six arrows in Fig. 1(b), i.e., the average distance from each reference point to the nearest point in the dominated region by the solution set. Let us assume that we have two solution sets A and B where A is better than B with respect to the Pareto dominance relation (i.e., B is dominated by A). In this case, B can be evaluated as being better than A by the IGD indicator as shown in Fig. 2(a). However, B cannot be evaluated as being better than A by the IGD$^+$ indicator. This is because the dominated region by B is included in the dominated region by A as shown in Fig. 2(b). Such an inclusion relation between the dominated regions by the two solution sets is directly related to the weak Pareto compliant property of the IGD$^+$ indicator.

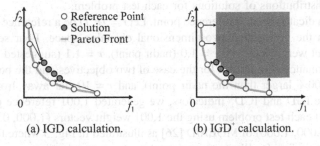

(a) IGD calculation.   (b) IGD$^+$ calculation.

Fig. 1. Illustration of the IGD and IGD$^+$ indicators for a minimization problem.

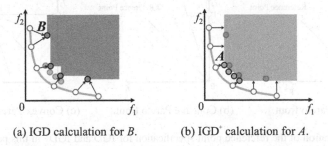

(a) IGD calculation for B.   (b) IGD$^+$ calculation for A.

Fig. 2. Comparison of two solution sets A and B by the IGD and IGD$^+$ indicators. The solution set A is correctly evaluated as being better than B in (b) by the IGD$^+$ indicator whereas B is evaluated as being better than A in (a) by the IGD indicator.

3   Optimal Distributions for Two-Objective Problems

The optimal distribution of solutions for HV maximization was theoretically derived
for two-objective problems in [1, 4, 10]. It was shown that the HV indicator is maximized by evenly distributed solutions when the Pareto front is linear. When the Pareto
front is nonlinear, it was shown that the optimal distribution of solutions depends on
the slope of the Pareto front. For multi-objective problems with three or more objectives, the optimal distribution was empirically discussed in [12, 13]. The optimal
distribution of solutions for IGD minimization was discussed only for the case of linear
Pareto fronts in [14]. The IGD$^+$ indicator has not been analyzed from the viewpoint of
the optimal distribution of solutions.

In this section, we show numerically obtained near-optimal distributions of solutions for optimizing each indicator for two-objective problems. We used the framework
of SMS-EMOA [3, 9] to optimize each indicator for three types of Pareto fronts: linear,
concave and convex. We used two-objective versions of DTLZ1 (linear Pareto front
[8]), DTLZ2 (concave [8]), and Minus-DTLZ2 (convex [16]). We normalized the
objective space of each test problem so that the nadir point and the ideal point are (1, 1)
and (0, 0) in the normalized objective space, respectively. This is for easy comparison
of obtained distributions of solutions for each test problem.

The HV indicator needs a reference point. Let $r = (r, r)$ be a reference point for HV
calculation in the normalized two-dimensional objective space. Four settings of the
reference point were examined: $r = 1.0$ (nadir point), $r = 1.1$ (suggested value in [13]
for fair performance comparison for the case of two objectives and the population size
11), $r = 2$ (100% larger than the nadir point), and $r = 10$ (far away from the Pareto
front). For the IGD and IGD$^+$ indicators, we generated 1,001 reference points on the
Pareto front of each test problem using the 1,001 weight vectors (1.000, 0.000), (0.999,
0.001), ..., (0.000, 1.000) of MOEA/D [26] as illustrated in Fig. 3 where the number of
weight vectors is five for illustration purposes.

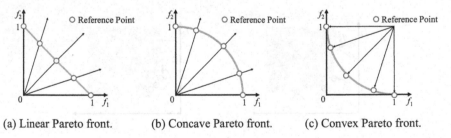

(a) Linear Pareto front.   (b) Concave Pareto front.   (c) Convex Pareto front.

Fig. 3. Illustration of the reference point specification for IGD and IGD$^+$ in this paper. Whereas
five points are generated in Fig. 3 for illustration purposes, the number of reference points in our
computational experiments is 1,001 for two-objective problems.

An indicator-based algorithm with the SMS-EMOA framework was applied to each test problem for optimizing each indicator under the following setting:

Number of distance variables: 0,
Number of position variables: $m - 1$ (m: Number of objectives),
Population size: 11,
Crossover: SBX with the index 20 (probability: 1.0),
Mutation: Polynomial mutation with the index 20
    (probability: $1/L$ where L is the string length),
Termination condition: 100,000 generations.

The number of distance variables (i.e., k in the DTLZ test suite) was specified as 0. This means that all feasible solutions are Pareto optimal. In this manner, we can focus on the optimization of the distribution of solutions on the Pareto front. That is, the role of the optimization algorithm under this setting is to adjust the location of each solution on the Pareto front for optimizing each indicator. The execution of the optimization algorithm was iterated 11 times for each test problem. The best result among those 11 runs with respect to the corresponding indicator is reported in this paper as a near-optimal distribution of solutions. Experimental results are shown in Figs. 4, 5 and 6.

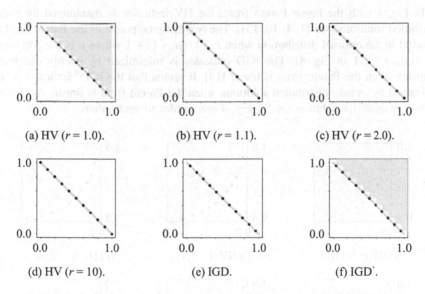

(a) HV ($r = 1.0$).   (b) HV ($r = 1.1$).   (c) HV ($r = 2.0$).

(d) HV ($r = 10$).   (e) IGD.   (f) IGD$^+$.

Fig. 4. Results on the normalized DTLZ1 problem with the linear Pareto front.

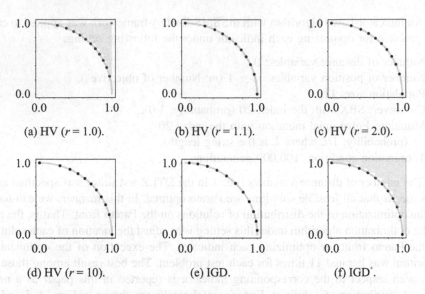

(a) HV ($r = 1.0$).   (b) HV ($r = 1.1$).   (c) HV ($r = 2.0$).

(d) HV ($r = 10$).   (e) IGD.   (f) IGD$^+$.

Fig. 5. Results on the normalized DTLZ2 problem with the concave Pareto front.

In Fig. 4 with the linear Pareto front, the HV indicator is maximized by evenly distributed solutions (see [1, 4, 10, 13]). The two extreme points of the Pareto front are included in the optimal distribution when $r \geq 1/(\mu - 1) + 1$ where μ is the population size (i.e., $r \geq 1.1$ in Fig. 4). The IGD indicator is minimized by evenly distributed solutions when the Pareto front is linear [14]. It seems that the IGD$^+$ indicator is also minimized by evenly distributed solutions when the Pareto front is linear. As a result, all the obtained six solution sets in Fig. 4 are similar to each other.

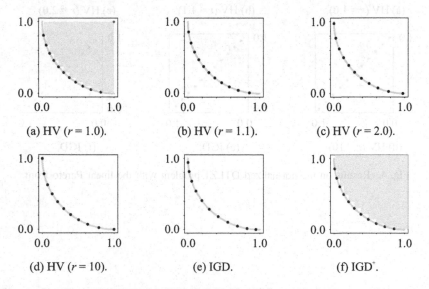

(a) HV ($r = 1.0$).   (b) HV ($r = 1.1$).   (c) HV ($r = 2.0$).

(d) HV ($r = 10$).   (e) IGD.   (f) IGD$^+$.

Fig. 6. Results on the normalized Minus-DTLZ2 problem with the convex Pareto front.

When the Pareto front is nonlinear in Figs. 5 and 6, the optimal distribution of solutions for HV maximization depends on the slope of the Pareto front. The highest density of solutions is at the region with the 45° slope. The density decreases as the slope decreases or increases from the 45° slope. Thus more solutions are obtained around the center of the Pareto front in Figs. 5(a)–(d) and 6(a)–(d). When the Pareto front is convex in Fig. 6, solutions around the two extreme points of the Pareto front cannot have large HV contributions. As a result, no solutions around the two extreme points were obtained even in Fig. 6(c) with $r = 2.0$. On the contrary, the two extreme points have large HV contributions in the case of the concave Pareto front in Fig. 5. Thus, they were obtained in Fig. 5(b)–(d). Independent of the shape of the Pareto front, the two extreme points cannot be obtained when the nadir point (i.e., $r = 1.0$) is used as the reference point for HV calculation. This is because their HV contributions are always zero when $r = 1.0$.

From Figs. 4, 5 and 6(e), we can see that similar distributions were obtained for IGD minimization independent of the shape of the Pareto front. IGD minimization is to minimize the average distance from each reference point on the Pareto front to the nearest solution. This problem can be viewed as a clustering problem of the given reference points. In Figs. 4, 5 and 6(e), the given 1,001 reference points are divided into 11 clusters. That is, the Pareto front in each figure is divided into 11 lines or curves. The nonlinearity of each segment (whose length is about 1/11 of the Pareto front) is weak. That is, each segment is similar to a line even when the Pareto front is nonlinear. As a result, similar distributions were obtained for IGD minimization in Figs. 4, 5 and 6(e).

An interesting observation in Figs. 5 and 6 is that the results in (f) for IGD$^+$ minimization are more similar to those in (b) for HV maximization with $r = 1.1$ than those in (e) for IGD minimization. This can be explained by the IGD$^+$ calculation mechanism illustrated in Fig. 2(b). From Fig. 2(b), we can see that the calculation of IGD$^+$ is to approximately evaluate the difference between the true Pareto front and the dominated region by a solution set. The minimization of this difference is the maximization of the HV of a solution set. That is, IGD$^+$ minimization is closely related to HV maximization. Thus, similar distributions of solutions were obtained in Figs. 5 and 6 from IGD$^+$ minimization in (f) and HV maximization in (b). This similarity can be also explained by the slope of the Pareto front. As shown in Figs. 5(f) and 6(f), the Pareto front is very close to the dominated region by the obtained solutions in the regions where the slope is close to 0 or 90°. Thus, many solutions are not needed for IGD$^+$ minimization in those regions. More solutions are needed around the center of the Pareto front with the 45° slope. The importance of the two extreme points for IGD$^+$ minimization depends on the shape of the Pareto front (i.e., concave or convex). These characteristics of the optimal distribution for IGD$^+$ minimization are the same as those for HV maximization.

4   Optimal Distributions for Three-Objective Problems

In this section, we show numerically obtained near-optimal distributions of solutions for optimizing each indicator for three-objective problems in the same manner as in the previous section. We used six test problems, which correspond to the six combinations of three curvature properties (linear, concave and convex) and two shape properties (triangular and inverted triangular) of the Pareto front as summarized in Table 1. All test problems were normalized so that the ideal point and the nadir point are $(0, 0, 0)$ and $(1, 1, 1)$ in the normalized objective space, respectively.

Table 1. Six three-objective test problems.

Problem	DTLZ1 [8]	Minus-DTLZ1 [16]	DTLZ2 [8]	Inverted DTLZ2 [17]	Convex DTLZ2 [7]	Minus-DTLZ2 [16]
Curvature	Linear	Linear	Concave	Concave	Convex	Convex
Shape	Triangular	Inverted triangular	Triangular	Inverted triangular	Triangular	Inverted triangular

As in the previous section, we used the framework of SMS-EMOA to optimize each indicator for the six three-objective problems. The population size was specified as 66. Four settings of the reference point $r = (r, r, r)$ were examined: $r = 1.0$ (nadir point), $r = 1.1$ (suggested value in [13] for fair performance comparison for the case of three objectives and the population size 66), $r = 2$ (100% larger than the nadir point), and $r = 10$ (far away from the Pareto front). For the IGD and IGD$^+$ indicators, we generated 5,151 reference points on the Pareto front of each test problem using the 5,151 weight vectors $(1.00, 0.00, 0.00)$, $(0.99, 0.01, 0.00)$, ..., $(0.00, 0.00, 1.00)$ of MOEA/D [26]. For each test problem with a triangular Pareto front, the weight vectors were used from the ideal point as in Fig. 3(a) and (b) to generate 5,151 reference points. The generated reference points were rotated to use them for the corresponding rotated test problem with an inverted triangular Pareto front.

Experimental results are shown in Figs. 7, 8, 9, 10, 11 and 12. In Fig. 7 with the linear triangular Pareto front, the effect of the location of the reference point for HV calculation is small. The effect of the location of the reference point is also small in Fig. 9 with the concave triangular Pareto front. Frequently-used test problems such as DTLZ1-4 [8] and WFG4-9 [11] have linear or concave triangular Pareto fronts. Thus, the importance of the location of the reference point for HV calculation has not been stressed in the EMO community. However, when the Pareto front is inverted triangular (Figs. 8, 10 and 12), the location of the reference point has a large effect on the optimal distribution of solutions for HV maximization. This means that HV-based comparison results strongly depend on the location of the reference point.

Among Figs. 7, 8, 9, 10, 11 and 12, only when the Pareto front is linear triangular in Fig. 7, similar results are obtained from the six settings of the indicators (i.e., HV with the four settings of the reference point, IGD and IGD$^+$). In Fig. 8 with the linear inverted triangular Pareto front, similar results are obtained from HV with $r = 1.1$, IGD and IGD$^+$. These observations suggest that similar comparison results are obtained from the three indicators when the Pareto front is linear (and the reference point for HV calculation is appropriately specified in the case of the inverted triangular Pareto front).

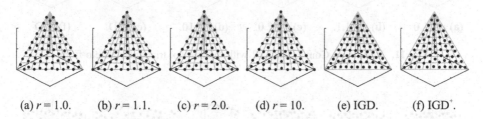

(a) $r = 1.0$.   (b) $r = 1.1$.   (c) $r = 2.0$.   (d) $r = 10$.   (e) IGD.   (f) IGD$^+$.

Fig. 7. Results on DTLZ1 with the linear triangular Pareto front.

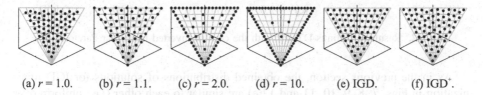

(a) $r = 1.0$.   (b) $r = 1.1$.   (c) $r = 2.0$.   (d) $r = 10$.   (e) IGD.   (f) IGD$^+$.

Fig. 8. Results on Minus-DTLZ1 with the linear inverted triangular Pareto front.

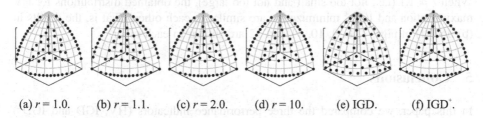

(a) $r = 1.0$.   (b) $r = 1.1$.   (c) $r = 2.0$.   (d) $r = 10$.   (e) IGD.   (f) IGD$^+$.

Fig. 9. Results on DTLZ2 with the concave triangular Pareto front.

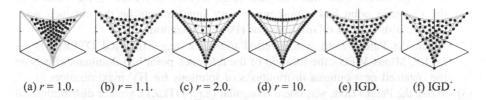

(a) $r = 1.0$.   (b) $r = 1.1$.   (c) $r = 2.0$.   (d) $r = 10$.   (e) IGD.   (f) IGD$^+$.

Fig. 10. Results on Inverted DTLZ2 with the concave inverted triangular Pareto front.

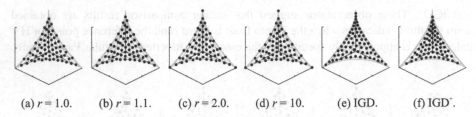

(a) $r = 1.0$.   (b) $r = 1.1$.   (c) $r = 2.0$.   (d) $r = 10$.   (e) IGD.   (f) IGD⁺.

Fig. 11. Results on Convex DTLZ2 with the convex triangular Pareto front.

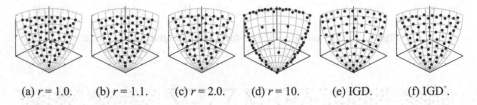

(a) $r = 1.0$.   (b) $r = 1.1$.   (c) $r = 2.0$.   (d) $r = 10$.   (e) IGD.   (f) IGD⁺.

Fig. 12. Results on Minus-DTLZ2 with the convex inverted triangular Pareto front.

As in the previous section, the obtained distributions of solutions for IGD minimization in Figs. 7, 8, 9, 10, 11 and 12(e) are similar to each other (i.e., uniform over the entire Pareto front). On the contrary, the obtained distributions of solutions for HV maximization and IGD⁺ distributions strongly depend on the shape of the Pareto front. When $r = 1.1$ (i.e., not too small and not too large), the obtained distributions for HV maximization and IGD⁺ minimization are similar to each other. That is, the results in (b) and (f) in Figs. 7, 8, 9, 10, 11 and 12 are similar to each other.

5   Conclusions

In this paper, we compared the three performance indicators (HV, IGD and IGD⁺) through computational experiments where near-optimal distributions of solutions were found for each indicator. We obtained the following observations:

(1)   When the Pareto front was triangular (e.g., DTLZ1-4, WFG4-9), the location of the reference point had almost no or very small effects on the obtained near-optimal distributions of solutions for HV maximization.

(2)   When the Pareto front was inverted triangular (e.g., Inverted DTLZ, Minus-DTLZ, Minus-WFG), the location of the reference point had dominant effects on the obtained near-optimal distributions of solutions for HV maximization.

(3)   When the Pareto front was linear triangular (e.g., DTLZ1), similar distributions of solutions were obtained for the three indicators.

(4)   Evenly distributed solutions were always obtained for IGD minimization independent of the shape of the Pareto front.

(5) When the reference point was not too small and not too large (i.e., when it was specified by the method in [13]), similar distributions of solutions were obtained for HV maximization and IGD$^+$ minimization independent of the shape of the Pareto front.

These observations suggest that IGD$^+$ can be used as a substitute for HV (i.e., as an overall performance indicator) when the use of HV is not easy due to the large computation load. It is also suggested that IGD$^+$ and HV are not always good indicators for evaluating the uniformity of solutions. Whereas IGD is not always a good overall performance indicator due to its Pareto incompliant property, IGD is a good indicator for evaluating the uniformity of solutions as shown in our computational experiments.

A future research topic is to examine the optimal distributions of solutions for many-objective problems, which are often counter-intuitive as shown in [14] for the IGD indicator on the 10-objective DTLZ1 problem. Another interesting research topic is to examine the optimal distribution of solutions for other performance indicators which were not examined in this paper. Theoretical discussions on the optimal distribution of solutions for the IGD$^+$ indicator are also an important future research topic.

Acknowledgments. This work was supported by National Natural Science Foundation of China (Grant No. 61876075), the Program for Guangdong Introducing Innovative and Entrepreneurial Teams (Grant No. 2017ZT07X386), Shenzhen Peacock Plan (Grant No. KQTD2016112514355 531), the Science and Technology Innovation Committee Foundation of Shenzhen (Grant No. ZDSYS201703031748284), and the Program for University Key Laboratory of Guangdong Province (Grant No. 2017KSYS008).

References

1. Auger, A., Bader, J., Brockhoff, D., Zitzler, E.: Hypervolume-based multiobjective optimization: theoretical foundations and practical implications. Theoret. Comput. Sci. **425**, 75–103 (2012)
2. Bader, J., Zitzler, E.: HypE: an algorithm for fast hypervolume-based many-objective optimization. Evol. Comput. **19**, 45–76 (2011)
3. Beume, N., Naujoks, B., Emmerich, M.: SMS-EMOA: multiobjective selection based on dominated hypervolume. Eur. J. Oper. Res. **181**, 1653–1669 (2007)
4. Brockhoff, D.: Optimal μ-distributions for the hypervolume indicator for problems with linear bi-objective fronts: exact and exhaustive results. In: Deb, K., et al. (eds.) SEAL 2010. LNCS, vol. 6457, pp. 24–34. Springer, Heidelberg (2010). https://doi.org/10.1007/978-3-642-17298-4_2
5. Coello Coello, C.A., Reyes Sierra, M.: A study of the parallelization of a coevolutionary multi-objective evolutionary algorithm. In: Monroy, R., Arroyo-Figueroa, G., Sucar, L.E., Sossa, H. (eds.) MICAI 2004. LNCS, vol. 2972, pp. 688–697. Springer, Heidelberg (2004). https://doi.org/10.1007/978-3-540-24694-7_71
6. Deb, K., Pratap, A., Agarwal, S., Meyarivan, T.: A fast and elitist multiobjective genetic algorithm: NSGA-II. IEEE Trans. Evol. Comput. **6**, 182–197 (2002)
7. Deb, K., Jain, H.: An evolutionary many-objective optimization algorithm using reference-point-based non-dominated sorting approach, part I: solving problems with box constraints. IEEE Trans. Evol. Comput. **18**, 577–601 (2014)

8. Deb, K., Thiele, L., Laumanns, M., Zitzler, E.: Scalable multi-objective optimization test problems. In: Proceedings of IEEE CEC 2002, pp. 825–830 (2002)
9. Emmerich, M., Beume, N., Naujoks, B.: An EMO algorithm using the hypervolume measure as selection criterion. In: Coello Coello, C.A., Hernández Aguirre, A., Zitzler, E. (eds.) EMO 2005. LNCS, vol. 3410, pp. 62–76. Springer, Heidelberg (2005). https://doi.org/10.1007/978-3-540-31880-4_5
10. Emmerich, M., Deutz, A., Beume, N.: Gradient-based/evolutionary relay hybrid for computing Pareto front approximations maximizing the S-metric. In: Bartz-Beielstein, T., et al. (eds.) HM 2007. LNCS, vol. 4771, pp. 140–156. Springer, Heidelberg (2007). https://doi.org/10.1007/978-3-540-75514-2_11
11. Huband, S., Hingston, P., Barone, L., While, L.: A review of multiobjective test problems and a scalable test problem toolkit. IEEE Trans. Evol. Comput. 10, 477–506 (2006)
12. Ishibuchi, H., Imada, R., Setoguchi, Y., Nojima, Y.: Hypervolume subset selection for triangular and inverted triangular Pareto fronts of three-objective problems. In: Proceedings of FOGA 2017, pp. 95–110 (2017)
13. Ishibuchi, H., Imada, R., Setoguchi, Y., Nojima, Y.: How to specify a reference point in hypervolume calculation for fair performance comparison. Evol. Comput. 26, 411–440 (2018)
14. Ishibuchi, H., Imada, R., Setoguchi, Y., Nojima, Y.: Reference point specification in inverted generational distance for triangular linear Pareto front. IEEE Trans. Evol. Comput. 22, 961–975 (2018)
15. Ishibuchi, H., Masuda, H., Tanigaki, Y., Nojima, Y.: Modified distance calculation in generational distance and inverted generational distance. In: Gaspar-Cunha, A., Henggeler Antunes, C., Coello, C.C. (eds.) EMO 2015. LNCS, vol. 9019, pp. 110–125. Springer, Cham (2015). https://doi.org/10.1007/978-3-319-15892-1_8
16. Ishibuchi, H., Setoguchi, Y., Masuda, H., Nojima, Y.: Performance of decomposition based many-objective algorithms strongly depends on Pareto front shapes. IEEE Trans. Evol. Comput. 21, 169–190 (2017)
17. Jain, H., Deb, K.: An improved adaptive approach for elitist nondominated sorting genetic algorithm for many-objective optimization. In: Purshouse, R.C., Fleming, P.J., Fonseca, C. M., Greco, S., Shaw, J. (eds.) EMO 2013. LNCS, vol. 7811, pp. 307–321. Springer, Heidelberg (2013). https://doi.org/10.1007/978-3-642-37140-0_25
18. Jiang, S.W., Ong, Y.-S., Zhang, J., Feng, L.: Consistencies and contradictions of performance metrics in multiobjective optimization. IEEE Trans. Cybern. 44, 2391–2404 (2014)
19. Jiang, S.W., Zhang, J., Ong, Y.-S., Zhang, A.N., Tan, P.S.: A simple and fast hypervolume indicator-based multiobjective evolutionary algorithm. IEEE Trans. Cybern. 45, 2202–2213 (2015)
20. Lopez, E.M., Coello Coello, C.A.: An improved version of a reference-based multi-objective evolutionary algorithm based on IGD$^+$. In: Proceedings of GECCO 2018, pp. 713–720 (2018)
21. Ravber, M., Mernik, M., Crepinkek, M.: The impact of quality indicators on the rating of multi-objective evolutionary algorithms. Appl. Soft Comput. 55, 265–275 (2017)
22. Sierra, M.R., Coello Coello, C.A.: A new multi-objective particle swarm optimizer with improved selection and diversity mechanisms. Technical report. CINVESTAV-IPN (2004)
23. Sun, Y., Yen, G.G., Yi, Z.: IGD indicator-based evolutionary algorithm for many-objective optimization problems. IEEE Trans. Evol. Comput. (Early Access Paper: Online Available)
24. Tian, Y., Cheng, R., Zhang, X.Y., Cheng, F., Jin, Y.C.: An indicator-based multiobjective evolutionary algorithm with reference point adaptation for better versatility. IEEE Trans. Evol. Comput. 22, 609–622 (2018)

25. Wagner, T., Beume, N., Naujoks, B.: Pareto-, aggregation-, and indicator-based methods in many-objective optimization. In: Obayashi, S., Deb, K., Poloni, C., Hiroyasu, T., Murata, T. (eds.) EMO 2007. LNCS, vol. 4403, pp. 742–756. Springer, Heidelberg (2007). https://doi.org/10.1007/978-3-540-70928-2_56

26. Zhang, Q., Li, H.: MOEA/D: a multiobjective evolutionary algorithm based on decomposition. IEEE Trans. Evol. Comput. **11**, 712–731 (2007)

27. Zitzler, E., Brockhoff, D., Thiele, L.: The hypervolume indicator revisited: on the design of Pareto-compliant indicators via weighted integration. In: Obayashi, S., Deb, K., Poloni, C., Hiroyasu, T., Murata, T. (eds.) EMO 2007. LNCS, vol. 4403, pp. 862–876. Springer, Heidelberg (2007). https://doi.org/10.1007/978-3-540-70928-2_64

28. Zitzler, E., Thiele, L.: Multiobjective optimization using evolutionary algorithms—a comparative case study. In: Eiben, A.E., Bäck, T., Schoenauer, M., Schwefel, H.-P. (eds.) PPSN 1998. LNCS, vol. 1498, pp. 292–301. Springer, Heidelberg (1998). https://doi.org/10.1007/BFb0056872

29. Zitzler, E., Thiele, L., Laumanns, M., Fonseca, C.M., Fonseca, V.G.: Performance assessment of multiobjective optimizers: an analysis and review. IEEE Trans. Evol. Comput. **7**, 117–132 (2003)

Diversity-Indicator Based
Multi-Objective Evolutionary Algorithm:
DI-MOEA

Yali Wang[✉], Michael Emmerich, André Deutz, and Thomas Bäck

Leiden Institute of Advanced Computer Science, Leiden University,
Niels Bohrweg 1, 2333 CA Leiden, The Netherlands
{y.wang,m.t.m.emmerich}@liacs.leidenuniv.nl
http://moda.liacs.nl

Abstract. In this paper we propose a Diversity-Indicator based Multi-Objective Evolutionary Algorithm (DI-MOEA) for fast computation of evenly spread Pareto front approximations. Indicator-based optimization has been a successful principle for multi-objective evolutionary optimization algorithm (MOEA) design. The idea is to guide the search for approximating the Pareto front by a performance indicator. Ideally, the indicator captures both convergence to the Pareto front and a high diversity, and it does not require a priori knowledge of the Pareto front shape and location. It is, however, so far difficult to define indicators that scale well in computation time for high dimensional objective spaces, and that distribute points evenly on the Pareto front. Moreover, the behavior of commonly applied indicators depends on additional information, such as reference points or sets. The proposed DI-MOEA adopts a hybrid search scheme for combining the advantages of Pareto dominance-based approaches to ensure fast convergence to the Pareto front, with indicator based approaches to ensure convergence to an evenly distributed, diverse set. In addition, it avoids the use of complex structure and parameters in decomposition-based approaches. The Euclidean distance-based geometric mean gap is used as diversity indicator. Experimental results show that the new algorithm can find uniformly spaced Pareto fronts without the involvement of any reference points or sets. Most importantly, our algorithm performs well on both the hypervolume indicator and IGD when comparing with state-of-the-art MOEAs (NSGA-II, SMS-EMOA, MOEA/D and NSGA-III).

Keywords: Multi-objective optimization · Diversity indicator ·
Evolutionary Algorithm · Indicator-based MOEAs

This work is part of the research programme Smart Industry SI2016 with project name CIMPLO and project number 15465, which is (partly) financed by the Netherlands Organisation for Scientific Research (NWO).

© Springer Nature Switzerland AG 2019
K. Deb et al. (Eds.): EMO 2019, LNCS 11411, pp. 346–358, 2019.
https://doi.org/10.1007/978-3-030-12598-1_28

1   Introduction

Many real-world problems require multiple objectives to be optimized, leading us to the so-called "Multi-objective Optimization Problems (MOPs)" [7]. It is usually difficult to find the optimal solutions for MOPs because their objectives are often conflicting with each other, and we are searching for a representative set of Pareto optimal solutions rather than for a single globally optimal solution because no single solution exists that can simultaneously optimize all objectives.

Classical Pareto dominance-based MOEAs, such as NSGA-II [2], use Pareto dominance as a first ranking criterion and use a second ranking criterion to maintain and increase diversity. Pareto dominance-based MOEAs have been a mainstream class for a long time in the field of evolutionary multi-objective optimization (EMO). They are very efficient on MOPs with two or three objectives. However, their performance degrades significantly on many-objective optimization problems (MaOPs), in which the number of objectives is greater than three, due to their ineffectiveness in distinguishing the quality of solutions when the number of objectives becomes large.

As the performance assessment of MOEAs reached a mature stage, performance measures (indicators) on the quality of Pareto front approximations were adopted to search for solutions. These indicators capture both convergence and diversity in a single value. Additionally for Pareto-compliant indicators, it can be shown that they obtain their maximum in a diversified set of solutions on the Pareto front. In general, Indicator-based Evolutionary Algorithms (IBEAs) [14] have strong theoretical support. However, the commonly used performance indicators lead to a convergence in distribution with a high density on the boundary of the Pareto front, as well as on knee regions [1]. If the aim is to obtain uniformly distributed and evenly spread solution sets, so far only indicators that employ an estimate of the true Pareto front as a reference set could be used.

Decomposition [6,13] is a search paradigm that was originally applied by EMO two decades ago [7] and recently regained prominence from the MOEA/D framework [13] and NSGA-III [3]. Decomposition-based MOEAs transform the original multi-objective problem into simpler, single-objective subproblems by means of scalarizations with different weights, therefore they can converge to a well defined, diverse set. However, the central issue in decomposition-based methods is how to select a set of weighting vectors that can provide a well distributed set of Pareto optimal points, given that the location and shape of the Pareto front are unknown a priori. Moreover, the number of weights required to sample a Pareto front with a sufficient resolution suffers a exponential growth from the objective space dimension [6].

Our paper suggests algorithms that combine principles from Pareto dominance-based approach and from indicator-based algorithms. Instead of requiring the indicator to take into account diversity and Pareto dominance, we propose to

- use dominance rank as a primary selection indicator, in order to ensure convergence to the Pareto front;

– use performance indicators that measure the diversity *of a set* of mutually non-dominated solutions.

However, as opposed to Pareto dominance-based approaches such as SPEA2 and NSGA-II that also maintain diversity, we decide the diversity of a set is measured by a scalar value, such that convergence to a maximum diverse set can be achieved and theoretically assessed.

The proposed diversity-indicator based MOEA (DI-MOEA) therefore takes advantage of Pareto dominance-based approaches, and excludes the complex structure and parameters in decomposition-based and contemporary indicator-based approaches. Most importantly, experimental results show that our algorithm can find well converged and evenly spaced Pareto front approximations without the involvement of any reference points and assumptions about the location and shape of the Pareto front.

The rest of this paper is organized as follows: First, in Sect. 2, we introduce the diversity indicator. Then, we describe the proposed algorithm in Sect. 3. Section 4 shows experimental results on benchmark problems. Section 5 concludes the work and outlines some possible future work.

2   Diversity Indicators and Gap Contribution

There exist many indicators that assess the diversity of a distribution of points in $\mathbb{R}^m$. Among these, the Weitzman indicator and discrepancy measures have excellent theoretical properties, but their computation is expensive. The Hausdorff distance and related measures are indicators that would require the knowledge of the set on which points should be distributed, which is typically not available in Pareto optimization. The Solow-Polasky indicator has been suggested in the context of diversity assessment due to its moderate computational effort and good theoretical properties [10]. However, it is sensitive to the choice of the correlation strength parameter of an exponential kernel function and it requires matrix inversion which might cause numerical instability. The gap indicators (or the averages of distances to nearest neighbours) have been suggested in [4]. They are very fast to compute and easy to implement diversity indicators. In addition, they have certain favorable theoretical properties and empirical results show that their maximization results in diversified, evenly spread approximation sets. These results were obtained for multimodal optimization [12] and evolutionary level set approximation [9] for a wide range of test problems.

Let A define a set of points in $\mathbb{R}^m$, $D(x, A \setminus \{x\}) = \min_{a \in A \setminus \{x\}}\{d(x, a)\}$ and d denote the Euclidean distance, then the gap indicators (GI) are defined as follows:

$$GI_{\min}(A) = \min_{x \in A}\{D(x, A \setminus \{x\})\} \qquad \text{Minimal gap}$$

$$GI_{\Sigma}(A) = \frac{1}{|A|} \sum_{x \in A} D(x, A \setminus \{x\}) \qquad \text{Arithmetic mean gap}$$

$$GI_{\Pi}(A) = (\prod_{x \in A} D(x, A \setminus \{x\}))^{\frac{1}{|A|}} \qquad \text{Geometric mean gap}$$

Note, that $GI_{\min}$ is the well known diversity indicator used in the max-min diversity problem [5]. One can leave out the exponent in GI_Π and this yields the product distance to the nearest neighbour (PDNN) indicator, considered by Wessing [12] in the context of multimodal optimization. Wessing [12] pointed out that GI_Π obtains the value of zero in case of duplicates in the set, a property that also holds for $GI_{\min}$. Besides, it can only be used for comparing sets of equal size. Since we are using the indicator contribution as a relative measure of performance of points, these two properties do not cause problems.

In indicator-based steady state selection [1], the aim is to optimize a quality indicator QI for a solution set. W.l.o.g. we assume the quality indicator is to be maximized. The selection strategy is to add a non-dominated solution x to an approximation set A of size μ and then retain the best subset $S \subset P$ with $|S| = \mu$ of the new set $P = A \cup \{x\}$. This can be achieved by removing the point that contributes the least to the quality indicator. The indicator contribution of a point $p \in P$ is defined as:

$$\Delta_{QI}(p, P) \leftarrow QI(P) - QI(P \setminus \{p\})$$

In our algorithm, the set-indicator contribution of the individual $p \in P$ is defined as the difference of the geometric mean gap indicator value of the set with the individual p minus the indicator-value of the set without it. The computation of the minimal contributor in case of the gap indicators can be solved by computing the solution to the all point nearest neighbour problem (APNN). The straightforward implementation, i.e. measuring distance between all pairs, requires a running time of $O(n^2)$. The APNN problem can be solved by Vaidya's algorithm [11] in optimal time $O(n \log n)$ for a fixed dimensional space and any Minkowski metric, including the Euclidean metric. We propose to choose the Euclidean distance due to its rotational invariance.

3   Proposed Algorithm

In the algorithm, we utilize a hybrid selection scheme: the $(\mu + \mu)$ generational selection operator and the $(\mu + 1)$ steady state selection operator. The algorithm consists of two components:

- The $(\mu + \mu)$ generational selection operator: When the population is layered to multiple (more than one) dominance ranks, it indicates that the population has not yet converged to the true Pareto front. In this case, the $(\mu + \mu)$ generational selection operator is used to explore the decision space for dominating solutions. In this stage, a strict consideration of the diversity indicator is not yet the key determinant factor. Rather the first priority should be to push the population quickly to the Pareto front. Still, diversity is considered as a secondary ranking criterion in order to bring the points in a good starting position for searching for a uniformly distributed population. Overall, the selection operator is using non-dominated sorting as a primary ranking criterion, then if more than μ solutions are obtained by adding a layer, we propose two alternative strategies to truncate: the crowding distance (*variant 1*) as in NSGA-II, and the diversity indicator contribution (*variant 2*), where points are successively removed in a greedy manner and the contributions are recomputed after each removal. Under the condition that the μ selected solutions are mutually non-dominated after an iteration, the algorithm switches to the $(\mu + 1)$ steady state selection operator.

- The $(\mu + 1)$ steady state selection operator: When the parent population consists of only one non-dominated set, it is likely that the population has already reached a region near the Pareto front. In this case, the indicator-based $(\mu+1)$ steady state selection operator is applied, as described in Sect. 2. It discards the least contributor to the quality indicator, here, the diversity indicator. The intent is to achieve a uniformly distributed set on the Pareto front, that is to converge to a maximum of the diversity indicator. If there are more than one dominance ranks in the resulting population, the algorithm switches back to a $(\mu + \mu)$ generational selection operator.

Besides the hybrid selection scheme, another important design choice is the quality indicator, to be specific, the Euclidean distance based geometric mean gap indicator is used to guide the search towards the uniformly distributed Pareto front approximations regardless of the shape of the Pareto front.

The proposed algorithm is presented as pseudo-code in Algorithm 1 and a MOEA-Framework implementation is made available on http://moda.liacs.nl.

4 Experimental Results and Discussion

4.1 Experimental Setup

In this section, simulations are conducted to demonstrate the performance of the proposed algorithm. Because two different diversity measures are employed in the $(\mu + \mu)$ generational selection operator, two variants of DI-MOEA are involved in the experiments: the crowding distance and the set-indicator contribution are chosen as the second measure in the generational $(\mu + \mu)$ selection operator in algorithm DI-1 and algorithm DI-2 respectively.

All experiments are implemented based on the MOEA Framework 2.1 (http://www.moeaframework.org/), which is a Java-based framework for multi-objective optimization. In the simulations, we use the SBX operator with an index of 15 (30 in NSGA-III and a differential evolution operator is used in MOEA/D.) and polynomial mutation with an index 20. The crossover and mutation probabilities are set to 1 and $1/N$ respectively and N is the number of variables. In NSGA-III, the number of subdivisions is 99 for bi-objective problems, and 12 for three objective problems. The number of evaluation (NE) is chosen to be dependent on the complexity of the test problem. 20000 NE is used for ZDT problems and 100000 NE for DTLZ problems. The population size is 100 for all problems.

Algorithm 1. DI-MOEA

1: $P_0 \leftarrow init()$;   //Initialize random population
2: $popsize \leftarrow |P_0|$;
3: $(R_1, ..., R_{\ell_0}) \leftarrow$ Partition P_0 into subsets of increasing dominance rank; //Non-dominated sorting
4: **for each** $i \in \{1, ..., \ell_0\}$ **do**
5:     calculate diversity indicator for all solutions based on the current front;
6: $t \leftarrow 0$;
7: **while** Stop criterion not satisfied() **do**
8:     **if** $\ell_t > 1 \;||\; t == 0$ **then**
9:         // $(\mu + \mu)$ selection operator
10:        $Q_t \leftarrow Gen(P_t)$; // Generate offspring with the size of $popsize$ by variation
11:        Evaluate Q_t;
12:        $P_t = P_t \cup Q_t$ // Combine offspring and parents
13:        $(R_1, ..., R_{\ell_t}) \leftarrow$ Partition P_0 into subsets of increasing dominance rank; //Non-dominated sorting
14:        $i \leftarrow 0$; $P_{t+1} \leftarrow \emptyset$;
15:        **while** $|P_{t+1}| < popsize$ **do**
16:            $P_{t+1} \leftarrow$ all solutions on i-th front R_i;
17:            $i \leftarrow i + 1$;
18:        **if** $|P_{t+1}| > popsize$ **then**
19:            $n \leftarrow |P_{t+1}| - popsize$
20:            **while** $n > 0$ **do**
21:                calculate diversity indicator for all solutions on the last front;
22:                remove the least contributor solution based on rank and diversity;
23:                $n \leftarrow n - 1$;
24:    **else**
25:        // $(\mu + 1)$ selection operator
26:        $q \leftarrow Gen(P_t)$;   // Generate only an offspring by variation
27:        $P_t \leftarrow P_t \cup \{q\}$;
28:        Rank P_t based on Pareto dominance rule; //Non-dominated sorting
29:        **for** each front **do**
30:            calculate set-indicator contribution for all solutions on the least ranked front $|R_{\ell_t}|$, if $|R_{\ell_t}| > 1$;
31:        remove the least contributor to diversity-indicator on the least ranked front;

4.2   Experiments on Bi-objective Problems

For bi-objective problems, algorithms are tested on ZDT1, ZDT2 and ZDT3 with 30 variables. Two new algorithms, DI-1 and DI-2, are compared with NSGA-II, SMS-EMOA, NSGA-III and MOEA/D. Tables 1 and 2 show the aggregate hypervolume and aggregate inverted generational distance (IGD) across 30 runs. The aggregate value is the value obtained when the Pareto solutions from all runs are combined into one. For each problem in the two tables, the upper row denotes the aggregate hypervolume/IGD. (The best value is highlighted in bold.) The lower row is the standard deviation (Std) of results from 30 runs. The Mann-Whitney U test is used to determine if the medians of different algorithms for the same problem are significantly indifferent. In the tables, we also highlight algorithms whose median performance is indifferent to the algorithm with the best aggregate performance. It can be observed that SMS-EMOA or NSGA-III can achieve the best hypervolume and the best IGD on all three problems, and the proposed DI-MOEA can obtain better hypervolume and IGD than NSGA-II and MOEA/D. In some instances, DI-MOEA can even get better hypervolume and IGD than NSGA-III or SMS-EMOA.

Table 1. The aggregate hypervolume on bi-objective problems

Hypervolume Std ↘	NSGA-II	SMS-EMOA	NSGA-III	MOEA/D	DI-1	DI-2
ZDT1	0.66399	**0.66602**	0.66428	0.66029	0.66473	0.66491
	4.8379e−04	7.2331e−05	3.9507e−04	0.0028	3.5973e−04	2.8447e−04
ZDT2	0.33002	0.33265	**0.33266**	0.32849	0.33073	0.33141
	4.7756e−04	8.7207e−05	0.0086	0.0030	4.9232e−04	5.8483e−04
ZDT3	0.51600	0.51718	**0.51720**	0.51582	0.51623	0.51634
	3.9954e−04	0.0013	0.0010	0.0011	4.1969e−04	2.7955e−04

Table 2. The aggregate IGD on bi-objective problems

IGD Std ↘	NSGA-II	SMS-EMOA	NSGA-III	MOEA/D	DI-1	DI-2
ZDT1	0.00163	**0.00039**	0.00168	0.00385	0.00116	0.00106
	2.6517e−04	1.9915e−05	8.2835e−04	0.0018	1.4110e−04	9.7026e−05
ZDT2	0.00202	0.00084	**0.00051**	0.00247	0.00159	0.00120
	2.1844e−04	1.0340e−04	0.0088	0.0014	2.1557e−04	2.4062e−04
ZDT3	0.00092	**0.00037**	0.00054	0.00190	0.00087	0.00092
	1.5809e−04	0.0100	0.0080	8.6720e−04	1.6713e−04	1.3157e−04

4.3 Experiments on Three Objective Problems

For three objective problems, DTLZ1 with 7 variables, DTLZ2 with 12 variables and DTLZ7 with 22 variables are tested. Both DI-1 and DI-2 behave very well, and they are indifferent on the statistical significance of median hypervolume and IGD. Statistical data averaging 10 runs per problem and algorithm are shown on Tables 3 and 4. DI-1 beats all the algorithms on the aggregate hypervolume on all problems, and DI-2 also behaves better than other algorithms except for SMS-EMOA on DTLZ1. For IGD, the new algorithms perform the best on DTLZ1 and DTLZ2 problems. NSGA-II obtains the best IGD on DTLZ7, while IGD

Table 3. The aggregate hypervolume on three objective problems

Hypervolume Std ↘	NSGA-II	SMS-EMOA	NSGA-III	MOEA/D	DI-1	DI-2
DTLZ1	0.80605	0.80732	0.78400	0.80198	**0.80806**	**0.80645**
	0.0062	1.8738e−04	0.0179	0.0015	0.0013	6.1716e−04
DTLZ2	0.44263	0.45269	0.41915	0.42907	**0.45511**	**0.45489**
	0.0070	5.8698e−05	5.1471e−04	0.0031	0.0033	0.0014
DTLZ7	0.31064	0.24694	0.30624	0.30164	**0.31227**	**0.31339**
	0.0034	0.0038	0.0328	0.0055	0.0051	0.0137

Table 4. The aggregate IGD on three objective problems

IGD Std ↘	NSGA-II	SMS-EMOA	NSGA-III	MOEA/D	DI-1	DI-2
DTLZ1	0.02149	0.02074	0.04266	0.02779	**0.01966**	**0.02381**
	0.0063	8.1450e−04	0.0159	0.0018	0.0017	0.0016
DTLZ2	0.02414	0.03415	0.05181	0.03902	**0.01799**	**0.01909**
	0.0047	0.0014	2.1056e−04	0.0026	0.0019	0.0030
DTLZ7	**0.01820**	0.09182	0.02381	0.041367	0.01826	0.02191
	0.0027	0.0020	0.2151	0.0867	0.0017	0.0944

values of DI-1 and DI-2 are only slightly lower than NSGA-II on DTLZ7, but better than all other algorithms.

To easily observe the results of algorithms, we visualized the results on the three objective problems. Figure 1 shows the Pareto front approximations of a typical run on DTLZ1. It can be observed that the solutions of NSGA-II and MOEA/D are not uniformly distributed, and there are several overlaps in the result of NSGA-III. While, SMS-EMOA and our algorithms can obtain evenly spaced solutions on the linear Pareto front.

Figure 2 shows the Pareto front approximations of a typical run on DTLZ2. For NSGA-III, we observed the same phenomenon: some solutions are overlapping or very close. The result of SMS-EMOA is distributed across the Pareto front with emphasis on the boundary and knee regions of the Pareto front. The results of the two DI-MOEA variants are uniformly distributed and evenly spaced on the Pareto front.

DI-MOEA also behaves well on the multimodal DTLZ7 problem, which has non-linear disconnected Pareto front regions. Figure 3 shows the results under 200 population size and 500000 NE.

When running the DI-MOEA, it can be observed that the population evolves towards the Pareto front at the initial stage (the first phase) using the generational selection operator. After a short period where the two selection operators alternate (the second phase), the steady state selection operator takes over and

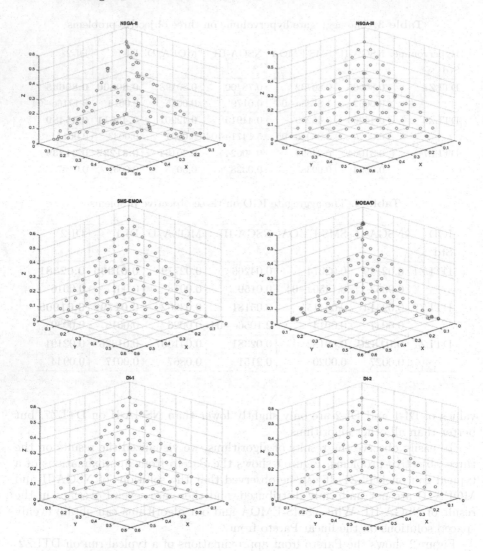

Fig. 1. Representative PF approximations on DTLZ1.

the population converges to a set with maximum diversity (the third phase). When the number of objectives becomes large, the third phase is more prominent than the previous two phases because it is more likely for solutions to be mutually non-dominated for a large objective number. In the runs conducted on three objective problems in this paper, the generational selection operator was applied around 100–200 iterations before it switched to the steady state selection operator for the first time. The intermittent alternating phase took about 20–50 iterations, and in most of the running time, the algorithm used the steady state selection operator and throughout this phase, only occasionally the algorithm switched back to generational selection operator for at most a single iteration.

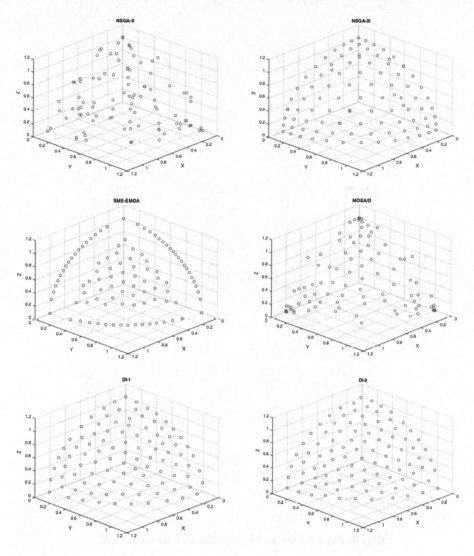

Fig. 2. Representative PF approximations on DTLZ2.

Overall, the first and the second phase took only a minor amount of the total running time.

It is worth noting that we observed Dominance Resistant Solutions (DRSs) [8] occasionally on the linear Pareto front of DI-2 on DTLZ1 three objective problem; these are points that have a large contribution to diversity, but dominate only a very narrow region exclusively. It might be necessary to keep these "special solutions", but on the other side, they make the Pareto front approximation less evenly distributed. We already tested a strategy to eliminate DRSs. Before the calculation of the set-indicator contribution for a front, each solution is checked

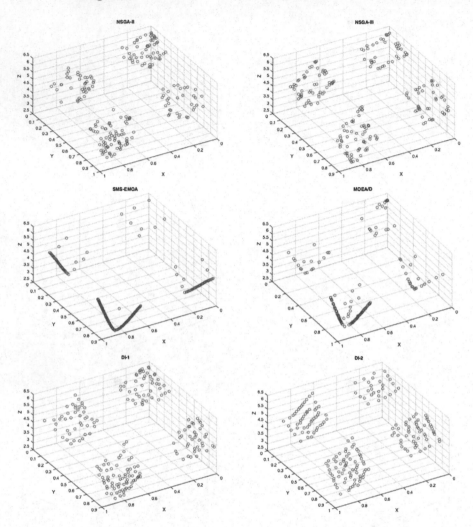

Fig. 3. Representative PF approximations on DTLZ7.

by comparing with all other solutions: the distances between two solutions in all dimensions are calculated, if the result of the minimal distance divided by the maximal distance is too small, the current solution will be removed from the front. Therefore, a shrinked front is created and diversity indicator can be calculated only in the new front. The underlying idea of this strategy is that for two solutions, if their distance is too close in one dimension and too large in another dimension, keeping both of them will result in an uneven distribution.

5   Conclusions and Further Work

The proposed DI-MOEA combines the advantage of Pareto dominance-based and indicator-based methods. Moreover, the achieved Pareto front approximations are excellent in both hypervolume indicator and IGD. Especially, the relative performance of our algorithms even gets better with increasing number of objectives. The set-indicator used in our algorithms is computationally simpler than the hypervolume indicator and only depend linearly on the number of objectives, making it possesses a potential advantage on MaOPs. Most importantly, the uniformly distributed, evenly spaced solution set can be achieved without the use of decomposition sets and the estimation of the location and shape of the true Pareto front.

In the current implementation of DI-MOEA, only a naive way of calculating the Euclidean distance based geometric mean gap is implemented. Although the computational time of the implemented algorithm is shorter than SMS-EMOA, it should be further improved, e.g., by using Vaidya's algorithm [11] and incremental updates of contributions. Besides, the new DI-MOEA holds the promise of performing well in many-objective optimization. To study this, its performance should be tested on MaOP benchmarks, paying special attention to effects that might occur in high dimensional objective spaces, such as distance concentration and the increasing number of non-dominated solutions. Also, more MOEAs can be involved in comparison, such as MACE-gD [6] and IBEA [14].

References

1. Beume, N., Naujoks, B., Emmerich, M.: SMS-EMOA: multiobjective selection based on dominated hypervolume. Eur. J. Oper. Res. **181**(3), 1653–1669 (2007)
2. Deb, K., Agrawal, S., Pratap, A., Meyarivan, T.: A fast and elitist multiobjective genetic algorithm: NSGA-II. IEEE Trans. Evol. Comput. **6**(2), 182–197 (2002)
3. Deb, K., Jain, H.: An evolutionary many-objective optimization algorithm using reference-point-based nondominated sorting approach, part I: solving problems with box constraints. IEEE Trans. Evol. Comput. **18**(4), 577–601 (2014)
4. Emmerich, M.T.M., Deutz, A.H., Kruisselbrink, J.W.: On quality indicators for black-box level set approximation. In: Tantar, E., et al. (eds.) EVOLVE - A Bridge Between Probability, Set Oriented Numerics and Evolutionary Computation. SCI, vol. 447, pp. 157–185. Springer, Heidelberg (2013). https://doi.org/10.1007/978-3-642-32726-1_4
5. Ghosh, J.B.: Computational aspects of the maximum diversity problem. Oper. Res. Lett. **19**(4), 175–181 (1996)
6. Giagkiozis, I., Purshouse, R.C., Fleming, P.J.: Generalized decomposition and cross entropy methods for many-objective optimization. Inf. Sci. **282**, 363–387 (2014)
7. Hajela, P., Lin, C.-Y.: Genetic search strategies in multicriterion optimal design. Struct. Optim. **4**(2), 99–107 (1992)
8. Hanne, T.: On the convergence of multiobjective evolutionary algorithms. Eur. J. Oper. Res. **117**(3), 553–564 (1999)

9. Liu, L.-Y., Basto-Fernandes, V., Yevseyeva, I., Kok, J., Emmerich, M.: Indicator-based evolutionary level set approximation: mixed mutation strategy and extended analysis. In: Ferrández Vicente, J.M., Álvarez-Sánchez, J.R., de la Paz López, F., Toledo Moreo, J., Adeli, H. (eds.) IWINAC 2017. LNCS, vol. 10337, pp. 146–159. Springer, Cham (2017). https://doi.org/10.1007/978-3-319-59740-9_15
10. Ulrich, T., Bader, J., Thiele, L.: Defining and optimizing indicator-based diversity measures in multiobjective search. In: Schaefer, R., Cotta, C., Kołodziej, J., Rudolph, G. (eds.) PPSN 2010. LNCS, vol. 6238, pp. 707–717. Springer, Heidelberg (2010). https://doi.org/10.1007/978-3-642-15844-5_71
11. Vaidya, P.M.: An o(n log n) algorithm for the all-nearest-neighbors problem. Discrete Comput. Geom. **4**(2), 101–115 (1989)
12. Wessing, S.: Two-stage methods for multimodal optimization. Ph.D. thesis. Universitätsbibliothek Dortmund (2015)
13. Zhang, Q., Li, H.: MOEA/D: a multiobjective evolutionary algorithm based on decomposition. IEEE Trans. Evol. Comput. **11**(6), 712–731 (2007)
14. Zitzler, E., Künzli, S.: Indicator-based selection in multiobjective search. In: Yao, X., et al. (eds.) PPSN 2004. LNCS, vol. 3242, pp. 832–842. Springer, Heidelberg (2004). https://doi.org/10.1007/978-3-540-30217-9_84

The Expected R2-Indicator Improvement for Multi-objective Bayesian Optimization

André Deutz[✉], Michael Emmerich, and Kaifeng Yang

LIACS, Leiden University, Niels Bohrweg 1, 2333 CA Leiden, The Netherlands
{a.h.deutz,m.t.m.emmerich,k.yang}@liacs.leidenuniv.nl

Abstract. In multi-objective Bayesian optimization, an infill criterion is an important part, as it is the indicator to evaluate how much good a new set of solutions is, compared to a Pareto-front approximation set. This paper presents a deterministic algorithm for computing the Expected R2 Indicator for bi-objective problems and studies its use as an infill criterion in Bayesian Global Optimization. The R2-Indicator was introduced in 1998 by M. Hansen and A. Jaszkiewicz for performance assessment in multi-objective optimization and is more recently also used in indicator-based multi-criterion evolutionary algorithms (IBEAs). In Bayesian Global Optimization, we propose the Expected R2-indicator Improvement (ER2I) as an infill criterion. It is defined as the expected decrease of the R2 indicator by a point that is sampled from a predictive Gaussian distribution. The ER2I can also be used as a pre-selection criterion in surrogate-assisted IBEAs. It provides an alternative to the Expected Hypervolume-Indicator Improvement (EHVI) that requires a reference point, bounding the Pareto front from above. In contrast, the ER2I works with a utopian reference point that bounds the Pareto front from below. In addition, the ER2I supports preference modelling with utility functions and its computation time grows only linearly with the number of considered weight combinations. It is straightforward to approximate the ER2I by Monte Carlo Integration, but so far a deterministic algorithm to solve the non-linear integral remained unknown. We outline a deterministic algorithm for the computation of the bi-objective ER2I with Chebychev utility functions. Moreover, we study monotonicity properties of the ER2I w.r.t. parameters of the predictive distribution and numerical simulations demonstrate fast convergence to Pareto fronts of different shapes and the ability of the ER2I Bayesian optimization to fill gaps in the Pareto front approximation.

Keywords: R2 indicator · Expected improvement ·
Surrogate models · Multiobjective Bayesian optimization ·
Chebychev utility function

1 Introduction

A central object of interest for multi-objective optimization is the Pareto Front (PF) for the minimization of a vector-valued function $\mathbf{f} : \mathbb{R}^d \rightarrow \mathbb{R}^n$, where

© Springer Nature Switzerland AG 2019
K. Deb et al. (Eds.): EMO 2019, LNCS 11411, pp. 359–370, 2019.
https://doi.org/10.1007/978-3-030-12598-1_29

$\mathbf{f} = (f_1, \ldots, f_n)$ denotes the objective functions[1]. Many algorithms have been developed in order to approximate PFs by finite sets. An approximation set to the PF is a finite, non-dominated set consisting of image points of the feasible set – a set is non-dominated with respect to the Pareto order if for each element of the set one cannot find an element of the set that Pareto dominates it.

The indicator based algorithms for multi-objective optimization problems are guided by performance indicators for the quality of approximation sets. So-called unary indicators are of particular interest, as they do not require the knowledge of the Pareto front. Among the unary indicators, besides the hypervolume indicator, the R2 indicator [5] attracted wide spread interest, also in many-objective optimization. It is relatively frugal in using computational resources as compared to other indicators such as the hypervolume indicator and requires a user-defined utopian (or ideal) point, instead of a reference point that bounds the Pareto front from above – as it is required by the hypervolume indicator. In many practical applications, such as resource or error minimization, a utopian point is easier to provide or more natural to the problem. The R2 indicator was introduced by Hansen and Jaszkiewicz [5] and it was proposed and studied for indicator-based evolutionary algorithm design by Brockhoff, Wagner and Trautmann [2]. Strictly speaking, the R2 indicator is a family of indicators that average a utility function over different choices of its weighting parameters. A common choice is the R2 indicator with the Chebychev utility function, which, as opposed to a linear weighting utility function, is also suitable for concave Pareto fronts.

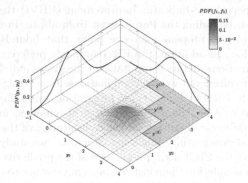

Fig. 1. Pareto front approximation and 2-D predictive distribution.

In the *optimization with expensive black box objective functions*, it is common to use statistical models fitted by data from past evaluations of the expensive objective functions (surrogate models) to guide the search. Surrogate-model assisted Evolutionary Algorithms (SEA) and Bayesian (global) optimization are two examples where this strategy is used. Bayesian (Global) Optimization

---

[1] Here we restrict ourselves to continuous optimization.

(BGO) originated by Žilinskas and Mockus [7,9] selects a single point to be evaluated in each iteration based on an expected improvement infill criterion that is defined on a Gaussian process model (GP) learned by all past evaluations. Recently, different generalization of infill criteria for multi-objective optimization were proposed. They use Gaussian process based predictions with uncertainty quantification (mean value $\mu(\mathbf{x})$, stddev. $\sigma(\mathbf{x})$ for $\mathbf{x} \in \mathbb{R}^d$) of vector valued function as illustrated in Fig. 1. Examples of such algorithms are ParEGO [6], S-Metric Selection EGO [8], or the Expected Hypervolume Improvement Algorithm [4]. In parallel, statistical models learned from past evaluations were used in surrogate-model assisted multi-criterion optimization. Besides, BGO, also surrogate-model assisted evolutionary algorithms use expected improvement criteria for selecting promising points [1].

In this paper, in Sect. 2 we propose the expected improvement indicator based on the R2 indicator improvement of an approximation set that can be computed by Monte Carlo methods, and for the biobjective case we provide an accurate deterministic procedure for its computation (Sect. 3). Moreover, we study its sensitivity with respect to changes of the mean and variance of the Gaussian process prediction, as well as its ability to guide BGO towards the Pareto front and to fill in the gaps in a Pareto front approximation (Sect. 4).

2   The R2 Indicator and Its Expected Improvement

From now on, we will consider solely Chebychev utility functions derived from weighted Chebyshev distance functions, i.e., functions of the form $d_{wc}(a,b) := \max_{i \in \{1,\cdots,n\}} \lambda_i |a_i - b_i|$ for some $\lambda \in \mathbb{R}^n_{\geq 0}, i = 1,\cdots,n$, i.e., λ_i is non-negative. A point in $\mathbb{R}^n$ is called *utopian* if and only if it is not dominated by an element of $\mathbf{f}(\mathcal{X})$. Technically the R2 indicator for finite approximation sets $P \subseteq \mathbb{R}^n$ to the Pareto front with respect to some utopian point $\mathbf{z}^* \in \mathbb{R}^n$ and $U = \{u_1, \cdots, u_s\}$ – $s \in \mathbb{N}$ – a finite set of utility functions each provided with a probability $p_i, p_1 + \cdots p_s = 1$ is defined as follows:

$$\text{R2}(P, U, \mathbf{z}^*) := \sum_{i=1}^{s} p_i(u_i(\mathbf{z}^*) - \max_{a \in P}\{u_i(a)\}).$$

For a uniform distribution - the one we will use exclusively – this boils down to:

$$\text{R2}(P, U, \mathbf{z}*) := \frac{1}{s} \sum_{i=1}^{s} (u_i(\mathbf{z}^*) - \max_{a \in A}\{u_i(a)\}).$$

By keeping the utopian point fixed this reduces to:

$$\text{R2}(P, U) = \frac{1}{s} \sum_{i=1}^{s} \min_{a \in P}\{u_i(a)\}, \text{ as } \frac{1}{s} \sum_{i=1}^{s} u_i(\mathbf{z}^*) \text{ is a constant.}$$

We can specialize U to a set of utility functions derived from a set of distance functions as follows: for each $u \in U$ there is a distance function d on $\mathbb{R}^m$ such that $u(a) = d(\mathbf{z}^*, a), a \in \mathbb{R}^n$. Note also if d is a distance function, then λd, where λ is a positive, real number, is a distance function as well.

Example 1. An example is given in Fig. 2. The set U exists of a singleton with $U = \{(\lambda_1, \lambda_2)\} = (\lambda_1 = 3/8, \lambda_2 = 5/8)$. The utopian point is given by $(z_1, z_2) = (1, 2)$. The approximation set is given by $P = (\mathbf{p}^{(1)}, \ldots, \mathbf{p}^{(4)})$ and $\mathbf{p}^{(3)}$ determines the minimal utility function value of all points in P, which is $\mathbf{d}_{wc} = 9/6$. In the case of a singleton U this value $(9/6)$ is also the resulting R2-indicator value. Note that the R2 indicator is defined as the average over all weight vectors in U and therefore it can be computed by means of a summation, when U contains more than one element.

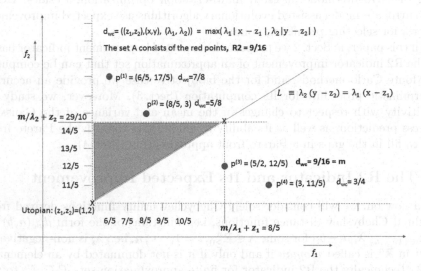

Fig. 2. Example R2 indicator for an approximation set P to the PF, U is a singleton, with $U = \{(\lambda_1, \lambda_2)\}$, $\lambda_1 = 3/8, \lambda_2 = 5/8$. (Color figure online)

2.1   Expected R2-Indicator Improvement

The *improvement* of the R2 indicator is defined as follows: the R2 indicator is evaluated on the given approximation set of the PF to which a point in the image of the feasible set is added. The resulting difference is the R2-improvement of the chosen point with respect to the given approximation set. The *expected improvement* is the mean of the improvement over dominated set by the utopian point in $\mathbb{R}^n$ with respect to the given $\text{pdf}_{\mathbf{x}}$.

$$\text{ER2I}(\mathbf{x}) = \int_{\mathbf{y} \in [\mathbf{z}, \infty)} (\text{R2}(P \cup \{\mathbf{y}\}) - \text{R2}(P)) \, \text{pdf}_{\mathbf{x}}(\mathbf{z}; \mu, \sigma, \mathbf{y}) d\mathbf{y} \qquad (1)$$

It is important to consider the truncated Gaussian probability density function, that is the predictive distribution conditioned by the knowledge that the utopian point dominates all potential solutions. Assuming conditional independence of outputs w.r.t. to a given input the function $\text{pdf}_\mathbf{x}$ is defined as:

$$\text{pdf}_\mathbf{x}(\mathbf{z}; \mu, \sigma, \mathbf{y}) := \prod_{i=1}^{n} \frac{1}{\sigma_i} \cdot \phi(\frac{y_i - \mu_i}{\sigma_i}) / \prod_{i=1}^{n} \left(1 - \Phi(\frac{z_i - \mu_i}{\sigma_i})\right)$$

3 Computation of the Expected R2-Indicator Improvement

In order to derive the formulas for the expected improvement for the 2-D R2-indicator (R2-indicator for biobjective optimization), we use the following variables.

1. $\mathbf{z} = (z_1, z_2)^\top$ is the utopian point.
2. m is the current value of the R2-indicator for some set, say some approximation set to the PF.
3. The Chebyshev weights are $\lambda = (\lambda_1, \lambda_2)^\top$
4. $a := \frac{m}{\lambda_1} + z_1$ and $b := \frac{m}{\lambda_2} + z_2$
5. The line $y_2 = \frac{\lambda_1}{\lambda_2} y_1 + z_2 - z_1 \frac{\lambda_1}{\lambda_2}$ through the utopian point: above this line the weighted Chebyshev distance to the utopian point is equal to $\lambda_2 |y_2 - z_2|$ for a point $(y_1, y_2)^\top \in \mathbb{R}^2$; similarly for a point below this line the weighted Chebyshev distance to the utopian point is equal to $\lambda_1 |y_1 - z_1| -$ we assume non-negative Chebyshev weights.
6. With the above notation: the m-Chebyshev ball with center the utopian point intersected with the dominated space of the utopian point is equal to $[\mathbf{z}, (a, b)^\top]$.
7. $d := z_2 - \frac{\lambda_1}{\lambda_2} \cdot z_1$

In the upper triangle (the blue triangle in Fig. 2), the R2-indicator improvement for a point $(y_1, y_2)^\top$ is equal to $\lambda_2(b - y_2)$ and in the lower triangle it is equal to $\lambda_1(a - y_1)$. Therefore, we split the integration for the expected improvement with respect to a predictive distribution in two parts (i.e., we integrate over the lower triangle and subsequently over the upper triangle). Since we assume that for the minimization problems we are thinking of, we know a priori that solution vectors in the decision space will map to points of the dominated space of the utopian point, we need to truncate the predictive distributions to this quadrant. Therefore we need to normalize. The normalization magnitude (the integral of the predictive distribution over the dominated space of the utopian point) is denoted by N.

The lower triangle:

$$\frac{1}{N} \frac{1}{\sqrt{2\pi\sigma_1^2}} \frac{1}{\sqrt{2\pi\sigma_2^2}} \int_{z_1}^{a} \int_{z_2}^{\frac{\lambda_1}{\lambda_2}y_1 + d} \lambda_1(a - y_1) e^{-\frac{1}{2}(\frac{y_2 - \mu_2}{\sigma_2})^2} e^{-\frac{1}{2}(\frac{y_1 - \mu_1}{\sigma_1})^2} \, dy_2 \, dy_1 \quad (2)$$

The upper triangle:

$$\frac{1}{N}\frac{1}{2\pi\sigma_1\sigma_2}\int\limits_{z_1}^{a}\int\limits_{\frac{\lambda_1}{\lambda_2}y_1+d}^{b}\lambda_2(b-y_2)e^{-\frac{1}{2}(\frac{y_2-\mu_2}{\sigma_2})^2}e^{-\frac{1}{2}(\frac{y_1-\mu_1}{\sigma_1})^2}dy_2dy_1 \qquad (3)$$

The sum of these two integrals is the expected improvement of the R2-indicator *at the weight vector* λ. For the R2-indicator, the expected improvement at each of the s weight vectors $\lambda^{(i)} : i = 1, \cdots, s$ is computed, then the expected improvement is the average of these numbers.

In future we will omit the factor $\frac{1}{N}\frac{1}{2\pi\sigma_1\sigma_2}$, and we will assume that the following integrals we compute need to be pre-multiplied by this factor. We continue with Expression (2):

$$\int\limits_{z_1}^{a}\lambda_1(a-y_1)e^{-\frac{1}{2}(\frac{y_1-\mu_1}{\sigma_1})^2}\int\limits_{z_2}^{\frac{\lambda_1}{\lambda_2}y_1+d}e^{-\frac{1}{2}(\frac{y_2-\mu_2}{\sigma_2})^2}dy_2dy_1 = \qquad (4)$$

$$\int\limits_{z_1}^{a}\lambda_1(a-y_1)e^{-\frac{1}{2}(\frac{y_1-\mu_1}{\sigma_1})^2}\left[\sqrt{\frac{\pi}{2}}\cdot\sigma_2\text{ erf}(\frac{y_2-\mu_2}{\sqrt{2}\sigma_2})\right]_{z_2}^{\frac{\lambda_1}{\lambda_2}y_1+d}dy_1 = \qquad (5)$$

$$\int\limits_{z_1}^{a}\lambda_1 ae^{-\frac{1}{2}(\frac{y_1-\mu_1}{\sigma_1})^2}\left[\sqrt{\frac{\pi}{2}}\cdot\sigma_2\text{ erf}(\frac{y_2-\mu_2}{\sqrt{2}\sigma_2})\right]_{z_2}^{\frac{\lambda_1}{\lambda_2}y_1+d}dy_1 - \qquad (6)$$

$$\int\limits_{z_1}^{a}\lambda_1 y_1 e^{-\frac{1}{2}(\frac{y_1-\mu_1}{\sigma_1})^2}\left[\sqrt{\frac{\pi}{2}}\cdot\sigma_2\text{ erf}(\frac{y_2-\mu_2}{\sqrt{2}\sigma_2})\right]_{z_2}^{\frac{\lambda_1}{\lambda_2}y_1+d}dy_1 = \qquad (7)$$

$$\lambda_1\sqrt{\frac{\pi}{2}}\cdot\sigma_2\int\limits_{z_1}^{a}ae^{-\frac{1}{2}(\frac{y_1-\mu_1}{\sigma_1})^2}\left[\text{erf}(\frac{\frac{\lambda_1}{\lambda_2}y_1+d-\mu_2}{\sqrt{2}\sigma_2})-\text{erf}(\frac{z_2-\mu_2}{\sqrt{2}\sigma_2})\right]dy_1 - \qquad (8)$$

$$\lambda_1\sqrt{\frac{\pi}{2}}\cdot\sigma_2\int\limits_{z_1}^{a}y_1 e^{-\frac{1}{2}(\frac{y_1-\mu_1}{\sigma_1})^2}\left[\text{erf}(\frac{\frac{\lambda_1}{\lambda_2}y_1+d-\mu_2}{\sqrt{2}\sigma_2})-\text{erf}(\frac{z_2-\mu_2}{\sqrt{2}\sigma_2})\right]dy_1. \qquad (9)$$

Unraveling Expression (8) and (9) we get the following 4 integrals:

Integral I

$$\lambda_1\sqrt{\frac{\pi}{2}}\cdot\sigma_2 a\int\limits_{z_1}^{a}e^{-\frac{1}{2}(\frac{y_1-\mu_1}{\sigma_1})^2}\text{ erf}\left(\frac{\frac{\lambda_1}{\lambda_2}y_1+d-\mu_2}{\sqrt{2}\sigma_2}\right)dy_1 + \qquad (10)$$

Integral II

$$- \lambda_1 \sqrt{\frac{\pi}{2}} \cdot \sigma_2 a \, \mathrm{erf}\left(\frac{z_2 - \mu_2}{\sqrt{2}\sigma_2}\right) \int\limits_{z_1}^{a} e^{-\frac{1}{2}\left(\frac{y_1 - \mu_1}{\sigma_1}\right)^2} dy_1 + \tag{11}$$

Integral III

$$\lambda_1 \sqrt{\frac{\pi}{2}} \cdot \sigma_2 \int\limits_{z_1}^{a} y_1 e^{-\frac{1}{2}\left(\frac{y_1 - \mu_1}{\sigma_1}\right)^2} \, \mathrm{erf}\left(\frac{\frac{\lambda_1}{\lambda_2} y_1 + d - \mu_2}{\sqrt{2}\sigma_2}\right) dy_1 + \tag{12}$$

Integral IV

$$- \lambda_1 \sqrt{\frac{\pi}{2}} \cdot \sigma_2 \, \mathrm{erf}\left(\frac{z_2 - \mu_2}{\sqrt{2}\sigma_2}\right) \int\limits_{z_1}^{a} y_1 e^{-\frac{1}{2}\left(\frac{y_1 - \mu_1}{\sigma_1}\right)^2} dy_1. \tag{13}$$

In turn we shall evaluate the Integrals I, II, III, and IV. For Integral I (Expression (10)) we apply the following substitution: $t := \frac{1}{\sqrt{2}}\frac{y_1 - \mu_1}{\sigma_1}$. Applying this substitution, Integral I is transformed into:

$$\lambda_1 \sqrt{\frac{\pi}{2}} \cdot \sigma_2 \int\limits_{\frac{1}{\sqrt{2}}\frac{z_1 - \mu_1}{\sigma_1}}^{\frac{1}{\sqrt{2}}\frac{a - \mu_1}{\sigma_1}} e^{-t^2} \, \mathrm{erf}\left(\frac{\lambda_1}{\lambda_2}\frac{\sigma_1}{\sigma_2} t + \frac{\lambda_1 \mu_1 + \lambda_2 d - \lambda_2 \mu_2}{\sqrt{2}\sigma_2 \lambda_2}\right) \sqrt{2}\sigma_1 dt = \tag{14}$$

$$\lambda_1 \sqrt{\pi} \cdot \sigma_2 \sigma_1 \int\limits_{\frac{1}{\sqrt{2}}\frac{z_1 - \mu_1}{\sigma_1}}^{\frac{1}{\sqrt{2}}\frac{a - \mu_1}{\sigma_1}} e^{-t^2} \, \mathrm{erf}\left(At + B\right) dt, \tag{15}$$

where $A := \frac{\lambda_1}{\lambda_2}\frac{\sigma_1}{\sigma_2}$ and $B := \frac{\lambda_1 \mu_1 + \lambda_2 d - \lambda_2 \mu_2}{\sqrt{2}\sigma_2 \lambda_2}$ As $\mathrm{erf}(x) = \frac{2}{\sqrt{\pi}}\sum_{i=0}^{\infty}\frac{(-1)^i x^{2i+1}}{i! \cdot (2i+1)}$ with a radius of convergence equal to ∞, we can approximate the error function to any degree of accuracy. Integral I is approximately equal to:

$$\lambda_1 \sqrt{\pi} \cdot \sigma_2 \sigma_1 \int\limits_{\frac{1}{\sqrt{2}}\frac{z_1 - \mu_1}{\sigma_1}}^{\frac{1}{\sqrt{2}}\frac{a - \mu_1}{\sigma_1}} e^{-t^2} \frac{2}{\sqrt{\pi}}\left(\frac{(At+B)^1}{0! \cdot 1} - \frac{(At+B)^3}{1! \cdot 3} + \frac{(At+B)^5}{2! \cdot 5}\right) dt. \tag{16}$$

The function $\frac{2}{\sqrt{\pi}}e^{-t^2}\left(\frac{(At+B)^1}{0! \cdot 1} - \frac{(At+B)^3}{1! \cdot 1} + \frac{(At+B)^5}{2! \cdot 5}\right)$ has a primitive, and it is computed in terms of A and B as follows.

$$\int \exp(-t^2) \cdot \frac{2}{\sqrt{\pi}} \sum_{i=0}^{2} \frac{(-1)^i}{i! \cdot (2i+1)}(At+B)^{2i+1} dt = \tag{17}$$

$$\frac{2}{\sqrt{\pi}} \left(-\frac{1}{2} A \exp(-t^2) + \frac{1}{2} B \sqrt{\pi} \operatorname{erf}(t) \right) - \frac{1}{6} \operatorname{erf}(t) B (3A^2 + 2B^2) + \tag{18}$$

$$- \frac{1}{6\sqrt{\pi}} \exp(-t^2)(-2A) \left[3B^2 + 3ABt + A^2(1 + t^2) \right] + \tag{19}$$

$$\frac{1}{40\sqrt{\pi}} \exp(-t^2)(-2A) \left[10B^4 + 2AB^3 t + 20A^2 B^2(1 + t^2) \right] + \tag{20}$$

$$\frac{1}{40\sqrt{\pi}} \exp(-t^2)(-2A)[5A^3 Bt(3 + 2t^2) + 2A^4(2 + 2t^2 + t^4)] + \tag{21}$$

$$\frac{1}{40} B \operatorname{erf}(t) \left[15A^4 + 20A^2 B^2 + 4B^4 \right]. \tag{22}$$

This shows the first part of our task. Next we compute Integral II (Expression (11)):

$$-\lambda_1 \sqrt{\frac{\pi}{2}} \cdot \sigma_2 a \operatorname{erf} \left(\frac{z_2 - \mu_2}{\sqrt{2}\sigma_2} \right) \int_{z_1}^{a} e^{-\frac{1}{2}\left(\frac{y_1 - \mu_1}{\sigma_1}\right)^2} dy_1 =$$

$$- \lambda_1 \sqrt{\frac{\pi}{2}} \cdot \sigma_2 a \operatorname{erf} \left(\frac{z_2 - \mu_2}{\sqrt{2}\sigma_2} \right) \sqrt{\frac{\pi}{2}} \sigma_1 \left[\operatorname{erf} \left(\frac{a - \mu_1}{\sqrt{2}\sigma_1} \right) - \operatorname{erf} \left(\frac{z_1 - \mu_1}{\sqrt{2}\sigma_1} \right) \right], \tag{23}$$

as $\int e^{-\frac{1}{2}\left(\frac{y_1 - \mu_1}{\sigma_1}\right)^2} dy_1 = \sqrt{\frac{\pi}{2}} \sigma_1 \operatorname{erf} \left(\frac{y_1 - \mu_1}{\sqrt{2}\sigma_1} \right)$.

Next we proceed to compute Integral III (Expression (12)). For this integral we apply the same substitution as for Integral I, obtaining:

$$\lambda_1 \frac{1}{\sqrt{2}} \sqrt{\pi} \sigma_2 \int_{\frac{1}{\sqrt{2}} \frac{z_1 - \mu_1}{\sigma_1}}^{\frac{1}{\sqrt{2}} \frac{a - \mu_1}{\sigma_1}} (\sqrt{2}\sigma_1 t + \mu_1) e^{-t^2} \operatorname{erf} \left(\frac{\lambda_1 \sigma_1}{\lambda_2 \sigma_2} t + \frac{\lambda_1 \mu_1 + \lambda_2 d - \lambda_2 \mu_2}{\sqrt{2}\lambda_2 \sigma_2} \right) \sqrt{2}\sigma_1 dt = \tag{24}$$

As before we rewrite this in terms of A and B:

$$\lambda_1 \sqrt{\pi} \sigma_1 \sigma_2 \int_{\frac{1}{\sqrt{2}} \frac{z_1 - \mu_1}{\sigma_1}}^{\frac{1}{\sqrt{2}} \frac{a - \mu_1}{\sigma_1}} (\sqrt{2}\sigma_1 t + \mu_1) e^{-t^2} \operatorname{erf} (At + B) \, dt = \tag{25}$$

$$\sqrt{\pi} \sigma_1^2 \sigma_2 \sqrt{2} \int_{\frac{1}{\sqrt{2}} \frac{z_1 - \mu_1}{\sigma_1}}^{\frac{1}{\sqrt{2}} \frac{a - \mu_1}{\sigma_1}} t e^{-t^2} \operatorname{erf} (At + B) \, dt + \tag{26}$$

$$\lambda_1 \sqrt{\pi} \sigma_1 \sigma_2 \mu_1 \int_{\frac{1}{\sqrt{2}} \frac{z_1 - \mu_1}{\sigma_1}}^{\frac{1}{\sqrt{2}} \frac{a - \mu_1}{\sigma_1}} e^{-t^2} \operatorname{erf} (At + B) \, dt. \tag{27}$$

The primitive of the first summand of Eq. 26 is equal to $\int t e^{-t^2} \operatorname{erf}(At + B)\, dt =$
$-\frac{1}{2} e^{-t^2} \operatorname{erf}(At + B) + \int \frac{1}{2} e^{-t^2} \frac{2A}{\sqrt{\pi}} e^{-(At+B)^2}\, dt$. Now we can apply the completing
of the square to find the primitive of $e^{-t^2} e^{-(At+B)^2} = e^{-((A^2+1)t + 2ABt + B^2)} =$
$e^{-(A^2+1)(t+\frac{AB}{A^2+1})^2} e^{\frac{(AB)^2}{A^2+1} - B^2} = e^{\frac{(AB)^2}{A^2+1} - B^2} \frac{\sqrt{\pi}}{2\sqrt{A^2+1}} \operatorname{erf}\left(\sqrt{A^2+1}(t + \frac{AB}{A^2+1})\right).$

Thus:

$$\lambda_1 \sqrt{\pi} \sigma_1^2 \sigma_2 \sqrt{2} \int_{\frac{1}{\sqrt{2}} \frac{z_1 - \mu_1}{\sigma_1}}^{\frac{1}{\sqrt{2}} \frac{a - \mu_1}{\sigma_1}} t e^{-t^2} \operatorname{erf}(At + B)\, dt = \tag{28}$$

$$\left[\left(-\frac{1}{2} e^{-t^2} \operatorname{erf}(At + B) + e^{\frac{(AB)^2}{A^2+1} - B^2} \frac{\sqrt{\pi}}{2D} \operatorname{erf}\left(Dt + \frac{AB}{D}\right) \right) \right]_{\frac{1}{\sqrt{2}} \frac{z_1 - \mu_1}{\sigma_1}}^{\frac{1}{\sqrt{2}} \frac{a - \mu_1}{\sigma_1}}. \tag{29}$$

which needs to be multiplied by $\lambda_1 \sqrt{\pi} \sigma_1^2 \sigma_2 \sqrt{2}$, and $D = \sqrt{A^2 + 1}$. Remains
Expression (27). With this integral we proceed as before, this integral is approximately equal to:

$$\lambda_1 \sqrt{\pi} \sigma_2 \sigma_1 \mu_1 \int_{\frac{1}{\sqrt{2}} \frac{z_1 - \mu_1}{\sigma_1}}^{\frac{1}{\sqrt{2}} \frac{a - \mu_1}{\sigma_1}} e^{-t^2} \sum_{i=0}^{2} \frac{(-1)^i (At + B)^{2i+1}}{i! \cdot (2i+1)}\, dt \tag{30}$$

for which we have the antiderivative which is the sum of the Expressions (18),
(19), (20), (21) and (22). Lastly we compute Integral IV:

$$-\lambda_1 \sqrt{\frac{\pi}{2}} \cdot \sigma_2 \operatorname{erf}\left(\frac{z_2 - \mu_2}{\sqrt{2}\sigma_2}\right) \int_{z_1}^{a} y_1 e^{-\frac{1}{2}\left(\frac{y_1 - \mu_1}{\sigma_1}\right)^2}\, dy_1 = \tag{31}$$

$$-\lambda_1 \sqrt{\frac{\pi}{2}} \cdot \sigma_2 \operatorname{erf}\left(\frac{z_2 - \mu_2}{\sqrt{2}\sigma_2}\right) \left[-\sigma_1^2 \cdot e^{-\frac{1}{2}\left(\frac{y_1 - \mu_1}{\sigma_1}\right)^2} + \mu_1 \sigma_1 \sqrt{\frac{\pi}{2}} \operatorname{erf}\left(\frac{y_1 - \mu_1}{\sqrt{2}\sigma_1}\right) \right]_{z_1}^{a}. \tag{32}$$

The integration over the upper triangle is done in a similar way. Sum over the
lower triangle and the upper triangle pre-multiplied with $\frac{1}{N} \frac{1}{2\pi\sigma_1\sigma_2}$ is the expected
improvement of the R2-indicator at a single weight vector. As said before, the
computation needs to be carried out for each of the s weight vectors and the
average of these numbers is the R2-indicator.

4   Numerical Results

In the numerical results we show the behavior for varying values of σ and μ. The
results are displayed in Fig. 3a (the monotonicity of the ER2I w.r.t. the σ) and
in Fig. 3b (the landscapes of the ER2I). Here the utopian point z is set to zero.
$P = [4.5\ 1; 3\ 3; 2\ 4; 1\ 6]$; $\mu = [3.5\ 0.5; 3.5\ 3.5]$; $\sigma = [1\ 1; 1\ 1] * \alpha$; $N_{iter} =$
10000; $\alpha = [0.1, 10]$ with a stepsize of 0.1.

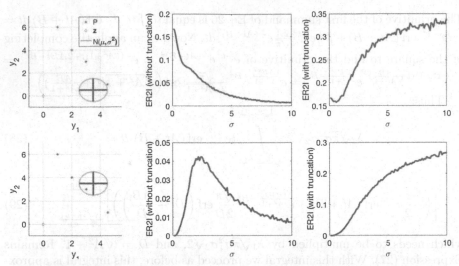

(a) Study of variance monotonicity of the non-truncated (middle) ER2I, and the truncated (right) ER2I. for two different positions of the predictive mean value (left).

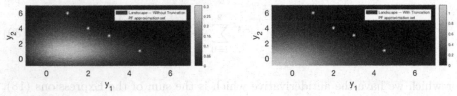

(b) Influence of the position of the mean value on the ER2I for the non-truncated ER2I and the truncated ER2I (right). The variance is a constant in the plots. Parameters for Fig 3b: $P = [4.5\ 1; 3\ 3; 2\ 4; 1\ 6]; \sigma = [1\ 1]; N_{iter} = 1000$. μ is a meshgrip with a stepsize of 0.1 for each coordinate from -1 to 7.

Fig. 3. Behavior of the ER2I.

Clearly, the truncation has a big influence on the result. Without it an increasing variance will first lead to a slight improvement of the ER2I but then the improvement will get worse, since points fall beyond the region of interest. If we integrate over the truncated expected improvement, increasing the variance will be rewarded. Moreover, in this case, a distribution that is centered around the utopian point yields the maximum ER2I value (see Fig. 3b (right)), as it should be. If truncation is not used, another region is more advantageous in terms of ER2I, see Fig. 3b (left).

The ER2I-MOBGO, which takes ER2I as the infill criterion in multi-objective Bayesian global optimization, are performed on three test problems, including the EYDWF problem [4], two generalized Schaffer problems (GSPs) [3] with parameter $\gamma = 0.4$, 1.8. All the parameters for all the experiments are the same. The size of the initial sampling set is 10, the initial sampling method is Latin Hypercube Sampling (LHS), the number of total function evaluations is

25, the optimizer is the grid search with spaced 30 points for each coordinate within the search range, and $N_{iter} = 200, \mathbf{z} = (0, 0)$. The search space is $\mathbf{x} = (x_1, x_2) \in [-2, 2] \times [-2, 2] \subset \mathbb{R}^2$ and $\mathbf{x} = (x_1, x_2) \in [0, 1] \times [0, 1] \subset \mathbb{R}^2$ for the EYDWF problem and the GSPs, respectively.

Figure 4 shows the experimental results, where blue stars and red diamond symbols represent the initial set generated by the LHS and the 15 "optimal" solutions are found during the sequential interactions of BGO by grid search to find the ER2I maximizer. Figure 4a shows the landscapes of the GSPs predictions for y_1, y_2 (first row), the landscapes of the standard deviations of the predictions σ_1, σ_2 (second row), the landscape of ER2I (left figure in the third row) and objective values of 25 evaluations in the objective space. Figure 4b shows the objective values of the GSPs over 25 evaluations. The experimental results show that the ER2I can find a good Pareto-front approximation set for liner, concave and convex problems.

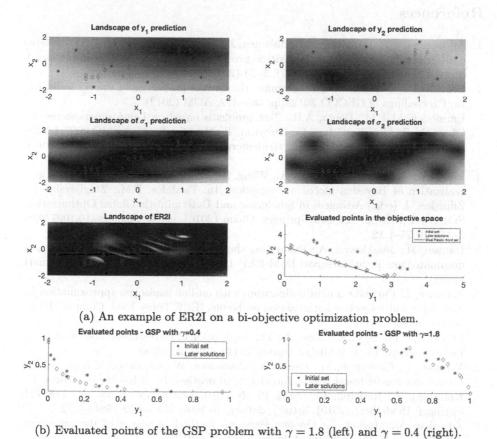

(a) An example of ER2I on a bi-objective optimization problem.

(b) Evaluated points of the GSP problem with $\gamma = 1.8$ (left) and $\gamma = 0.4$ (right).

Fig. 4. Experimental results on GSPs. (Color figure online)

5   Conclusions and Outlook

We discussed the numerical computation of the bi-objective Expected R2-Indicator improvement. To solve certain integrals of the erf(.)-functions Taylor expansions were used. The ER2I was proposed as an indicator in Bayesian Global optimization. In the 2-D case, and it was shown empirically, that it rewards a high variance and a better mean value of the distribution. For Pareto fronts of different geometry, it is able to guide the search towards the Pareto front and to fill in the gaps of a Pareto front approximation. Future work will have to compare it to other infill criteria and on challenging real-world problems and extend the deterministic computation scheme to more than two objective functions.

MATLAB and Mathematica implementations and data used in this study will be made available by the authors on http://moda.liacs.nl.

References

1. Allmendinger, R., Emmerich, M., Hakanen, J., Jin, Y., Rigoni, E.: Surrogate-assisted multicriteria optimization: complexities, prospective solutions, and business case. J. Multi-Criteria Decis. Anal. **24**(1–2), 5–24 (2017)
2. Brockhoff, D., Wagner, T., Trautmann, H.: On the properties of the R2 indicator. In: Proceedings of GECCO 2012, pp. 465–472. ACM (2012)
3. Emmerich, M.T.M., Deutz, A.H.: Test problems based on Lamé superspheres. In: Obayashi, S., Deb, K., Poloni, C., Hiroyasu, T., Murata, T. (eds.) EMO 2007. LNCS, vol. 4403, pp. 922–936. Springer, Heidelberg (2007). https://doi.org/10.1007/978-3-540-70928-2_68
4. Emmerich, M., Yang, K., Deutz, A., Wang, H., Fonseca, C.M.: A multicriteria generalization of Bayesian global optimization. In: Pardalos, P.M., Zhigljavsky, A., Žilinskas, J. (eds.) Advances in Stochastic and Deterministic Global Optimization. SOIA, vol. 107, pp. 229–242. Springer, Cham (2016). https://doi.org/10.1007/978-3-319-29975-4_12
5. Hansen, M., Jaszkiewicz, A.: Evaluating the quality of approximations to the non-dominated set. Technical report IMM-REP-1998-7. Technical University of Denmark (1998)
6. Knowles, J.: ParEGO: a hybrid algorithm with on-line landscape approximation for expensive multiobjective optimization problems. IEEE Trans. Evol. Comput. **10**(1), 50–66 (2006)
7. Mockus, J., Tiesis, V., Žilinskas, A.: The application of Bayesian methods for seeking the extremum. Towards Global Optim. **2**(117–129), 2 (1978)
8. Wagner, T., Emmerich, M., Deutz, A., Ponweiser, W.: On expected-improvement criteria for model-based multi-objective optimization. In: Schaefer, R., Cotta, C., Kołodziej, J., Rudolph, G. (eds.) PPSN 2010. LNCS, vol. 6238, pp. 718–727. Springer, Heidelberg (2010). https://doi.org/10.1007/978-3-642-15844-5_72
9. Žilinskas, A., Mockus, J.: On one Bayesian method of search of the minimum. Avtomatica i Vychislitel'naya Teknika **4**, 42–44 (1972)

Innovization and Surrogates

Trust-Region Based Multi-objective Optimization for Low Budget Scenarios

Proteek Chandan Roy$^{(\boxtimes)}$ (iD), Rayan Hussein, Julian Blank (iD),
and Kalyanmoy Deb (iD)

Michigan State University, East Lansing, MI 48824, USA
{royprote,husseinr,blankjul,kdeb}@egr.msu.edu
http://www.coin-laboratory.com/

Abstract. In many practical multi-objective optimization problems, evaluation of objectives and constraints are computationally time-consuming, because they require expensive simulation of complicated models. Researchers often use a comparatively less time-consuming surrogate or metamodel (model of models) to drive the optimization task. Effectiveness of the metamodeling method relies not only on how it manages the search process (to find infill sampling) but also how it deals with associated error uncertainty between metamodels and the true models during an optimization run. In this paper, we propose a metamodel-based multi-objective evolutionary algorithm that adaptively maintains regions of trust in variable space to make a balance between error uncertainty and progress. In contrast to other trust-region methods for single-objective optimization, our method aims to solve multi-objective expensive problems where we incorporate multiple trust regions, corresponding to multiple non-dominated solutions. These regions can grow or shrink in size according to the deviation between metamodel prediction and high-fidelity computed values. We introduce two performance indicators based on hypervolume and achievement scalarization function (ASF) to control the size of the trust regions. The results suggest that our proposed trust-region based methods can effectively solve test and real-world problems using a limited budget of solution evaluations with increased accuracy.

Keywords: Surrogate modeling · Metamodel · Trust-region method · Multi-objective optimization

1 Introduction

Most real-world problems involve time-consuming experiments and simulations that cause optimization to be increasingly expensive. To face this challenge and to reduce the computational cost, metamodels as approximations of exact or high-fidelity based computational models are used for the optimization task. There are a few challenges and decision factors in metamodel-based multi-objective optimization. First, given a multi-objective optimization problem with

© Springer Nature Switzerland AG 2019
K. Deb et al. (Eds.): EMO 2019, LNCS 11411, pp. 373–385, 2019.
https://doi.org/10.1007/978-3-030-12598-1_30

M number of objectives and J number of constraints, one can model each objective and constraint separately thus having a total of $(M + J)$ metamodels. Also, one can combine all objectives using a scalarization method, e.g., weighted-sum, ϵ-constraint, Tchebychev, or achievement scalarization function (ASF) [16,26] and metamodel them separately, thereby reducing the total number of metamodels to $(1 + J)$. The choice of metamodeling methodologies are discussed in recent papers [2,3,5,9,13,19,21,22] by the authors. Second, a great deal of research has been done to formulate the criteria for finding infill or subsequent points for high-fidelity evaluation during optimization. For example, Emmerich et al. [11] has generalized the concept of probability of improvement and the expected improvement to find infill solutions. Next, computational cost of constructing surrogates is a practical issue that prohibits us to build a large number of metamodels. Finding the best metamodel or approximation method is another concern for metamodel-based optimization. There is a wide variety of metamodels, such as Kriging, neural network, support vector regression, polynomial approximation and others, used in past studies [14]. Interestingly, the choice of metamodeling method may vary according to early, intermediate or late stage of the optimization process and is certainly not known a priori. Therefore, researchers have attempted to use multiple surrogate models in few efforts [25].

Although most existing methods are directed towards proposing more accurate metamodels or introducing efficient search schemes, there is a need for managing error uncertainty of one particular under-performing metamodel during optimization. A better management of a metamodel can, not only restrain the model from becoming worse, but also boost the performance by recognizing the inherent complexity of search regions. In this paper, we introduce a trust region concept for multi-objective optimization to reduce model uncertainty during metamodel-based optimization. This may allow a continuous convergence to the Pareto-front in some cases. Therefore, we don't completely rely on the assumptions made by the metamodel from the first iteration on.

The rest of this paper is organized as follows. Section 2 presents the previous works that are relevant to the trust region, uncertainty of metamodeling and overall metamodel-based algorithms. Section 3 discusses the new concepts introduced in this paper. Based on those concepts, the algorithm is presented in Sect. 4. Experimental settings and results are presented in Sect. 5. Section 6 concludes our study and suggests future work.

2   Related Studies

There have been several studies in metamodel-based multi-objective evolutionary algorithms for constrained and unconstrained problems. ParEGO [15], MOEA/D-EGO [27], SMS-EGO [18] and KRVEA [4] use scalarization methods (e.g., Tchebycheff) to combine multiple objectives into one and solve multiple scalarized versions of them to find a trade-off set of solutions. While these methods are mostly useful for unconstrained problems, they need to be modified for constrained scenarios. Hypervolume-based expected improvement [10] and maximum hypervolume contribution [18] are used as a performance criteria for infill points. Few recent studies [4,19] outperformed standard evolutionary multi-objective optimization methods for unconstrained test problems.

Trust region methods are an effective mechanism to identify new infill points with a specific certainty. A few researchers have suggested using metamodel-based optimization with a trust region concept [1,17]. They proposed a trust region framework for using approximation models with varying fidelity. Their approach is based on the trust region concept from nonlinear programming literature and was shown to be provably convergent for some of the original high-fidelity problems. A sequential quadratic approximation model was used in their study. In [17], a global version of the trust region method—Global Stochastic Trust Augmented Region (G-STAR) was proposed. The trust region was used to focus on simulation effort and balance between exploration and exploitation. They used Kriging as a metamodel for unconstrained single-objective optimization problems only. Few recent studies have considered for bi-objective [24] and multi-objective [12] problems with a convergence guarantee under mild conditions.

3   Trust Region Method for Single-Objective Optimization

The classical trust region method for single-objective optimization proceeds by building a metamodel $\hat{f}(.)$ for the original objective function $f(.)$. The prediction of the metamodel $\hat{f}(.)$ is minimized to obtain new infill points [1]:

$$\text{Minimize}_q \ \hat{f}(q), \quad \text{Subject to } \|q - p\| \leq \delta_k. \tag{1}$$

Here p is the current iterate (solution) and q is the new predicted point that can replace p in the next iteration. Typically, a quadratic model is used as $\hat{f}(.)$. The search is restricted within a radius δ_k from the current point p so that the metamodel approximates f well. The distance $\|q - p\|$ can be calculated using any norm. Without loss of generality, we use the Euclidean norm here. The trust region is updated by comparing the exact and the predicted value of the new point $(f(q)$ and $\hat{f}(q))$ with respect to the old point p by the following eq. [1]:

$$r = \frac{f(p) - f(q)}{f(p) - \hat{f}(q)}. \tag{2}$$

Depending on the performance indicator r, the trust region might increase, decrease or remain the same. To decide what operation should be performed, two constants r_1 and r_2 are defined and the trust region is adapted as follows:

– If the model fails to improve objective value (that is, $r < r_1$), we reduce the trust region by multiplying existing δ_k with c_1 (<1) and do not replace p with the new point q.

- If the model performs good in predicting function improvement from previous solution (that is, $r > r_2$), we increase δ_k for the next iteration by multiplying existing δ_k with c_2 (>1) and we replace the old point p by new point q.
- Otherwise, we leave the trust region size δ_k as it was before.

We replace the old point p with the new point q, whenever q is a better point. The current point (p or q) is always associated with the updated trust radius. Suitable values of c_1 and c_2 are used.

3.1 Challenges and Motivation for Multi-objective Optimization

The main challenges for applying the trust region concept in multi-objective evolutionary algorithms (MOEA) are handling multiple objectives and constraints. In addition, since MOEAs are population based methods, we also need to deal with multiple solutions and their individual trust regions. Moreover, there is a need for a meaningful performance metric to adapt trust radii of multiple high-fidelity solutions.

4   Proposed Trust Region in Metamodel-Based Multi-objective Evolutionary Algorithm

A multi-objective optimization problem can be formulated as follows. Here, we omit the vector notation of $\{x, p, q\}$ and F to denote a multi-dimensional point or objective vector.

$$\text{Minimize } F(x) = (f_1(x), f_2(x), \ldots, f_M(x))$$
$$\text{Subject to } g_j(x) \geq 0, \ \forall j \in \{1, \ldots, J\} \tag{3}$$
$$x \in \Omega \subseteq \mathbb{R}^n \text{ and, } F \in \Lambda \subseteq \mathbb{R}^M$$

Here, feasible variable space and respective feasible objective space are defined by Ω and Λ, respectively. The goal of this optimization is to find the best trade-off hyper-surface.

4.1 Proposed Trust Region Concept

We propose several modifications on the classical trust region method in order to make it applicable to metamodel-based multi-objective evolutionary algorithms:

1. We store all high fidelity solutions in an archive A, instead of replacing them with better solutions.
2. We maintain an independent trust region in the variable space for each solution. The regions may overlap with each other. They can either grow or shrink in size independently during optimization according to the quality of prediction. The algorithm restricts its search within the combined trust regions of A.

3. To compare a newly predicted point q with the neighbor point p (q is within trust region of p), we define two performance indicators PI that calculate r (analogous to Eq. 2) for a multi-objective problem. Moreover, we propose a novel scheme to compare between feasible and infeasible solutions.
4. If the new point q is within the trust regions of multiple points $P \subseteq A$, then we update the trust radius δ_k for each of them using pair-wise performance metric (PI). The trust radius of point q will be the minimum of trust radii of P.

Thus, we optimize the following metamodel-based optimization to obtain a set of new infill points:

$$\text{Minimize}_{q \in \Omega} \, \hat{f}_1(q), \dots, \hat{f}_M(q)$$
$$\hat{g}_j(x) \geq 0, \ \forall j \in \{1, \dots, J\} \tag{4}$$
$$\text{Subject to } \|q - p\| \leq \delta_k^p, \ \exists p \in A$$

Here $p \in A$ are the exactly evaluated solutions from the current archive. Figure 1 illustrates the population based extension of the trust region method. Five exactly evaluated points $\{P_1, P_2, P_3, P_4, P_5\}$ with their trust regions (regions within the circles) are shown. Say, a new point P_{new} is predicted by the algorithm after optimizing on the model space. Note that P_{new} is inside the trust regions of P_1 and P_2. Assuming that the performance indicator reports an improvement of P_{new} over P_2, but no improvement over P_1. Then we reduce the size of the trust region of P_2 and increase that of P_1. The trust radius of the new point will be the smaller of the trust radii of P_1 and P_2.

4.2   Performance Indicators for Updating Trust Radius

To update the trust radius of solutions, we propose two performance indicators (PI).

Scalarization based Performance Indicator(PI_{ASF}): Scalarization method is used to convert a multi-objective problem into a number of parameterized single-objective optimization problems. We use the achievement scalarization function (ASF) [26] as a performance indicator. The scalarization is based on a weight vector w and a reference point z. The ASF formulation is given below:

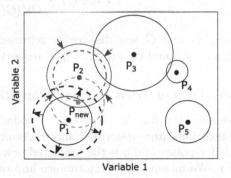

Fig. 1. Adaptive trust region concept for multiple solutions.

$$\text{ASF}(x) = \max_{i=1}^{M} \frac{f_i(x) - z_i}{w_i}. \tag{5}$$

The proposed performance criteria using ASF function for trust radius update is presented as follows:

$$PI_{ASF}(q) = \frac{ASF(p) - ASF(q)}{ASF(p) - \widehat{ASF}(q)}. \tag{6}$$

Here $\widehat{ASF}$ is obtained from predicted objectives. The estimated improvement may differ for different reference directions.

Hypervolume based Performance Indicator (PI_{HV}): Hypervolume [10] is a widely used indicator in multi-objective optimization. It takes a set of solutions and a reference point, and computes the dominated region (in objective space) enclosed by the set and the reference point. In order to find the improvement of a new point over old point, we calculate the difference of their absolute hypervolume measures. We include archive points (A) as a common ground for computation. We then compute the ratio between actual improvement and predicted improvement and adjust the trust radii of old points. The predicted hypervolume is calculated by the objective values evaluated in model space using $\hat{F}(.)$. Since larger values indicate better hypervolume, we use negative of the hypervolume:

$$PI_{HV}(q) = \frac{HV(F(A) \cup F(q)) - HV(F(A))}{HV(F(A) \cup \hat{F}(q)) - HV(F(A))}. \tag{7}$$

Performance Indicator for Constrained Problems: We use constrained violation CV function [7], by accumulating violation of each constraint function ($g_j(x) \geq 0$), given as: $CV(x) = \sum_{j=1}^{J} \langle \bar{g}_j(x) \rangle$, where the bracket operator $\langle \alpha \rangle$ for g is $-\alpha$ if $\alpha < 0$ and zero, otherwise. The functions $\bar{g}_j$ are the normalized version of constraint functions g_j [7].

$$PI_{CV}(q) = \frac{CV(G(p)) - CV(G(q))}{CV(G(p)) - CV(\hat{G}(q))} \tag{8}$$

Here, G and $\hat{G}$ are the vector representations of constraint functions $G = (g_1, \ldots, g_J)$ and $\hat{G} = (\hat{g}_1, \ldots, \hat{g}_J)$, respectively.

4.3   Overall Trust Region Adaptation

We now describe the procedure of updating the trust regions using the performance indicators described above. Assume that solution p is one of the high-fidelity points and q is the predicted new point which is within the trust region of p. We measure the performance improvement by the following equation.

$$r = \begin{cases} PI_{HV}(q) \text{ or } PI_{ASF}(q), & \text{if both } p \text{ and } q \text{ feasible}, \\ r_2 + \epsilon, & \text{if } p \text{ infeasible}, q \text{ feasible}, \\ r_1 - \epsilon, & \text{if } p \text{ feasible}, q \text{ infeasible}, \\ PI_{CV}(q), & \text{otherwise}. \end{cases} \tag{9}$$

Here $\epsilon > 0 \in \mathcal{R}$ is a small positive number. The pre-defined positive constants $0 < r_1 < r_2 < 1$ are the hyper-parameters that regulate expansion and contraction of the trust regions. After estimating performance indicator PI of a new point q with respect to old point p we update trust radius of p by the following rule.

$$\delta^p_{k+1} = \begin{cases} c_1 \delta^p_k & \text{if } r < r_1 \\ \min\{c_2 \delta^p_k, \Delta_{max}\} & \text{if } r > r_2 \\ \delta^p_k & \text{otherwise} \end{cases} \tag{10}$$

The positive constants $0 < c_1 < 1$ and $c_2 > 1$ controls the size of subsequent trust radius. As mentioned earlier, we assign the trust radius of q to be the smaller of the trust radii of all neighboring solutions of which q is inside their trust regions. The parameter Δ_{max} is the largest allowed trust radius for the solutions.

5   Proposed Overall Algorithm

We now present trust region based algorithm for multi-objective optimization for low-budget problems. We refer our algorithm to TR-NSGA-II.

The overall procedure is described in Algorithm 1, the metamodeling algorithm starts with an archive of ρ initial population members created using the Latin hypercube sampling (LHS) method on the entire search space. The trust radii of initial solutions are then set to a predefined initial value δ_{init}. Thereafter, these solutions are evaluated exactly (high-fidelity) and metamodels are constructed for all M objectives $(\hat{f}_i(x); i = 1, \ldots, M)$ and J constraints $(\hat{g}_j(x); j = 1, \ldots, J)$. Then, a multi-objective evolutionary algorithm NSGA-II [6] with faster non-dominated sorting algorithm [20,23] is run for τ generations starting with μ initial random solutions in model space. The NSGA-II algorithm returns $min(\mu, E - e)$ solutions where e is the current number of high-fidelity solution evaluations. The solutions are then evaluated using high-fidelity simulation and included in the archive (line 13). Then, new metamodels are then build from scratch and the process is repeated until termination. The trust radii are updated after each NSGA-II run, for new and old points according to the update rules discussed before. We have used both Hypervolume based and ASF based performance indicator alternatively for updating trust radius. ASF values are computed using reference point set W. PI_{ASF} is calculated using the best ASF values for the new solutions.

Algorithm 1. Trust Region Based Algorithm or TR-NSGA-II

Input : Obj: $[f_1, \ldots, f_m]^T$, Constr: $[g_1, \ldots, g_J]^T$, n (vars), ρ (sample
size), E (max. high-fidelity SEs), NSGA-II (multi-obj EA)
with pop-size μ, number of generation for model optimization
τ, other parameters of NSGA-II Γ, Constraint violation
function **CV**, Trust region parameters $\delta_{init}, \Delta_{max}, c_1, c_2, r_1$
and r_2

Output: Solution set P_T

1 $t, e \leftarrow 0$;
2 $P_t, F_t, G_t \leftarrow \varnothing$;
3 $P_{new} \leftarrow \text{LHS}(\rho, n)$// Initial solutions
4 $\delta^\ell \leftarrow \delta_{init}, \forall \ell \in \{1, \ldots, \rho\}$;
5 **while** *True* **do**
6 $\quad$ $F^i_{new} \leftarrow f_i(P_{new}), \forall i \in \{1, \ldots, M\}$// eval obj.
7 $\quad$ $G^j_{new} \leftarrow g_j(P_{new}), \forall j \in \{1, \ldots, J\}$// eval constr.
8 $\quad$ **if** $t > 0$ **then**
9 $\quad\quad$ $\widehat{F}^i_{new} \leftarrow \widehat{f}^i_t(P_{new}), \forall i \in \{1, \ldots, M\}$// predicted
10 $\quad\quad$ $\widehat{G}^j_{new} \leftarrow \widehat{g}^j_t(P_{new}), \forall j \in \{1, \ldots, J\}$// predicted
11 $\quad\quad$ $\delta \leftarrow \text{UPDATE\_TRUSTREGION}(F_t, \widehat{F}_{new}, G_t, \widehat{G}_{new}, \delta)$
12 $\quad$ **end**
13 $\quad$ $P_{t+1}, F_{t+1}, G_{t+1} \leftarrow (P_t \cup P_{new}), (F_t \cup F_{new})$ and $(G_t \cup G_{new})$;
14 $\quad$ $e \leftarrow e + |P_{new}|$;
15 $\quad$ **break if** $e \geq E$;
16 $\quad$ $\widehat{f}^i_{t+1} \leftarrow \text{METAMODEL}(F^i_{t+1}), \forall i \in \{1, \ldots, M\}$// metamodel obj.
17 $\quad$ $\widehat{g}^j_{t+1} \leftarrow \text{METAMODEL}(G^j_{t+1}), \forall j \in \{1, \ldots, J\}$// metamodel constrt.
18 $\quad$ $P_{new} \leftarrow \text{NSGA-II}(\widehat{f}_{t+1}, \widehat{g}_{t+1}, \mu, \tau, \Gamma, E - e, \textbf{CV}, \delta)$; // Optimize model
$\quad$ space
19 $\quad$ $t \leftarrow t + 1$;
20 **end**
21 **return** $P_T \leftarrow$ filter the best solutions from P_{t+1}

In one epoch $|W|$ solutions are returned directly by running NSGA-II while in
another epoch we choose $|W|$ solutions (after running NSGA-II) such that they
minimizes ASF according to $w \in W$ reference directions. In the end, trust regions
are updated according to Hypervolume and ASF respectively. The major steps
of this method are outlined in Algorithm 1.

6   Results

We present experimental results obtained by running four different optimization
algorithms. We refer our algorithm as TR-NSGA-II. We compare the proposed
algorithm with three other baseline algorithms: (a) M1-2 [9] which works similar to our TR-NSGA-II (Algorithm 1) but without the trust region, and (b)
state-of-the-art multi-objective evolutionary method NSGA-II [8] and recently
proposed K-RVEA [4]. We got source code of K-RVEA from the authors. The
code currently doesn't handle constraints, thus we don't apply it to constrained
problems. In NSGA-II, we use the binary tournament selection operator, simulated binary crossover (SBX), and polynomial mutation with parameters as
follows: population size $= 10n$, where n is a number of variables, number of
generations $= 100$, crossover probability $= 0.95$, mutation probability $= 1/n$,
distribution index for SBX operator $= 15$, and distribution index for polynomial mutation operator $= 20$. The NSGA-II procedure, wherever used, uses the

same parameter values. Initial trust radius is $\delta_{init} = 0.75\Delta_{max}$ for all problems, where $\Delta_{max} = \sqrt{n}$ is the largest diagonal of an n-dimensional unit hypercube. We take $c_1 = 0.75, c_2 = 1.10, r_1 = 0.9, r_2 = 1.05$ for all the problems. All the distances calculated here are in the normalized space. We perform 11 runs for each algorithm on all test and engineering design problems.

For NSGA-II, we have used population size 20 to maximize the evolution effect and that provided the best results for these low-budget problems. Other parameters are kept identical across all algorithms to provide a representative performance of each algorithm. Median IGD values and p-values of Wilcoxon rank sum test are provided in Table 1.

Table 1. IGD values for 11 test problems are computed. Best algorithm and other statistically similar methods are marked in bold.

Problem/Method	NSGA-II		M1-2		TR-NSGA-II		K-RVEA	
	IGD	GD	IGD	GD	IGD	GD	IGD	GD
ZDT1	0.27131	0.34582	0.01161	0.01091	**0.00121**	**0.00122**	0.07964	0.03715
	p=1.852e-05	p=1.852e-05	p=7.7801e-04	p=7.4613e-04	-	-	p=1.852e-05	p=1.852e-05
ZDT2	0.98265	0.61637	0.00975	0.00755	**0.00057**	**0.00081**	0.03395	**0.00080**
	p=1.852e-05	p=1.852e-05	p=1.852e-05	p=1.852e-05	-	p=0.2851	p=1.852e-05	-
ZDT3	0.32080	0.38940	0.01251	0.00761	**0.00870**	**0.00230**	0.02481	0.00650
	p=1.852e-05	p=1.852e-05	p=1.852e-05	p=1.852e-05	-	-	p=1.852e-05	p=1.802e-04
ZDT4	25.24040	34.43350	7.11881	10.10851	6.97620	12.92170	**4.33221**	**4.50901**
	p=1.852e-05	p=1.852e-05	p=0.7928	p=0.1007	p=0.8955	p=0.2934	-	-
ZDT6	5.00571	4.80922	1.55861	2.27535	**0.31070**	2.84941	0.65462	**1.50551**
	p=1.852e-05	p=1.852e-05	p=1.852e-05	p=0.001	-	p=0.0151	p=1.852e-05	-
BNII	0.78981	0.19842	0.45272	0.13606	**0.09651**	**0.09092**	-	-
	p=1.852e-05	p=1.852e-05	p=1.852e-05	p=1.852e-05	-	-		
SRN	1.66162	2.11235	**0.67285**	**0.75337**	1.44045	1.74951	-	-
	p=1.852e-05	p=1.852e-05	-	-	p=1.852e-05	p=1.852e-05		
TNK	0.04182	0.01341	0.01543	0.01008	**0.00141**	**0.00201**	-	-
	p=1.852e-05	p=1.852e-05	p=1.852e-05	p=1.852e-05	-	-		
OSY	35.80211	27.43991	4.78063	0.59202	**0.16731**	**0.25063**	-	.
	p=1.059e-05	p=1.331e-05	p=1.832e-05	p=1.852e-05	-	-		
Welded Beam	1.10272	**0.21092**	0.92692	1.68806	**0.07681**	1.72811	-	-
	p=1.852e-05	-	p=1.8267e-04	p=0.0042	-	p=0.0012		
C2DTLZ2	0.13733	0.04792	**0.03355**	**0.02373**	0.06411	0.02991	-	-
	p=1.852e-05	p=1.852e-05	-	-	p=1.8267e-04	p=1.8267e-04		

6.1 Two-Objective Unconstrained Problems

First, we apply our proposed method to two-objective unconstrained problems ZDT1, ZDT2, ZDT3, ZDT4 and ZDT6 with ten ($n = 10$) variables, $|W| = 21$ reference directions, and with a maximum of only $E = 500$ high-fidelity solution evaluations. The obtained non-dominated solutions are shown in Fig. 2(a)–(e). It is evident from the figure that trust region method with hypervolume perform better than method M1-2 without trust region. Because of the lack of enough solution evaluations, NSGA-II could not converge enough to these problems. On the contrary, trust region based methods provide increased accuracy (for example TR-NSGA-II has IGD 0.00121 compared to 0.01161 of M1-2 for ZDT1) for these test problems. K-RVEA performed the third best in ZDT1 with IGD = 0.07964. TR-NSGA-II performs the best for all ZDT problems both in terms of GD and IGD. For ZDT4, all the methods find it hard to converge and perform equivalently (with p-value 0.05) except NSGA-II. For ZDT6, K-RVEA has better GD value but TR-NSGA-II has a better distribution (IGD = 0.31070).

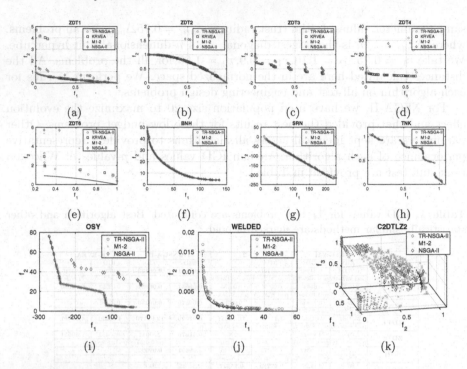

Fig. 2. Obtained non-dominated solutions of median run for 11 test problems using four different algorithms are shown.

6.2 Two-Objective Constrained Problems

Next, we apply our algorithms to two-objective constrained problems: BNH, SRN, TNK, and OSY [8], For each problem, we use $|W| = 21$ reference directions and a total of 500 solution evaluations. The obtained non-dominated solutions are shown in Fig. 2(f)–(i). With the trust region method, we find better convergence as well as diversity for OSY and TNK, although no extra effort has been made to maintain diversity. BNH, SRN and TNK have only two variables and two constraints. NSGA-II, along with all other methods, performs well in BNH and SRN. We have achieved increased accuracy (IGD 0.00141 compared to 0.01543 of M1-2) for TNK problem. OSY is a difficult problem with six variables and six constraints. But our proposed method is able to find a good distribution on the true Pareto-front with only 500 solution evaluations with better IGD and GD. In SRN, method M1-2 performs the best in terms of both GD and IGD.

6.3 Three-Objective and Real-World Problems

We have applied three methods (except K-RVEA) to three-objective constrained problem C2DTLZ2 (Fig. 2(k)). For C2DTLZ2, M1-2 without trust region performs the best. Trust region based method TR-NSGA-II performs the second best. Due to restricted search region in 3-dimensional space, our method suffers from premature convergence. We also apply our algorithm to a real-world welded beam design problem. Surprisingly, with the optimum population size, NSGA-II performs better in terms of GD, whereas our method has the best IGD.

Median IGD values of 11 runs of 11 test problems are presented in Table 1 for all the algorithms. The table demonstrates that trust region methods perform usually better than non-trust region based methods whenever solutions reach to near Pareto-optimal front. It would be interesting to incorporate our method to other recently proposed multi-objective evolutionary algorithms including K-RVEA.

6.4 Dynamics of Trust Region Adaptation

In Fig. 3, we investigate the dynamics of trust region adaptation for the evolving population in different test problems. The value of δ starts from the $\sqrt{n}$ where n is number of variables. The maximum value remains the same for most ZDT problems because the obtained non-dominated solutions go beyond these regions after some epochs. In contrast, minimum, median and mean values are always decreasing throughout the optimization process. As discussed before, based on the improvement of the neighboring solutions, the regions are either expanded or contracted. In order to increase the trust region, the evolving population has to maintain $r_2 = 1.05$ or 5% Hypervolume improvement over previous generations. In general, this condition is hard to meet when solutions are converged in the end. Therefore, our method focus more on exploitation in the last stage of optimization.

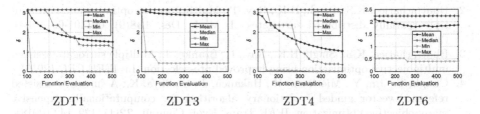

ZDT1   ZDT3   ZDT4   ZDT6

Fig. 3. Trust region adaptation for evolving population is presented. Minimum, median, average and maximum δ values during the optimization are shown.

7 Conclusions

In this paper, we have presented an adaptive trust region concept for multi-objective optimization with a low budget of solution evaluations. Trust regions are used as a constraint in the variable space during optimization to deal with uncertainties of metamodels. This study makes three main contributions: First, we have proposed two performance indicators based on scalarization and hyper-volume to adapt appropriate trust regions. Second, a constraint handling scheme is presented in order to handle the trust region adaptation in the presence of constraints. Third, since multi-objective optimization aims to find a set of Pareto-optimal solutions, we need to manage multiple trust regions with multiple trade-off solutions compare to single best solution, as proposed in the classical literature. Our results on several test and one engineering design problems have shown that we can achieve better convergence using the proposed method than that without a trust region. While other MOEAs spend thousands of function evaluations, our trust region based method can solve test and real-world problems with limited budget yet with increased accuracy.

The current study has introduced some new parameters, such as the initial trust radius and their updating factors. Although our experiments are based on reasonable parameter settings, a detailed parameter study is a good starting point for future research. Moreover, other distance metrics besides the Euclidean norm can be used to define a trust region. Also, it needs to be ensured that the trust region concept scales up for problems with high-dimensional variable, objective, and constraint spaces. Nevertheless, this pilot first study has made one aspect of metamodeling task for multi-objective optimization clear – a balance between a trust of metamodels around high-fidelity points and progress of the overall search is essential for an efficient application.

References

1. Alexandrov, N.M., Dennis, J.E., Lewis, R.M., Torczon, V.: A trust-region framework for managing the use of approximation models in optimization. Struct. Optim. **15**(1), 16–23 (1998)
2. Bhattacharjee, K.S., Singh, H.K., Ray, T.: Multi-objective optimization with multiple spatially distributed surrogates. J. Mech. Des. **138**(9), 091401-091401-10 (2016)
3. Bhattacharjee, K.S., Singh, H.K., Ray, T., Branke, J.: Multiple surrogate assisted multiobjective optimization using improved pre-selection. In: IEEE CEC (2016)
4. Chugh, T., Jin, Y., Miettinen, K., Hakanen, J., Sindhya, K.: A surrogate-assisted reference vector guided evolutionary algorithm for computationally expensive many-objective optimization. IEEE Trans. Evol. Comput. **22**(1), 129–142 (2018)
5. Deb, K., Hussein, R., Roy, P.C., Toscano, G.: A taxonomy for metamodeling frameworks for evolutionary multi-objective optimization. IEEE Trans. Evol. Comput. (in Press)
6. Deb, K., Pratap, A., Agarwal, S., Meyarivan, T.: A fast and elitist multiobjective genetic algorithm: NSGA-II. IEEE Trans. Evol. Comput. **6**(2), 182–197 (2002)
7. Deb, K.: An efficient constraint handling method for genetic algorithms. Comput. Methods Appl. Mech. Eng. **186**, 311–338 (2000)
8. Deb, K.: Multi-Objective Optimization Using Evolutionary Algorithms. Wiley, New York (2001)
9. Deb, K., Hussein, R., Roy, P., Toscano, G.: Classifying metamodeling methods for evolutionary multi-objective optimization: first results. In: Trautmann, H., et al. (eds.) EMO 2017. LNCS, vol. 10173, pp. 160–175. Springer, Cham (2017). https://doi.org/10.1007/978-3-319-54157-0_12

10. Emmerich, M.T.M., Deutz, A.H., Klinkenberg, J.W.: Hypervolume-based expected improvement: monotonicity properties and exact computation. In: 2011 IEEE Congress of Evolutionary Computation, CEC, pp. 2147–2154 (2011)

11. Emmerich, M.T.M., Giannakoglou, K.C., Naujoks, B.: Single- and multiobjective evolutionary optimization assisted by Gaussian random field metamodels. IEEE Trans. Evol. Comput. **10**(4), 421–439 (2006)

12. Fliege, J., Vaz, A.I.F.: A method for constrained multiobjective optimization based on SQP techniques. SIAM J. Optim. **26**(4), 2091–2119 (2016)

13. Hussein, R., Deb, K.: A generative kriging surrogate model for constrained and unconstrained multi-objective optimization. In: GECCO 2016. ACM Press (2016)

14. Jin, Y.: Surrogate-assisted evolutionary computation: recent advances and future challenges. Swarm Evol. Comput. **1**(2), 61–70 (2011)

15. Knowles, J.: ParEGO: a hybrid algorithm with on-line landscape approximation for expensive multiobjective optimization problems. IEEE Trans. Evol. Comput. **10**, 50–66 (2006)

16. Miettinen, K.: Nonlinear Multiobjective Optimization. Kluwer, Boston (1999)

17. Pedrielli, G., Ng, S.: G-STAR: a new kriging-based trust region method for global optimization. IEEE Press, United States, January 2017

18. Ponweiser, W., Wagner, T., Biermann, D., Vincze, M.: Multiobjective optimization on a limited budget of evaluations using model-assisted S-metric selection. In: Rudolph, G., Jansen, T., Beume, N., Lucas, S., Poloni, C. (eds.) PPSN 2008. LNCS, vol. 5199, pp. 784–794. Springer, Heidelberg (2008). https://doi.org/10.1007/978-3-540-87700-4_78

19. Roy, P., Deb, K.: High dimensional model representation for solving expensive multi-objective optimization problems. In: IEEE CEC, pp. 2490–2497 (2016)

20. Roy, P.C., Deb, K., Islam, M.M.: An efficient nondominated sorting algorithm for large number of fronts. IEEE Trans. Cyber. 1–11 (2018)

21. Roy, P., Hussein, R., Deb, K.: Metamodeling for multimodal selection functions in evolutionary multi-objective optimization. In: GECCO 2017. ACM Press (2017)

22. Roy, P.C., Blank, J., Hussein, R., Deb, K.: Trust-region based algorithms with low-budget for multi-objective optimization. In: GECCO, pp. 195–196. ACM (2018)

23. Roy, P.C., Islam, M.M., Deb, K.: Best order sort: a new algorithm to non-dominated sorting for evolutionary multi-objective optimization. In: GECCO (2016)

24. Ryu, J.H., Kim, S.: A derivative-free trust-region method for biobjective optimization. SIAM J. Optim. **24**(1), 334–362 (2014)

25. Viana, F.A.C., Haftka, R.T., Watson, L.T.: Efficient global optimization algorithm assisted by multiple surrogate techniques. J. Global Optim. **56**, 669–689 (2013)

26. Wierzbicki, A.P.: The use of reference objectives in multiobjective optimization. In: Fandel, G., Gal, T. (eds.) Multiple Criteria Decision Making Theory and Application. LNE, vol. 177, pp. 468–486. Springer, Heidelberg (1980). https://doi.org/10.1007/978-3-642-48782-8_32

27. Zhang, Q., Liu, W., Tsang, E., Virginas, B.: Expensive multiobjective optimization by MOEA/D With Gaussian process model. IEEE Trans. Evol. Comput. **14**(3), 456–474 (2010)

Approximating Pareto Set Topology by Cubic Interpolation on Bi-objective Problems

Yuri Marca[1,5(✉)], Hernán Aguirre[1,5], Saúl Zapotecas Martinez[2,5], Arnaud Liefooghe[3,5], Bilel Derbel[3,5], Sébastien Verel[4,5], and Kiyoshi Tanaka[1,5]

[1] Faculty of Engineering, Shinshu University, Nagano, Japan
yurimarca@gmail.com, {ahernan,ktanaka}@shinshu-u.ac.jp
[2] Universidad Autónoma Metropolitana Cuajimalpa, Mexico City, Mexico
szapotecas@correo.cua.uam.mx
[3] University of Lille, CNRS, Centrale Lille, UMR 9189 – CRIStAL,
Inria Lille – Nord Europe, 59000 Lille, France
{arnaud.liefooghe,bilel.derbel}@univ-lille1.fr
[4] Univ. Littoral Côte d'Opale, LISIC, 62100 Calais, France
verel@univ-littoral.fr
[5] International Associated Laboratory LIA-MODO, Nagano, Japan

Abstract. Difficult Pareto set topology refers to multi-objective problems with geometries of the Pareto set such that neighboring optimal solutions in objective space differ in several or all variables in decision space. These problems can present a tough challenge for evolutionary multi-objective algorithms to find a good approximation of the optimal Pareto set well-distributed in decision and objective space. One important challenge optimizing these problems is to keep or restore diversity in decision space. In this work, we propose a method that learns a model of the topology of the solutions in the population by performing parametric spline interpolations for all variables in decision space. We use Catmull-Rom parametric curves as they allow us to deal with any dimension in decision space. The proposed method is appropriated for bi-objective problems since their optimal set is a one-dimensional curve according to the Karush-Kuhn-Tucker condition. Here, the proposed method is used to promote restarts from solutions generated by the model. We study the effectiveness of the proposed method coupled to NSGA-II and two variations of MOEA/D on problems with difficult Pareto set topology. These algorithms approach very differently the Pareto set. We argue and discuss their behavior and its implications for model building.

Keywords: Evolutionary algorithm · Multi-objective optimization · Interpolation · Difficult Pareto set topology

1 Introduction

Multi-objective Evolutionary Algorithms (MOEA) are metaheuristic methods based on natural evolution principles that have attracted a lot of attention due

© Springer Nature Switzerland AG 2019
K. Deb et al. (Eds.): EMO 2019, LNCS 11411, pp. 386–398, 2019.
https://doi.org/10.1007/978-3-030-12598-1_31

to their good performance to deal with multi-objective optimization problems (MOP) [3]. Indeed, with the development of different MOEAs, many methodologies to improve their performance have been proposed [1,3–5,7,17,18].

Despite successful results obtained by MOEAs, studies have shown that their performance can deteriorate significantly when facing problems with difficult Pareto set (PS) topology [8]. Okabe et al. [11] observed that the PS topology of most artificial test problems, such as DTLZ [6], have an oversimplified geometry, arguing that we should not expect such simplification on real world problems. Since then, new test problems have been developed with some challenging PS topologies [7,11], along with new approaches to solve these problems. Special sessions and competitions dedicated to solve problems with difficult PS topology [16] have served to promote research in this area and to improve some multiobjective algorithms. Nonetheless, efficiency and scalability remain an open question for improved algorithms, such as enhanced versions of decomposition-based algorithms. For other classes of algorithms, such as those based on Pareto dominance, performance in terms of convergence and diversity in both decision and objective space is still poor. Overall, besides final results, there is still not a clear understanding of how various classes of algorithms work on these classes of difficult PS topology problems.

On the other hand, learning and model assisted optimization is gaining attention to enhance evolutionary search, where models are built to capture properties of the landscape, learn dependencies between variables, identify variables for recombination, and so on. The models are in turn used to guide the evolutionary algorithm aiming to improve the overall efficiency and effectiveness of the search. Some recent works have tried to incorporate learning models for better solutions [9,10,12] when optimizing problems with difficult PS topology. These models try to learn certain regions in decision space where good solutions are more likely to be found, restricting and guiding the evolutionary search.

From this standpoint, in this paper we present a method that learns a model of the topology of the solutions in the population by performing parametric spline interpolations for all variables in decision space, aiming to assist multi-objective evolutionary algorithms on bi-objective problems with difficult PS topology. To build the model, we use Catmull-Rom parametric curves as they allow us to deal with any dimension in decision space. The proposed method is appropriated for bi-objective problems since their optimal set is a one-dimensional curve according to the Karush-Kuhn-Tucker condition. The model allows to identify and query regions in decision space that are under represented in the population of the evolutionary algorithm. That is, based on these polynomial interpolations, the model can be used to generate new candidate solutions well distributed in decision space aiming to guide the search towards approximations of the Pareto set with better distribution in decision and objective space.

In this work, the proposed method is used to promote restarts from solutions generated by the model. We study the effectiveness of the proposed method embedded in three algorithms: NSGA-II [5], MOEA/D [7], and MOEA/D-DRA [15]. These algorithms are good representatives of Pareto-dominance and decomposition based approaches to multi-objective optimization. These algorithms also show quite different behavior approaching the Pareto optimal solutions. While MOEA/D have a fast approach to the optimal for some of its weights, NSGA-II slowly moves a better distributed set of solutions towards the optimal front. Since models to guide the evolutionary search are mostly built from the solution in the population, it is important to understand how the behavior of the algorithms affect the quality of the model. We test the performance of the modified algorithms using problems with difficult PS topology proposed in [16]. Simulation results presented in this work clarifies the correlation between the way an algorithm approaches the Pareto optimal set and the quality of the model, showing that the proposed method can help evolutionary algorithms to find better distributed solutions depending on algorithm's evolutionary behavior.

The rest of the paper is organized as follows. Firstly, Sect. 2 elucidates the meaning of a MOP with difficult PS topology. Section 3 describes the learning model to enhance result of evolutionary algorithms on such problems. In Sect. 4, we present the experimental design of our comparative study, and Sect. 5 presents a discussion on the results obtained. Finally, Sect. 6 presents our conclusion and future work.

2   Difficult PS Topology

Solving a MOP consists in maximizing or minimizing simultaneously m objective functions subject to constrains and bounds of a set of n decision variables. Often, there is no single solution to these problems, instead a set of optimal solutions that captures the trade-offs between solutions are demanded. This set is called Pareto set (PS) in decision space, and Pareto front (PF) in objective space.

Pareto set topology refers to the geometry created by optimal solutions of a multi-objective problem in the decision space. According to the Karush-Kuhn-Tucker condition, it can be induced under certain assumptions that the PS of a continuous MOP defines a piecewise continuous $(m - 1)$-dimensional manifold in the decision space [7]. In such case, the PS would be a piecewise continuous one-dimensional curve in $\Re^n$ for bi-objective optimization problems, a two-dimensional curve for three-objective problems, and so on. Considering this property of continuous MOPs, Okabe et al. [11] observed that PS topologies were oversimplified for most artificial problems, arguing that we should not expect such simplification on real-world problems. For example, in DTLZ2 [6] the optimal solutions lies on the interval between 0 and 1 for the variables related to diversity, and 0.5 for all other variables related to convergence. Therefore, it might be simple to find a well-spread set of solutions in objective space on this problem since after finding one optimal point, changing only one variable would create another optimal solution. On the other hand, a more difficult case would demand a change in multiple decision variables. Such curves have received distinct denominations on different studies such as complicated PS shapes [7] and difficult PS topology [14]. In this paper, we refer to it as difficult PS topology (Fig. 1).

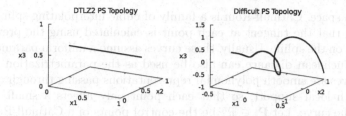

Fig. 1. DTLZ2 PS topology (left) compared to difficult PS topology (right).

Evolutionary algorithms are a powerful tool to find good solutions in multi-objective problems, but they lack proper distribution of solutions sometimes, particularly in problems with difficult topology [7,8]. Figure 2 illustrates how MOEAs can guide solutions towards optimum values by mixing them with evolutionary operators, but evolution gets stuck at some point before finding a good representation of the PS. In this case, the decision maker would have fewer options to choose from in decision space, and most likely some trade-offs in objective space will be missing.

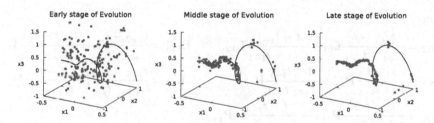

Fig. 2. Example of solutions found by NSGA-II (in red) and the PS (in black). (Color figure online)

3   Interpolation of PS Topology

In this work, we propose a model that builds polynomial interpolations of the decision variables from the data contained in the population and use these polynomials to generate new candidate solutions to update the population. If the interpolation is close to the true PS topology, we can distribute solutions across decision space hoping to produce non-dominated solutions in objective space and give more options to the decision maker in both spaces. In the following, we describe the proposed method in detail.

For bi-objective optimization problems, decision space topology is a one-dimensional curve presented in hyper-dimensional space as illustrated in Fig. 2. Thus, we use parametric Catmull-Rom curves [2] to perform the interpolation of decision space. Catmull-Rom is a family of cubic interpolating splines formulated such that the tangent at each point is calculated using the previous and next point on the spline. Usually, these curves assume a uniform parameter spacing, but Euclidean distance can also be used as the parametrization space [2]. These curves are smooth polynomial representations passing through all control points with local support, so that each point only affects a small neighborhood on the curve. Let $\mathbf{P}_i \in \Re^n$ be the control points of a Catmull-Rom curve, $i = 1, 2, \ldots, n_{cp}$, and t_i its associated parametric value. A Catmull-Rom curve is composed of $n_{cp} - 1$ polynomial segments between consecutive control points. Let $\mathbf{Q}_{i,i+1}$ be the polynomial interpolation between control points $\mathbf{P}_i$ and $\mathbf{P}_{i+1}$, associated to parameters t_i and t_{i+1}. The polynomial segment $\mathbf{Q}_{i,i+1}$ is influenced by both adjacent control points $\mathbf{P}_{i-1}$ and $\mathbf{P}_{i+2}$. Note that for extreme segments $\mathbf{Q}_{1,2}$ and $\mathbf{Q}_{n_{cp}-1,n_{cp}}$, there is no $\mathbf{P}_0$ and $\mathbf{P}_{n_{cp}+1}$, so we define them as $\mathbf{P}_1 - 0.5(\mathbf{P}_2 - \mathbf{P}_1)$ and $\mathbf{P}_{n_{cp}} - 0.5(\mathbf{P}_{n_{cp}} - \mathbf{P}_{n_{cp}-1})$, respectively.

The $\mathbf{Q}_{i,i+1}$ segment is defined by:

$$\mathbf{Q}_{i,i+1} = \frac{t_{i+1} - t}{t_{i+1} - t_i}\mathbf{L}_{012} + \frac{t - t_i}{t_{i+1} - t_i}\mathbf{L}_{123} \tag{1}$$

where:

$$\mathbf{L}_{012} = \frac{t_{i+1} - t}{t_{i+1} - t_{i-1}}\mathbf{L}_{01} + \frac{t - t_{i-1}}{t_{i+1} - t_{i-1}}\mathbf{L}_{12}, \qquad \mathbf{L}_{123} = \frac{t_{i+2} - t}{t_{i+2} - t_i}\mathbf{L}_{12} + \frac{t - t_i}{t_{i+2} - t_i}\mathbf{L}_{23},$$

$$\mathbf{L}_{01} = \frac{t_i - t}{t_i - t_{i-1}}\mathbf{P}_{i-1} + \frac{t - t_{i-1}}{t_i - t_{i-1}}\mathbf{P}_i, \qquad \mathbf{L}_{12} = \frac{t_{i+1} - t}{t_{i+1} - t_i}\mathbf{P}_i + \frac{t - t_i}{t_{i+1} - t_i}\mathbf{P}_{i+1},$$

$$\mathbf{L}_{23} = \frac{t_{i+2} - t}{t_{i+2} - t_{i+1}}\mathbf{P}_{i+1} + \frac{t - t_{i+1}}{t_{i+2} - t_{i+1}}\mathbf{P}_{i+2}$$

For Catmull-Rom curves, it is common to define the parametrization from its geometric embedding in Euclidean space. Therefore, we can define t_{i+1} as the Euclidean distance between consecutive control points by:

$$t_{i+1} = |\mathbf{P}_{i+1} - \mathbf{P}_i|^\alpha + t_i \tag{2}$$

Here, centripetal parametrization ($\alpha = 0.5$) has been chosen since it guarantees no intersections within curve segments [13].

Figure 3 illustrates the parametrization of decision space, where red dots are examples of control points. In this case, control points are in fact solutions found during the evolutionary process. Therefore, it is possible to describe the PS topology with Catmull-Rom method if solutions are good control points, i.e. well converged in some regions. However, this method requires a proper ordering of solutions in decision space.

The clustering method k-means is used to sample control points from the population of the evolutionary algorithm. Rather than using all solutions, it is reasonable to select few of them since it can be redundant to do interpolation between too close points. By applying k-means in objective space, solutions can

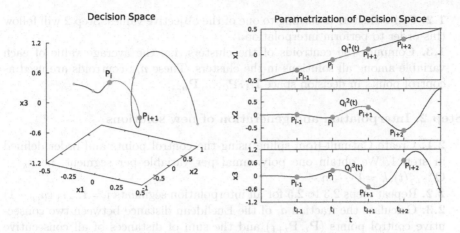

Fig. 3. Illustration of a hypothetical PS and its Catmull-Rom parametric curves. (Color figure online)

be clustered in groups, from which we can get their centroids as control points for the interpolation. Figure 4 presents an illustration of solutions being divided in objective space, and their respective values in decision space.

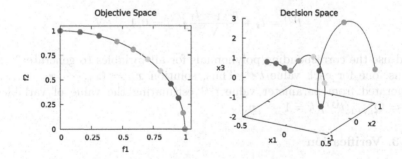

Fig. 4. k-means can distinguish solutions in different groups.

There are several ways in which the proposed method can be used within the evolutionary algorithm. One is to use it as a restart mechanism, where the solutions generated from the interpolation polynomials replace all solutions in the population. Another way is to allow competition between solutions in the population with those generated using the polynomials. To test our proposed method, in this work we use it to perform restarts during the evolutionary search.

The pseudocode of the proposed method is as follows:

Step 1. Sample and ordering

1.1. Apply k-means to distribute solutions in n_{cp} different clusters in objective space.

1.2. Order clusters according to one of the objective values. Step 2 will follow this order to perform interpolation.

1.3. Compute the centroids of the clusters, i.e. the average value of each variable among all solutions in the clusters. These n_{cp} centroids are used as control points in decision space $\mathbf{P}_1, \mathbf{P}_2, ..., \mathbf{P}_{n_{cp}}$.

Step 2. Interpolation and generation of new solutions

2.1. Create Catmull-Rom spline using the control points and order defined in step 1. We obtain one polynomial per variable per segment, i.e $x_k = \mathbf{Q}_{i,i+1}^k(t), k = 1 \cdots, n$.

2.2. Repeat steps 2.3 to 2.5 for all interpolation segments $i = 1, ..., (n_{cp} - 1)$

2.3. Calculate the fraction d_i of the Euclidean distance between two consecutive control points $(\mathbf{P}_i, \mathbf{P}_{i+1})$ and the sum of distances of all consecutive control points.

$$d_i = \frac{dist(\mathbf{P}_i, \mathbf{P}_{i+1})}{\sum_{k=1}^{n_{cp}-1} dist(\mathbf{P}_k, \mathbf{P}_{k+1})} \tag{3}$$

2.4. Define the number of solutions $\bar{N}_i$ to be generated in the ith interval as $\bar{N}_i = round(d_i \times N)$, where N is the population size.

2.5. Generate new solutions based on Catmull-Rom spline. Here, calculate $\bar{N}_i$ values of the parameter $t \in [t_i, t_{i+1}]$ by

$$t^{(j)} = t_i + \frac{t_{i+1} - t_i}{\bar{N}_i} j, \; j = 0, 1, ..., \bar{N}_i \tag{4}$$

and use the corresponding polynomials for all variables to generate $\bar{N}_i$ solutions, one for each value $t^{(j)}$. Thus, solution $x_j = (x_1, \cdots, x_k, \cdots, x_n)$ is generated from parameter value $t^{(j)}$ estimating the value of variables by $x_k = \mathbf{Q}_{i,i+1}^k(t^{(j)}), k = 1 \cdots, n$.

Step 3. Verification

3.1. Discard solutions off boundaries.

3.2. Check if the number of new solutions is equal to population size. If there are fewer solutions, we randomly include individuals from the current population. In case we have more solutions, we randomly throw away individuals so we can have same population size in the restart.

4   Experimental Setup

In total, five bi-objective unimodal CEC09 competition problems [16] were used, namely UF1, UF2, UF3, UF4, and UF7, setting number of variables **n = 30**.

To test the proposed method with different evolutionary methodologies following the CEC09 competition parameters setting, we implement our model in three algorithms: NSGA-II [5], MOEA/D [7], and an improved version of

MOEA/D to solve CEC 2009 competition problems denominated MOEA/D-DRA [15]. Differential Evolution (DE) crossover operator and polynomial mutation were used, since it produces better results than SBX operator [7]. Crossover rate is $pc = 1.0$, and mutation rate per variable is $pm = 1/n$. DE operator parameter is set to $F = 0.5$, and the distribution exponent of polynomial mutation is set to $\eta_m = 20$. All algorithms perform a total number of function evaluations equals to 300000 with population size $N = 600$. For MOEA/D, Tchebycheff approach and neighborhood size of $T = 60$ were used in both versions. Here, we tested different numbers of control points $n_{cp} = 150, 300, 500$, and $restarts = 2, 5, 10, 20$. Restarts are equally spaced in generations. We run each algorithm 30 times using the same set of seeds. Finally, IGD [3] metric was used to compare results.

5 Experimental Results and Discussion

To compare the original evolutionary algorithm against the version coupled with the model, Table 1 presents IGD results for all tested problems. To illustrate, we present results for the model with 10 restarts and 300 samples of control points to build the interpolations which overall produce good results. For each problem, we present the average and standard deviation values of IGD among all 30 runs, together with the p-value of t-tests on the IGD sets obtained with the original algorithm and its improved version. A p-value smaller than 0.05 indicates

Table 1. Results of IGD for 300 control points and 10 restarts.

		NSGA-II		MOEA/D		MOEA/D-DRA	
		Original	Model	Original	Model	Original	Model
UF1	Average	0.018806	**0.012445**	0.001479	0.001171	0.001427	**0.001117**
	SD	0.001136	0.003395	0.025630	0.020608	0.000324	0.000078
	p-value	3.561984e−13		0.958135		0.000141	
UF2	Average	0.015440	**0.009106**	0.005799	0.007126	0.003703	0.003718
	SD	0.001315	0.001615	0.020539	0.003534	0.001798	0.001606
	p-value	4.992806e−25		0.120524		0.601648	
UF3	Average	0.145545	**0.082388**	0.009209	0.007392	0.006490	0.005320
	SD	0.011775	0.007450	0.011240	0.006558	0.008079	0.005899
	p-value	1.340732e−33		0.215363		0.342249	
UF4	Average	0.039056	0.038930	0.064486	**0.064192**	0.059472	0.060421
	SD	0.001765	0.001742	0.000028	0.000015	0.004593	0.004190
	p-value	0.548976		7.984198e−36		0.730446	
UF7	Average	0.008453	**0.007527**	0.003074	0.003415	0.001409	**0.001250**
	SD	0.001826	0.002527	0.159385	0.139367	0.000314	0.000165
	p-value	0.001213		0.600833		0.019249	

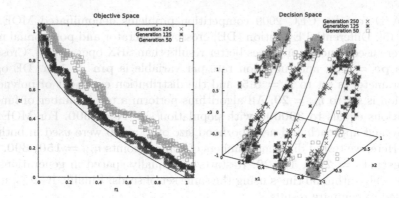

Fig. 5. Behavior of solutions found by NSGA-II (without model) on UF1. (Color figure online)

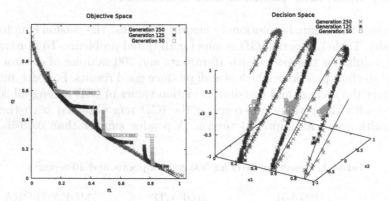

Fig. 6. Behavior of solutions found by MOEA/D (without model) on UF1. (Color figure online)

with 95% confidence that the averages are statistically different. Statistically better IGD averages are shown in bold, i.e. smaller average IGD value and p-value smaller than 0.05. According to these results, the proposed method could find approximations with better IGD for some problems. Besides UF4, IGD values improved for all problems when using NSGA-II. In case of the decomposition algorithm, the model could improve IGD results for problem UF4 when using MOEA/D, and problems UF1 and UF7 when using its improved version MOEA/D-DRA. Note that the model did not deteriorate results in any case.

Since the proposed model is built based on solutions found by the algorithms, whether or not the model can improve results by performing interpolation would depend on the distribution of algorithm's population. In other words, if the population is a good representation of the PS topology, the model can be effective by creating solutions from the interpolation. In this case, a good representation would include solutions in the inflection points of the curve in decision space, where the topology changes concavity. Thus, by looking at where solutions are placed through generations, we can understand whether the interpolation would properly approximate the PS topology.

Figures 5 and 6 present solutions found by NSGA-II and MOEA/D in distinct generations, presented in different colors. While NSGA-II's population steadily approaches the optimal front with large coverage as the evolution proceeds, MOEA/D converges solutions very fast in some regions of objective space at first, and distribute solutions on those regions' neighborhood. These different approaches have an effect on decision space, where the Pareto-dominance based algorithm seems to produce solutions better suitable to the proposed model. As NSGA-II's solutions are placed on the inflection points of the PS topology since early generations, it offers to the learning model a better representation of the topology. In contrast, the decomposition algorithm finds at first solutions in fewer regions far from the inflection points, misleading the learning model. For example, the interpolation on solutions from generation 50 by MOEA/D would produce something close to a strait line in decision space.

Figure 7 presents IGD values by NSGA-II and MOEA/D-DRA on problem UF1 with different sampling size and number of restarts. Note that IGD steadily improves as the number of restarts increases. Also, note that a sample of control points half of the population size (300) gives overall good results.

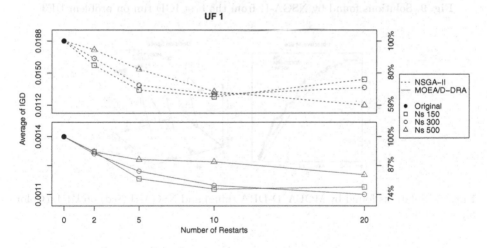

Fig. 7. Average of IGD for 30 runs including all non-dominated solutions found during the search.

Results for NSGA-II have improved significantly according to IGD metric. Figures 8 and 9 illustrates for UF1 and UF3 problems all non-dominated solutions found by the original algorithm and the one coupled with the model in their best IGD run. Both figures show that distribution of solutions have improved in both spaces when using the proposed method. Note that for UF3

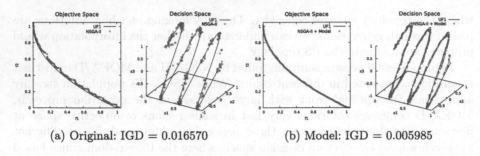

(a) Original: IGD = 0.016570   (b) Model: IGD = 0.005985

Fig. 8. Solutions found by NSGA-II from the best IGD run on problem UF1.

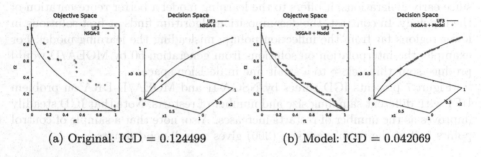

(a) Original: IGD = 0.124499   (b) Model: IGD = 0.042069

Fig. 9. Solutions found by NSGA-II from the best IGD run on problem UF3.

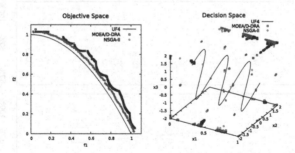

Fig. 10. Solutions found by MOEA/D-DRA (blue) and NSGA-II (red) on UF4. (Color figure online)

the improvement in convergence is more significant. These figures illustrate well the main idea behind our learning method, where interpolation takes advantage that NSGA-II can provide a set of control points to build a good interpolation to generate new candidate solutions and improve quality of results.

In contrast to previous problems, Fig. 10 shows that restarts could not improve results for problem UF4. In this case, both NSGA-II and MOEA/D fail to converge solutions close to the optimal Pareto set, so when our method tries to perform interpolation based on EA's population, it fails to represent PS topology. Therefore, this problem shows that the proposed method with restarts may not improve algorithm's performance when the population is far from the Pareto set.

6 Conclusion

In this paper, we presented a method that learns a model of the solutions' topology in the population by performing parametric spline interpolation for all variables in decision space. Here, Catmull-Rom parametric curves were used to perform interpolation, which allow us to deal with any dimension in decision space, but limited to bi-objective problems. We coupled the model with NSGA-II and two version of MOEA/D to perform restarts from solutions generated by the model. We showed that the proposed model could improve distribution and convergence of solutions for most problems in the case of NSGA-II, and for some problems in the case of MOEA/D-DRA. Also, we showed that the effectiveness of the interpolation depends on the behavior of the algorithm.

In the future, we would like to study other methods to perform interpolation to solve problems with more objectives. Also, we want to study other ways to couple this model with evolutionary algorithms. For instance, allowing competition of solutions created by the model with the current population, instead of totally replacing the population. Another aspect that we would like to investigate is the scalability of the model in decision space.

References

1. Aguirre, H., Yazawa, Y., Oyama, A., Tanaka, K.: Extending AεSεH from many-objective to multi-objective optimization. In: Dick, G., et al. (eds.) SEAL 2014. LNCS, vol. 8886, pp. 239–250. Springer, Cham (2014). https://doi.org/10.1007/978-3-319-13563-2_21
2. Catmull, E., Rom, R.: A class of local interpolating splines. In: Barnhill, R.E., Riesenfeld, R.F. (eds.) Computer Aided Geometric Design, pp. 317–326. Academic Press (1974)
3. Coello, C.A.C., Lamont, G.B., Veldhuizen, D.A.V.: Evolutionary Algorithms for Solving Multi-Objective Problems. Genetic and Evolutionary Computation. Springer, Secaucus (2006). https://doi.org/10.1007/978-0-387-36797-2
4. Corne, D.W., Knowles, J.D., Oates, M.J.: The Pareto envelope-based selection algorithm for multiobjective optimization. In: Schoenauer, M., et al. (eds.) PPSN 2000. LNCS, vol. 1917, pp. 839–848. Springer, Heidelberg (2000). https://doi.org/10.1007/3-540-45356-3_82
5. Deb, K., Pratap, A., Agarwal, S., Meyarivan, T.: A fast and elitist multiobjective genetic algorithm: NSGA-II. IEEE Trans. Evol. Comput. 6(2), 182–197 (2002)
6. Deb, K., Thiele, L., Laumanns, M., Zitzler, E.: Scalable multi-objective optimization test problems. In: Congress on Evolutionary Computation, CEC 2002, pp. 825–830. IEEE Press (2002)
7. Li, H., Zhang, Q.: Multiobjective optimization problems with complicated Pareto sets, MOEA/D and NSGA-II. IEEE Trans. Evol. Comput. 13(2), 284–302 (2009)

8. Marca, Y., et al.: Pareto dominance-based MOEAs on problems with difficult Pareto set topologies. In: Proceedings of the Genetic and Evolutionary Computation Conference Companion, GECCO 2018, pp. 189–190. ACM, New York (2018)
9. Mo, L., Dai, G., Zhu, J.: The RM-MEDA based on elitist strategy. In: Cai, Z., Hu, C., Kang, Z., Liu, Y. (eds.) ISICA 2010. LNCS, vol. 6382, pp. 229–239. Springer, Heidelberg (2010). https://doi.org/10.1007/978-3-642-16493-4_24
10. Morgan, D., Waldock, A., Corne, D.: MOPC/D: a new probability collectives algorithm for multiobjective optimisation. In: 2013 IEEE Symposium on Computational Intelligence in Multi-Criteria Decision-Making, MCDM, pp. 17–24 (2013)
11. Okabe, T., Jin, Y., Olhofer, M., Sendhoff, B.: On test functions for evolutionary multi-objective optimization. In: Yao, X., et al. (eds.) PPSN 2004. LNCS, vol. 3242, pp. 792–802. Springer, Heidelberg (2004). https://doi.org/10.1007/978-3-540-30217-9_80
12. Schütze, O., Mostaghim, S., Dellnitz, M., Teich, J.: Covering Pareto sets by multi-level evolutionary subdivision techniques. In: Fonseca, C.M., Fleming, P.J., Zitzler, E., Thiele, L., Deb, K. (eds.) EMO 2003. LNCS, vol. 2632, pp. 118–132. Springer, Heidelberg (2003). https://doi.org/10.1007/3-540-36970-8_9
13. Yuksel, C., Schaefer, S., Keyser, J.: Parameterization and applications of Catmull-Rom curves. Comput. Aided Des. **43**(7), 747–755 (2011)
14. Zapotecas-Martínez, S., Coello, C.A.C., Aguirre, H.E., Tanaka, K.: A review of features and limitations of existing scalable multi-objective test suites. IEEE Trans. Evol. Comput. (2018). https://doi.org/10.1109/TEVC.2018.2836912
15. Zhang, Q., Liu, W., Li, H.: The performance of a new version of MOEA/D on CEC09 unconstrained MOP test instances. In: 2009 IEEE Congress on Evolutionary Computation, pp. 203–208, May 2009. https://doi.org/10.1109/CEC.2009.4982949
16. Zhang, Q., Zhou, A., Zhao, S., Suganthan, P.N., Liu, W., Tiwari, S.: Multiobjective optimization test instances for the CEC 2009 special session and competition. Technical report. University of Essex and Nanyang Technological University (2008)
17. Zitzler, E., Künzli, S.: Indicator-based selection in multiobjective search. In: Yao, X., et al. (eds.) PPSN 2004. LNCS, vol. 3242, pp. 832–842. Springer, Heidelberg (2004). https://doi.org/10.1007/978-3-540-30217-9_84
18. Zitzler, E., Laumanns, M., Thiele, L.: SPEA2: improving the strength Pareto evolutionary algorithm. Technical report TIK-Report 103. ETH Zurich, Switzerland (2001)

Linear Search Mechanism for Multi- and Many-Objective Optimisation

Heiner Zille[✉] and Sanaz Mostaghim

Institute for Intelligent Cooperative Systems, Otto von Guericke University,
Magdeburg, Germany
{heiner.zille,sanaz.mostaghim}@ovgu.de
http://www.ci.ovgu.de

Abstract. This article proposes a search mechanism based on linear combinations of population members to increase the solution quality of multi-objective and many-objective optimisation algorithms. Our approach makes use of the inherent knowledge in the solution population at a given time step, and forms new solutions through linear combinations of the existing ones. A population of coefficient vectors is formed and optimised by a metaheuristic to explore and exploit promising areas of the search space. In addition, our proposed method provides a reduction of dimensionality for large search spaces. The concept is formally introduced and implemented into a generic algorithm structure to be used in arbitrary metaheuristics. The experimental evaluation uses four multi- and many-objective algorithms (NSGA-II, GDE3, NSGA-III and RVEA) and is performed on a total of 60 test instances from three benchmark families with 2 to 5 objective functions and 30 to 514 decision variables. The results indicate that the performance of existing methods can be significantly improved by the proposed search strategy, especially in high-dimensional search spaces and for many-objective problems.

Keywords: Multi-objective optimisation ·
Many-objective optimisation · Large-scale optimisation ·
Evolutionary algorithm · Exploration · Linear combination ·
Dimensionality reduction

1 Introduction

The search for a well-spread non-dominated front in multi- and many-objective optimisation is still an ongoing challenge. This is especially true in large-scale optimisation which contains a very large number of decision variables or many objective functions. Previous methods try to balance the trade-off between convergence and diversity in different ways. The research in the last years has led to a variety of many-objective optimisation methods (e.g. [2,6,10]), as well as a number of methods that can deal with hundreds or thousands of decision variables (e.g. [11,17,18,20]). Concepts that can be found in this area are dimensionality

© Springer Nature Switzerland AG 2019
K. Deb et al. (Eds.): EMO 2019, LNCS 11411, pp. 399–410, 2019.
https://doi.org/10.1007/978-3-030-12598-1_32

reduction (e.g. [14]), variable interaction analyses or the division of design variables into convergence-related and diversity-related parameters (e.g. [11,17]), some of which involve increased computational costs.

In this work, we propose a method to increase exploration in multi- and many-objective optimisation by searching in subspaces defined by the current populations' design variables. The approach is based on the assumption that the evolutionary process finds promising areas of the search space (i.e. areas where good solutions are located) and adjusts the variables in the population accordingly to cover and explore these areas. Once certain promising results have been found, a recombination of these solutions through linear combinations is subject to optimisation in order to create new solutions which benefit from the inherent information in the population. Our newly proposed exploration method can easily be included into any metaheuristic optimisation algorithm and can also lead to a dimensionality reduction without the need for dividing variables into subcomponents. In this work, the mathematical concept is introduced and analysed, and an experimental evaluation shows its benefits when embedded into multi- and many-objective algorithms. The equipped algorithms are tested on a total of 60 different benchmark function instances from the literature with 2 to 5 objective functions and 30 to 514 decision variables.

The remainder of this article is structured as follows. In Sect. 2 the basic principles of multi-objective optimisation are outlined briefly and a short overview about related work on multi-objective and large-scale approaches is given. In Sect. 3 the proposed linear-combination approach is introduced. The mathematical concept is explained first before describing the inclusion of the concept into existing algorithms. The experimental evaluation in Sect. 4 equips a number of well-known metaheuristics with the proposed exploration method and compares their performance on a variety of benchmark functions. Finally, a summary and outlook on future work directions is given in Sect. 5.

2   Multi-objective Optimisation and Related Work

Problems in nature and science often contain multiple conflicting goals which need to be optimised simultaneously. These problems are called multi-objective problems (MOPs) and can be formulated as:

$$Z: \quad min \quad \boldsymbol{f}(\boldsymbol{x}) = (f_1(\boldsymbol{x}), f_2(\boldsymbol{x}), ..., f_m(\boldsymbol{x}))^T$$
$$s.t. \quad \boldsymbol{x} \in \Omega \subseteq \mathbb{R}^n \tag{1}$$

where $m \geq 2$. This kind of MOP maps the decision space $\Omega = \{\boldsymbol{x} \in \mathbb{R}^n | \boldsymbol{g}(\boldsymbol{x}) \leq 0\}$ of dimension n to the objective space of dimension m. In most problems, some of the constraints define a domain for each variable with lower and upper bounds, i.e. $x_i \in [x_{i,min}, x_{i,max}]$, $i = 1, .., n$. For such problems, a single optimal solution can often not be determined, since there is usually a trade-off between the objective functions. Modern problem solving methods instead concentrate on finding an approximation of a Pareto-optimal solution set [4,6,9].

Two key challenges in finding such a set of solutions are convergence and diversity of the solution set. Convergence refers to the search for better, non-dominated solutions, which improve all objective functions from the current solution set, and therefore bring the whole set closer to the true Pareto-optimal solutions. Diversity on the other hand is necessary to obtain a widely spread set of solutions. The output of a metaheuristic algorithm should provide a diverse set of different solutions which represent different trade-offs among the objective functions and cover the whole Pareto-optimal front as good as possible. Finding a non-dominated set of solutions that is as close to the true Pareto-set and as diverse as possible is an ongoing challenge, especially when many-objective and large-scale problems are involved [11,17,18,20].

In the area of large-scale (i.e. many-variable) optimisation, the topic of reducing the dimensionality of a problem is often of importance. Concepts like Cooperative Coevolution [1] aim to optimise smaller subspaces of the n-dimensional search space by dividing the variables into groups or subcomponents based on different criteria. Approaches to achieve a better balance between convergence and diversity have been used e.g. in [11,17]. These works carry out an analysis to identify variables which influence the diversity of the solution set before starting the optimisation process. In addition, both methods utilize an interaction detection to from groups of variables. A major drawback of these and similar approaches is that the analysis of variables and formation of variable groups requires an additional and very large computational budget for this pre-processing step, while the actual benefit compared to a random assignment of variables to the groups or less expensive methods is not always guaranteed [13]. Another method called WOF [19,20] shows a superior convergence behaviour in large-scale problems with up to thousands of variables [21]. WOF aims to balance diversity and convergence through the selection of certain solutions from the population, which are used in a fast-converging transformation and optimisation step of the algorithm.

In the area of many-objective optimisation, a variety of algorithms has been developed in recent years, among them many who adapt the concept of reference vectors like NSGA-III [6], MOEA/DD [10] or RVEA [2]. Reference vectors or directions are a common concept that is used to solve the problem of decreasing selection pressure when Pareto-dominance-based approaches are used for many-objective problems. Such methods have increased the capabilities of metaheuristics to solve many-objective problems. Due to that, an area that might draw increased focus in the future is solving problems with many objectives and a large number of decision variables at the same time, while keeping computational budget as low as possible. This work therefore aims to propose a mechanism that can be used to reduce the dimensionality of such problems and by that improve the solution quality of existing many-objective methods.

3   Proposed Approach

In this section, we propose a search strategy that can be used to enhance exploration of the search space and at the same time reduce the dimensionality of a problem without using variable groups. In metaheuristic evolutionary optimisation, one general assumption is that the encoding of the problem ensures that promising solutions can be generated from a combinations of other good solutions. By extension, Pareto-optimal solutions might share certain characteristics, i.e. decision variable values, which might be similar throughout the whole Pareto-optimal set (as for instance the convergence-related variables of common benchmark families [3,8]), and which might be approximated by an optimiser. The main goal of this approach is to exploit this information that is inherent in the population of an evolutionary algorithm at a given time, i.e. the information which (sub-)vector-space of the n-dimensional search space contains the (at that point) best or most promising solutions. Based on this, a search inside this subspace in conducted. This concept is also related to that of "innovization" from the literature ([5,7]), which aims to extract information or design principles from the outcome or during the process of optimisation. In the following, the concept of the proposed search strategy is explained.

3.1 Concept

Suppose we have an optimisation problem with n real-valued decision variables and m objectives as given in Eq. 1. Let the population of an algorithm be P and its size be $s := |P|$. At each given time of the optimisation process the population consists of s solution vectors each of dimensionality n: $P = \{x^{(1)}, x^{(2)}, ..., x^{(s)}\}$. Each solution is a vector $\in \mathbb{R}^n$:

$$x^{(i)} = (x_1^{(i)}\ x_2^{(i)}\ ...\ x_n^{(i)}) \tag{2}$$

and the set of solutions P defines a vector (sub)space. The dimensionality of this subspace is given by the rank of the matrix of the spanning vectors. We therefore compose the matrix $\hat{X} \in \mathbb{R}^{s \times n}$ which contains in each row one solution of the population.

$$\hat{X} = \begin{pmatrix} x_1^{(1)} & x_2^{(1)} & \cdots & x_n^{(1)} \\ x_1^{(2)} & x_2^{(2)} & \cdots & x_n^{(2)} \\ \vdots & \vdots & \ddots & \vdots \\ x_1^{(s)} & x_2^{(s)} & \cdots & x_n^{(s)} \end{pmatrix} \tag{3}$$

An evolutionary algorithm (EA) combines the solutions in the current population by using crossover operators. However, instead of classical crossover methods, it is also possible to combine the existing solutions linearly. This can be done through convex, conical or arbitrary linear combinations. In the following we focus on general linear combinations as they include the convex and conical combinations as subsets. A linear combination of the solutions in the population is defined as follows:

$$x' = y\hat{X} = y_1 x^{(1)} + y_2 x^{(2)} + ... + y_s x^{(s)} \tag{4}$$

where y is the vector of coefficients of the combination:

$$y = (y_1\ y_2\ ...\ y_s) \tag{5}$$

With this concept it is possible to search the subspace spanned by the s vectors in the population, and thereby find improved solutions from combinations of the existing ones. The actual dimension of this subspace is defined by the rank of the matrix $\hat{X}$ which is bounded between $1 \leq rank(\hat{X}) \leq min\{n, s\}$.

In a way, this method can be seen as a s-parent crossover method. In contrast to a crossover, in our proposed method the parameters of the vector y are subject to an optimisation process. While a multi-parent crossover would produce a random combination and let the evolutionary process judge whether this was a good one, the proposed procedure uses an evolutionary process to find "optimal" combinations.

In order to find a good linear combination, the values of the vector y are optimised instead of the original variables. As a consequence, this newly formed optimisation problem has only s decision variables compared to the original one with n variables. This means, in case a problem with $n = 30$ variables is optimised with a population size of $s = 100$, there might be redundancy in the newly formed linear-combination-problem, and the algorithm now searches in 100 dimensions, even though the actual space in which the solutions are created is still 30-dimensional, and some of the vectors of the linear combinations are not independent in this case. However, the situation differs when applied to a high-dimensional problem with for instance $n = 500$ variables. The s solutions can at most define a s-dimensional subspace. If all solutions are randomly created in the beginning, it is not guaranteed that good solutions actually lie in the defined subspace. However, when the algorithm is allowed a certain progress to find a preliminary approximation of the optimal areas, we can assume that promising parameter combinations might have been found already, and the spanned subspace might contain additional good solutions. In that case, optimising the linear-combination-solutions can also be regarded as a dimensionality reduction technique, as it enables the algorithm to search in a 100-dimensional subspace instead of the 500-dimensional original search space. This makes the method not only promising for multi- and many-objective problems, but also for the area of large-scale optimisation.

3.2 Inclusion into Other Algorithms

The proposed concept can be used inside arbitrary metaheuristic optimisation algorithms. To do so, we define a population Q of y-vectors, where each vector in the population defines one linear combination of the members of P as described above. By this, we can use any metaheuristic optimiser on this newly formed population to find suitable linear combinations of the underlying original solutions in the population P. Since the optimisation of the population Q relies on the assumption, see above, that the population P defines a promising subspace of Ω, the proposed method is included into other metaheuristics as an additional search step. In particular, we apply the original (arbitrary) metaheuristic in turns with the proposed linear-combination-search. As a further step to concentrate on promising solutions, the linear combinations are only performed on the non-dominated solutions in the population.

Let $\hat{X}$ be the matrix of the decision variable values of all non-dominated solutions in P as seen above, where each row in $\hat{X}$ corresponds to one non-dominated solution in P. As a result, $\hat{X}$ is an $s' \times n$ matrix, where s' is the number of non-dominated solutions. In the same way, let $\hat{Y}$ be the matrix of the decision variable values (i.e. coefficients of linear combinations) of the solutions in Q. The population size of Q is t, therefore $\hat{Y} \in \mathbb{R}^{t \times s'}$. The original objective function evaluation can be applied to the new population by simply multiplying $\hat{X}$ with $\hat{Y}$ and computing $f(\hat{Y}\hat{X})$, i.e. applying f to each row in $\hat{Y}\hat{X}$. For practical reasons and to limit the search space of the newly found problem, the variables y_i are also equipped with lower and upper bounds, i.e. $y_i \in [y_{i,min}, y_{i,max}]$, $i = 1, .., s'$.

The outline of the resulting algorithm looks as follows:

1. Optimise the population P with any multi-objective method for a specified time.
2. Use the first non-dominated front of the current population P to build the matrix $\hat{X}$ out of its decision variables' values.
3. Create a random population Q of linear-combination-vectors.
4. Optimise Q for a certain time using an arbitrary optimisation method. Store all evaluated solutions in an Archive A.
5. Merge the population P with A and proceed with the normal optimisation process (Step 1).

4 Evaluation

To evaluate the proposed method, we have included it into several well-known optimisation algorithms from the areas of multi- and many-objective optimisation. These algorithms are NSGA-II [4], GDE3 [9] as representatives of traditional evolutionary methods, both classical and differential evolution, and NSGA-III [6] and RVEA [2] to represent many-objective methods. The aim of the experiments is not to show the superiority of one of these methods over one another, but rather to show that the proposed exploration method has a positive influence when applied to existing algorithms. Due to space limitations, an inclusion into dedicated large-methods like LMEA [17], MOEA/DVA [11] or WOF [20], and the analysis of this methods' capabilities as a dimensionality reduction mechanism, is subject to future work. Each of the four used algorithms has been equipped with the proposed method by applying in terms 100 generations of the original algorithm and after that 100 generations of the search in the formed subspace as described above. The created solutions are then merged back into the original population using the usual selection method of the respective algorithm. For the optimisation of the population Q, the NSGA-II algorithm is used in all cases. This is done so that all algorithms use the same exploration mechanism for searching the formed subspace. Future research might deal with different mechanisms in this regard, as the NSGA-II mechanism might not be the optimal choice for many-objective problems. This procedure is repeated until the maximum amount of function evaluations is reached. The modified versions of the algorithms are denoted with an "x" in front of their names, i.e. xNSGA-II, xGDE3, xNSGA-III and xRVEA. To test the performance, we use a total of 60 test problem instances from three common benchmark families with 2 to 5 objectives and 30 to 514 decision variables. The used problems are as follows:

- Six problems from the LSMOP (large-scale multi- and many-objective test problems) family [3]: LSMOP1-6. Each of them is tested with 2, 3, 4 and 5 objective functions, resulting in 206, 307, 413 and 514 decision variables.
- Six problems from the popular WFG family [8]: WFG2-5, WFG7 and WFG8. All of them are tested with 2 and 3 objectives, in combination with both 40 and 400 decision variables. The WFG problems were chosen by an analysis done in [11], where WFG2 and 3 represent problems with a sparse number of interacting variables, WFG4 and 5 have no interactions and WFG7 and 8 have a high number of interacting variables.
- Six problems from the CEC 2009 unconstrained benchmarks: UF1-3 are 2-objective problems, UF8-10 are 3-objective problems. All of them are tested with 30 and 300 variables.

For implementation, the PlatEMO framework [15] version 1.5 is used. For each experiment we perform 31 independent runs and report the median and interquartile range (IQR) values of the relative hypervolume (HV) indicator [16]. The relative HV is the hypervolume obtained by a solution set in relation to the hypervolume obtained by a sample of the Pareto-front of the problem, consisting of 10,000 solutions as provided by the PlatEMO framework. The used reference point for the indicator is obtained by using the nadir point of our Pareto-front sample (i.e. the point in the objective space containing the worst value in each dimension throughout the sample) and multiply it by 2.0 in each dimension. Statistical significance is tested using a two-sided Mann-Whitney-U Test with the null hypothesis that the tested samples have equal medians. Statistical significance is assumed for a value of $p < 0.01$.

4.1 Parameter Settings

The maximum number of function evaluations for all algorithms and problem instances is set to 100,000. The number of position-related variables for the WFG problems has been set to $n/4$ and the parameter n_k in the LSMOP benchmarks was set to 5. The population size is set to 40 in all instances of NSGA-II and GDE3. The population sizes of NSGA-III and RVEA are set to 40, 36, 35 and 40 for $m = 2, 3, 4$ and 5 objectives respectively, due to the uniform generation of reference vectors. All algorithms use polynomial mutation with a distribution index of 20.0 and a probability of $1/n$. All algorithms, except GDE3, use the simulated binary crossover with a distribution index of 20.0 and a probability of 1.0. In GDE3 are $CR = 1$ and $F = 0.5$. In RVEA, $\alpha = 2$ and $f_r = 0.1$ as in the original work. The bounds of the coefficients for the linear combination are set to $y_{i,min} = -10.0$, and $y_{i,max} = 10.0$.

4.2 Results

The results of the experiments are shown in Tables 1 and 2, where each algorithm is compared with its respective linear-combination-enhanced counterpart. Overall, the proposed method is beneficial for the performance of all algorithms in most problem instances.

First, we take a look at the two traditional evolutionary algorithms. The xNSGA-II performs significantly better (based on the used Mann-Whitney-U Test) compared to the original NSGA-II in 44 out of 60 problem instances, and achieves an equal result in 9 cases. xGDE3 is significantly better or equal to its counterpart in 52 out of 60 cases. Notable is that both original algorithms can perform better just in a few 2-objective and 3-objective instances, while in all higher dimensional problems with 4 and 5 objectives, they lack the ability to even achieve any solution beyond the reference point for the HV calculation, resulting in a HV of zero (denoted as dashes in the tables). The linear combination technique enables these algorithms to achieve significantly better results in even high-dimensional problems with 5 objectives and over 500 decision variables.

Next, we examine the two many-objective algorithms NSGA-III and RVEA. Also in these methods the proposed approach is able to improve the performance of both algorithms significantly. In NSGA-III, the modified version with linear combination performs significantly better in 49 problem instances and performed equally well in another 6. xRVEA outperforms its original version in 49 instances as well, with 5 more draws. An interesting observation is that even though both algorithms are originally designed to work with many-objective problems, their enhanced versions increase their performances even in these instances to a great extent. It is worth to note that the performance on the many-objective instances is significantly increased, even though the subspace of linear combinations is searched with the NSGA-II mechanism, which is usually not designed for many-objective problems. A possible explanation for this fact might be that NSGA-III and RVEA do not posses a mechanism for dealing with high-dimensional search spaces. Since the LSMOP problems do not only contain many objective function but also high-dimensional search spaces, this might turn out a challenge for these algorithms. The positive influence of the linear-combination-search might be, at least partly, due to the inherent reduction of dimensionality. This is also supported by the fact that for all the four algorithms, the original version did only perform better than the x-versions in low-dimensional problems, almost exclusively in UF and WFG problems with only 30 or 40 variables.

Another interesting observation concerns the type of problem where the linear-combination-search seems to work less effectively. Among the few instances where the modified algorithms do not perform best are the (low-dimensional) WFG4 and WFG5 problems, both with 2 and with 3 objectives. WFG4 and WFG5 are both fully separable. Furthermore, NSGA-II and NSGA-III outperform their modified counterparts on the 2-objective LSMOP5 problem, which is also fully separable. The separability of variables suggests that an algorithm can reach optimal solutions by altering variables completely independent of each

Table 1. Obtained median and IQR values of the relative Hypervolume for NSGA-II and xNSGA-II as well as GDE3 and xGDE3. An asterisk in the column of an x-Algorithm indicates statistical significance to the respective original version of that algorithm. Best performances are marked in bold and shaded where significant.

	m	n	NSGA-II	xNSGA-II	GDE3	xGDE3
LSMOP1		206	0.75134 (1.13E-1)	0.85510 * (9.07E-2)	0.56263 (4.52E-2)	0.86759 * (2.47E-2)
LSMOP2		206	0.94481 (4.97E-3)	0.96903 * (6.71E-3)	0.94321 (2.31E-3)	0.98270 * (1.41E-3)
LSMOP3	2	206	— (—)	0.23474 * (1.38E-1)	— (—)	0.03916 * (3.23E-3)
LSMOP4		206	0.90440 (8.53E-3)	0.94727 * (4.55E-3)	0.94008 (9.01E-3)	0.96784 * (6.26E-3)
LSMOP5		206	0.79282 (6.00E-2)	0.62224 * (—)	0.30990 (9.27E-2)	0.81408 * (3.48E-2)
LSMOP6		206	0.46222 (1.79E-2)	0.57904 * (5.69E-3)	0.50444 (3.28E-2)	0.58816 * (1.95E-2)
LSMOP1		307	— (—)	0.80958 * (2.78E-2)	— (—)	0.71837 * (2.01E-2)
LSMOP2		307	0.97074 (2.47E-3)	0.97026 (2.91E-3)	0.97153 (1.59E-3)	0.97553 * (1.70E-3)
LSMOP3	3	307	— (—)	0.51111 * (—)	— (—)	0.51111 * (—)
LSMOP4		307	0.90478 (7.82E-3)	0.92602 * (4.70E-3)	0.89693 (6.83E-3)	0.93835 * (3.50E-3)
LSMOP5		307	— (—)	0.85655 * (9.56E-4)	— (—)	0.87305 * (1.33E-2)
LSMOP6		307	— (—)	0.29197 * (6.70E-3)	— (—)	0.26725 * (6.36E-3)
LSMOP1		413	— (—)	0.76046 * (1.25E-2)	— (—)	0.66323 * (1.91E-2)
LSMOP2		413	0.98192 (1.77E-3)	0.98272 * (1.29E-3)	0.98080 (1.59E-3)	0.98570 * (1.03E-3)
LSMOP3	4	413	— (—)	0.02216 * (9.49E-4)	— (—)	0.02356 * (3.31E-4)
LSMOP4		413	0.96053 (2.76E-3)	0.96187 (2.80E-3)	0.95663 (3.94E-3)	0.96756 * (2.80E-3)
LSMOP5		413	— (—)	0.91945 * (4.68E-2)	— (—)	0.93033 * (5.17E-3)
LSMOP6		413	0.33750 (3.50E-3)	0.66594 * (2.55E-2)	0.33652 (5.55E-3)	0.67133 * (5.52E-2)
LSMOP1		514	— (—)	0.63753 * (1.49E-2)	0.50034 (—)	0.63655 * (2.09E-2)
LSMOP2		514	0.99283 (7.79E-4)	0.99313 (5.96E-4)	0.99323 (4.98E-4)	0.99416 * (5.24E-4)
LSMOP3	5	514	— (—)	0.50034 * (—)	— (5.00E-1)	0.50034 * (—)
LSMOP4		514	0.96323 (2.79E-3)	0.97878 * (1.84E-3)	0.96271 (3.12E-3)	0.98035 * (2.37E-3)
LSMOP5		514	— (—)	0.90318 * (5.62E-2)	— (—)	0.69514 * (3.07E 1)
LSMOP6		514	— (—)	0.21220 * (1.02E-1)	— (—)	0.26520 * (1.19E-1)
UF1		30	0.91887 (4.11E-2)	0.96229 * (2.89E-2)	0.97379 (1.39E-2)	0.96932 (1.53E-2)
UF2		30	0.95267 (3.99E-2)	0.98183 * (8.40E 3)	0.07080 (9.59E-3)	0.97173 * (8.26E-3)
UF3	2	30	0.71558 (5.38E-2)	0.95430 * (6.00E-3)	0.96908 (1.45E-2)	0.95301 * (1.46E-2)
UF1		300	0.90110 (2.48E-2)	0.90481 (6.03E-2)	0.44383 (9.25E-2)	0.29760 * (4.25E-2)
UF2		300	0.89138 (2.24E-2)	0.88332 (1.39E-2)	0.86877 (5.74E-3)	0.88162 * (4.62E-3)
UF3		300	0.73023 (8.34E-3)	0.95459 * (5.81E-3)	0.73585 (1.42E 2)	0.04091 * (1.92E-3)
UF8		30	0.80694 (3.35E-2)	0.84659 * (9.42E-2)	0.24162 (2.42E-1)	0.64186 * (6.77E-2)
UF9		30	0.72903 (9.67E-2)	0.82914 * (8.45E-2)	0.31863 (1.46E-1)	0.59303 * (5.85E-2)
UF10	3	30	0.38003 (1.29E-1)	0.82597 * (1.22E-2)	— (—)	0.54190 * (3.09E-2)
UF8		300	0.79119 (2.05E-2)	0.85501 * (9.30E-4)	0.65592 (4.89E-2)	0.82907 * (8.35E-3)
UF9		300	0.63539 (2.04E-2)	0.62206 (2.44E-2)	0.50443 (2.73E-2)	0.62197 * (1.26E-2)
UF10		300	0.03827 (5.22E-2)	0.84973 * (2.89E-3)	— (—)	0.73813 * (8.86E-2)
WFG2		41	0.85891 (7.60E-3)	0.97933 * (1.27E-2)	0.97452 (5.03E-2)	0.97187 (2.02E-2)
WFG3		41	0.84987 (2.91E-3)	0.84869 (6.62E-2)	0.84444 (1.00E-2)	0.83283 * (7.49E-3)
WFG4		40	0.99235 (1.06E-3)	0.99019 * (2.60E-3)	0.94351 (6.56E-3)	0.94394 (9.17E-3)
WFG5		40	0.97948 (2.92E-3)	0.97841 * (2.40E-3)	0.95348 (1.31E-2)	0.93875 * (2.60E-2)
WFG7		40	0.99345 (1.02E-3)	0.99262 * (8.23E-4)	0.96279 (3.02E-2)	0.95901 (1.44E-2)
WFG8	2	40	0.89218 (5.87E-3)	0.89593 * (6.48E-3)	0.85652 (1.44E-2)	0.83074 * (1.64E-2)
WFG2		401	0.72887 (3.64E-2)	0.86591 * (2.49E-2)	0.76243 (6.42E-3)	0.87080 * (2.68E-3)
WFG3		401	0.63592 (7.54E-3)	0.78669 * (7.04E-2)	0.66454 (1.52E-2)	0.73419 * (6.82E-3)
WFG4		400	0.66487 (1.36E-2)	0.83624 * (1.42E-2)	0.79988 (2.20E-2)	0.82247 * (1.51E-2)
WFG5		400	0.64103 (1.90E-2)	0.85848 * (1.42E-2)	0.82362 (6.03E-3)	0.85548 * (8.87E-3)
WFG7		400	0.67641 (1.19E-2)	0.86891 * (2.39E-2)	0.74902 (8.54E-3)	0.78553 * (9.29E-3)
WFG8		400	0.57725 (1.06E-2)	0.78999 * (1.48E-2)	0.57692 (2.33E-2)	0.78252 * (1.79E-2)
WFG2		40	0.90095 (9.05E-3)	0.97075 * (8.38E-3)	0.94492 (1.92E-2)	0.93262 * (3.45E-2)
WFG3		40	0.93396 (2.31E-2)	0.93407 (2.58E-2)	0.84310 (3.27E-2)	0.86879 * (4.69E-2)
WFG4		40	0.96941 (3.62E-3)	0.95785 * (3.62E-3)	0.88129 (2.56E-2)	0.85787 * (2.53E-2)
WFG5		40	0.94920 (4.28E-3)	0.94571 * (2.34E-3)	0.88469 (2.12E-2)	0.88316 (2.18E-2)
WFG7		40	0.96833 (2.17E-2)	0.97117 (3.13E-2)	0.86183 (3.23E-2)	0.86343 (2.79E-2)
WFG8	3	40	0.92742 (6.25E-3)	0.91145 * (8.30E-3)	0.80365 (1.82E-2)	0.80627 (1.82E-2)
WFG2		400	0.68917 (9.54E-3)	0.85622 * (5.57E-3)	0.69053 (5.17E-3)	0.84083 * (5.10E-3)
WFG3		400	0.57700 (2.56E-2)	0.79436 * (4.23E-2)	0.57939 (1.48E-2)	0.67991 * (4.62E-3)
WFG4		400	0.59041 (1.47E-2)	0.71919 * (6.97E-2)	0.70378 (1.63E-2)	0.71675 * (2.00E-2)
WFG5		400	0.55374 (1.23E-2)	0.75534 * (4.74E-2)	0.69452 (1.47E-2)	0.72009 * (1.29E-2)
WFG7		400	0.59516 (1.35E-2)	0.71240 * (8.34E-3)	0.68170 (9.85E-3)	0.69439 * (9.10E-3)
WFG8		400	0.52326 (1.77E-2)	0.72452 * (1.80E-2)	0.59209 (1.95E-2)	0.72812 * (1.85E-2)

other. These results imply that for such problems, at least in low-dimensional search spaces, a combination of solutions, which actually alters all variables at the same time through the linear coefficients, might not be suitable.

Table 2. Obtained median and IQR values of the relative Hypervolume for NSGA-III and xNSGA-III as well as RVEA and xRVEA. An asterisk in the column of an x-Algorithm indicates statistical significance to the respective original version of that algorithm. Best performances are marked in bold and shaded where significant.

	m	n	NSGA-III	xNSGA-III	RVEA	xRVEA
LSMOP1		206	0.64220 (1.53E-1)	0.79579 * (1.15E-1)	0.06873 (5.66E-1)	0.83933 * (1.61E-2)
LSMOP2		206	0.95088 (2.79E-3)	0.98200 * (4.04E-3)	0.94387 (4.98E-3)	0.97949 * (2.83E-3)
LSMOP3		206	— (—)	0.08521 * (9.18E-2)	0.56963 (9.96E-3)	0.01974 * (3.53E-4)
LSMOP4	2	206	0.92227 (5.51E-3)	0.95692 * (3.62E-3)	0.89610 (5.73E-3)	0.95196 * (6.33E-3)
LSMOP5		206	0.80093 (5.06E-2)	0.62224 * (—)	— (—)	0.62224 * (7.05E-4)
LSMOP6		206	0.45949 (8.70E-2)	0.57262 * (7.48E-3)	0.49474 (5.00E-2)	0.53730 * (2.34E-1)
LSMOP1		307	0.07558 (1.15E-1)	0.82378 * (2.30E-2)	0.53715 (5.74E-2)	0.76242 * (1.24E-2)
LSMOP2		307	0.97954 (4.31E-4)	0.98317 * (5.45E-4)	0.97803 (5.73E-4)	0.98107 * (7.71E-4)
LSMOP3		307	— (—)	0.51112 * (1.37E-2)	— (—)	0.51081 * (5.01E-4)
LSMOP4	3	307	0.93995 (2.68E-3)	0.96222 * (1.53E-3)	0.93470 (3.10E-3)	0.96314 * (2.83E-3)
LSMOP5		307	— (—)	0.85486 * (9.61E-4)	0.53581 (6.39E-6)	0.91034 * (2.29E-2)
LSMOP6		307	— (—)	0.28687 * (4.61E-3)	— (—)	0.16058 * (1.06E-1)
LSMOP1		413	— (—)	0.79609 * (2.13E-2)	0.56234 (1.34E-1)	0.76333 * (1.34E-2)
LSMOP2		413	0.98752 (2.75E-4)	0.99087 * (3.96E-4)	0.98672 (4.39E-4)	0.99004 * (7.35E-4)
LSMOP3		413	— (—)	0.02076 * (1.47E-2)	0.02149 (3.03E-1)	0.02041 (2.93E-3)
LSMOP4	4	413	0.97775 (1.31E-3)	0.98127 * (1.63E-3)	0.97015 (3.45E-3)	0.98206 * (1.20E-2)
LSMOP5		413	— (—)	0.93971 * (8.78E-4)	0.51144 (3.22E-4)	0.92817 * (5.73E-3)
LSMOP6		413	0.36714 (7.75E-3)	0.70352 * (8.04E-3)	0.40710 (2.81E-2)	0.74121 * (8.87E-2)
LSMOP1		514	— (—)	0.69758 * (4.51E-2)	0.67646 (1.28E-1)	0.74755 * (1.11E-2)
LSMOP2		514	0.99601 (8.07E-5)	0.99644 * (6.37E-5)	0.99566 (2.15E-3)	0.99625 * (1.14E-4)
LSMOP3		514	— (—)	0.50033 * (2.10E-5)	— (7.30E-3)	0.50033 * (3.31E-2)
LSMOP4	5	514	0.98389 (1.12E-3)	0.98972 * (6.96E-4)	0.98235 (2.83E-3)	0.98890 * (1.14E-3)
LSMOP5		514	— (—)	0.98680 * (1.82E-3)	0.50400 (5.49E-6)	0.98609 * (2.35E-3)
LSMOP6		514	— (—)	0.53851 * (1.35E-2)	0.19527 (2.39E-1)	0.31759 * (1.33E-1)
UF1		30	0.90857 (5.92E-2)	0.97172 * (2.62E-2)	0.86021 (7.86E-2)	0.94097 * (1.38E-2)
UF2		30	0.95574 (3.03E-2)	0.98049 * (1.00E-2)	0.95095 (3.30E-2)	0.95587 (1.82E-2)
UF3	2	30	0.70910 (4.33E-2)	0.94851 * (1.07E-2)	0.69021 (1.81E-2)	0.90003 * (2.32E-2)
UF1		300	0.88442 (6.09E-2)	0.88944 (4.52E-2)	0.70355 (5.68E-2)	0.65673 * (4.88E-2)
UF2		300	0.88425 (1.64E-2)	0.88747 (6.78E-3)	0.85611 (9.08E-3)	0.87460 * (4.22E-3)
UF3		300	0.70306 (1.05E-2)	0.96051 * (1.68E-3)	0.69537 (8.73E-3)	0.95204 * (4.75E-3)
UF8		30	0.84654 (6.85E-3)	0.85433 * (1.79E-3)	0.84549 (1.19E-3)	0.84531 (1.13E-1)
UF9		30	0.72999 (3.78E-3)	0.86360 * (7.90E-3)	0.68531 (2.59E-2)	0.86237 * (1.95E-1)
UF10	3	30	0.46645 (2.00E-1)	0.85371 * (1.78E-3)	0.44133 (8.21E-2)	0.84551 * (2.97E-4)
UF8		300	0.82401 (6.90E-3)	0.85408 * (1.07E-3)	0.75309 (2.85E-2)	0.84802 * (7.03E-4)
UF9		300	0.56338 (1.33E-2)	0.56357 (1.11E-2)	0.57018 (1.40E-2)	0.56862 (9.94E-3)
UF10		300	0.49573 (1.36E-1)	0.85499 * (9.58E-4)	0.41166 (3.40E-1)	0.84215 * (3.51E-1)
WFG2		41	0.85728 (1.06E-2)	0.97518 * (1.18E-2)	0.84715 (1.23E-2)	0.95977 * (1.02E-2)
WFG3		41	0.84681 (6.36E-3)	0.84961 (6.54E-3)	0.83505 (1.19E-2)	0.84317 * (8.32E-3)
WFG4		40	0.99153 (9.83E-3)	0.98916 (8.35E-3)	0.98032 (1.09E-2)	0.97351 * (9.98E-3)
WFG5		40	0.97941 (3.86E-3)	0.97447 * (4.24E-3)	0.98076 (4.17E-3)	0.97430 * (3.69E-3)
WFG7	2	40	0.93748 (1.22E-2)	0.99385 * (5.10E-4)	0.92685 (1.41E-2)	0.98565 * (4.19E-3)
WFG8		40	0.86666 (2.03E-2)	0.89541 * (1.08E-2)	0.83284 (1.65E-2)	0.84491 * (1.49E-2)
WFG2		401	0.73729 (4.17E-2)	0.86429 * (1.96E-2)	0.71967 (1.06E-2)	0.85805 * (4.48E-3)
WFG3		401	0.63240 (1.17E-2)	0.74807 * (3.22E-2)	0.60861 (8.36E-3)	0.75809 * (8.93E-3)
WFG4		400	0.66360 (2.09E-2)	0.84435 * (1.72E-2)	0.62383 (1.44E-2)	0.79269 * (2.27E-2)
WFG5		400	0.64081 (9.68E-3)	0.85923 * (1.75E-2)	0.58413 (1.36E-2)	0.84903 * (1.56E-2)
WFG7		400	0.67987 (1.46E-2)	0.87752 * (1.64E-2)	0.63490 (1.33E-2)	0.80371 * (2.21E-2)
WFG8		400	0.56587 (1.33E-2)	0.79794 * (3.07E-2)	0.52654 (1.04E-2)	0.75307 * (2.12E-2)
WFG2		40	0.89508 (9.06E-3)	0.97233 * (1.06E-2)	0.88068 (1.72E-2)	0.95820 * (1.28E-2)
WFG3		40	0.89644 (3.21E-2)	0.91990 * (1.88E-2)	0.92193 (3.11E-2)	0.90139 * (1.91E-2)
WFG4		40	0.97227 (3.95E-3)	0.96489 (5.08E-3)	0.96010 (9.14E-3)	0.96054 (1.07E-2)
WFG5		40	0.95600 (3.73E-3)	0.95345 * (1.85E-3)	0.96749 (2.00E-3)	0.95277 * (1.38E-3)
WFG7	3	40	0.98000 (5.03E-2)	0.98470 * (3.55E-3)	0.96277 (3.84E-2)	0.98642 * (1.04E-3)
WFG8		40	0.94130 (4.94E-3)	0.92604 * (7.27E-3)	0.84503 (9.21E-2)	0.91875 * (8.59E-3)
WFG2		400	0.68287 (9.66E-3)	0.82952 * (7.41E-3)	0.65750 (5.27E-3)	0.83219 * (1.32E-2)
WFG3		400	0.54703 (2.47E-2)	0.65643 * (7.34E-2)	0.33447 (4.64E-2)	0.67913 * (3.18E-2)
WFG4		400	0.46450 (1.38E-2)	0.61422 (2.85E-1)	0.46569 (2.93E-2)	0.50973 * (3.65E-2)
WFG5		400	0.53172 (4.44E-2)	0.64958 * (1.86E-1)	0.49506 (2.40E-2)	0.60250 * (2.47E-2)
WFG7		400	0.48076 (1.48E-2)	0.68523 * (4.92E-2)	0.49942 (6.12E-2)	0.71808 * (4.20E-2)
WFG8		400	0.39701 (5.85E-2)	0.66022 * (2.87E-2)	0.36715 (1.04E-2)	0.72648 * (1.53E-2)

In summary, we conclude that the proposed approach of optimising linear combination of the population members is able to increase the performance of multi- and many-objective algorithms in most cases. This is especially true for higher numbers of decision variables and higher numbers of objective functions. The authors further tested the method on the remaining problems from the WFG, UF and LSMOP families, which were not reported here due to page limitations, and obtained similar superior performance of the proposed method.

5 Conclusion and Future Work

This article proposed a new mechanism for exploration and solution creation in multi- and many-objective optimisation. The mathematical concept is able to focus the search on relevant areas and at the same time reduce the dimensionality of the original problem without using (possibly expensive) variable grouping methods. After we introduced the mathematical concept, we described how this approach can be incorporated into existing metaheuristic algorithms and explored its capabilities on a variety of benchmark functions with different characteristics and dimensionality. The results indicate that this linear-combination approach can improve the performance of existing methods in both large-scale and many-objective optimisation.

Future work in this area involves exploring the possibilities of this approach further. It can be included into specific large-scale metaheuristics like the WOF as a dimensionality reduction technique. Another possible application might be in constrained problems. Linear combinations have been applied to particle swarm optimisation in [12] to preserve the feasibility of individuals. The approach described in this article can easily be adapted to only allow certain linear combinations, for instance convex ones. If the search is restrained in this way to convex combinations, the algorithm can by definition only create feasible solutions out of existing feasible ones, provided that constraints are linear. Therefore, this can be a promising direction for constraint handling in (large-scale) many-objective optimisation.

References

1. Antonio, L.M., Coello Coello, C.A.: Use of cooperative coevolution for solving large scale multiobjective optimization problems. In: IEEE Congress on Evolutionary Computation (CEC), pp. 2758–2765 (2013)
2. Cheng, R., Jin, Y., Olhofer, M., Sendhoff, B.: A reference vector guided evolutionary algorithm for many-objective optimization. IEEE Trans. Evol. Comput. **20**(5), 773–791 (2016). https://doi.org/10.1109/TEVC.2016.2519378
3. Cheng, R., Jin, Y., Olhofer, M., Sendhoff, B.: Test problems for large-scale multiobjective and many-objective optimization. IEEE Trans. Cybern. **47**(12), 4108–4121 (2017)
4. Deb, K., Pratap, A., Agarwal, S., Meyarivan, T.: A fast and elitist multiobjective genetic algorithm: NSGA-II. IEEE Trans. Evol. Comput. **6**(2), 182–197 (2002)
5. Deb, K., Srinivasan, A.: Innovization: innovating design principles through optimization. In: Proceedings of the 8th Annual Conference on Genetic and Evolutionary Computation, pp. 1629–1636. ACM (2006)

6. Deb, K., Jain, H.: An evolutionary many-objective optimization algorithm using reference-point-based nondominated sorting approach, part i: solving problems with box constraints. IEEE Trans. Evol. Comput. **18**(4), 577–601 (2014)
7. Gaur, A., Deb, K.: Effect of size and order of variables in rules for multi-objective repair-based innovization procedure. In: 2017 IEEE Congress on Evolutionary Computation (CEC), pp. 2177–2184. IEEE (2017)
8. Huband, S., Hingston, P., Barone, L., While, L.: A review of multiobjective test problems and a scalable test problem toolkit. IEEE Trans. Evol. Comput. **10**(5), 477–506 (2006)
9. Kukkonen, S., Lampinen, J.: GDE3: the third evolution step of generalized differential evolution. In: IEEE Congress on Evolutionary Computation (CEC), vol. 1, pp. 443–450. IEEE (2005). https://doi.org/10.1109/CEC.2005.1554717
10. Li, K., Deb, K., Zhang, Q., Kwong, S.: An evolutionary many-objective optimization algorithm based on dominance and decomposition. IEEE Trans. Evol. Comput. **19**(5), 694–716 (2015). https://doi.org/10.1109/TEVC.2014.2373386
11. Ma, X., et al.: A multiobjective evolutionary algorithm based on decision variable analyses for multiobjective optimization problems with large-scale variables. IEEE Trans. Evol. Comput. **20**(2), 275–298 (2016)
12. Mostaghim, S., Halter, W., Wille, A.: Linear multi-objective particle swarm optimization. In: Abraham, A., Grosan, C., Ramos, V. (eds.) Stigmergic Optimization. Studies in Computational Intelligence, vol. 31, pp. 209–238. Springer, Heidelberg (2006). https://doi.org/10.1007/978-3-540-34690-6_9
13. Sander, F., Zille, H., Mostaghim, S.: Transfer strategies from single- to multi-objective grouping mechanisms. In: Proceedings of the Genetic and Evolutionary Computation Conference, GECCO 2018, pp. 729–736. ACM, New York (2018). https://doi.org/10.1145/3205455.3205491
14. Singh, H.K., Isaacs, A., Ray, T.: A pareto corner search evolutionary algorithm and dimensionality reduction in many-objective optimization problems. IEEE Trans. Evol. Comput. **15**(4), 539–556 (2011). https://doi.org/10.1109/TEVC.2010.2093579
15. Tian, Y., Cheng, R., Zhang, X., Jin, Y.: Platemo: a MATLAB platform for evolutionary multi-objective optimization. CoRR abs/1701.00879 (2017). http://arxiv.org/abs/1701.00879
16. While, L., Hingston, P., Barone, L., Huband, S.: A faster algorithm for calculating hypervolume. IEEE Trans. Evol. Comput. **10**(1), 29–38 (2006)
17. Zhang, X., Tian, Y., Jin, Y., Cheng, R.: A decision variable clustering-based evolutionary algorithm for large-scale many-objective optimization. IEEE Trans. Evol. Comput. **22**(1), 97–112 (2018). https://doi.org/10.1109/TEVC.2016.2600642
18. Zille, H., Ishibuchi, H., Mostaghim, S., Nojima, Y.: Mutation operators based on variable grouping for multi-objective large-scale optimization. In: 2016 IEEE Symposium Series on Computational Intelligence (SSCI) (2016)
19. Zille, H., Ishibuchi, H., Mostaghim, S., Nojima, Y.: Weighted optimization framework for large-scale multi-objective optimization. In: Companion of Genetic and Evolutionary Computation Conference - GECCO. ACM (2016)
20. Zille, H., Ishibuchi, H., Mostaghim, S., Nojima, Y.: A framework for large-scale multi-objective optimization based on problem transformation. IEEE Trans. Evol. Comput. **22**(2), 260–275 (2018). http://ieeexplore.ieee.org/document/7929324/
21. Zille, H., Mostaghim, S.: Comparison study of large-scale optimisation techniques on the LSMOP benchmark functions. In: 2017 IEEE Symposium Series on Computational Intelligence (SSCI), November 2017

Estimating Relevance of Variables
for Effective Recombination

Taishi Ito[1,4]($\boxtimes$), Hernán Aguirre[1,4], Kiyoshi Tanaka[1,4], Arnaud Liefooghe[2],
Bilel Derbel[2], and Sébastien Verel[3]

[1] Faculty of Engineering, Shinshu University, Nagano, Japan
{17w2006d,ahernan,ktanaka}@shinshu-u.ac.jp
[2] University of Lille, CNRS, Centrale Lille, UMR 9189 – CRIStAL,
Inria Lille – Nord Europe, 59000 Lille, France
{arnaud.liefooghe,bilel.derbel}@univ-lille.fr
[3] Univ. Littoral Cote d'Opale, LISIC, 62100 Calais, France
verel@lisic.univ-littoral.fr
[4] International Associated Laboratory LIA-MODO, Nagano, Japan

Abstract. Dominance, extensions of dominance, decomposition, and indicator functions are well-known approaches used to design MOEAs. Algorithms based on these approaches have mostly sought to enhance parent selection and survival selection. In addition, several variation operators have been developed for MOEAs. We focus on the classification and selection of variables to improve the effectiveness of solution search. In this work, we propose a method to classify variables that influence convergence and increase their recombination rate, aiming to improve convergence of the approximation found by the algorithm. We incorporate the proposed method into NSGA-II and study its effectiveness using three-objective DTLZ and WFG functions, including unimodal, multimodal, separable, non-separable, unbiased, and biased functions. We also test the effectiveness of the proposed method on a real-world bi-objective problem. Simulation results verify that the proposed method can contribute to achieving faster and better convergence in several kinds of problems, including the real-world problem.

Keywords: Evolutionary multi-objective optimization ·
Variables classification · Variables selection · Recombination operators

1 Introduction

Multi-objective evolutionary algorithms [2,4] (MOEAs) have been used to solve multi-objective optimization problems on all kinds of application domains. Due to their success, MOEAs are being applied to real-world problems of increased complexity. Scalability in decision and objective spaces, epistasis, effectiveness on problems with difficult topologies of the Pareto optimal set, and a limited budget of evaluations due to computationally expensive fitness functions are some of the challenges the new generation of MOEAs have to face.

© Springer Nature Switzerland AG 2019
K. Deb et al. (Eds.): EMO 2019, LNCS 11411, pp. 411–423, 2019.
https://doi.org/10.1007/978-3-030-12598-1_33

The enhancement of MOEAs performance is an active research subject. Dominance, extensions of dominance, decomposition, and indicator functions are well-known approaches used to design MOEAs [1,5,9–11]. Algorithms based on these approaches have mostly sought to enhance parent selection and survival selection. In addition, several variation operators have been developed and incorporated within MOEAs.

Evolutionary multi-objective algorithms commonly select individuals for variation based on their fitness. However, the operators of variation are commonly applied to variables randomly chosen. Typically, an operator rate per variable controls the expected number of variables that will be subject to variation, but the decision of what variables will be modified is left to chance. Modifying a variable of a solution in a multi-objective problem can have one of the following effects. The modification improves one or several objectives without worsening others. This would be the case if the solution subject to the modification is suboptimal, which is commonly observed in random initial populations and during early stages of the optimization. The modification improves one or more objectives but worsens others. This will typically be observed if the solution being modified is Pareto optimal or if it belongs to a local front. Multi-objective evolutionary algorithms aim to find an approximation of the Pareto optimal set, commonly with good qualities in terms of convergence and diversity in objective space. If the effects a variable has on convergence and diversity can be learned or estimated during the optimization [8], the effectiveness of the search could be enhanced by targeting particular variables for variation to find better approximations of the Pareto optimal set.

We focus on the classification and selection of variables for variation aiming to improve the ability of solution search. In this work, we propose a method to identify variables that influence convergence and increase their selection probabilities, so that recombination can select them more frequently to improve convergence of the approximation found by the algorithm. The proposed method selects randomly a solution from the instantaneous Pareto set and creates variations of it mutating one variable at the time. Variables are classified into `influential` and `uninfluential` based on whether there is a dominance relation or not between the original solution and the corresponding one-variable mutants. The method estimates that `influential` variables affect convergence of solutions in objective space and increase their recombination rate.

In this paper, we incorporate the proposed method into NSGA-II [5] and study its effectiveness using three-objective DTLZ [3] and WFG [6] functions, including unimodal, multimodal, separable, non-separable, unbiased, and biased functions. We also verify the effectiveness of the proposed method on a real-world bi-objective problem [7]. Three ways to determine the trial values of variables to create the mutants are investigated. Simulation results verify that the proposed method can contribute to achieving faster and better convergence in several kinds of problems.

2   Method

In this section, we describe the proposed variable selection method, applied every generation after front sorting before truncation selection. The method first classifies variables that can influence convergence and then update their probabilities so that recombination can select them more frequently.

The procedure to classify variables is illustrated in the pseudocode of Fig. 1. First, we assign a label 0 to each variable, $L_i = 0$, $i = 1, \cdots, n$. Next, we obtain trial values for each variable $x^{\mathbf{trial}} = (x_1^{\mathrm{trial}}, \cdots, x_n^{\mathrm{trial}})$ from the non-dominated solution set F_1 in the population P_t at generation t. We randomly select one solution $x = (x_1, \cdots, x_n)$ from F_1. Then, for each variable i, we generate a solution y duplicate of x and modify the i-th variable with its corresponding trial value, $y_i = x_i^{trial}$. Evaluate y and calculate the dominance relation between solutions y and x. If either y dominates x or x dominates y, we update the corresponding label to 1, $L_i = 1$. This procedure returns the vector of labels L assigned to the variables, where $L_i = 1$ if the change in the i-th variable induced a dominance relation ($\succ$) between the randomly sampled solution and its one-variable mutant. $L_i = 0$ otherwise.

```
 1: procedure VARIABLECLASSIFICATION(F₁)
 2:     L = (L₁, L₂, ···, Lₙ) = (0, 0, ···, 0)
 3:     x^trial = (x₁^trial, x₂^trial, ···, xₙ^trial) = ReferenceValues(F₁)
 4:     x = xʲ ∈ F₁, j = rand(1, |F₁|)
 5:     for i = 1 to n do
 6:         y = (y₁, y₂, ···, yₙ) = x = (x₁, x₂, ···, xₙ)
 7:         yᵢ = xᵢ^trial
 8:         Evaluate(y)
 9:         if x ≻ y OR y ≻ x then
10:             Lᵢ = 1
11:         end if
12:     end for
13:     return L
14: end procedure
```

Fig. 1. Classification of variables

We explore three procedures called `random`, `far`, and `near` to set trial values for variables.

`random` sets x_i^{trial} to the value of the i-th variable of a solution randomly selected from F_1. A different solution j is randomly chosen for each variable i,

$$x_i^{trial} = z_i^j \mid z^j \in F_1 \wedge j = rand(1, |F_1|). \tag{1}$$

far sets x_i^{trial} to the farthest value of x_i in F_1. That is,

$$k = \arg\max_{j=1...|F_1|}(|x_i - z_i^j|), \qquad z^j \in F_1, x_i \neq z_i^j \tag{2}$$
$$x_i^{trial} = z_i^k.$$

On the other hand, **near** sets x_i^{trial} to the nearest value of x_i in F_1. That is,

$$k = \arg\min_{j=1...|F_1|}(|x_i - z_i^j|), \qquad z^j \in F_1, x_i \neq z_i^j \tag{3}$$
$$x_i^{trial} = z_i^k.$$

In **near** and **far**, if all solutions in F_1 have the same value in x_i, then $x_i^{trial} = x_i$ (no change).

After variables have been classified, a recombination rate $p'_{cv,i}$ for each variable i is computed as follows

$$p'_{cv,i} = p_{cv}, \qquad if \quad N_L = 0. \tag{4}$$

$$p'_{cv,i} = \begin{cases} 1, & if \quad L_i = 1 \\ \frac{E-N_L}{n-N_L}, & otherwise, \end{cases}, \qquad if \quad 0 < N_L \leq E. \tag{5}$$

$$p'_{cv,i} = \begin{cases} \frac{E}{N_L}, & if \quad L_i = 1 \\ 0, & otherwise. \end{cases}, \qquad if \quad N_L > E. \tag{6}$$

where $L_i \in \{0, 1\}$ is the label assigned to variables, $N_L = \sum_{i=1}^n L_i$ is the number of variables classified with label $L_i = 1$, $E = n \times p_{cv}$ is the expected number of crossed variables using the default rate. If after exploring n one-variable mutants no changes in dominance are observed, $N_L = 0$, the recombination rate $p'_{cv,i}$ is set to the default rate p_{cv} for each variable. If there are changes in dominance and these are less than the expected number of crossed variables when the default crossover rate is used, $0 < N_L \leq E$, then the recombination rate $p'_{cv,i}$ is set to 1 for variables labeled $L_i = 1$ and to $\frac{E-N_L}{n-N_L} < p_{cv}$ for variables labeled $L_i = 0$. Otherwise, if $N_L > E$, $p'_{cv,i}$ is set to $\frac{E}{N_L} \geq p_{cv}$ for variables labeled $L_i = 1$ and to 0 for variables labeled $L_i = 0$. Note that the expected number of recombined variables with p_{cv} and p'_{cv} is the same.

In this work, we use the variable classification procedure to update recombination probabilities of variables. However, this method can be easily extended to influence probabilities for mutation or other variation operators.

3   Test Problems

DTLZ [3] and WFG [6] benchmark multi-objective optimization problems are used to evaluate the proposed method. These problems are scalable in the number of variables and the number of objective functions. From the DTLZ family of problems we use DTLZ2 and DTLZ3, whereas from the WFG family we use WFG1-WFG9. Some properties of these problems are summarized in Table 1.

Table 1. Features of test problems. Separability: separable S, non-separable NS. Modality: unimodal U, multimodal M.

Problem	Sep.	Modality	Bias	Other Features
DTLZ2	S	U	-	
DTLZ3	S	M	-	
WFG1	S	U	Yes	Polynomial bias $\alpha = 0.02$. Bias variables towards 1
WFG2	NS	U,M	-	U: $f_1, \cdots, f_{M-1}$, M: f_M, f_M. Discontinuous front
WFG3	NS	U	-	Degenerancy constants $A_1 = 1, A_{2:M-1} = 0$, front reduces to two dimensions
WFG4	S	M	-	
WFG5	S	M	-	
WFG6	NS	U	-	
WFG7	S	U	Yes	Parameter dependent bias: $z_{i=1:k} \leftarrow z_{i+1}, \ldots, z_n$
WFG8	NS	U	Yes	Parameter dependent bias: $z_{i=k+1:n} \leftarrow z_1, \ldots, z_{i-1}$
WFG9	NS	M	Yes	Parameter dependent bias: $z_{i=1:n-1} \leftarrow z_{i+1}, \ldots, z_n$

We include separability, modality and bias. Separable problems are marked with S and non-separable with NS. Unimodal problems are marked U and multimodal problems with M.

We also test the effectiveness of the proposed method on a bi-objective real-world problem [7]. The problem is to design a platform with a motor mounted on it. The machine setup is simplified as a pin-pin supported beam carrying a weight (motor). A vibratory disturbance is imparted from the motor onto the beam, which is of length L, width b, and symmetrical about its mid-plane. The beam is made of three layers of material. Variables d_1 and d_2, respectively, locate the contact of materials 1 and 2, and 2 and 3. Variable d_3 locates the top of the beam. The values of d_1, d_2, and d_3 are measured from the mid-plane of the beam. M_i refers to the material type for layer i ($i = 1, 2, 3$). The mass density (ρ), Young's modulus of elasticity (E), and cost per unit volume (c) for each material type is given [7]. The objective functions are the fundamental frequency, f_1, to be maximized and the cost of the set up, f_2, to be minimized. The complete formulation is as follows,

$$f_1(d_1, d_2, d_3, b, L) = (\pi/2L^2)(EI/\mu)^{1/2},$$
$$EI = (2b/3)[E_{M_1} d_1^3 + E_{M_2}(d_2^3 - d_1^3) + E_{M_3}(d_3^3 - d_2^3)], \qquad (7)$$
$$\mu = 2b[\rho_{M_1} d_1 + \rho_{M_2}(d_2 - d_1) + \rho_{M_3}(d_3 - d_2)],$$

$$f_2(d_1, d_2, d_3, b) = 2b[c_{M_1}d_1 + c_{M_2}(d_2 - d_1) + c_{M_3}(d_3 - d_2)], \qquad (8)$$

subject to $\mu L - 2800 < 0$, $d_2 - d_1 \leq 0.01$, $d_3 - d_2 \leq 0.01$, $0.35 \leq b \leq 0.5$, $3 \leq L \leq 6$, and $d_1, d_2, d_3 \in [0.01, 0.6]$.

4   Experimental Setup and Performance Measures

In this paper, we use DTLZ2, DTLZ3, WFG1-WFG9 as benchmark problems, setting the number of objectives $M = 3$ and the number of variables $n = 12$ for all problems. The number of position variables is $M - 1$ in DTLZ. Similarly, we set the number of position variables to $k = M - 1$ in WFG. Thus, the number of position variables is 2 and the number of distance variables is 10 in all benchmark problems dealt with in this paper.

We use five algorithms to study and verify the performance of the variable selection method proposed in this work. The base algorithm is conventional NSGA-II [5]. NSGA-II randomly selects variables for recombination with the same probability p_{cv} per variable. In the following, NSGA-II is named org. We also use three variations of NSGA-II using the proposed variable selection method with one of the procedures to set trial variables, as explained in Sect. 2. These variations are named random, far, and near. The fifth algorithm, named ideal, knows in advance the correct classification of distance and position variables to compute the recombination rate per variable. Thus, org provides a baseline for performance comparison, whereas ideal provides the performance reference of an algorithm with a perfect classification of variables.

The algorithms are run for 2000 generations setting the number of individuals to 100, the recombination rate per individual to 1.0, and the mutation rate to $1/n$. The recombination rate per variable is set to $p_{cv} = 0.5$ in org and to p'_{cv} in random, far, near, and ideal, computed with Eqs. (4–6) as explained in Sect. 2. The number of runs is 30 in all experiments.

To verify improvements on convergence we use generational distance GD in the case of DTLZ problems and the value of variable x_M in WFG problems. GD is computed analitycally. The smaller the value of GD, the better the convergence of the set of obtained nondominated solutions. In WFG distance variables are finally aggregated into variable $x_M \in [0.0, 1.0]$. The smaller the value of x_M is, the closer to the true optimum the solution is. When $x_M = 0$, the obtained solution is the true optimal solution. We also use the hypervolume metric to evaluate performance in the real world problem.

5   Simulation Results and Discussion

5.1   Three-Objective Benchmark Problems

First, we apply each algorithm to DTLZ2 and DTLZ3 functions. Figure 2(a) shows the transition of the average GD value in 30 runs over the number of fitness evaluations for DTLZ2. Similarly, Fig. 2(b) shows results for DTLZ3. In this experiment, population size is set to 100 and the number of generations to 2,000. Therefore, the total number of evaluations in `org` and `ideal` is 200,000 per run. In the case of `random`, `far`, and `near` the algorithms evaluate an additional trial solution per variable. Since the number of variables is 12, `random`, `far`, and `near` evaluate 112 solutions in each generation and 224,000 evaluations per run.

Comparing GD by the algorithms in Fig. 2(a) and (b), it can be seen that `random`, `far`, and `near` approach `ideal` and obtain significantly smaller GD than `org` in both separable problems, the unimodal DTLZ2 and in the multimodal DTLZ3. In DTLZ2, no significant difference is observed among `random`, `far`, and `near`. In the multimodal DTLZ3, during the latest stages of the search, `far` seems to perform better than `random` and `near`, in that order. Note that there is one order of magnitude difference between `far` and `org` in DTLZ2 and two orders of magnitude difference in DTLZ3.

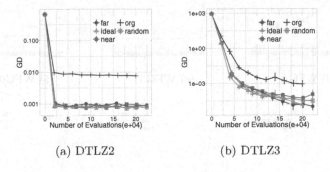

(a) DTLZ2   (b) DTLZ3

Fig. 2. Transition of the median value of GD over the number of fitness evaluations in DTLZ2 and DTLZ3 problems

Next, we apply each algorithm to WFG1-WFG9 problems. Figure 3 shows the transition of the median value of the distance variable x_M in the set of non-dominated solutions over the number of evaluations. Comparing the algorithms, it can be seen that `random`, `far`, and `near` approach `ideal` and achieve smaller x_M values than `org` in WFG1, WFG2, WFG4, WFG5, WFG6, and WFG8, improving significantly convergence. In WFG3, WFG7, and WFG9 x_M value is similar for all algorithms.

Table 2. Rate of absolute classification of all variables.

	DTLZ2	DTLZ3	WFG1	WFG2	WFG3	WFG4	WFG5	WFG6	WFG7	WFG8	WFG9
far	99.6	76.6	38.0	99.9	34.6	99.7	99.9	100	4.06	98.1	0.805
near	71.6	43.1	14.9	92.3	34.6	98.3	69.5	99.2	6.41	91.3	0.485
random	86.9	53.6	13.3	86.7	30.8	89.6	83.5	89.6	2.78	86.7	0.240

Table 2 shows the average rate of absolute classification of variables. A classification is counted as absolute when all position and distance variables are

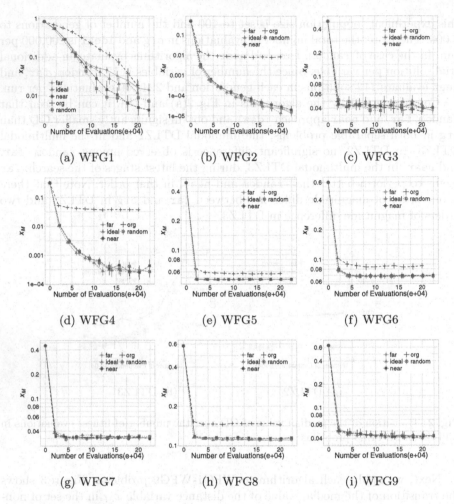

Fig. 3. Transition of the median value of the distance variable x_M over the number of fitness evaluations in WFG problems

correctly classified. From Table 2, it can be seen that the rates of absolute classification are high in problems DTLZ2, WFG2, WFG4-6, and WFG8, where convergence improves in the algorithms with the proposed method as shown in Figs. 2 and 3. Absolute classification rate is low in problems WFG3, WFG7, and WFG9, were convergence did not improve. Note that absolute classification is low in DTLZ3 and WFG1, although the proposed method significantly improves convergence.

To analyze with more detail the classification of variables and its impact on performance, Fig. 4 shows the classification rate per variables over the generations for some problems. For each variable, we compute the percentage it was assigned label 1 in every 100 generations.

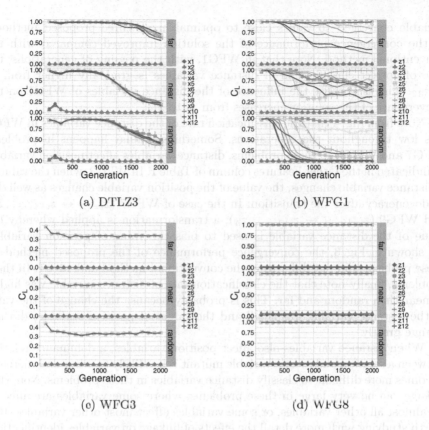

(a) DTLZ3 (b) WFG1

(c) WFG3 (d) WFG8

Fig. 4. Classification rate per variables over the generations

In all DTLZ and WFG problems, the classification rates of position variables x_1, x_2 and z_1, z_2 are 0 in nearly all generations. In WFG8, the classification rate of distance variables $x_3, \ldots, x_{12}$ and $z_3, \ldots, z_{12}$ are 1 in almost all generations. A similar situation occurs in problems DTLZ2, WFG2, and WFG4-6 (not shown here). As shown in Figs. 2 and 3, the convergence performance improved in the proposed methods compared with the conventional NSGA-II in problems where classification rates of the distance variable are high. In these problems, it is considered that the convergence performance has improved since the distance variables are estimated correctly and they are searched intensively.

Looking at DTLZ3 and WFG1 in Fig. 4, distance variables were correctly classified in the early generations, but their classification rate decreases in later generations. This is because, in these problems, the value of distance variable tends to converge to the same value, i.e. solutions are trapped in local optima. When this happens, the trial value of the variable in the mutant is the same as the value of the variable in the original solution and therefore a dominance relation does to occur between them. From Figs. 2 and 3, DTLZ3 and WFG1, in which distance variables were estimated correctly in the early generations, distance variable converged to values close to optimal in all three proposed methods, so the convergence performance of the solution improved compared with the conventional method. Note that in WFG1, with the passing of generations, the rate of correct classification of distance variables is gradually higher from z_3 to z_{12}. This is because the influence of the distance variables of WFG1 on the convergence of the solution increases from z_3 to z_{12}.

Note from Fig. 4 that the classification rate of the distance variables in WFG3 was low throughout the generations. Something similar happens in problems WFG7 and WFG9. In these problems, distance variables affect position variables as indicated in the Other Features column of Table 1. In WFG3, when the value of a distance variable changes, the value of the position variable changes as well due to degeneracy after the transition. In the case of WFG7 ($z_{i=1:k} \leftarrow z_{i+1}, \ldots, z_n$) and WFG9 ($z_{i=1:n-1} \leftarrow z_{i+1}, \ldots, z_n$), a transformation is applied whereby the value of the distance variable is used to bias the value of position variables. As shown in Fig. 3, the convergence performance of the proposed method on these problems did not differ from the conventional method and ideal. In these problems, finally note that the classification rate of distance variable was higher in near than random and far. This is probably because the change of the value of the distance variable was small and the values of position variables did not change greatly.

When distance variables also affect position variables, a dominance relation between a solution and its one-variable mutant is less likely to occur and therefore becomes more difficult to classify distance variables in these problems. Note that linkage can be very large in these problems, where some variables are affected by almost all other variables, or some variables affect most other variables. It is worth studying with more detail the effects of linkage on variables identification.

On the other hand, note that variables are correctly classified when position variables affect distance variables ($z_{i=k+1:n} \leftarrow z_1, \ldots, z_{i-1}$), and convergence can be improved as shown by the results on WFG8.

Modality and non-separability of a function seem not to affect the correct classification of variables. Note that there is a high classification rate for uni-modal and multi-modal problems, separable and non-separable, when distance variables do not affect the position of solutions. Also, linkage between distance variables and linkage from position to distance variables are not an issue for correct classification.

In this work, the proposed method favors recombination of variables that can improve convergence. However, in addition to convergence, the aim of a multi-objective optimizer is to also achieve a set of well-distributed solutions. To verify whether the proposed method has a negative impact on diversity we also compute the hypervolume of the solutions found by the algorithms, which measures both convergence and diversity. Figure 5 shows the hypervolume for WFG3, WFG5, and WFG8. In general, where there is a clear improvement in convergence we also observe an improvement in hypervolume, as illustrated in Fig. 3(h) and Fig. 5(c) for WFG8. In cases where there is no difference in convergence or is very small, hypervolume is also similar, as shown in Fig. 3(c)

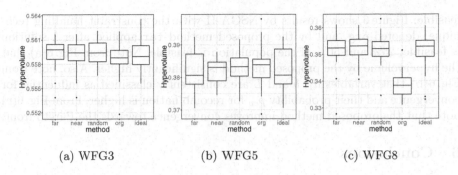

(a) WFG3   (b) WFG5   (c) WFG8

Fig. 5. Hypervolume

and Fig. 5(a) for WFG3 and Fig. 3(e) and Fig. 5(b) for WFG5. These results suggest that there is not a serious detriment to diversity of solutions. However, emphasizing variation of variables that improve diversity would enhance further the performance of the multi-objective algorithm. In the future, we would like to extend the method to focus on diversity as well.

Summarizing, the convergence performance of the proposed method on benchmark problems with **random**, **far**, or **near** procedures to set the trial variables improves compared to **org**, because distance variables are correctly estimated and the frequency of recombining them is increased.

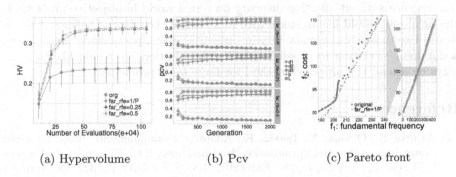

(a) Hypervolume   (b) Pcv   (c) Pareto front

Fig. 6. Results on the vibrating beam problem

5.2 Vibrating Beam Problem

This bi-objective problem consists of 5 variables and 3 constraints. A random initialization of the population in this problem leads to all solutions being unfeasible. Approximately, 3 in every 10.000 randomly created solutions is feasible. However, a simple constraint handling method allows the algorithm fo find feasible solutions after 5 or 6 generations (500–600 function evaluations) and accumulate them thereafter. Around generation 20 all solutions in the population are feasible. Figure 6 shows results by NSGA-II with the constraint handling technique, denoted `org`, and by the proposed method `far` applied after 1 solution is feasible, 25% and 50% of the population is feasible. Note from Fig. 6(a) that the hypervolume by the proposed method is significantly higher. Also, note from Fig. 6(b) that variables d_1, b and L are consistently classified as influential for convergence and their probability p_{cv} for recombination is higher. From Fig. 6(c) note that the proposed method improves convergence towards the Pareto front.

6   Conclusions

In this work we have proposed a method to classify variables that influence convergence and increased their selection probabilities to recombine them more often. The classification procedure is based on whether there is a Pareto dominance relation between one-variable mutants. The proposed method was tested on DTLZ and WFG functions, including unimodal, multimodal, separable, non-separable, unbiased and biased functions. Our experimental results show that the proposed method can improve significantly the performance of the well known NSGA-II algorithm in most instances for 3 objective functions. Modality and non-separability of a function seem not to affect the correct classification of variables. Also, linkage between distance variables and linkage from position to distance variables are not an issue for correct classification. However, performance could not be improved in problems where linkage from distance to position variables is very large. We also verified that the proposed method can improve convergence without affecting diversity on a real world bi-objective problem. In the future we would like to study with more detail the effect of linkage. Also, we would like to study the scalability in objective space, particularly since we use a dominance relation to classify variables. In addition, we would like to study ways to extend the proposed method for large scale problems.

References

1. Aguirre, H., Oyama, A., Tanaka, K.: Adaptive ε-sampling and ε-hood for evolutionary many-objective optimization. In: Purshouse, R.C., Fleming, P.J., Fonseca, C.M., Greco, S., Shaw, J. (eds.) EMO 2013. LNCS, vol. 7811, pp. 322–336. Springer, Heidelberg (2013). https://doi.org/10.1007/978-3-642-37140-0_26
2. Coello, C.C., Lamont, G., van Veldhuizen, D.: Evolutionary Algorithms for Solving Multi-Objective Problems. Genetic and Evolutionary Computation, 2nd edn. Springer, Heidelberg (2002). https://doi.org/10.1007/978-1-4757-5184-0
3. Deb, K., Thiele, L., Laumanns, M., Zitzler, E.: Scalable multi-objective optimization test problems. In: Congress on Evolutionary Computation, pp. 825–830. IEEE Service Center (2002)
4. Deb, K.: Multi-Objective Optimization using Evolutionary Algorithms. Wiley, Hoboken (2001)
5. Deb, K., Agrawal, S., Pratap, A., Meyarivan, T.: A fast elitist non-dominated sorting genetic algorithm for multi-objective optimization: NSGA-II. In: Schoenauer, M., et al. (eds.) PPSN 2000. LNCS, vol. 1917, pp. 849–858. Springer, Heidelberg (2000). https://doi.org/10.1007/3-540-45356-3_83

6. Huband, S., Hingston, P., Barone, L., While, R.: A review of multi-objective test problems and a scalable test problem toolkit. IEEE Trans. Evol. Comput. **10**(5), 477–506 (2007)
7. Narayanan, S., Azarm, S.: On improving multiobjective genetic algorithms for design optimization. Struct. optim. **18**(2), 146–155 (1999)
8. Sagawa, M., et al.: Learning variable importance to guide recombination. In: IEEE SSCI, pp. 1–7 (2016)
9. Zhang, Q., Li, H.: MOEA/D: a multiobjective evolutionary algorithm based on decomposition. IEEE Trans. Evol. Comput. **11**(6), 712–731 (2007)
10. Zitzler, E., Laumanns, M., Thiele, L.: SPEA2: improving the strength pareto evolutionary algorithm for multiobjective optimization. In: Evolutionary Methods for Design Optimization and Control with Applications to Industrial Problems, pp. 95–100. International Center for Numerical Methods in Engineering (2001)
11. Zitzler, E., Künzli, S.: Indicator-based selection in multiobjective search. In: Yao, X., et al. (eds.) PPSN 2004. LNCS, vol. 3242, pp. 832–842. Springer, Heidelberg (2004). https://doi.org/10.1007/978-3-540-30217-9_84

sParEGO – A Hybrid Optimization Algorithm for Expensive Uncertain Multi-objective Optimization Problems

João A. Duro[1(✉)], Robin C. Purshouse[1(✉)], Shaul Salomon[1,2],
Daniel C. Oara[1], Visakan Kadirkamanathan[1], and Peter J. Fleming[1]

[1] University of Sheffield, Sheffield, UK
{j.a.duro,r.purshouse}@sheffield.ac.uk
[2] ORT Braude College of Engineering, Karmiel, Israel
shaulsal@braude.ac.il

Abstract. Evaluations of candidate solutions to real-world problems are often expensive to compute, are characterised by uncertainties arising from multiple sources, and involve simultaneous consideration of multiple conflicting objectives. Here, the task of an optimizer is to find a set of solutions that offer alternative robust trade-offs between objectives, where robustness comprises some user-defined measure of the ability of a solution to retain high performance in the presence of uncertainties. Typically, understanding the robustness of a solution requires multiple evaluations of performance under different uncertain conditions – but such an approach is infeasible for expensive problems with a limited evaluation budget. To overcome this issue, a new hybrid optimization algorithm for expensive uncertain multi-objective optimization problems is proposed. The algorithm – sParEGO – uses a novel uncertainty quantification approach to assess the robustness of a candidate design without having to rely on expensive sampling techniques. Hypotheses on the relative performance of the algorithm compared to an existing method for deterministic problems are tested using two benchmark problems, and provide preliminary indication that sParEGO is an effective technique for identifying robust trade-off surfaces.

Keywords: Expensive optimization · Surrogate-based optimization ·
Robust optimization · Multi-objective optimization

1   Introduction

The ability of simulations to predict the performance of a candidate design is constantly increasing. While some simulations can produce high-fidelity outputs relatively quickly, a typical mesh-based simulation can run for several hours, and even days. Even if a design team has access to supercomputing resources, the extensive run-time still implies that perhaps only a few hundred candidate designs can be explored using high-fidelity modelling resources. Unfortunately,

© Springer Nature Switzerland AG 2019
K. Deb et al. (Eds.): EMO 2019, LNCS 11411, pp. 424–438, 2019.
https://doi.org/10.1007/978-3-030-12598-1_34

conventional multi-objective optimization algorithms implemented in commercial packages typically require tens of thousands of function evaluations to converge on a high quality solution [1]. Therefore, the search for a promising design using expensive evaluation functions on a limited computational budget poses a great challenge.

To exacerbate this problem, optimizing for a robust solution is itself a computationally demanding task. In order to gain confidence over the robustness of a solution to uncertainties, the statistical properties of the expected solution's performance must be quantified. In a world where the complexity of the high-fidelity models essentially produces a black-box mapping of inputs to outputs, such statistical properties would typically be found through repeated evaluation of the same solution using those high-fidelity models. However, repeatedly sampling a single candidate design is computationally expensive.

To address the above, we propose a framework for expensive uncertain multi-objective optimization problems (MOPs). The key aims are to: (i) exploit expensive, black-box evaluation function for a candidate design; (ii) account for multiple sources of uncertainty, such as fidelity of evaluation functions and manufacturing tolerances; and (iii) provide an understanding of the risk and opportunity trade-offs between candidate designs with respect to a given robustness metric. To achieve this the framework leverages ParEGO [2], an algorithm for multi-objective optimization, which has been demonstrated to provide good results for optimization runs limited to a small number of function evaluations. ParEGO itself is a multi-objective extension to Jones et al.'s [3] seminal efficient global optimization (EGO) algorithm for single-objective problems. The main limitation of ParEGO is that it has not been designed to handle problems featuring uncertainty (although there is some evidence that it can perform favourably in noisy environments [4]). Therefore a fundamental part of the framework is how ParEGO can be extended to consider evaluation functions as samples of random variates. We refer to this new algorithm as *stochastic ParEGO* or *sParEGO*.

In the remainder of this paper, first the robustness metric used is described in Sect. 2, and the proposed framework is presented in Sect. 3. The hypotheses on the relative performance of the algorithm are introduced in Sect. 4. The experimental settings and findings are in Sect. 5. The paper concludes with Sect. 6.

2   Threshold-Based Robustness Metric

A general single-objective robust optimisation problem can be formulated as:

$$\min_{\mathbf{x} \in \Omega} S = \mathbf{f}(\mathbf{x}, \mathbf{U}). \tag{1}$$

Here, $\mathbf{x} = [x_1, \ldots, x_{n_x}]$ is a vector of n_x decision variables in a feasible domain Ω, $\mathbf{U}$ is a vector of random variables that includes all the uncertainties associated with the optimisation problem. These uncertainties may be an outcome of manufacturing tolerances, a noisy environment, evaluation inaccuracies etc. A single scenario of the variate $\mathbf{U}$ is denoted as $\mathbf{u}$. Since uncertainties are involved, the scalar objective S is also a random variate, where every scenario of the uncertainties, $\mathbf{u}$, is associated with an objective value s.

In a robust optimisation scheme, the random objective value is replaced with a robustness criterion, denoted by the indicator $I[S]$. Several criteria are commonly used in the literature, which can be broadly categorised into three main approaches:

1. **Worst-Case Scenario.** The worst objective vector, considering a bounded domain in the neighbourhood of the nominal values of the uncertain variables.
2. **Aggregated Value.** An integral measure of robustness that amalgamates the possible values of the uncertain variables (e.g. mean value or variance).
3. **Threshold Probability.** The probability for the objective function to be better than a defined threshold.

In our framework the third approach, suggested by Beyer and Sendhof [5], is used. A threshold q is considered as a satisficing performance for the objective value s. When s is uncertain, denoted by the random variable S, the probability for S to satisfy the threshold level can be seen as a confidence level c. For a minimization problem this can be written as:

$$c(S, q) = \Pr(S < q). \tag{2}$$

A robustness indicator used in this paper is based on minimization of the threshold q for a pre-defined confidence level c, meaning that the confidence in the resulting performance can be specified (e.g. by a decision-maker).

A stochastic unconstrained multi-objective optimization problem (MOP), which is the focus of this study, can be formulated as:

$$\min_{\mathbf{x} \in \Omega} \mathbf{Z} = \mathbf{f}(\mathbf{x}, \mathbf{U}). \tag{3}$$

where $\mathbf{Z}$ is a multivariate random vector of n_z performance criteria, and $\mathbf{f}$ is a set of functions mapping from decision-space to objective-space. Due to uncertainties over the problem parameters or the mapping functions themselves, every evaluation of the same decision vector may result in a different realisation of the objective vector $\mathbf{z} = [z_1, \ldots, z_{n_z}]$.

3   The Framework of the sParEGO Algorithm

sParEGO is a surrogate-based multi-objective optimization algorithm for dealing with stochastic MOPs. The algorithm shares many similarities with ParEGO including the ability to approximate expensive MOPs over a realistically small number of function evaluations. The main idea is that the uncertain distribution in objective space of every candidate solution is not quantified through uncertainty quantification methods (e.g. Monte Carlo sampling). Instead, every solution is evaluated once, and the distribution is approximated based on the performance of nearby solutions. A pseudo-code of sParEGO is presented in Algorithm 1 and a general description of its working principles is as follows.

The decision variables and objectives are normalised to non-dimensional units in the following manner:

$$\tilde{x}_i = (x_i - x_i^l)/(x_i^u - x_i^l), \; i = 1,\ldots,n_x, \tag{4}$$

$$\tilde{z}_j = (z_j - z_j^*)/(z_j^n - z_j^*), \; j = 1,\ldots,n_z, \tag{5}$$

where x_i^u and x_i^l are the upper and lower boundaries of the ith decision variable, z_j^n and z_j^* are the jth components of the estimated nadir and ideal vectors, and the tilde accent represents a normalised, dimensionaless variable. The normalised values are used for all operations within the algorithm. Before a candidate design is evaluated, it is re-scaled to the natural dimensions.

sParEGO decomposes the overall MOP into a number of single-objective problems by using a set of (reference) direction vectors to guide the search towards different regions of the Pareto front[1]. The set of all direction vectors is denoted by $\mathcal{D}$ (Line 1). The direction vectors are picked (one at the time) based on their sequence in the set $\mathcal{D}$. Once all direction vectors have been traversed by the optimizer the vectors in the set $\mathcal{D}$ are shuffled (Line 5). This prevents any bias that might arise due to repeatedly using the same sequence of direction vectors during the entire optimization process.

The procedure used to generate the initial set of solutions ($\mathcal{X}$) is described in Sect. 3.2 (Line 2). Following this, all solutions in the set $\mathcal{X}$ are evaluated and their performance is stored in the set $\mathcal{Z}$ (Line 3). The ideal and nadir vectors are then updated (Line 7). A scalar fitness value is obtained for each solution by using a scalarising function as mentioned in Sect. 3.1 (Line 8). The robustness indicator values of the solutions are estimated and stored in the set $\mathcal{I}$ (Line 9), and these are used to construct a surrogate model (Line 10). A search procedure is then conducted over the model to find a solution $\mathbf{x}^{\text{new}}$ that optimizes the given robustness indicator based on the concept of expected improvement (Line 11). A new solution $\mathbf{x}^{\text{pert}}$ is generated by applying a perturbation to $\mathbf{x}^{\text{new}}$ (Line 12). This ensures that all generated solutions have at least one nearby solution. The new solutions are added to $\mathcal{X}$ (Line 13) and, once evaluated, their performance is stored in $\mathcal{Z}$ (Line 14). The algorithm goes back to Line 4 and the procedure repeats itself until a stopping criteria is satisfied.

The robustness indicator values of the solutions are estimated based on the procedure in Line 17. The first step is to identify, for each solution, all the nearby solutions. For this, we define the concept of neighbourhood and consider that two solutions are neighbours if their distance in normalised decision-space is within a user-defined neighbourhood distance δ (Line 19). The statistical properties of the performance of a solution is approximated from the other neighbouring solutions (Line 23). Finally, the robustness indicator values of the solutions are estimated for a given robustness criterion I (Line 24).

---

[1] More details about the decomposition strategy are provided in Sect. 3.1.

Algorithm 1. sParEGO Pseudo-code

Parameters: initial set size n_{init}, surrogate model maximum set size n_{max}, maximum distance between newly generated solutions δ_{pert}, robustness criterion I, neighbourhood distance δ

1: $\mathcal{D} \leftarrow$ set of all reference direction vectors   ▷ Eq. 6 (Sect. 3.1)
2: $\mathcal{X} \leftarrow$ generate initial set of solutions using n_{init} and δ_{pert} ▷ Sect. 3.2
3: $\mathcal{Z} \leftarrow \mathbf{f}(\mathcal{X})$ ▷ evaluate the initial set
4: **while** stopping criteria not satisfied **do**
5:      Shuffle the set $\mathcal{D}$
6:      **for all** $\mathbf{d} \in \mathcal{D}$ **do**
7:          update ideal and nadir vectors
8:          $\mathcal{S} \leftarrow$ calculate scalar fitness value of all solutions   ▷ Eq. 7 (Sect. 3.1)
9:          $\mathcal{I} \leftarrow$ RobustnessApproximation$(\mathcal{X}, \mathcal{S}, \delta)$ ▷ Sects. 3.3 and 3.4
10:          $model \leftarrow$ fit a Surrogate model to the indicator values $\mathcal{I}$ using n_{max} ▷ Sect. 3.5
11:          $\mathbf{x}^{\text{new}} \leftarrow$ maximize the expected improvement based on $model$
12:          $\mathbf{x}^{\text{pert}} \leftarrow$ add a neighbour to $\mathbf{x}^{\text{new}}$ using δ_{pert} ▷ Sect. 3.5
13:          $\mathcal{X} \leftarrow \mathcal{X} \cup \{\mathbf{x}^{\text{new}}, \mathbf{x}^{\text{pert}}\}$
14:          $\mathcal{Z} \leftarrow \mathcal{Z} \cup \{\mathbf{f}(\mathbf{x}^{\text{new}}), \mathbf{f}(\mathbf{x}^{\text{pert}})\}$ ▷ evaluate the new solutions
15:      **end for**
16: **end while**

17: **procedure** RobustnessApproximation$(\mathcal{X}, \mathcal{S}, \delta)$
18:      **for all** $\mathbf{x}_i \in \mathcal{X}$ **do**
19:          update the neighbourhood $\mathcal{N}(\mathbf{x}_i)$ for a given δ ▷ Eq. 10 (Sect. 3.3)
20:      **end for**
21:      $\mathcal{I} \leftarrow \emptyset$
22:      **for all** $\mathbf{x}_i \in \mathcal{X}$ **do**
23:          approximate the distribution of S_i ▷ Sect. 3.3
24:          calculate robustness indicator $I[S_i]$ ▷ Sect. 3.4
25:          $\mathcal{I} \leftarrow \mathcal{I} \cup I[S_i]$
26:      **end for**
27:      **return** $\mathcal{I}$
28: **end procedure**

3.1 Decomposition

A decomposition-based algorithm decomposes the MOP into a number of single-objective problems, each approaching the global trade-off surface from a different direction. The ith sub-problem is associated with a reference direction vector $\mathbf{d}_i$ which is taken from the set $\mathcal{D}$. The set is constructed by using a Simplex Lattice design:

$$\mathcal{D} = \left\{ \mathbf{d} = [d_1, \ldots, d_{n_z}] \mid \sum_{j=1}^{n_z} d_j = 1 \wedge d_j \in \left\{ \frac{0}{h}, \frac{1}{h}, \ldots, \frac{h}{h} \right\} \text{ for all } j \right\}, \quad (6)$$

where h is a parameter that defines the number of divisions for each objective.

Each sub-problem assigns a scalar fitness value to each solution. This is achieved by using a scalarising function $f(\mathbf{z}, \mathbf{w})$ that maps an objective vector $\mathbf{z}$ into a scalar value according to a vector of weights $\mathbf{w} = [w_1, \ldots, w_{n_z}]$. The scalarising function used is the weighted Tchebycheff, which is given by:

$$s = \max_{1 \leq i \leq n_z} \{w_i z_i\}. \quad (7)$$

For a given direction vector $\mathbf{d}$ there is a corresponding weighting vector that minimizes the scalarising function [6]. The optimal weighting vector $\mathbf{w}$ for the scalarising function in (7) is defined as:

$$w_i = t_i \Big/ \sum_{i=1}^{n_z} t_i, \quad \text{where } t_i = (d_i + \epsilon)^{-1}, \quad i = 1, \ldots, n_z, \tag{8}$$

where ϵ is a small number to prevent division by zero, and the normalisation enforces the weighting vector's elements to sum up to one.

3.2 Initialisation

In sParEGO, the robustness assessment of a candidate design relies on the determination of its statistical properties, which in turn depends on the information of the neighbouring solutions. Hence, in order to support the robustness assessment from the beginning of the optimization process, for any solution in the initial set there is at least one nearby solution in desision-space. Let n_{init} denote the size of the initial set, then the procedure is as follows:

1. To provide a good coverage, a space-filling design technique (Latin Hypercube sampling) is used to generate a fraction of the total n_{init}. We suggest this fraction to be a quarter.
2. For every existing solution in $\mathcal{X}$, another solution is generated by applying a random perturbation where the upper bound is within a hypersphere with a radius of δ_{pert} which is smaller than δ.
3. The rest of the solutions are generated by randomly selecting an existing solution from $\mathcal{X}$ and applying a perturbation as in the previous step. This step is repeated until the number of solutions in $\mathcal{X}$ is equal to n_{init}.

The second step enforces that every solution has at least one neighbour. The third step seeds the initial population with neighbourhoods of different sizes.

3.3 Uncertainty Quantification

The most important difference between sParEGO and ParEGO is that the former assumes that the outcome of an evaluation function is a realization of a random variate. Therefore, the scalarised function value cannot be used directly to construct the surrogate model, and a utility indicator value is used instead. For every direction vector, the surrogate model is constructed to search for a design that will optimize a given robustness indicator (described in Sect. 3.4). The guiding principle is to avoid having to repeatedly sample every candidate design to assess its statistical properties in objective-space. Instead, these properties (specifically, measures of central tendency and dispersion) are approximated from the available information of other candidate design evaluations.

Approximation of the Central Tendency. The stochasticity of the problem might originate from a variety of sources, including variations in decision-space.

For this type of uncertainty, two designs with similar nominal values can be identical when realised. Therefore, the performance of a candidate design should be calculated from the performance of neighbouring designs as well. Two solutions $\mathbf{x}_i$ and $\mathbf{x}_j$ are considered as neighbours if their Euclidean distance in normalised decision-space is smaller than or equal to δ, that is:

$$\|\mathbf{x}_i - \mathbf{x}_j\|_2 \leq \delta. \tag{9}$$

For a solution $\mathbf{x}_i$ with a scalar fitness given by s_i, the statistical properties of the scalar fitness are approximated from the neighbouring solutions as follows: First, the neighbourhood $\mathcal{N}(\mathbf{x}_i)$ of the solution is defined[2]:

$$\mathcal{N}(\mathbf{x}_i) = \left\{ \mathbf{x}_j \in \mathcal{X} \mid \|\mathbf{x}_i - \mathbf{x}_j\|_2 \leq \delta \right\}. \tag{10}$$

Next, the approximated mean function value, μ_s, is derived from the neighbourhood. Members that are closer to $\mathbf{x}_i$ are given a larger weight, denoted as v in Eq. (11), in approximating its properties. Since the weight of most solutions in the neighbourhood is smaller than 1, the overall "neighbourhood size" ς_i is smaller than $|\mathcal{N}(\mathbf{x}_i)|$:

$$v_j = \frac{\delta - \|\mathbf{x}_i - \mathbf{x}_j\|_2}{\delta}, \ \forall \mathbf{x}_j \in \mathcal{N}(\mathbf{x}_i), \tag{11}$$

$$\varsigma_i = \sum_{\mathbf{x}_j \in \mathcal{N}(\mathbf{x}_i)} v_j, \tag{12}$$

$$\mu_{s,i} = \frac{1}{\varsigma_i} \sum_{\mathbf{x}_j \in \mathcal{N}(\mathbf{x}_i)} v_j s_j, \tag{13}$$

where $\mu_{s,i}$ is the approximated mean of the scalar fitness function for $\mathbf{x}_i$.

Approximation of the Dispersion. Once the expected mean is known, the expected value for the variance is calculated:

$$\sigma_{s,i}^2 = \frac{1}{\varsigma_i} \sum_{\mathbf{x}_j \in \mathcal{N}(\mathbf{x}_i)} v_j (s_j - \mu_{s,i})^2. \tag{14}$$

An example is shown in Fig. 1(a) for an optimization problem with a single decision variable where 5 solutions are divided into two neighbourhoods. The mean and variance of the scalar fitness function is estimated based on their scalar fitness values and their decision-space distance within the neighbourhood.

3.4   Estimating the Robustness Indicator Value

Once the statistical properties of the scalar fitness function have been estimated, the random variable $S(\mathbf{x})$ is assumed to follow a normal distribution with the

---

[2] Note that according to (10), $\mathbf{x}_i$ is included in the neighbourhood $\mathcal{N}(\mathbf{x}_i)$.

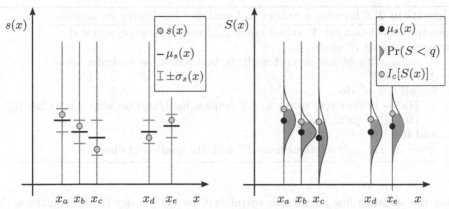

(a) Estimation of the mean value and the variance from the neighbourhood

(b) Assuming a normal distribution for S according to μ_s and σ_s. The shaded area represents a confidence of 80%.

Fig. 1. Approximation of the statistical properties, and estimation of robustness indicator $I_c[\bullet]$.

estimated mean and variance. The robustness indicator is calculated for this distribution with respect to a desired confidence level c (assuming $c \in [0, 100]$). The indicator, denoted $I_c[S]$, is then equal to the cth percentile of the normal distribution with mean μ_s (Eq. 13) and variance σ_s^2 (Eq. 14).

An example for $I_c[S]$ is given in Fig. 1(b) where $I_c[S]$ value corresponds to the 80th percentile of the normal distribution. Following this, the indicator value $I_c[S]$ is considered as the solution's fitness at the current iteration.

3.5   Fitting a Surrogate Model to the Fitness

Now that every solution is associated with a scalar fitness value based on the robust indicator, the algorithm proceeds in a similar fashion to EGO and ParEGO [3,7]. A surrogate model is fitted to the fitness values, and the expected improvement function is constructed from the model. Above a certain size (approximately 50 solutions), the surrogate model becomes prohibitively expensive to construct. When the number of evaluated solutions exceed this size, a subset of size n_{max} is chosen according to Algorithm 2.

The first step in Algorithm 2 is to select $n_{\text{max}}/2$ solutions from the population set $\mathcal{X}$ with the best robustness indicator value, and to add these to the set $\mathcal{X}'$ (Line 1). The next step is to select from the remaining solutions those that are closer to the current direction vector $\mathbf{d}$. For this, the normalised objective vectors are projected to the $n_z - 1$ simplex (Line 4). The Euclidean norm between the vectors $\hat{\mathbf{z}}(\mathbf{x})$ and $\mathbf{d}$ is given by the operator Δ (Line 5). Finally, the $n_{\text{max}}/2$ solutions from $\mathcal{X}''$ with the smallest Δ distance are added to $\mathcal{X}'$ (Line 7).

To use the expected improvement function we need to estimate its variance ($\hat{\sigma}^2$). For this, we use the density of the solutions in decision-space by knowing

---

Algorithm 2. Choosing a Subset to Construct the Surrogate Model

---

Require: population set $\mathcal{X}$, subset size $n_{\max}$, current direction vector $\mathbf{d}$
Ensure: a subset $\mathcal{X}'$ of size $n_{\max}$
1: $\mathcal{X}' \leftarrow n_{\max}/2$ solutions from $\mathcal{X}$ with the best robustness indicator value
2: $\mathcal{X}'' \leftarrow \mathcal{X} \setminus \mathcal{X}'$
3: **for all** $\mathbf{x} \in \mathcal{X}''$ **do**
4: $\quad \hat{\mathbf{z}}(\mathbf{x}) \leftarrow$ project $\mathbf{z}(\mathbf{x})$ to the $n_z - 1$ simplex, implying that $\hat{\mathbf{z}}(\mathbf{x}) = \mathbf{z}(\mathbf{x}) \|\mathbf{z}(\mathbf{x})\|_1^{-1}$
5: $\quad \Delta(\mathbf{x}, \mathbf{d}) \leftarrow \|\hat{\mathbf{z}}(\mathbf{x}) - \mathbf{d}\|_2$
6: **end for**
7: $\mathcal{X}' \leftarrow \mathcal{X}' \cup n_{\max}/2$ solutions from $\mathcal{X}''$ with the smallest Δ distance

---

that the variance has an inverse correlation to the density of the solutions. A suitable way to estimate the density at a given point $\mathbf{x}$ is to use a non-parametric statistical approach, and in this case we use a kernel density model given by:

$$p(\mathbf{x}) = \frac{1}{n_{\max}} \sum_{\mathbf{x}_i \in \mathcal{X}'} \frac{1}{(2\pi h_b^2)^{n_x/2}} \exp\left\{ -\frac{\|\mathbf{x} - \mathbf{x}_i\|^2}{2h_b^2} \right\}, \tag{15}$$

where h_b is the bandwidth. We suggest setting the bandwidth to be equal to one hundredth of the mean span of all solutions, that is:

$$h_b = \frac{1}{100 \times n_{\max}} \sum_{\mathbf{x}_i \in \mathcal{X}'} (\max(\mathbf{x}_i) - \min(\mathbf{x}_i)). \tag{16}$$

Based on experimental results we have observed that the kernel density model can be very sensitive to any changes in the density, thus we have used a smoothing function (in this case the arctan function), and the estimated variance at $\mathbf{x}$ is:

$$\hat{\sigma}^2(\mathbf{x}) = \left(\frac{1}{\pi/2} \arctan\left(\frac{1}{p(\mathbf{x})} \right) \right)^2. \tag{17}$$

After fitting a surrogate model to the solutions from $\mathcal{X}'$, the next task is to find the solution that maximizes the expected improvement function. For this, any suitable off-the-shelf single-objective optimizer can be used, and we have chosen ACROMUSE [8]. The identified solution, denoted by $\mathbf{x}^{\mathrm{new}}$, is added to the population together with a neighbouring solution $\mathbf{x}^{\mathrm{pert}}$, generated using the same perturbation as that described in Sect. 3.2.

4   Hypothesis Testing

We employ a hypothesis testing approach to study the performance of sParEGO compared with ParEGO in dealing with MOPs on a limited computational budget. We postulate two hypotheses, each relating to anticipated pathological behaviour of one of the algorithms:

1. If the problem is deterministic, and the region close to the Pareto front is highly multi-modal, sParEGO will incorrectly interpret the multimodality as stochasticity, and converge on seemingly 'robust' solutions that are actually non-optimal. However ParEGO's convergence will be unaffected.
2. If the problem is highly stochastic, and the region close to the Pareto front is smooth, ParEGO will identify seemingly high-performance solutions that are actually non-robust. However sParEGO's convergence will be unaffected.

To test these hypotheses, we use two variants of the WFG4 problem [9]. Both variants have two objectives and five decision variables. The first two decision variables are position parameters and the last three are distance parameters. For the first problem, namely P1, we have modified WFG4 to increase the density and the number of local optima in the periphery of the global optimum. This simulates the effect that stochasticity can have when approaching the Pareto-optimal Front (PF). The second problem, namely P2, is characterised by having a more smooth landscape with no local minima surrounding the global optimum, and stochasticity is added by the toolkit from [10].

The modification applied to WFG4 is as follows. The original formulation of WFG4 applies a transformation to each input parameter (y) given by:

$$s\_multi(y, a, b, c) = \left(1 + \cos(r_2) + b(r_1)^2\right) / (b + 2),$$
$$r_1 = |y - c| / (\lfloor c - y \rfloor + c), \tag{18}$$
$$r_2 = (4a + 2)\pi(0.5 - 0.5r_1),$$

where a controls the number of minima, b controls the magnitude of the "hill sizes" of the multi-modality, and c is the value for which y is mapped to zero. The number of minima increases up-to $2a + 1$ which includes the global optimum at c. We propose a modification to Eq. 18 as follows:

$$s\_multi^*(y, a, b, c, d, e) = \left(1 + \cos(r_2 r_3) + b|r_1|^e\right) / (b + 2),$$
$$r_3 = (1 - |r_1|)^{2d}, \tag{19}$$

where d controls the density of the hills around the optimum, and e specifies the polynomial order of the base curve. The effect of these parameters is shown in Fig. 2, in that: the density of hills around the optimum increases with an increase in d as shown in Fig. 2(a), and; the proximity of local minima from the value zero decreases with an increase in e as shown in Fig. 2(b).

The toolkit from [10] is used here to transform the objective vectors of WFG4 into random vectors. The parameters have been chosen to ensure that uncertainty increases towards more optimal regions. The uncertainty also decreases up to a point when moving away from the Pareto region, and then starts increasing again for regions that are further away from the PF. The perturbation is applied to the objective vector by using only its radial component, implying that the perturbation radius is set to zero. This means that an objective vector $\mathbf{z}$ is perturbed only along one direction, defined by the $n_z - 1$ simplex and given by $\hat{\mathbf{z}} = \mathbf{z}/\sum z_i$, for $i = 1, \ldots, n_z$. In practice, instead of using the deterministic value of the distance term in WFG4, we consider it as a random variate

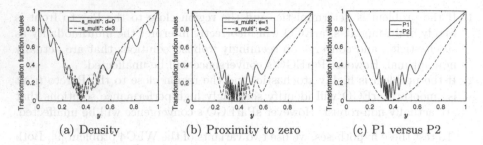

(a) Density   (b) Proximity to zero   (c) P1 versus P2

Fig. 2. Transformation function in WFG4 as a function of the input parameter (y). For (a) and (b) the parameters are $a = 5$, $b = 10$, and $c = 0.35$. The difference between P1 and P2 is shown in (c). Moreover, $e = 1$ for (a) and $d = 3$ for (b).

with a uniform distribution. The lower bound of the distribution is situated at the deterministic value from the test problem, and the upper bound increases as solutions approach the Pareto region. As a result, for a given robustness criterion (say, the worst-case scenario as mentioned in Sect. 2), the worst performance of the Pareto-optimal solutions can be worse than that of some of the non-Pareto-optimal solutions. This gives rise to the term *Robust Pareto-optimal Set* (RPS), which is defined as the set of solutions with the best performance with respect to the given robustness indicator.

Following the above, we have chosen $c = 0.35$ for all test instances. The remaining parameters are: $a = 5$, $b = 10$, $d = 3$, and $e = 1$ for P1; and $a = 0$, $b = 8$, $d = 0$, and $e = 2$ for P2. The transformation function values for these settings are shown in Fig. 2(c). The PF for P1 corresponds to a quadrant with extremes of 2 and 4 for objectives f_1 and f_2, respectively, and it is shown in Fig. 3(a). The PF has been obtained by uniformly generating points along the quadrant. Figure 3(b) shows the performance of the RPS with respect to $I_c[S]$ for difference confidence levels c. The RPS has been obtained by using an enumeration where the uniform distribution over the distance term of WFG4 has been replaced by the cth percentile of the same distribution.

5   Experimental Results

5.1   Experimental Settings

For both ParEGO and sParEGO the number of direction vectors is set to 10. Other common parameters are: $n_{init} = 10$, $n_{max} = 50$ and the optimization budget is set to 5000 function evaluations. For sParEGO: $\delta = 0.1\sqrt{n_x}$, $\delta_{pert} = \delta/2$, and the confidence level c of the robustness indicator $I_c[S]$ is set to 90%. Inverted Generational Distance (IGD) [6] is used to measure the quality of the obtained sets by the optimizers. The 10 solutions that are marked with a filled circle in Fig. 3 are used as the reference sets for IGD, and these solutions correspond to the best optimal solutions for the chosen direction vectors.

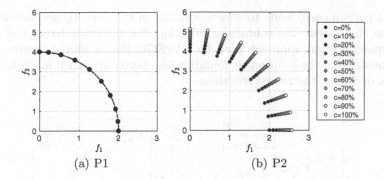

Fig. 3. Pareto-optimal front is shown for P1 in (a), and in (b) it is shown for P2 the performance of the RPS with respect to the robustness indicator $I_c[S]$ for difference confidence levels c.

The optimizers report only one solution per direction vector, implying that only 10 solutions are identified at the end of the optimization process. For each direction vector the solution with the minimum scalarised fitness value is chosen. For this, ParEGO uses the scalarised fitness values determined directly by Eq. 7, while sParEGO uses the fitness attributed by the robustness indicator.

5.2   Findings

This section presents the experimental results for problems P1 and P2. The results shown in Fig. 4 provide both a visual and a analytical assessment of the quality of the solutions obtained by the optimizers in terms of their convergence to and diversity across the PF. The objective vectors for P2 have been determined by evaluating 100 times each decision vector, and the performance of each objective is equal to the 90th percentile of its marginal distribution.

For P1, ParEGO's approximation to the PF is slightly better than for sParEGO as shown in Fig. 4(a) and (c). This indicates that the multi-modality in P1, close to the vinicity of the PF, is interpreted by sParEGO as a region of high uncertainty. The performance of the solutions with respect to the robustness indicator in this region is captured as being poor according to the statistical inferences made by sParEGO. Hence, most sParEGO solutions are just outside the region where the magnitude of the hill sizes of the multi-modality become relatively large. Nevertheless, it is expected for sParEGO to improve its convergence to the PF with more function evaluations, since the statistical assessment made about the true performance of the solutions that are on the PF is also expected to improve.

For P2, sParEGO approximation to the PF obtained with respect to the robustness indicator is better than ParEGO as shown in Fig. 4(b) and (d). The convergence of ParEGO deteriorates along the optimization run as shown in Fig. 4(d), while the convergence of sParEGO improves. This indicates that the selection criterion used by ParEGO that promotes solutions with a better nominal performance, leads to a deterioration in the convergence towards the solutions that satisfy the robustness criterion. On the other hand, the uncertainty quantification approach used by sParEGO that is used to estimate the true robustness of the solutions is found to be a better approach in dealing with the task of finding the robust solutions.

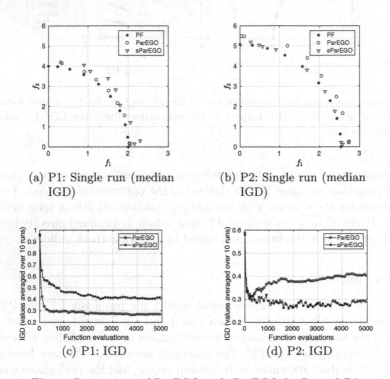

(a) P1: Single run (median IGD)

(b) P2: Single run (median IGD)

(c) P1: IGD

(d) P2: IGD

Fig. 4. Comparison of ParEGO and sParEGO for P1 and P2.

6   Conclusion

This paper has proposed a new multi-objective optimization algorithm for dealing with expensive uncertain MOPs, namely sParEGO. The comparative analysis with the existing algorithm ParEGO has demonstrated that the statistical inferences made by sParEGO are better equipped to assess the robustness of the candidate solutions for a given robustness criterion. However, we have also shown that the existence of a not-so-well-behaved problem landscape can mislead the uncertainty quantification approach when (as is necessary) working on a limited budget of evaluations. Given this, the assumptions made within the framework described in this paper, and their associated risks, are as follows:

1. *The landscape is well-behaved (i.e. smooth, continuous).* The uncertainty distributions are approximated according to available information for other candidate solutions. The underlying assumption for approximating in this way

is that similar solutions have similar performance. If the functions are highly ragged and discontinuous, the surrogate models cannot accurately predict their behaviour.

2. *The problem dimensionality is small to medium.* The search is conducted on a surrogate model fitted to the existing evaluated solutions. The surrogate model used in this framework typically produces good estimates for problems with up to 20 design variables.

3. *The maximum distance between solutions to be considered as neighbours, specified by δ, affects the variance of solutions and the convergence rate.* For smooth and continuous functions, a tight neighbourhood is likely to result in smaller variance, but also uses less information from other solutions, which reduces the convergence rate.

Further benchmarking of sParEGO[3] is now needed to confirm its capabilities across a wider set of problem instances. This includes conducting a comparative analysis with other multi-objective robust optimization algorithms, such as those described in the survey in [11]. Other future research directions include: how to approximate the statistical inferences of isolated solutions; how to incorporate constraints; and incorporation of alternative robustness criteria.

Acknowledgements. This work was supported by Jaguar Land Rover and the UK-EPSRC grant EP/L025760/1 as part of the jointly funded Programme for Simulation Innovation. The open-source version of Liger was also supported by the Advanced Propulsion Centre UK (grant J14921). The authors acknowledge Joshua Knowles for useful discussions during the design process of sParEGO, and thank all reviewers for their insightful comments and suggestions.

References

1. Zhou, A., et al.: Multiobjective evolutionary algorithms: a survey of the state of the art. Swarm Evol. Comput. **1**(1), 32–49 (2011)
2. Knowles, J.: ParEGO: a hybrid algorithm with on-line landscape approximation for expensive multiobjective optimization problems. IEEE Trans. Evol. Comput. **10**(1), 50–66 (2006)
3. Jones, D., et al.: Efficient global optimization of expensive black-box functions. J. Glob. Optim. **13**(4), 455–492 (1998)
4. Knowles, J., Corne, D., Reynolds, A.: Noisy multiobjective optimization on a budget of 250 evaluations. In: Ehrgott, M., Fonseca, C.M., Gandibleux, X., Hao, J.-K., Sevaux, M. (eds.) EMO 2009. LNCS, vol. 5467, pp. 36–50. Springer, Heidelberg (2009). https://doi.org/10.1007/978-3-642-01020-0_8
5. Beyer, H.G., et al.: Robust optimization - a comprehensive survey. Comput. Methods Appl. Mech. Eng. **196**(33–34), 3190–3218 (2007)
6. Van Veldhuizen, D., et al.: On measuring multiobjective evolutionary algorithm performance. In: Proceedings of the CEC 2000, vol. 1, pp. 204–211 (2000)

---

[3] The source code of the sParEGO algorithm, as well as the test problems, is found within the Liger software: https://github.com/ligerdev/liger.

7. Knowles, J., Hughes, E.J.: Multiobjective optimization on a budget of 250 evaluations. In: Coello Coello, C.A., Hernández Aguirre, A., Zitzler, E. (eds.) EMO 2005. LNCS, vol. 3410, pp. 176–190. Springer, Heidelberg (2005). https://doi.org/10.1007/978-3-540-31880-4_13

8. Ginley, B.M., et al.: Maintaining healthy population diversity using adaptive crossover, mutation, and selection. IEEE Trans. Evol. Comput. **15**(5), 692–714 (2011)

9. Huband, S., et al.: A review of multiobjective test problems and a scalable test problem toolkit. IEEE Trans. Evol. Comput. **10**(5), 477–506 (2006)

10. Salomon, S., et al.: A toolkit for generating scalable stochastic multiobjective test problems. In: Proceedings of the GECCO (2016)

11. Jin, Y., et al.: Evolutionary optimization in uncertain environments-a survey. IEEE Trans. Evol. Comput. **9**(3), 303–317 (2005)

Knowledge Discovery in Scheduling Systems Using Evolutionary Bilevel Optimization and Visual Analytics

Julian Schulte[✉], Niclas Feldkamp, Sören Bergmann, and Volker Nissen

Ilmenau University of Technology, Ilmenau, Germany
{julian.schulte, niclas.feldkamp, soeren.bergmann, volker.nissen}@tu-ilmenau.de

Abstract. Scheduling systems are subject to a variety of influencing factors, some of which (e.g. number of vehicles or employees) can be determined by the company itself. Since these framework conditions can have a major impact on the scheduling system's performance, their determination is an important management task. The difficulty of this task increases when conflicting objectives have to be considered, such as costs and performance. Even though evolutionary bilevel optimization can be used to solve this kind of strategic multi-objective problems, it remains hard to gain deeper insights into the scheduling system's behavior by only analyzing the obtained set of Pareto optimal solutions. In this paper, we propose an approach for knowledge discovery in scheduling systems by applying visual analytics on the whole set of evaluated individuals during the evolutionary algorithm. The proposed concept of bilevel innovization is demonstrated by using a nested NSGA-II to solve a strategic personnel planning problem and subsequently applying visual analytics to support decision making regarding the number of employees and implemented shifts. The results show that bilevel innovization can be used to get a better understanding of a scheduling system's behavior and to support the decision making process in a strategic planning context.

Keywords: Innovization · Evolutionary bilevel optimization · Visual analytics · Scheduling · Staffing

1 Introduction

Scheduling problems in general deal with complex resource allocation tasks and arise in a variety of domains, such as manufacturing (e.g. shop scheduling), transportation (e.g. vehicle routing), management (e.g. personnel scheduling) or computer science (e.g. task scheduling). As these problems are challenging itself, the difficulty increases when the framework conditions of the considered scheduling system have to be determined. This becomes relevant when looking at a scheduling problem from a strategic rather than an operative perspective, for example by finding an adequate number of vehicles for a public transportation company in order to simultaneously minimize fleet size and passenger travel time [13]. Other examples could be deciding

© Springer Nature Switzerland AG 2019
K. Deb et al. (Eds.): EMO 2019, LNCS 11411, pp. 439–450, 2019.
https://doi.org/10.1007/978-3-030-12598-1_35

on appropriate data placement policies in a data center considering energy savings on the one side and server performance on the other side [23] or determining size and structure of a company's workforce while minimizing labor costs and maintaining high quality personnel schedules [15].

To deal with this kind of strategic problems, evolutionary bilevel optimization can be used (see [18] for a comprehensive review). Bilevel optimization in general can be considered as a form of hierarchical optimization problem, whose hierarchical relationship is closely related to the problem of Stackelberg [19]. Here, a follower (lower-level optimization problem) optimizes his objective based on the parameters determined by the leader (upper-level optimization problem). The leader, in turn, optimizes his own objective under consideration of the follower's possible reactions [4]. In the context of bilevel innovization, the upper-level problem determines the framework conditions (e.g. policies, number of vehicles or set of available employees) of the lower-level scheduling problem. A popular approach to solve bilevel problems is using an evolutionary algorithm at the upper-level and any kind of optimization algorithm at the lower-level, resulting in a nested evolutionary algorithm [18]. As the upper-level algorithm faces at least two conflicting objectives, a multi-objective optimization problem has to be solved.

Evolutionary multi-objective optimization supports the decision making process by providing a set of Pareto optimal solutions. The final solution to be selected by the decision maker will therefore be a trade-off among the considered objectives [2]. To support this trade-off decision and to gain a deeper understanding of the considered problem, Deb and Srinivasan [6] introduced the concept of innovization. Here, a post-optimality analysis is conducted by applying data mining methods on the approximated Pareto front to identify relationships among input variables and objective values and consequently finding new design principles. Having its origin in the domain of engineering, innovization was successfully applied in other areas, such as manufacturing [1]. However, analyzing only Pareto optimal solutions still reveals limited insights into the overall problem structure. Hence, the concept of innovization was extended in the context of simulation-based optimization of manufacturing respectively production systems, so called simulation-based innovization [7, 14]. Here, data mining and visualization techniques are applied not only on the set of Pareto optimal solutions, but on all evaluated solutions during the optimization procedure in order to discover unknown relationships and patterns for the optimal design of production systems. A similar approach, also in the domain of manufacturing simulation, was proposed by Feldkamp et al. [9, 10]. Instead of using optimization, the authors fully enumerate a predefined parameter space of input variables. Subsequently, visual analytics is applied on the obtained data sets to uncover relationships in the considered model.

In this paper, we transfer the idea of simulation-based innovization and knowledge discovery in manufacturing simulations to the domain of scheduling. For demonstration, the proposed concept of bilevel innovization is applied to a strategic workforce planning problem with the conflicting objectives to minimize labor costs and to maximize scheduling quality (measured by penalty functions). At the upper-level, a nested NSGA-II is used to determine the number of employees in different categories as well as the shift patterns to be implemented by the company. At the lower-level, an elitist Genetic Algorithm (GA) is used to solve the resulting personnel scheduling problem, due to the widespread usage and successful application of GA to this type of problem [22].

The remainder of this paper is structured as follows: In Sect. 2 the concept of bilevel innovization is presented. Section 3 describes the considered workforce planning problem. The applied algorithms are described in Sect. 4. In Sect. 5 the bilevel innovization process is applied to the investigated problem. Finally, conclusions and suggestions for further research are presented in Sect. 6.

2 Bilevel Innovization

2.1 Visual Analytics and Knowledge Discovery

In general, visualization is a useful and important tool for the interpretation of data [21]. Therefore, in the context of simulation, simulation based optimization and related applications, the analysis of generated output and its relation to the corresponding input data sets often relies on visual inspection. Frequently applied visualizations are for example confidence intervals on certain output metrics, histograms for distribution analysis, scatter plots for finding causation in correlated structures, as well as animated visualization of dynamic processes [24]. Visual analytics (VA) goes beyond those commonly applied visualization techniques by defining a research discipline of its own. VA is defined as "an iterative process that involves information gathering, data pre-processing, knowledge representation, interaction and decision making" [12]. By combining automated data analysis and interactive visualizations, it also combines the strengths of both machine and human capabilities. On the one hand, patterns from large amounts of data can be extracted and processed through data mining with statistical and mathematical models. This is commonly referred to as knowledge discovery in databases [8]. On the other hand, visualizations of the processed data can be explored by making use of the human capabilities to perceive, relate, and recognize visual patterns and draw conclusions, encouraged by a high degree of user interaction. Figure 1 shows the visual analytics process. The user is in a constant loop of refining data mining hyperparameters and interacting with the visualization of the data, for example zooming, filtering or applying different visualization schemes. Through the combination of both model building and visualization, conclusions can be drawn that lead to a better understanding of the underlying data and ultimately to a creation of knowledge. Implementing the findings through a feedback loop may restart the process entirely if needed.

2.2 Bilevel Innovization Process

The here proposed concept of bilevel innovization for knowledge discovery in scheduling systems is based on the ideas of simulation-based innovization and knowledge discovery in manufacturing simulations. In general, the bilevel innovization process (see Fig. 2) can be divided into two parts: data generation and data analysis. For data generation, evolutionary bilevel optimization is used. The part of data analysis is based on the visual analytics process.

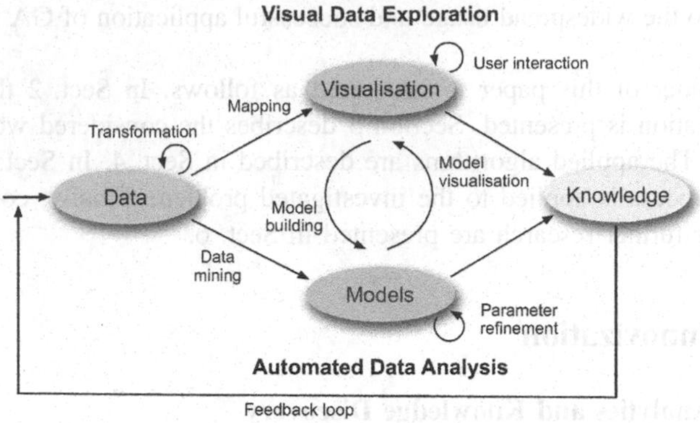

Visual Data Exploration

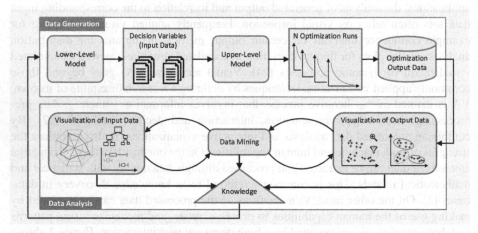

Fig. 1. Visual analytics process [11].

Fig. 2. Bilevel innovization process [16].

Data Generation. The first part of the bilevel innovization process serves the purpose of generating the data set for the subsequent analysis by solving the investigated bilevel optimization problem.

Starting point is the actual scheduling system to be analyzed and lower-level optimization model, respectively. The system behavior will be represented by the lower-level objective value. The optimization algorithm here is freely selectable depending on the considered scheduling problem.

The next step is to determine the framework conditions that want to be investigated (e.g. policies or set of available employees). These will serve as decision variables for the upper-level problem.

Subsequently, the resulting upper-level problem has to be modeled. For solving the upper-level problem an evolutionary multi-objective algorithm is used. In the context of strategic decision making, at least two conflicting objectives are assumed to be optimized within the bilevel optimization problem.

Thereafter, a large number of independent runs of the upper-level algorithm should be conducted in order to obtain as many different solutions as possible for the subsequent analysis. The number of runs depends, among others, on the problem structure, the number and characteristics of the decision variables at the upper-level and the required computation time.

The last step of this stage is to prepare the obtained data sets, i.e. evaluated individuals, from the optimization runs in a suitable manner for the following visual analytics task. Furthermore, duplicates, i.e. individuals with identical decision vectors, should be removed to avoid bias.

Data Analysis. In the context of bilevel innovization, each data record is composed by the objective values of an evaluated individual at the upper-level problem (output data) and the corresponding decision variables (input data).

As common in visual analytics, the whole data analysis can be seen as an iterative process. The first step is to visualize the output data and to identify an area of interest for deeper analysis by zooming into the data (e.g. selected range around the Pareto optimal front). Subsequently, the filtered data can be visualized again to explore shape and distribution of the objective values and possible linkages. This may lead to first insights regarding the system's behavior.

Now, data mining methods (e.g. clustering, classification or decision trees) can be applied on the filtered data set both on the output and the input data. Thereafter, the data mining results should be visualized to uncover interesting patterns and to get a better understanding of the analyzed system's behavior. Suitable diagrams are, among others, scatter plots, parallel coordinate plots, radar charts, pie charts or box plots.

Each of the previously mentioned steps may lead to knowledge, which in turn could be used to start the data generation process at an arbitrary step (e.g. adjust the lower-level model, add or remove decision variables, pick a new algorithm at the upper level or conduct more optimization runs).

Although it is assumed that the optimization models on both levels are verified and perform correctly, the bilevel innovization process allows the decision maker to further verify the optimization models if unexpected behavior is identified that may arise from modeling issues.

3 Problem Description

In this section, a strategic workforce planning problem of a midsized inbound call center of a utility is described, which is an extended version of the problem presented in [15]. Due to internal restrictions of the utility, the presented problem as well as the specific setting in Sect. 5.1 are derived and abstracted from a real world problem commonly found in strategic workforce planning.

In a strategic context, the purpose of staffing is to determine the adequate future number of employees needed in different categories. Scheduling, as an operative task, is concerned with getting the right people to the right place at the right time. In the here presented problem, the decision making task not only concerns staffing but also determining the shift patterns to be implemented by the company. Hence, scheduling quality not only depends on the staffing decision, but also on the available shifts.

3.1 Planning Problem

The considered workforce planning problem is modeled as bilevel problem [20], with x representing the planning decision regarding staffing and shift design. For each x, the scheduling problem $f(x, y)$ will be optimized yielding personnel schedules for the considered planning horizon (represented by y).

The call center of the investigated problem has a need of k different skill types $s \in S$, with $S = \{s_1, s_2, \ldots, s_k\}$. The skills are meant to be categorical, i.e. they determine the tasks each employee can perform. However, it is possible to cross-train employees so they can perform more than one type of task [3]. The qualification of an employee can therefore be seen as set of different skill combinations $q \subseteq S$. The contract type $t \in T$ of an employee determines his average weekly working time.

Within its staffing decision, the company has to predefine feasible employee types $\bar{E}$. Each employee type $\bar{e}_{qt} \in \bar{E}$ is defined by its qualification q and contract type t. Moreover, each employee type $\bar{e}_{qt}$ is linked to costs $c_{\bar{e}_{qt}}$ that arise for employing one employee of this type over the considered planning horizon. The number of employees of each type is represented by the decision variable $x_{\bar{e}_{qt}}$. Regarding the shift design decision, the company furthermore has to predefine a set of feasible shifts. Each shift pattern $\bar{m} \in \bar{M}$ is constrained by the operating times of the call center and the minimum and maximum shift length. The number of shift patterns ϑ to be implemented is defined by (1c). Whether a shift pattern is implemented or not is represented by the decision variable $x_{\bar{m}}$. The specific planning problem setting will be described in Sect. 5.1.

The objective of the upper-level problem is to minimize the overall staffing costs (1a) subject to constraints (1b)–(1c) and the optimized scheduling decision y at the lower-level problem (1d). The decision vector passed to the lower-level is defined as $x = (x_{\bar{e}_{qt}}, x_{\bar{m}})$. Since $x_{\bar{m}}$ has no direct impact on the upper-level objective value, it is left out of the objective function.

$$\min_{x \in X, y \in Y} F\left(\sum_{\bar{e}_{qt} \in \bar{E}} x_{\bar{e}_{qt}} c_{\bar{e}_{qt}}, y \right) \tag{1a}$$

$$\text{s.t.} \quad x_{\bar{e}_{qt}} \geq 0 \text{ and integer} \quad \forall \bar{e}_{qt} \in \bar{E} \tag{1b}$$

$$\sum_{\bar{m} \in \bar{M}} x_{\bar{m}} = \vartheta \tag{1c}$$
$$x_{\bar{m}} \in \{0, 1\} \quad \forall \bar{m} \in \bar{M}$$

$$y \in \text{argmin}_{y \in Y} \{result\ scheduling\ problem\} \tag{1d}$$

3.2   Scheduling Problem

The lower-level scheduling problem considers the daily staff scheduling of a call center over a planning horizon $W = \{1, 2, \ldots, w_{max}\}$. Each week of the planning horizon $w \in W$ is partitioned into periods $p \in P = \{1, 2, \ldots, p_{max}\}$, representing the operating days of the call center. Moreover, each operating day again is segmented into time intervals $i \in I = \{i_1, i_2, \ldots, i_{max}\}$. The set of shift patterns M and set employees E are determined by the staffing decision at the upper-level, with a concrete employee for each $x_{\bar{e}qt}$. An employee can be assigned to one shift each day and only if he has the required skill.

For each time interval i on day p in week w and each skill s a certain staffing level has to be satisfied. If a deviation arises from the staffing target, penalty points are generated. To provide an equal workload distribution and to ensure that employees are staffed according to their contract types, further penalty is calculated based on how far employees exceeded or fell below their average weekly working time. The objective is to minimize the overall penalty points over the considered planning horizon.

For a more detailed description of the scheduling problem as well as the mathematical formulation, we refer to [15]. The specific scheduling problem setting will be described in Sect. 5.1.

4   Algorithmic Approach

Following the taxonomy given by Talbi [20], the algorithm used to solve the considered strategic planning problem can be defined as a nested constructing approach with metaheuristics on both levels. In this type of bilevel model, an upper-level metaheuristic calls a lower-level metaheuristic during its fitness assessment. In doing so, the upper-level heuristic determines the decision vector x (here the number of employees of each type and the set of implemented shift patterns) as input for the lower-level algorithm, which in turn determines the decision vector y (optimized schedules). Both decision vectors are subsequently used to solve the bilevel problem at the upper-level.

At the upper-level, the NSGA-II [5] is used to solve the multi-objective problem of minimizing staffing costs and scheduling penalty. The chromosomes of the upper-level individuals are composed of two one-dimensional integer vectors, each corresponding to one subproblem. The staffing vector $u = (u_1, u_2, \ldots, u_\varrho)$ contains $\varrho = |\bar{E}|$ integer values corresponding to the number of employee types. Each value u_ζ represents the number of employees to be employed for this type. However, $u_\varrho \in \mathbb{N}$ should be limited in a reasonable manner to reduce the search space. The shift vector is denoted by $v = (v_1, v_2, \ldots, v_\vartheta)$, with $v_\varphi \in \{0, 1, \ldots, |\bar{M}| - 1\}$ representing one integer encoded shift pattern $\bar{m} \in \bar{M}$ and ϑ (see constraint 2c) corresponding to the total number of shifts to be implemented.

As u and v have different search spaces (metric and categorical) as well as different value ranges, they are handled independently during reproduction. However, for both vectors one-point, uniform and n-point crossover are applied, randomly selected for each reproduction process. Furthermore, intermediate crossover and random walk mutation (both as described in [17]) are used for the reproduction of $\boldsymbol{u}$. For mutation of $\boldsymbol{v}$ random integer mutation is applied. Because the values of $\boldsymbol{v}$ have to be unique, a repair procedure is implemented replacing duplicates with the nearest feasible value.

At the lower-level, an elitist GA with the same elitism principle as NSGA-II is used. For a more detailed description of the lower-level algorithm we refer to [15].

5 Experimental Study

5.1 Problem Setting

Within the investigated scenario, the call center is planning its workforce for a 12 week planning horizon $W = \{1, 2, \ldots, 12\}$. Furthermore, the operating days are Monday till Friday $P = \{1, 2, \ldots, 5\}$ from 8 a.m. to 6 p.m. $I = \{1, 2, \ldots, 10\}$ (due to the strategic context, hourly scheduling intervals were chosen). Shifts are allowed in the range between 4 and 8 h, resulting in 25 possible shift patterns. Moreover, due to organizational regulations the company plans to implement $\vartheta = 7$ different shifts. Regarding its staffing decision, the company has a need of two different skills $S = \{agent, support\}$. The forecasted demand of agents is highly volatile both during the day and across the planning horizon. The demand of support employees is calculated based on a staffing ratio of one support for each four agents. Table 1 shows the predetermined employee types with related costs. The costs of each employee type are represented by a relative factor summing up annual wages, payroll taxes, overhead and training costs.

Table 1. Employee types and related costs.

Contract type	Qualification	Costs
20 h/40 h	Agent	0.6/1
20 h/40 h	Support	0.65/1.1
20 h/40 h	Agent - support	0.75/1.3

5.2 Data Generation

The first three steps of the bilevel innovization data generation part were described in Sects. 3 and 4. The next step deals with the execution of the optimization runs. The parameters of both GA were set based upon preliminary studies. For the upper-level GA, a population size of 30, a generation number of 100 and n = 30 restarts were chosen, with each restart having a random initial population. For the lower-level GA a trade-off regarding computation time, solution quality and solution noise had to be made. Therefore, the GA was configured with a population size of 40 and a generation number of 60. On both levels, the mutation rate was set to 1/v, with v being the number of bits of the encoded individual, and the application of the available crossover operators was uniformly distributed. The fitness at the upper-level was evaluated by Eq. (1a), for the fitness evaluation at the lower-level Eq. (1d) was used.

After removing individuals with identical upper-level decision vectors, the optimization runs yielded 88,762 unique data records.

5.3 Data Analysis

In the first step of the bilevel innovization data analysis part, the output data was visualized with respect to the two upper-level objectives scheduling penalty (i.e. scheduling quality) and staffing costs. Subsequently, an area of interest was selected for more detailed investigations (see Fig. 3). The selected subset still contains more than 60% (57,531) of the explored solutions and shows a high density around to the Pareto optimal front. We see that the scheduling quality increases with higher staffing costs.

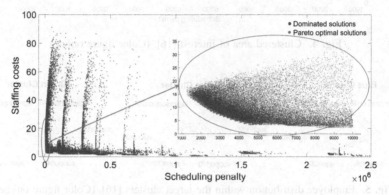

Fig. 3. Objective space and selected area of interest.

Prior to a more detailed examination of the input data, further target areas within the selected area of interest were identified by clustering the output data regarding overall penalty and costs. In doing so, we calculated an optimal clustering structure for this data set according to the silhouette coefficient. The best structuring was found with five clusters, the k-means clustering algorithm and a cosine-based distance measure (see Fig. 4). In the further process, we focus on the three clusters along the Pareto front: blue (16,563 solutions/2,451 avg. penalty), green (10,521/2,966) and violet (13,378/3,985).

First, the clusters were investigated regarding the six discrete input variables affecting workforce size and structure by using radar charts (see Fig. 5). The dashed lines show the upper and lower quartiles, the solid line the median of the employee distribution in one solution. The gap between lower quartile and median in the blue cluster reflects its noisiness. It becomes apparent that solutions in the blue cluster have significantly more staff involved, especially flexible and part-time workers. The green and violet clusters form a similar shape, differing mainly in the number of employees. However, for going from violet to green, investing in agent40 and agentsupport40 seems to be the most efficient way to increase scheduling quality.

In the next step, the 25 binary input variables regarding shift design were analyzed. The pie charts (see Fig. 6) show the proportional amount of shift patterns (24-h clock)

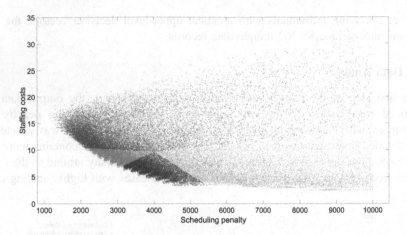

Fig. 4. Clustered area of interest [16]. (Color figure online)

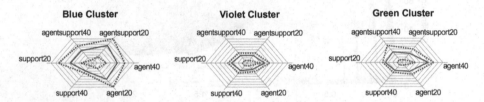

Fig. 5. Employee distribution within the target clusters [16]. (Color figure online)

in each cluster. Variables with a high proportion have a positive effect on solutions to appear in the corresponding cluster, variables with low proportion may inhibit solutions from appearing in this cluster. It has to be considered that, because of seven unique shifts per solution, about 14% is the maximal proportion to be reached. This means, for example, that in the green cluster about 91% of all solutions contain shift 10 to 18. Interestingly, the shift structure in the three clusters is very similar.

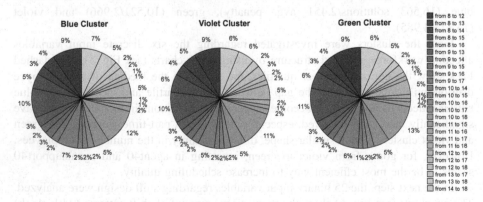

Fig. 6. Proportional amount of shift patterns. (Color figure online)

6 Conclusions and Future Research

A new approach to support the decision making process for determining framework conditions of a scheduling system in a strategic planning context was presented in this paper. The proposed concept of bilevel innovization was demonstrated by solving a multi-objective problem commonly found in strategic workforce planning. In doing so, a nested NSGA-II was used to determine size and structure of a call center's workforce as well as shifts to be implemented, considering the conflicting objectives staffing costs and scheduling quality. In a next step, not only Pareto optimal but all explored solutions during multiple restarts of this bilevel model were analyzed by iteratively applying suitable visualization techniques and data mining methods.

The results of the conducted study show, that the application of bilevel innovization enables the decision maker to get a better understanding of the investigated scheduling problem and system, respectively, and its behavior under different framework conditions. Those insights comprise both the behavior regarding output parameters (i.e. objective values) as well as input parameters (i.e. decision variables).

However, especially due to the combinatorial nature of both the staffing and the shift design problem, interaction effects and dependencies of the decision variables have to be investigated. Therefore, data mining methods should be applied not only to the output parameters (as done in this study) but also to the input parameters. Further research should also be conducted by evaluating the concept of bilevel innovization in the context of other scheduling problems as well as bilevel optimization problems in general.

References

1. Bandaru, S., Aslam, T., Ng, A.H.C., Deb, K.: Generalized higher-level automated innovization with application to inventory management. Eur. J. Oper. Res. **243**(2), 480–496 (2015)
2. Branke, J. (ed.): Multiobjective optimization: Interactive and evolutionary approaches. Lecture Notes in Computer Science, vol. 5252. Springer, Berlin (2008). https://doi.org/10.1007/978-3-540-88908-3
3. de Bruecker, P., van den Bergh, J., Beliën, J., Demeulemeester, E.: Workforce planning incorporating skills. State Eur. J. Oper. Res. **243**(1), 1–16 (2015)
4. Colson, B., Marcotte, P., Savard, G.: An overview of bilevel optimization. Ann. Oper. Res. **153**(1), 235–256 (2007)
5. Deb, K., Pratap, A., Agarwal, S., Meyarivan, T.: A fast and elitist multiobjective genetic algorithm: NSGA-II. IEEE Trans. Evol. Computat. **6**(2), 182–197 (2002)
6. Deb, K., Srinivasan, A.: Innovization. In: Cattolico, M. (ed.) Proceedings of the 8th annual conference on Genetic and evolutionary computation - GECCO 2006, p. 1629. ACM Press, New York (2006)
7. Dudas, C., Hedenstierna, P., Ng, A.H.C.: Simulation-based innovization for manufacturing systems analysis using data mining and visual analytics. In: Proceedings of the 4th Swedish Production Symposium, Lund, Schweden, pp. 374–382 (2011)
8. Fayyad, U., Piatetsky-Shapiro, G., Smyth, P.: From data mining to knowledge discovery in databases. AI Mag. **17**(3), 37–54 (1996)

9. Feldkamp, N., Bergmann, S., Strassburger, S.: Knowledge discovery in manufacturing simulations. In: Taylor, S.J.E., Mustafee, N., Son, Y.-J. (eds.) Proceedings of the 3rd ACM Conference on SIGSIM-Principles of Advanced Discrete Simulation - SIGSIM-PADS 2015, pp. 3–12. ACM Press, New York (2015)
10. Feldkamp, N., Bergmann, S., Strassburger, S.: Visual analytics of manufacturing simulation data. In: 2015 Winter Simulation Conference (WSC), pp. 779–790. IEEE (2015)
11. Keim, D. (ed.): Mastering the Information Age - Solving Problems with Visual Analytics. Eurographics Association, Goslar (2010)
12. Keim, D.A., Mansmann, F., Schneidewind, J., Thomas, J., Ziegler, H.: Visual analytics: scope and challenges. In: Simoff, S.J., Böhlen, M.H., Mazeika, A. (eds.) Visual Data Mining. LNCS, vol. 4404, pp. 76–90. Springer, Heidelberg (2008). https://doi.org/10.1007/978-3-540-71080-6_6
13. Liu, Z., Shen, J.: Regional bus operation bi-level programming model integrating timetabling and vehicle scheduling. Syst. Eng. Theory Pract. 27(11), 135–141 (2007)
14. Ng, A.H.C., Dudas, C., Nießen, J., Deb, K.: Simulation-Based innovization using data mining for production systems analysis. In: Wang, L., Ng, A.H.C., Deb, K. (eds.) Multi-objective Evolutionary Optimisation for Product Design and Manufacturing, pp. 401–429. Springer, London (2011). https://doi.org/10.1007/978-0-85729-652-8_15
15. Schulte, J., Günther, M., Nissen, V.: Evolutionary bilevel approach for integrated long-term staffing and scheduling. In: Proceedings of the 8th Multidisciplinary International Conference on Scheduling: Theory and Applications (MISTA 2017), Kuala Lumpur, Malaysia, pp. 144–157 (2017)
16. Schulte, J., Feldkamp, N., Bergmann, S., Nissen, V.: Bilevel innovization. In: Takadama, K., Aguirre, H. (eds.) Proceedings of the Genetic and Evolutionary Computation Conference Companion on - GECCO 2018, pp. 197–198. ACM Press, New York (2018)
17. Luke, S.: Essentials of Metaheuristics. Lulu, Vancouver (2013)
18. Sinha, A., Malo, P., Deb, K.: A review on bilevel optimization: from classical to evolutionary approaches and applications. IEEE Trans. Evol. Comput. 22(2), 276–295 (2018). https://doi.org/10.1109/TEVC.2017.2712906
19. Stackelberg, H.: The Theory of Market Economy. Oxford University Press, Oxford (1952)
20. Talbi, E.-G. (ed.): Metaheuristics for Bi-level Optimization. Studies in Computational Intelligence, vol. 482. Springer, Heidelberg (2013). https://doi.org/10.1007/978-3-642-37838-6
21. Thomas, J.J. (ed.): Illuminating the Path: The Research and Development Agenda for Visual Analytics. IEEE Computer Society, Los Alamitos (2005)
22. van den Bergh, J., Beliën, J., de Bruecker, P., Demeulemeester, E., de Boeck, L.: Personnel scheduling: a literature review. Eur. J. Oper. Res. 226(3), 367–385 (2013)
23. Wang, X., Wang, Y., Cui, Y.: A new multi-objective bi-level programming model for energy and locality aware multi-job scheduling in cloud computing. Future Gener. Comput. Syst. 36, 91–101 (2014)
24. Wenzel, S., Bernhard, J., Jessen, U.: A taxonomy of visualization techniques for simulation in production and logistics. In: Proceedings of the 2003 International Conference on Machine Learning and Cybernetics (IEEE Cat. No.03EX693), 2003 Winter Simulation Conference, pp. 729–736. IEEE (2003)

Pareto Optimal Set Approximation by Models: A Linear Case

Aimin Zhou[1($\boxtimes$)], Haoying Zhao[1], Hu Zhang[2], and Guixu Zhang[1]

[1] Department of Computer Science and Technology, East China Normal University, Shanghai, China
{amzhou,gxzhang}@cs.ecnu.edu.cn, 52184506010@stu.ecnu.edu.cn
[2] Beijing Electro-Mechanical Engineering Institute, Beijing, China
jxzhanghu@126.com

Abstract. The optimum of a *multiobjective optimization problem (MOP)* usually consists of a set of tradeoff solutions, called *Pareto optimal set*, that balances different objectives. In the community of evolutionary computation, an internal or external population with a limited size is usually used to approximate the Pareto optimal set. Since the Pareto optimal set forms a manifold in both the decision and objective spaces under mild conditions, it is possible to use a model as well as a population of solutions to approximate the Pareto optimal set. Following this idea, the paper proposes to use a set of linear models to approximate the Pareto optimal set in the decision space. The basic idea is to partition the manifold into different segments and use a linear model to approximate each segment in a local area. To implement the algorithm, the models are incorporated in the *multiobjective evolutionary algorithm based on decomposition (MOEA/D)* framework. The proposed algorithm is applied to a test suite, and the comparison study demonstrates that models can help to improve the performance of algorithms that only use solutions to approximate the Pareto optimal set.

Keywords: Evolutionary multiobjective optimization ·
Regularity model · MOEA/D

1 Introduction

This paper considers the following continuous *multiobjective optimization problems (MOPs)*:

$$\text{minimize } F(x) = (f_1(x), \ldots, f_m(x)) \qquad (1)$$
$$\text{subject to } \qquad x \in \Omega$$

This work is supported by the National Natural Science Foundation of China under Grant Nos. 61673180, 61731009, and 61703382, and the Fundamental Research Funds for the Central Universities.

© Springer Nature Switzerland AG 2019
K. Deb et al. (Eds.): EMO 2019, LNCS 11411, pp. 451–462, 2019.
https://doi.org/10.1007/978-3-030-12598-1_36

where $\Omega \subseteq R^n$ defines the feasible region of the decision space, x is an n-D decision vector, $F : \Omega \to R^m$ consists of m real-valued continuous objective functions $f_1, \ldots, f_m$, and R^m is the objective space.

Due to the conflicting nature of the objectives, there usually does not exist a single solution that can optimize all the objectives simultaneously. Instead, the best tradeoff solutions among the objectives, called Pareto optimal solutions, are of interest to decision makers. The set of all the Pareto optimal solutions is called *Pareto set (PS)* in the decision space, and its image in the objective space is called *Pareto front (PF)* [1,2]. The PS and PF can provide decision makers with a good understanding of tradeoff relationships among different objectives and help them to make a final choice.

Multiobjective evolutionary algorithms (MOEAs) have been accepted as a major methodology for approximating the PS and PF [3]. Most, if not all, MOEAs approximate the PS and PF by using a finite number of solutions. In other words, these algorithms conduct a 0-order approximation to the PS and PF. It is a very natural and convenient choice since MOEAs work with a population of candidate solutions. However, model-based approximations of the PS and PF may be interesting and even necessary since they can provide a deeper understanding of the problem to decision makers [4], and help to improve the search efficiency [5].

Some efforts have been made to approximate the PF to guide the selection [6–8] or to model the mapping between the PS and the PF to reduce the function evaluations [9–11]. Some more efforts have been applied to approximate the PS to sample new trial solutions. These algorithms fall into the category of *estimation of distribution algorithm (EDA)* in which a probabilistic model is built to model the population distribution and to sample new trial solutions [12]. Some models, such as mixture Gaussian [13–16], Boltzmann machine [17], B-spline basis function [18], Bayesian network [19], Gaussian neural network [20], Gaussian process [21], have been considered in multiobjective optimization.

It is reasonable that to improve algorithm performance, EAs should utilize more problem-specific knowledge. Under mild conditions, it can be induced from the Karush-Kuhn-Tucker condition that the PS (PF) of a continuous MOP forms a piecewise continuous $(m-1)$-D manifold [22]. By considering this regularity property, we have proposed a *regularity model-based multiobjective estimation of distribution algorithm (RM-MEDA)* [23], in which a manifold learning method was applied to detect the PS manifold. A variety of work has been done to improve the algorithm performance from different aspects [24–31].

This paper considers approximating the PS by a set of linear models as well as a population of candidate solutions. The *Multiobjective optimization evolutionary algorithm based on decomposition (MOEA/D)* [32,33] is used as a basic framework. The basic idea is as follows: since each subproblem (each solution) in MOEA/D is linked with a weight vector, a Pareto optimal solution x in the PS can be regarded as a function of weight vector λ as done in [34]. Therefore, the solutions in the current population with their associated weight vectors can be treated as a set of noisy (λ, x) pairs, and these pairs can be used for building an approximation model for the PS at each generation. The model thus built can be helpful for generating new solutions as in EDAs.

The rest of the paper is organized as follows. Section 2 introduces the local linear model used in the paper. Section 3 presents the algorithm framework. Section 4 gives the experimental studies. Finally, the paper is concluded in Sect. 5 with some suggestions for future work.

2   Local Linear Model

We assume that the PS of (1) is a piecewise continuous $(m - 1)$-D manifold. It is reasonable to approximate the PS locally by an $(m - 1)$-D plane. Let $x^{i_1}, \ldots, x^{i_K} \in R^n$ be K solutions to the neighboring subproblems of subproblem i in MOEA/D. Solution x^{i_j} is for subproblem i_j and is associated with the weight vector λ^{i_j}. These K solutions will hopefully converge to a small part of the PS. Therefore, these solutions are scattered around an $(m - 1)$-D plane and can be modeled as follows:

$$x = Bw + \epsilon \tag{2}$$

where $x \in R^n$, B is an $n \times m$ constant matrix, $w \in R^m$ is a latent variable vector with a 1 in its first position, and ϵ is an n-D Gaussian noise vector. For the sake of simplicity, we assume that $\epsilon \sim N(0, \sigma^2 I)$ where I is an $n \times n$ unity matrix.

Let w^{i_j} be the value of w corresponding to x^{i_j} in (2), $X = (x^{i_1}, \ldots, x^{i_K})$ and $W = (w^{i_1}, \ldots, w^{i_K})$. To estimate B, we consider:

$$\min_{B,W} ||X - BW||^2, \tag{3}$$

where $|| \cdot ||$ is the 2-norm, and both B and W are unknown. We adopt the widely-used alternating optimization technique for solving (3), which is given in Algorithm 1.

Algorithm 1. Procedure to optimize B and W

---
1 Initialize W^0 by the corresponding weight vectors, and set $B^0 = 0$;
2 Set $t = 0$;
3 **repeat**
4   Set $B^{t+1} = \arg\min_{B} ||X - BW^t||^2$;
5   Set $W^{t+1} = \arg\min_{W} ||X - B^{t+1}W||^2$;
6   Set $t = t + 1$;
7 **until** $||B^t W^t - B^{t-1}W^{t-1}||^2 < \varepsilon$;

In Algorithm 1, W^0 is initialized by the weight vectors that are associated with the corresponding neighboring solutions, i.e., $w^{i_j 0} = \begin{pmatrix} 1 \\ \lambda^{i_j} \end{pmatrix}$, in MOEA/D since the weight vectors are predefined in a simplex. ε is a predefined threshold that determines the termination condition of the procedure. According to our experiments, W^0 is well initialized by this way and Algorithm 1 usually stops in less than 5 iterations. In *Lines* 4 & 5, the least squares method is applied to solve the optimization problems.

Given the optimal estimations of the basis and coefficients, B and W, we can estimate the variance of the noise as

$$\sigma^2 = ||X - BW||^2.$$

With B and σ^2, it is possible to generate a new trial solution x from (2). However, the corresponding coefficient w of the trial solution is unknown. Therefore, we firstly generate a coefficient and then sample a new solution. The detailed procedure is given in Algorithm 2, where $rand()$ is a function that returns a random number in $[0, 1]$.

Algorithm 2. Procedure to sample a trial solution

---

1 Randomly select three different indices a, b, and c from $\{1, 2, \cdots, K\}$;
2 Generate a new coefficient vector

$$w = w^a + (w^b - w^c)rand();$$

3 Sample a noise vector

$$\epsilon \sim N(0, \sigma^2 I);$$

4 Generate a new solution

$$y = Bw + \epsilon;$$

---

3   Algorithm Framework

In this paper, we use the following Tchebycheff technique to define subproblem:

$$\min g(x|\lambda^i, z^*) = \max_{1 \le j \le m} \lambda_j^i |f_j(x) - z_j^*| \tag{4}$$

where $\lambda^i = (\lambda_1^i, \cdots, \lambda_m^i)^T$ is a weight vector, $z^* = (z_1^*, \cdots, z_m^*)^T$ is a reference point, i.e., z_j^* is the minimal value of f_j in the objective space. For simplicity, we use $g^i(x)$ to denote $g(x|\lambda^i, z^*)$. In most cases, subproblems with close weight vectors will have similar optimal solutions. Based on the distances among the weight vectors, MOEA/D defines neighborhood relations among the subproblems.

In MOEA/D, the ith ($i = 1, \cdots, N$) subproblem maintains:

- its objective function $g^i(x)$, which is defined in (4),
- its current solution x^i and the objective vector of x^i, i.e. $F^i = F(x^i)$, and
- the index set of its neighboring subproblems, B^i.

MOEA/D also maintains a reference point $z^* = (z_1^*, \cdots, z_m^*)^T$. The main framework of MOEA/D with linear models, called MOEA/D-LM, is given in Algorithm 3.

MOEA/D-LM is generally a variant of MOEA/D-DE [33], and some comments on MOEA/D-LM are given as follows.

Algorithm 3. Main Framework of MOEA/D-LM

---

1 Initialize a set of subproblems, i.e., x^i, F^i, B^i, g^i for $i = 1, \cdots, N$;
2 Initialize the reference point z^* as $z_j^* = \min\limits_{i=1,\cdots,N} f_j(x^i)$ for $j = 1, \cdots, m$;
3 **while** *not terminate* **do**
4   **foreach** $i \in perm(\{1, \cdots, N\})$ **do**
5   Build a linear model by using solutions with indices in B^i through Algorithm 1;
6   Sample a new solution y through Algorithm 2;
7   **foreach** $j \in \{1, \cdots, m\}$ **do**
8   **if** $f_j(y) < z_j^*$ **then**
9   Set $z_j^* = f_j(y)$;
10   **end**
11   **end**
12   Update the population by the new trial solution y;
13   **end**
14 **end**

- N is the number of subproblems (the population size), K is the neighborhood size for local linear model building, $perm(\cdot)$ randomly permutes the input values, and $rand()$ generates a random real number in $[0, 1]$.
- *Line 1:* The initial solutions for the subproblems are uniformly randomly sampled from Ω. The weight vectors with the subproblems are uniformly distributed, and the details are referred to [33].
- *Line 3:* A maximum number of generations is used as the termination condition.
- *Line 4:* In each iteration, a subproblem is randomly selected.
- *Lines 7–11:* The reference point z^* is updated by the newly generated solution. Since z^* is not fixed, the subproblem objectives g^i, $i = 1, \cdots, N$, are changing during the run.
- *Line 12:* The population is updated by the new trial solution, which is the same as in [33].

More details about the MOEA/D framework and its variants could be found in [33, 35].

4   Experimental Study

This section is devoted to empirically study the performance of MOEA/D-LM. RM-MEDA [23] and MOEA/D-DE [33] are used as baseline algorithms for comparison, and the 10 problems from [23] are used as benchmark problems.

In this paper, we use the inverted general distance (IGD) [36] and hypervolume difference (I_H^-) [37] metrics to measure the algorithm performance. In the

Table 1. The mean (std.) metric values obtained by three algorithms over 60 runs

Instance	Algorithm	IGD	I_H^-
F1	MOEA/D-DE	$0.0041_{0.0001}(-)$	$0.0064_{0.0002}(-)$
	RM-MEDA	$0.0043_{0.0001}(-)$	$0.0064_{0.0003}(-)$
	MOEA/D-LM	$\mathbf{0.0039_{0.0000}}$	$\mathbf{0.0057_{0.0001}}$
F2	MOEA/D-DE	$0.0040_{0.0000}(\sim)$	$0.0061_{0.0002}(-)$
	RM-MEDA	$0.0042_{0.0001}(-)$	$0.0064_{0.0005}(-)$
	MOEA/D-LM	$\mathbf{0.0039_{0.0000}}$	$\mathbf{0.0055_{0.0001}}$
F3	MOEA/D-DE	$0.2305_{0.0245}(-)$	$0.2333_{0.0227}(-)$
	RM-MEDA	$0.0073_{0.0044}(\sim)$	$0.0162_{0.0107}(\sim)$
	MOEA/D-LM	$\mathbf{0.0041_{0.0011}}$	$\mathbf{0.0096_{0.0031}}$
F4	MOEA/D-DE	$0.0356_{0.0005}(\sim)$	$-0.0502_{0.0006}(\sim)$
	RM-MEDA	$0.0424_{0.0008}(-)$	$-0.0337_{0.0018}(-)$
	MOEA/D-LM	$\mathbf{0.0352_{0.0005}}$	$\mathbf{-0.0526_{0.0011}}$
F5	MOEA/D-DE	$0.0052_{0.0004}(-)$	$0.0086_{0.0006}(-)$
	RM-MEDA	$0.0052_{0.0008}(-)$	$0.0093_{0.0023}(-)$
	MOEA/D-LM	$\mathbf{0.0041_{0.0001}}$	$\mathbf{0.0061_{0.0002}}$
F6	MOEA/D-DE	$0.0495_{0.1677}(+)$	$0.0435_{0.1173}(+)$
	RM-MEDA	$\mathbf{0.0139_{0.0169}}(+)$	$\mathbf{0.0279_{0.0256}}(+)$
	MOEA/D-LM	$0.7875_{0.1966}$	$0.5025_{0.1196}$
F7	MOEA/D-DE	$0.2732_{0.3374}(-)$	$0.2914_{0.0989}(-)$
	RM-MEDA	$0.1981_{0.2791}(-)$	$0.1061_{0.1177}(-)$
	MOEA/D-LM	$\mathbf{0.0212_{0.0008}}$	$\mathbf{0.0479_{0.0028}}$
F8	MOEA/D-DE	$0.1054_{0.1097}(-)$	$0.0105_{0.0683}(-)$
	RM-MEDA	$0.0568_{0.0028}(-)$	$0.0037_{0.0062}(-)$
	MOEA/D-LM	$\mathbf{0.0417_{0.0019}}$	$\mathbf{-0.0385_{0.0039}}$
F9	MOEA/D-DE	$0.0138_{0.0098}(\sim)$	$0.0257_{0.0166}(\sim)$
	RM-MEDA	$\mathbf{0.0081_{0.0047}}(\sim)$	$\mathbf{0.0150_{0.0079}}(+)$
	MOEA/D-LM	$0.0092_{0.0047}$	$0.0211_{0.0094}$
F10	MOEA/D-DE	$3.3843_{2.2331}(-)$	$1.0961_{0.0382}(-)$
	RM-MEDA	$128.5240_{19.5941}(-)$	$1.1066_{0.0000}(-)$
	MOEA/D-LM	$\mathbf{0.5304_{0.0159}}$	$\mathbf{0.4917_{0.0160}}$

Table 2. The t-test results of MOEA/D-LM vs MOEA/D-DE and RM-MEDA

		IGD	I_H^-
MOEA/D-LM vs MOEA/D-DE	~	3	2
	+	1	1
	−	6	7
MOEA/D-LM vs RM-MEDA	~	2	1
	+	1	2
	−	7	7

experiments, 100,000 evenly distributed points are selected from the PF to be the reference PF P^*, and $(1.2, 1.2)^T$ and $(1.2, 1.2, 1.2)^T$ are the nadir points for bi-objective and tri-objective problems respectively.

The parameters are as follows:

- The number of decision variables: It is set to be 30 for all instances.
- The population size: It is 101 for bi-objective problems, and 253 for tri-objective problems.
- The maximum generation: It is 200 for F1, F2, F4, F5, F6, and F8, and 1000 for F3, F7, F9, and F10.
- The number of runs: It is 60 for all algorithms on all test instances.
- Parameters in RM-MEDA: The number of clusters is 5 for all instances.
- Parameters in MOEA/D-DE: The update size is $n_r = 2$, the neighborhood size $T = 30$, and the neighborhood search probability $\delta = 0.9$. The parameters are the same as in [33].
- Parameters in MOEA/D-LM: The neighborhood size is $K = 30$, the threshold $\varepsilon = 10^{-5}$, and the other parameters are the same as in MOEA/D-DE.

4.1   Comparison Study

In this section, the proposed MOEA/D-LM is compared with RM-MEDA and MOEA/D-DE on the given test suite. The IGD and I_H^- metric values obtained by the three algorithms over 60 runs are presented in Table 1. The Student's t-test is applied to compare MOEA/D-DE and RM-MEDA with MOEA/D-LM with a significance level of 95%. In the tables, ~, +, and − denote that the results obtained by MOEA/D-LM are similar, worse, and better than those obtained by MOEA/D-DE or RM-MEDA according to the Student's test.

The results in Table 1 clearly show that MOEA/D-LM obtained the best results on 8 out of 10 test problems according to both IGD and I_H^- metrics. The Student's test results in Table 2 indicate that MOEA/D-LM performs significantly better than MOEA/D-DE on 6 problems according to the IGD metric and on 7 problems according to the I_H^- metric. A similar result can be obtained by comparing MOEA/D-LM and RM-MEDA.

The comparison study suggests that MOEA/D-LM outperforms MOEA/D-DE and RM-MEDA on most of the given test problems. The reasons might be explained as follows.

- The major difference between MOEA/D-LM and MOEA/D-DE is on the procedure to generate offspring solutions. The reason might be as discussed in Sect. 1 that the linear models can successfully capture the PS manifold and thus help to improve the approximation quality of the population.
- The major differences between MOEA/D-LM and RM-MEDA are on two folds: firstly, MOEA/D-LM is a decomposition based MOEA and RM-MEDA is a Pareto domination based MOEA; secondly, they use different methods to build local linear models. In MOEA/D-LM, the model building process uses more information from the algorithm framework than that in RM-MEDA. Therefore, the model quality in MOEA/D-LM is higher than that in RM-MEDA. This might be the reason why MOEA/D-LM outperforms RM-MEDA.

4.2   Sensitivity to Control Parameters

In MOEA/D-LM, there are two control parameters that may influence the performance, the threshold ε that determines the stop condition in Algorithm 1, and the neighborhood size K that determines the size of points to build the model in Algorithm 1.

According to the preliminary study, MOEA/D-LM is not sensitive to the threshold ε. Actually, Algorithm 1 will return good approximations to B and W even if the main loop is only repeated for several iterations. Therefore, this section mainly focuses on the neighborhood size K. MOEA/D-LM with $K = 5, 10, \cdots, 40$ is applied to the test suite. The other control parameters are the same as in the previous section.

The mean IGD metric values versus different K values obtained by MOEA/D-LM after 20%, 60%, and 100% function evaluations on F1-F5 are drawn in Fig. 1. The mean I_H^- metric values versus different K values obtained by MOEA/D-LM after 20%, 60%, and 100% function evaluations on F6-F10 are drawn in Fig. 2. The figures show that except on $F6$, the performance increases as K increases in MOEA/D-LM. This suggests that a big neighborhood is useful. It is reasonable because a big neighborhood means a big training dataset to build a linear model, and a relatively big training dataset is generally helpful to build an accurate model. Balancing the cost of model building and the accuracy of the built model, $K = 30$ might be a tradeoff choice.

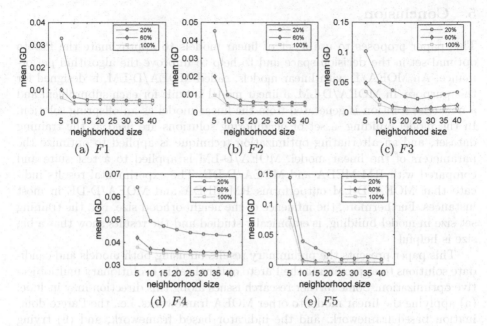

Fig. 1. The average IGD values obtained by MOEA/D-LM with different neighborhood sizes and percentages of function evaluations over 60 runs on F1-F5.

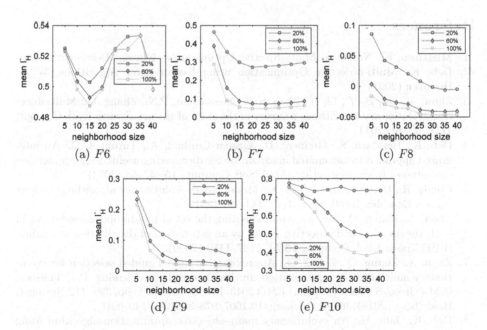

Fig. 2. The average I_H^- values obtained by MOEA/D-LM with different neighborhood sizes and percentages of function evaluations over 60 runs on F6-F10.

5   Conclusion

This paper proposes to use a set of linear models to approximate the Pareto optimal set in the decision space and to help to improve the algorithm performance. An MOEA/D with linear models, called MOEA/D-LM, is designed for this purpose. In MOEA/D-LM, a linear model is built for each subproblem and a new trial solution is generated from the linear model as an offspring solution. In the model building, a set of neighboring solutions are used as the training dataset, and an alternating optimization technique is applied to optimize the parameters of the linear model. MOEA/D-LM is applied to a test suite and compared with RM-MEDA and MOEA/D-DE. The experimental results indicate that MOEA/D-LM outperforms RM-MEDA and MOEA/D-DE in most instances. Furthermore, the influence of the neighborhood size, i.e., the training set size in model building, is empirically studied and the results show that a big size is helpful.

This paper presents the preliminary results on using both models and candidate solutions to approximate the Pareto optimal set in evolutionary multiobjective optimization. Some further research issues along this direction may include: (a) applying the linear model to other MOEA frameworks, i.e., the Pareto domination based framework, and the indicator based framework, and (b) trying some global models, instead of local linear models, to approximate the Pareto optimal set.

References

1. Miettinen, K.: Nonlinear Multiobjective Optimization. Kluwer, Dordrecht (1999)
2. Deb, K.: Multi-objective Optimization using Evolutionary Algorithms. Wiley, Hoboken (2001)
3. Zhou, A., Qu, B.-Y., Li, H., Zhao, S.-Z., Suganthan, P.N., Zhang, Q.: Multiobjective evolutionary algorithms: a survey of the state of the art. Swarm Evol. Comput. 1(1), 32–49 (2011)
4. Deb, K., Bandaru, S., Greinerc, D., Gaspar-Cunhad, A., Tutum, C.C.: An integrated approach to automated innovization for discovering useful design principles: case studies from engineering. Appl. Soft Comput. 15, 42–56 (2014)
5. Cheng, R., He, C., Jin, Y., Yao, X.: Model-based evolutionary algorithms: a short survey. Complex Intell. Syst. 4(4), 283–292 (2018)
6. Zhou, A., Zhang, Q., Jin, Y.: Approximating the set of pareto-optimal solutions in both the decision and objective spaces by an estimation of distribution algorithm. IEEE Trans. Evol. Comput. 13(5), 1167–1189 (2009)
7. Zhou, A., Zhang, Q., Zhang, G.: Approximation model guided selection for evolutionary multiobjective optimization. In: Purshouse, R.C., Fleming, P.J., Fonseca, C.M., Greco, S., Shaw, J. (eds.) EMO 2013. LNCS, vol. 7811, pp. 398–412. Springer, Heidelberg (2013). https://doi.org/10.1007/978-3-642-37140-0_31
8. Deb, K., Jain, H.: An evolutionary many-objective optimization algorithm using reference-point-based nondominated sorting approach, Part I: solving problems with box constraints. IEEE Trans. Evol. Comput. 18(4), 577–601 (2014)

9. Zhang, Q., Liu, W., Tsang, E., Virginas, B.: Expensive multiobjective optimization by MOEA/D with Gaussian process model. IEEE Trans. Evol. Comput. **14**(3), 456–474 (2010)

10. Deb, K., Hussein, R., Roy, P., Toscano, G.: Classifying metamodeling methods for evolutionary multi-objective optimization: first results. In: Trautmann, H., et al. (eds.) EMO 2017. LNCS, vol. 10173, pp. 160–175. Springer, Cham (2017). https://doi.org/10.1007/978-3-319-54157-0_12

11. Volz, V., Rudolph, G., Naujoks, B.: Surrogate-assisted partial order-based evolutionary optimisation. In: Trautmann, H., et al. (eds.) EMO 2017. LNCS, vol. 10173, pp. 639–653. Springer, Cham (2017). https://doi.org/10.1007/978-3-319-54157-0_43

12. Pelikan, M., Sastry, K., Goldberg, D.E.: Multiobjective estimation of distribution algorithms. In: Pelikan, M., Sastry, K., CantúPaz, E. (eds.) Scalable Optimization via Probabilistic Modeling. Studies in Computational Intelligence, vol. 33, pp. 223–248. Springer, Berlin, Heidelberg (2006). https://doi.org/10.1007/978-3-540-34954-9_10

13. Bosman, P.A., Thierens, D.: Multi-objective optimization with diversity preserving mixture-based iterated density estimation evolutionary algorithms. Int. J. Approx. Reason. **31**(3), 259–289 (2002)

14. Zapotecas-Martínez, S., Derbel, B., Liefooghe, A., Brockhoff, D., Aguirre, H.E., Tanaka, K.: Injecting CMA-ES into MOEA/D. In: Proceedings of the Annual Conference on Genetic and Evolutionary Computation (GECCO), pp. 783–790. ACM (2015)

15. Wang, T.-C., Liaw, R.-T., Ting, C.-K.: MOEA/D using covariance matrix adaptation evolution strategy for complex multi-objective optimization problems. In: IEEE Congress on Evolutionary Computation (CEC), pp. 983–990 (2016)

16. Li, H., Zhang, Q., Deng, J.: Biased multiobjective optimization and decomposition algorithm. IEEE Trans. Cybern. **47**(1), 52–66 (2017)

17. Shim, V.A., Tan, K.C., Cheong, C.Y., Chia, J.Y.: Enhancing the scalability of multi-objective optimization via restricted Boltzmann machine-based estimation of distribution algorithm. Inf. Sci. **248**, 191–213 (2013)

18. Bhardwaj, P., Dasgupta, B., Deb, K.: Modelling the pareto-optimal set using B-spline basis functions for continuous multi-objective optimization problems. Eng. Optim. **46**(7), 912–938 (2014)

19. Ahn, C.W., Ramakrishna, R.S.: Multiobjective real-coded Bayesian optimization algorithm revisited: diversity preservation. In: Proceedings of the Annual Conference on Genetic and Evolutionary Computation (GECCO), pp. 593–600 (2007)

20. Martí, L., García, J., Berlanga, A., Molina, J.M.: Multi-objective optimization with an adaptive resonance theory-based estimation of distribution algorithm. Ann. Math. Artif. Intell. **68**(4), 247–273 (2013)

21. Cheng, R., Jin, Y., Narukawa, K., Sendhoff, B.: A multiobjective evolutionary algorithm using Gaussian process-based inverse modeling. IEEE Trans. Evol. Comput. **19**(6), 838–856 (2015)

22. Hillermeier, C.: Nonlinear Multiobjective Optimization: A Generalized Homotopy Approach. Birkhauser, Basel (2001)

23. Zhang, Q., Zhou, A., Jin, Y.: RM-MEDA: a regularity model-based multiobjective estimation of distribution algorithm. IEEE Trans. Evol. Comput. **12**(1), 41–63 (2008)

24. Dai, G., Wang, J., Zhu, J.: A hybrid multi-objective algorithm using genetic and estimation of distribution based on design of experiments. In: IEEE International Conference on Intelligent Computing and Intelligent Systems (ICIS), vol. 1, pp. 284–288 (2009)
25. Liu, Y., Xiao, B., Dai, G.: Hybrid multi-objective algorithm based on probabilistic model. J. Comput. Appl. **31**(9), 2555–2558 (2011)
26. Yang, D., Jiao, L., Gong, M., Feng, H.: Hybrid multiobjective estimation of distribution algorithm by local linear embedding and an immune inspired algorithm. In: IEEE Congress on Evolutionary Computation (CEC), pp. 463–470 (2009)
27. Qi, Y., Liu, F., Liu, M., Gong, M., Jiao, L.: Multi-objective immune algorithm with Baldwinian learning. Appl. Soft Comput. **12**(8), 2654–2674 (2012)
28. Li, Y., Xu, X., Li, P., Jiao, L.: Improved RM-MEDA with local learning. Soft Comput. **18**(7), 1383–397 (2014)
29. Wang, H., Jiao, L., Shang, R., He, S., Liu, F.: A memetic optimization strategy based on dimension reduction in decision space. Evol. Comput. **18**(1), 69–100 (2015)
30. Mo, L., Dai, G., Zhu, J.: The RM-MEDA Based on elitist strategy. In: Cai, Z., Hu, C., Kang, Z., Liu, Y. (eds.) ISICA 2010. LNCS, vol. 6382, pp. 229–239. Springer, Heidelberg (2010). https://doi.org/10.1007/978-3-642-16493-4_24
31. Wang, Y., Xiang, J., Cai, Z.: A regularity model-based multiobjective estimation of distribution algorithm with reducing redundant cluster operator. Appl. Soft Comput. **12**(11), 3526–3538 (2012)
32. Zhang, Q., Li, H.: MOEA/D: a multiobjective evolutionary algorithm based on decomposition. IEEE Trans. Evol. Comput. **11**(6), 712–731 (2007)
33. Li, H., Zhang, Q.: Multiobjective optimization problems with complicated Pareto sets, MOEA/D and NSGA-II. IEEE Trans. Evol. Comput. **13**(2), 284–302 (2009)
34. Eichfelder, G.: Adaptive Scalarization Methods in Multiobjective Optimization. Springer, Heidelberg (2008). https://doi.org/10.1007/978-3-540-79159-1
35. Trivedi, A., Srinivasan, D., Sanyal, K., Ghosh, A.: A survey of multiobjective evolutionary algorithms based on decomposition. IEEE Trans. Evol. Comput. **21**(3), 440–462 (2017)
36. Zhou, A., Zhang, Q., Jin, Y., Tsang, E.P.K., Okabe, T.: A model-based evolutionary algorithm for bi-objective optimization. In: IEEE Congress on Evolutionary Computation (CEC), pp. 2568–2575 (2005)
37. Zitzler, E., Thiele, L., Laumanns, M., Fonseca, C.M., da Fonseca, V.G.: Performance assessment of multiobjective optimizers: an analysis and review. IEEE Trans. Evol. Comput. **7**(2), 117–132 (2003)

On Dealing with Uncertainties from Kriging Models in Offline Data-Driven Evolutionary Multiobjective Optimization

Atanu Mazumdar[1](✉), Tinkle Chugh[2], Kaisa Miettinen[1],
and Manuel López-Ibáñez[3]

[1] University of Jyvaskyla, Faculty of Information Technology, P.O. Box 35 (Agora),
FI-40014 University of Jyvaskyla, Finland
`atanu.a.mazumdar@student.jyu.fi`
[2] Department of Computer Science, University of Exeter, Exeter, UK
[3] Alliance Manchester Business School, University of Manchester, Manchester, UK

Abstract. Many works on surrogate-assisted evolutionary multiobjective optimization have been devoted to problems where function evaluations are time-consuming (e.g., based on simulations). In many real-life optimization problems, mathematical or simulation models are not always available and, instead, we only have data from experiments, measurements or sensors. In such cases, optimization is to be performed on surrogate models built on the data available. The main challenge there is to fit an accurate surrogate model and to obtain meaningful solutions. We apply Kriging as a surrogate model and utilize corresponding uncertainty information in different ways during the optimization process. We discuss experimental results obtained on benchmark multiobjective optimization problems with different sampling techniques and numbers of objectives. The results show the effect of different ways of utilizing uncertainty information on the quality of solutions.

Keywords: Machine learning · Gaussian process · Pareto optimality · Metamodelling · Surrogate

1 Introduction

Sometimes in real applications, multiple conflicting objectives should be optimized, but there is no mathematical or simulation model of the objectives involved. Instead, there is data, e.g., obtained via physical experiments. In such cases, surrogate models can be built using the given data and optimization is then performed with the surrogate models. In the literature, surrogate models such as Kriging [8], neural networks [18] and support vector regression [16] have been typically used for solving computationally expensive optimization problems [6,10]. If we may conduct new (expensive) function evaluations when needed, this process is called online data-driven optimization [20]. When we do not have

© Springer Nature Switzerland AG 2019
K. Deb et al. (Eds.): EMO 2019, LNCS 11411, pp. 463–474, 2019.
https://doi.org/10.1007/978-3-030-12598-1_37

access to additional data during the optimization, we call it *offline data-driven optimization* [11].

In using surrogate models, the main challenge is to manage the models for improving convergence and diversity without too much sacrifice in the accuracy of models. In online data-driven optimization problems, an infill criterion [6] is maximized or minimized for updating the models iteratively during the optimization process. However, this is not applicable for offline data-driven optimization when no further data is available during the optimization process. So far, little research has been conducted on solving optimization problems, where no new data is available for managing the surrogates [4, 11, 20]. In such case, the quality of the solutions obtained after using the surrogate models is entirely dependent on the accuracy of the models and optimizer used.

When solving an offline data-driven problem with multiple conflicting objectives, one can fit models using all the data available for each objective function. Then an evolutionary multiobjective optimization (EMO) algorithm can be used on these models to find a set of approximated nondominated solutions. Essentially, in that case, an offline data-driven multiobjective optimization problem (MOP) can be divided into two major parts: model building and using an EMO algorithm.

Some surrogate models, like Kriging, provide uncertainty information (or standard deviation) about the predicted values. A low standard deviation implies that the actual objective function value has a higher chance of being close to the predicted value (though the actual function may remain unknown and the only information is the data available). Therefore, one possible way to improve the accuracy of the model is to utilize uncertainty in the fitted model as an additional objective to be optimized.

In this article, we study different ways to deal with the uncertainty information provided by the Kriging models in offline data-driven multiobjective optimization. Moreover, we consider the effect of using different initial sampling techniques on some benchmark test problems. In this study, we simulate offline problems by generating data for problems with known optimal solutions to be able to analyze the results. The results show the effect of utilizing uncertainty information in the quality of solutions.

The rest of this article is organized as follows. We summarize the basic concepts of data-driven optimization and Kriging model in Sect. 2. In Sect. 3, we present different approaches of incorporating uncertainty information in the optimization problem and present and analyze the results in Sect. 4. Finally, we draw conclusions in Sect. 5.

2   Background

2.1   Generic Offline Data-Driven EMO

We consider MOPs of the following form:

$$
\begin{aligned}
&\text{minimize } \{f_1(\mathbf{x}), \dots, f_k(\mathbf{x})\}, \\
&\quad \text{subject to } \mathbf{x} \in S,
\end{aligned}
\tag{1}
$$

with k (≥ 2) objective functions and the feasible set S is a subset of the decision space $\mathbb{R}^n$. For any feasible decision vector $\mathbf{x}$ we have a corresponding objective vector $f(\mathbf{x}) = (f_1(\mathbf{x}), \ldots, f_k(\mathbf{x}))$.

MOPs that are offline in nature can generally be solved by the approach given in Fig. 1. In what follows, we refer to it as a *generic approach*. As described in [11,21], the solution process can be split into three major components: (1) data collection, (2) model building and management, and (3) EMO method utilized. The collection of data may also incorporate data pre-processing, if it is required. Once the data has been obtained, the objectives and constraints of the MOP are formulated. The next stage is to build surrogate models (also known as meta-models) e.g. for each objective function using the available data. Finally, an EMO method is used to find nondominated solutions utilizing the surrogates as objective functions. As objectives to be optimized in (1) we have for $i = 1, \ldots, k$ the predicted means $\hat{f}_i$ of the surrogate of objective f_i and our objective vector is denoted by:

$$\hat{\mathbf{f}} = (\hat{f}_1(\mathbf{x}), \ldots, \hat{f}_k(\mathbf{x})). \tag{2}$$

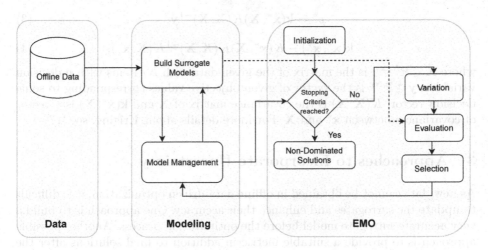

Fig. 1. Flowchart of a generic offline data-driven evolutionary multiobjective optimization approach.

Selecting proper surrogate models is a challenging task in model management. In online data-driven EMO, the quality of the surrogate models can be accessed and updated as new data becomes available during the optimization process. However, for offline data-driven EMO this is not possible. It becomes even more challenging with the data being noisy [22], skewed [23], time-varying [2] or heterogeneous [3]. Thus, it is crucial to build, before optimization, surrogates that are as good approximations as possible of the "true" objective functions. One way to improve the accuracy of the surrogates is to enhance the quality of the data. In this research, our consideration is on a general level and we do not go into the characteristics of the data.

In offline data-driven EMO, the possible ways to improve the accuracy of the surrogate models are to have an effective data pre-processing for noise removal [4], creating synthetic data [23], transferring knowledge [15] or applying advanced machine leaning techniques [19,20]. However, it is quite possible that the surrogate models are not good representations of the true objectives. It may even happen that the solutions obtained are actually worse than the data used for fitting the models.

2.2   Kriging

Kriging or Gaussian process regression has been widely used as a surrogate model for solving expensive optimization problems [6]. The main advantage of using Kriging is its ability to provide uncertainty information of the predicted values. Given a Kriging model, the approximated mean value y^* and its variance s^2 for a sample (or decision variable value) $\mathbf{x}^*$ are as follows:

$$y^* = \mathbf{k}(\mathbf{x}^*, \mathbf{X}) K(\mathbf{X}, \mathbf{X})^{-1} \mathbf{y}, \tag{3}$$

$$s^2 = \mathbf{k}(\mathbf{x}^*, \mathbf{x}^*) - K(\mathbf{x}^*, \mathbf{X}) K(\mathbf{X}, \mathbf{X})^{-1} K(\mathbf{X}, \mathbf{x}^*), \tag{4}$$

where $\mathbf{X} \in \mathbb{R}^{N_I \times n}$ is the matrix of the given data with N_I items with n decision variables, $\mathbf{y} \in \mathbb{R}^{N_I}$ is the vector of given objective values corresponding to some decision vector, $K(\mathbf{X}, \mathbf{X})$ is the covariance matrix of $\mathbf{X}$ and $\mathbf{k}(\mathbf{x}^*, \mathbf{X})$ is a vector of covariances between $\mathbf{x}^*$ and $\mathbf{X}$. For more details about Kriging, see [17].

3   Approaches to Incorporate Uncertainty

As new data cannot be obtained in offline data-driven optimization, it is difficult to update the surrogates and enhance their accuracy. One approach is to build a very accurate surrogate model before the optimization process. Another possible approach is to provide a suitable metric in addition to final solutions after the optimization process, which can be used to measure the accuracy of solutions obtained. This approach can be beneficial when the surrogate models cannot provide a very exact representation of the true objective functions. One such instance can be when the data consists of optimal solutions. In such a case, the surrogate might not be a good representation of the actual objectives, which might lead to degraded final solutions. Providing a set of solutions together with the uncertainty information of predicted final solutions can be helpful in the decision making process.

As previously discussed, the two major components in offline data-driven optimization are building a surrogate model and using an EMO algorithm. In this research we have limited ourselves by focusing on a few variations of the optimization problem which try to minimize the uncertainty in the final solutions. As shown in Fig. 2, the uncertainties in the predicted value of the Kriging models are utilized as additional objective functions. By considering uncertainties in this way, the EMO method tries to minimize the predicted mean values from the fitted Kriging models by subsequently minimizing the standard deviations in the prediction. Thus, the final set of nondominated solutions will consist of solutions with different levels of uncertainty.

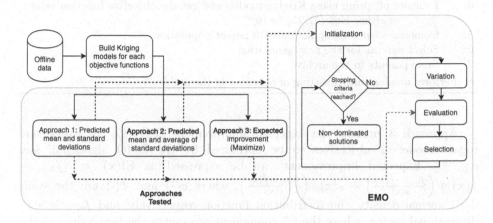

Fig. 2. Flowchart of offline data-driven optimization with uncertainty.

We have tested three different approaches for utilizing uncertainties in the optimization. Approach 1 uses all the standard deviations given by each surrogate model as additional objectives. The resulting objective vector in Approach 1 is:

$$\hat{\mathbf{f}} = (\hat{f}_1(\mathbf{x}), \ldots, \hat{f}_k(\mathbf{x}), s_1(\mathbf{x}), \ldots, s_k(\mathbf{x})), \tag{5}$$

where $\hat{f}_i(\mathbf{x})$ and $s_i(\mathbf{x})$ and are the predicted mean and the standard deviation values for the ith objective. Final solutions are obtained by performing a nondominated sort on the archive of predicted solutions (predicted mean values and standard deviations) stored while optimization. It might be possible that the solutions have different uncertainties for different objectives. We double the number of objectives which may increase the complexity of solving the resulting optimization problem.

Approach 2 utilizes the average of the standard deviations given by each of the surrogate models as an additional objective and the resulting objective vector is:

$$\hat{\mathbf{f}} = (\hat{f}_1(\mathbf{x}), \ldots, \hat{f}_k(\mathbf{x}), \bar{s}(\mathbf{x})), \tag{6}$$

where $\bar{s}(\mathbf{x})$ is the average of the standard deviations from Kriging models built for each objective function. This method has fewer objectives when compared to Approach 1, however, either of the approaches provide solutions with a range of uncertainty values. Both Approaches 1 and 2 can provide an option for filtering solutions based on the uncertainty information.

Algorithm 1. Uncertainties as additional objective functions

    Input: k Kriging models, one for each objective function and an empty archive
    Output: Final nondominated approximated solutions from the archive

1: Generate parent population
2: **while** *Stopping criteria are not reached* **do**
3:     Generate offspring with crossover and mutation
4:     Evaluate offspring using Kriging models and get the objective function values
        of either Eqs. (4), (5) or (6)
5:     Combine offspring population with parent population
6:     Select parents for the next generation
7:     Store parents in the archive
8: Perform nondominated sorting of solutions in the archive

Approach 3 utilizes the expected improvement (EI) [12] for every surrogate model as objectives to be optimized by the EMO algorithm, see, e.g. [9]. Expected improvement can be expressed as $\mathrm{EI}(\mathbf{x}) = (f_{min} - \hat{f}(\mathbf{x}))\Phi\left(\frac{f_{min}-\hat{f}(\mathbf{x})}{s(\mathbf{x})}\right) + s(\mathbf{x})\phi\left(\frac{f_{min}-\hat{f}(\mathbf{x})}{s(\mathbf{x})}\right)$, where $\phi(\cdot)$ and $\Phi(\cdot)$ are the standard normal density and distribution function respectively, and f_{min} is a k-dimensional vector, where the ith component represents the best values of the ith objective function in the given data. The objective vector in this case is:

$$\hat{\mathbf{f}} = (\mathrm{EI}_1(\mathbf{x}), \ldots, \mathrm{EI}_k(\mathbf{x})), \tag{7}$$

where $\mathrm{EI}_i(\mathbf{x})$ is the expected improvement value for the ith objective. The EI criterion takes the predicted mean value and the standard deviation into account.

Now we have introduced three approaches for incorporating uncertainty information. Algorithm 1 shows the process of applying any of them in the offline optimization process, where k is the number of objectives and we can use the maximum number of evaluations using surrogate models as a stopping criterion.

4   Experimental Results

We compare the three different approaches to each other and also to a generic approach (as (2) in Subsect. 2.1), using test problems DTLZ2, DTLZ4–DTLZ7 with 2, 3 and 5 objectives. As said, we generate data for these problems and fit Kriging models there. The dimension of the decision variable space n is fixed to 10.

The size of the data set used is 109 (corresponds to the $11n - 1$ [5,13,24]). The sampling techniques for creating the data sets were *Latin hypercube sampling* (LHS), *uniform random sampling* and a special case of sampling which we call *optimal-random sampling*. In the latter, 50% of the data are nondominated solutions and the remaining 50% are uniform random samples. This kind of hypothetical sampling might resemble a special case where most of the samples in the given data set are close to optimal, and thus the optimization process could no longer improve the solutions further. However, in such a scenario the offline optimization technique should not compute final solutions which are worse than the provided samples. A total of 31 independent runs from each sampling were performed for each case.

We used indicator based evolutionary algorithm (IBEA) [25] as the EMO method as it has been demonstrated to perform well in [1] even for problems with a higher number of objectives. The selection criterion was $I_{\epsilon+}$ (Step 6 in Algorithm 1) with κ parameter values 0.51, 0.87 and 0.48 for $k = 2, 3$ and 5, respectively, and κ value of 0.5 for any other number of objectives. The population size was 100 and the maximum number of function evaluations was 40 000 according to [1]. We used Matlab implementation of Kriging models with first order polynomial functions and a Gaussian kernel function.

For measuring the performance of different approaches, we first performed a nondominated sort on the archive (also including the additional objective(s)). These nondominated solutions were then evaluated with the real objective function. After obtaining their true objective function values, dominated solutions were removed producing the final nondominated set. For comparing the quality of solutions for all the approaches, inverted generational distance (IGD) metric was utilized with 5000 points in the reference set for all problems.

Table 1 shows the comparison between the mean and standard deviation values of the IGD for all the three approaches and the generic approach. It was observed that Approaches 1 and 2 performed better than the generic approach for LHS and uniform random sampling for all the problems with various numbers of objectives with the exception of DTLZ6 and DTLZ7. However, while using optimal-random sampling, Approaches 1 and 2 performed better than the generic approach for DTLZ2, DTLZ4-5 and better for DTLZ6 and DTLZ7 for few of the objectives. Approach 3 did not produce good results for any of the problems, objectives or sampling technique.

Adding uncertainties as additional objectives pose a major problem in explaining the effect of optimization as the fitness landscape of the uncertainties is mostly unknown. A possible explanation that no noticeable performance improvement is observed in DTLZ6 when using Approaches 1 and 2 is because the problem consists has a non-uniform (or biased) [7] degenerated Pareto front. Adding additional uncertainty objectives makes the problem even harder to solve and fewer nondominated solutions are obtained. For DTLZ7, a possible explanation for the worse performance of Approaches 1 and 2 is that the objective functions are completely separable [14]. Thus, the additional objectives added by Approaches 1 and 2 only make the problem more difficult than the generic approach.

For optimal-random sampling the advantage of Approaches 1 and 2 was clearly visible. Despite the initial sampling including also nondominated solutions, the generic approach failed to provide good solutions. This is because the surrogate models do not provide a perfect representation of the true objectives. While utilizing EIs as objectives in Approach 3, the solutions were actually worse (comparing mean IGD values) for most of the cases. This is because EI tries to balance between convergence and diversity. Therefore, it can select a solution with a high uncertainty for achieving its goal.

Figure 3 shows the root mean square error (RMSE) of the final solutions obtained by different approaches with LHS sampling on problems with two objectives. It can be observed that the solutions obtained by Approaches 1 and 2 are more accurate in most of the cases. This means that using uncertainty as additional objective(s) helps to find solutions with a low approximation error. Therefore, using uncertainty in the optimization process can be considered as an advantage in solving an offline data-driven EMO problem where there is no possibility for updating the surrogate models. An illustration of solutions obtained after evaluating them with real objectives for the DTLZ2 problem with LHS and optimal-random sampling is shown in Fig. 4. Due to space limitations, further analysis is available at http://www.mit.jyu.fi/optgroup/extramaterial.html as additional material. The performance of the proposed approaches on other test problems (i.e., DTLZ1, DTLZ3, WFG1-WFG3, WFG5 and WFG9) can also be found at the above-mentioned website.

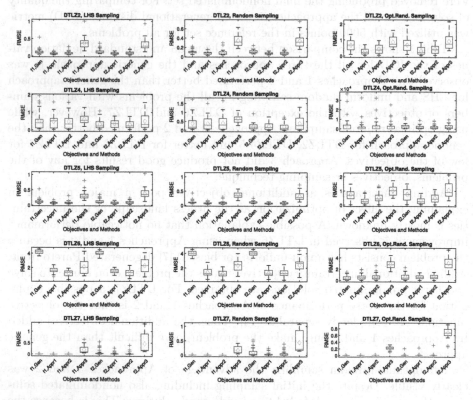

Fig. 3. RMSE of the final solutions for bi-objective problems. Here f1 and f2 are the objectives and "Gen", "Appr1", "Appr2" and "Appr3" are the generic and Approaches 1, 2 and 3, respectively. Opt.Rand is optimal-random sampling.

Table 1. Means and standard deviations of IGD values of the final archive, evaluated on the true objective functions, obtained by each approach, for various problems and sampling techniques. (Best values are in bold)

Sampling	Problems	k	Generic		Approach 1		Approach 2		Approach 3	
			Mean	Std.Dev.	Mean	Std.Dev.	Mean	Std.Dev.	Mean	Std.Dev.
LHS	DTLZ2	2	0.0989	0.1260	**0.0722**	**0.0431**	0.0770	0.0651	0.3377	0.0477
		3	0.2027	0.0910	0.1787	0.0530	**0.1665**	0.0539	0.3471	**0.0365**
		5	0.2708	0.0873	0.2689	**0.0343**	**0.2574**	0.0396	0.3993	0.0395
	DTLZ4	2	0.6311	**0.1619**	**0.3951**	0.1935	0.4919	0.1852	0.6467	0.2098
		3	0.7306	0.2021	**0.5309**	0.1413	0.5867	0.1467	0.7166	**0.1162**
		5	0.6929	0.0766	**0.5640**	0.0653	0.6062	0.0545	0.7173	**0.0514**
	DTLZ5	2	0.1030	0.1326	0.1032	0.0905	**0.0814**	**0.0570**	0.3716	0.0580
		3	0.1191	0.0982	**0.0684**	**0.0315**	0.0701	0.0452	0.2676	0.0388
		5	0.0934	0.0606	**0.0655**	**0.0277**	0.0805	0.0453	0.1486	0.0387
	DTLZ6	2	**0.1570**	**0.1078**	1.6188	0.7635	2.4518	0.5797	3.5210	1.1369
		3	**0.9871**	**0.2737**	1.7564	0.7308	1.5561	0.7159	3.2847	1.1907
		5	**0.8207**	**0.2158**	2.3859	0.4822	1.3725	0.3734	2.8157	1.0211
	DTLZ7	2	**0.0023**	**0.0049**	0.0292	0.0095	0.0095	0.0086	0.6157	0.1767
		3	**0.0549**	**0.0120**	0.1791	0.1721	0.0956	0.1449	0.6529	0.1016
		5	**0.2800**	**0.0541**	0.5254	0.2175	0.3675	0.1234	0.7169	0.0888
Random	DTLZ2	2	0.0947	0.0893	0.0879	0.0468	**0.0828**	0.0493	0.3673	**0.0395**
		3	0.2315	0.0712	0.1907	0.0534	**0.1692**	**0.0316**	0.3591	0.0433
		5	0.2843	0.0790	0.2593	**0.0268**	**0.2514**	0.0335	0.4188	0.0289
	DTLZ4	2	0.5986	0.1857	**0.4461**	0.1850	0.4665	**0.1735**	0.4935	0.2243
		3	0.7885	0.1465	**0.5354**	0.1474	0.5682	**0.1320**	0.7680	0.1544
		5	0.7064	0.1731	**0.5487**	0.1021	0.6034	0.1127	0.7391	**0.0697**
	DTLZ5	2	0.1144	0.1211	0.0949	0.0495	**0.0889**	0.0506	0.3590	**0.0481**
		3	0.1114	0.0367	**0.0610**	0.0291	0.0615	**0.0283**	0.2823	0.0350
		5	0.0644	0.0447	**0.0498**	**0.0169**	0.0542	0.0254	0.1521	0.0319
	DTLZ6	2	**0.2826**	**0.3739**	1.8949	1.0420	2.6166	0.7696	4.6779	1.2463
		3	**1.2833**	**0.2710**	2.9273	0.4893	1.2966	0.4552	3.0290	0.9259
		5	**0.7897**	**0.2869**	2.5206	0.6990	1.6732	0.6577	2.9527	1.1470
	DTLZ7	2	**0.0081**	**0.0113**	0.0444	0.0254	0.0260	0.0382	0.5942	0.1295
		3	**0.0500**	**0.0261**	0.1635	0.1030	0.0853	0.0443	0.6159	0.0980
		5	**0.2821**	**0.0235**	0.5763	0.2356	0.4916	0.3096	0.7254	0.0781
Optimal-Random	DTLZ2	2	0.4220	0.2079	**0.0053**	**0.0020**	0.0090	0.0029	0.1244	0.1827
		3	0.3152	0.2285	**0.0517**	**0.0101**	0.0554	0.0120	0.2088	0.1247
		5	0.1619	0.0604	0.1582	0.0143	**0.1404**	0.0253	0.2758	**0.0078**
	DTLZ4	2	0.8335	0.8480	**0.0194**	**0.0160**	0.0526	0.0351	0.5851	0.4683
		3	0.7853	0.1831	**0.2662**	**0.0738**	0.2966	0.0857	0.5575	0.1704
		5	0.5789	0.1020	**0.4319**	0.1062	0.4730	0.0904	0.6047	**0.0801**
	DTLZ5	2	0.7489	0.4255	**0.0086**	**0.0024**	0.0094	0.0032	0.2086	0.2516
		3	0.3323	0.3085	**0.0064**	0.0018	0.0076	**0.0017**	0.1010	0.0845
		5	0.1890	0.2090	**0.0049**	**0.0019**	0.0055	0.0021	0.0251	0.0232
	DTLZ6	2	**0.0064**	**0.0031**	0.0077	**0.0013**	0.0081	0.0019	0.0147	0.0019
		3	0.0556	0.0868	**0.0075**	**0.0021**	0.0085	0.0029	0.0198	0.0104
		5	0.0396	0.0986	**0.0069**	0.0014	0.0085	**0.0012**	0.0171	0.0078
	DTLZ7	2	**0.0005**	0.0004	0.0013	**0.0003**	0.0020	0.0007	0.0177	0.0033
		3	0.0397	0.0093	**0.0365**	**0.0043**	0.0388	0.0058	0.1012	0.0124
		5	0.1910	0.0179	**0.1855**	**0.0141**	0.1825	0.0220	0.3404	0.0367

LHS Sampling

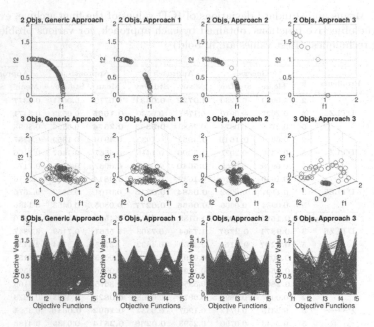

Optimal-Random Sampling

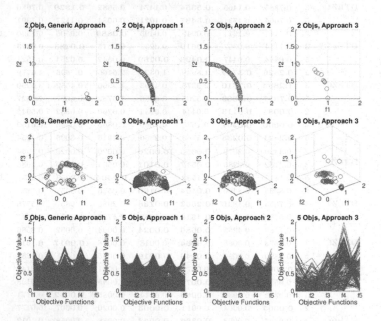

Fig. 4. Final solutions obtained of the run with the median IGD value using different approaches for LHS sampling (top three rows) and optimal-random sampling (bottom three rows) for the DTLZ2 problem.

5   Conclusions

We have considered offline data-driven optimization with evolutionary multiobjective optimization. We used Kriging to fit surrogate models to data and proposed and tested three approaches to utilize uncertainty information from Kriging models in the optimization. A comparison was done with several benchmark problems, sampling techniques and varying the number of objectives in solving offline data-driven multiobjective optimization problems. Adding uncertainty as one or more objectives showed improvements in the final solutions for certain problems in our benchmark testing. However, utilizing expected improvements as objectives (in Approach 3) did not seem to be effective in solving this kind of problems. The analysis also revealed that the solutions obtained in Approaches 1 and 2 are more accurate compared to the ones obtained using a generic approach (without uncertainty information).

Future work will include comparing the performance of the proposed approaches with bigger initial sample sizes, higher number of decision variables and higher number of objectives. Aiding the decision making process by giving a decision maker an option to select a final solution using the uncertainty information is another direction to work on. Moreover, filtering techniques can be applied to remove solutions with higher uncertainties. Testing on real-world data sets and exploring different ways to deal with uncertainties using other surrogate models will also be future research topics.

Acknowledgements. This research is related to the thematic research area Decision Analytics utilizing Causal Models and Multiobjective Optimization (DEMO) at the University of Jyvaskyla. This work was partially supported by the Natural Environment Research Council [NE/P017436/1].

References

1. Bezerra, L.C.T., López-Ibáñez, M., Stützle, T.: A large-scale experimental evaluation of high-performing multi- and many-objective evolutionary algorithms. Evol. Comput. **26**, 621–656 (2018)
2. Blackwell, T., Branke, J.: Multiswarms, exclusion, and anti-convergence in dynamic environments. IEEE Trans. Evol. Comput. **10**(4), 459–472 (2006)
3. Castano, S., Antonellis, V.D.: Global viewing of heterogeneous data sources. IEEE Trans. Knowl. Data Eng. **13**(2), 277–297 (2001)
4. Chugh, T., Chakraborti, N., Sindhya, K., Jin, Y.: A data-driven surrogate-assisted evolutionary algorithm applied to a many-objective blast furnace optimization problem. Mater. Manuf. Process. **32**(10), 1172–1178 (2017)
5. Chugh, T., Jin, Y., Miettinen, K., Hakanen, J., Sindhya, K.: A surrogate-assisted reference vector guided evolutionary algorithm for computationally expensive many-objective optimization. IEEE Trans. Evol. Comput. **22**(1), 129–142 (2018)
6. Chugh, T., Sindhya, K., Hakanen, J., Miettinen, K.: A survey on handling computationally expensive multiobjective optimization problems with evolutionary algorithms. Soft Comput. (to appear). https://doi.org/10.1007/s00500-017-2965-0

7. Coello, C., Lamont, G., Veldhuizen, D.: Evolutionary Algorithms for Solving Multi-Objective Problems, 2nd edn. Springer, New York (2007). https://doi.org/10.1007/978-0-387-36797-2

8. Forrester, A., Sobester, A., Keane, A.: Engineering Design via Surrogate Modelling. Wiley, Hoboken (2008)

9. Jeong, S., Obayashi, S.: Efficient global optimization (EGO) for multi-objective problem and data mining. In: 2005 IEEE Congress on Evolutionary Computation, vol. 3, pp. 2138–2145 (2005)

10. Jin, Y.: Surrogate-assisted evolutionary computation: recent advances and future challenges. Swarm Evol. Comput. **1**, 61–70 (2011)

11. Jin, Y., Wang, H., Chugh, T., Guo, D., Miettinen, K.: Data-driven evolutionary optimization: an overview and case studies. IEEE Trans. Evol. Comput. (to appear). https://doi.org/10.1109/TEVC.2018.2869001

12. Jones, D.R., Schonlau, M., Welch, W.J.: Efficient global optimization of expensive black-box functions. J. Global Optim. **13**(4), 455–492 (1998)

13. Knowles, J.: ParEGO: a hybrid algorithm with on-line landscape approximation for expensive multiobjective optimization problems. IEEE Trans. Evol. Comput. **10**(1), 50–66 (2006)

14. Li, K., Omidvar, M.N., Deb, K., Yao, X.: Variable interaction in multi-objective optimization problems. In: Handl, J., Hart, E., Lewis, P.R., López-Ibáñez, M., Ochoa, G., Paechter, B. (eds.) PPSN 2016. LNCS, vol. 9921, pp. 399–409. Springer, Cham (2016). https://doi.org/10.1007/978-3-319-45823-6_37

15. Pan, S.J., Yang, Q.: A survey on transfer learning. IEEE Trans. Knowl. Data Eng. **22**(10), 1345–1359 (2010)

16. Pilat, M., Neruda, R.: Aggregate meta-models for evolutionary multiobjective and many-objective optimization. Neurocomputing **116**, 392–402 (2013)

17. Rasmussen, C., Williams, C.: Gaussian Processes for Machine Learning (Adaptive Computation and Machine Learning). The MIT Press, Cambridge (2005)

18. Regis, R.G.: Evolutionary programming for high-dimensional constrained expensive black-box optimization using radial basis functions. IEEE Trans. Evol. Comput. **18**(3), 326–347 (2014)

19. Sun, X., Gong, D., Jin, Y., Chen, S.: A new surrogate-assisted interactive genetic algorithm with weighted semisupervised learning. IEEE Trans. Cybern. **43**(2), 685–698 (2013)

20. Wang, H., Jin, Y., Jansen, J.O.: Data-driven surrogate-assisted multiobjective evolutionary optimization of a trauma system. IEEE Trans. Evol. Comput. **20**(6), 939–952 (2016)

21. Wang, H., Jin, Y., Sun, C., Doherty, J.: Offline data-driven evolutionary optimization using selective surrogate ensembles. IEEE Trans. Evol. Comput. (to appear). https://doi.org/10.1109/TEVC.2018.2834881

22. Wang, H., Zhang, Q., Jiao, L., Yao, X.: Regularity model for noisy multiobjective optimization. IEEE Trans. Cybern. **46**(9), 1997–2009 (2016)

23. Wang, S., Minku, L.L., Yao, X.: Resampling-based ensemble methods for online class imbalance learning. IEEE Trans. Knowl. Data Eng. **27**(5), 1356–1368 (2015)

24. Zhang, Q., Liu, W., Tsang, E., Virginas, B.: Expensive multiobjective optimization by MOEA/D with Gaussian process model. IEEE Trans. Evol. Comput. **14**(3), 456–474 (2010)

25. Zitzler, E., Künzli, S.: Indicator-based selection in multiobjective search. In: Yao, X., et al. (eds.) PPSN 2004. LNCS, vol. 3242, pp. 832–842. Springer, Heidelberg (2004). https://doi.org/10.1007/978-3-540-30217-9_84

Convergence Acceleration for Multiobjective Sparse Reconstruction via Knowledge Transfer

Bai Yan[1]([⊠]), Qi Zhao[2], J. Andrew Zhang[3], Yonghui Li[4], and Zhihai Wang[5]

[1] Institute of Laser Engineering, Beijing University of Technology, Beijing, China
yanbai@emails.bjut.edu.cn
[2] College of Economics and Management, Beijing University of Technology,
Beijing, China
qzhao@emails.bjut.edu.cn
[3] Global Big Data Technologies Centre, University of Technology Sydney,
Sydney, Australia
Andrew.Zhang@uts.edu.au
[4] School of Electrical and Information Engineering, University of Sydney,
Sydney, Australia
yonghui.li@sydney.edu.au
[5] Key Laboratory of Optoelectronics Technology, Ministry of Education,
Beijing University of Technology, Beijing, China
wangzhihai@bjut.edu.cn

Abstract. Multiobjective sparse reconstruction (MOSR) methods can potentially obtain superior reconstruction performance. However, they suffer from high computational cost, especially in high-dimensional reconstruction. Furthermore, they are generally implemented independently without reusing prior knowledge from past experiences, leading to unnecessary computational consumption due to the re-exploration of similar search spaces. To address these problems, we propose a sparse-constraint knowledge transfer operator to accelerate the convergence of MOSR solvers by reusing the knowledge from past problem-solving experiences. Firstly, we introduce the deep nonlinear feature coding method to extract the feature mapping between the search of the current problem and a previously solved MOSR problem. Through this mapping, we learn a set of knowledge-induced solutions which contain the search experience of the past problem. Thereafter, we develop and apply a sparse-constraint strategy to refine these learned solutions to guarantee their sparse characteristics. Finally, we inject the refined solutions into the iteration of the current problem to facilitate the convergence. To validate the efficiency of the proposed operator, comprehensive studies on extensive simulated signal reconstruction are conducted.

Keywords: Sparse reconstruction ·
Multiobjective evolutionary algorithm · Learning · Knowledge transfer

© Springer Nature Switzerland AG 2019
K. Deb et al. (Eds.): EMO 2019, LNCS 11411, pp. 475–487, 2019.
https://doi.org/10.1007/978-3-030-12598-1_38

1   Introduction

In the compressed sensing (CS) theory [2,7], a sparse reconstruction problem is often considered:

$$\min_{\mathbf{x}} \|\mathbf{x}\|_0, \ s.t. \ \mathbf{b} = \mathbf{A}\mathbf{x} + \mathbf{e}, \tag{1}$$

where $\mathbf{x} \in \Re^n$ is a k-sparse signal (i.e., there are k nonzero values ($k < n$) in the signal), $\mathbf{A} \in \Re^{m \times n} (m \leq n)$ is the measurement matrix, $\mathbf{b} \in \Re^{m \times 1}$ is the measurement vector, and $\mathbf{e} \in \Re^{m \times 1}$ denotes the noise vector.

The problem (1) can be rewritten as an unconstrained optimization problem:

$$\arg\min_{\mathbf{x}} \ \lambda\|\mathbf{x}\|_p + \frac{1}{2}\|\mathbf{b} - \mathbf{A}\mathbf{x}\|_2^2, \ p \in [0,1] \tag{2}$$

where λ is a pre-chosen positive regularization parameter being introduced to balance the two conflicting objective terms (the regularization term and measurement error). Unfortunately, there is no optimal rule for determining λ. Some heuristics methods are used, e.g., the Homotopy continuation methods [6,12], and the cross validation method [17].

The regularization methods can be naturally solved by the multiobjective evolutionary algorithms (MOEAs) [5,15,22]. MOEAs can simultaneously optimize all the objectives and obtain a number of nondominated solutions (termed as Pareto front, PF). In this regards, (2) is transformed into a MOSR problem:

$$f(\mathbf{x}) = \min_{\mathbf{x}}(\|\mathbf{x}\|_0, \|\mathbf{A}\mathbf{x} - \mathbf{b}\|_2^2). \tag{3}$$

The first solver to the MOSR problem is the soft-thresholding evolutionary multiobjective (StEMO) algorithm [14]. StEMO is based on the NSGA-II framework [5] and it uses the IST method [4] for local search. It was observed that the knee region can provide the best trade-off solution. To enhance the reconstruction precision, the LBEA is proposed in [21], which employed the improved linear Bregman-based local search operator in the differential evolution paradigm to accelerate the convergence. A two-phase evolutionary approach for sparse reconstruction is proposed in [23]. In phase 1, the statistical features of the nondominated solutions from MOEA/D [22] were extracted to generate new solutions. In phase 2, a forward-based selection method was designed for better locating the nonzero entries. An improved MOEA/D equipped with sparse preference-based local search, denoted as SPLS, was proposed in [13]. The knee region was exploited with preference. In [20], an adaptive decomposition-based evolutionary approach (ADEA) is proposed. With the guidance of reference vectors, more search effort on the approximating knee region was executed by adaptively adding the reference vectors.

Although these MOSR solvers can achieve better reconstruction performance than conventional algorithms, they suffer from high computational cost, especially in high-dimensional reconstruction scenarios. When the signal is less sparse or there are fewer measurements, more iterations are needed and the computational cost is further increased.

In addition, many optimization solvers, including MOSR, are implemented independently without reusing the previous problem-solving knowledge. This causes unnecessary computational complexity due to the re-exploration of similar search spaces [10]. In fact, for practical artificial systems, the problems to be solved are rarely isolated, but may be repetitive or share domain-specific similarities. Some studies on evolutionary optimizers [8,11,16] have demonstrated the accelerated effect of reusing the prior information. This finding motivates us to exploit the solution search knowledge from past solved problems for the MOSR problem, where similarities to the evolutionary optimizers exist in the problem form and the solution search process.

In this paper, we propose a sparse-constraint knowledge transfer operator to accelerate the convergence of MOSR solvers by exploiting the knowledge from a previous problem-solving process. Firstly, we introduce the deep nonlinear feature coding (DNFC) [19] to extract the feature mapping for the searching process between the current and a previously solved MOSR problem. Through the mapping, we learn a set of knowledge-induced solutions which contain the search experience of the past solved problem. We then propose a sparse constraint strategy to refine these learned solutions to guarantee their sparsity characteristics. Finally, we inject the refined solutions into the iteration process for solving the current problem to facilitate the convergence.

The rest of this paper is organized as follows. Section 2 gives a brief introduction to the background techniques. Section 3 details the proposed sparse-constraint knowledge transfer operator. In Sect. 4, the performance of the proposed operator is examined using two baseline MOSR algorithms StEMO and ADEA. Finally, conclusions are described in Sect. 5.

2   Preliminaries

2.1   Maximum Mean Discrepancy

The maximum mean discrepancy (MMD) [9] is a distance estimation method, which measures the discrepancy between two distributions by comparing the difference of the mean values. Specifically, let $\mathbf{X}_s = [\mathbf{x}_1, \mathbf{x}_2, \ldots, \mathbf{x}_{n_s}]$ and $\mathbf{Y}_t = [\mathbf{y}_1, \mathbf{y}_2, \ldots, \mathbf{y}_{n_t}]$ denote the samples of two distributions on a domain χ, and Ω is a function: $\chi \to \Re$, then the MMD can be formulated as

$$\| \frac{1}{n_s} \sum_{i=1}^{n_s} \Omega(\mathbf{x}_i) - \frac{1}{n_t} \sum_{i=1}^{n_t} \Omega(\mathbf{y}_i) \|_\chi. \tag{4}$$

Further, the study [18] performed the MMD method in a Reproducing Kernel Hilbert Space (RKHS) to capture the nonlinear divergence between $\mathbf{X}_s$ and $\mathbf{Y}_t$. The function Ω is replaced by a kernel-induced feature map $\Phi : \chi \to \mathcal{H}$ with $\mathbf{K}(\mathbf{x}_i, \mathbf{x}_j) = \Phi(\mathbf{x}_i)^T \Phi(\mathbf{x}_j)$ as the kernel of $\mathcal{H}$, (4) can be written as

$$\| \frac{1}{n_s} \sum_{i=1}^{n_s} \Phi(\mathbf{x}_i) - \frac{1}{n_t} \sum_{i=1}^{n_t} \Phi(\mathbf{y}_i) \|_\mathcal{H}. \tag{5}$$

2.2   Marginalized Denoising Autoencoder

The marginalized denoising autoencoder (mDA) [3] aims at learning a linear mapping $\mathbf{W}$ to reconstruct the original data from its corrupted versions. Assume $\overline{\mathbf{X}} = [\mathbf{X}, \mathbf{X}, \ldots \mathbf{X}]$ is the union of r-times repeated $\mathbf{X} = [\mathbf{x}_1, \mathbf{x}_2, \ldots, \mathbf{x}_{n_X}]$; and $\widetilde{\mathbf{X}} = [\widetilde{\mathbf{X}^1}, \widetilde{\mathbf{X}^2}, \ldots, \widetilde{\mathbf{X}^r}]$ is the combination of different r-times corrupted versions of $\mathbf{X}$ by random feature removal (i.e., each feature of $\mathbf{X}$ is corrupted to 0 with probability p), then the objective function of mDA can be modeled as

$$\min L(\mathbf{W}) = \frac{1}{2rn_X} tr[(\overline{\mathbf{X}} - \mathbf{W}\widetilde{\mathbf{X}})^T(\overline{\mathbf{X}} - \mathbf{W}\widetilde{\mathbf{X}})] \tag{6}$$

where $tr(\cdot)$ is a function to compute the trace of a matrix.

3   MOSR via Transfer Operator

In this section, we propose a novel sparse-constraint knowledge transfer operator to speed up the convergence of MOSR solvers. The motivation is that, although sparse reconstruction problems vary from each other, they are not isolated and may be repetitive or have some domain-specific similarities. Therefore, we design this operator to reuse the structural knowledge from previous search experiences, accelerating the convergence.

3.1   Framework

The workflow of proposed operator in MOSR solvers is provided in Fig. 1, with the corresponding pseudo-code provided in Algorithm 1. For convenience, we denote the current MOSR problem as the *target problem* and name the previously solved problem which we want to transfer knowledge from as the *source problem*. $\mathbf{P}^t$ and $\mathbf{PS}^t$ are the solution sets to the target and source problems at generation t, respectively, and $\mathbf{PS}^{t_{max}}$ is the set of the optimized solutions to the source problem. As shown in Fig. 1, in each iteration of the target problem, the recombination (i.e., crossover and mutation), local search, and selection steps of MOSR solvers are firstly executed. Subsequently, the proposed sparse-constraint knowledge transfer operator, depicted in the dotted box, is implemented. This operator will be detailed in the next subsection. Finally, $\mathbf{P}^t$ and the new obtained population $\mathbf{T}^t$ by knowledge transfer will undergo the selection process, and a final solution would be identified.

3.2   Sparse-Constraint Knowledge Transfer Operator

The pseudo-code of sparse-constraint knowledge transfer operator is given in Algorithm 2, which includes four steps: (a) feature mapping extraction: it aims to learn a mapping $\mathbf{W}$ that provides the connection between the source and target problem; (b) knowledge-induced solutions acquisition: with this mapping, the most valuable search experience from the source problem $\mathbf{PS}^{t_{max}}$ is injected for improving the target problem-solving ability; (c) solution sparsification: it can ensure the sparse characteristics of solutions; (d) selection: a number of N better solutions are selected. Next, we will introduce these steps in detail.

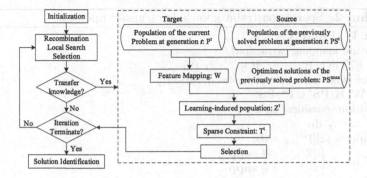

Fig. 1. Workflow of MOSR via knowledge transfer

Algorithm 1. Pseudo-code of MOSR via knowledge transfer

Require: A, b, $\mathbf{PS}^t$, N
1: $\mathbf{P}^0 \leftarrow$ Initialization;
2: **while** the stopping criterion is not met **do**
3:    $\mathbf{Q}^t =$Recombination$(\mathbf{P}^t)$;
4:    $\mathbf{L}^t =$Local_Search$(\mathbf{P}^t \cup \mathbf{Q}^t)$;
5:    $\mathbf{P}^t =$Selection$(\mathbf{P}^t \cup \mathbf{Q}^t \cup \mathbf{L}^t)$;
6:    **if** knowledge transfer **then**
7:        $\mathbf{P}^{t+1} =$Knowledge Transfer$(\mathbf{P}^t, \mathbf{PS}^t, \mathbf{PS}^{t_{max}})$;
8:    **end if**
9:    $t = t + 1$;
10: **end while**
11: $\mathbf{x} \leftarrow$Final_ Solution_Identification$(\mathbf{P}^t)$

Feature Mapping Extraction. It aims at finding a "connective bridge" **W** for the searching process between the source and target problems. DNFC [19] is an effective method for domain adaptation and provides a closed-form solution. Therefore, we employ the single-layer form of DNFC (named NFC) to predict the feature mapping matrix **W**. NFC incorporates the MMD (refer to Sect. 2.1) and kernelization into the mDA (refer to Sect. 2.2), in which the MMD enables the extracted features from the source and target problem to have a small distribution discrepancy, the kernelization ensures the nonlinearity relationship between domains to be well exploited, and mDA is for extracting deep features.

Now let us consider how to obtain **W** by the NFC method. Firstly, we define $\mathbf{X} = \mathbf{PS}^t \cup \mathbf{P}^t$, $\overline{\mathbf{X}} = [\mathbf{X}, \mathbf{X}, \dots \mathbf{X}]$ as the union of the r-times copies of **X**, $\widetilde{\mathbf{X}} = [\widetilde{\mathbf{X}^1}, \widetilde{\mathbf{X}^2}, \dots \widetilde{\mathbf{X}^r}]$ as the union of the r-times corrupted versions of **X**, n_s and n_t denote the population size of $\mathbf{PS}^t$ and $\mathbf{P}^t$ respectively. Then, the objective function to obtain the feature mapping matrix **W** can be formulated as

---

Algorithm 2. Sparse-constraint knowledge transfer operator

Require: $\mathbf{P}^t$, $\mathbf{PS}^t$, $\mathbf{PS}^{t_{max}}$, θ
1: /*Feature mapping extraction*/
2: $\mathbf{W} = E[\mathbf{R}1](E[\mathbf{R}2] + \theta E[\mathbf{R}3])^{-1}$; // (9)
3: /*Knowledge-induced solutions acquisition*/
4: $\mathbf{Z}^t = \mathbf{W}_k \mathbf{K}(\mathbf{PS}^t \cup \mathbf{P}^t, \mathbf{PS}^{t_{max}})$;
5: /*Solution sparsification*/
6: **for** $i = 1 : N$ **do**
7:     $\mathbf{supp} = \{z|[\mathbf{P}^t]_{i,z} = 0\}$;
8:     $[\mathbf{T}^t]_{i,z} = \begin{cases} [\mathbf{Z}^t]_{i,z}, & z \notin \mathbf{supp} \\ 0, & z \in \mathbf{supp} \end{cases}$;
9: **end for**
10: /*Selection*/
11: $\mathbf{P}^{t+1} = \text{Selection}(\mathbf{P}^t \cup \mathbf{T}^t)$

$$\Gamma(\mathbf{W}) = tr[(\overline{\mathbf{X}} - \mathbf{W}\widetilde{\Phi(\mathbf{X})})^T(\overline{\mathbf{X}} - \mathbf{W}\widetilde{\Phi(\mathbf{X})})]+$$

$$\|\frac{1}{n_s}\sum_{i=1}^{n_s}\Phi(\widetilde{\mathbf{X}_i^r}) - \frac{1}{n_t}\sum_{i=n_s+1}^{n_t}\Phi(\widetilde{\mathbf{X}_i^r})\|^2$$

$$= \underbrace{tr[(\overline{\mathbf{X}} - \mathbf{W}_k\widetilde{\mathbf{K}})^T(\overline{\mathbf{X}} - \mathbf{W}_k\widetilde{\mathbf{K}})]}_{\text{mDA}} + \underbrace{\theta\, tr(\mathbf{W}_k\widetilde{\mathbf{K}}\widetilde{\mathbf{G}}\widetilde{\mathbf{K}}^T\mathbf{W}_k^T)}_{\text{MMD}} \quad (7)$$

where $\mathbf{W} = \mathbf{W}_k\Phi(\mathbf{X})^T$, $\Phi(\mathbf{X})$ is the mapped $\mathbf{X}$ in the RKHS; $\mathbf{K} = \Phi(\mathbf{X})^T\Phi(\mathbf{X})$ is the corresponding kernel matrix; $\widetilde{\mathbf{K}}$ is the corrupted kernel matrix with a corruption probability p; $\mathbf{G} = [\mathbf{G}_{i,j}]_{(n_s+n_t)\times(n_s+n_t)}$ with $G_{i,j} = 1/n_s^2$ if $\mathbf{X}_{i,j} \in \mathbf{PS}^t$, $G_{i,j} = 1/n_t^2$ if $\mathbf{X}_{i,j} \in \mathbf{P}^t$, $G_{i,j} = -1/(n_sn_t)$ otherwise; θ is the balancing parameter. Applying the weak law of large numbers and computing the expectations when $r \to \infty$, a closed-form solution for $\mathbf{W}_k$ that minimizes (7) can be obtained as

$$\mathbf{W}_k = E[\mathbf{R}1](E[\mathbf{R}2] + \theta E[\mathbf{R}3])^{-1} \quad (8)$$

with

$$E[\mathbf{R}1] = (1-p)\mathbf{X}\mathbf{K}^T$$

$$E[\mathbf{R}2]_{i,j} = \begin{cases} (1-p)^2\mathbf{K}\mathbf{K}^T, & i \neq j \\ (1-p)\mathbf{K}\mathbf{K}^T, & i = j \end{cases}$$

$$E[\mathbf{R}3]_{i,j} = \begin{cases} (1-p)^2\mathbf{K}\mathbf{G}\mathbf{K}^T, & i \neq j \\ (1-p)^2\mathbf{K}\mathbf{G}\mathbf{K}^T + p(1-p)\mathbf{K}\mathbf{F}\mathbf{K}^T, & i = j, \end{cases} \quad (9)$$

where $\mathbf{F}$ is a diagonal matrix having the same diagonal elements with $\mathbf{G}$. For detailed derivation process of Eqs. (7)–(9), please refer to [19].

Acquirement of Knowledge-Induced Solutions. With $\mathbf{W}_k$, the search experience from the optimized solutions to the source problem $\mathbf{PS}^{t_{max}}$ can be transferred into the current iterations for resolving the target problem to improve the solution quality. As $\mathbf{W}_k$ is a connective mapping between $\mathbf{PS}^t$ and $\mathbf{P}^t$, we can obtain the knowledge-induced solution set $\mathbf{Z}^t$ as

$$
\begin{aligned}
\mathbf{Z}^t &= \mathbf{W}\Phi(\mathbf{PS}^{t_{max}}) \\
&= \mathbf{W}_k\Phi(\mathbf{X})^T\Phi(\mathbf{PS}^{t_{max}}) \\
&= \mathbf{W}_k\mathbf{K}(\mathbf{X}, \mathbf{PS}^{t_{max}}).
\end{aligned}
\tag{10}
$$

Sparse Constraint. To guarantee the sparsity characteristics of the acquired knowledge-induced solutions, we propose a sparse constraint strategy by an example in Fig. 2, where $\mathbf{T}^t$ is the new obtained solution set based on $\mathbf{P}^t$ and $\mathbf{Z}^t$, the white and uncolored lattices denote zero and nonzero entries, respectively. Specifically, we firstly find the locations of all zero elements in the i-th solution of $\mathbf{P}^t$: $\mathbf{supp} = \{z|[\mathbf{P}^t]_{i,z} = 0\}$, where $[\cdot]_{i,z}$ represents the element in the i-th row, z-th column. Then, the updated solution set $\mathbf{T}^t$ can be obtained as

$$
[\mathbf{T}^t]_{i,z} = \left\{ \begin{array}{l} [\mathbf{Z}^t]_{i,z}, z \notin \mathbf{supp}, \\ 0, \quad\ z \in \mathbf{supp}. \end{array} \right\}, \quad i = \{1, 2, \ldots N\}.
\tag{11}
$$

Therefore, $\mathbf{T}^t$ can not only possess the valuable knowledge extracted from the search experience for the past problem, but also inherit the sparse structure.

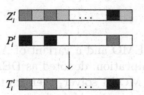

Fig. 2. Illustration of sparse constraint.

Selection. The selection operator of the MOSR solver is implemented in $\mathbf{P}^t \cup \mathbf{T}^t$ to select N elitism solutions for the next generation. If the knowledge transfer is beneficial, some of the sparsified learning-induced solutions survive in the selection procedure; otherwise, solutions in $\mathbf{T}^t$ will not be sent to the next generation, which avoids negative transfer.

4   Experiments and Discussions

4.1   Experimental Settings

Test Problems. We artificially generate a series of simulated signals as test problems. Firstly, a k-sparse signal $\mathbf{x}$ is produced, in which the nonzero elements are sampled from a Gaussian distribution $\mathcal{N}(0, 1)$. Then, a Gaussian matrix $\mathbf{A}$

is yielded and the measurement vector $\mathbf{b}$ is obtained by $\mathbf{b} = \mathbf{Ax}$. Lastly, the measurement $\mathbf{b}$ is corrupted by additive white Gaussian noise with elements from the normal distribution $N(0, 0.01)$. Each test problem involves three key parameters: (n, m, k). To better explore the effects of knowledge transfer, six complex test problems are randomly generated, as shown in Table 1.

Table 1. A list of test problems

Problem	(n, m, k)	Problem	(n, m, k)
P1	(1000, 200, 60)	P4	(1200, 250, 60)
P2	(1000, 300, 60)	P5	(1200, 350, 60)
P3	(1000, 300, 100)	P6	(1200, 350, 120)

Settings of the Sparse-Constraint Transfer Learning Operator. The corruption probability p is set between 0.5 and 0.9 with an interval of 0.1 by doing the cross-validation on the population of a past problem in the first generation. The balancing parameter θ and the kernel function are suggested to be 10^3 and 'RBF' respectively according to [19]. This operator is executed every five generations.

MOSR Solvers. Here, StEMO and a variant of ADEA (i.e., the ADEA without the reference vector adaptation, denoted as DEA) are employed as baseline solvers. We denote Δ as a MOSR solver, then its three versions are compared in this paper: the first version is the original solver Δ; the second and third versions are both equipped with the sparse-constraint knowledge transfer operator but receive the past experience from the source problems "P1–P3" and "P4–P6" respectively. We use different settings in the source problems because "P1–P3" and "P4–P6" have different dimensions and are heterogeneous. For convenience, we denote the second and third version as Δ-tr1 and Δ-tr2, respectively. For these solvers, their basic parameters are set as suggested in their original versions [14,20]. The population size of StEMO and ADEA is set to 50.

Terminate Criterion. For a fair comparison, all methods stop running when the maximum function evaluations reach 5000 times. Each algorithm runs 15 times in each test case.

Evaluation Criterion. All the versions of MOSR solvers are evaluated by hypervolume (HV) [1], which is the only parameter to measure the quality of a solution set. The larger HV values, the better the reconstruction quality achieved. For all scenarios, the reference sets for HV computation are all set to $(1000, 1.2)$, and the obtained HV results are normalized. In addition, the average reconstruction error (RE) and the RE variance under each test case are also compared, where $RE = \|\mathbf{x} - \mathbf{x}_G\|_2/\|\mathbf{x}_G\|_2$, $\mathbf{x}$ and $\mathbf{x}_G$ are the estimated and ground-truth signals. Smaller RE represents better reconstruction quality.

4.2   Experimental Results and Discussions

To evaluate the convergence performance of the proposed knowledge transfer operator, the HV values obtained by different versions of DEA and StEMO, across 15 independent runs with 5000 function evaluations are depicted in Figs. 3 and 4. In these two figures, the sub-figures (a)–(f) correspond to the six target problems P1–P6 respectively; the red, blue and black curve in each sub-figure indicates the original MOSR solver, the solvers with knowledge transfer from "P1–P3" and "P4–P6" respectively.

As seen from Fig. 3, DEA-tr1, and DEA-tr2 achieve higher accuracy or faster convergence than its original version in most problems. For P1, P3 and P4, DEA-tr1 and DEA-tr2 obtain much larger HV values than DEA when the function evaluations arrive at 5000. For P5 and P6, compared with DEA, the convergence of DEA-tr1 and DEA-tr2 is significantly faster, more than 500 function evaluations ahead. Except that DEA-tr1 converges slightly slower than DEA when solving P2, DEA-tr2 spends only 1500 function evaluations to generate the same HV values with those by DEA which takes about 2500 function evaluations.

Figure 4 shows similar observations when StEMO is employed as a baseline solver. StEMO-tr1 and StEMO-tr2 have larger HV values than StEMO when solving P1 and P5. For P3, P4 and P6 they converge much faster than StEMO, with a save of at least 500 function evaluations. When solving P2, three versions of StEMO have almost the same convergence performance.

The corresponding average REs and the variances of all solver versions for P1–P6 within 5000 function evaluations are given in Tables 2 and 3 respectively. In these tables, the symbols "≈", "+" and "−" represent that the average RE of the corresponding solver is similar, smaller and larger than that of its baseline solver respectively. As can be observed from Table 2, DEA-tr1 and DEA-tr2 have better or comparable reconstruction quality with respect to DEA in 10 out of 12 scenarios. Furthermore, the mean RE values of DEA-tr1 and DEA-tr2 are much smaller than that of DEA.

Similar results can be seen in Table 3 which presents the comparison results for the StEMO algorithms. The reconstruction quality of StEMO-tr1 and StEMO-tr2 is higher or similar to that of StEMO in 11 out of 12 test cases, thanks to the knowledge transfer. Besides, in terms of mean RE results for all problems, StEMO-tr1 and StEMO-tr2 achieve more satisfying reconstruction performance compared with StEMO.

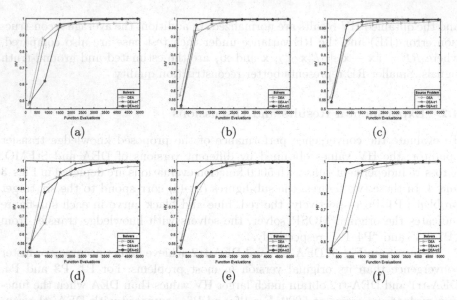

Fig. 3. Mean HV values obtained by three versions of DEA. (a)–(f) are the target problems to be solved: (a) P1. (b) P2. (c) P3. (d) P4. (e) P5. (f) P6. (Color figure online)

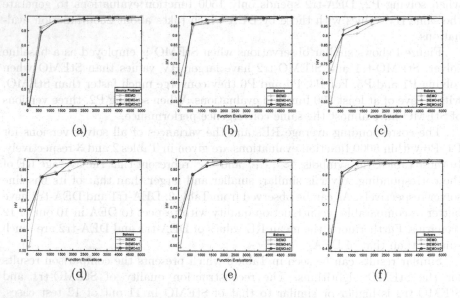

Fig. 4. Mean HV values obtained by three versions of StEMO. (a)–(f) are the target problems to be solved: (a) P1. (b) P2. (c) P3. (d) P4. (e) P5. (f) P6. (Color figure online)

Table 2. Comparisons of average REs and variance obtained by three versions of DEA

Problems	DEA	DEA-tr1	DEA-tr2
P1	0.3421 (±1.76E−2)	0.2249 (±3.46E−3) +	0.2359 (±2.00E−3) +
P2	0.1307 (±1.43E−5)	0.1302 (±1.56E−5) ≈	0.1418 (±1.63E−5) −
P3	0.5261 (±7.67E−4)	0.4248 (±7.83E−4) +	0.4319 (±9.03E−4) +
P4	0.2725 (±7.77E−2)	0.2287 (±3.82E−3) +	0.2265 (±2.97E−3) +
P5	0.2027 (±1.87E−4)	0.2016 (±2.00E−4) ≈	0.2124 (±1.85E−4) −
P6	0.3031 (±5.07E−3)	0.2889 (±8.98E−4) +	0.2968 (±8.23E−4) +
Mean	0.2962(±1.67E−2)	0.2498 (±1.53E−2) +	0.2575 (±1.15E−2) +

Table 3. Comparisons of average REs and variance obtained by three versions of StEMO

Problems	StEMO	StEMO-tr1	StEMO-tr2
P1	0.7201 (±1.73E−2)	0.5665 (±1.01E−2) +	0.5923 (±9.76E−3) +
P2	0.4754 (±9.07E−3)	0.4304 (±3.90E−4) +	0.4699 (±3.01E−4) ≈
P3	0.6632 (±9.82E−3)	0.6201 (±1.30E−2) +	0.6329 (±9.93E−3) +
P4	0.6995 (±6.10E−2)	0.5794 (±9.82E−3) +	0.6280 (±8.82E−3) +
P5	0.5791 (±2.85E−4)	0.5227 (±1.19E−4) +	0.5921 (±1.97E−3) −
P6	0.5997 (±6.11E−3)	0.5706 (±5.23E−3) +	0.5992 (±7.84E−3) ≈
Mean	0.6228 (±1.72E−2)	0.5483 (±6.36E−3) +	0.5586 (±6.44E−2) +

5 Conclusion

We presented a scheme for accelerating the convergence of MOSR solvers by introducing a sparse-constraint knowledge transfer operator that reuses the search experience from a previously solved problem. We employ the NFC technique to extract the feature mapping between the source and target problems, and then apply it to generate a set of knowledge-induced solutions for the target problem. A sparse constraint strategy is then proposed for sparsifying the obtained knowledge-induced solutions to reserve the sparse characteristics. Using StEMO and ADEA as the baseline MOSR solvers for several experimental problems, we demonstrate that the proposed operator can improve the convergence speed of MOSR solvers with high probability by transferring knowledge across either homogeneous or heterogeneous problems, without causing degradation in reconstruction accuracy.

Acknowledgments. This work was supported by the China Scholarship Council under Grant 201706540025.

References

1. Bader, J., Zitzler, E.: HypE: an algorithm for fast hypervolume-based many-objective optimization. Evol. Comput. **19**(1), 45–76 (2011)
2. Candès, E.J., Wakin, M.B.: An introduction to compressive sampling. IEEE Signal Process. Mag. **25**(2), 21–30 (2008)
3. Chen, M., Xu, Z., Weinberger, K., Sha, F.: Marginalized denoising autoencoders for domain adaptation. In: Proceedings of the 29th International Conference on Machine Learning (2012)
4. Combettes, P.L., Wajs, V.R.: Signal recovery by proximal forward-backward splitting. Multiscale Model. Simul. **4**(4), 1168–1200 (2005)
5. Deb, K., Pratap, A., Agarwal, S., Meyarivan, T.: A fast and elitist multiobjective genetic algorithm: NSGA-II. IEEE Trans. Evol. Comput. **6**(2), 182–197 (2002)
6. Dong, Z., Zhu, W.: Homotopy methods based on l_0-norm for compressed sensing. IEEE Trans. Neural Netw. Learn. Syst. **29**(4), 1132–1146 (2018)
7. Donoho, D.L.: Compressed sensing. IEEE Trans. Inf. Theory **52**(4), 1289–1306 (2006)
8. Feng, L., Ong, Y.S., Jiang, S., Gupta, A.: Autoencoding evolutionary search with learning across heterogeneous problems. IEEE Trans. Evol. Comput. **21**(5), 760–772 (2017)
9. Gretton, A., Borgwardt, K.M., Rasch, M., Schölkopf, B., Smola, A.J.: A kernel method for the two-sample-problem. In: Proceedings of the Conference on Neural Information Processing System, pp. 513–520 (2007)
10. Gupta, A., Ong, Y.S., Feng, L.: Insights on transfer optimization: because experience is the best teacher. IEEE Trans. Emerg. Topics Comput. Intell. **2**(1), 51–64 (2018)
11. Jiang, M., Huang, Z., Liming, Q., Huang, W., et al.: Transfer learning based dynamic multiobjective optimization algorithms. IEEE Trans. Evol. Comput. **22**(4), 501–514 (2017)
12. Jiao, Y., Jin, B., Lu, X.: Iterative soft/hard thresholding with homotopy continuation for sparse recovery. IEEE Signal Process. Lett. **24**(6), 784–788 (2017)
13. Li, H., Zhang, Q., Deng, J., Xu, Z.B.: A preference-based multiobjective evolutionary approach for sparse optimization. IEEE Trans. Neural Netw. Learn. Syst. **29**(5), 1716–1731 (2018)
14. Li, L., Yao, X., Stolkin, R., Gong, M., He, S.: An evolutionary multiobjective approach to sparse reconstruction. IEEE Trans. Evol. Comput. **18**(6), 827–845 (2014)
15. Liu, C., Zhao, Q., Yan, B., Elsayed, S., Ray, T., Sarker, R.: Adaptive sorting-based evolutionary algorithm for many-objective optimization. IEEE Trans. Evol. Comput. (in press). https://doi.org/10.1109/TEVC20182848254
16. Pan, S.J., Tsang, I.W., Kwok, J.T., Yang, Q.: Domain adaptation via transfer component analysis. IEEE Trans. Neural Netw. **22**(2), 199–210 (2011)
17. Sentelle, C.G., Anagnostopoulos, G.C., Georgiopoulos, M.: A simple method for solving the SVM regularization path for semidefinite kernels. IEEE Trans. Neural Netw. Learn. Syst. **27**(4), 709–722 (2016)
18. Steinwart, I.: On the influence of the kernel on the consistency of support vector machines. J. Mach. Learn. Res. **2**(Nov), 67–93 (2001)
19. Wei, P., Ke, Y., Goh, C.K.: Deep nonlinear feature coding for unsupervised domain adaptation. In: IJCAI, pp. 2189–2195 (2016)

20. Yan, B., Zhao, Q., Wang, Z., Zhang, J.A.: Adaptive decomposition-based evolutionary approach for multiobjective sparse reconstruction. Inf. Sci. **462**, 141–159 (2018)
21. Yan, B., Zhao, Q., Wang, Z., Zhao, X.: A hybrid evolutionary algorithm for multiobjective sparse reconstruction. Signal Image Video P. **11**, 993–1000 (2017)
22. Zhang, Q., Li, H.: MOEA/D: a multiobjective evolutionary algorithm based on decomposition. IEEE Trans. Evol. Comput. **11**(6), 712–731 (2007)
23. Zhou, Y., Kwong, S., Guo, H., Zhang, X., Zhang, Q.: A two-phase evolutionary approach for compressive sensing reconstruction. IEEE Trans. Cybern. **47**(9), 2651–2663 (2017)

Combinatorial EMO

Evolving Generalized Solutions for Robust Multi-objective Optimization: Transportation Analysis in Disaster

Keiki Takadama[1]([✉]), Keiji Sato[2], and Hiroyuki Sato[1]

[1] The University of Electro-Communications,
1-5-1, Chofugaoka, Chofu, Tokyo, Japan
keiki@inf.uec.ac.jp, h.sato@uec.ac.jp
[2] National Maritime Research Institute,
6-38-1, Shinkawa, Mitaka, Tokyo, Japan
sato-k@nmri.go.jp

Abstract. This paper proposes the multi-objective evolutionary algorithm (MOEA) that can evolve the *generalized* individuals, which include *many* solutions that can be applied into different situations with the *minimal change*. The intensive simulations on the waterbus route optimization problem as the real world problem have revealed the following implications: (1) the proposed MOEA cannot only optimize the solutions like general MOEAs but also can evolve the generalized individuals; and (2) the proposed MOEA can analyze the feature of the river transportation in the waterbus route optimization.

Keywords: Generalization ·
Evolutionary multi-objective optimization · Don't care symbol ·
Route optimization

1 Introduction

When disaster occurs in a center of city, a large number of persons are difficult to go home due to suspension of transportation service. In such a situation, bus (including waterbus) transportation attracts attention as the solution of this problem because the bus route can be changed flexibly according to road conditions (*e.g.*, traffic jam, road repair). However, many changed routes make passengers be confused. For this problem, it is necessary to develop a *robust* routes which can cope with *many* situations with a *minimal* route modification.

To tackle this issue, we focus on the multi-objective evolutionary algorithm (MOEA) [3] and propose a new MOEA that can evolve the *generalized* individuals, which include *many* solutions that can be applied into different situations with the *minimal change*, by employing the concept of the *generalization* in the context of Learning Classifier Systems (LCSs) [5, 7]. In detail, a *generalized* individual is represented by the chromosome including *don't care symbol* # which can

© Springer Nature Switzerland AG 2019
K. Deb et al. (Eds.): EMO 2019, LNCS 11411, pp. 491–503, 2019.
https://doi.org/10.1007/978-3-030-12598-1_39

be changed into any other symbol. Considering the binary chromosome represented by 0011##, for example, this chromosome means one of {001100, 001101, 001110, 001111}. From the viewpoint of MOEA, the proposed MOEA does not only evolve Pareto optimal solutions (POS) of the *specific* chromosomes (such as 001100) but also evolves POS of the *generalized* chromosomes (such as 0011##). More importantly, the solutions (even POS) evolved by general MOEAs cannot be applied in the route optimization in disaster while the generalized solutions evolved by the proposed MOEA can be applied, because the former solutions are fixed and mostly different each other while the latter solutions are flexible and similar as a set of individuals, which corresponds to many/small route change in the route optimization. For the implementation of the proposed MOEA, most MOEAs can be employed, but this paper employs NSGA-II (Non-dominated Sorting Genetic Algorithm II) [4] as one of major MOEAs and extends it to evolve the generalized individuals. Here, the proposed MOEA is called as the *generalization-based MOEA* (G-MOEA in short).

This paper is organized as follows. Section 2 introduce the *generalization* in the context of LCSs and proposes G-MOEA. Section 3 explains the water-bus route optimization problem as the real world problem. Section 4 shows the experimental result and discusses it. Finally, our conclusion is given in Sect. 5.

2   Generalization-Based MOEA

2.1   Learning Classifier System and Its Generalization

Learning classifier system (LCS) [5] is a machine learning system that learns a set of if-then rules, called *classifiers*. LCS aims at generalizing the classifiers by employing the don't care symbol "#" which represents any symbol such as 0 or 1 in the binary problem. By employing the "#" symbol, LCS can generate the classifiers which can be applied into many situations.

2.2   Swap-Based Generalization

The generalized individual represented by the "#" symbols can keep the system performance (*i.e.*, fitness) even if the value in the "#" locus changes. In this paper, we call such a generalized individual as s *robust* individual. To understand it in detail, let's focus on the multi-objective knapsack problem of 10 items, which determines whether the item is selected or not in the knapsack to maximize the total value of the selected items but not to exceed the prefixed maximum weight of the selected items. For example, the individual "#001#00111" in Fig. 1 means that the 4th, 8th, 9th, and 10th items are selected, the 2nd, 3rd, 6th, and 7th items are not selected, and the 1st and 5th items are either selected or not selected. If the rank of this individual can be kept in spite of a selection or not-selection of the 1st and 5th items, the individual is regarded as the generalized individual.

However, it is generally difficult to find such a generalized individual. Considering that the individual "#001#00111" in Fig. 1 covers four solutions, *a*, *b*, *i*

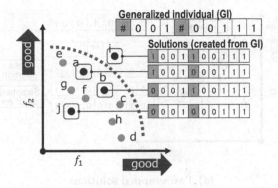

Fig. 1. Generalization in multi-objective problem

and j, in which the solution i exceeds the maximum weight of the knapsack while the fitness of the solution j is lower than that of a and b. This indicates that the fitness of four solutions differ, meaning that the individual "#001#00111" does not have the solutions located in the same rank. In contrast, if we assume that the individual "#001#00111" covers only two solutions, a and b, then the fitness of two solutions are similar because they are located in the same Pareto front. In this case, we can call it as the robust individual.

To obtain such robust individuals, this paper proposes the *swap-based generalization* where its individual is represented by (i) the base solution and (ii) the swapped solutions derived from the base solution by swapping the values in the "s#" (swap #) locus as shown in Fig. 2. As a function of "s#", "s#" should be "1" when the locus of the base solution is "0" and vice versa to create the swapped solution. In Fig. 2(a), the swap-based generalized individual is composed of three solutions, a, b, and c, where a is the base solution while b and c are the solutions created from a by changing to the opposite values of a in the 1st and 5th locus of b and in the 3rd and 4th locus of c. As shown in Fig. 2(b), any number of the swapped solutions can be created according to the number of a set of "s#" (*e.g.*, the four sets of "s#" in this figure), and any number of "s#" can be inserted in one set (*e.g.*, three "s#" are inserted in the 4th set of "s#" in this figure). Note that "1-0" (meaning that "1" and "0" in the "s#" locus), "0-0", "1-1", and "1-0-0" in the base solution respectively change to "0-1","1-1", "0-0", and "0-1-1" in the swapped solutions in Fig. 2(b).

2.3 Swap-Based Generalization Mechanism

Figure 3 shows how the individual is generalized. At the first cycle, (1) one of individuals (A in this figure) is selected from the population; (2) the base solution ("1001111000") in A searches the other individuals (from B to X) which are *exactly* matched with the base solution; (3) the base solution in A searches the other individuals (from B to X) which are *approximately* matched with the base solution under the condition that the hamming distance between the base

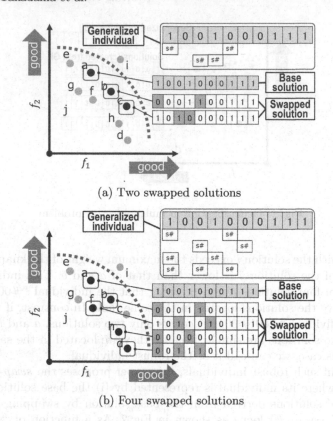

(a) Two swapped solutions

(b) Four swapped solutions

Fig. 2. Representation of swap-based generalization

solution in A and the base solution in one of the individuals from B to X is less than ($maxS\#$). In this figure, the base solution in A is matched with the solution ("1101101000") in one of the individuals from B to X, which values at the 2nd and 6th locus differ; (4) two "s#" are created at the 2nd and 6th locus in A. If the base and/or exactly matched solutions ("1001111000") have the "s#", the set of "s#" is added to A (no "s#" is added in this case); and (5) return to (1) to generalize the remaining individuals. Through this cycle, most of all individuals are generalized by adding the sets of "s#". At the X-th cycle, the steps from (1) to (3) are the same, but most of all solutions have the sets of "s#". In this figure, the base solution in A is matched with the solution ("1100110000") in one of the individuals from B to X, which values at the 2nd, 4th, and 7th locus differ; (4) three "s#" are created at the 2nd, 4th, and 7th locus in A. Furthermore, the set of "s#" (located at the 2nd and 6th locus) in the base solution and the set of "s#" (located at the 3rd and 6th locus) in the exactly matched solutions are also added to A; and (5) return to (1).

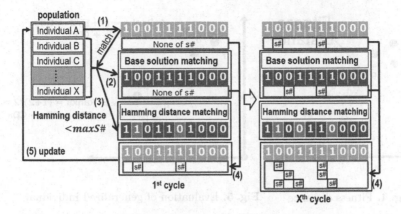

Fig. 3. Swap-based generalization

2.4 Fitness Distance

The swap-based generalization mechanism contributes to generalizing the individuals by adding "s#" but does not guarantee to have *similar* fitness. To address this issue, the *fitness distance* is proposed to calculate the distance between the base solution and each swapped solution in the solution space. Concretely, the fitness distance between them is calculated (normalized) by Eq. (1), where the #*object*, $basefit_i$, $swapfit_i$, $maxfit_i$, and $minfit_i$ indicate the number of objectives, the fitness of the *base*, *swapped*, *maximum*, and *minimum* solutions in the *i*-th objective, respectively. Figure 4 shows an example of the fitness distance between the base solution b and the swapped solution h, both of which are covered by the generalized individual A in the 2 objectives solution space.

$$FitnessDistance = \sqrt{\sum_{i=1}^{\#object} \left(\frac{basefit_i - swapfit_i}{maxfit_i - minfit_i} \right)^2} \tag{1}$$

To evolve the generalized individual composed of the solutions with the similar fitness, the swapped solution is eliminated from the generalized individual when the fitness distance of the swapped solution is larger than the threshold θ_{SD}. Note that (1) θ_{SD} depends on tasks (*i.e.*, how much similar fitness should be accepted in tasks), meaning that decision makers are required to determine θ_{SD}; and (2) the fitness distance ignores the rank of solutions, but it is not problem because the fitness distance is employed to remove the swapped solution which is far from the base solution (but not the swapped solution which is lower than the base solution).

2.5 Evaluation of Generalized Individual

Since the generalized individual is composed of many solutions (*i.e.*, the base and swapped solutions), its rank and its crowding distance should be designed to

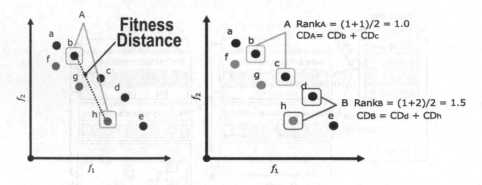

Fig. 4. Fitness distance   **Fig. 5.** Evaluation of generalized individual

consider them. In detail, the *rank* of the generalized individual is calculated by the averaged rank of all composed solutions, while the *crowding distance* of the generalized individual is calculated by the summation of all composed solutions. Figure 5 gives an example, which shows that the generalized individual A is composed of two solutions b and c while the generalized individual B is composed of two solutions d and h. As shown in this figure, the rank of A is calculated by $(1 + 1)/2 = 1$ as the averaged rank because the rank of b and c is 1, while the rank of B is calculated by $(1+2)/2 = 1.5$ as the averaged rank because the rank of d and h is 1 and 2, respectively. The crowding distance of A (*i.e.*, CD_A) is calculated by $(CD_b + CD_c)$ as the summation of the crowding distance, where CD_b and CD_c are the crowding distances of b and c, respectively. Note that the rank and crowding distance of B is calculated as the same manner of A.

2.6   Algorithm of Generalization-Based MOEA (G-MOEA)

The algorithm of G-MOEA shown in Fig. 6 is described as follows. Note that two red colored mechanisms (*i.e.*, duplicated individual deletion and generalization) and red colored rank and crowding distance in the generalized individuals are newly employed in NSGA-II.

Step 1: The population $R_t(= P_t + Q_t)$ at the t-th generation is formed after generating the offspring Q_t from P_t as the same as NSGA-II.

Step 2: The population R_t' is created from R_t by eliminating the duplicated individuals. This mechanism is needed to avoid filling a lot of generalized individuals.

Step 3: The non-dominated sorting and crowding distance are executed according to new rank and new crowding distance, and the half of the top individuals are selected.

Step 4: All individuals are generalized as described in Sect. 2.3. Note that the swapped solution is eliminated from the generalized individual when the fitness distance of the swapped solution is larger than the threshold θ_{SD}.

Step 5: Return to step 1 with increasing t and continue this cycle until reaching the maximum generation.

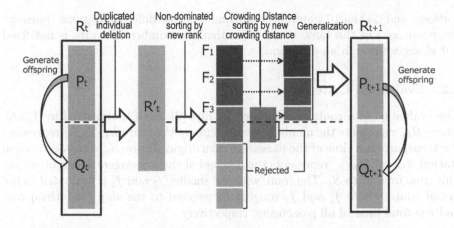

Fig. 6. Overview of generalization-based MOEA (Color figure online)

3   Waterbus Route Optimization Problem

3.1   Problem Description

As the real world problem, Sumida River in Tokyo is modeled to consist of the 11th stations, where the station distance and the number of passengers (*i.e.*, OD (origin-destination) matrix that represents the number of passengers from the origin to the destination) are based on the actual data [6]. Figure 7 shows the OD matrix, where the origin is represented in the row, while the destinations is represented in the column. For example, $E \to D$ in OD represents that 500 passengers move from the station E to D.

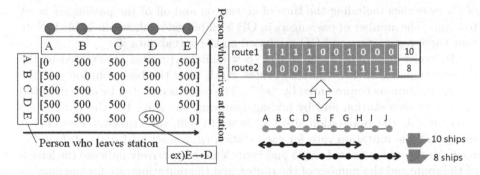

Fig. 7. Origin-destination matrix   **Fig. 8.** Representation of the route network

As the route model, Fig. 8 shows the two routes, each of which represents the anchored (stopped) stations in the river and the number of the waterbuses. The route 1 indicates that (1) the waterbus anchors (stops) in A, B, C, D, and G stations, and (2) the 10 waterbuses starting from the different stations transport the passengers in the same route. Note that the number of routes is not fixed but changes through an evolution.

3.2   Evaluation Criteria

The evaluation of a route is calculated by Eqs. (2) and (3) employed in [1,2,8], where B_{L_k} represents the number of waterbuses in the route L_k, T_{S_i,S_j} represents the transportation time of the passenger from origin station S_i to the destination station S_j, and D_{S_i,S_j} represents the demand of the passengers which occur per unit time from S_i to S_j. The route with the smaller f_1 and f_2 is evaluated as the better route, where f_1 and f_2 roughly correspond to the ship (waterbus) cost and the total time of all passengers, respectively.

$$f_1 = \sum_{L_k} B_{L_k} \tag{2}$$

$$f_2 = \sum_{S_i \neq S_j} T_{S_i,S_j} \cdot D_{S_i,S_j} \tag{3}$$

4   Experiment

4.1   Waterbus Route Optimization Problem

To investigate the effectiveness of G-MOEA for environmental changes, this paper applies it into Sumida river waterbus route optimization problem as the real world problem. The parameters related to the waterbus are set as follows: (1) the speed of the waterbus is set to 10 knot; (2) the capacity of the waterbus is set to 50 passengers; (3) the total time of arriving and leaving alongside pier of the waterbus including the time of getting on and off of the passengers is set to 3 min. The number of passengers in OD is employed as shown in Table 1. Note that these parameters and OD are based on the actual data [6].

To evolve the routes in G-MOEA, new (children) routes are created by the 2-point crossover of two (parent) routes which have the base solution and the swapped solutions (represented by "s#"). The route is mutated by changing stop or pass in each station and by adding/removing one ship. For the parameters of G-MOEA, the population size (N) is set to 300, the crossover rate (P_c) is set to 1.0, the mutation rate for each station (μ_r) is set to $1.0/(routeLength \cdot routeNum)$, where $routeLength$ and $routeNum$ respectively indicate the length of the route and the number of the routes, and the mutation rate for the number of ship (μ_s) is set to $1.0/routeLength$. When mutating the number of ship, one ship is added or removed with the 0.5 probability. For an evaluation criteria, the hypervolume (HV) and the number of the solutions (i.e., the base and swapped solutions) are employed.

In this experiment, the following methods are compared. In particular, $\theta_{SD} < 0.05$ in G-MOEA allows the only swapped solutions which have very similar

Table 1. OD matrix of Sumida river

	Hinode	Shin-kawa	Hama-machi	Ryo-goku	Azuma-bashi	Sumidaku-tyosyamae	Sakur-bashi	Senju	Higa-shiogu	Araka-wayuen	Kamiya
Hinode	0	464.4	610.2	200	62	9.1	7.3	10.2	0	0.8	0.6
Shinkawa	412.8	0	0	40.6	0	5.7	1.7	0	0	0	0
Hamamachi	1116.6	0	0	0	46.4	64.8	43.5	39.3	0	0.4	0.2
Ryogoku	252.4	94.4	0	0	198.8	539.7	17.7	21.7	0.2	3.3	2.8
Azumabashi	34.4	13.5	219.7	500	0	0	0	10	0	0.2	0.2
Sumidaku-tyosyamae	51.5	15.3	173.3	337.3	0	0	0	30.4	0	0.2	0.2
Sakurbashi	100.8	59.3	334.2	102.6	0	0	0	3.7	0	6.6	1.5
Senju	95.3	36.6	359.7	300.4	162.4	20.6	8	0	1.7	54.8	5.5
Higasiogu	3.7	2.2	14.4	14.4	14.4	3.1	0.4	7.1	0	0	2.8
Arakawayuen	8.4	3.3	36.2	28.8	6.2	2.2	3.3	16.8	0	0	54.4
Kamiya	8.4	4.8	23.5	8.4	6.6	3.7	1.1	13.7	1.1	71.1	0

fitness (*i.e.*, the other routes which have the mostly same number of ships and transportation time) to keep the mostly same service even in disaster situations. Note that the different results can be obtained by setting the different θ_{SD}, but the appropriate θ_{SD} depends on how much number of ships can be provided and how much transportation time can be extended in given situations.

- NSGA-II
- G-MOEA ($\theta_{SD} < 0.05$, $maxS\# = 2, 3, 4, 5$).

4.2 Experimental Results

Figure 9 shows the hypervolume of NSGA-II and G-MOEA, where the horizontal axis indicates NSGA-II and G-MOEA with the different $maxS\#$ while the vertical axis indicates the hypervolume (averaged from 10 runs) in the last generation. From this figure, the hypervolume of G-MOEA except for $maxS\# = 4$ is better than that of NSGA-II, even though G-MOEA evolves not only the base solutions but also their swapped solutions (while NSGA-II only evolves the solutions corresponding to the base solutions). Note that the hypervolume of G-MOEA with $maxS\# = 4$ is discussed later.

Figure 10 shows the number of the solutions (*i.e.*, the base and swapped solutions) in the population, where the horizontal and vertical axes indicate $maxS\#$ and the number of the solutions (averaged from 100 runs), respectively. Figure 10 indicates that the number of the solutions increases as $maxS\#$ increases, meaning that the generalized individual has the many alternative solutions as $maxS\#$ increases. Such generalized individuals are robust to environmental changes because of many alternative solutions.

4.3 Discussion

To analyze the effect of the generalized individual acquired by G-MOEA for environmental changes, Table 2 shows one of the generalized individual with

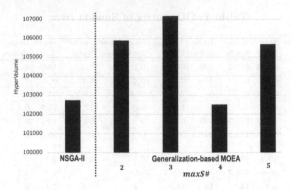

Fig. 9. Hypervolume of NSGA-II and G-MOEA

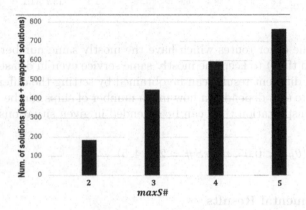

Fig. 10. Number of individuals (base and swapped solutions)

$maxS\# = 5$. The column in this table indicates the solution number (No), f_1 (the total number of the waterbuses), f_2 (the total time), the gene (Station No.), and the number of the waterbuses in each route.

From this table, the generalized individual is composed of one base set of the routes (*i.e.*, the base solution) and the 17 alternative set of the routes (*i.e.*, the swapped solutions), where these sets are evolved with the different number of waterbuses. For example, the base solution can transport all passengers by 32 waterbuses in 2304 min, which is divided into two routes, "00110111111" with 11 waterbuses and "11111110000" with 21 waterbuses. Note that "1" means to stop at the station while "0" means to pass the station. The colored cells of the stations in the alternative set of the routes have the opposite values from the base set of the routes by $s\#$. For example, the waterbus of the first route in the base set passes the 5th station but stops at the 6th station, while the waterbus of the first route in the No. 1 alternative set stops at the 5th station but passes the 6th station. This indicates that the first route in the No. 1 alternative set can be employed instead of the base set by stopping at the 5th station instead of the 6th station when the 6th station cannot be stopped due to disaster. In this

Table 2. Acquired generalized individual

No	f1 Ship num.	f2 Time[min]	1	2	3	4	5	6	7	8	9	10	11	Route ship num.
base	32	2304	0	0	1	1	0	1	1	1	1	1	1	11
			1	1	1	1	1	1	1	0	0	0	0	21
1	32	2304	0	0	1	1	1	0	1	1	1	1	1	11
			1	1	1	1	1	1	1	0	0	0	0	21
2	32	2304	0	0	1	1	1	0	0	1	1	1	1	11
			1	1	1	1	1	1	1	0	0	0	0	21
3	32	2304	0	0	1	1	1	1	1	1	1	1	1	11
			1	1	1	1	1	1	1	0	0	0	0	21
4	32	2304	0	1	1	1	1	1	0	1	1	1	1	11
			1	1	1	1	1	1	1	0	0	0	0	21
5	32	2304	1	1	1	1	1	1	0	1	1	1	1	11
			1	1	1	1	1	1	1	0	0	0	0	21
6	32	2304	0	0	1	1	1	1	0	1	1	1	1	11
			1	1	1	1	1	1	1	0	0	0	0	21
7	32	2304	0	1	1	1	1	1	1	1	1	1	1	11
			1	1	1	1	1	1	1	0	0	0	0	21
8	32	2304	1	1	1	1	1	1	0	1	1	1	1	11
			1	1	1	1	1	1	1	0	0	0	0	21
9	31	2345	1	1	1	1	0	1	1	1	1	1	1	10
			1	1	1	1	1	1	1	0	0	0	0	21
10	32	2339	0	1	1	1	0	1	0	1	1	1	1	15
			1	1	1	1	1	1	1	0	0	0	0	17
11	33	2297	0	0	1	1	1	1	1	1	1	1	1	11
			1	1	1	1	0	1	1	0	0	0	0	22
12	32	2345	0	1	1	1	1	0	0	1	1	1	1	11
			1	1	1	1	1	1	1	0	0	0	0	21
13	31	2341	1	0	1	1	0	1	0	1	1	1	1	10
			1	1	1	1	1	1	1	0	0	0	0	21
14	33	2297	0	1	1	1	0	1	1	1	1	1	1	11
			1	1	1	1	1	1	1	0	0	0	0	22
15	33	2297	1	0	1	1	1	1	1	1	1	1	1	11
			1	1	1	1	1	1	0	0	0	0	0	22
16	33	2297	1	1	1	1	1	0	1	1	1	1	1	11
			1	1	1	1	1	1	1	0	0	0	0	22
17	34	2304	1	0	1	1	1	0	1	1	1	1	1	12
			1	1	1	1	1	1	1	0	0	0	0	22

case, the number of waterbuses and the transportation time do not change by the route change. Among all alternative solutions, the number of waterbuses and the transportation time do not change drastically, which suggests that the set of routes can be changed without affecting the results. As mentioned in Sect. 1, general MOEAs cannot evolve such solutions because solutions (even POS) in general MOEAs are mostly different each other.

From the total viewpoint, the stations 1, 2, 5, 6, and 7 in the black solid box shown in Table 2 can change by environmental changes as the same reason above. On the other hand, the station 3 and 4 are all "1 (stopped)" in the generalized individual (*i.e.*, all waterbuses should stop at both stations), meaning that both are core stations in Sumida river transportation, which should be protected for future disaster. From the above generalized individual analysis, G-MOEA cannot only evolve the robust individuals, but also can analyze the feature of the Sumida river transportation by clarifying which stations (*e.g.*, the stations 1, 2, 5, 6, and 7 in this case) can change to be passed from to be stopped or vice versa and which stations (*e.g.*, the stations 3 and 4 in this case) should be protected because of no alternative set of routes which pass both stations.

Finally, the black solid box shown in Table 2 provides the reason why the hypervolume of G-MOEA with $maxS\# = 4$ is lower than NSGA-II. Both of the number of the waterbuses and the transportation time in the solutions from No. 1 to 8 can keep the best (*i.e.*, 32 and 2304) with one to five stations change (mostly two or three stations change), while the solution No. 12 (*i.e.*, 32 and 2345) calculated by four stations change is worse than (can be dominated by) the solutions from No. 1 to 8. This means that there are not so many good solutions with four stations change in the Sumida river transportation, which decreases the hypervolume of G-MOEA with $maxS\# = 4$.

5   Conclusions

This paper proposed the generation-based MOEA (G-MOEA) that can evolve the *generalized* individuals, which include *many* solutions that can be applied into different situations with the *minimal change*. In particular, the *generalized* individual is robust to environmental changes because it is composed of not only (i) the base solution but also (ii) the similar solutions derived from the base one by introducing the *don't care symbol* # (representing any situation). By this extension, the generalized solutions evolved by G-MOEA can be applied in the route optimization in disaster while the solutions (even POS) evolved by general MOEAs cannot be applied, because the former solutions are flexible and similar as a set of individuals while the latter solutions are fixed and mostly different each other. The intensive simulations on the waterbus route optimization problem as the real world problem have revealed the following implications: (1) G-MOEA cannot only optimize the solutions like general MOEAs but also can evolve the generalized individuals; and (2) G-MOEA can analyze the feature of the river transportation in the waterbus route optimization.

What should be noted here is that the above implications have only been obtained from one problem, which suggests that further careful qualifications and justifications (such as an application of G-MOEA into other domains) are needed to generalize our results. As described in Sect. 4.1, the different analysis can be done by setting the different parameter θ_{SD}, which should be investigated. Such important directions must be pursued in the near future in addition to an application of the concept of generalization to other MOEAs such as MOEA/D.

References

1. Baaj, M.H., Mahmassani, H.S.: An AI-based approach for transit route system planning and design. J. Adv. Transp. **25**(2), 187–210 (1991)
2. Ceder, A., Wilson, N.H.M.: Bus network design. Transp. Res. **20B**(4), 331–344 (1986)

3. Deb, K.: Multi-objective Optimization Using Evolutionary Algorithms. Wiley, Hoboken (2001)
4. Deb, K., Pratap, A., Agarwal, S., Meyarivan, T.: A fast elitist multi-objective genetic algorithm: NSGA-II. IEEE Trans. Evol. Comput. **6**, 182–197 (2002)
5. Holland, J.H.: Adaptation in Natural and Artificial Systems. MIT Press, Cambridge (1992)
6. Japan Association of Marine Safety: Research practical on use of river transportation by networking of bases for the main wide disaster prevention (2006, in Japanese)
7. Wilson, S.W.: Classifier fitness based on accuracy. Evol. Comput. **3**(2), 149–175 (1995)
8. Zhao, F., Zeng, X.: Optimization of user and operator cost for large-scale transit network. J. Transp. Eng. **133**(4), 240–251 (2007)

Runtime Analysis of Evolutionary Multi-objective Algorithms Optimising the Degree and Diameter of Spanning Trees

Wanru Gao(✉), Mojgan Pourhassan, Vahid Roostapour, and Frank Neumann

Optimisation and Logistics, School of Computer Science, The University of Adelaide, Adelaide, Australia
wanru.gao@adelaide.edu.au

Abstract. Motivated by the telecommunication network design, we study the problem of finding diverse set of minimum spanning trees of a certain complete graph based on the two features which are maximum degree and diameter. In this study, we examine a simple multi-objective EA, GSEMO, in solving the two problems where we maximise or minimise the two features at the same time. With a rigorous runtime analysis, we provide understanding of how GSEMO optimize the set of minimum spanning trees in these two different feature spaces.

Keywords: Evolutionary multi-objective optimisation ·
Algorithm analysis

1 Introduction

Evolutionary algorithms (EAs) have wide application in solving complex problems in various areas such as combinatorial optimization, bioinformatics and engineering. In EA research, the algorithm works with a set of solutions which is called the population and is evolved during the optimization process to cover a so-called Pareto front. Most evolutionary algorithms incorporate certain diversity mechanisms which ensure that the population consists of a diverse set of individuals [3,16]. By presenting a set of different solutions with acceptable quality to the decision maker, EAs with diversity maximisation provide a better exploration and understanding of the search space. In recent years, EAs with diversity optimisation mechanism have been proposed and examined in both theoretical and practical aspects [8,9,14].

There have been many EAs that are applied in solving multi-objective optimisation problems and have gained significant success. Evolutionary multi-objective optimization (EMO) aims at achieving a set of solutions which is used to approximate the so-called Pareto front. The solutions are evaluated based on two or more conflicting objective functions and EAs are suitable in computing several trade-off during a single process. There have been many well-known

© Springer Nature Switzerland AG 2019
K. Deb et al. (Eds.): EMO 2019, LNCS 11411, pp. 504–515, 2019.
https://doi.org/10.1007/978-3-030-12598-1_40

multi-objective evolutionary algorithms (MOEAs) which include MOEA/D [17], IBEA [18] and NSGA-II/III [5,6].

In this paper, we consider a simple MOEA which finds a diverse set of Minimum spanning trees (MSTs) with different features for an undirected unweighted complete graph and analyse the algorithm theoretically. Minimum spanning tree problem is a fundamental problem with diverse applications including network design and approximation algorithms design of NP-hard problems [1,4,12]. A spanning tree of a graph refers to a subgraph that contains all the vertices in the graph and is a tree. A graph may have many spanning trees. When all edges are assigned weights or lengths, the minimum spanning tree of a graph is the one with the minimum sum of weights. For an unweighted complete graph, all spanning trees are MSTs which have different structures. Although they have the same total weights, they have various features which make them different to the decision makers. There exist many different features other than the total weight that researchers use to evaluate a MST including the maximum degree, diameter and depth. The features examined in this paper are the maximum degree and diameter, which evaluate different structural characteristics.

Finding MSTs with different maximum degree and diameter is important for real-world applications such as telecommunication network design with certain connection requirement. When designing a telecommunication network, there are a lot of factors that affect the choice of the decision makers. The degree of each node indicates the number of descendants which is proportional to the workload of that certain node. It is essential to control the maximum degree of all nodes in the tree which ensures the amount of work that each node has to do is under control [15]. In order to guarantee the communication speed, a MST with low diameter is preferred [11]. The diameter is also important in forcing the reliability constraints which should be taken into consideration of the designer.

Although finding a minimum spanning tree in a given graph is solvable in polynomial time, achieving a MST with certain maximum degree requirement is NP-hard [2]. There have been studies into the problem of approximating the search space of diversifying MSTs based on feature values [7,13].

In this research, we focus on optimizing these two features in MSTs which are the maximum degree and diameter at the same time. Since maximising or minimising maximum degree leads to a MST with minimum or maximum diameter, it is suitable to consider the problem in a multi-objective space.

The paper is organized as follows. First, we introduce the background of the problem in Sect. 2. Then in Sects. 3 and 4, we examine the MOEA on two multi-objective problems about MSTs. Finally, we finish the paper with some conclusions in Sect. 5.

2 Preliminaries

In our research, we focus on the multi-objective optimization problem of finding a population containing MSTs of a complete graph with various feature values. Let $G = (V, E)$ be an undirected graph, where V and E denote the set of nodes and set of edges respectively. Define $|V| = n$ and $|E| = m$. A spanning tree of G is defined as a connected subgraph containing all vertices in V without cycles. In this study, we represent a spanning tree as a set of edges and use a bitstring of size m where each bit shows the existence of a certain edge in the subgraph to denote the spanning tree.

We characterize MSTs by two feature values which are the maximum degree and the diameter of an MST. The maximum degree $d(s)$ of an MST s is defined as the maximum value of the degrees of all nodes in V. The diameter $l(s)$ of an MST s is defined as the length of the longest path in s. We also define the number of longest paths in an MST s as $p(s)$.

Considering these two features as objectives, we examine the Global Simple Evolutionary Multi-objective Optimiser (GSEMO) [10] which is presented in Algorithm 1 in optimizing the problem. For the concept of dominance, we use the following definition.

Definition 1 (Dominance). *In multi-objective optimization, there exists a fitness function that maps each solution in the search space X to a vector of real values, i.e. $f : X \rightarrow \mathbb{R}^k$. Assume all k objectives should be minimised. For two solutions $s, s' \in X$, s is said to weakly dominate s' iff $f_i(s) \leqslant f_i(s')$, where $1 \leqslant i \leqslant k$. S is said to (strictly) dominate s' iff s weakly dominates s' and $f(s) \neq f(s')$.*

The definition of dominance can be adapted to problems where one or more objectives should be maximised.

Definition 2 (Pareto optimality). *A solution s is Pareto optimal if it is not dominated by any other solution in the search space. The set of all Pareto-optimal solutions is called the Pareto set. The set of all Pareto optimal objective vectors is called the Pareto Front.*

Algorithm 1. GSEMO

1: Choose an initial MST $x \in \{0,1\}^m$ uniformly at random for a certain complete graph G with n vertices and m edges.
2: Let $P := x$
3: **while** stopping criteria not met **do**
4:     Pick s from P uniformly at random.
5:     Create an offspring s' by flipping each bit in s with probability $1/m$.
6:     **if** s' is a tree and is not dominated by any individual in P **then**
7:         Add s' to P, and remove all individuals weakly dominated by s' from P.
8:     **end if**
9: **end while**

We focus our analysis on the simple multi-objective EA which is GSEMO proposed by Giel [10] because of its simplicity and suitability for the theoretical analysis. The algorithm starts with an MST which is selected uniformly at random for the complete graph G. Before the stopping criteria is reached, the algorithm selects a solution s uniformly at random from population P and an offspring s' is generated by flipping each bit of s with probability $1/m$. In the case where s' is not dominated by any solution in P, it is added to P. The new population contains only non-dominated solutions.

The algorithm is examined in terms of the number of generations until it has achieved a population that covers the whole Pareto front which is equivalent to the number of evaluations. The expected optimisation time refers to the expected number of iterations to reach this goal.

3   The Max-Max Problem

We look into two multi-objective problems considering these two features. In the first problem we aim at maximising both the diameter and the maximum degree at the same time, which is referred to as the MAX-MAX problem in this paper. The dominance definition for the MAX-MAX problem is defined as follows.

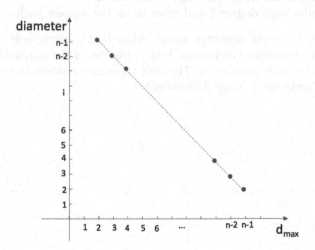

Fig. 1. The Pareto front for the multi-objective problem of maximising the diameter and maximising the max degree of a MST.

Definition 3 (Domination for Max-Max Problem). *For two MSTs s and s' of an unweighted complete graph G, in the* MAX-MAX *problem, s dominates s' iff $d(s) \geq d(s')$ and $l(s) \geq l(s')$.*

Lemma 1. *Let s be a Pareto optimal solution of the* MAX-MAX *problem, then $d(s) + l(s) = n + 1$, where n denotes the number of nodes in the graph.*

Proof. Assume the minimum spanning tree with the maximum degree is s and its diameter and maximum degree are represented as $l(s)$ and $d(s)$. In MST s, the longest path has length $l(s)$ which has $l(s) + 1$ nodes on it. Then there are another $n - (l(s) + 1)$ nodes which are not on the path. In order to maximise $d(s)$, these nodes should be connected to one of the nodes on the path except the tailing ones. Hence,

$$d(s) = 2 + n - (l(s) + 1) = n - l(s) + 1.$$

The sum of the diameter and the maximum degree equals to $l(s) + n - l(s) + 1 = n + 1$. □

According to Lemma 1, the Pareto front of the MAX-MAX problem is as shown in Fig. 1. It is easy to see that each Pareto solution consists of a star node with degree d and a longest path of length l as shown in Fig. 2. Note that for a specific degree and diameter, the Pareto solution is not unique. However, a solution is Pareto optimal if and only if

1. It has at most one node with degree more than 2.
2. All the nodes with degree 2 and more lie on the longest path.

Moreover, for each diameter value, Algorithm 1 keeps only one solution because of the dominance definition. Hence, the size of the population produced by the algorithm is at most $n - 2$. The next theorem considers the expected time to find the Pareto front using Algorithm 1.

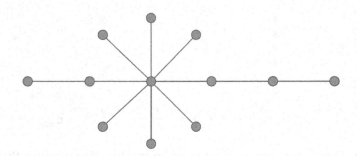

Fig. 2. A Pareto solution with degree 8 and diameter 5

Theorem 1. *Algorithm 1 finds all the Pareto optimal solutions of the* MAX-MAX *problem in expected time* $O(n^2 m^2)$.

Proof. Let $s_P \in P$ denote a Pareto optimal solution in the population with diameter $l(s_P)$. We consider the proof in the following two phases. The first phase is to maintain P such that it contains at least one Pareto optimal solution. The second phase is to find other Pareto optimal solutions starting from s_P. We prove that each phase needs expected time $O(n^2 m^2)$ to be completed.

Now let $s' \in P$ be a solution with the highest diameter $l(s') < n - 1$. The mutation step on s' that detaches a leaf from a node with degree more than 2 and attaches it to one side of the longest path increases $l(s')$ by one. Since the probability of any single bit flip is $\frac{1}{m}$, the probability of such a mutation is at least $\frac{1}{em^2}$. On the other hand, the size of P is upper bounded by n and the probability of selecting s' for mutation is at least $\frac{1}{n}$. Hence, after expected time $O(nm^2)$, the diameter of s' is increased by one. Furthermore, the diameter of s' is at least 2. Therefore, we need at most $n - 2$ such mutations to obtain the Pareto optimal solution with maximum degree 2 and diameter $n - 1$. It implies that Algorithm 1 needs expected time $O(n^2m^2)$ to complete the first phase.

Now we analyse the second phase and assume that there is at least one Pareto optimal solution s_P in the population. Assume that the Pareto optimal solution with diameter $l(s_P) - 1$ is not included in P yet. In this case, the mutation step on s_P that removes a leaf from one side of the longest path and connects it to the node with the highest degree will produce a new Pareto optimal solution with diameter $l(s_P) - 1$ and maximum degree $d(s_P) + 1$. Similar to the argument in the first part of the proof, the algorithm needs expected time $O(nm^2)$ to perform this mutation. Furthermore, from the first phase, it is known that the solution s_P with diameter $n - 1$ exists in P. Hence, the algorithm is able to produce all the Pareto optimal solutions gradually, starting from s_P. Since the size of the Pareto set is $n - 2$, Algorithm 1 finds all the Pareto optimal solution in expected time $O(n^2m^2)$. □

4   The Min-Min Problems

In this section, we investigate the second problem, in which both feature values are minimised at the same time. The minimum diameter happens when the MST has a star structure where the diameter is 2 and the node in the centre has the maximum degree $n - 1$. The minimum maximum degree happens when the graph is a single path. In this case, the maximum degree is 2 and the diameter is $n - 1$. The general dominance definition is adapted for the MIN-MIN problem as follows.

Definition 4 (Dominance for the Min-Min Problem). *For two MSTs s and s' of an unweighted complete graph G, in the* MIN-MIN *problem, a solution s is said to dominate solution s' iff $d(s) \leq d(s')$ and $l(s) \leq l(s')$.*

Based on the fact that the diameter is either even or odd, the Pareto optimal MSTs with diameter l have different structures. Figure 3 shows the structure of an optimal MST with odd diameter. The Pareto optimal MST with even diameter only contains multiple subtrees with the same depth $l/2$.

The Pareto front for this problem is not as simple as the Pareto front for the MAX-MAX problem. Having a solution s in the Pareto front, the adjacent solution with a smaller diameter, named s', can have $d(s') = d(s) + i$ and $l(s') = l(s) - j$ for some $i \geq 1$ and $j \geq 1$. Therefore, it is not always possible to find the solution s' by means of a 2-bit flip on s.

In order to overcome this problem we use a different definition of dominance in analysing the MIN-MIN problem, which still leads to a population of linear size. The new definition of dominance is presented in Definition 5, where $p(s)$ is the number of longest paths. Furthermore, it should be noted that we consider Algorithm 1 with the new definition of dominance (instead of weak dominance in lines 6 and 7).

Definition 5 (Extended dominance for the Min-Min Problem). *In the* MIN-MIN *problem, for two MSTs s and s' of a complete graph, s dominates s' iff $l(s') = l(s) \wedge d(s') = d(s) \wedge p(s') \leq p(s)$ or $l(s') < l(s) \wedge d(s') < d(s)$.*

We define the *Extended Pareto optimal solution* and the *Extended Pareto front* to be the Pareto optimal solution and the Pareto front with the new definition of dominance in Definition 5. Then in Lemma 2 we prove that the Extended Pareto front set is a superset of the original Pareto front, which is defined by the original definition of dominance.

Lemma 2. *The Extended Pareto front is a super set of the Pareto front for the* MIN-MIN *problem.*

Proof. According to Definition 4, a Pareto optimal solution s of the MIN-MIN problem should fulfil the requirement that $\nexists s'$ dominates s where s' is any other MST of the same graph. This indicates that $\nexists s'$, where $d(s') \leq d(s)$ and $l(s') \leq l(s)$.

Assume there exists a MST s'' that dominates s according to Definition 5. Then either $l(s) = l(s'') \wedge d(s) = d(s'') \wedge p(s) \leq (s'')$ or $l(s) < l(s'') \wedge d(s) < d(s'')$ is true. For maximum degree and diameter, it should fulfil that $l(s) \leq l(s'') \wedge d(s) \leq d(s'')$, which is contradict to the fact that $\nexists s'$ in the search space, where $d(s') \leq d(s)$ and $l(s') \leq l(s)$.

Therefore, the MST s is not dominated by any other solutions in the extended Pareto front which means it should be included in the Extended Pareto set.   □

In the following, we analyse the performance of the algorithm in finding the whole Extended Pareto front. Since this set is a super set for the original Pareto front, we are also analysing the performance of the algorithm in finding the original Pareto front. Lemma 3 proves an upper bound on the size of the population during the optimisation process.

Lemma 3. *The population size is upper bounded by $2n$.*

Proof. Here we prove that the maximum size of the population is $2n - 5 < 2n$. According to Definition 5, solution s does not dominate solution s' if $d(s) = d(s')$ and $l(s) \leq l(s')$. Similarly, s' is not dominated by s when $d(s) \leq d(s')$ and $l(s) = l(s')$. Moreover, for each specific combination of diameter and maximum degree, the algorithm keeps only one solution.

Let P be the population of an arbitrary iteration during the process. We partition P to at most $n-2$ subsets $P^i = \{s_1^i, \cdots, s_{k_i}^i\}$, $2 \leq i \leq n-1$, such that

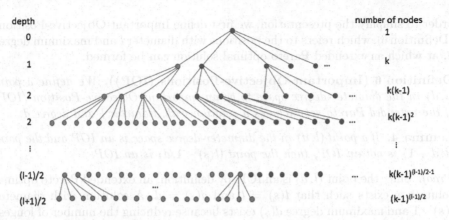

Fig. 3. The MST with odd diameter and maximum degree minimised, where l denotes the diameter and k denotes the maximum degree in the MST. The nodes coloured in orange are from a single subtree of the root. (Color figure online)

for any $s \in P^i$, $d(s) = i$. Moreover, for any $j_1 < j_2 \le k_i$ we have $l(s^i_{j_1}) > l(s^i_{j_2})$. For each subset, we have $|P^i| = k_i \le l(s^i_1) - l(s^i_{k_i}) + 1$. Without loss of generality, let all the subsets have at least one solution. Since P is the set of non-dominated solutions, for any $2 \le i \le n - 2$ we have $l(s^i_{k_i}) \ge l(s^{i+1}_1)$. Otherwise, $s^i_{k_i}$ dominates s^{i+1}_1. Hence, for any subsets P^i and P^{i+1}, we have

$$|P^i \cup P^{i+1}| \le l(s^i_1) - l(s^i_{k_i}) + 1 + l(s^{i+1}_1) - l(s^{i+1}_{k_{i+1}}) + 1 \le l(s^i_1) - l(s^{i+1}_{k_{i+1}}) + 2.$$

With the same argument we have

$$|P| = |\bigcup_{j=2}^{n-1} P^j| \le l(s^2_1) - l(s^{n-1}_{k_{n-1}}) + (n - 2)$$

$$\le (n - 1) - 2 + (n - 2)$$

$$\le 2n - 5$$

□

In any tree with n nodes, there is only one path between any two nodes. Hence the total number of paths in a tree, which is an upper bound for the number of paths with length d, is

$$\binom{n}{2} \le n^2.$$

Therefore, in a solution with diameter d, the number of paths with length d is upper bounded by n^2.

In Lemmas 4 and 5 we give some properties about Extended Pareto optimal solutions and show how they are produced in $O(n^3m^2)$ for each diameter size. In order to simplify the presentation, we first define Important-Objective-Positions (Definition 6), which refers to the positions with diameter l and maximum degree d, at which an extended Pareto optimal solution can be formed.

Definition 6 (Important-Objective-Positions (IOP)). *We define a point (l, d) in the diameter-degree space to be an Important-Objective-Position (IOP) if the extended Pareto set includes a solution with diameter l and degree d.*

Lemma 4. *If a point (l, d) in the diameter-degree space is an IOP and the point $(l, d + 1)$ is not an IOP, then the point $(l(s) - 1, d)$ is an IOP.*

Proof. Since the point (l, d) is an IOP, by definition, an extended Pareto optimal solution s exists such that $l(s) = l$ and $d(s) = d$. A solution with diameter $l(s) - 1$ and maximum degree $d(s)$ exists because reducing the number of longest paths in solution s either results in a solution with a larger maximum degree (which we have assumed that does not belong to the extended Pareto set), or a solution with smaller diameter. Moreover, a solution with diameter $l(s) - 1$ and maximum degree $d(s)$ can only be dominated by a solution with the same maximum degree and diameter, or a solution s' with maximum degree $d(s') < d(s)$ and diameter $l(s') < l(s) - 1$ (Definition 5). If solution s' exists, then it would have dominated solution s as well, which contradicts with the assumption that s is an extended Pareto optimal solution. Therefore, a solution with diameter $l(s) - 1$ and maximum degree $d(s)$ can only be dominated with a solution with the same maximum degree and diameter, which implies that the point $(l(s) - 1, d)$ is an IOP.   □

Lemma 5. *Assume that points $(l, d + i)$, for $0 \leq i < k$ and $k > 1$, are IOPs and, a solution s with $l(s) = l$ and $d(s) = d$, is in the population. In expected time $O(n^3 m^2)$, all Pareto optimal solutions with diameter l and also a solution s' with $l(s') = l - 1$ and $d(s') = d + k - 1$ are added to the population.*

Proof. Since the position $(l(s), d(s))$ is an IOP, the solution s can only be removed from the population if a solution with the same diameter and maximum degree and a smaller number of longest paths is found (Definition 5).

In a solution s, there always exists at least one pair of 2-bit flips that reduces the number of longest paths, $p(s)$. This can be done by disconnecting a leaf of one of these paths and connecting it to an inner node. At each step, with probability $\frac{1}{|P|}$ the assumed solution is selected for mutation, where $|P|$ is the size of the population. Moreover, while there exist inner nodes with degree less than $d(s)$, with probability $\frac{1}{e \cdot m^2}$ a proper 2-bit flip happens, which reduces $p(s)$ without increasing the maximum degree of the solution. Due to Definition 5, solution s is dominated and replaced by the new solution. The same process with reducing $p(s)$ continues until the algorithm reaches a solution s^0 that belongs to the extended Pareto front and stays in the population. Denoting the total reduction on the number of longest paths by $\Delta_{s^0} = p(s) - p(s^0)$, we can observe that the expected time until reaching the solution s^0 is $O(|P| m^2 \Delta_{s^0})$.

We define solutions s^i, $i < k$, to be extended Pareto optimal solutions of diameter $l(s)$ and degree $d(s) + i$. We also define $\Delta_{s^i} = p(s^{i-1}) - p(s^i)$,

$1 \leq i < k - 1$ as the total difference on the number of longest paths between solutions s^{i-1} and s^i. With similar analysis we can show that after reaching the solution s^i, $i < k - 1$, at each step with probability $\frac{1}{|P|m^2}$ a solution with degree $d(s^i) + 1$, diameter $l(s^i)$ and number of longest paths $p(s^i) - 1$ is produced, which is, due to Definition 5, either accepted by the algorithm, or dominated by a solution with the same degree and diameter, but a smaller number of longest paths. This process continues until reaching a solution with minimum number of longest paths, which implies that a solution s^{i+1} is reached by the algorithm in expected time $O(|P|m^2 \Delta_{s^1})$. This means that all extended Pareto optimal solutions with diameter $l(s)$ can be found in expected time

$$\sum_{i=0}^{k-1} O(|P|m^2 \Delta_{s^i}).$$

Moreover, since the solution s^{k-1} is the extended Pareto optimal solution with diameter $l(s)$ that maximises $d(s)$, it only contains one longest path. Therefore, moving an edge from it results in obtaining the solution s' with $l(s') = l(s) - 1$ and $d(s') = d(s^{k-1})$. This would also happen in expected time $O(|P|m^2 p(s^{k-1}))$. Together with the expected time of finding extended Pareto optimal solutions with diameter $l(s)$, the total expected time of finding all k extended Pareto optimal solutions with diameter $l(s)$ and also a solution s' with diameter $l(s') = l(s) - 1$ and maximum degree $d(s) + k - 1$ would be

$$\sum_{i=0}^{k-1} O(|P|m^2 \Delta_{s^i}) + O(|P|m^2 p(s^{k-1})) = O(|P|m^2 p(s)).$$

The equality holds because the total number of longest paths that have been reduced in the process is $\sum_{i=0}^{k-1} \Delta_{s^i} + p(s^{k-1}) = p(s)$. Since the number of longest paths in solution s is upper bounded by n^2 and the population size is upper bounded by $2n$ (Lemma 3), the obtained expected time is upper bounded by $O(n^3 m^2)$. □

Now we present the main theorem of this section, in which, starting from a solution with maximum degree of 2 (a path), the expected time until finding all Pareto front set is analysed.

Theorem 2. *Starting with a population that contains a solution s with $d(s) = 2$, Algorithm 1 finds the Pareto set of the* MIN-MIN *problem in expected time* $O(n^4 m^2)$.

Proof. Firstly, we prove that Algorithm 1 finds the extended Pareto set in expected time $O(n^4 m^2)$.

Since the maximum degree of a minimum spanning tree on a graph of at least three nodes cannot be less than 2, solution s belongs to the extended Pareto front. This solution is a path of length $n - 1$, which implies that $l(s) = n - 1$ and the corresponding IOP is $(n - 1, 2)$.

Having a solution s at IOP position $(l(s), d(s))$, from Lemma 5, we know that in expected time $O(n^3 m^2)$, all k extended Pareto optimal solutions with diameter $l(s)$ are added to the population in addition to a solution s' with diameter $l(s) - 1$ and degree $d(s) + k - 1$. The largest maximum degree among solutions with diameter $l(s)$ would be $d(s)+k-1$, which implies that a diameter-degree position $(l(s), d(s) + k)$ is not an IOP. Therefore, by Lemma 4 we know that the position $(l(s) - 1, d(s) + k - 1)$ is an IOP. Since solution s' is placed at this position, it can be used for Lemma 5 and diameter size $l(s) - 1$. We can use similar argument for smaller diameter sizes. Since we start with a diameter size of $n - 1$, all extended Pareto optimal solutions for all diameter sizes are found in expected time $O(n^4 m^2)$.

Since the extended Pareto front is a superset of the Pareto front, the dominated solutions according to Definition 4 should be eliminated before the Pareto set of the MIN-MIN problem is achieved. As the population size is upper bounded by $2n$, getting rid of all dominated solutions takes expected $O(n^2)$ time. Hence, the statement of the theorem is proved.

□

5   Conclusions

The MOEAs, which are used to optimise several objective functions, always involve a set of solutions which approximates the so-called Pareto front. These algorithms are suitable in dealing with conflicting objective functions. In this paper, we examine a simple multi-objective optimisor on two bi-objective optimisation problems about MSTs of a complete graph. Inspired by the real-world application in telecommunication, we focus on the MAX-MAX and MIN-MIN problems which provide insights in dealing with the trade-off between optimising the features of maximum degree and diameter. With a rigorous runtime analysis, we provide a better understanding of the search space and the computational complexity of such problems.

Acknowledgements. This work has been supported by Australian Research Council (ARC) grants DP160102401.

References

1. Abuali, F.N., Schoenefeld, D.A., Wainwright, R.L.: Designing telecommunications networks using genetic algorithms and probabilistic minimum spanning trees. In: Proceedings of the 1994 ACM Symposium on Applied Computing, SAC 1994, pp. 242–246. ACM, New York (1994). https://doi.org/10.1145/326619.326733
2. Bui, T.N., Zrncic, C.M.: An ant-based algorithm for finding degree-constrained minimum spanning tree. In: Proceedings of the 8th Annual Conference on Genetic and Evolutionary Computation, GECCO 2006, pp. 11–18. ACM, New York (2006). https://doi.org/10.1145/1143997.1144000
3. Chaiyaratana, N., Piroonratana, T., Sangkawelert, N.: Effects of diversity control in single-objective and multi-objective genetic algorithms. J. Heuristics **13**(1), 1–34 (2007)

4. Cormen, T.H., Leiserson, C.E., Rivest, R.L., Stein, C.: Introduction to Algorithms, 3rd edn. The MIT Press, Cambridge (2009)
5. Deb, K., Jain, H.: An evolutionary many-objective optimization algorithm using reference-point-based nondominated sorting approach, part i: solving problems with box constraints. IEEE Trans. Evol. Comput. **18**(4), 577–601 (2014). https://doi.org/10.1109/TEVC.2013.2281535
6. Deb, K., Pratap, A., Agarwal, S., Meyarivan, T.: A fast and elitist multiobjective genetic algorithm: NSGA-II. IEEE Trans. Evol. Comput. **6**(2), 182–197 (2002). https://doi.org/10.1109/4235.996017
7. Dekker, A., Pérez-Rosés, H., Pineda-Villavicencio, G., Watters, P.: The maximum degree & diameter-bounded subgraph and its applications. J. Math. Model. Algorithms **11**(3), 249–268 (2012). https://doi.org/10.1007/s10852-012-9182-8
8. Gao, W., Nallaperuma, S., Neumann, F.: Feature-based diversity optimization for problem instance classification. In: Handl, J., Hart, E., Lewis, P.R., López-Ibáñez, M., Ochoa, G., Paechter, B. (eds.) PPSN 2016. LNCS, vol. 9921, pp. 869–879. Springer, Cham (2016). https://doi.org/10.1007/978-3-319-45823-6_81
9. Gao, W., Neumann, F.: Runtime analysis for maximizing population diversity in single-objective optimization. In: Genetic and Evolutionary Computation Conference, GECCO 2014, Vancouver, BC, Canada, 12–16 July 2014, pp. 777–784 (2014). https://doi.org/10.1145/2576768.2598251
10. Giel, O.: Expected runtimes of a simple multi-objective evolutionary algorithm. In: The 2003 Congress on Evolutionary Computation 2003, CEC 2003, vol. 3, pp. 1918–1925, December 2003. https://doi.org/10.1109/CEC.2003.1299908
11. Gouveia, L., Magnanti, T.L.: Network flow models for designing diameter-constrained minimum-spanning and steiner trees. Networks **41**(3), 159–173. https://doi.org/10.1002/net.10069
12. Khan, M., Pandurangan, G., Kumar, V.S.A.: Distributed algorithms for constructing approximate minimum spanning trees in wireless sensor networks. IEEE Trans. Parallel Distrib. Syst. **20**(1), 124–139 (2009). https://doi.org/10.1109/TPDS.2008.57
13. Könemann, J., Levin, A., Sinha, A.: Approximating the degree-bounded minimum diameter spanning tree problem. Algorithmica **41**(2), 117–129 (2005). https://doi.org/10.1007/s00453-004-1121-2
14. Neumann, A., Gao, W., Doerr, C., Neumann, F., Wagner, M.: Discrepancy-based evolutionary diversity optimization. In: Proceedings of the Genetic and Evolutionary Computation Conference, GECCO 2018, Kyoto, Japan, 15–19 July 2018, pp. 991–998 (2018). https://doi.org/10.1145/3205455.3205532
15. Robins, G., Salowe, J.S.: Low-degree minimum spanning trees. Discrete Comput. Geom. **14**, 151–165 (1999)
16. Ursem, R.K.: Diversity-guided evolutionary algorithms. In: Guervós, J.J.M., Adamidis, P., Beyer, H.-G., Schwefel, H.-P., Fernández-Villacañas, J.-L. (eds.) PPSN 2002. LNCS, vol. 2439, pp. 462–471. Springer, Heidelberg (2002). https://doi.org/10.1007/3-540-45712-7_45
17. Zhang, Q., Li, H.: MOEA/D: a multiobjective evolutionary algorithm based on decomposition. IEEE Trans. Evol. Comput. **11**(6), 712–731 (2007). https://doi.org/10.1109/TEVC.2007.892759
18. Zitzler, E., Laumanns, M., Bleuler, S.: A tutorial on evolutionary multiobjective optimization. In: Gandibleux, X., Sevaux, M., Sörensen, K., T'kindt, V. (eds.) Metaheuristics for Multiobjective Optimisation, vol. 535, pp. 3–37. Springer, Heidelberg (2004). https://doi.org/10.1007/978-3-642-17144-4_1

Bi-objective Orienteering: Towards a Dynamic Multi-objective Evolutionary Algorithm

Jakob Bossek[1], Christian Grimme[1(✉)], Stephan Meisel[1], Günter Rudolph[2], and Heike Trautmann[1]

[1] Department of Information Systems, University of Münster, Leonardo-Campus 3, 48149 Münster, Germany
{bossek,grimme,meisel,trautmann}@wi.uni-muenster.de
[2] Department of Computer Science, TU Dortmund University, Otto-Hahn-Str. 14, 44227 Dortmund, Germany
guenter.rudolph@tu-dortmund.de

Abstract. We tackle a bi-objective dynamic orienteering problem where customer requests arise as time passes by. The goal is to minimize the tour length traveled by a single delivery vehicle while simultaneously keeping the number of dismissed dynamic customers to a minimum. We propose a dynamic Evolutionary Multi-Objective Algorithm which is grounded on insights gained from a previous series of work on an a-posteriori version of the problem, where all request times are known in advance. In our experiments, we simulate different decision maker strategies and evaluate the development of the Pareto-front approximations on exemplary problem instances. It turns out, that despite severely reduced computational budget and no oracle-knowledge of request times the dynamic EMOA is capable of producing approximations which partially dominate the results of the a-posteriori EMOA and dynamic integer linear programming strategies.

Keywords: Multi-objective optimization · Metaheuristics ·
Vehicle routing · Combinatorial optimization · Dynamic optimization

1 Introduction

Bi-objective orienteering belongs to the class of vehicle routing problems. It differs from classical Traveling Salesperson Problems (TSP) in that the number of cities resp. customers is not fixed but rather a certain number of dynamic customer requests have to be handled on the way from the start to the end depot. Naturally, both the overall tour length as well as the number of unvisited customers are desired to be minimized and we would like to dynamically react to new customer requests so that previously optimized tours can be adjusted in an efficient and optimal way. The design of an appropriate optimization algorithm given this scenario is not trivial, especially as, additionally, decision makers' preferences regarding the importance of both objectives have to be taken

© Springer Nature Switzerland AG 2019
K. Deb et al. (Eds.): EMO 2019, LNCS 11411, pp. 516–528, 2019.
https://doi.org/10.1007/978-3-030-12598-1_41

into account which might vary in the course of the operation time of the whole tour. This paper introduces such a real-time expert system in terms of a specific dynamic evolutionary multi-objective algorithm (EMOA) integrating local search strategies via inexact TSP solvers. The algorithm was designed by relating to the detailed problem insights gained by previous studies which approached the problem in a retrospective, offline way leading to a Pareto-front approximation exploiting the full information about the dynamic problem characteristics.

Experimental studies provide a proof-of-concept analysis of the proposed approach. It will be shown that it has the potential of outperforming competitive integer linear programming (ILP) strategies in terms of solution quality. Moreover, the algorithm is capable of generating Pareto-front approximations which come very close to and even partially dominate the solutions which resulted from the offline approach. First results show that clustered instances are more challenging compared to random ones. As purely numerical performance assessment is not trivial due to a lack of an appropriate performance indicator capturing all requirements stated above, sophisticated visualizations illustrate algorithm characteristics.

The paper is organized as follows: Sect. 2 gives an overview on related work, followed by a detailed description of our proposed dynamic multi-objective evolutionary algorithm in Sect. 3. Experimental results are provided in Sect. 4 and summarized in Sect. 5, supplemented by an outlook on promising further research building on the consolidated findings.

2   Background and Related Work

2.1   Static Multi-objective Optimization Problems

Let $\mathbb{X}$ and Θ be nonempty sets and $f(x;\theta) = (f_1(x;\theta),\ldots,f_d(x;\theta))^\mathsf{T}$ a vector-valued mapping with $d \geq 2$ functions $f_i : \mathbb{X} \times \Theta \to \mathbb{R}$ for $i = 1,\ldots,d$, where x is variable and $\theta \in \Theta$ a tuple of fixed parameters. If these functions are to be minimized simultaneously, they are called *objective functions* of the *multi-objective optimization problem* $\min\{f(x;\theta) : x \in X\}$ with *decision set* $X \subseteq \mathbb{X}$. The optimality of a multi-objective optimization problem (MOP) is defined by the concept of *dominance*.

Let $u, v \in F \subseteq \mathbb{R}^d$ where F is equipped with the partial order $\preceq$ defined by $u \preceq v \Leftrightarrow \forall i = 1,\ldots d : u_i \leq v_i$. If $u \prec v \Leftrightarrow u \preceq v \wedge u \neq v$ then v is said to be *dominated by* u. An element u is termed *non-dominated* relative to $V \subseteq F$ if there is no $v \in V$ that dominates u. The set $\mathsf{ND}(V, \preceq) = \{u \in V \mid \nexists v \in V : v \prec u\}$ is called the *non-dominated set* relative to V.

If $F = f(X;\theta)$ is the *objective set* of some MOP with decision set $X \subseteq \mathbb{R}^n$ and objective function $f(\cdot)$ then the set $F^* = \mathsf{ND}(f(X;\theta), \preceq)$ is called the *Pareto-front* (PF). Elements $x \in X$ with $f(x) \in F^*$ are termed *Pareto-optimal* and the set X^* of all Pareto-optimal points is called the *Pareto set* (PS).

2.2   The Dynamic Multi-objective Vehicle Routing Problem

The dynamic vehicle routing problem we consider in this work consists of one vehicle that visits customer locations over time. The set of customers $C \backslash \{1, N\} = C^m \cup C^o$ resolves into C^m, the subset of initially known customers and the set C^o of additional locations, which become known randomly while the vehicle is en route. The vehicle starts its tour at a given location 1 (start depot) and ends at a different location N (end depot). Locations that are known initially must be visited by the vehicle (including depots), whereas locations that become known in the course of time are optional. We refer to the set of optional customers that have arrived until time t as $C^o_{\leq t}$.

Clearly, a static MOP (as defined above) has to be adapted as the Pareto-front and Pareto-set now depend on dynamic parameters θ, i.e., in general we have F^*_θ and X^*_θ. In a dynamic MOP the parameters are no longer constant but variable over time. As a consequence, a dynamic MOP (DMOP) at time step $t \geq 0$ is given by $\min\{f(x; \theta_t) : x \in X\}$ where $(\theta_t)_{t \geq 0}$ is a sequence of parameter tuples with time index $t \geq 0$. For each point in time $t \geq 0$ we could solve a static MOP with solution $F^*_{\theta_t}$ and $X^*_{\theta_t}$ and might regard the sequences of both sets as the final solution. For a general survey on dynamic MOO, see [1].

However, this solution concept has little practical relevance. Instead, we specify a closed time interval Δ_t and monitor (the quality of) the best solutions that can be achieved within the time interval. A similar solution concept can be found in [15]. This is repeated multiple times, where at the end of each so-called *era*, a decision maker (DM) is provided with the best solutions of that era. The quantitative assessment of the sequence of best solutions found within the time interval heavily depends on the application scenario.

Specifically, our VRP is dynamic in the sense that decisions about the vehicle's route (which of the customer locations known so far to visit, and how to sequence these locations) are made repeatedly over time by a decision maker. Although dynamic decision making has been an important research topic in the field of vehicle routing (see, e.g., [9,14]), and although static variants of bi-objective orienteering problems have been considered by a number of authors (e.g., [2,5,8,10]), the research on dynamic bi-objective orienteering problems still is in a very early stage. So far, only few authors work on dynamic multi-objective vehicle routing problems, most of them proposing solution approaches in terms of methodological frameworks that rely on evolutionary computation (e.g., [6,13]).

Over the past decade a number of authors have solved (single-objective) dynamic orienteering problems by combining integer linear programming with waiting strategies (see [11] for an overview). The idea is to maximize the number of visited customers over a given fixed time horizon by using linear programming for calculation of a routing plan at each decision time. Therefore, simple waiting strategies[1] are used, i.e., the vehicle remains idle at locations in these plans, hoping for a close-by customer request to occur in the near future.

This approach can be transformed into an a-posteriori benchmark solution for dynamic bi-objective optimization algorithms by selecting the best waiting strategy and by then solving the problem several times, each time with a different bound of the maximum tour length in the linear program. In Sect. 4 we use the waiting strategies and the linear program described in [11] as benchmark for the dynamic multi-objective evolutionary algorithm introduced in the following Sect. 3.

3   The Dynamic Multi-objective Evolutionary Algorithm

Our dynamic EMOA for the considered orienteering problem is based on the a-posteriori EMOA introduced in [10] with refined adjustments—in particular in initialization and mutation—to meet the requirements of the dynamic setting.

Algorithm 1. Dynamic EMOA

Require: Instance $I = (C^m, C^o)$,
    time resolution Δ_t, nr. of time slots n_t
1: $t := 0$
2: tour := LOCALSEARCH(C^m)   ▷ No dynamic
    customers, i.e., solve single-obj. problem
3: $t := t + \Delta_t$
4: $P = $ NIL
5: driven.tour := FINDDRIVENTOUR(tour, t)
6: **for** i in 1 to n_t **do**
7:    $(P, F(P)) := $ EMOA(I, driven.tour, t, P)
8:    tour := DECIDE($P, F(P)$)
9:    $t := t + \Delta_t$
10:    driven.tour := FINDDRIVENTOUR(tour, t)

Algorithm 2. EMOA

Require: Instance $I = (C^m, C^o)$, driven.tour,
    time t, population of previous era Q, popu-
    lation size μ
1: **for** i in 1 to μ **do**
2:    $P_i := $ INITINDIVIDUAL(I, driven.tour, t,
        Q_i)   ▷ Q_i is NIL on start
3: $F(P) := $ EVALUATEFITNESS(P)
4: **while** stopping condition not met **do**
5:    $O := $ MUTATE(P)
6:    $O := $ LOCALSEARCH(O)
7:    $(P, F(P)) := $ SELECT($P \cup O$)
8: **return** $(P, F(P))$

Algorithm 3. initIndividual

Require: instance $I = (C^m, C^o)$, driven.tour,
    current time t, template individual y
1: **if** not y is NIL $\wedge$ y is feasible **then**
2:    **return** y
3: $C^o_{\leq t} := $ Dyn. customers arrived so far
4: $D := C^o_{\leq t} \setminus$ driven.tour
5: $x.b, x.p, x.t$ are vectors of length $N - 2$
6: $x.b_i := 1, x.p_i := 0 \, \forall \, i \in C^m \vee$
        $i \in$ driven.tour
7: $x.t := $ CONCAT(driven.tour,
        RANDPERM($C \setminus$ driven.tour))
8: $x.p_i := 1/|D| \, \forall i \in D$
9: $u := \mathcal{R}(1, \ldots, |D|)$ ▷ Rnd. number
10: Set $x.b_i := 1$ for u rnd. customers from D
11: **if** not y is NIL **then**
12:    $x := $ TRANSFER(x, y)
13: **return** x

Algorithm 4. mutate

Require: Population P, swap prob. p_{swap},
    nr. of swaps σ_{swap}
1: **for** $x \in P$ **do**
2:    flip $x.b_i$ with probability $x.p_i$
3:    $t_{\text{active}} := $ seq. of active customers in $x.t$
4:    **if** $r \sim R(0, 1) \leq p_{\text{swap}}$ **then**
5:        **for** 1 to σ_{swap} **do**
6:            swap two random pos. in t_{active}
7: **return** P

---

[1] Two prominent strategies used also in this work for comparison reasons are Drive First (DF) and Distributed Waiting (DW). While in DF the vehicle only waits at its current customer location if both waiting time is available and the planned route only contains the end depot, the latter strategy distributes the amount of available waiting time equally among all customer locations of the current planned route.

We start with a high-level description of the dynamic EMOA framework accompanied by a example first and discuss the more complex solution encoding scheme and mutation later on. The dynamic EMOA (see Algorithm 1) is basically a wrapper around the static version introduced in [7] which uses NSGA-II [4] as the encapsulating meta-heuristic (see Algorithm 2). It is started at time $t = 0$. Note, that at this point in time only mandatory customers C^m are available. Since no subset selection is necessary in this special case the problem is of single-objective nature and we simply apply local search[2] to approximate the optimal tour serving all mandatory customers (see Fig. 1 left) and the first *era* ends. Here, the DM is given only a single choice and there is nothing left to do. In subsequent eras $j = 1, \ldots, n_t$ however, already time $j \cdot \Delta_t$, Δ_t being the adjustable time resolution, has passed and hence more and more dynamic customers request for service. To be precise, in era j dynamic customers with request times $r_i \in ((j-1) \cdot \Delta_t, j \cdot \Delta_t]$ arrive. In each such era the static EMOA is started feeding in the partial tour already driven by the vehicle (as time goes by, the vehicle already may have served some of the mandatory and/or dynamic customers). After termination, the resulting approximations are handed over to the DM who needs to choose exactly one solution (see line 8 in Algorithm 1 and Fig. 1 middle and right for example).

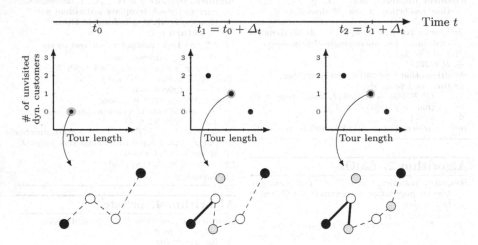

Fig. 1. Exemplary progress of the dynamic EMOA. The scatter plots show the Pareto-front approximations with selected solutions highlighted (●). Below the decision maker choices are depicted (depots ●, mandatory customers ○ and dynamic customers ◔). A dashed path indicates the tour chosen by the decision maker while the thick solid prefix path highlights the partial tour already driven.

---

[2] We adopt EAX [12] as the local search procedure with focus on tour length minimization. Note, that we need to solve a shortest Hamiltonian path problem, but EAX is a TSP solver. Thus, before application of the local search procedure, the problem is transformed into a TSP by a sequence of modifications to the distance matrix (see [10] for details).

The encoding of candidate solutions needs to account for both the subset selection of customers and the minimization of the Hamiltonian path serving all selected customers. Thus, three essential vectors of length $N - 2$ are maintained: (1) a permutation vector holds the sequence of customers, i.e., the actual tour, (2) vector $b = (b_2, \ldots, b_{N-1}) \in \{0,1\}^{N-2}$ indicates whether a customer $i \in \{2, \ldots, N-1\}$ is active ($b_i = 1$) or inactive ($b_i = 0$) and (3) vector $p \in [0,1]^{N-2}$ holds the flip probabilities for the mutation operator (see below). Building the initial population (see Algorithm 3) is a complex process since several dynamic aspects need to be considered: (1) we need to ensure, that both mandatory and already visited (potentially dynamic) customers are active and cannot be removed by mutation (line 6). Hence, $b_i = 1$ and $p_i = 0$ for those customers. (2) the partial tour already driven must not be changed and hence the first positions of the permutation vector correspond to this sequence (line 7). (3) We transfer knowledge from the final population of the previous era in order to not start from scratch. This is achieved by simply copying the individual if it is still feasible (line 2). Otherwise, we transfer as much information by keeping active customers active and maintaining the tour as far as possible (line 12). The initialization procedure guarantees feasibility of initial solutions.

Mutation is twofold to account for both objectives (see Algorithm 4). First, available customers are added or removed by flipping each bit b_i independently with probability p_i. Next, with probability $p_{\text{swap}} \in (0,1]$ some random position exchanges in the permutation vector are performed limited to active customers not yet visited, i.e., which are not part of the already driven part of the tour. Note, that mutation is non-destructive and hence feasibility is maintained. Finally, mutated solutions are subject to local search at certain generations. Here, we apply EAX [12] with the last customer of the already driven tour as the start node and the end depot as the destination node omitting already visited customers. It is important to stress, that the local search operator is focused on tour length minimization only, since we consider this objective to be more difficult. Furthermore, take notice that EAX does not take request times into consideration. Hence, the length of the resulting tour is a lower bound on the true tour length. We take the solid foundations and results laid down in [3,10] as a justification for this approach.

4 Computational Experiments

Experimental Setup: In order to evaluate the dynamic EMOA introduced in Sect. 3, we perform proof-of-concept experiments. We select 5 instances with $N = 100$ customers (including depots) each: one instance with locations distributed uniformly at random in the Euclidean plane and 4 instances with $2, 3, 5$ and 10 clusters respectively form the instances introduced in [10]. The proportion of dynamic customers is chosen to be 75% for all instances, in order to specifically analyze the working principles of our approach.

We fix the time resolution $\Delta_t = 100$ and determine the number of eras as $\lceil \max_{i \in C}(r_i)/\Delta_t \rceil + 1$, where $r_i \geq 0$ is the request time of customer $i \in C$. The final parametrization of the dynamic EMOA is gathered in Table 1. These settings deserve further explanation: Preliminary experiments were performed testing different parameter settings. More precisely, we varied local search (on/off), transfer of knowledge of previous eras (on/off), the swap-mutation probability $p_{swap} \in \{0.2, 0.4, \ldots, 1\}$ and the way available dynamic customers are being distributed in the initial solutions of each era (uniform/binomial). Unsurprisingly, local search (see our a-posteriori study in [3]) and knowledge transfer are beneficial settings to not discard progress already being made. The latter two varied parameters, p_{swap} and the distribution of dynamic customers in initial solutions, however, show strong interaction with local search. It turns out, that a high swap probability with binomial distribution leads to poor front coverage in areas with a high number of unvisited customers. This can be explained as follows: Local search pushes solutions to the left (focus on tour length minimization). Now assume, we are given a very good solution with respect to tour length and apply mutation with high swap probability. Assume further, that mutation deactivates some customers. Clearly, since the tour can only become even shorter, this step pushes the solution to the top left area of the Pareto-front approximation. Since the tour is already close to optimal, the subsequent swaps introduce edge crossings and have a destructive effect with overwhelming probability. Consequently, the mutated individual shifts to the right (larger tour length) and is likely to be dismissed by the following survival selection. In case of binomial distribution each available dynamic customer is activated with probability $1/2$. Hence, the number of activated dynamic customers is binomially distributed with expected value $N_t^d/2$ where N_t^d is the number of dynamic available customers at time $t > 0$. The probability that the actual number deviates from the expectation is rather low and hence is concentrated heavily around it. Thus, this type of initialization in combination with activated local search and high swap probability tends to produce the above mentioned poor coverage. We bypass this problem by adopting a uniform distribution of dynamic customers, i.e., each number of active available customers is active with equal probability.

Table 1. Dynamic EMOA parameterization.

Parameter	Setting
Generations per era	65.000
μ, λ	100
p_{swap}	0.6
σ_{swap}	$N/10 = 10$
LS application in generations	initial, half-time, last
Cutoff time for LS	1s
Transfer knowledge from last era	on
Distribution of dynamic customers	Uniform

In this study, we simulate different decision maker strategies which are based on order ranking of the first objective (tour length). In a nutshell, the n solutions of the EMOA are ordered in ascending order of tour length[3] and the DM decides for the $\lfloor \text{rank} \cdot n \rfloor$-ranked solution with rank $\in \{0.25, 0.5, 0.75\}$ in each era. Clearly, in real world scenarios, the DM can make different decisions in each era to adapt to different situations and we are aware of the limitations of our DM policies. However, for a first study and for an automated evaluation of the approach, we consider these fixed three strategies a good starting point.

We performed 10 independent runs on each instance. The implementation of our dynamic EMOA is available at a public repository[4].

Results: On the one hand, the following results contribute to the understanding of the working principle of the dynamic evolutionary approach. On the other hand, they show the applicability and provide a feeling for the potential of such an approach.

Figure 2 comprises two representative series of depictions of the intermediate Pareto-front approximations generated in each era of the algorithm run, for uniform (top) and clustered (bottom) topologies of customers. Each era bases on decisions made during previous process. For the decision making process three ranks were fixed. In each plot, the Pareto-fronts of the dynamic approach are colored per era from dark blue (first era) to light green (last era). For visual comparison, Pareto-front approximations of the a-posteriori EMOA recently proposed in [3] and of an ε-constrained-based ILP approach using the dynamic single-objective strategies [10] described in Sect. 2 are shown. Note, that – for comparison reasons – the results of all eras have been transformed to the a-posteriori solution space. Additionally, the sub-figures contain horizontal lines colored according to the eras. Those lines define a true upper bound of available unvisited customers for that era. It is clear that depending on the current era and previous actions of the DM, the upper bound decreases.

The first interesting finding is, that our approach is capable of outperforming the ILP-based a-posteriori strategy directly and the MOEA-based a-posteriori approach on the long run. Although the a-posteriori approaches possess complete information on the (virtually) dynamic service requests, the dynamic approach is able to generate comparable or even better solutions without foresight - especially for uniform topologies. For clustered topologies, the approach often outperforms the ILP-based strategy in its final era and sometimes even becomes comparable to the a-posteriori EMOA solutions. This is especially true, when the (higher) decision maker rank favors the second objective (number of unvisited customers).

The at a first glance surprising superiority over the a-posteriori approach is rooted in the fact, that the search space for the a-posteriori problem is much larger than the restricted dynamic scenario, in which previous decisions and a

---

[3] Note that in the bi-objective case this leads to a sorting in descendant order of the number of unvisited customers.

[4] Repository: https://github.com/jakobbossek/dynvrp/.

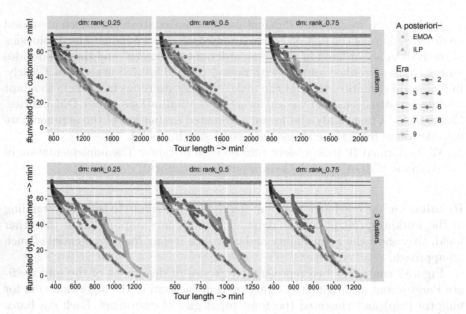

Fig. 2. Scatter plots of representative Pareto-front approximations for three different decision maker strategies on uniform (top) and clustered (bottom) topologies. Points are colored by era. Colored horizontal lines indicate a true upper bound for the number of unvisited available dynamic customers w.r.t. the era. For comparison front approximations based on complete a-posteriori knowledge obtained in [3] and [10] are shown. (Color figure online)

fixed partial tour reduce the search space dramatically. While in the a-posteriori case for selecting an optimal subset of visited customers, all customers are eligible, the dynamic approach can narrow the subset selection to still available customers w.r.t. the already fixed partial tour.

The analysis of the representative results in Fig. 2 for uniform and clustered topologies[5] shows that era results for uniform topologies are closer to the a-posteriori results than era results for clustered topologies. To provide a more detailed insight into this aspect, we show respective embeddings of found (intermediate) solution tours for both topologies in Fig. 3. For both settings, the DM selected solutions of era 1, era 4 and era 9 are plotted including the path to already visited customers (bold) and the plan for the remaining tour considering currently available customers. In the top row of Fig. 3, the tour starts with the mandatory customers and successively integrates new appearing customers into the tour. As customers are uniformly distributed in search space, later appearing customers can easily be integrated in the not yet fixed part of the tour.

In contrast to this, for clustered instances like in Fig. 3 (bottom), new customers appear over time in different clusters. Here, the mutation operator

---

[5] We find similar behavior for all investigated (but not shown) topologies for multiple repetitions.

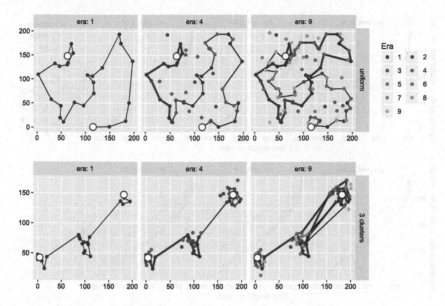

Fig. 3. Embedding of the actual tours the decision maker (rank 0.75) decides for at the end of eras 1, 4 and 9 respectively. The bold part of the tour is already fixed/visited and is hence not subject to change in subsequent eras.

(i.e. random activation of customers) and preferences of the DM potentially have major impact on the quality of the solution. On the one hand, mutation may include customers from a distant cluster. On the other hand, strong DM preference on maximizing the number of visited customers (set to 0.75 for example shown in Fig. 3) may force the algorithm to select newly available customers from a distant cluster. Both will lead to long traveled distances in the resulting tour and as such deteriorate the overall trade-off solution compared to the a-posteriori results. This suggests, that future work should deal with elaborated mutation mechanisms that try to avoid (or alternatively repair) multiple long distance travels between clusters.

In order to evaluate the process of decision making and to test our approach for stability w.r.t. multiple runs, we plot the intermediate decision results leading to the final realized tour in Fig. 4. According to our standard color scheme, we show picked solutions of the parametrized DM for all eras and over all runs. Additionally, the centroid of the final realizations is shown as black-framed dot. The solid black line connects the centroids of the intermediate decisions and shows the decision path. For the representative results in Fig. 4 we can conclude two aspects: (1) The dynamic approach is stable over multiple runs, i.e. the variance in produced solutions is low. (2) Compared to the a-posteriori approximated Pareto-front, the final decisions made under the dynamic evolutionary scheme quite perfectly reflect the parametrized ranking set up for the DM.

Note, that all qualitative results presented here also hold for the investigated topologies (different number of clusters) in the same way.

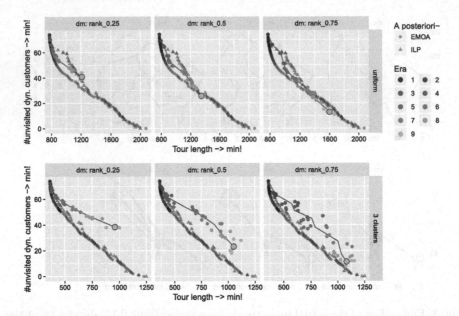

Fig. 4. Paths of decisions taken by different decision maker policies. Colored points represent the decisions made in the corresponding eras for all 5 independent runs. Paths run through the centroids of per era decisions. The centroid of final decisions is highlighted as black-framed dot at the end of the decision path. (Color figure online)

5   Conclusions and Outlook

Our previous studies on bi-objective orienteering from an offline perspective resulted in detailed insights into problem characteristics and challenges for respective multi-objective (evolutionary) algorithm design. This paper proposes an online-approach for multi-objective dynamic optimization, which is required in practice for adjusting a currently active vehicle tour to new customer requests. A crucial feature of the new real-time optimization system is the possibility of incorporating user preferences regarding both objectives, which can either be given as a fixed a-priori rule or interactively adjusted along the algorithm run whenever an adjustment decision of the current tour has to be made.

Initial proof-of-concept experiments indicate that ILP strategies are outperformed by our approach in terms of solution quality and efficiency. The latter point is especially important regarding scalability w.r.t. the instance size. With increasing instance size ILP strategies will become infeasible in terms of the real-time system requirement. Moreover, ILP methods are based on a-priori fixed waiting strategies in contrast to flexible preference incorporation. Additionally, in our settings, the dynamic approach comes close or even dominates certain parts of the Pareto-front approximation gained by the retrospective offline EMOA. We find, however, that dynamic optimization becomes more challenging on clustered instances due to higher probability of long distances travels between customers.

Next steps will include a comprehensive benchmark study on a large set of representative instances in terms of proportion of dynamic customers (different from here considered 75% optional customers), degree of clustering as well as instance sizes. Realistically, the current instance size is already quite large in terms of one vehicle serving 100 customers a day. From the hybridization point of view, the influence of local search has to be investigated for the online case. A straightforward extension will be allowing for more than one vehicle, which increases practical relevance but poses additional challenges onto dynamic EMOA design. For a systematic validation, a suitable performance indicator simultaneously incorporating the quality of the final Pareto-front approximation, the any-time performance along the EMOA run, robustness across multiple runs, and the degree of user preference fulfillment, has to be derived.

Acknowledgments. J. Bossek, C. Grimme, S. Meisel and H. Trautmann acknowledge support by the European Research Center for Information Systems (ERCIS).

References

1. Azzouz, R., Bechikh, S., Ben Said, L.: Dynamic multi-objective optimization using evolutionary algorithms: a survey. In: Bechikh, S., Datta, R., Gupta, A. (eds.) Recent Advances in Evolutionary Multi-objective Optimization. ALO, vol. 20, pp. 31–70. Springer, Cham (2017). https://doi.org/10.1007/978-3-319-42978-6_2
2. Berube, J.-F., Gendreau, M., Potvin, J.-Y.: An exact ∈-constraint method for bi-objective combinatorial optimization problems: application to the traveling salesman problem with profits. Eur. J. Oper. Res. **194**(1), 39–50 (2009)
3. Bossek, J., Grimme, C., Meisel, S., Rudolph, G., Trautmann, H.: Local search effects in bi-objective orienteering. In: Proceedings of the Genetic and Evolutionary Computation Conference, GECCO 2018, pp. 585–592. ACM, New York (2018)
4. Deb, K., Pratap, A., Agarwal, S., Meyarivan, T.: A fast and elitist multiobjective genetic algorithm: NSGA-II. IEEE Trans. Evol. Comput. **6**(2), 182–197 (2002)
5. Filippi, C., Stevanato, E.: Approximation schemes for bi-objective combinatorial optimization and their application to the TSP with profits. Comput. Oper. Res. **40**(10), 2418–2428 (2013)
6. Ghannadpour, S.F., Noori, S., Tavakkoli-Moghaddam, R.: A multi-objective vehicle routing and scheduling problem with uncertainty in customers' request and priority. J. Comb. Optim. **28**, 414–446 (2014)
7. Grimme, C., Meisel, S., Trautmann, H., Rudolph, G., Wölck, M.: Multi-objective analysis of approaches to dynamic routing of a vehicle. In: ECIS 2015 Completed Research Papers. Paper 62. AIS Electronic Library (2015)
8. Jozefowiez, N., Glover, F., Laguna, M.: Multi-objective meta-heuristics for the traveling salesman problem with profits. J. Math. Model. Algorithms **7**(2), 177–195 (2008)
9. Meisel, S.: Anticipatory Optimization for Dynamic Decision Making. Operations Research/Computer Science Interfaces Series, vol. 51. Springer, New York (2011). https://doi.org/10.1007/978-1-4614-0505-4
10. Meisel, S., Grimme, C., Bossek, J., Wölck, M., Rudolph, G., Trautmann, H.: Evaluation of a multi-objective EA on benchmark instances for dynamic routing of a vehicle. In: Proceedings of the Genetic and Evolutionary Computation Conference, GECCO 2015, pp. 425–432. ACM, New York (2015)

11. Meisel, S., Wölck, M.: Evaluating idle time policies for real-time routing of a service vehicle. In: ECIS 2015 Completed Research Papers. Paper 132. AIS Electronic Library (2015)
12. Nagata, Y., Kobayashi, S.: A powerful genetic algorithm using edge assembly crossover for the traveling salesman problem. INFORMS J. Comput. **25**(2), 346–363 (2013)
13. Nahum, O.E., Hadas, Y.: A framework for solving real-time multi-objective VRP. In: Żak, J., Hadas, Y., Rossi, R. (eds.) EWGT/EURO -2016. AISC, vol. 572, pp. 103–120. Springer, Cham (2018). https://doi.org/10.1007/978-3-319-57105-8_5
14. Pillac, V., Gendreau, M., Guéret, C., Medaglia, A.L.: A review of dynamic vehicle routing problems. Eur. J. Oper. Res. **225**(1), 1–11 (2013)
15. Raquel, C., Yao, X.: Dynamic multi-objective optimization: a survey of the state-of-the-art. In: Yang, S., Yao, X. (eds.) Evolutionary Computation for Dynamic Optimization Problems, pp. 85–106. Springer, Heidelberg (2013). https://doi.org/10.1007/978-3-642-38416-5_4

A Formal Model for Multi-objective Optimisation of Network Function Virtualisation Placement

Joseph Billingsley[1]($\boxtimes$), Ke Li[1]($\boxtimes$), Wang Miao[1]($\boxtimes$), Geyong Min[1]($\boxtimes$), and Nektarios Georgalas[2]($\boxtimes$)

[1] Department of Computer Science, University of Exeter, Exeter, UK
{jb931,k.li,wang.miao,g.min}@exeter.ac.uk
[2] Research and Innovation, British Telecom, Martlesham, UK
nektarios.georgalas@bt.com

Abstract. Ranging from web caches to firewalls, network functions play a critical role in modern networks. Network function virtualisation (NFV) has gained significant interests from both industry and academia, thus making the study of their placement an active research topic. Due to multiple criteria that must be considered by stake holders, e.g. the minimisation of the end-to-end latency and overall energy consumption, the NFV placement problem is in principle a multi-objective optimisation problem. This paper develops a formal model for the NFV placement problem based on queuing theory. By using the popular NSGA-II as the optimiser, the effectiveness of the proposed model is validated through a series of proof-of-concept experiments. In particular, some genetic operators have been developed to match the characteristics of the problem.

Keywords: Network function virtualisation ·
Multi-objective optimisation · Telecommunications · Queueing theory

1 Introduction

Virtualisation has transformed data centres in recent years. Despite the installed server base in data centres increasing by an estimated 6 million since 2007, the energy consumption of data centres has remained relatively flat [12]. A key step towards this improved energy consumption was the introduction of elastic scaling through virtualisation. Whilst observations of server workload show the peak workload can exceed the average by factors of 2 to 10 [1], elastic scaling allows the data centre to use only the resources requested at any time.

The next step of this transformation is network function virtualisation (NFV) which targets a key part of data centre infrastructure. Traditional data centres are composed of purpose-built computers called 'middleboxes' that perform a single network function such as deep packet inspection, encryption or analytics.

Supported by EPSRC Industrial CASE and British Telecom under grant 16000177.

© Springer Nature Switzerland AG 2019
K. Deb et al. (Eds.): EMO 2019, LNCS 11411, pp. 529–540, 2019.
https://doi.org/10.1007/978-3-030-12598-1_42

Traditional middleboxes have several drawbacks, the most notable of which is their high specialisation. This leads to high cost and inflexibility in the data centre, making the deployment/redeployment of services challenging and time consuming. Without redeployment of services, any placement will becomes inefficient or unsustainable as demand inevitably changes over time.

NFV is an application of virtualisation technologies to middleboxes. Virtual network functions (VNFs) can be run on off-the-shelf hardware and, as in cloud computing, the resources allocated to the VNFs can easily be scaled in accordance with demand. Furthermore, as no physical components are required, VNFs can easily be moved around in the data centre and virtual network structures can be formed to connect them allowing for services to be optimised over time to meet changing demand. NFV is a powerful technology and has been identified as a key component of 5G [10], the Internet of Things [3] and future data centres [7] all of which have the potential to be multi-billion dollars industries [11,12].

A major challenge of NFV is the optimal placement of VNFs in data centres which meet various criteria from stake holders, such as the demand on services and the energy consumption of placements. Note that it is not uncommon that data centres have tens of thousands of servers, each of which may run many virtual machines (VMs), leading to a problem of a tremendous scale. Dependencies (also know as interactions) among components make it yet more challenging and it is important to find solutions robust enough to handle fluctuations in demand. The need to construct virtual networks reflects the virtual network embedding problem [6] so that the NFV placement problem is NP-hard in principle.

Although many efforts have been devoted to the NFV placement problem, there has been no consensus on the problem formulation, especially the corresponding objective functions. Some researchers have opted to test their solution using actual hardware [13] or to simulate the network/hardware by using a discrete event simulator [9]. Note that both hardware- and simulation-based approaches require time to achieve a stable status. Hence whilst these approaches are useful for validating the effectiveness of some algorithms or heuristics, it is difficult to consider them within the actual placement optimisation process. Another alternative is to use some particular heuristics to evaluate solutions [2]. Although simple heuristics are responsive enough to be used in optimisation, designing appropriate heuristics is no-free-lunch. Furthermore, the final solutions may not be reliable if the heuristic is made with improper assumptions. Besides, it is difficult to compare the effectiveness of two algorithms which separately use different heuristics.

In this work we propose a general purpose model based on queuing theory that balances the speed of a heuristic approach against the accuracy of more expensive simulation or hardware approaches, allowing it to be used in the optimisation process. One of the merits of using queueing theory is its ability to handle complex dependencies among components. Further, it provides additional considerations to the number of network functions needed for a service and the arrival and service rates of VNFs, all of which have been ignored by the existing approaches. To validate the effectiveness of the proposed model, it is incorporated into a classic evolutionary multi-objective optimisation (EMO) algorithm, i.e. NSGA-II [5], in proof-of-concept experiments.

The remainder of the paper is organised as follows. Section 2 provides a formal definition of the NFV placement problem and develops a formal model to evaluate the end-to-end latency and energy consumption. Section 3 applies the model to a particular network topology and derives genetic operators using information from the model. Section 4 examines the effectiveness of the model through proof-of-concept experiments. Finally, Sect. 5 concludes this paper and outlines some potential future directions.

2   Problem Formulation and Model Building

A service is composed of several network functions that must be visited in a particular order. In traditional data centres, network functions are provided by middleboxes whereas with NFV these are provided by VNFs. In the NFV placement problem, the network can be considered as a graph consisting of servers running VMs connected by an arrangement of switches, as in Fig. 1. We assume that the resources of each server are evenly divided into several slices where each VM is allocated to a slice. A VNF can then be placed on to one of these VMs.

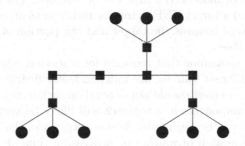

Fig. 1. An example graph with three servers, supporting three VMs (denoted as filled circles) connected by switches (denoted as filled squares).

Each service has a typical arrival rate, which is the amount of traffic it receives over some unit of time on average; while each VNF has a service rate which is the number of packets it can process over the same unit of time when it is placed on a VM slice. If the arrival rate at a switch or VNF is equal to or greater than its service rate, the length of queues at VNFs or switches will tend towards infinity and any dependant services will be inoperable. Similarily if a service requires a particular VNF but no instances of it exist, the service will be inoperable. These characteristics naturally form two constraints on the solution space.

The quality of a particular VNF placement solution can be evaluated by various metrics. In this initial implementation of the model we consider two essential but conflicting objectives, i.e. end-to-end latency and energy consumption. Specifically, the latency for a service is defined as the expected time taken for a request to visit each VNF in the service whilst the energy consumption can be measured by the number of switches/servers used and the traffic they received. These two objectives should be minimised but they are conflicting with each other. If we consider each objective in isolation, the best solution for latency will use as many servers as possible so as to widely distribute the load, whereas the best solution for energy consumption will lead to the usage of only as many servers as is necessary to obtain a feasible solution.

Before deriving the analytical objective functions, we must consider how packets will be routed through the data centre network since this determines the arrival rate at each switch. To this end, we first need to choose one or more VNFs to forward traffic towards and decide the portion of traffic that each one will receive on average. We also need to determine the path of switches the packets will take to their destination. Generally speaking, the NFV placement problem consists of three inter-connected problems: (1) VNF placement; (2) VNF selection; and (3) packet routing. Depending on the network, the packet routing component may be handled by an existing network protocol and hence not be a part of the optimisation problem. Here we propose a formal model that allows for heuristics or some optimisation techniques to be used to solve for any part of the problem. The model takes two user defined functions, named *selection* and *step*. Specifically, the selection function takes several candidate VNFs as inputs and returns the one or more VNFs that will be selected. The step function takes the current VNF and a target VNF as inputs and returns an object that contains the next possible steps towards the target and the portion of traffic that should be sent down each step.

It is reasonable to assume that requests for a service, which may come from different users or different sources, are independently distributed. Similarly, the time taken to serve a request should not depend on earlier requests. Moreover, the distance between components in a network will likely be very small so that the time spent in flight will be negligible. As a consequence, the end-to-end latency is given by the summation of waiting or processing time at VNFs, servers and switches. Each component in the network contains a packet buffer with a certain capacity, while packet loss occurs when the buffer exceeds this capacity. Bearing these considerations in mind, it is natural to consider using queueing theory as the baseline model. Although finite queues have been well studied in queueing theory, they introduced several additional complexities. If we assume queues have an effectively infinite length, instead of packet loss, the time a packet spends in a queue will increase with the length of the queue. Hence considering infinite queues and optimising for latency should in turn favour solutions that would minimise packet loss. A more thorough analysis of the impact of finite queues and packet loss is planned for future work.

Following the above reasoning, the following assumptions are made with regards to the construction of the network:

1. Every switch and server processes traffic according to a Poisson process with a mean rate of μ_{sw}, while each VNF has a particular service rate with respect to that VNF.

2. For each service there are a set of VMs that contain the first VNF in the service. Each of these VNFs produces traffic according to a Poisson process with a mean rate calculated as the arrival rate of the service divided by the number of VMs in the set.
3. Queues at each network component have an infinite capacity.

Based on these assumptions, we can represent each switch and VNF as an M/M/1 queue. The average time a packet will take to be served is given by [8]:

$$f_w(\mu, \lambda) = \frac{1}{\mu - \lambda} \tag{1}$$

where packets arrive at an average rate λ and are served at an average rate μ.

Subsequently, to determine the latency, we need to first calculate the arrival rate at each component and then calculate the expected latency considering the probability of taking each path. The set of paths can be deduced by using a simple recursive process of selection and step functions as illustrated in Fig. 2. Once the set of paths is determined, the total arrival rate at each node can also be calculated. Once all services have been considered we can calculate the expected latency at each node. This is simply the summation of the waiting time at each node in each path, where the waiting time is given by Eq. (1), then the summation is multiplied by the portion of traffic that is sent down that path. We must also consider that the arrival rate at a node may exceed the service rate and hence make the solution infeasible. In this situation we return the overall constraint violations, allowing a means to evaluate the usefulness of infeasible solutions.

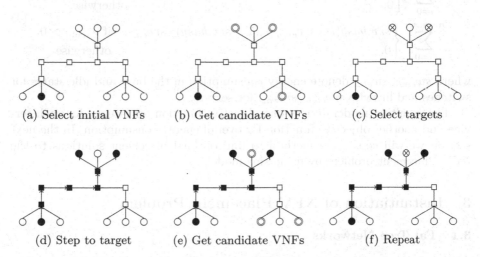

(a) Select initial VNFs   (b) Get candidate VNFs   (c) Select targets

(d) Step to target   (e) Get candidate VNFs   (f) Repeat

Fig. 2. The paths used by a service can be identified with a simple recursive process.

The energy consumption of the network only depends on the arrival rate at each component. Here we propose a three-level energy model that is suitable for a range of problems. Each component can either be one of three states, i.e. *off*, *idle* or *active*. A component is off only if it has an average arrival rate of zero. In this case, it is never used. A component is idle when it is not serving a request, otherwise the component is active. In particular, the time a physical switch is active is calculated as the length of its busy period:

$$sw\_busy(i) = \lambda_{sw[i]}/\mu_{sw}; \tag{2}$$

where $\lambda_{sw[i]}$ is the arrival rate at the switch i and μ_{sw} is its service rate. A server is busy if it is serving a request, or if any of the VMs running on the server is serving a request:

$$srv\_busy(i) = 1 - P(server\_idle \cap vms\_idle)$$
$$= 1 - (1 - (\lambda_{srv[i]}/\mu_{sw})) \cdot \prod_{j=0}^{k_{vm}} (1 - (\lambda_{vm[i][j]}/\mu_{vm[i][j]})) \tag{3}$$

where $\lambda_{srv[i]}$ and $\lambda_{vm[i][j]}$ is the arrival rate at the server i and the arrival rate of its jth VM, and μ_{sw} and $\mu_{vm[i][j]}$ are the corresponding service rates.

As services may share components in a data centre, the arrival rate at a component depends on the production rate of each service. Hence we propose to calculate the arrival rate and the expected latency in two steps. Finally the expected energy cost only depends on the arrival rate at each node:

$$\sum_{i=0}^{num\_sw} \begin{cases} sw\_busy(i) \cdot sw_{en\_b} + (1 - sw\_busy) \cdot sw_{en\_i}, & \text{if } \lambda_{sw[i]} > 0 \\ 0, & \text{otherwise} \end{cases}$$
$$+ \sum_{i=0}^{num\_srv} \begin{cases} srv\_busy(i) \cdot srv_{en\_b} + (1 - srv\_busy) \cdot srv_{en\_i}, & \text{if } \lambda_{srv[i]} > 0 \\ 0, & \text{otherwise} \end{cases} \tag{4}$$

where sw_{en_b}, sw_{en_i} denote energy consumption in the busy and idle states for switches and likewise srv_{en_b}, srv_{en_i} for servers.

The resulting model derives one objective function for latency for each service and another objective function for overall energy consumption. In the next section we will consider a method to find optimal placement solutions to the NFV Placement problem by using this model.

3   Instantiation of NFV Placement Problem

3.1   Fat Tree Networks

Although the model proposed in Sect. 2 is flexible, it is still difficult to define useful step and selection functions for an arbitrary graph. In this paper, for proof-of-concept purposes, we implement these functions for the case of a fat tree network. Fat tree networks are widely used in industry [4] and the underlying principles presented here can be extended to other network structures.

Fat tree networks are described by the number of ports at each switch. We define k as the number of ports for each physical switch and k_{vm} as the number of slices in each VM. In a fat tree topology, there are $(k/2)^2$ core switches. Each core switch connects to one switch in each of k pods. Each pod contains two layers (aggregation and edge) of $k/2$ switches. Each edge switch is connected to each of the $k/2$ aggregation switches of the pod. Each edge switch is also connected to $k/2$ servers. Each server contains a virtual switch connected to k_{vm} VMs. This topology results in $n = (k^3/4) \cdot k_{vm}$ VMs (Fig. 3).

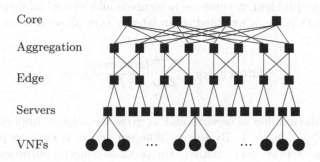

Fig. 3. An example NFV enabled fat tree network with four ports for each hardware switch and space for three VNFs per server.

Based on this definition, a NFV placement solution can be represented as a string of VNFs of length n. As some slices may not be used we also introduce the None character to represent an unused slice. To determine the arrival rates we will need to define the selection and routing functions. In this implementation we will use simple heuristics for both. A fat tree network can be efficiently traversed by stepping upwards to the parent node until a common ancestor between the initial and target nodes is reached. Due to the arrangement of aggregate and core switches, there can be several closest common ancestors which lead to equally efficient paths to the target node. In this implementation, the step function is constructed to distribute traffic evenly over each efficient path. As for the selection criteria, we will simply select all candidate VNFs from the closest server.

3.2 Optimisation Algorithm

Having decided on the representation and objective function implementations, we now have enough information to be able to find optimal NFV placement solutions for a fat tree network. Due to its multi-objective nature, we use the most popular EMO algorithm, NSGA-II [5], as the baseline optimiser. However, considering the characteristics of the NFV placement problem and its constraints, some modifications on NSGA-II are developed as follows.

NSGA-II. From the problem formulation introduced in Sect. 2, a solution that violates any constraint will not have meaningful objective values in the NFV placement problem. In this case, we propose to assign all infeasible solutions the lowest possible crowding distance of zero. When comparing two infeasible solutions, the one having the lower constraint violation is preferred.

In principle, when considering the end-to-end latency, each service has its own latency to optimise. In this case, it may end up with a multi-objective optimisation problem with as many objectives as services in the network. However, curse-of-dimensionality is always the Achilles' heel of an optimisation algorithm. To simplify the problem, we propose to combine all services' latencies into a new objective function, i.e. a weighted mean latency over all services as:

$$latency\_agg = \sum_{i=0}^{N_s} \frac{latencies[i]}{N_s} \cdot w_i \tag{5}$$

where N_s is the number of services and w_i gives the relative importance of each service and $\sum_{i=1}^{N_s} w_i = 1$. By setting different w_i, it is easy to prioritise the latency of one service over another. For proof-of-concept purposes, this paper assumes that all services are equally important, i.e. $w_i = 1/N_s, i \in \{1, \cdots, N\}$.

Initialisation. In the original NSGA-II, the initial population is generated by a random sampling over the solution space. Given the existence of constraints, it is highly likely to generate infeasible placement solutions.

To remedy this issue, we need to make sure that at least one instance of each VNF in every service is placed. However, this does not guarantee a feasible solution is made. It is possible that we may need more than one VNF to fulfil a service if the arrival rate exceeds the service rate of a single VNF. In fact, it may be necessary to have multiple instances of earlier VNFs as well, e.g. if the selection function only ever selects one VNF. Whilst these problems mean that we cannot guarantee that all initialised solutions will be feasible, we can determine a lower bound for the required number of instances of a VNF by using the following equation:

$$min\_num(vnf) = \lceil \mu_{vnf}/\lambda_{vnf} \rceil \tag{6}$$

This information can be used to increase the chance of generating feasible solutions. In order to generate a range of solutions, some multiple of the minimum number of each VNF can be placed in the solution. In our proposed initialisation procedure, we opt to randomise the order of these VNFs in the solution but condense the VNFs towards one end of the data centre. This is motivated by the fact that the energy efficiency of a solution is not greatly affected by the placement of VNFs while latency can be significantly harmed if the next VNF in a service is further away. The pseudo code of the initialisation procedure is given in Algorithm 1.

Algorithm 1. Initialisation

$vnfs \leftarrow []$
  for all $(i, service)$ in $services$ **do**   ▷ Find the minimum number of each VNF
    for all vnf in $service.vnfs$ **do**
       $min\_vnfs \leftarrow ceil(\lambda_{service}/\mu_{vnf})$
       $append(vnfs, [vnf \mid min\_vnfs])$
    end for
  end for
  $max\_copies \leftarrow floor(n/len(vnfs))$ ▷ Find the maximum copies that could fit
  $copies \leftarrow rand(0..max\_copies)$
  $vnfs \leftarrow append(vnfs, [vnfs \mid copies])$ ▷ Copy the VNFs
  $solution \leftarrow shuffle(vnfs)$ ▷ Shuffle the VNFs
  return $append(solution, [None \mid n - len(vnfs)])$ ▷ Pad the output with None

---

Reproduction Operators. In order to generate new candidate placement solutions, we use crossover and mutation, which are the normal practice in genetic algorithm, to serve as reproduction operators. As for the crossover operation, this paper uses the vanilla uniform crossover without any modification. Due to the consideration of constraint violation, the mutation operator needs some further development to guarantee the feasibility of the mutated solutions. When a mutation occurs, the new value is chosen from the set of VNF's and the empty character, weighted by Eq. (6) – the minimum number of instances of the VNF that is required for a feasible solution.

4 Proof-of-Concept Experiments

In this section, we present some proof-of-principle results to demonstrate the effectiveness of the proposed model to serve as objective functions. We also aim to identify some interesting properties of the NFV placement problem that could not have been found with existing models. Due to the page limit, the proof-of-principle experiments here focus on the key results that differentiate the model from previous approaches.

Except where otherwise stated the following parameters are used: $k = 4$, the number of slices on each server $k_{vm} = 3$, 4 services with length 3 are deployed, the service arrival rate for each service is $service_i.arr = 5$, the service rate at each switch $\mu_{srv} = 20$, the service rate at each VNF $\mu_{vnf} = 3$, finally the idle and busy energy consumptions for the servers are $srv_{en\_b} = 2.0$, $sw_{en\_i} = 0.2$, with each switch using $sw_{en\_b} = 1.0$ and $sw_{en\_i} = 0.1$.

To ensure that the model is accurate we compared the results from the model against those from a discrete event simulation. First the NSGA-II algorithm with a population size of 100 is run for 500 generations to produce a set of non-dominated solutions. Then three representative solutions were selected: one from each objective, and a third solution close to the mean. The objectives calculated by the model where compared against those from a discrete event simulator for the same solutions over a range of arrival rates and are plotted in Figs. 4a and b.

From these two subfigures, we can see that the model and simulation agree to a high level of accuracy up to the point that the queues become saturated. At this point, where the arrival rate exceeds the service rate, the waiting time at a queue will approach infinity and the assumptions the model is constructed with are no longer valid.

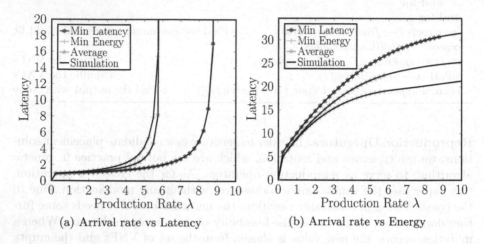

(a) Arrival rate vs Latency

(b) Arrival rate vs Energy

Fig. 4. Model *versus* simulation results for different arrival rate λ settings.

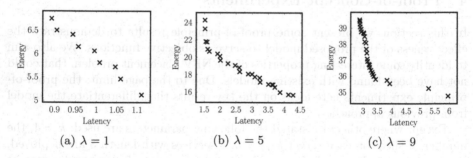

(a) $\lambda = 1$

(b) $\lambda = 5$

(c) $\lambda = 9$

Fig. 5. Non-dominated solutions from 30 runs of 500 generations for different production rate λ settings.

To gain some insight into the properties of the problem under different parameters, the non-dominated solutions from 30 runs of NSGA-II were gathered and plotted in Figs. 5a to c for different production rates. As expected, we can easily see a trade-off between latency and energy. Less obvious however is the relationship between the two objectives. In Fig. 5a the relationship appears to be linear. However as the production rate increases, improvements in latency require increasingly more resources. This may be due to the choice of *selection* algorithm which only uses VNFs from the nearest server. If a whole server is required for each VNF to provide enough resources for a feasible solution, groups of VNFs must be deployed to distribute the load. It is unclear at this stage whether this relationship remains under permutations of other parameters but further research could lead to more informed heuristics.

Finally for the sake of future comparison, the hypervolume for different parameter settings was calculated over the course of 30 runs. The nadir point for each setting of α was estimated using the worst objective values for the non-dominated points considering all runs and is used as the reference point. All relevant data including reference points is presented in Table 1 and Fig. 6.

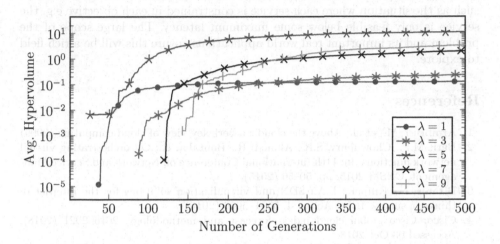

Fig. 6. Average hypervolume for different production rate λ settings.

Table 1. Mean and standard deviation of the hypervolume for different production rates λ and different generations

λ	Ref. point (latency, energy)	Generation				
		100	200	300	400	500
1	(1.111, 6.750)	0.072 ± 0.048	0.114 ± 0.048	0.119 ± 0.048	0.126 ± 0.049	0.129 ± 0.047
3	(1.451, 15.770)	0.004 ± 0.013	0.074 ± 0.085	0.151 ± 0.141	0.206 ± 0.168	0.239 ± 0.164
5	(4.279, 24.280)	0.892 ± 1.436	5.908 ± 2.713	8.613 ± 2.534	10.232 ± 2.705	11.292 ± 2.688
7	(3.421, 32.110)	0 ± 0	0.481 ± 0.797	1.115 ± 1.096	1.560 ± 1.211	2.043 ± 1.286
9	(5.954, 39.520)	0 ± 0	0.250 ± 0.756	1.522 ± 1.858	2.563 ± 2.627	3.415 ± 3.193

5 Conclusion

This paper developed a formal model for the multi-objective telecommunications problem of NFV placement. Whilst this model has been studied previously the proposed model is more thorough than existing approaches and is able to consider important aspects such as the production and arrival rates that existing models cannot. Additionally through the development of new genetic operators we propose an initial algorithm that demonstrates the utility of the model.

There are many practical extensions to the model that could be considered for future work, including the derivation of other objectives using queuing theory. Objectives such as packet loss and service bandwidth have been well studied in queueing theory [8]. Additionally variants of the problem could be considered such as the situation where each service is constrained in each objective e.g. the service is only feasible below some maximum latency. The large scope of the problem and its important real world applications ensure this will be a rich field to explore.

References

1. Armbrust, M., et al.: Above the clouds: a berkeley view of cloud computing (2009)
2. Bari, M.F., Chowdhury, S.R., Ahmed, R., Boutaba, R.: On orchestrating virtual network functions. In: 11th International Conference on Network and Service Management, CNSM 2015, pp. 50–56 (2015)
3. Bizanis, N., Kuipers, F.A.: SDN and virtualization solutions for the internet of things: a survey. IEEE Access 4, 5591–5606 (2016)
4. Cisco: Cisco global cloud index: forecast and methodology, 2016–2021 (2018). Accessed 03 Oct 2018
5. Deb, K., Agrawal, S., Pratap, A., Meyarivan, T.: A fast and elitist multiobjective genetic algorithm: NSGA-II. IEEE Trans. Evol. Comput. 6(2), 182–197 (2002)
6. Fischer, A., Botero, J.F., Beck, M.T., de Meer, H., Hesselbach, X.: Virtual network embedding: a survey. IEEE Commun. Surv. Tutor. 15(4), 1888–1906 (2013)
7. Han, B., Gopalakrishnan, V., Ji, L., Lee, S.: Network function virtualization: challenges and opportunities for innovations. IEEE Commun. Mag. 53(2), 90–97 (2015)
8. Kleinrock, L.: Queueing Systems: Theory, vol. 1. Wiley, Hoboken (1975)
9. Mijumbi, R., Serrat, J., Gorricho, J., Bouten, N., Turck, F.D., Davy, S.: Design and evaluation of algorithms for mapping and scheduling of virtual network functions. In: Proceedings of the 1st IEEE Conference on Network Softwarization, NetSoft 2015, London, United Kingdom, 13–17 April 2015, pp. 1–9 (2015)
10. Pei, X., et al.: Network functions virtualisation - white paper on NFV priorities for 5G (2017)
11. Reichert, C.: 5G industry to be worth $1.2 trillion by 2026: ericsson. ZDNet, February 2017. Accessed 03 Oct 2018
12. Shehabi, A., et al.: United States data center energy usage report (2016)
13. Xu, J., Fortes, J.A.B.: A multi-objective approach to virtual machine management in datacenters. In: 8th International Conference on Autonomic Computing, ICAC 2011, pp. 225–234 (2011)

NSGA-II for Solving Multiobjective Integer Minimum Cost Flow Problem with Probabilistic Tree-Based Representation

Behrooz Ghasemishabankareh$^{(\boxtimes)}$, Melih Ozlen, and Xiaodong Li

School of Science, RMIT University, Melbourne, Australia
{behrooz.ghasemishabankareh,melih.ozlen,xiaodong.li}@rmit.edu.au

Abstract. Network flow optimisation has many real-world applications. The minimum cost flow problem (MCFP) is the most common network flow problem, which can also be formulated as a multiobjective optimisation problem, with multiple criteria such as time, cost, and distance being considered simultaneously. Although there exist several multiobjective mathematical programming techniques, they often assume linearity or convexity of the cost functions, which are unrealistic in many real-world situations. In this paper, we propose to use the non-dominated sorting genetic algorithm, NSGA-II, to solve this sort of Multiobjective MCFPs (MOMCFPs), because of its robustness in dealing with optimisation problems of linear as well as nonlinear properties. We adopt a probabilistic tree-based representation scheme, and apply NSGA-II to solve the multiobjective integer minimum cost flow problem (MOIM-CFP). Our experimental results demonstrate that the proposed method has superior performance compared to those of the mathematical programming methods in terms of the quality as well as the diversity of solutions approximating the Pareto front. In particular, the proposed method is robust in handling linear as well as nonlinear cost functions.

Keywords: Multiobjective optimisation ·
Minimum cost flow problem · Genetic algorithm

1 Introduction

Minimum cost flow problem (MCFP) is the most general case of a network optimisation problem where a commodity is transferred through the network to satisfy a demand and minimise/maximise objective function(s). There are different applications of the MCFP such as distribution and manufacturing problems, optimal loading of a Hopping aeroplane, or human resource management [1].

For MCFP, sometimes it is necessary to consider multiple criteria such as time, cost, and distance. In this case, we can formulate the MCFP as a multiobjective MCFP (MOMCFP) [14]. This will allow the decision maker to consider

© Springer Nature Switzerland AG 2019
K. Deb et al. (Eds.): EMO 2019, LNCS 11411, pp. 541–552, 2019.
https://doi.org/10.1007/978-3-030-12598-1_43

various conflicting objective criteria and select the most appropriate solution from a set of Pareto-optimal solutions.

There are various types of MOMCFP: linear MOMCFP, integer multiobjective MCFP (MOIMCFP), and nonlinear integer multiobjective MCFP [14]. The linear MOMCFP has continuous decision variables with linear cost functions and constraints, where the solution set consists of only supported non-dominated points [14]. In contrast, the MOIMCFP has integer decision variables with linear cost functions and constraints, and the solution set consists of the supported and a large number of unsupported non-dominated points [5]. Supported non-dominated solutions can be found by applying weighted sum methods, but finding an unsupported non-dominated solution is a challenging task for mathematical programming methods [3,17]. The supported non-dominated solutions are located on the convex hull of the feasible region, while the unsupported non-dominated solutions are found inside the feasible region [4]. The supported and unsupported non-dominated solutions for a bi-objective MCFP are shown in Fig. 1 (See Sect. 2 for the definition of supported and unsupported points). Finally, the nonlinear integer multiobjective MCFP has integer decision variables and employs nonlinear cost functions. The solution set for nonlinear integer multiobjective MCFP generally consists of both supported and unsupported points.

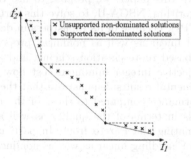

Fig. 1. Supported and unsupported non-dominated solutions in the objective space, for a bi-objective MCFP.

The approaches for solving MOMCFP and MOIMCFP can be categorised into exact and approximation methods [9]. A pseudo polynomial approximation algorithm was proposed in [15] to solve MOMCFP by approximating the optimal value function. In [6], a piecewise linear convex curve of a MOMCFP was approximated by following the trade-off curve. Other approximation methods make use of upper and lower bounds which "sandwich" the Pareto-front [9].

To solve MOMCFP using exact methods, a generalisation of the *out-of-kilter* method can be used [12]. However, this algorithm dealt with only small-sized networks, and the extreme non-dominated solution in the objective space could not be found. A modified out-of-kilter algorithm was developed in [10] to overcome the drawbacks of the generalised out-of-kilter method. In [16], a method was used to identify all the efficient extreme points in the objective space [16]. On the other hand, a branch-and-bound method [18] can be used to produce a representative set of Pareto-optimal solutions for MOIMCFP. All the above mentioned methods considered only small to medium-sized linear MOMCFPs or MOIMCFPs.

The main challenge in mathematical programming for solving MOIMCFPs is to find the unsupported non-dominated points and ultimately the entire set of non-dominated solutions efficiently, without violating the MCFP's constraints [5]. Although algorithms exist to generate the entire set of non-dominated solutions, e.g., using the ϵ-constraint and branch-and-bound method to generate a set of non-dominated solutions for bi-objective integer MCFP [3], or using a two-phase parametric simplex algorithm [13], these algorithms can only solve small and medium-sized bi-objective MCFPs with reasonable efficiency. However, when dealing with a much larger MCFP, e.g., a network consisting of 50 nodes and 870 arcs, the computational time can be as high as 14,000 s.

Mathematical programming methods make a strong assumption that the cost function is linear or convex. However, most real-world problems are nonlinear and non-convex in their more realistic settings. It would be desirable to have a method that does not make these assumptions and can handle both linear and nonlinear cost functions effectively.

The above identified limitations motivated us to employ NSGA-II to solve MOIMCFP with a probabilistic tree-based representation scheme. NSGA-II is able to handle MOIMCFP using either linear or nonlinear cost functions. The algorithm can also control the number of non-dominated solutions to be generated. We compare the NSGA-II method with the state-of-the-art mathematical programming method Bensolve [11] on a set of small to medium-sized MOIM-CFP instances. Our results suggest the superiority of NSGA-II over Bensolve in at least two aspects: (1) NSGA-II can find a set of non-dominated solutions (both supported and unsupported) with control on its size, but Bensolve can only generate one solution depending on its previous solution in a sequential manner, with no control over the total number of solutions. Furthermore, it produces only supported non-dominated points; (2) NSGA-II can handle both linear and nonlinear cost functions, but Bensolve is confined to handling just linear cost functions.

The rest of the paper is structured as follows: Sect. 2 provides the preliminaries and the definition of MOIMCFP. The NSGA-II method using the probabilistic tree-based representation is presented in Sect. 3. Section 4 presents the computational results, and finally Sect. 5 concludes the paper.

2 Problem Formulation

Let $G = (\mathcal{N}, A)$ be a set $\mathcal{N}$ which consists of n nodes and a set A of m arcs. Each arc (i, j) has a capacity of u_{ij} and lower bound of l_{ij} which denote the maximum and minimum amount that can be sent on the arc (i, j), respectively. Each node $i \in \mathcal{N}$ is associated with an integer value $b(i)$. If $b(i)$ is positive, it shows that node i is a supply node, if $b(i)$ is negative, node i is a demand node with demand of $| b(i) |$ and finally $b(i) = 0$ shows the transshipment node i. A decision variable in MOIMCFP is an *integer* flow and denoted by x_{ij}. $F(\mathbf{x}) = (f_1(x), \ldots, f_N(x))$ defines N objective functions for the MOIMCFP. Figure 2 shows an example of the bi-objective MCFP (with $n = 5$ nodes and $m = 7$ arcs), which has a supplier node ($b(1) = 10$) and a demand node ($b(5) = -10$). There are two different costs associated to each arc: $f_1(x_{ij})$ denotes the time that takes to send a flow from node i to node j; $f_2(x_{ij})$ denotes the cost of sending a flow on the arc (i, j).

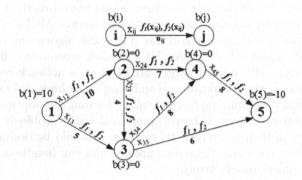

Fig. 2. An example of a bi-objective MCFP (time and cost) ($n = 5$, $m = 7$).

Generally in MOIMCFP, we aim to send flows through the network to satisfy all demands by minimising the objective functions. The formulation of the MOIMCFP is as follows [1]:

$$Minimise : F(\mathbf{x}) = \big(f_1(\mathbf{x}), \ldots, f_N(\mathbf{x})\big) \tag{1}$$

$$\text{s.t.} \sum_{\{j:(i,j)\in A\}} x_{ij} - \sum_{\{j:(j,i)\in A\}} x_{ji} = b(i) \quad \forall \ i \in \mathcal{N}, \tag{2}$$

$$0 \leq x_{ij} \leq u_{ij} \ \forall \ (i,j) \in A, \tag{3}$$

$$x_{ij} \in Z \ \forall \ (i,j) \in A, \tag{4}$$

where Eq. 1 minimises multiple (N) objective functions through the network. Equation 2 is a flow balance constraint which states the difference between the total outflow (first term) and the total inflow (second term). The flow on each arc should be between an upper bound and zero (Eq. 3), and finally all the flow values are integer numbers (Eq. 4). In this paper we consider the following assumptions for the MCFP: (1) the network is directed; (2) there are no two or more arcs with the same tail and head in the network; (3) the total demands and supplies in the network are equal, i.e., $\sum_{i=1}^{n} b(i) = 0$.

Since we are dealing with the multiobjective optimisation problem (MOIMCFP), it is necessary to declare the following definitions. The feasible region in the variable space is defined by $X^I = \{\mathbf{x} = (x_{i_1 j_1}, \ldots, x_{i_n j_n}) \in \mathbb{Z}^n : \mathbf{x}$

satisfies Eqs. 2–4}. The feasible region in the objective space is defined by $Y^I = F(X^I) = \{\mathbf{y} = (y_1, \ldots, y_N) : y_1 = f_1(\mathbf{x}), \ldots, y_N = f_N(\mathbf{x}), \mathbf{x} \in X^I\}$. Let $X = conv(X^I)$ and $Y = conv(Y^I)$ be the convex hull of the sets X^I and Y^I, respectively. Let $y', y'' \in \mathbb{R}^N$, the following notation $y' \leqq y''$ denotes that $y'_q \leq y''_q, \forall q = 1, \ldots, N$.

Definition 2.1. Consider two feasible vectors y' and y'' in $Y^I(Y)$, y' *dominates* y'' if $y' \leqq y''$ and $y' \neq y''$ $(y'_q \leq y''_q)$ with at least one strict inequality. The vector y' is said to be a *non-dominated* solution if there does not exist another y in Y^I which satisfies $y \leqq y'$ and $y \neq y'$. The set of all non-dominated points in Y^I is denoted by $ND(Y^I)$.

Definition 2.2. There are two categories of non-dominated solutions in MOIM-CFP called supported and unsupported non-dominated solutions. Let $Y^{\geqq} = conv(ND(Y^I) + \mathbb{R}^N_{\geqq})$ where $\mathbb{R}^N_{\geqq} = \{y \in \mathbb{R}^N | y \geq 0\}$ and $ND(Y^I) + \mathbb{R}^N_{\geqq} = \{y \in \mathbb{R}^N : y = y' + y'', y' \in ND(Y^I)$ and $y'' \in \mathbb{R}^N_{\geqq}\}$. If y is on the boundary of the $Y^{\geqq}$, y is referred to as a *supported* non-dominated solution, otherwise y is an *unsupported* non-dominated solution. Note that y is always denoted here as a non-dominated solution in the objective space.

Definition 2.3. A solution $x' \in X^I$ (in the variable space) is called *efficient* if it is not possible to find another solution $(x \in X^I)$ with a better objective function value without deteriorating the value of at least another objective value.

3   NSGA-II and the Probabilistic Tree-Based Representation for MCFP

As aforementioned, when dealing with MOIMCFP, there are supported and unsupported non-dominated solutions in the objective space. Supported non-dominated solutions can be found by applying weighted sum methods [3,17], while finding an unsupported non-dominated solution is a challenging task by exact methods. Several such exact methods rely on a mechanism that must find a large number of non-dominated solutions one by one across the entire Pareto-front. However, there is no control of the number of non-dominated solutions to be produced. The excessive number of solutions is unnecessary to the decision maker (DM) and may also require a very high computational cost [5]. In this paper, we propose to use NSGA-II to solve the MCFP with a probabilistic tree-based representation, in order to find a controllable set of non-dominated solutions including both supported and unsupported points.

3.1   Representation

The most popular representation method for solving MCFP using genetic algorithm (GA) is priority-based representation (PbR) [7]. However, the PbR method has serious drawbacks. To counteract the limitations of PbR, the PTbR (Probabilistic Tree based Representation) is introduced in [8]. The PTbR chromosome has $n-1$ sub-chromosomes (Sub.Ch) and the value of each gene is a random number between 0 and 1 which is then accumulated to 1 in each sub-chromosome. In order to obtain a feasible solution from PTbR, in phase I, a path is first constructed, and then a feasible flow is sent through the constructed path in phase II. An example of PTbR and its feasible solution (for the network in Fig. 2) are shown in Fig. 3.

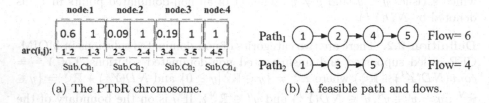

(a) The PTbR chromosome.   (b) A feasible path and flows.

Fig. 3. PTbR and its corresponding feasible solution for the network in Fig. 2.

3.2   NSGA-II for Solving MOIMCFP

After describing the PTbR, we now present how to adopt PTbR and apply NSGA-II to solve MOIMCFP.

Initialisation: First a population with *pop_size* individuals (chromosomes) is generated (P_0). To initialise the population, we can either generate the whole population randomly or use a heuristic technique to seed the initial population. As aforementioned, the mathematical programming techniques can efficiently generate the supported non-dominated points for MOIMCFP with linear objective function. One state-of-the-art mathematical solver that can create the supported non-dominated points is Bensolve [11].

To initialise the population, we apply two different approaches. The first approach is to generate individuals (chromosomes) randomly, namely NSGA-II(R). The second approach is to initialise $\alpha\%$ of the individuals using the supported point obtained from a mathematical programming method (e.g., Bensolve) and the remaining individuals are generated randomly (namely NSGA-II(H)). After initialising the population the binary tournament and genetic operators are applied to create the population Q_0.

NSGA-II Procedure: In each iteration It, we combine the P_{It} with Q_{It} to form the R_{It} population, which is then sorted based on the non-dominance definition. We then calculate the crowding distance, and by sorting the population in each front we select the best-fit *pop_size* individuals in the population as P_{It+1} [2].

Crossover and Mutation: The new population P_{It+1} is now used for binary tournament selection, crossover, and mutation to form a new population Q_{It+1}. A two-point crossover operation is applied, where two blocks (sub-chromosomes) of the selected chromosome (parents) are first randomly selected. Then, two parents swap the selected sub-chromosomes to generate new offspring. To perform mutation, a random parent is selected and the randomly chosen sub-chromosome is regenerated to create a new offspring [8].

Termination Criteria: After genetic operators are applied, we first check the stopping criteria: (1) if the maximum number of function evaluations (NFEs) is reached; or (2) no improvement in the average of the objective function values on the Pareto-front for β successive iterations. If any of the criteria is satisfied the algorithm is terminated; otherwise we go back to the NSGA-II step.

4   Computational Results

In the following subsection, we first describe the cost functions, followed by the network instances which we have adopted for MOIMCFP. We tackle the test instances using two variants of NSGA-II, i.e., NSGA-II(R) and NSGA-II(H), as well as the state-of-the-art mathematical solver Bensolve. Since Bensolve is not able to solve the MOIMCFP using nonlinear cost functions, we only present results in comparing NSGA-II(H) and NSGA-II(R) for MOIMCFP using non-linear cost functions. Note that for NSGA-II(H) using nonlinear cost functions, the *fmincons()* in MATLAB is employed to generate the initial solutions.

4.1   Test Instances

This paper considers MOIMCFP using linear and nonlinear cost functions. Note that, two objectives considered here are time (f_1) and cost (f_2). Our aim is to compare the performance of NSGA-II and Bensolve on bi-objective MCFPs. Equation 1 can be rewritten as follows by employing the linear cost functions [9]:

$$Minimise : F(\mathbf{x}) = \{f_1(\mathbf{x}) = \sum_{(i,j)\in A} c_{ij}^1 x_{ij}, \ f_2(\mathbf{x}) = \sum_{(i,j)\in A} c_{ij}^2 x_{ij}\}, \quad (5)$$

where c_{ij}^1 and c_{ij}^2 are non-negative integer time and cost associated with one unit of flow on arc (i, j) respectively. To consider nonlinearity, the following concave nonlinear cost functions are adopted [20]:

$$Minimise : F(\mathbf{x}) = \{f_1(\mathbf{x}) = \sum_{(i,j)\in A} c_{ij}^1 \sqrt{x_{ij}}, \ f_2(\mathbf{x}) = \sum_{(i,j)\in A} c_{ij}^2 \sqrt{x_{ij}}\}. \quad (6)$$

For our experiments, a set of 30 MOIMCFP instances with different number of nodes ($n = \{5, 10, 20, 40, 60, 80\}$) is randomly generated and presented in

Table 1 (*No.* denotes the instance number, and each instance has *n* nodes and *m* arcs). Note that, for each node size (n), five different networks are randomly generated. The number of supply/demand for nodes $1/n$ are set to $q = 20/-20$ in the test instances up to 20 nodes and for all other test problems supply/demand are set to $q = 30/-30$ [8].

Table 1. A set of 30 randomly generated MOIMCFP instances.

No.	n	m	No.	n	m	No.	n	m	No.	n	m	No.	n	m	No.	n	m
1		6	6		25	11		86	16		287	21		697	26		1322
2		7	7		28	12		81	17		336	22		721	27		1298
3	5	8	8	10	25	13	20	87	18	40	370	23	60	635	28	80	1356
4		8	9		28	14		74	19		334	24		693	29		1250
5		6	10		27	15		92	20		358	25		695	30		1140

4.2   Results and Analysis

NSGA-II and PTbR are implemented in MATLAB on a PC with Intel(R) Core(TM) i7-6500U 2.50 GHz processor with 8 GB RAM, and we run 30 times for each problem instance. To solve MOIMCFP instances using a mathematical solver, we use the MATLAB version of Bensolve[1].

The parameter settings for NSGA-II are as follows: maximum number of iterations ($It_{max} = 200$), population size ($pop\_size = min\{n \times 10, 300\}$), crossover rate ($P_c = 0.95$), mutation rate ($P_m = 0.3$), maximum number of function evaluations ($NFEs = 100{,}000$) and the termination criterion $\beta = 30$ [8]. For NSGA-II(H) only $\alpha = 10\%$ of the initial individuals are generated using the heuristic method explained in Sect. 3.2 and the rest are generated randomly.

To evaluate the performance of a multiobjective optimisation algorithm, two aspects need to be measured: convergence and distribution of the solutions approaching the Pareto front [2]. We adopt Hypervolume (HV) (or S metric), a widely-used metric for evaluating the performance of a multiobjective optimisation algorithm [21]. Hypervolume computes how close the solutions are to the Pareto-front as well as the spread of the solutions across the Pareto-front [19].

We compare the performance of NSGA-II(R), NSGA-II(H) with Bensolve on all network instances using linear cost functions (Eq. 5). The results are presented in Table 2 (where t, nPF and HV denote the average running time in second, average number of solutions on the Pareto-front and average of the hypervolume metric, respectively over 30 runs). Since Bensolve cannot solve the nonlinear MOIMCFP, we only present the results of the NSGA-II(R) and NSGA-II(H) for solving the network instances using concave nonlinear cost functions (Eq. 6) in Table 3, including the average time (t) and average hypervolume value (HV). Note that for NSGA-II(H) using nonlinear cost functions (Table 3), we employ

---
[1] Bensolve MATLAB version is available on: http://bensolve.org/.

Table 2. Results for solving MOIMCFP using linear objective functions.

No.	n	m	Bensolve			NSGA-II(R)			NSGA-II(H)		
			t	nPF	HV	t	nPF	HV	t	nPF	HV
1		6	1	2	4248	16	3.0	**4274.4**	17	3.0	**4274.4**
2		7	1	3	2326	14	11.0	**2385.7**	14	11.0	**2385.7**
3	5	8	1	2	2910	13	6.0	**3030.0**	14	6.0	**3030.0**
4		8	1	2	1570	16	3.0	**1605.0**	16	3.0	**1605.0**
5		6	1	2	2114	13	6.0	**2153.6**	14	6.0	**2153.6**
6		25	1	3	7373	59	17.9	7624.9	66	18.0	**7643.4**
7		28	1	3	10155	59	2.9	9058.7	66	6.3	**11047.8**
8	10	25	1	2	4368	66	4.0	**4394.6**	83	4.0	**4394.6**
9		28	1	2	4347	53	5.0	**4509.3**	58	5.0	**4509.3**
10		27	1	3	5872	55	5.8	5629.6	63	8.7	**5994.7**
11		86	1	2	3866	105	9.8	4712.1	102	10.1	**4760.3**
12		81	1	2	2722	113	12.3	3003.3	105	12.8	**3010.0**
13	20	88	1	3	2894	96	4.0	2850.8	106	4.4	**2913.6**
14		74	1	2	2700	103	3.3	2821.0	94	5.7	**2999.5**
15		92	1	3	9872	98	7.3	10345.8	109	7.9	**10408.7**
16		287	1	1	**1093**	108	1.0	**1093.0**	106	1.0	**1093.0**
17		336	1	2	3168	107	2.1	2669.9	108	2.2	**3189.1**
18	40	370	1	3	8269	111	2.2	7915.6	111	3.1	**8280.4**
19		334	1	3	4423	105	2.1	3685.8	105	3.0	**4426.8**
20		358	1	1	**1215**	103	1.0	**1215.0**	105	1.0	**1215.0**
21		697	1	5	**5898**	192	2.0	5057.3	223	5.7	**5898.0**
22		721	1	6	18194	217	10.5	14139.5	213	27.4	**19755.7**
23	60	635	1	5	16731	216	4.0	14576.1	214	10.5	**16920.3**
24		693	1	2	7203	231	5.5	7730.7	224	5.9	**7795.5**
25		695	1	3	7643	228	3.6	6807.9	226	5.5	**7889.3**
26		1322	1	3	6484	269	19.0	**6723.6**	277	19.0	**6723.6**
27		1298	1	4	8692	227	13.0	8714.5	217	14.0	**8961.7**
28	80	1356	1	5	25518	317	22.2	25141.1	293	28.0	**26632.6**
29		1250	1	4	11130	271	6.0	10838.6	262	8.0	**11178.7**
30		1140	1	3	9494	243	6.0	8674.7	258	12.0	**9548.9**

fmincons() in MATLAB to seed the initial population by converting the bi-objective problem to the single objective problem using a weighted sum method (i.e., considering equal weights for all objective functions). *fmincons()* uses an interior point method by default to solve constrained nonlinear single objective problems.

As shown in Table 2, NSGA-II(H) has greater or equal HV value than those of the Bensolve and NSGA-II(R). Note that, in Tables 2 and 3 the algorithm with the best HV value for each instance is highlighted in boldface. It is noticeable that Bensolve terminates after 1 s on all instances since they cannot find any other solutions. Although our NSGA-II variants took longer to generate the non-dominated solutions, it can converge to a better set of non-dominated solutions with a better diversity as indicated by the HV metric. As can be seen in Fig. 4, Bensolve is only capable of finding supported non-dominated points, and NSGA-II(R) is able to find just one part of the Pareto-front, however NSGA-II(H) is able

Table 3. Results for solving MOIMCFP using nonlinear objective functions.

Algorithms	No.	1	2	3	4	5	6	7	8	9	10
NSGA-II(R)	t	20	17	16	18	15	71	63	75	64	72
	HV	21970.0	7975.4	14794.0	10521.8	12478.0	28017.9	38407.0	28573.1	28265.2	37078.8
NSGA-II(H)	t	20	18	14	17	15	69	62	69	57	64
	HV	21970.0	7975.4	14794.0	10521.8	12478.0	28017.9	38653.6	28573.1	28616.9	38624.5
Algorithms	No.	11	12	13	14	15	16	17	18	19	20
NSGA-II(R)	t	129	113	128	129	131	139	147	145	139	141
	HV	9801.6	8184.8	8947.6	7944.3	14962.6	5096.2	10519.8	12607.4	10436.2	4791.2
NSGA-II(H)	t	111	111	99	101	103	104	104	107	109	112
	HV	9801.6	8184.8	8973.3	7944.3	14962.6	5130.8	10804.1	13023.9	10982.4	4791.2
Algorithms	No.	21	22	23	24	25	26	27	28	29	30
NSGA-II(R)	t	254	275	264	264	264	304	264	289	337	324
	HV	4471.0	8195.8	19618.1	25460.3	30509.0	283.3	6072.9	13885.3	3122.5	10808.9
NSGA-II(H)	t	238	235	222	244	248	333	232	309	312	322
	HV	13799.6	27278.5	30252.9	31030.9	34056.3	24882.2	17830.5	35182.3	21993.3	31489.4

to find a set of non-dominated solutions with better diversity and convergence. This shows the superiority of NSGA-II(H) over NSGA-II(R) and Bensolve.

As shown in Fig. 4a, for network instance No. 22, Bensolve can find only 6 points on the Pareto-front, which are all supported non-dominated points with $HV = 18,194$, while NSGA-II(R) can find on average 10.5 non-dominated points with $HV = 14, 139.5$ and NSGA-II(H) can obtain on average 27.4 non-dominated points with $HV = 19, 755$. It shows that NSGA-II(H) provided not only a better quality of non-dominated solutions, but also better solution diversity (Fig. 4b). This pattern is observed on all instances in Table 2, indicating that NSGA-II(H) has better performance than Bensolve and NSGA-II(R).

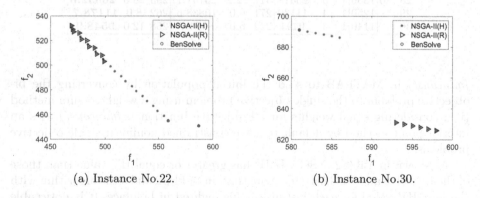

(a) Instance No.22.   (b) Instance No.30.

Fig. 4. Results for solving MOIMCFP using linear objective functions.

Table 3 shows the results on the MOIMCFP instances in Table 1 using concave nonlinear cost functions (Eq. 6) using NSGA-II(H) and NSGA-II(R). On all instances NSGA-II(H) has equal or better performance than the NSGA-II(R). For example, Fig. 5 shows that NSGA-II(H) can converge to a better non-dominated solution set as compared to NSGA-II(R) on instances No.26 and 27. It is consistent with the results of MOIMCFP using linear cost functions and it suggests that using heuristic initialisation (or seeding) can dramatically improve the performance of NSGA-II in dealing with MOIMCFPs.

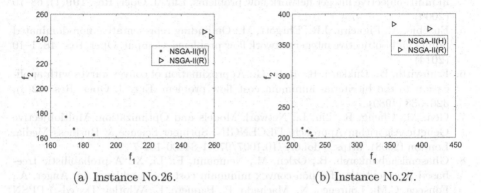

(a) Instance No.26.   (b) Instance No.27.

Fig. 5. Results for solving MOIMCFP using nonlinear objective functions (Bensolve is not included since it cannot handle nonlinear objective functions).

5 Conclusion

In this paper we have adopted the probabilistic tree-based representation for handling the MCFPs, and apply NSGA-II to solve the MOIMCFP using linear and nonlinear cost functions. Unlike the mathematical solvers which are unable to handle nonlinear cost functions, NSGA-II is more robust in dealing with various types of cost functions. The performance of the two variants of NSGA-II (i.e., NSGA-II(H) and NSGA-II(R)) algorithms are evaluated on a set of 30 MOIMCFP instances and compared with that of the state-of-the-art mathematical solver Bensolve. The experimental results demonstrate that NSGA-II(H) has superior performance than that of the Bensolve and NSGA-II(R) in terms of the quality of solutions as well as the diversity of solutions in the objective space. As can be seen in Fig. 4, Bensolve only managed to generate a limited number of solutions (i.e., supported non-dominated solutions) and it cannot find the unsupported non-dominated solutions. However, NSGA-II does not have such a limitation and is able to generate a controllable set of non-dominated solutions. Furthermore, Bensolve cannot handle nonlinearity. It is also worth noting that using a heuristic initialisation procedure (i.e., seeding with solutions found by an exact method) can improve the performance of NSGA-II for solving MOIMCFP.

References

1. Ahuja, R.K., Magnanti, T.L., Orlin, J.B.: Network Flows: Theory, Algorithms, and Applications, pp. 4–6. Prentice hall, Upper Saddle River (1993)

2. Deb, K., Pratap, A., Agarwal, S., Meyarivan, T.: A fast and elitist multiobjective genetic algorithm: NSGA-II. IEEE Trans. Evol. Comput. **6**(2), 182–197 (2002)
3. Eusébio, A., Figueira, J.R.: Finding non-dominated solutions in bi-objective integer network flow problems. Comput. Oper. Res. **36**(9), 2554–2564 (2009)
4. Eusébio, A., Figueira, J.R.: On the computation of all supported efficient solutions in multi-objective integer network flow problems. Eur. J. Oper. Res. **199**(1), 68–76 (2009)
5. Eusébio, A., Figueira, J.R., Ehrgott, M.: On finding representative non-dominated points for bi-objective integer network flow problems. Comput. Oper. Res. **48**, 1–10 (2014)
6. Fruhwirth, B., Bukkard, R., Rote, G.: Approximation of convex curves with application to the bicriterial minimum cost flow problem. Eur. J. Oper. Res. **42**(3), 326–338 (1989)
7. Gen, M., Cheng, R., Lin, L.: Network Models and Optimization: Multiobjective Genetic Algorithm Approach. DECENGIN. Springer Science & Business Media, London (2008). https://doi.org/10.1007/978-1-84800-181-7
8. Ghasemishabankareh, B., Ozlen, M., Neumann, F., Li, X.: A probabilistic tree-based representation for non-convex minimum cost flow problems. In: Auger, A., Fonseca, C.M., Lourenço, N., Machado, P., Paquete, L., Whitley, D. (eds.) PPSN 2018. LNCS, vol. 11101, pp. 69–81. Springer, Cham (2018). https://doi.org/10.1007/978-3-319-99253-2_6
9. Hamacher, H.W., Pedersen, C.R., Ruzika, S.: Multiple objective minimum cost flow problems: a review. Eur. J. Oper. Res. **176**(3), 1404–1422 (2007)
10. Lee, H., Pulat, P.S.: Bicriteria network flow problems: continuous case. Eur. J. Oper. Res. **51**(1), 119–126 (1991)
11. Löhne, A., Weißing, B.: The vector linear program solver Bensolve-notes on theoretical background. Eur. J. Oper. Res. **260**(3), 807–813 (2017)
12. Malhotra, R., Puri, M.: Bi-criteria network problem. Cahiers du Centre d'études de recherche opérationnelle **26**(1–2), 95–102 (1984)
13. Raith, A., Ehrgott, M.: A two-phase algorithm for the biobjective integer minimum cost flow problem. Comput. Oper. Res. **36**(6), 1945–1954 (2009)
14. Raith, A., Sedeño-Noda, A.: Finding extreme supported solutions of biobjective network flow problems: an enhanced parametric programming approach. Comput. Oper. Res. **82**, 153–166 (2017)
15. Ruhe, G.: Algorithmic Aspects of Flows in Networks. MAIA, vol. 69. Springer Science & Business Media, Dordrecht (1991). https://doi.org/10.1007/978-94-011-3444-6
16. Sedeño-Noda, A., González-Martın, C.: The biobjective minimum cost flow problem. Eur. J. Oper. Res. **124**(3), 591–600 (2000)
17. Steuer, R.: Multiple Criteria Optimization: Theory, Computation, and Application. Willey, New York (1986)
18. Sun, M.: A branch-and-bound algorithm for representative integer efficient solutions in multiple objective network programming problems. Networks **62**(1), 56–71 (2013)
19. While, L., Hingston, P., Barone, L., Huband, S.: A faster algorithm for calculating hypervolume. IEEE Trans. Evol. Comput. **10**(1), 29–38 (2006)
20. Yan, S., Shih, Y., Wang, C.: An ant colony system-based hybrid algorithm for square root concave cost transhipment problems. Eng. Optim. **42**(11), 983–1001 (2010)
21. Zitzler, E.: Evolutionary algorithms for multiobjective optimization: methods and applications, vol. 63. Citeseer (1999)

Opposition-Based Multi-objective Binary Differential Evolution for Multi-label Feature Selection

Azam Asilian Bidgoli[1,2(✉)] ⓘ, Shahryar Rahnamayan[2] ⓘ,
and Hessein Ebrahimpour-Komleh[1] ⓘ

[1] Department of Computer and Electrical Engineering,
University of Kashan, Kashan, Iran
ebrahimpour@kashanu.ac.ir

[2] Nature Inspired Computational Intelligence (NICI) Lab,
Department of Electrical, Computer, and Software Engineering,
University of Ontario Institute of Technology (UOIT), Oshawa, Canada
{azam.asilianbidgoli,shahryar.rahnamayan}@uoit.ca

Abstract. Multi-label learning problem is a data analytic task in which every sample is associated with more than single label. The complexity of such problems declares the importance of feature selection task as a preprocessing step prior for multi-label learning. Feature selection can make a better learning performance both in terms of reducing computational complexity and increasing classification accuracy. Selecting the best subset of features with two objectives, the smaller number of features and higher accuracy of classification can be treated as a binary multi-objective optimization problem. Since feature selection is inherently a binary optimization problem, applying continuous metaheuristic algorithms to solve this problem decreases the diversity of solutions in the optimal Pareto-front, because of many-to-one mapping and low exploration power, accordingly. This paper proposed a binary version of Generalized Differential Evolution (BGDE3) for multi-label feature selection based on majority voting of solutions and opposition-based learning (OBL). Experimental results show that the proposed algorithm outperforms the continuous GDE3 for multi-label feature selection.

Keywords: Multi-objective optimization ·
Generalized Differential Evolution · Opposition-based learning ·
Binary differential evolution · Feature selection ·
Multi-label classification

1 Introduction

Classification is one of the main tasks in machine learning which is defined as constructing a model on some training data to predict the label of unseen samples. In traditional machine learning, each sample has only one label while in many

ⓒ Springer Nature Switzerland AG 2019
K. Deb et al. (Eds.): EMO 2019, LNCS 11411, pp. 553–564, 2019.
https://doi.org/10.1007/978-3-030-12598-1_44

real-world datasets, there is more than one label for each sample which is known as multi-label data [1]. Applications such as text categorization, gene functional classification, and object recognition are examples of multi-label learning problems. For instance in text categorization, every text sample can belong to different categories such as sport, politics, and economy. So learning a model able to classify such samples is more sophisticated than single label data. The algorithms which classify multi-label data are divided into two main categories [2]. Some of the algorithms transform the multi-label problem into the single label. For example Binary Relevance algorithm [3] is one of the transformation methods which constructs a model for every label. The trained model classifies the samples that belong to every class from those doesn't. Another category of multi-label learning algorithms adapts existing single label classifiers to classify multi-label data. One of the most known adaptive algorithms is Multi-label K-Nearest Neighbors (ML-KNN) [4] that is based on single label KNN.

The quality of features influences the performance of multi-label learning. The importance of feature selection as a preprocessing step of learning task has been clarified in many related studies [5–8]. Feature selection is the process of reducing irrelevant and redundant features which aim to improve classification performance. However removing such features can increase learning accuracy, elimination of useful and relevant features reverse the classification efficiency. So feature selection is a challenging crucial task mainly due to a massive search space, where the total number of possible solutions is 2^n for a dataset with n features [9]. Since feature selection algorithms try to choose the best combination of features, the task could be considered as an optimization problem. Evolutionary computation (EC) techniques as one of the most efficient groups of search techniques are applied for feature selection because of their popularity for global search ability.

Generally, feature selection is a multi-objective problem. Two conflicting objective functions are the number of features which is desirable to be decreased and the accuracy of classification which should be maximized. Although feature selection can increase the accuracy of the classification task, the excessive reduction of relevant features will reduce accuracy. Therefore, the main goal of feature selection is to minimize the number of features while maintaining an acceptable classification accuracy. Most recently, some feature selection methods based on multi-objective optimization algorithms have also been proposed [10–12]. Since feature selection is basically a binary optimization problem, applying metaheuristic algorithm with binary representation and operators is well-suited to solve it.

This paper proposes a novel binary mutation for Differential Evolution (DE) [13,14] for multi-label feature selection. This operator works based on majority voting of selected solutions in population and opposition-based learning (OBL) [15,16]. The operator is utilized in Generalized Differential Evolution (GDE3) [17] to generate new solutions. Results show that applying metaheuristic algorithms with continuous scheme degrades the diversity of obtained candidate solutions for a binary optimization problem such as feature selection. So, the proposed binary method supports the well-distribution of resulted Pareto front solutions. The remaining parts of this paper are organized as follows. Section 2 presents a background on GDE3 and definition of objective functions. Section 3 details a description of the proposed method. The performance of the proposed method is investigated for various multi-label datasets in Sect. 4. This paper is concluded in Sect. 5.

2   Background Review

2.1   Generalized Differential Evolution (GDE3)

The DE is an evolutionary algorithm originally for solving continuous optimization problems which improves initial population using the crossover and mutation operations. Creation of new generation is done by a mutation and a crossover operator. The mutation operator for a gene, j, is defined as follows.

$$v_{j,i} = x_{j,i_1} + F.(x_{j,i_2} - x_{j,i_3}) \tag{1}$$

Applying this operator generates a new D dimensional vector, v_i, using three randomly selected individuals, x_{j,i_1}, x_{j,i_2}, and x_{j,i_2} from the current population. Parameter F, mutation factor, scales difference between two vectors. The crossover operator changes some or all of genes of parent solution based on Crossover Rate (CR) parameter as follows.

$$j_{rand} = floor(rand_i[0,1).D) + 1$$
$$for(j = 1; j \leq D; j = j + 1)$$
$$\{$$
$$\quad if(rand_j[0,1) < CR) \tag{2}$$
$$\quad\quad u_{j,i} = v_{j,i}$$
$$\quad else$$
$$\quad\quad u_{j,i} = x_{j,i}$$
$$\}$$

Similar to other population-based algorithms, the single objective version of DE starts with a uniform randomly generated population. Next generation is created using mentioned mutation and crossover operations; then best individual (between parent and new individual) is selected based on objective value; which is called a greedy selection. It continues to meet a stopping criterion such as a predefined number of generations.

There are also several variants of DE based algorithms for multi-objective optimization. The first version of Generalized Differential Evolution (GDE) [18] changed the selection mechanism for producing the next generation. The idea in the selection was based on constraint-domination. The new vector is selected if it dominates the old vector. Other sections of the algorithm including mutation and cross-over operations remain as the single objective version of DE. GDE2 [19], the next version of multi-objective DE algorithm, added the crowding distance measure to decide selection between old and newly generated vectors. If both vectors are non-dominating each other, the vector with a higher crowding distance will be selected.

The third version of GDE (GDE3) extends DE algorithm for multi-objective optimization problems with M objectives and K constraints. DE operators are applied using three randomly selected vectors to produce an offspring per parent in each generation. The selection strategy is similar to the GDE2 except in two parts: 1. Applying constraints during selection process. 2. The non-dominating case of two candidate solutions. Selection rules in GDE3 are as follows: when old and new vectors are infeasible solutions, each solution that dominates other in constraint violation space is selected. In the case that one of them is feasible vector, feasible vector is selected. If both vectors are feasible, then one is selected for the next generation that dominates other. In non-dominating case, both vectors are selected. Therefore, the size of the population generated may be larger than the population of the previous generation. If this is the case, it is then decreased back to the original size. Selection strategy for this step is similar to NSGA-II algorithm [20]. It sorts individuals in the population, based on the non-dominated sorting algorithm and crowding distance measure. The population is sorted into a hierarchy of sub-populations based on the ordering of Pareto dominance. Then the selection of solutions starts from first ranks until the number of the selected solutions exceeds the size of the initial population. The remaining members of the population are chosen from subsequent non-dominated fronts in the decreasing order of crowding distance. Similar to other population-based multi-objective algorithms, the selected individuals go to the next generation to continue optimization processing.

2.2   Objective Functions for Multi-label Feature Selection

The goal of feature selection as a preprocessing method in machine learning is the selection of best features to enhance of classification error and computational cost. However, without enough relevant features, the learning algorithm will fail to classify samples. So the rate of selected features and classification accuracy are two conflicting objectives for multi-objective optimization feature selection problems. So the number of selected features divided by the total number of features, a real value in the interval (0, 1], should be minimized as the first objective. The second objective function should be a measure to evaluate the quality of selected features in the classification task. Hamming loss is the most well-known measures for this purpose. It is defined as follow:

$$hloss(h) = \frac{1}{p} \sum \frac{1}{q} |h(x_i) \Delta Y_i| \tag{3}$$

where p and q are the number of labels and the number of samples, respectively. $h(x_i)$ represents the result of classification for ith sample (x_i) in the dataset and Y_i shows the actual labels of that sample. Hamming loss computes for every sample the differences (Δ) between predicted labels and actual labels and then averages over the obtained differences for total samples of the dataset. The goal of the optimization algorithm of feature selection is the minimization of hamming loss as the second objective function. Since two mentioned objective functions conflict with each other, similar to other multi-objective optimization problems, the proposed algorithm results a set of non-dominated solutions (feature subsets).

3   Binary GDE3 for Multi-label Feature Selection

The main version of GDE3 was designed for continuous multi-objective optimization. Feature selection is inherently a binary optimization problem. The problem dimension is equal to the number of features. So every gene of representation indicates the selection (1) or non-selection (0) of each feature. A straightforward method to use continuous operators to solve feature selection problem can be explained as follows. The initial population is produced uniform randomly in continuous search space so that every gene is a real value in the interval [0, 1]. To calculate objective functions, the scheme should be converted to a binary number with a simple mapping. The value of variable smaller (greater) than 0.5 is changed to 0 (1) to indicate non-selection (selection) of the corresponding feature. So a temporary binary vector is produced just before the evaluation step. Although this method is simple, it causes many-to-one mapping problem for the search process. All values of variables in interval [0, 0.5) are mapped to 0 and values in [0.5, 1] are mapped to 1. So there are not any differences between the two solutions even with the non-equal value of one variable which are in the same interval (for example 0.25 and 0.45). This case maybe occurs especially when new variable produced by continues mutation operator of DE remained in the same interval of parent solution. Therefore this issue decreases exploration power of an evolutionary algorithm for the binary optimization problem.

In this paper, a binary version of GDE3 (BGDE3) for multi-label feature selection is proposed. Binarization can be performed during the initialization and crossover without any changes. Similar to the main GDE3, the algorithm starts with a uniform randomly initialized population but using binary vectors in the length of the number of total features. Since in the crossover processing, the values of genes are exchanged, this operator also cannot disturb the binary structure of vectors. The only problem is with mutation operation which is addressed in this paper. In each generation for every target vector, a candidate solution is produced using the proposed binary mutation operator. The operator is based on the majority voting between the target vector and two randomly selected vectors from the population. The addition and subtraction arithmetic operators in DE mutation are replaced with majority voting discrete operator. Every variable of new candidate solution is obtained using the voting of values (0, 1) of voters variables. In feature selection problem, it is expected to gather best features of voters using majority voting. The proposed mutation operator is defined as follows:

$$u_{j,i} = vote(x_i, x_i, x_{r_1}, \breve{x}_{r_2})$$
$$if(rand_j[0,1] < OPR)$$
$$u_{j,i} = \breve{u}_{j,i} \qquad\qquad (4)$$
$$else$$
$$u_{j,i} = u_{j,i}$$

where OPR (opposition rate) is a random number between 0 and 1 which determines the rate of applying voting or opposite of voting for producing new vectors. $u_{j,i}$ and x_i are new produced individual and parent individual respectively. x_{r_1} and x_{r_2} are also two randomly selected vectors from population. The parent is duplicated to emphasize the role of the current solution. Since the number of voters is even (4 votes), in the event of a tie, the value of parent gene is considered. $\breve{}$ indicates the opposite of a point which is defined based on opposition based learning (OBL). The idea of opposition based operators confirms that searching both a random direction and its opposite simultaneously gives a higher chance to find the promising regions [15, 21]. The mathematical definition of opposition of D dimensional binary point is defined as [22]:

Let $X(x_1, x_2, \ldots, x_D)$ be a binary vector in D-dimensional space where $x_i \in \{0, 1\}, i = 1, 2, \ldots, D$. The opposite point of X is defined by $\breve{X}(\breve{x}_1, \breve{x}_2, \ldots, \breve{x}_3)$ where $\breve{x}_i = 1 - x_i, i = 1, 2, \ldots, D$.

It works similar to *Not* logical operator for a binary vector. Switching between voting and opposite voting, with defined OPR rate, leads to a better diversity in producing candidate solutions because especially in large-scale datasets, applying majority voting alone creates sparse vectors (candidate solutions with a lot of zero). The reason for this issue can be explained as follows: as it is mentioned before, one of the objective functions for feature selection is the number of features. This objective is a stronger objective (i.e., high impact objective) in comparison with classification error because decreasing in classification error obtained harder than changes the number of features. Therefore candidate solutions with few numbers of features can dominate other individuals in the population. As the algorithm progress, vectors with the fewer number of features (more genes with zero value) are produced, and majority voting also accelerates creating such vectors. To promote search regions of solutions with more features (more genes with value 1) and less classification error, applying opposition based voting operator is necessary. Figure 1 shows an example for mutating parent and two randomly selected solutions using the proposed operator. As it is presented, the parent solution is duplicated and the opposition of one of the two randomly selected solutions is computed. Then voting or opposite voting of four vectors (parent, duplicated parent, first selected solution, opposition of second selected solution) are considered as the new solution according to OPR.

The remaining of the algorithm including selection strategy is similar to GDE3. Of course, feature selection is an optimization problem with any constraint, so selection between new and old individuals is made based on non-dominating sorting method and crowding distance measure as mentioned in the previous section.

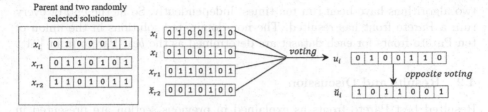

Fig. 1. An example of BGDE3 mutation operator.

4 Experiments

4.1 Datasets and Settings

To evaluate the performance of the proposed method, standard multi-label datasets are considered. Table 1 presents the description of datasets in terms of the domain, number of features, number of training and test sets and the number of labels. These datasets are benchmark datasets from various application domains including image, biology, audio, and text. The datasets are originally split into training and test set based on MULAN library [23] which datasets are taken from. The optimization process is done on the training set, and then its performance is reported on the test set.

Table 1. Multi-label datasets used in the experiments.

Datasets	Domain	#Training instances	#Test instances	#Labels	#Features
Emotions	Music	391	202	6	72
Scene	Image	1211	1196	6	294
Yeast	Biology	1500	917	14	103
Birds	Audio	322	323	19	260
Genbase	Biology	463	199	27	1186
Medical	Text	645	333	45	1449
Enron	Text	1123	579	53	1001

The BGDE3 algorithm is compared with GDE3 to show that binary version can achieve competitive performance compare to the continues version. For evaluating every combination of selected features, a classifier algorithm is required. ML-KNN is the most well-known multi-label classifier which is adopted from the single label KNN classifier. It determines the label of unseen data based on the majority of K neighboring samples. K is set to 10 in experiments according to reference [4]. The population size and maximum iteration number are set to 100 and 300, respectively. BGDE3 is implemented with the parameter of OPR = 0.8 and GDE3 with parameters of CR = 1 and F = 0.5. For each dataset, two algorithms have been run ten times, independently. So at the end of every run, a Pareto front has resulted. The non-dominated solutions in the union of ten Pareto fronts for each dataset are determined as the *best* Pareto front [24].

4.2   Results and Discussion

Resulted best Pareto fronts as explained in previous section are presented in Fig. 2. In each chart, the horizontal axis indicates the number of selected features and the vertical axis represents the classification error rate, the Hamming Loss. On the top of each chart, the number in the bracket shows the classification error rate using all features. As it is presented, BGDE3 achieved more diversity of non-dominated solutions than GDE3 in most cases. Applying continuous operator on binary optimization problem limits search space, while the proposed binary operator empowers exploration of the multi-objective algorithm. Although the deep search (fine-tuning) of GDE3 in a limited PF region leads to decreasing classification error rate, the results generally suggest that BGDE3 obtain more feature subsets with a fewer number of selected features and better classification performance than GDE3.

In order to test the performance of BGDE3, it is compared with GDE3 in terms of three important evaluation metrics of multi-objective optimization algorithms: Set coverage (SC) [25], Hypervolume (HV) indicator [26], and Pure diversity (PD) [27]. Set coverage of two algorithms indicates the rate of solutions on Pareto front of one algorithm which are dominated using solutions of another algorithm. For example SC(A, B) = 0.25 means that the obtained final solutions of algorithms A dominate 25% of the solutions resulted from algorithm B. The comparison of two algorithm on set coverage measure is presented in Table 2. The results are computed based on best Pareto fronts of the algorithms. As seen in the table BGDE3 has achieved higher set coverage than GDE3 in five out of seven datasets.

Table 2. Comparison of BGDE3 and GDE3 on set coverage metric.

Datasets	Genbase	Emotions	Birds	Enron	Yeast	Medical	Scene
SC(BGDE3, GDE3)	**0.650**	0.250	**0.615**	**0.529**	0.250	**0.231**	**0.438**
SC(GDE3, BGDE3)	0.167	**0.400**	0.231	0.148	**0.348**	0.105	0.179

HV indicator is desired to measure the closeness of the estimated solutions to the true Pareto front. It gives the volume of the dominated portion of the objective space bounded from below by a reference point. In a minimization problem, the maximum value for each objective function could be considered as a reference point. Table 3 shows the results of worst, best, mean and standard deviation of HV indicator for BGDE3 and GDE3 for 30 independent runs. Since the binary operator gives a more distributed solutions than continuous GDE3's

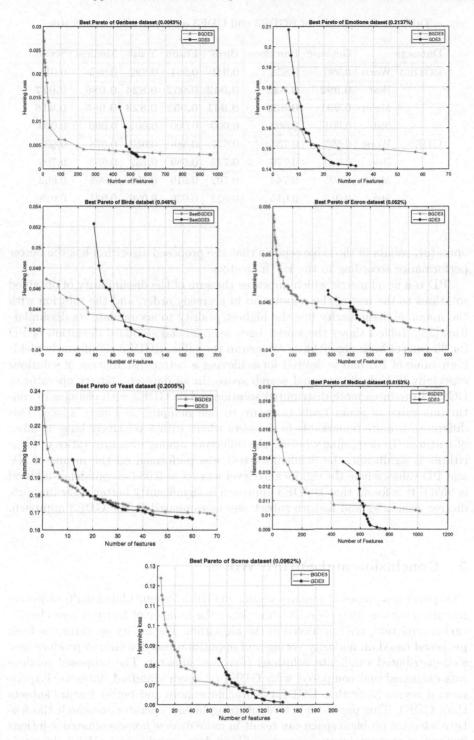

Fig. 2. Best Pareto front for every dataset.

Table 3. Comparison of BGDE3 and GDE3 on hyper volume indicator.

Datasets		Genbase	Emotions	Birds	Enron	Yeast	Medical	Scene
BGDE3	Worst	0.990	0.835	0.941	0.951	0.822	0.985	0.915
	Best	**0.992**	**0.842**	**0.942**	**0.952**	**0.824**	**0.986**	**0.917**
	Mean	**0.991**	**0.839**	**0.941**	**0.952**	**0.823**	**0.985**	**0.916**
	Std	0.001	0.002	0.000	0.000	0.001	0.000	0.000
GDE3	Worst	0.593	0.731	0.677	0.586	0.653	0.578	0.621
	Best	0.630	0.776	0.743	0.689	0.732	0.663	0.709
	Mean	0.605	0.744	0.702	0.619	0.701	0.608	0.662
	Std	0.013	0.015	0.022	0.036	0.026	0.031	0.028

operator, results in the table reports that the proposed algorithm has the better performance according to the HV indicator.

PD is a novel metric which indicates the sum of the dissimilarity of obtained solutions to the rest of the population in a greedy order, and the solution with the maximal dissimilarity has the highest priority to accumulate its dissimilarities [28]. Table 4 shows the worst, best, average and standard deviation of PD for 30 runs of both algorithms. As seen in the table, BGDE3 significantly yields high value of PD and is desired for achieving a better distribution of solutions especially in high dimensional search space. In all datasets, binary operator of BGDE3 produces more distributed solutions while GDE3 with changes in continuous value of genes leads to many to one mapping in binary space. This difference is more remarkable in datasets where with a relatively large number of features. To determine whether the difference among measures ratios of algorithms is significant, the Student's t-test was performed on the means of HV and PD values where the significance level was set as 0.05 (or confidence interval is 95%). It is found that BGDE3 approach is significantly better regarding predictive accuracies and feature subset size as compared to the GDE3 approach.

5   Conclusion and Future Work

This paper has proposed a binary version of GDE3 for multi-label multi-objective feature selection. Two objective functions, the number of features and classification error rate, are considered in the algorithm. The binary operator has been proposed based on majority voting and opposition based voting to produce new well-distributed candidate solutions (features subsets). The proposed method was examined and compared with GDE3 on seven standard datasets. Experimental results show that BGDE3 can achieve more and better feature subsets than GDE3. This paper finds that utilizing binary operator to search the feature selection problem space can result in more diverse non-dominated solutions instead of the continuous operator. Since deep searching of GDE3 decreases

Table 4. Comparison of BGDE3 and GDE3 on pure diversity.

Datasets		Genbase	Emotions	Birds	Enron	Yeast	Medical	Scene
BGDE3	Worst	104.99	174.05	80.76	92.63	175.41	81.70	160.08
	Best	**248.63**	**291.73**	**160.59**	**167.16**	**241.26**	**126.28**	**283.52**
	Mean	**168.09**	**215.59**	**119.09**	**137.42**	**208.63**	**103.12**	**220.94**
	Std	45.13	35.65	28.81	22.83	20.08	14.34	38.99
GDE3	Worst	12.63	98.67	39.48	15.49	87.55	24.20	50.65
	Best	51.10	183.64	75.15	41.41	157.59	40.31	92.30
	Mean	32.48	124.32	55.82	32.08	127.47	31.73	63.25
	Std	10.80	27.12	11.34	8.48	23.84	5.05	14.45

error classification in some limited space (in other words, it converts many-to-one mapping to a one-to-one mapping), in the future, we intend to investigate the use of local search for improving exploitation power of the proposed method.

References

1. Gibaja, E., Ventura, S.: Multi-label learning: a review of the state of the art and ongoing research. Wiley Interdisc. Rev. Data Min. Knowl. Discov. 4(6), 411–444 (2014)
2. Zhang, M.L., Zhou, Z.H.: A review on multi-label learning algorithms. IEEE Trans. Knowl. Data Eng. 26(8), 1819–1837 (2014)
3. Godbole, S., Sarawagi, S.: Discriminative methods for multi-labeled classification. In: Dai, H., Srikant, R., Zhang, C. (eds.) PAKDD 2004. LNCS (LNAI), vol. 3056, pp. 22–30. Springer, Heidelberg (2004). https://doi.org/10.1007/978-3-540-24775-3_5
4. Zhang, M.L., Zhou, Z.H.: ML-KNN: a lazy learning approach to multi-label learning. Pattern Recogn. 40(7), 2038–2048 (2007)
5. Chandrashekar, G., Sahin, F.: A survey on feature selection methods. Comput. Electr. Eng. 40(1), 16–28 (2014)
6. Zhang, Y., Gong, D.W., Sun, X.Y., Guo, Y.N.: A PSO-based multi-objective multi-label feature selection method in classification. Sci. Rep. 7(1), 376 (2017)
7. Li, F., Miao, D., Pedrycz, W.: Granular multi-label feature selection based on mutual information. Pattern Recogn. 67, 410–423 (2017)
8. Gu, S., Cheng, R., Jin, Y.: Feature selection for high-dimensional classification using a competitive swarm optimizer. Soft Comput. 22(3), 811–822 (2018)
9. Xue, B., Zhang, M., Browne, W.N., Yao, X.: A survey on evolutionary computation approaches to feature selection. IEEE Trans. Evol. Comput. 20(4), 606–626 (2016)
10. Xue, B., Zhang, M., Browne, W.N.: Particle swarm optimization for feature selection in classification: a multi-objective approach. IEEE trans. Cybern. 43(6), 1656–1671 (2013)
11. Huang, B., Buckley, B., Kechadi, T.M.: Multi-objective feature selection by using NSGA-II for customer churn prediction in telecommunications. Expert Syst. Appl. 37(5), 3638–3646 (2010)

12. Mierswa, I., Wurst, M.: Information preserving multi-objective feature selection for unsupervised learning. In: Proceedings of the 8th Annual Conference on Genetic and Evolutionary Computation, pp. 1545–1552. ACM (2006)

13. Storn, R., Price, K.: Differential evolution-a simple and efficient heuristic for global optimization over continuous spaces. J. Glob. Optim. **11**(4), 341–359 (1997)

14. Price, K.V., Storn, R.M., Lampinen, J.A.: Differential Evolution: A Practical Approach to Global Optimization. NCS. Springer Science & Business Media, Heidelberg (2005). https://doi.org/10.1007/3-540-31306-0

15. Rahnamayan, S., Tizhoosh, H.R., Salama, M.M.: Opposition versus randomness in soft computing techniques. Appl. Soft Comput. **8**(2), 906–918 (2008)

16. Rahnamayan, S., Wang, G.G., Ventresca, M.: An intuitive distance-based explanation of opposition-based sampling. Appl. Soft Comput. **12**(9), 2828–2839 (2012)

17. Kukkonen, S., Lampinen, J.: GDE3: the third evolution step of generalized differential evolution. In: 2005 IEEE Congress on Evolutionary Computation, vol. 1, pp. 443–450. IEEE (2005)

18. Lampinen, J., et al.: DE's selection rule for multiobjective optimization. Technical report, Lappeenranta University of Technology, Department of Information Technology, pp. 03–04 (2001)

19. Kukkonen, S., Lampinen, J.: An extension of generalized differential evolution for multi-objective optimization with constraints. In: Yao, X., et al. (eds.) PPSN 2004. LNCS, vol. 3242, pp. 752–761. Springer, Heidelberg (2004). https://doi.org/10.1007/978-3-540-30217-9_76

20. Deb, K., Pratap, A., Agarwal, S., Meyarivan, T.: A fast and elitist multiobjective genetic algorithm: NSGA-II. IEEE Trans. Evol. Comput. **6**(2), 182–197 (2002)

21. Mahdavi, S., Rahnamayan, S., Deb, K.: Opposition based learning: a literature review. Swarm Evol. Comput. **39**, 1–23 (2018)

22. Seif, Z., Ahmadi, M.B.: Opposition versus randomness in binary spaces. Appl. Soft Comput. **27**, 28–37 (2015)

23. Tsoumakas, G., Katakis, I., Vlahavas, I.: Mining multi-label data. In: Maimon, O., Rokach, L. (eds.) Data Mining and Knowledge Discovery Handbook, pp. 667–685. Springer, Boston (2009). https://doi.org/10.1007/978-0-387-09823-4_34

24. Cervante, L., Xue, B., Shang, L., Zhang, M.: A multi-objective feature selection approach based on binary PSO and rough set theory. In: Middendorf, M., Blum, C. (eds.) EvoCOP 2013. LNCS, vol. 7832, pp. 25–36. Springer, Heidelberg (2013). https://doi.org/10.1007/978-3-642-37198-1_3

25. Zitzler, E., Thiele, L.: Multiobjective optimization using evolutionary algorithms—a comparative case study. In: Eiben, A.E., Bäck, T., Schoenauer, M., Schwefel, H.-P. (eds.) PPSN 1998. LNCS, vol. 1498, pp. 292–301. Springer, Heidelberg (1998). https://doi.org/10.1007/BFb0056872

26. While, L., Hingston, P., Barone, L., Huband, S.: A faster algorithm for calculating hypervolume. IEEE Trans. Evol. Comput. **10**(1), 29–38 (2006)

27. Wang, H., Jin, Y., Yao, X.: Diversity assessment in many-objective optimization. IEEE Trans. Cybern. **47**(6), 1510–1522 (2017)

28. Parsana, S., et al.: Machining parameter optimization for EDM machining of Mg-RE-Zn-Zr alloy using multi-objective passing vehicle search algorithm. Arch. Civ. Mech. Eng. **18**(3), 799–817 (2018)

Configuration of a Dynamic MOLS Algorithm for Bi-objective Flowshop Scheduling

Camille Pageau[1]([✉]), Aymeric Blot[1], Holger H. Hoos[2],
Marie-Eléonore Kessaci[1], and Laetitia Jourdan[1]

[1] Université de Lille, CNRS, UMR 9189 – CRIStAL, Lille, France
{camille.pageau,aymeric.blot,mkessaci,laetitia.jourdan}@univ-lille.fr
[2] LIACS, Leiden University, Leiden, The Netherlands
hh@liacs.nl

Abstract. In this work, we propose a dynamic multi-objective local search (MOLS) algorithm whose parameters are modified while it is running and a protocol for automatically configuring this algorithm. Our approach applies automated configuration to a static pipeline that sequentially runs multiple configurations of the MOLS algorithm. In a series of experiments for well-known benchmark instances of the bi-objective permutation flowshop scheduling problem, we show that our dynamic approach produces substantially better results than static MOLS, and that longer pipeline (with a higher number of parameters) outperform shorter ones.

Keywords: Algorithm configuration ·
Multi-objective combinatorial optimisation · Local search

1 Introduction

Many metaheuristic algorithms for solving multi-objective optimisation problems have parameters that highly affect their performance, and that should be set to different values to achieve good performance for various types of problem instances. The problem of configuring such parameters for optimised performance can be approached in an off-line or on-line manner. Static algorithm configuration approaches can handle many parameters but provide configurations that can be highly specific to a given set or distribution of problem instances (see, e.g., [10,14]). Dynamic configuration approaches adapt parameters during the run of a given algorithm but generally consider only one or two parameters (see, e.g., [11]); they can, in principle, achieve robust performance over a broad range of problem instances. In this work, we leverage the advantages of both types of approaches by considering a framework in which we switch between different configurations of a multi-objective optimisation algorithm while it is running on a given problem instances. We determine these configurations, and

© Springer Nature Switzerland AG 2019
K. Deb et al. (Eds.): EMO 2019, LNCS 11411, pp. 565–577, 2019.
https://doi.org/10.1007/978-3-030-12598-1_45

the static schedule we use for switching between them, using a general-purpose, static algorithm configurator. Our approach thus represents a simple mechanism for dynamically changing many algorithm parameters in a way that optimises overall performance on a given type of problem instances.

The bi-objective permutation flowshop scheduling problem (bPFSP), in which makespan and total flowtime are to be minimised, is a prominent and widely studied combinatorial multi-objective optimisation problem. The bPFSP can be solved effectively by multi-objective local search (MOLS) algorithms [6,12]; in the design of these algorithms, multiple design choices are encountered, and when using them, several parameters have to be set. Therefore, MOLS algorithms for the bPFSP provide an excellent test bed for our dynamic configuration approach.

The remainder of this article is organised as follows. First, in Sect. 2, we introduce our dynamic algorithm framework and a protocol to automatically configure it. Then, in Sect. 3, we describe the multi-objective local search algorithm. Sections 4 and 5 detail the setup of our experimental study and the results obtained from it, respectively. Finally, Sect. 6 provides some conclusions and perspectives on future work.

2   Automatic Design of a Dynamic Algorithm

2.1   Static *vs* Dynamic Design Approaches

Over the last decade, automatic algorithm configuration (AAC) techniques have been increasingly exploited in the off-line design of high-performance heuristic algorithms, such as metaheuristics. These algorithms present design choices, such as strategy components, and tunable parameters that heavily affect their performance. In the following, we will assume that all design choices have been exposed as parameters.

Given a parametrised *target algorithm* $\mathcal{A}$, a *configuration* θ is a specific setting of all the parameters of $\mathcal{A}$. The configuration space Θ of $\mathcal{A}$ is the set of all valid configurations. Automated algorithm configuration (AAC) can be seen as an optimisation problem, where the objective is to determine one or more configurations that lead to the best performance for a given set or distribution of problem instances. AAC can be seen as a supervised, off-line learning process, in which training instances are used to learn and determine the best configurations of the given target algorithm. This configuration is then fixed and used, in a completely static manner, whenever $\mathcal{A}$ is run on new problem instances. Prominent AAC procedures include irace [14] and ParamILS [10], which optimise a single configuration objective, and MO-ParamILS, a recent extension of ParamILS that handles multiple configuration objectives [1].

In parallel with AAC procedures, *dynamic* algorithm design techniques have been proposed [11] to permit the modification of strategy components or numerical parameters of a given target algorithm $\mathcal{A}$ while it is running. These so-called parameter control approaches use techniques such as multi-armed bandits [8] or adaptive pursuit [19] to dynamically determine good parameter settings in response to observations made while trying to solve a given problem instance. However, the number of configurations of $\mathcal{A}$ that can be handled by such approaches is very limited.

In this work, we are interested in algorithms that expose several design choices, in the form of categorical parameters. This scenario falls outside of most dynamic design scenarios, as they usually deal with a single numerical parameter or very few categorical choices. Nevertheless, we want to be able to dynamically modify parameters while running our target algorithm, and to this end, we introduce a framework that successively runs several configurations, in the form of a static pipeline, which we configure using a standard, general-purpose AAC procedure.

2.2   A Dynamic Algorithm Framework

Given a configurable algorithm $\mathcal{A}$ and its configuration space Θ, we use $\mathcal{A}_{\theta,T}$ to denote $\mathcal{A}$ under configuration $\theta \in \Theta$ with cut-off time T. Then, we define the dynamic algorithm $\mathcal{F}^{\mathcal{A}}_{(\theta_i,T_i)^k}$ as a pipeline with k stages, which sequentially runs $\mathcal{A}_{\theta_1,T_1}, \mathcal{A}_{\theta_2,T_2}, \ldots, \mathcal{A}_{\theta_k,T_k}$. Specifically, when applied to a multi-objective optimisation problem, we first run $\mathcal{A}$ under configuration θ_1, starting from a initial set of solutions, up to time T_1. At that point, we switch to configuration θ_2 and continue our computation from the current set of solutions, with a cut-off time of T_2. We note that $\mathcal{A}$ is *not* restarted when switching between configurations. Overall, the maximum running time of the dynamic algorithm is then $T = \sum_{i=1}^{k} T_i$.

Figure 1 depicts two examples of dynamic algorithms $\mathcal{F}^{\mathcal{A}}$ and $\mathcal{F}'^{\mathcal{A}}$. While $\mathcal{F}$ uses $k = 3$ configurations to divide the total time budget into three intervals of equal duration, $\mathcal{F}'$ uses $k = 4$ configurations, of which two are run quickly in the beginning, after which more time is allocated to last two configurations.

The configuration space of our framework comprises the Cartesian product Θ^k, the time budgets $T_1, \ldots, T_k$ and the integer $k \geq 1$. For $k = 1$, our framework degenerates to the original, static target algorithm $\mathcal{A}$.

2.3   Automatic Configuration of Our Framework

The purpose of this work is to assess the performance gains that can be obtained by switching between different configurations of an algorithm $\mathcal{A}$ while it is running. Towards this end, we use a general-purpose, static algorithm configurator to configure the framework introduced in the previous section. Since the size of the configuration space exponentially increases with the maximum number of pipeline stages, K, we only consider a fixed number s_k of different cut-off times for each stage, where k is the number of actual pipeline stages used in a specific instantiation of our framework. This leads to a configuration space of size $\sum_{k=1}^{K} s_k \cdot |\Theta|^k$. Using this approach, we can also assess the influence of K and s_k (for $k = 1, \ldots, K$) on the performance achieved by automatically configuring our dynamic algorithm framework.

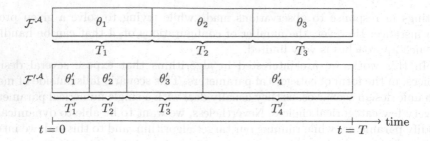

Fig. 1. Two examples of dynamic algorithms, $\mathcal{F}^{\mathcal{A}}$ and $\mathcal{F'}^{\mathcal{A}}$

2.4 Related Work

In addition to being conceptually related to adaptive algorithms or hyper-heuristics, since it enables modifications of the configuration of an algorithm while it is running, our approach also bears resemblance to per-instance algorithm scheduling [13]. There are, however, several major differences. Firstly, per-instance algorithm scheduling uses instance features to determine which of a given set of distinct algorithms to run, one after the other, on a given problem instance; in contrast, our approach uses different configurations of a single algorithm and does not require instance features. Secondly, in per-instance algorithm scheduling, results are not passed from one stage of the schedule to the next, while in our pipeline approach, each stage continues from the result of the previous stage – as explained previously, it can thus be seen as a single algorithm whose parameter configuration changes while running on a given problem instance. Finally, the primary goal of per-instance algorithm scheduling is *robustness* resulting from performance complementarity between the algorithms in the schedule; the goal of our approach is to achieve improvements over the performance of the static version of the given target algorithm, which uses a single configuration for the entire run, based on the idea that different configurations are best suited for different phases of solving a given problem instance.

3 Multi-objective Local Search

In the following, we consider a Pareto optimisation approach to solve the bi-objective permutation flowshop scheduling problem. More precisely, we focus on multi-objective local search algorithms, since they are known to provide good solutions to classical multi-objective permutation problems [2,7,12].

3.1 The MOLS Framework

Stochastic local search (SLS) algorithms are widely used for solving a broad range of NP-hard problems, including many single-objective optimisation problems [9]. The key idea is to iteratively improve a candidate solutions, by choosing, in each step, a neighbouring solution to move to, making use of randomisation to balance intensification and diversification. SLS algorithms have also been developed for multi-objective optimisation problems, where they operate on a set of non-dominated candidate solutions dubbed an *archive*. Among the most widely used SLS methods for multi-objective optimisation problems we find Pareto Local Search (PLS) [17] and its numerous variants, such as the stochastic PLS [5], the iterated PLS [16] and the anytime PLS [7], and the Dominance-based Multi-Objective local search [12].

Recently, Blot *et al.* have proposed a generic local search framework that encompasses most of the multi-objective local search (MOLS) algorithms of the literature as well as many new variants [3]. A MOLS algorithm iterates over several phases: *selection* of solutions within the current archive, *exploration* of these solutions, and *archiving* of the neighbouring solutions that have been visited. Similarly to single-objective local search algorithms, iterated local search (ILS) approaches have been developed, which add a perturbation phase designed to more effectively explore of the underlying search space [15]. Within the generic MOLS framework, different strategies can be selected for each of these phases in order to optimise performance for a given set or distribution of benchmark instances.

3.2   MOLS Component Strategies

In the following, we explain the different components of the MOLS algorithms and describe the strategies available for instantiating them in our experiments (see Sect. 5). Since our investigation is focussed on these strategies, all numerical parameters have been set to values determined in previous work [2].

Initialisation. First step of MOLS, in which one or more solutions are generated from which the search process is started. Here, 10 solutions are generated uniformly at random; these form the initial archive.

Selection. Solutions are chosen within the current archive according to strategy `select_strat`. One option is to select `all` solutions in the archive; alternatively, a subset of s solutions can be selected uniformly at `random`, or according to their age (*i.e.*, the time they have been in the archive), among the `newest` or the `oldest`. In our experiments (see Sect. 5), s has been set to 1.

Exploration. The neighbourhood of each solution that has been selected in the previous step is explored, and an archive of candidate solutions is created, containing some of the visited neighbours. The strategy for exploring the neighbourhood (`explor_strat`) can either involve exploring it *entirely* or *partially*, using different techniques for comparing new candidate solutions with those in the current archive. In the first case, the `all` and `all_imp` strategies evaluate all the neighbours of the selected solution and consider as candidates either all non-dominated or all dominating neighbours, respectively. On the other hand,

Table 1. Configuration space of MOLS considered in our experiments.

Parameter	Values
select_strat	{rand, all, new, old}
explor_strat	{imp, ndom, imp_ndom, all, all_imp}
perturb_strat	{restart, kick, kick_all}

the exploration may end before all the neighbours have been visited, when r non-dominated neighbours have been evaluated (ndom), or when r dominating neighbours have been found. In this last case, either only the dominating neighbours are kept (imp), or dominating neighbours as well as all visited non-dominated neighbours (imp_ndom) are considered as candidate solutions. In the following experiments (see Sect. 5), r has been set to 5.

Archiving. All candidate solutions identified in the exploration phase are added to the current archive; then, dominated solutions are removed from the archive.

Perturbation. In order to facilitate exploration of the search space, the perturbation strategy (perturb_strat) can either *restart* the search process, or merely *kick* (*i.e.*, remove) solutions from the current archive. A restart is performed by forming a new archive, as in the initialisation phase. The *kick* strategy replaces one or more solutions by neighbours selected uniformly at random. It can be applied to either r solutions in the current archive (kick), or to all the solutions in the archive (kick_all). In the following experiments (see Sect. 5), r has been set to 1.

Table 1 shows all strategies we considered when configuring our MOLS framework; these jointly give rise to 60 ($4 \times 5 \times 3$) different configurations of MOLS.

4   Experimental Setup

Benchmark Sets for the bPFSP. As previously mentioned, we are considering a bi-objective version of the classical Permutation Flowshop Scheduling Problem (PFSP), which involves scheduling a set of n jobs $\{J_1, \ldots, J_n\}$ on a set of m machines $\{M_1, \ldots, M_m\}$. In the PFSP, each machine can only process one job at a time, and each job J_i is sequentially processed on each of the m machines, with fixed processing times $\{p_{i,1}, \ldots, p_{i,M}\}$. Furthermore, the jobs are processed in the same order on every machine. Therefore, each solution of a PFSP instance (called the schedule) can be represented by a permutation of jobs of size n. In the bi-objective PFSP (bPFSP), two objectives are considered: the makespan and the flowtime of the schedule, where makespan is the total completion time, and flowtime is the sum of the individual completion times of the n jobs. We use a widely studied set of benchmark instances proposed by Taillard [18]. It is known that the difficulty of these instances increases with the number of jobs. We evaluated our approach on 6 sets of 10 Taillard instances each, with 20 jobs and 20 machines, 50 jobs and 5 machines, 50 jobs and 10 machines, 50 jobs and 20 machines, 100 jobs and 10 machines and 100 jobs and 20 machines, respectively.

Dynamic MOLS for the bPFSP. We used the implementation of MOLS for the bPFSP provided by Blot *et al.* [2] and considered two instantiations of our dynamic algorithm framework described in Sect. 2.2, with up to $K = 2$ and $K = 3$ pipeline stages, respectively, and three ways of dividing the overall running times between the pipeline stages: For $K = 2$, we used $(T_1, T_2) = (1/4, 3/4) \cdot T$, $(1/2, 1/2) \cdot T$ and $(3/4, 1/4) \cdot T$, where T is the overall cut-off time, while for $K = 3$, we considered $(T_1, T_2, T_3) = (1/3, 1/3, 1/3) \cdot T$, $(1/4, 1/4, 1/2) \cdot T$ and $(1/2, 1/4, 1/4) \cdot T$. Therefore, whilst the basic MOLS algorithm has 60 distinct configurations, the dynamic MOLS algorithm, dubbed D-MOLS, has $60 + 3 \cdot 60^2 \approx 1.1 \cdot 10^4$ configurations for $K = 2$, and $60 + 3 \cdot 60^2 + 3 \times 60^3 = 6.6 \cdot 10^5$ configurations for $K = 3$ stages. We note that this configuration space is very large compared to on-line algorithms from the literature, which typically involve only very few configurations. In our experiments, we chose an overall cut-off time of $T = n^2 \dot{m}/1000$ for D-MOLS.

Automatic Configuration of D-MOLS. Blot et al. [2] showed that a multi-objective AAC is the best approach to automatically configure multi-objective algorithms such as MOLS. Therefore, in order to configure D-MOLS, we used the state-of-the-art multi-objective algorithm configurator MO-ParamILS [1], with two performance indicators: unary hypervolume [20], a volume-based convergence performance indicator, and Δ spread [4], a distance-based distribution metric. In order to simplify the use of MO-ParamILS and interpretation of results, we used a variant of hypervolume, denoted $1 - HV$, in which after normalisation to the interval $[0, 1]$, the hypervolume values are subtracted from 1, so that both indicators ($1 - HV$ and Δ) need to be minimised.

To obtain training sets to be used as the basis for automatic configuration, we generated uniformly at random a set 100 instances for each the six instance size we considered, following the same protocol as Taillard [18]. Since MO-ParamILS is a stochastic algorithm, we performed 20 independent runs for each configuration scenario, each with 1000 and 10000 runs of D-MOLS for $K = 2$ and $K = 3$, respectively. Then, the best of the 20 resulting D-MOLS configurations (according to performance on the respective training set) was evaluated on the 10 Taillard instances in each of our testing sets, based on 15 independent runs. The performance indicators – hypervolume and spread – reported for a single D-MOLS configuration were obtained by averaging the respective values over the 15 independent runs and the 10 instances per set.

5   Experimental Results

First, we present results for D-MOLS, our dynamic version of MOLS, for the bPFSP for $K = 2$ and 3 pipeline stages, *i.e.*, one or two changes in configuration during each run. Next, we compare the results for D-MOLS with those for static MOLS.

5.1   Evaluation of Dynamic MOLS

Table 2 shows the number of D-MOLS configurations in the Pareto-optimal sets obtained from automatic configuration using MO-ParamILS; specifically, for $K = 2$ and $K = 3$, we report the number of non-dominated configurations with $k = 1$, 2 and 3 pipeline stages. For example, for the 20x20 scenario and $K = 3$, we obtained 1 configuration for static MOLS, 7 for dynamic MOLS with $K = 2$ and 9 for dynamic MOLS with $K = 3$ pipeline stages. For 8 of the 11 benchmark sets considered, all non-dominated D-MOLS configurations obtained from MO-ParamILS had at least 2 pipeline stages $k \geq 2$, which clearly indicates the performance advantage gained by switching between configurations during a single run of MOLS.

Figure 2 shows the Pareto fronts of D-MOLS configurations obtained in our experiments with $K = 2$ (left) and 3 (right), respectively, for the benchmark instances with 20 jobs and 20 machines. For $K = 2$, static MOLS ($k = 1$) achieves better hypervolume, while D-MOLS(2) obtains better spread; for $K = 3$, on the other hand, D-MOLS(2) and D-MOLS(3) yield better results w.r.t. both indicators. Figure 3 shows the our results for benchmark instances with 50 jobs and 20 machines. As also seen in Table 2, no configurations from static MOLS are found in the final Pareto sets; furthermore, the sets of configurations from both D-MOLS scenarios are well distributed over the Pareto front.

5.2   Performance of the Dynamic *vs* Static MOLS

In this section, we further assess the performance of our dynamic MOLS algorithm against static MOLS. Since there are only 60 configurations of static

Table 2. Number of non-dominated D-MOLS configurations determined through automatic configuration (see text for details).

Instances	$K = 1$	$K = 2$		$K = 3$		
		$k = 1$	$k = 2$	$k = 1$	$k = 2$	$k = 3$
20×20	20	5	4	1	7	9
50×5	9	-	7	-	5	9
50×10	9	-	7	-	10	8
50×20	11	-	12	-	3	8
100×10	8	1	9	-	5	5
100×20	8	-	13	N/A	N/A	N/A

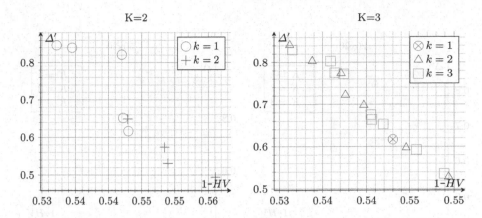

Fig. 2. Performance of Pareto-optimal D-MOLS configurations for the **20x20** benchmarks.

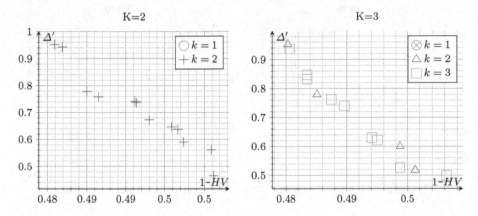

Fig. 3. Performance of Pareto-optimal D-MOLS configurations for the **50x20** benchmarks.

MOLS, we were able to evaluate all of them. Figure 4 shows the Pareto fronts of configurations for static MOLS ($K = 1$) *vs* dynamic MOLS for $K = 2$ and $K = 3$. Only few of the 60 configurations of MOLS ended up in the Pareto-optimal sets for each of our benchmarks. We further note that for each instance size, the Pareto fronts obtained for $K = 2$ and $K = 3$ are of roughly similar size. For 50x10, 50x20, 100x10 and 100x20, the configurations obtained for D-MOLS are better distributed along the respective fronts. For 100x10 and 100x20, the fronts obtained by static MOLS ($K = 1$) are very poorly distributed. Most of the configurations are tightly clustered; this is particularly pronounced for 100x20, where there are two types of configurations that obtain either good hypervolume or good spread, but never both. The configurations for dynamic MOLS ($K \geq 2$), on the other hand, are well distributed and cover a broad range of tradeoffs between the objectives. Furthermore, the configurations for $K = 1$

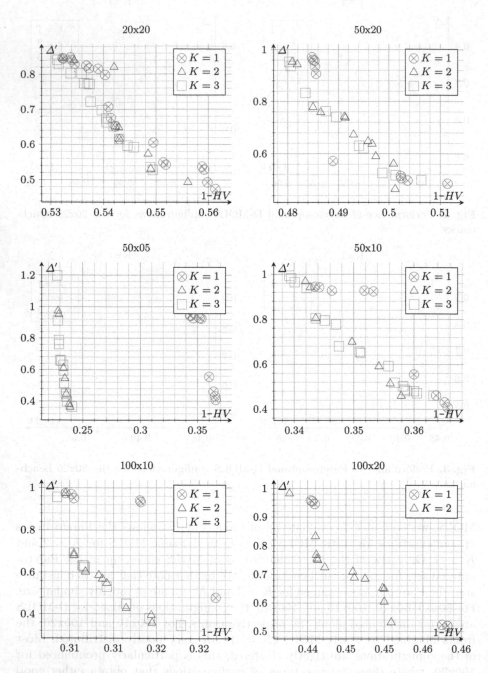

Fig. 4. Pareto fronts of static ($K = 1$) and dynamic ($K \geq 2$) MOLS configurations; for details see text.

are all dominated by those for $K \geq 2$. For the smallest instance size, 50x5, we observed a large improvement in hypervolume, while spread remains comparable; this effect is less obvious for the 20x20 and 50x20 instances. For 50x20, static MOLS dominates parts of the fronts for dynamic MOLS, likely as a result of the large configuration spaces for $K \geq 2$; nevertheless, for $K \geq 2$, more homogeneous Pareto fronts of configurations are obtained. For 20x20, all three fronts are quite close to each other and reasonably well distributed, with the configurations of dynamic MOLS ($K \geq 2$) filling some of the gaps in the front obtained for static MOLS. We note that, even though the fronts for $K = 2$ and $K = 3$ are roughly similar in size, the one for $K = 3$ contains more configurations and is overall preferable.

6   Conclusions and Future Work

In this work, we have investigated the use of automatic algorithm configuration techniques for generating dynamic algorithms that modify their parameters while solving a given problem instance. Specifically, we proposed a dynamic algorithm framework that can be automatically configured with a standard, general-purpose algorithm configurator. Given a parameterised static algorithm, using our approach, it is easy to automatically construct a dynamic version of the algorithm whose parameter configuration is adjusted, according to an optimised, static schedule, while it is running.

We evaluated this approach by applying it to a multi-objective local search (MOLS) algorithm for the bi-objective permutation flowshop scheduling problem. Our experiments show that the dynamic MOLS algorithm obtained using our approach shows better performance than the underlying static MOLS procedure on the widely studied Taillard instances.

In future work, we plan to analyse the behaviour of our dynamic MOLS algorithm to further understand how the optimised configurations used by it contribute to its overall performance. We also intend to apply our approach to single- and multi-objective metaheuristic algorithms for other challenging combinatorial problems.

References

1. Blot, A., Hoos, H.H., Jourdan, L., Kessaci-Marmion, M.É., Trautmann, H.: MO-ParamILS: a multi-objective automatic algorithm configuration framework. In: Festa, P., Sellmann, M., Vanschoren, J. (eds.) LION 2016. LNCS, vol. 10079, pp. 32–47. Springer, Cham (2016). https://doi.org/10.1007/978-3-319-50349-3_3
2. Blot, A., Jourdan, L., Kessaci, M.E.: Automatic design of multi-objective local search algorithms: case study on a bi-objective permutation flowshop scheduling problem. In: Proceedings of the Genetic and Evolutionary Computation Conference, GECCO 2017, pp. 227–234. ACM, New York (2017)

3. Blot, A., Kessaci, M.É., Jourdan, L., De Causmaecker, P.: Adaptive multi-objective local search algorithms for the permutation flowshop scheduling problem. In: Battiti, R., Brunato, M., Kotsireas, I., Pardalos, P.M. (eds.) LION 12 2018. LNCS, vol. 11353, pp. 241–256. Springer, Cham (2019). https://doi.org/10.1007/978-3-030-05348-2_22
4. Deb, K., Pratap, A., Agarwal, S., Meyarivan, T.: A fast and elitist multiobjective genetic algorithm: NSGA-II. IEEE Trans. Evol. Comput. **6**(2), 182–197 (2002)
5. Drugan, M.M., Thierens, D.: Stochastic pareto local search: pareto neighbourhood exploration and perturbation strategies. J. Heuristics **18**(5), 727–766 (2012)
6. Dubois-Lacoste, J., López-Ibáñez, M., Stützle, T.: A hybrid TP + PLS algorithm for bi-objective flow-shop scheduling problems. Comput. Oper. Res. **38**(8), 1219–1236 (2011)
7. Dubois-Lacoste, J., López-Ibáñez, M., Stützle, T.: Anytime pareto local search. Eur. J. Oper. Res. **243**(2), 369–385 (2015)
8. Fialho, Á., Da Costa, L., Schoenauer, M., Sebag, M.: Dynamic multi-armed bandits and extreme value-based rewards for adaptive operator selection in evolutionary algorithms. In: Stützle, T. (ed.) LION 2009. LNCS, vol. 5851, pp. 176–190. Springer, Heidelberg (2009). https://doi.org/10.1007/978-3-642-11169-3_13
9. Hoos, H.H., Stützle, T.: Stochastic Local Search: Foundations and Applications. Elsevier, Amsterdam (2004)
10. Hutter, F., Hoos, H.H., Leyton-Brown, K., Stützle, T.: ParamILS: an automatic algorithm configuration framework. J. Artif. Intell. Res. **36**(1), 267–306 (2009)
11. Karafotias, G., Hoogendoorn, M., Eiben, Á.E.: Parameter control in evolutionary algorithms: trends and challenges. IEEE Trans. Evol. Comput. **19**(2), 167–187 (2015)
12. Liefooghe, A., Humeau, J., Mesmoudi, S., Jourdan, L., Talbi, E.G.: On dominance-based multiobjective local search: design, implementation and experimental analysis on scheduling and traveling salesman problems. J. Heuristics **18**(2), 317–352 (2012)
13. Lindauer, M., Bergdoll, R.-D., Hutter, F.: An empirical study of per-instance algorithm scheduling. In: Festa, P., Sellmann, M., Vanschoren, J. (eds.) LION 2016. LNCS, vol. 10079, pp. 253–259. Springer, Cham (2016). https://doi.org/10.1007/978-3-319-50349-3_20
14. López-Ibáñez, M., Dubois-Lacoste, J., Cáceres, L.P., Birattari, M., Stützle, T.: The irace package: Iterated racing for automatic algorithm configuration. Oper. Res. Perspect. **3**, 43–58 (2016)
15. Lourenço, H.R., Martin, O.C., Stützle, T.: Iterated local search. In: Glover, F., Kochenberger, G.A. (eds.) Handbook of Metaheuristics. ISOR, vol. 57, pp. 320–353. Springer, Boston (2003). https://doi.org/10.1007/0-306-48056-5_11
16. Olson, R.S., Moore, J.H.: TPOT: a tree-based pipeline optimization tool for automating machine learning. In: Hutter, F., Kotthoff, L., Vanschoren, J. (eds.) Proceedings of the Workshop on Automatic Machine Learning. Proceedings of Machine Learning Research, vol. 64, pp. 66–74. PMLR, New York (2016)
17. Paquete, L., Chiarandini, M., Stützle, T.: Pareto local optimum sets in the biobjective traveling salesman problem: an experimental study. In: Gandibleux, X., Sevaux, M., Sörensen, K., T'kindt, V. (eds.) Metaheuristics for Multiobjective Optimisation. LNE, vol. 535, pp. 177–199. Springer, Heidelberg (2004). https://doi.org/10.1007/978-3-642-17144-4_7
18. Taillard, É.D.: Benchmarks for basic scheduling problems. Eur. J. Oper. Res. **64**(2), 278–285 (1993). Project Management anf Scheduling

19. Thierens, D.: An adaptive pursuit strategy for allocating operator probabilities. In: Proceedings of the 7th Annual Conference on Genetic and Evolutionary Computation, pp. 1539–1546. ACM (2005)
20. Zitzler, E., Thiele, L.: Multiobjective evolutionary algorithms: a comparative case study and the strength pareto approach. IEEE Trans. Evol. Comput. $3(4)$, 257–271 (1999)

MCDM and Interactive EMO

Reliable Biobjective Solution
of Stochastic Problems Using Metamodels

Marius Bommert$^{(\boxtimes)}$ and Günter Rudolph

TU Dortmund University, 44221 Dortmund, Germany
marius.bommert@tu-dortmund.de

Abstract. For many real world optimization problems, the objective
function is stochastic. When optimizing a stochastic function f, one has
to deal with the problem of varying outputs $f(x, C)$ for the same input x
due to the effects of a random variable C. One possibility for optimizing
f is considering the expectation and standard deviation of $f(x, C)$ and
choosing x such that the expected value of $f(x, C)$ is optimal, e.g. minimal and the standard deviation of $f(x, C)$ is minimal. This turns the
optimization of f into a biobjective optimization problem. We investigate the optimization of expensive stochastic black box functions $f(x, C)$
with $x \in \mathbb{R}$ and C being a one dimensional random variable. Because f is
an expensive function, we want to evaluate it seldom. Therefore, we use a
surrogate model $\hat{f}$ of f and numerical integration to estimate the expectation $\mathrm{E}(f(x, C))$ and the standard deviation $\mathrm{S}(f(x, C))$. We perform a
simulation study to analyze how well our approach works and compare
it to a classic method. Our approach enables us to estimate $\mathrm{E}(f(x, C))$
and $\mathrm{S}(f(x, C))$ for each feasible x-value with a comparably high quality
and yields a good approximation of the true Pareto set at the cost of
requiring that C is observable.

Keywords: Biobjective optimization · Stochastic black box function ·
Metamodel

1 Introduction

Many optimization problems are influenced by random effects. Therefore, the
respective objective functions are stochastic. Such a stochastic function can be
denoted as $f(x, C)$ where x is a decision vector and C is a random variable. In
this paper, we consider the case of f with real-valued output, $x \in \mathbb{R}$ and C being
a one dimensional random variable.

The evaluation of f in any point x does not yield a deterministic output. Instead, $f(x, C)$ is a random variable whose distribution depends on x.
The expectation $\mathrm{E}(f(x, C))$ provides information about the central location of
the distribution when f is evaluated in x. It describes the expected output of
$f(x, C)$. If f is evaluated in x many times, the mean output value will be close
to $\mathrm{E}(f(x, C))$ (law of large numbers). A single evaluation of f in x does not

© Springer Nature Switzerland AG 2019
K. Deb et al. (Eds.): EMO 2019, LNCS 11411, pp. 581–592, 2019.
https://doi.org/10.1007/978-3-030-12598-1_46

necessarily yield a value close to $E(f(x, C))$. The variance $V(f(x, C))$ gives the expected quadratic deviation of an evaluation of $f(x, C)$ from $E(f(x, C))$. It is a measure of the spread of the distribution and can be used to assess the uncertainty about how far $f(x, C)$ will likely deviate from $E(f(x, C))$. As the variance uses a quadratic scale, the standard deviation $S(f(x, C)) = \sqrt{V(f(x, C))}$ is often used instead.

Our aim is the optimization of f by simultaneously optimizing the expectation $E(f(x, C))$ and the standard deviation $S(f(x, C))$. Depending on the context, $E(f(x, C))$ is minimized or maximized. $S(f(x, C))$ is always minimized to achieve small uncertainty. So, instead of the single-objective stochastic function f, we optimize the biobjective deterministic function $(E(f(x, C)), S(f(x, C)))$.

For many optimization tasks, the analytic form of the objective function is not known and the budget for the optimization is small. Therefore, we assume that f is an expensive stochastic black box function.

In the field of portfolio optimization the approach of optimizing a stochastic black box function by the maximization of the expected value and the minimization of a risk measure is very common. The risk is often assessed by the variance or the standard deviation of f or domain specific quantities like the value at risk. A very popular approach is the mean-variance model introduced by Markowitz [8]. In this model, expectation and variance are scalarized to a single objective function and then optimized. The portfolio optimization has also been investigated as a multi-objective problem, see for example [10]. The expectation and risk measure are estimated using empirical quantities.

Several methods for optimizing stochastic functions like the two-stage stochastic multi-objective optimization are explained in [6]. The set of possible approaches for the optimization of stochastic functions presented in [6] also includes the simultaneous optimization of $E(f(x, C))$ and a risk measure like $S(f(x, C))$. The authors do not suggest a method for estimating $E(f(x, C))$ and $S(f(x, C))$.

Paenke et al. [9] search for robust solutions of optimization problems. They look at functions f where the design variables are disturbed by some random effects. They estimate expectation $E(f(x, C))$ and variance $V(f(x, C))$ using local approximation models and Monte Carlo integration.

In our approach, we estimate $E(f(x, C))$ and $S(f(x, C))$ which are defined as integrals over terms that only consist of f and the probability density function p_C. We estimate $E(f(x, C))$ and $S(f(x, C))$ by building a metamodel for f, estimating p_C and using numerical integration. Our approach requires C to be observable. Our approach is similar to [9] but we use Kriging instead of a local model and we estimate the probability density function p_C. To the best of our knowledge, nobody else has tried this specific approach before.

The remainder of this article is organized as follows: In the next section, the fundamental concepts and methods Kriging, Pareto optimality and attainment functions are explained. In Sect. 3, our and the classic approach for estimating expectation and standard deviation of f are described. The design of our experiments for comparing the two approaches is explained in Sect. 4. In Sect. 5, we evaluate our experiments. Plots of the empirical attainment functions and the Pareto sets are compared to assess the quality of the approximations of the Pareto frontier and Pareto set. Concluding remarks and future work are presented in Sect. 6.

2 Concepts and Methods

2.1 Kriging

Let $x_1, \ldots, x_d$ denote d points which are evaluated with a deterministic function f. Kriging is an interpolating method to build a metamodel $\hat{f}$ of f using the d given points. For that purpose, it is assumed that $f(x_1), \ldots, f(x_d)$ are realizations of a gaussian random field. A numerical optimization is performed to fit the model to the data. For more information on Kriging see [12].

2.2 Pareto Frontier and Pareto Set

Let $f : X \to \mathbb{R}^m, f(x) = (f_1(x), \ldots, f_m(x))$ denote an objective function where all components should be minimized. $x \in X$ weakly dominates $y \in X$ (notation: $x \preceq y$ or $f(x) \preceq f(y)$) if

$$\forall i \in \{1, \ldots, m\} : f_i(x) \leq f_i(y).$$

The point x dominates the point y if it weakly dominates it, and

$$\exists i \in \{1, \ldots, m\} : f_i(x) < f_i(y).$$

$x^\star \in X$ is Pareto optimal if there is no $x \in X$ which dominates it. The set of all Pareto optimal points is called Pareto set and the corresponding image is the Pareto frontier. For more information on Pareto frontiers and Pareto sets see [3].

2.3 Attainment Function

Let $\mathcal{X}_1, \ldots, \mathcal{X}_k$ denote k approximations of the same Pareto frontier resulting from k optimization runs. Each approximation can be seen as a realization of a random non dominated point set $\mathcal{X}^\star$. $\mathcal{X}^\star = \{X_1^\star, \ldots, X_N^\star\}$ is a random set of vectors in $\mathbb{R}^m$. The attainment function allows analyzing the distribution of this random set with respect to its location. A point $z \in \mathbb{R}^m$ is attained by the set $\mathcal{X}^\star$ if

$$X_1^\star \preceq z \vee \ldots \vee X_N^\star \preceq z =: \mathcal{X}^\star \trianglelefteq z.$$

The symbol $\preceq$ denotes weak Pareto dominance. So, a point is attained by a set if at least one element of the set weakly dominates the point. For each $z \in \mathbb{R}^m$ the attainment function is defined as the probability that z is attained by $\mathcal{X}^\star$:

$$a(z) = P(\mathcal{X}^\star \trianglelefteq z).$$

The empirical attainment function estimates the attainment function. It is defined as

$$e(\mathcal{X}_1,\ldots,\mathcal{X}_k;z) = \frac{1}{k}\sum_{i=1}^{k} I(\mathcal{X}_i \trianglelefteq z)$$

with $I(\cdot) : \mathbb{R}^m \mapsto \{0,1\}$ denoting the indicator function and $\mathcal{X}_1,\ldots,\mathcal{X}_k$ the k approximations of the true Pareto frontier.

For visualizing the empirical attainment function in the two dimensional case, contour lines are plotted. These lines display the tightest set of points which are attained for a given percentage of the k approximations of the true Pareto frontier. These and further information on the attainment function can be found in [4,5].

3   Approaches for Estimating Expectation and Standard Deviation of f

3.1   General Idea

Let $f(x,C)$ be a stochastic function where $x \in \mathbb{R}$ is a controllable parameter and C is a one dimensional random variable. Our aim is optimizing $f(x,C)$ by minimizing the expectation

$$\mathrm{E}(f(x,C)) := \int_{-\infty}^{\infty} f(x,c) \cdot p_C(c)dc$$

and the standard deviation $\mathrm{S}(f(x,C)) = \sqrt{\mathrm{V}(f(x,C))}$ where

$$\mathrm{V}(f(x,C)) := \int_{-\infty}^{\infty} (f(x,c) - \mathrm{E}(f(x,C)))^2 \cdot p_C(c)dc$$

is the variance of the function f in x. $p_C(c)$ denotes the probability density function of C at the point c. It is assumed that the distribution of C is continuous. For a discrete distribution a summation over the support of C is needed instead of the integration. If the expectation should be maximized, this can be transformed into a minimization problem by multiplying the expectation with -1.

For optimizing expectation and standard deviation, it is necessary to calculate or at least approximate $\mathrm{E}(f(x,C))$ and $\mathrm{V}(f(x,C))$ for different values of x. The parameter x should be bounded by an interval $[x_l, x_u] \subset \mathbb{R}$.

If the function f is given in an analytic form and the probability density function p_C is known, it is possible to calculate $\mathrm{E}(f(x,C))$ and $\mathrm{V}(f(x,C))$ exactly. We assume that f is an expensive black box function. This means that the analytic form is not known and the evaluation of f requires many resources. Because we look at a black box function, it is not possible to calculate $\mathrm{E}(f(x,C))$ and $\mathrm{V}(f(x,C))$ exactly so they have to be estimated. Because each function evaluation is expensive, we want to evaluate f seldom.

3.2 Description of Our Approach

Our approach is to build a metamodel $\hat{f}$ for the function f. This metamodel can be used to estimate the expectation $E(f(x,C))$ and the standard deviation $S(f(x,C))$ by numerical integration. To estimate $E(f(x,C))$ and $S(f(x,C)) = \sqrt{V(f(x,C))}$ the following integrals are calculated:

$$\hat{E}(f(x,C)) = \int_{c_l}^{c_u} \hat{f}(x,c) \cdot \hat{p}_C(c) dc$$

and

$$\hat{V}(f(x,C)) = \int_{c_l}^{c_u} (\hat{f}(x,c) - \hat{E}(f(x,C)))^2 \cdot \hat{p}_C(c) dc.$$

c_l and c_u denote the lower and upper integration limits and $\hat{p}_C(c)$ the estimated probability density function of C. For good estimations of $E(f(x,C))$ and $S(f(x,C))$, a good model $\hat{f}(x,c)$, a good choice of c_l and c_u and a good estimation $\hat{p}_C(c)$ are required. It is assumed that C is observable so that the probability density function p_C can be estimated. It is possible to use $c_l = -\infty$ and $c_u = \infty$ like for the true expectation and variance but preliminary experiments have shown that other choices can lead to better results.

To build the metamodel $\hat{f}(x,c)$, a data set with values for x, c and $f(x,c)$ is needed. This data set should contain different values of x in the interval $[x_l, x_u]$ and different values of c in the interval $[c_l, c_u]$. For the model it is best to have a space-filling design on $[x_l, x_u] \times [c_l, c_u]$ like a latin hypercube design. The values of x can be chosen in a controlled and smart way whereas the values c are given as realizations of the random variable C and hence cannot be chosen as desired. The values for x could for example be placed equidistantly and with repetitions. We choose all values for x at the beginning. Since C is a random variable, the realizations of C could be good or bad with regard to a space-filling design. The distribution class of C has a big influence on the quality of the data set which is used to build a metamodel.

The metamodel enables us to use the information of the evaluations of f very well. It is not only possible to estimate $E(f(x,C))$ and $S(f(x,C))$ for the x-values in which f is evaluated but for every $x \in [x_l, x_u]$. This is a big advantage in comparison to the following classic approach where this is not possible.

3.3 Description of the Classic Method

In this section, we explain a classic method which could be applied in our setting, too. Our approach is compared to this method in Sect. 5.

For estimating $E(f(x,C))$ and $S(f(x,C))$, we take the same observations as in the data set which is used to build the metamodel in Subsect. 3.2. For each value of x, we estimate the expectation with the arithmetic mean and the standard deviation with the empirical standard deviation of the corresponding observations $f(x, C)$. For calculating the empirical standard deviation, at least two values are needed. Because of this, it is not possible to estimate the standard deviation using this approach if f is evaluated in x only once whereas this is possible with our approach. Also, it is only possible to estimate $E(f(x, C))$ and $S(f(x, C))$ for x-values where $f(x, C)$ is evaluated.

4   Experiments

In this section, we describe the design of our experiments for comparing the two approaches and some computational aspects.

4.1   Design of Experiments

Analyzing the optimization of expensive stochastic black box functions entails two problems for performing experiments. Firstly, for such a function the true expectation $E(f(x, C))$ and standard deviation $S(f(x, C))$ are not known. But it is necessary to know them for comparing the two approaches explained in Sect. 3. Therefore, we use functions where the analytic form is known. Secondly, it is desired to perform a large number of experiments. To reach this goal, we use a function which can be evaluated fast instead of an expensive black box function. But we limit the budget of function evaluations like it is common for expensive black box optimization.

We consider a simple version of the newsvendor model [1] as objective function f. This model is motivated by selling some product like newspapers which loses value very fast. It is not possible to sell the newspapers after a short period of time because they are outdated then. Given a stock of x newspapers and a random demand C, the minimum of x and C is sold for a price p. The cost for purchase or production for x products is lx. So the profit is given as

$$f(x, C) = p \min(x, C) - lx$$

with $p > l$. If p is not larger than l it does not make sense to sell any products. The profit should be maximized and the uncertainty minimized. Therefore $-E(f(x, C))$ and $S(f(x, C))$ can be minimized. We use $p = 5, l = 3$ and $x \in [0, 100]$.

For C we draw realizations c from a normal distribution. It is possible to calculate the expectation $E(f(x, C))$ and the standard deviation $S(f(x, C))$ for this example. For $C \sim \mathcal{N}(\mu, \sigma^2)$ with $x, \mu \in \mathbb{R}$ and $\sigma^2 > 0$ the expectation of $\min(x, C)$ is

$$E(\min(x, C)) = x + (\mu - x)\Phi\left(\frac{x - \mu}{\sigma}\right) - \sigma\phi\left(\frac{x - \mu}{\sigma}\right)$$

where Φ and ϕ denote the cumulative distribution function and the probability density function of the standard normal distribution. This leads to

$$E(f(x, C)) = pE(\min(x, C)) - lx.$$

The variance $V(f(x, C))$ is

$$p^2 \left(x^2 + (\mu^2 + \sigma^2 - x^2)\Phi\left(\frac{x - \mu}{\sigma}\right) - (x\sigma + \mu\sigma)\phi\left(\frac{x - \mu}{\sigma}\right) - E(\min(x, C))^2 \right).$$

We generate the values for C by drawing from random variables. Because of this, C is observable, the probability density function is independent of x and the distribution class is known. We use C normally distributed with expectation $E(C) = 50$ and $V(C) = 49$. If the variance of C is small, it is very easy to achieve good estimations of $E(f(x, C))$ and $S(f(x, C))$. But the greater the variance of C is, the more difficult it is to estimate expectation and standard deviation because the variance of the estimators increases for a higher variance of C.

In our experiments we analyze different numbers of evaluations of f. Let n denote the number of different x-values and r the number of repetitions of each x-value. We look at each combination of $n \in \{11, 21, 51, 101\}$ and $r \in \{1, 2, 5, 10\}$.

We use the maximum likelihood method to estimate the parameters for the probability density function p_C. We choose the integration limits c_l and c_u as the 10^{-8}- and $(1 - 10^{-8})$-quantile of the estimated distributions. This choice of parameters has been good in preliminary experiments. Because the results are influenced by random effects, we repeat every configuration 100 times.

Beyond the scope of this paper, we performed a large number of experiments with several objective functions, several distributions of C and several levels of variance of C to compare the two approaches. Due to the page limit we decided to report only the results of one representative example in detail here. The results for our other experiments do not differ a lot and the conclusions are identical.

4.2 Computational Aspects

For our experiments we use the software R [11]. As metamodel we apply Kriging which is implemented in the package DiceKriging [12]. The maximum likelihood estimation of the parameters for the Kriging model is not deterministic. Because of this, we build five models per experiment and choose the one which approximates the true function f best regarding the set of evaluated points. It is possible to exchange Kriging with another type of metamodel but preliminary studies have shown that Kriging leads to the best results for our situation. For calculating the integrals $\hat{E}(f(x, C))$ and $\hat{V}(f(x, C))$, the standard method integrate is used for numerical integration. To determine the Pareto sets and Pareto frontiers, the package ecr [2] is employed. For generating the empirical attainment functions, the package eaf [7] is used.

5 Evaluation of the Experiments

In this section, we compare our approach and the classic approach regarding the approximation quality of the Pareto frontiers and Pareto sets.

5.1   Pareto Frontiers

First, we analyze the quality of the approximations of the true Pareto frontier generated with our and the classic approach. Figures 1 and 2 show the empirical attainment functions for our and the classic approach. With our approach, $\hat{E}(f(x,C))$ and $\hat{S}(f(x,C))$ are available for every $x \in [0,100]$. For the classic approach, $E(f(x,C))$ and $S(f(x,C))$ can only be estimated for the x-values in which f has been evaluated. In order to conduct a fair comparison between the two approaches, we decided to consider $\hat{E}(f(x,C))$ and $\hat{S}(f(x,C))$ only for the x-values in which f has been evaluated. If $(\hat{E}(f(x,C)),\hat{S}(f(x,C)))$ are only available for a small number of x-values, the approximation of the true Pareto frontier is rough. With our approach a smoother approximation of the true Pareto frontier than the one presented in Fig. 1 is possible when considering $(\hat{E}(f(x,C)),\hat{S}(f(x,C)))$ for more x-values.

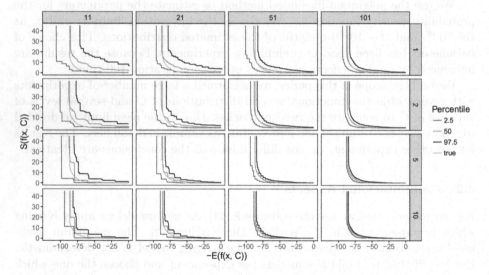

Fig. 1. Empirical attainment functions for the approximations of the true Pareto frontiers with our approach. In the rows are the plots with $r \in \{1,2,5,10\}$ and in the columns $n \in \{11,21,51,101\}$. The colors display the used percentiles of the empirical attainment function as well as the true Pareto frontier. (Color figure online)

For the empirical attainment functions, it is desired that the lines for the 2.5-, 50- and 97.5-percentile are very close to the line for the true Pareto frontier. If the line for the 50-percentile is close to the true Pareto frontier, this means that the median approximation is quite accurate. If the lines of the 2.5- and 97.5-percentile are close to each other, this means that the variation of the approximated Pareto frontiers and hence the uncertainty in the estimation of the true Pareto frontier is quite low.

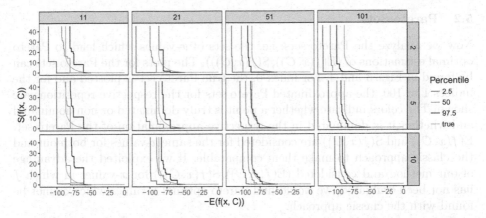

Fig. 2. Empirical attainment functions for the approximations of the true Pareto frontiers with the classic approach. In the rows are the plots with $r \in \{1, 2, 5, 10\}$ and in the columns $n \in \{11, 21, 51, 101\}$. The colors display the used percentiles of the empirical attainment function as well as the true Pareto frontier. (Color figure online)

Looking at the results of our approach, it can be seen that quite good approximations of the true Pareto frontier can already be achieved by building a metamodel based on 101 different x-values each repeated once or 51 different x-values and two repetitions. If more evaluations of f are allowed like for example with 51 different x-values and 5 repetitions, it is possible to improve the approximation of the true Pareto frontier a bit but this costs much more resources.

Figure 2, which displays the results for the classic approach, contains no plots for the configuration where each x-value is only evaluated once. In this situation, it is not possible to estimate the standard deviation $S(f(x, C))$ with the classic approach. In many of the plots, the expectation $E(f(x, C))$ is overestimated. In these cases, all three lines of the empirical attainment function are left of the true Pareto frontier. A design with a larger number of different x-values leads to a worse result for the classic approach if the number of repetitions is the same. The reason for this seems to be that there is a higher probability that there is at least one point which falsely dominates all other points because the expectation $E(f(x, C))$ is overestimated. Furthermore, the estimation of a single point $(E(f(x, C)), S(f(x, C)))$ is not improved by evaluating a larger number of different x-values here, because only function evaluations with the same x-value are used for the estimation. A higher number of repetitions is the only possibility to improve the estimations using the classic approach. The reason for this is that the variance of the estimations decreases for a larger number of observations. For the classic method it could be better to distribute the $n \cdot r$ evaluations of f in another way than in our experiments and use more repetitions instead of more different x-values. But for a fair comparison of the two approaches it is necessary to use the same values of x and c for evaluating f.

5.2 Pareto Sets

Now we analyze the Pareto sets, i.e. the sets of x-values which lead to Pareto optimal estimations of $(\mathrm{E}(f(x,C)), \mathrm{S}(f(x,C)))$. The plots for the Pareto sets can be found in Figs. 3 and 4. For index 0, the true Pareto set is plotted and for the indices 1 to 100, the approximated Pareto sets for the respective repetitions are shown. The colors indicate whether a point is truly dominated or non dominated and whether it is dominated in the respective approximation of the Pareto set. $\hat{\mathrm{E}}(f(x,C))$ and $\hat{\mathrm{S}}(f(x,C))$ are considered for the same x-values for both our and the classic approach to make them comparable. If we exploited the advantage of our method and considered $(\hat{\mathrm{E}}(f(x,C)), \hat{\mathrm{S}}(f(x,C)))$ for x-values in which f has not been evaluated, we would find Pareto optimal x-values which cannot be found with the classic approach.

For our approach, most of the truly dominated points are dominated in the approximations as well. It seems to be a much more difficult task to find all non dominated points. It can be seen that especially small x-values are often truly non dominated but dominated in the approximations. This can be explained by the fact that there is only a small slope for small x-values in the true Pareto frontier. Because of this, a small underestimation of the standard deviation can result in a lot of truly non dominated points being dominated in the approximation. For a larger number of repetitions, the quality of the approximations of the true Pareto set improves a bit but not very much.

Looking at the classic approach, only a small percentage of the truly non dominated points is non dominated as well in the approximations. The approxi-

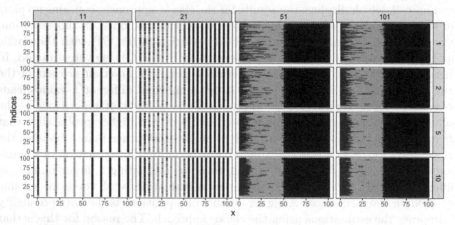

- truly non dominated but approximation dominated
- truly non dominated and approximation non dominated
- truly dominated but approximation non dominated
- truly dominated and approximation dominated

Fig. 3. Approximated Pareto sets with our approach. In the rows are the plots with $r \in \{1, 2, 5, 10\}$ and in the columns $n \in \{11, 21, 51, 101\}$. The colors display if the assignments are correct. Index 0 is used to display the true memberships of the x-values and indices 1 to 100 for the approximations. (Color figure online)

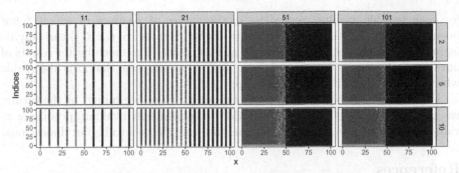

● truly non dominated but approximation dominated   ● truly dominated but approximation non dominated
● truly non dominated and approximation non dominated   ● truly dominated and approximation dominated

Fig. 4. Approximated Pareto sets with the classic approach. In the rows are the plots with $r \in \{1, 2, 5, 10\}$ and in the columns $n \in \{11, 21, 51, 101\}$. The colors display if the assignments are correct. Index 0 is used to display the true memberships of the x-values and indices 1 to 100 for the approximations. (Color figure online)

mations seem to be worse with a larger number of different x-values like for the Pareto frontiers. Furthermore, it can be seen that more truly dominated points are classified as non dominated in the approximations than with our approach. A higher number of repetitions does not improve the approximation quality very much.

6   Conclusion

In this paper, we analyzed the optimization of expensive stochastic black box functions $f(x, C)$ with a parameter $x \in \mathbb{R}$ and a one dimensional random variable C. We transformed the problem of optimizing the stochastic function f into the biobjective optimization of the deterministic expectation $\mathrm{E}(f(x, C))$ and standard deviation $\mathrm{S}(f(x, C))$.

$\mathrm{E}(f(x, C))$ and $\mathrm{S}(f(x, C))$ are defined as integrals over terms which only consist of f and the probability density function p_C. Our approach is estimating $\mathrm{E}(f(x, C))$ and $\mathrm{S}(f(x, C))$ by building a metamodel for f, estimating p_C and using numerical integration. Our approach requires C to be observable.

We performed a simulation study to compare our approach to the classic approach where $\mathrm{E}(f(x, C))$ and $\mathrm{S}(f(x, C))$ are estimated using the arithmetic mean and the empirical standard deviation. The results of our experiments show that the approximation quality of the Pareto frontiers of our approach is much better than for the classic approach. We also compared the approximation qualities of the Pareto sets. The Pareto sets resulting from our approach are often very close to the true Pareto set. With the classic approach, only a small percentage of the truly non dominated points is found.

In conclusion, our approach performs much better than the classic approach. Moreover, for our approach it is possible to estimate $\mathrm{E}(f(x, C))$ and $\mathrm{S}(f(x, C))$

for each feasible x-value, regardless of the choice of x-values in which f has been evaluated. For the classic approach, $E(f(x, C))$ and $S(f(x, C))$ can only be estimated if f has been evaluated in x at least twice. The much better results of our approach come at the cost of the additional assumption that C is observable which is not required for the classic approach.

In our approach we use all evaluations of f in the beginning. In the future, we will employ model based optimization to sequentially evaluate promising x-values. In addition to this, we will generalize our method to higher dimensions.

References

1. Arrow, K.J., Harris, T., Marschak, J.: Optimal inventory policy. Econometrica **19**(3), 250–272 (1951). https://www.jstor.org/stable/1906813
2. Bossek, J.: ecr: Evolutionary Computation in R (2017). https://CRAN.R-project.org/package=ecr. R package version 2.1.0
3. Branke, J., Deb, K., Miettinen, K., Słowiński, R. (eds.): Multiobjective Optimization. LNCS, vol. 5252. Springer, Heidelberg (2008). https://doi.org/10.1007/978-3-540-88908-3
4. Fonseca, C.M., Guerreiro, A.P., López-Ibáñez, M., Paquete, L.: On the computation of the empirical attainment function. In: Takahashi, R.H.C., Deb, K., Wanner, E.F., Greco, S. (eds.) EMO 2011. LNCS, vol. 6576, pp. 106–120. Springer, Heidelberg (2011). https://doi.org/10.1007/978-3-642-19893-9_8
5. Grunert da Fonseca, V., Fonseca, C.M., Hall, A.O.: Inferential performance assessment of stochastic optimisers and the attainment function. In: Zitzler, E., Thiele, L., Deb, K., Coello Coello, C.A., Corne, D. (eds.) EMO 2001. LNCS, vol. 1993, pp. 213–225. Springer, Heidelberg (2001). https://doi.org/10.1007/3-540-44719-9_15
6. Gutjahr, W.J., Pichler, A.: Stochastic multi-objective optimization: a survey on non-scalarizing methods. Ann. Oper. Res. **236**(2), 475–499 (2016). https://doi.org/10.1007/s10479-013-1369-5
7. López-Ibáñez, M., Paquete, L., Stützle, T.: Exploratory analysis of stochastic local search algorithms in biobjective optimization. In: Bartz-Beielstein, T., Chiarandini, M., Paquete, L., Preuss, M. (eds.) Experimental Methods for the Analysis of Optimization Algorithms, pp. 209–222. Springer, Heidelberg (2010). https://doi.org/10.1007/978-3-642-02538-9_9
8. Markowitz, H.: Portfolio selection. J. Finance **7**(1), 77–91 (1952). https://doi.org/10.1111/j.1540-6261.1952.tb01525.x
9. Paenke, I., Branke, J., Jin, Y.: Efficient search for robust solutions by means of evolutionary algorithms and fitness approximation. IEEE Trans. Evol. Comput. **10**(4), 405–420 (2006). https://doi.org/10.1109/TEVC.2005.859465
10. Ponsich, A., Jaimes, A.L., Coello Coello, C.A.: A survey on multiobjective evolutionary algorithms for the solution of the portfolio optimization problem and other finance and economics applications. IEEE Trans. Evol. Comput. **17**(3), 321–344 (2013). https://doi.org/10.1109/TEVC.2012.2196800
11. R Core Team: R: a language and environment for statistical computing. R Foundation for Statistical Computing, Vienna (2018). https://www.R-project.org/
12. Roustant, O., Ginsbourger, D., Deville, Y.: DiceKriging, DiceOptim: two R packages for the analysis of computer experiments by kriging-based metamodeling and optimization. J. Stat. Softw. **51**(1), 1–55 (2012). https://doi.org/10.18637/jss.v051.i01. http://www.jstatsoft.org/v51/i01/

Mutual Rationalizability in Vector-Payoff Games

Erella Eisenstadt-Matalon[1,2] and Amiram Moshaiov[1(✉)]

[1] Tel-Aviv University, 69978 Tel-Aviv, Israel
erella@braude.co.il, moshaiov@eng.tau.ac.il
[2] ORT Braude College of Engineering, Karmiel, Israel

Abstract. This paper deals with vector-payoff games, which are also known as Multi-Objective Games (MOGs), multi-payoff games and multi-criteria games. Such game models assume that each of the players does not necessarily consider only a scalar payoff, but rather takes into account the possibility of self-conflicting objectives. In particular, this paper focusses on static non-cooperative zero-sum MOGs in which each of the players is undecided about the objective preferences, but wishes to reveal tradeoff information to support strategy selection. The main contribution of this paper is the introduction of a novel solution concept to MOGs, which is termed here as Multi-Payoff Mutual-Rationalizability (MPMR). In addition, this paper provides a discussion on the development of co-evolutionary algorithms for solving real-life MOGs using the proposed solution concept.

Keywords: Game theory · Non-cooperative games · Set-based optimization · Set domination · Multi-criteria decision-analysis

1 Introduction

Game theoretic studies have proven to be significant to making strategic decisions in both cooperative and non-cooperative situations of relevance to many application areas including economy, engineering, biology and sociology. Most such studies concern games in which each player has a scalar payoff to maximize, such as a monetary payoff. However, in many real-life game situations decision-makers are interested in more than one objective which are often incomparable and conflicting. Blackwell [1], was the first to study such vector-payoff games. Since his early work, vector-payoff games, which have also been termed as Multi-Objective Games (MOGs), multi-payoff games and multi-criteria games, have been studied by many others. As evident from reviews such as in [2, 3], the majority of studies on MOGs employed either an interpretation of Nash-equilibrium or interpretations of the MiniMax solution concept to such games. While providing possible solutions to MOGs, the focus of most of the traditional studies has neither been on finding tradeoff information nor on how to utilize such information for making a strategic decision.

In contrast to most previous studies on MOGs, several recent publications, including [4–6], suggested a method that is based on a novel solution concept to MOGs, which accounts for the performance tradeoffs. Focusing on pure strategies,

© Springer Nature Switzerland AG 2019
K. Deb et al. (Eds.): EMO 2019, LNCS 11411, pp. 593–604, 2019.
https://doi.org/10.1007/978-3-030-12598-1_47

zero-sum, static, non-cooperative MOGs, the method of [4–6] is based on the notion of rationalizable strategies, which was suggested for solving scalar games by Bernheim [7] and independently by Pearce [8]. To account for tradeoff information when solving MOGs, the suggested method of [4–6] involves two stages. In the first stage, a Set of Rationalizable Strategies (SRS) is found for each player using set domination relations and worst-case considerations. In addition, the first stage provides, for each rationalizable strategy, an associated set of payoff vectors, which results from the most harmful responses by the opponent. In the second stage, the associated sets of payoff vectors are used to analyze the performance tradeoffs in support of strategy selection out of the SRS. For the latter stage, the study in [4] provides two novel multi-criteria decision-analysis procedures for strategy selection that are based on set parameters.

The current paper suggests a revision to the first stage of the procedure in [4–6], which amounts to the introduction of a new solution concept to MOGs. In addition, this paper provides a discussion on the consequences of this suggestion to the development of co-evolutionary algorithms for finding good representations of rationalizable strategies.

In contrast to the rationalizability approach of [4–6], which assumes the possibility of an irrational opponent, the proposed revision aims to account for a rational opponent. In non-cooperative situations players are faced with the problem of uncertainties about the opponent. Traditionally, game theoretic studies have assumed that the players are rational. Yet, multi-payoff rationalizability, as suggested in [4–6], is based on the worst-case approach in which the opponent is assumed to play in the most harmful way. Namely, irrationalizable strategies of the opponent are considered by the player when taking the multi-payoff rationalizability approach that has been suggested in [4–6]. However, a player may have beliefs or even intelligence about the opponent, which may lead her to the assumption that the opponent is rational. In such a case, the approach of [4–6] should be considered as inconsistent with the aforementioned assumption.

The current study assumes that the opponent is rational. In the context of the considered game, it means that the opponent plays according to her best replies, rather than her most harmful replies, without a-priori decision on her objective preferences. Moreover, it is assumed that both players view their opponent as rational. In addition, it is assumed that each player knows it, and knows that the opponent knows it, and so on in the sense of Aumann's common knowledge of rationality [9], and in accordance with the original rationalizability approach of Bernheim and Pearce to scalar games [7, 8]. To distinguish between the rationalizability of [4–6] and the current approach, the latter is hereby termed Multi-Payoff Mutual-Rationalizability (MPMR), whereas the former is hereby termed as one-sided rationalizability.

The rest of this paper is organized as follows. First, in Sect. 2, the considered game is described and the MPMR solution concept is presented. Next, in Sect. 3, the proposed approach is demonstrated. In Sect. 4, a discussion is provided on the past and future development of evolutionary algorithms based on the suggested MPMR approach and the one-sided rationalizability approach of [4–6]. Finally, Sect. 5 summarizes this paper and provides suggestions for future research.

2 The Game and the Proposed MPMR Solution Concept

In this study, the considered MOG involves two players, each with self-conflicting objectives. The game is zero-sum with respect to each of the components of the payoff vector. The game is pure strategy, single-act (static), non-cooperative and with imperfect information. It is also a game of incomplete information since it concerns no a-priori declaration or determination of objective preferences by the players. The MOG and the proposed solution concept are described in the following.

2.1 Problem Formulation

The two players are P_{min} (minimizer) and P_{max} (maximizer). The maximizer aims at maximizing all the components of the payoff vector while the minimizer aims at minimizing these same components. Let S_{min} and S_{max} be the sets of all possible strategies for P_{min} and P_{max} respectively, such that $s^i_{min} \in S_{min}$ and $s^j_{max} \in S_{max}$. Note that $s^i_{min} \in \mathbb{R}^{N_{min}}$ and $s^j_{max} \in \mathbb{R}^{N_{max}}$ and the mth component of such a strategy is denoted as $s^{i(m)}_{min}$ and $s^{j(m)}_{max}$.

The interaction between the ith strategy and the jth strategy played by P_{min} and P_{max}, respectively, results in the following payoff vector:

$$f_{i,j} = \left[f^{(1)}_{i,j}, \ldots, f^{(k)}_{i,j}, \ldots, f^{(K)}_{i,j} \right] \in \mathbb{R}^K \tag{1}$$

where K is the number of objectives that the players consider. The set of all the interactions between strategy s^i_{min} of P_{min} and all the J strategies of P_{max} is the set of payoff vectors that represents the performances of strategy s^i_{min}:

$$F_{s^i_{min}} := \left\{ f_{i,j} \in \mathbb{R}^K \mid \forall j \in \{1, \ldots, J\} \bigwedge s^i_{min} \in S_{min} \right\}. \tag{2}$$

Similarly, for strategy s^j_{max} of player P_{max}:

$$F_{s^j_{max}} := \left\{ f_{i,j} \in \mathbb{R}^K \mid \forall i \in \{1, \ldots, I\} \bigwedge s^j_{max} \in S_{max} \right\}. \tag{3}$$

The multi-objective game is defined as:

$$G = \left(\{P_{min}, P_{max}\}, S_{min}, S_{max}, (f_{i,j}) \begin{matrix} i \in \{1, \ldots, I\} \\ j \in \{1, \ldots, J\} \end{matrix} \right). \tag{4}$$

2.2 The Proposed Solution Concept

According to the solution concept of rationalizability there is no single optimal strategy [7, 8]. Instead, a set of rationalizable strategies is sought for each of the players. In the context of the considered MOGs, a strategy is considered irrational if it is Never a Best Response (NBR) (never best reply) under any possible objective preferences. On the other hand, a strategy is considered rationalizable if it is a Best Response (BR) in some preference circumstances. Solving the game means finding, for each player, all the rationalizable strategies and their associated sets of payoff vectors.

In scalar games, obtaining the set of rationalizable strategies, under the assumption of common knowledge of rationality, involves an iterative elimination process, in which the players eliminate their strictly dominated strategies [7, 8]. To extend the concept of rationalizable strategies from scalar games to MOGs, in which the players are undecided about their objective preferences, there is a need to clarify what constitutes a strictly dominated strategy under such a condition. In studies such as [4–6], it has been suggested to employ worst-case set domination to obtain the sought sets of strategies. Yet, as explained in the introduction, such studies do not consider the assumption of common knowledge of rationality.

To account for the common knowledge of rationality, obtaining the set of rationalizable strategies involves an iterative elimination process, in which the players eliminate their strictly dominated strategies, which are inferior strategies under any objective preferences. During the iterative process, following elimination during a previous iteration, the players re-evaluate their strategies according to the updated sets of players' strategies and look for "new" irrational strategies, and so on. The following describes in details the proposed iterative process for the MPMR solution concept.

$S_{min}(\tau)$ and $S_{max}(\tau)$ are the sets of all possible $I(\tau)$ and $J(\tau)$ strategies, in the τth iteration, for the minimizer P_{min} and the maximizer P_{max}, respectively. Here, $\tau = 0$ for the initial iteration before eliminating any of the strategies. When evaluating the ith strategy s^i_{min} of P_{min} in the τth iteration, there is a need to consider all possible strategies of P_{max} in this iteration.

Let $F_{s^i_{min}}(\tau)$ be the set of all payoff vectors from all the interactions of the ith strategy of P_{min} with the remaining strategies of P_{max} in the τth iteration. Given that the objective preferences of P_{max} are undecided, then there is a set of non-dominated payoff vectors (in a maximization problem), which corresponds to all possible BRs of P_{max} (at the current iteration) to the ith strategy of P_{min}. This set is termed as the anti-optimal front of the ith strategy of P_{min} in the τth iteration. It is defined as:

$$F^{-*}_{s^i_{min}}(\tau) := \left\{ f_{i,j} \in F_{s^i_{min}}(\tau) \mid \nexists f_{i,j'} \in F_{s^i_{min}}(\tau) : f_{i,j'} \succ^{max} f_{i,j} \right\} \tag{5}$$

where $a \succ^{max} b$ stands for a dominates b in the maximization problem.

In the same way, the anti-optimal front of a strategy of the maximizer P_{max}, in the τth iteration, is:

$$F^{-*}_{s^j_{max}}(\tau) := \left\{ f_{i,j} \in F_{s^j_{max}}(\tau) \mid \nexists f_{i',j} \in F_{s^j_{max}}(\tau) : f_{i',j} \succ^{min} f_{i,j} \right\} \tag{6}$$

where $a \succ^{min} b$ stands for a dominates b in the minimization problem. Note that in this paper the superscript $(-*)$ specifies the fact that the front is a result of the inverse optimization problem of the player.

The set of the BRs of P_{max}, to the ith strategy of P_{min} in the τth iteration, is:

$$S_{s_{min}^i}^{BR}(\tau) := \left\{ s_{max}^j \in S_{max}(\tau) \mid \nexists s_{max}^{j'} \in S_{max}(\tau) : f_{i,j'} \succ^{max} f_{i,j} \right\}. \tag{7}$$

Any strategy of P_{max} that belongs to the BR set, $s_{max}^j \in S_{s_{min}^i}^{BR}(\tau)$, is related to one of the payoff vectors that form the anti-optimal front of strategy s_{min}^i, that is $f_{i,j} \in F_{s_{min}^i}^{-*}$. Similarly, the set of the BRs of P_{min}, to the jth strategy of P_{max} in the τth iteration, is:

$$S_{s_{max}^j}^{BR}(\tau) := \left\{ s_{min}^i \in S_{min}(\tau) \mid \nexists s_{min}^{i'} \in S_{min}(\tau) : f_{i',j} \succ^{min} f_{i,j} \right\}. \tag{8}$$

Any strategy of P_{min} that belongs to the set of best response, $s_{min}^i \in S_{s_{max}^j}^{BR}(\tau)$, is related to one of the payoff vectors that form the anti-optimal front of strategy s_{max}^j, that is $f_{i,j} \in F_{s_{max}^j}^{-*}$.

The set of all the anti-optimal fronts of P_{min} and the set of all the anti-optimal fronts of P_{max}, at the τth iteration, are the following sets of sets:

$$F_{min}^{-*}(\tau) := \left\{ F_{s_{min}^i}^{-*}(\tau) \mid \forall s_{min}^i \in S_{min}(\tau) \right\} \tag{9}$$

$$F_{max}^{-*}(\tau) := \left\{ F_{s_{max}^j}^{-*}(\tau) \mid \forall s_{max}^j \in S_{max}(\tau) \right\}. \tag{10}$$

To obtain the irrationalizable strategies, the following definitions of set domination relations are used. Set F dominates set H in a maximization problem, $F \succ^{max} H$, if $\forall h \in H \, \exists f \in F$ such that $f \succ^{max} h$. Also, Set F dominates set H in a minimization problem, $F \succ^{min} H$, if $\forall h \in H \, \exists f \in F$ such that $f \succ^{min} h$. Using these definitions, the set of *irrational* strategies of the minimizer, at the τth iteration, are the strategies with the anti-optimal front that dominates in the maximization problem (in the inverse problem) at least one other anti-optimal front. Namely:

$$S_{min}^{irr}(\tau) := \left\{ s_{min}^i \in S_{min}(\tau) \mid \nexists s_{min}^{i'} \in S_{min}(\tau) : F_{s_{min}^i}^{-*}(\tau) \succ^{max} F_{s_{min}^{i'}}^{-*}(\tau) \right\} \tag{11}$$

and the maximizer's set of *irrational* strategies at the τth iteration is:

$$S_{max}^{irr}(\tau) := \left\{ s_{max}^j \in S_{max} \mid \nexists s_{max}^{j'} \in S_{max} : F_{s_{max}^j}^{-*}(\tau) \succ^{min} F_{s_{max}^{j'}}^{-*}(\tau) \right\}. \tag{12}$$

The irrational strategies of the considered iteration are the worst-case dominated strategies using the available strategies at the iteration. These strategies are strictly inferior when comparing to the rest of the strategies of the iteration since there is at least one other strategy that yields preferred outcome in any objective preference. This means that an irrational strategy is NBR for any possible objective preference. This is equivalent to a strictly dominated strategy in scalar games.

The τth iteration is completed by the elimination of the irrational strategies of both players, as found in that iteration. The set of the remaining strategies of P_{min} is the relative complement of $S_{min}(\tau)$ and $S_{min}^{irr}(\tau)$:

$$S_{min}(\tau+1) = S_{min}(\tau) \backslash S_{min}^{irr}(\tau) \; for \; 0 \le \tau \qquad (13)$$

and the set of the remaining strategies of P_{max} is the relative complement of $S_{max}(\tau)$ and $S_{max}^{irr}(\tau)$:

$$S_{max}(\tau+1) = S_{max}(\tau) \backslash S_{max}^{irr}(\tau) \; for \; 0 \le \tau. \qquad (14)$$

The iterative deletion, of irrationalizable strategies, is terminated when there are no more irrational strategies to delete. Namely, at the last iteration, τ_{final}:

$$S_{min}^{irr}(\tau_{final}) = \emptyset \quad and \quad S_{max}^{irr}(\tau_{final}) = \emptyset. \qquad (15)$$

Therefore, the set of rationalizable strategies of the minimizer and the maximizer are the sets of strategies in τ_{final}:

$$S_{min}^{R} := S_{min}(\tau_{final}) \qquad (16)$$

$$S_{max}^{R} := S_{max}(\tau_{final}). \qquad (17)$$

Each of these strategies is represented in the objective space by its related anti-optimal front as resulting from the interactions with the rationalizable strategies of the opponent. The union of all the anti-optimal fronts, of the rationalizable strategies, forms the rationalizable layer of P_{min} and P_{max} respectively:

$$F_{min}^{R} := \left\{ F_{s_{min}^{i}}^{-*} \in F_{min}^{-*}(\tau_{final}) \mid s_{min}^{i} \in S_{min}(\tau_{final}) \right\} \qquad (18)$$

$$F_{max}^{R} := \left\{ F_{s_{max}^{j}}^{-*} \in F_{max}^{-*}(\tau_{final}) \mid s_{max}^{j} \in S_{max}(\tau_{final}) \right\}. \qquad (19)$$

The obtained sets of rationalizable strategies, which are based on mutual rationalizability, and their associated anti-optimal fronts, allow strategy selection. This can be done by the same procedures that have been suggested in [4] with respect to the onesided rationalizability approach.

2.3   Irrational Strategies and Set Domination

The proposed mutual rationalizability approach and also the one-sided rationalizability approach assume that irrational strategies can be found by set domination. While not providing any proof for such an assertion, the following aims to clarify the justification for using set domination as done here and in [4–6].

To illustrate the idea of finding irrational strategies by way of set domination, consider a bi-objective game in which the minimizer has strategies A, B, and C. Each of the three panels of Fig. 1 shows the three anti-optimal fronts of these strategies. The dashed-black, solid-black and gray lines mark the anti-optimal front of strategies A, B and C, respectively. In this case, the anti-optimal front of strategy C dominates the anti-optimal front of strategy B in a maximization problem. In view of the procedure above, C is an irrational strategy. The three panels depict three different objective preferences, as follows. The solid line, marked I, II and II in panel (a), (b) and (c), respectively, represents a preference of objectives of the minimizer. In the case of panel (a), the minimizer assigns higher weight to $f^{(2)}$. Each of the dashed thin lines depicts the maximal projection of a front on the preference line. Each of the projections is the maximal weighted-sum value, according to the minimizer preference, which can be achieved by the maximizer. Namely, this is the worst value that can be expected by the minimizer if she chooses the strategy associated with this front.

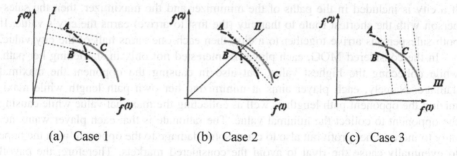

(a) Case 1   (b) Case 2   (c) Case 3

Fig. 1. Anti-optimal fronts of irrational and rationalizable strategies

Clearly, the best strategy for the minimizer in the case of panel (a) is strategy B which yields the minimal value of the maximal projections. By similar considerations, in the case of panel (b) the minimizer will also choose strategy B and in the case of panel (c) she will choose strategy A. Note that C is not chosen in any of these cases. The proposition is that strategy C will never be chosen under any objective preferences, which corresponds to the fact that C is declared irrational by the set domination procedure.

3 TSP-MOG Example

The following example is based on the competing traveling salespersons MOG, which has been introduced in [6] and used in [4]. As noted in [4], the classical TSP is NP-hard, let alone the considered TSP-MOG. The game arena is presented as a graph. The considered graph contains N vertices (cities), where each vertex represents a city from the set of cities $C = \{c(1), c(2), \ldots, c(n), \ldots c(N)\}$. Each city $c(n)$ has a value $v(c(n))$. This value represents the profit of the first salesperson that arrives to that city. The arcs of the graph represent the roads between the cities. The arc value is the road length. The game is between two competing salespersons (players), which are denoted by P_{min} and P_{max}. A strategy of a player, which is a chosen route, is defined as a partial permutation of the cities' set C. Each player may visit a city no more than once and returns to her first city at the end of the path. The routes of the players are described as the ordered sets $Path_{min} = \left\{ c_{min}^{(1)}, c_{min}^{(2)}, \ldots, c_{min}^{(N_{min})}, c_{min}^{(1)} \right\}$ and $Path_{max} = \left\{ c_{max}^{(1)}, c_{max}^{(2)}, \ldots, c_{max}^{(N_{max})}, c_{max}^{(1)} \right\}$, where $1 \leq N_{min}, N_{max} \leq N$ for the first and second player, respectively. Each element $c_{min}^{(q)}$ or $c_{min}^{(p)}$ is associated with a member city in the set C. Namely, $c_{min}^{(q)}, c_{min}^{(p)} \in C \, \forall q, p$. Also $c_{min}^{(q)} \neq c_{min}^{(q')} \, \forall q, q', q \neq q'$ and $c_{max}^{(p)} \neq c_{max}^{(p')} \, \forall p, p', p \neq p'$. It is noted that the subscripts min and max indicate which player visited that city. On the other hand, the superscripts indicate the order by which the player visited it. Each selected path has a length that is calculated as the sum of the distances between all successive cities of the path. When considering a path, each player takes into account not only the path length but also the value of the chosen route, which is in general a sum of the values of the cities of the chosen path, which meets the following criterion. If a city is included in the paths of the minimizer and the maximizer, then the salesperson with the shortest route to that city (the first to arrive) earns the city's value. If both salespersons arrive together to a city, then each one earns half of the city value.

In the considered MOG, each player is interested not only in shortening her path, while collecting the highest value, but also in causing the opponent the maximal damage. Namely, each player aims at minimizing her own path length while maximizing the opponent path length as well as collecting the maximal value while causing the opponent to collect the minimal value. The rationale is that each player wants not only to increase her profit but also to cause some damage to the opponent with the hope to eventually cause the rival to avoid the considered markets. Therefore, the payoff vector components are defined as:

$$f^{(1)} = L_{min} - L_{max} \quad and \quad f^{(2)} = V_{max} - V_{min} \tag{20}$$

where L_{min} and V_{min} denote the length and the value of the path (strategy) of the minimizer, and L_{max} and V_{max} are those of the maximizer.

In the considered problem with N cities, the number of possible paths of one salesperson is:

$$\Omega = \sum_{n=1}^{N} \frac{(N-1)!}{(N-n)!}. \tag{21}$$

As the problem presented here is a game, then the number of all possible interactions is Ω^2. Considering a small problem with $N = 10$ the number of all possible interactions is $\Omega^2 = 9.73 \cdot 10^{11}$.

For illustration purposes, the following example is planned such that full sorting, according to the proposed procedure of Sect. 2, is achievable. The left side of Fig. 2 represents the arena of the considered example. In the arena there are six cities that are marked by black dots. The bold numbers above the cities indicate the values of the cities, whereas a numbers in brackets, below the cities indicates the number of the city. In the considered games, the maximizer starts from City no. 4 (marked by a diamond) and the minimizer from city no. 5 (marked by a triangle). In Fig. 2, the links between cities are not shown. The lengths of the roads between the cities are taken according to the scale of the straight lines between the cities in the figure.

The right side of Fig. 2 presents the resulting lists of strategies for each player, as obtained at the first iteration. The strategies are presented as ordered sets of the visited cities. The four strategies of the minimizer, which are marked by rectangles, are found to be irrational under the assumption of common knowledge of rationality. Namely, these strategies did not survive the second iteration, while all the rest survived the iterative process.

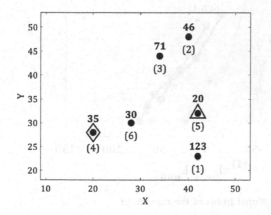

Fig. 2. The example arena (left) and the resulting strategies (right)

Clearly, as seen from the top items of the resulting lists, staying at the starting city is rationalizable. This happens when a player prefers just minimizing the path length (the travelling effort). It can also be seen that path $\{5, 3, 5\}$ of the minimizer was found to be irrational in the second iteration. When comparing the irrational path $\{5, 3, 5\}$ to the rationalizable path $\{5, 1, 5\}$, it might not be clear at first sight why path $\{5, 3, 5\}$ was suspected to be rationalizable at the first iteration. In fact, path $\{5, 1, 5\}$, yields both a higher value and shorter path for the minimizer, as compared with $\{5, 3, 5\}$ (the distance between City no. 5 and City no. 1 is slightly smaller than the distance between City no. 5 and City no. 3). Yet, one should remember that the elements of the payoff vectors are $f^{(1)} = L_{min} - L_{max}$ and $f^{(2)} = V_{max} - V_{min}$. Hence, the minimizer is aiming at maximizing her collected value V_{min} while minimizing the collected value of the maximizer V_{max}. In light of these objectives, and with the understanding that at the first iteration irrational strategies of the opponent are involved, then path $\{5, 3, 5\}$ could be suspected to be rationalizable at first. Similar argumentation can explain why other paths are rational in the first iteration and becomes irrational in the second.

Figure 3 depicts the objective space. The light gray dots mark all the 326×326 payoff vectors of all possible interactions between the strategies of the minimizer and the maximizer. The two black lines mark the two anti-optimal fronts of the rationalizable strategies of the maximizer and the two dark gray lines mark two typical anti-optimal fronts of the irrational strategies of the maximizer. In this figure it can be observed, as expected, that the two anti-optimal fronts of the rationalizable strategies of the maximizer are non-dominated by each other and that each of them is dominated by the two typical anti-optimal fronts of the irrational strategies, in a minimization problem.

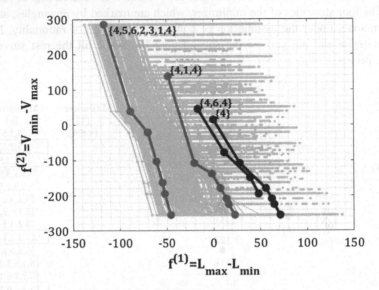

Fig. 3. Anti-optimal fronts of the maximizer

4 Aspects of Algorithm Development

It is beyond the scope of this paper to present details of an algorithm for solving MOGs, nor its evaluation. Rather, the goal here is to highlight a few issues that are worth mentioning in view of the discussion in Sect. V.A of [5]. In particular, the discussions here and in [5] refer to both the co-evolutionary algorithm of [10] and the one-sided evolutionary algorithm of [6]. These algorithms follow the lexicographic selection idea of NSGA-II [11]. Namely, strategy selection is based on a primary rank and a secondary grade. The primary rank aims to produce a selection pressure towards the rationalizable strategies. When two strategies are of the same rank, they are further evaluated according to the secondary grade within that rank. The aim of the secondary grade is to produce tangential pressure, namely to obtain well spread anti-optimal fronts within the rationalizable layers (see Eqs. 18 and 19). In [5], a revision has been suggested to the procedure of the secondary grade of [6] and [10].

As a result of the work towards the current solution concept, it has been clarified that it was wrongly suggested in [10] and in [5] that a co-evolutionary algorithm, as in [10], will generally solve MOGs according to the one-sided rationalizability approach of [4–6]. It appears that a similar wrong suggestion has been made in [12]. A possible reason for not spotting earlier this apparent mistake is that in both references [10] and [12] the algorithms have been tested on a simple demonstration case for which there is no need for an iterative process. This is due to the fact that, in the tug-of-war MOG of [10] and [12], the most harmful responses of the opponent are also her best replies.

In fact, as our unreported recent numerical studies show, inherent to co-evolution is the tendency to converge to the strategies that follow the proposed solution concept of mutual rationalizability and not to those of the one-sided rationalizability approach, which does not assume common knowledge of rationality. In fact, this phenomenon is quite intuitive, since that normally the competing populations have no memory of irrational strategies.

On the other hand, the one-sided evolutionary approach of [6] has the tendency to converge to the strategies that are based on the one-sided rationalizability approach, rather than to the strategies that result from mutual rationalizability.

5 Summary and Conclusions

Similar to the studies in [4–6], this paper deals with a rationalizability approach to MOGs. In [4–6] a worst-case consideration is taken, assuming that the opponent applies the most harmful responses, which are not necessarily her best responses. In contrast, here multi-payoff mutual rationalizability is considered. Namely, common knowledge of rationality is assumed, which leads to an iterative procedure for finding the rationalizable strategies of the player. Comparing the two solution concepts and the possibilities of using evolutionary algorithms to solve multi-objective games according to these approaches, two suggestions are made. First, a one-sided evolutionary algorithm, as in [6], appears suitable for finding the strategies according to the one-sided rationalizability approach of [4–6]. Yet, for solving MOGs according to mutual rationalizability, as presented here, a co-evolutionary algorithm is needed.

Studies as in [4–6] and in this paper deviate from two traditional approaches to solving MOGs including Shapley's equilibrium [13] and the Pareto Optimal Security Strategy (POSS) approach [14, 15]. Future studies may include elaborate comparisons between such approaches, and the approach that is proposed here. Finally, it should be noted that the proposed approach and also the approach of [4–6] assume that irrational strategies can be found by set domination. Future publications should aim at providing proofs concerning such assertions.

Acknowledgment. The authors would like to thank E. Solan for referring them to the work of Bernheim [7] and of Pearce [8], and to also thank the anonymous reviewers of this paper.

References

1. Blackwell, D.: An analog of the minimax theorem for vector payoffs. Pac. J. Math. **6**(1), 1–8 (1956)
2. Anand, L., Herath, G.: A survey of solution concepts in multicriteria games. J. Indian Inst. Sci. **75**(2), 141–174 (1995)

3. Nishizaki, I.: Nondominated equilibrium solutions of multiobjective two-person nonzero-sum games in normal and extensive forms. In: Proceedings of Fourth International Workshop on Computational Intelligence & Applications, pp. 13–22 (2008)

4. Eisenstadt, E., Moshaiov, A.: Decision-making in non-cooperative games with conflicting self-objectives. J. Multi-Criteria Decis. Anal. **25**, 1–12 (2018)

5. Eisenstadt, E., Moshaiov, A.: Novel solution approach for multi-objective attack-defense cyber games with unknown utilities of the opponent. IEEE Trans. Emerg. Top. Comput. Intell. **1**, 16–26 (2017)

6. Matalon-Eisenstadt, E., Moshaiov, A., Avigad, G.: The competing travelling salespersons problem under multi-criteria. In: Handl, J., Hart, E., Lewis, P.R., López-Ibáñez, M., Ochoa, G., Paechter, B. (eds.) PPSN 2016. LNCS, vol. 9921, pp. 463–472. Springer, Cham (2016). https://doi.org/10.1007/978-3-319-45823-6_43

7. Bernheim, B.D.: Rationalizable strategic behavior. Econometrica **52**(4), 1007–1028 (1984)

8. Pearce, D.: Rationalizable strategic behavior and the problem of perfection. Econometrica **52**(4), 1029–1050 (1984)

9. Aumann, R.J.: Backward induction and common knowledge of rationality. Games Econ. Behav. **8**(1), 6–19 (1995)

10. Eisenstadt, E., Moshaiov, A., Avigad, G.: Co-evolution of strategies for multi-objective games under postponed objective preferences. In: Proceedings of IEEE Conference Computational Intelligence and Games, pp. 461–468 (2015)

11. Deb, K., Pratap, A., Agarwal, S., Meyarivan, T.: A fast and elitist multiobjective genetic algorithm: NSGA-II. IEEE Trans. Evol. Comput. **6**(2), 182–197 (2002)

12. Żychowski, A., Gupta, A., Mańdziuk, J., Ong, Y.S.: Addressing expensive multi-objective games with postponed preference articulation via memetic co-evolution. Knowl.-Based Syst. **154**, 17–31 (2018)

13. Shapley, L.S.: Equilibrium points in games with vector payoffs. Naval Res. Logistics Q. **6**(1), 57–61 (1959)

14. Ghose, D., Prasad, U.R.: Multicriterion differential games with applications to combat problems. Comput. Math Appl. **18**(1–3), 117–126 (1989)

15. Ghose, D., Prasad, U.R.: Solution concepts in two-person multicriteria games. J. Optim. Theory Appl. **63**(2), 167–189 (1989)

Trend Mining: A Visualization Technique to Discover Variable Trends in the Objective Space

Sunith Bandaru$^{(\boxtimes)}$ and Amos H. C. Ng

Department of Production and Automation Engineering,
School of Engineering Science, University of Skövde, 541 28 Skövde, Sweden
{sunith.bandaru,amos.ng}@his.se

Abstract. Practical multi-objective optimization problems often involve several decision variables that influence the objective space in different ways. All variables may not be equally important in determining the trade-offs of the problem. Decision makers, who are usually only concerned with the objective space, have a hard time identifying such important variables and understanding how the variables impact their decisions and vice versa. Several graphical methods exist in the MCDM literature that can aid decision makers in visualizing and navigating high-dimensional objective spaces. However, visualization methods that can specifically reveal the relationship between decision and objective space have not been developed so far. We address this issue through a novel visualization technique called *trend mining* that enables a decision maker to quickly comprehend the effect of variables on the structure of the objective space and easily discover interesting variable trends. The method uses moving averages with different windows to calculate an *interestingness score* for each variable along predefined reference directions. These scores are presented to the user in the form of an interactive heatmap. We demonstrate the working of the method and its usefulness through a benchmark and two engineering problems.

Keywords: Visualization · Data mining ·
Multi-criteria decision making · Decision space · Trend analysis ·
Objective space

1 Introduction

Multi-objective optimization problems (MOOPs) are generally formulated as,

$$\text{Minimize} \quad \mathbf{f}(\mathbf{x}) = [f_1(\mathbf{x}), f_2(\mathbf{x}), \ldots, f_M(\mathbf{x})]^T$$
$$\text{Subject to} \quad g_j(\mathbf{x}) \geq 0 \ \forall \ j = 1, 2, \ldots, J$$
$$h_k(\mathbf{x}) = 0 \ \forall \ k = 1, 2, \ldots, K \tag{1}$$
$$\mathbf{x}^{(L)} \leq \mathbf{x} \leq \mathbf{x}^{(U)},$$

© Springer Nature Switzerland AG 2019
K. Deb et al. (Eds.): EMO 2019, LNCS 11411, pp. 605–617, 2019.
https://doi.org/10.1007/978-3-030-12598-1_48

where $\mathbf{f}(\mathbf{x})$ is the vector of M conflicting objectives, $g_j(\mathbf{x})$ and $h_k(\mathbf{x})$ represent J inequality and K equality constraints respectively, and $\mathbf{x}$ is a vector of n variables to be optimized within the bounds $\left[\mathbf{x}^{(L)}, \mathbf{x}^{(U)}\right]$. Thus, $\mathbf{f}(\mathbf{x})$ is a mapping from the *decision space* $\mathbb{R}^n$ to the *objective space* $\mathbb{R}^M$. The presence of conflicting objectives means that there exist multiple optimal solutions that provide a trade-off among the objectives. Together, these solutions form the Pareto-optimal set. However, in practice usually only a single optimal solution is desired for implementation and it can only be identified through a higher-level process, known as multi-criteria decision making (MCDM). The decision maker (DM) is a person (or group of persons) that provides the information required in this process.

1.1   A Brief Summary of MCDM Methods

There are four broad classes of MCDM methods [11]: (i) *no preference methods* do not involve a DM and therefore the optimal solution is obtained by making assumptions about a "reasonable" compromise between the objectives, (ii) *a priori methods* in which the DM's preferences are already known and the search involves finding a Pareto-solution that best satisfies those preferences, (iii) *a posteriori methods* where a representative Pareto-optimal set is generated first, which is then analyzed by the DM to choose a solution that best fits his/her preferences, (iv) *interactive methods* use the DM's preferences iteratively to guide candidate solutions towards a desirable region of the Pareto-optimal front.

This paper concerns a posteriori approaches that use multi-objective evolutionary algorithms (MOEAs) to obtain Pareto-optimal solutions. Because of their ability to generate multiple trade-off solutions simultaneously and ease of handling black-box objective functions, nonlinearities and mixed variable types, MOEAs are often the popular choice for solving practical MOOPs. However, as they typically evaluate a lot more candidate solutions than classical a posteriori MCDM techniques, the number of alternatives to be analyzed by the DM can be huge. This is especially true when we consider the fact that DMs may sometimes also be interested in non-Pareto-optimal (dominated) solutions due to hidden or qualitative objectives based on subjective knowledge that are not reflected in the MOOP formulation.

1.2   Decision Making and Decision Space

Traditionally, decision making is seen as an activity involving comparison of objective function values among different solutions. Therefore, DMs have always been associated with the objective space and most MCDM methods are built around this concept. However, in their search for the preferred solutions, DMs may also be interested in knowing various aspects of the decision space that affect the objective space. For example, (i) which variables are important in the preferred regions, (ii) how will a specific variable change when the preferences are changed, (iii) what implicit preferences does the DM impose on the variables with his/her decisions, or more generally, (iv) what makes a solution Pareto-optimal, and (v) which variables define the overall structure of the objective space. Such questions cannot be addressed with a sole view of the objective space. Standard graphical tools like bar charts, scatter plots and box plots can help to some extent, but are cumbersome to use for a large number of variables. On the other hand, it is worth noting that DMs are generally upper-level managers or business heads who may not be inclined to delve into statistical analysis to find answers to such questions.

Based on the discussion above, we argue that when several alternatives are to be analyzed for a MOOP with many objectives and variables, higher quality decisions can be facilitated by providing an intuitive decision support to the DM that can reveal the relationship between decision and objective spaces. To this end, we propose a novel visualization technique, called *trend mining*, that enables a DM to quickly understand the effect of variables on the structure of the objective space and to easily discover interesting variable trends.

The paper is organized as follows. In Sect. 2 we review existing visualization methods from the MCDM literature. The proposed trend mining procedure is laid out in Sect. 3. The working of trend mining is demonstrated in Sect. 4. The paper concludes with a few directions for future development of the technique.

2   Visualization in MCDM

Visualization methods in MCDM can be broadly classified as generic and specific. A generic method is any multivariate visualization tool that originated outside the area of MCDM or EMO. They include basic graphical methods like scatter plots, pie charts and bar plots, histograms, box-whisker plots, violin and bean plots, spider/radar/star/polar plots and parallel coordinate plots. They also include more advanced techniques like biplots, coplots, glyph plots (Chernoff faces, Andrew's curves, etc.), mosaic and spine plots, treemaps, dimensional stacking and radial coordinate visualization. Descriptions of generic methods and their variants can be found in review articles related to visual data mining [4, 9].

Dimensionality reduction and clustering techniques can also be categorized under generic methods of visualization when the corresponding reduced dimensions or clusters are visualized in 2D or 3D. Among dimensionality reduction techniques, principal component analysis, multidimensional scaling, Sammon mapping, neuroscale, isomaps, locally linear embedding, self-organizing maps and generative topographic maps have been applied for visualization in MCDM. Among clustering techniques, k-means, hierarchical and density based techniques have been used. A detailed discussion on these works can be found in [2, 7].

Generic methods do not differentiate between the "variates", meaning that plots can involve only objectives, or only variables, or a combination of them. On the other hand, specific methods recognize this difference as they have been developed within the MCDM/EMO field. They are mainly aimed at visualizing the Pareto-optimal front. Examples include, (i) distance and distribution charts (ii) value paths (iii) star coordinate system (iv) petal diagrams (v) Pareto race (vi) interactive decision maps (vii) Pareto shells (viii) level diagrams (ix) two-stage mapping (x) hyperspace diagonal counting (xi) heatmaps (xii) prosection method (xiii) aggregation trees. These methods are described with schematics and references to original works in [2,7].

Ironically, most methods developed specifically to aid the *decision* maker do not take the *decision* space into account! The reasoning often given is the fact that a DM's preferences are mainly related to objective values, and not decision variables [10]. We argue, especially in practical decision making processes, that the better informed a DM is about the impact of variables on decisions and vice versa, the higher are the confidence and quality of decisions. After all, it is the *decision* variables in terms of which the *decisions* of the DM shall be implemented.

3   Trend Mining Procedure

The main goal of trend mining is to give the DM a quick intuition about the effect of variables on the structure of the objective space and to help easily discover interesting variable trends, irrespective of the number of solutions (N), the number of objectives (M) and the number of variables (n). Given N solutions generated by an MOEA, the proposed procedure involves the following five steps: (i) creation of reference points and reference vectors, (ii) projection of solutions onto reference vectors, (iii) generation of variable trend lines, (iv) calculation of interestingness scores, and (v) heatmap visualization of interestingness scores. These are described in detail in the following subsections.

3.1   Creation of Reference Points and Reference Vectors

Different regions of the objective space are most likely affected in different ways by the variables. Hence, the first step is to define a region of interest. However, asking the DM to provide this information defeats the whole purpose of trend mining, which is decision support. Instead, trend mining uses a set of reference points to represent multiple initial regions of interest. The reference points are created uniformly on a standard $(M - 1)$-simplex[1] using the simplex-lattice design proposed in [5]. This method, described next, is commonly employed in many decomposition-based MOEAs to generate uniformly distributed weight vectors [12].

Simplex-lattice design (SLD) generates reference points $\boldsymbol{\lambda} = [\lambda_1, \lambda_2, \ldots, \lambda_M]^T$ in M dimensions using H equal divisions in $[0, 1]$ along each dimension. Thus, there are $H + 1$ possible values for each λ_i. However, for $\boldsymbol{\lambda}$ to lie on the $(M - 1)$-simplex, we also require that $\sum \lambda_i = 1$ and $\lambda_i \geq 0\ \forall i$. Therefore, there are $(M-1)$ free λ_i values to choose from $(H+1)$ coordinates *with repetition*. The total number of possible reference points is given by $\left(\binom{H+1}{M-1} \right)$, read as "$(H+1)$ multichoose $(M-1)$", which can be simplified as $\binom{H+M-1}{M-1}$. The reference points are generated sequentially in the following steps:

---

[1] A standard $(M - 1)$-simplex has M vertices in $\mathbb{R}^M$, each of which is one unit from the origin along each axis.

1. Set $j \leftarrow 1$.
2. Select $\lambda_1^{(j)}$ from $\left\{0, \frac{1}{H}, \frac{2}{H}, \ldots, 1\right\}$.
3. Select $\lambda_i^{(j)}$ from $\left\{0, \frac{1}{H}, \frac{2}{H}, \ldots, (1 - \sum_{t=1}^{i-1} \lambda_t^{(j)})\right\}$ for $i = 2, \ldots, M - 1$.
4. Set $\lambda_M^{(j)} = 1 - \sum_{t=1}^{M-1} \lambda_t^{(j)}$.
5. If $j < \binom{H+M-1}{M-1}$, then set $j \leftarrow j + 1$ and go to Step 2.

The main limitation of this approach is that the number of points needed to satisfactorily cover the $(M - 1)$-simplex grows rapidly with M. Since decomposition-based MOEAs use population sizes that are in proportion to this number, it becomes impractical to use SLD for $M > 20$. In practice, it is convenient to specify the maximum number of desired reference points, say P, from which H can be calculated such that $\binom{H+M-1}{M-1} \leq P$. In this work, a large value of P does not drastically affect trend mining because the runtime is linear in P.

In order to mimic change of preferences of a DM or simulate navigation through the objective space, we also define P different reference vectors from the origin to each of the reference points created above. These reference vectors will serve as representative *decision paths* along which the variable trends will be identified. Since they emanate from the origin, the set of reference vectors can be denoted as $\{\boldsymbol{\lambda}^{(1)}, \boldsymbol{\lambda}^{(2)}, \ldots, \boldsymbol{\lambda}^{(P)}\}$.

3.2 Projection of Solutions onto Reference Vectors

Given a set of N solutions evaluated by an MOEA, the next step is to normalize each objective in $[0, 1]$ to account for differences in their magnitudes. This is easily accomplished with, $f_i^{nr} = (f_i - f_i^{min})/(f_i^{max} - f_i^{min}) \ \forall i$, where f_i^{min} and f_i^{max} are, respectively, the minimum and maximum values of f_i among the N solutions. The *ideal point* of the solutions now lies at $[0, 0, \ldots, 0]^T$. Therefore, the reference vectors created above can be used for navigating the normalized objective space, $\mathbf{f}^{nr}$.

For identifying variable trends along each reference vector, the solutions also need to be totally ordered in the objective space. This can be done by projecting all solutions onto the reference vector and ordering them by their distance from the ideal point. In other words, the solutions are ordered by their distance from the ideal point, measured along the reference vector under consideration. For a given solution $\mathbf{x}^{(t)}$ and reference vector $\boldsymbol{\lambda}^{(j)}$, this distance is given by $d_{tj} = \frac{\boldsymbol{\lambda}^{(j)T} \mathbf{f}^{nr}(\mathbf{x}^{(t)})}{\|\boldsymbol{\lambda}^{(j)}\|}$. The solutions can be ordered either in ascending or descending order of this distance. We choose to sort them in ascending order of d_{tj}.

3.3 Generation of Variable Trend Lines

Let D_j be the ordered set of distances d_{tj} along reference vector $\boldsymbol{\lambda}^{(j)}$, and I_j be the corresponding ordered set of original indices of the solutions. Then I_{jk} represents the original index of the k-th closest solution to the ideal point along $\boldsymbol{\lambda}^{(j)}$. The trend of a variable x_i along a reference vector $\boldsymbol{\lambda}^{(j)}$ can be interpreted as the variation in x_i values of solutions as one moves from the ideal point along the vector. This trend can be shown graphically as a line plot of $x_i^{(I_{jk})}$ versus k for $k = 1, 2, \ldots, N$. We refer to this line plot as the *trend line* $x_i^{(I_j)}$. Trend mining involves analyzing these trend lines for *interesting* features. The definition of interestingness used in this work is presented in the next section. Note that, the sequence of values $x_i^{(I_{jk})}$ $\forall k$ can be treated as a time series, and methods developed within the field of time series analysis can be used to extract interesting patterns. We choose a much simpler approach in Sect. 3.4.

Illustration. Figure 1 illustrates the procedure for generating a variable trend line on a MOOP with $M = 2$ objectives and $n = 2$ variables. Given N solutions to the MOOP that need to be analyzed by a DM, the first step is to choose the number of reference points. Here, we choose $P = 7$ points and create the corresponding reference vectors. Next, the objective values are normalized to $[0, 1]$ for all objectives, so that the ideal point is at $(0, 0)$. Let's say we want to generate the variable trend for x_2 along $\boldsymbol{\lambda}^{(3)}$. All N solutions are first projected onto $\boldsymbol{\lambda}^{(3)}$. For clarity, Fig. 1 shows this process for five solutions. Next, the distances d_{t3}, measured along $\boldsymbol{\lambda}^{(3)}$ from the ideal point to each projected solution, are calculated. For example, the figure shows that d_{43} is this distance for solution $\mathbf{x}^{(4)}$. The calculated distances are then sorted in ascending order, which gives $D_3 = \{d_{23}, d_{43}, d_{13}, d_{33}, d_{53}\}$, and the corresponding order of original solution indices, $I_3 = \{2, 4, 1, 3, 5\}$. The trend of variable x_2 along $\boldsymbol{\lambda}^{(3)}$ is denoted by $x_2^{(I_3)}$, where I_3 defines the order in which solutions should appear on the trend line. Thus, the k-th solution on the trend line should be I_{3k} (k-th element of I_3). The corresponding values of variable x_2, denoted as $x_2^{(I_{3k})}$, are plotted on the Y-axis.

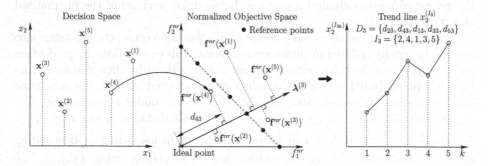

Fig. 1. Illustration of procedure for generating a variable trend line on a MOOP with $M = 2$ objectives and $n = 2$ variables.

Algorithm 1. Calculation of Interestingness Score, S_{ij}

Input: Variable trend line $x_i^{(I_{jk})}$ for $k = 1, 2, \ldots, N$
Output: S_{ij}

1:   $S_{ij} \leftarrow 0$
2:   **for** $s \leftarrow 1$ to $\lfloor \log_2 N \rfloor$ **do**
3:      $WindowSize \leftarrow \lfloor N/2^s \rfloor$
4:      $y_k \; \forall k \leftarrow \texttt{SimpleMovingAverage}\left(x_i^{(I_{jk})} \; \forall k, WindowSize\right)$
5:      $UpTicks \leftarrow 0, \; DownTicks \leftarrow 0$
6:      **for** $k \leftarrow 1$ to $N-1$ **do**
7:         **if** $y_{k+1} > y_k$ **then** $UpTicks \leftarrow UpTicks + 1$
8:         **if** $y_{k+1} < y_k$ **then** $DownTicks \leftarrow DownTicks + 1$
9:      $S_{ij} \leftarrow S_{ij} + |UpTicks - DownTicks| \times 100/(N-1)$
10: $S_{ij} \leftarrow S_{ij}/\lfloor \log_2 N \rfloor$ %% Average over $\lfloor \log_2 N \rfloor$ window sizes

---

3.4   Calculation of Interestingness Scores

The total number of trend lines is $n \times P$, n being the number of variables and P being the number of reference vectors. In real-world MOOPs with high n and M (hence P), it can be difficult for a DM to visually analyze the trends. Moreover, without a way to quantify the strength of a variable trend, any interpretation about the effect of that variable is prone to subjectivity. We therefore define a simple metric called the *interestingness score* that can be calculated for each pair of variable and reference vector.

In this paper, we consider monotonically increasing or decreasing values of variables (along the reference vectors) as characteristic features that can aid in decision making. The interestingness score, S_{ij}, measures how close the variable trend $x_i^{(I_j)}$ is to being monotonic. It is defined as a function of the number of $UpTicks$ and $DownTicks$ in the trend line. An uptick occurs every time consecutive points on the trend line show an increase in the variable value, and a downtick occurs when they show a decrease. For example, the trend line generated in Fig. 1 has $UpTicks = 3$ and $DownTicks = 1$. A higher value of the ratio $|UpTicks - DownTicks|/(N-1)$ means that the trend is closer to being monotonic. In order to account for possible non-uniformity of solutions in the objective space and the corresponding fluctuations in the trend line, S_{ij} is obtained by aggregating $|UpTicks - DownTicks|/(N-1)$ over several copies of the trend line at different levels of smoothing. Any appropriate smoothing method may be used. In this paper, we use simple moving average with a range of window sizes chosen such that the number of non-overlapping windows increases in powers of two. Thus, the largest window size will have 2^1 non-overlapping windows and the smallest window size will have $2^{\lfloor \log_2 N \rfloor}$ non-overlapping windows, which corresponds to no smoothing. The complete pseudocode is shown in Algorithm 1. Note that S_{ij} is expressed as a percentage of monotonicity.

3.5   Heatmap Visualization of Interestingness Scores

The interestingness scores are presented to the DM in the form of an interactive heatmap grid of size $n \times P$. The use of colors makes it easy to identify pairs of variables and reference vectors with high interestingness scores when n or P are high. The DM can choose to investigate a specific $\{x_i, \boldsymbol{\lambda}^{(j)}\}$ pair by clicking the color cell (i, j), which shows the corresponding trend line $x_i^{(I_j)}$. A smoothed version of $x_i^{(I_j)}$, with $WindowSize$ that maximizes $|UpTicks - DownTicks|$, is overlaid on the original trend line to highlight the monotonicity of x_i. For MOOPs with $M = 2$ or 3, a scatter plot of the objective space is also displayed with the solutions color-mapped to values of x_i. The corresponding reference vector, $\boldsymbol{\lambda}^{(j)}$, can also be shown, if desired. For $M > 3$, any of the visualization techniques discussed in Sect. 2 may be used.

4   Results and Discussion

We now demonstrate the working of trend mining on three MOOPs, namely (i) WFG2, (ii) Clutch-brake design problem (CLUTCH), and (iii) Flexible machining cell design problem (FMC). WFG2 is one of the nine scalable test problems proposed in [8]. CLUTCH is a mechanical design problem that has analytical formulation and FMC is a real-world stochastic simulation based MOOP. These problems have been specifically chosen because they have interesting structural features in the objective space that are not directly apparent from their problem formulations. We use $P = 10$ reference vectors for all problems.

WFG2. The WFG2 problem with $M = 3$ and $n = 12$ is solved using NSGA-III [6] with a population of 100 and maximum number of evaluations set to 10,000. All other parameters take recommended values.

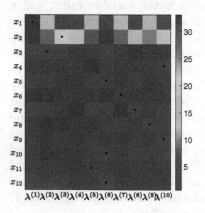

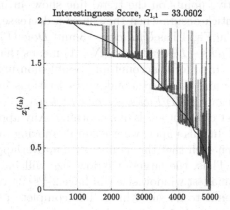

Fig. 2. Heatmap of interestingness scores for WFG2.

Fig. 3. Trend line $x_1^{(I_1)}$ for WFG2 along with smoothed version (black line). (Color figure online)

Trend mining is applied to the evaluated solutions. Figure 2 shows the obtained heatmap. For each variable, the black dots indicate the reference vector along which interestingness is maximum. The figure shows that variables x_1 and x_2 have relatively high interestingness scores along $\lambda^{(1)}$ and $\lambda^{(3)}$, respectively, while all other variables have similar low scores in all directions. Selecting cell (1,1) generates the trend line $x_1^{(I_1)}$, shown in Fig. 3, and the scatter plot of the objective space in Fig. 4 where the solutions are color-mapped to x_1. The figures together show how x_1 influences the structure of the objective space.

Clutch-Break Design Problem (CLUTCH). The problem involves two objectives for minimizing the system mass (f_1) and the stopping time (f_2) and five discrete variables namely, (i) x_1: inner disk radius, (ii) x_2: outer disk radius, (iii) x_3: disk thickness, (iv) x_4: actuating force, and (v) x_5: number of disks. The complete problem formulation can be found in [1]. The problem is solved using NSGA-II with recommended parameter settings.

Trend mining generates the heatmap shown in Fig. 5 which indicates that, due to its high interestingness score, the most important variable with respect to the structure of the objective space is x_5. Selecting cell (5, 2) generates the trend line $x_5^{(I_2)}$ and the scatter plot of the objective space. The latter, shown in Fig. 6, reveals that indeed x_5 defines various regions of the objectives space.

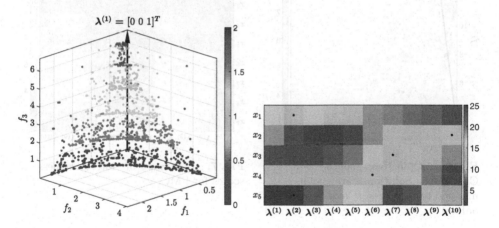

Fig. 4. Objective space for WFG2. Solutions are color-mapped to x_1 values. (Color figure online)

Fig. 5. Heatmap of interestingness scores for CLUTCH.

Flexible Machining Cell Design Problem (FMC). This problem from production engineering domain concerns multi-objective optimization of a flexible machining cell using a stochastic discrete event simulation model. The cell involves two workstations, the first of which performs operation $Op1$ and the second performs one of three operations $Op2A$, $Op2B$ and $Op2C$ depending on the product variant being machined. There are a total of nine variables, (i) x_1 and x_2 are buffer levels available for the two workstations that take integer values in $[5, 10]$, (ii) x_3 is a categorical variable which decides whether to use two slow machines ($x_3 = 1$) or one fast machine ($x_3 = 2$) for $Op1$, (iii) x_9 is also a categorical variable which decides whether to use dedicated machines ($x_9 = 1$) or flexible machines ($x_9 = 2$) for $Op2A$, $Op2B$ and $Op2C$, (iv) x_4, x_5, x_6, x_7 and x_8 are the availabilities (in percent) of the two slow and the three dedicated machines. They can each take one of the values in $\{90, 91, 92, 93, 94, 95\}$. The objectives are, (i) minimize the total investment cost (f_1), (ii) maximize the throughput of the cell (f_2), and (iii) minimize the total number of buffers used (f_3). The investment cost is calculated as: $f_1(\mathbf{x}) = \sum_{i=1}^{2} 10000(x_i - 5) + 100000(x_3 - 1) + \sum_{i=4}^{8} 10000(x_i - 5) + 90000(x_9 - 1)$. The throughput is obtained using discrete event simulation. The total number of buffers is simply $f_3(\mathbf{x}) = x_1 + x_2$. The complete description of the problem and the simulation model can be found in [3].

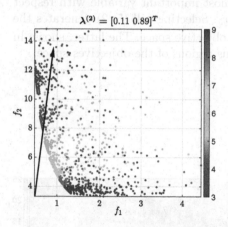

Fig. 6. Objective space for CLUTCH. Solutions are color-mapped to x_5 values. (Color figure online)

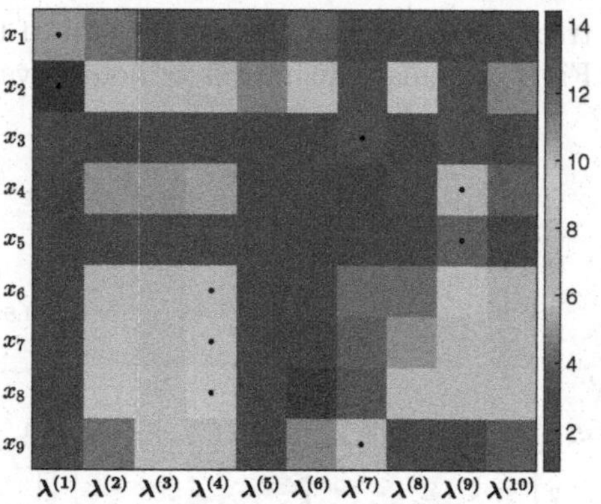

Fig. 7. Heatmap of interestingness scores for FMC.

Trend mining generates the heatmap shown in Fig. 7. According to the figure, trend line $x_2^{(I_1)}$ has the highest interestingness. Selecting cell (2,1), generates the trend line and scatter plot of the objective space as shown in Figs. 8 and 9, respectively. The objective space has different "layers" for different values of f_3. Even though, it is known that $f_3 = x_1 + x_2$, trend mining shows us that these layers in the objective space are primarily caused by the variation of x_2.

According to the heatmap, variables x_6, x_7 and x_8 have similar interestingness scores along $\boldsymbol{\lambda}^{(4)}$. This indicates that the three variables have similar effect on the objective space. Selecting cell (8, 4) generates Fig. 10 which shows that the

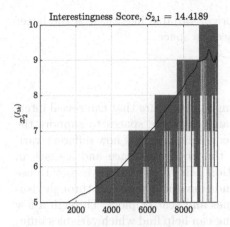

Fig. 8. Trend line $x_2^{(I_1)}$ for FMC along with smoothed version (black line). (Color figure online)

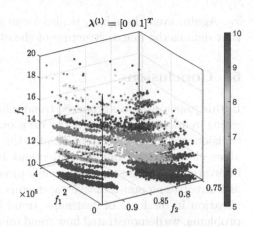

Fig. 9. Objective space for FMC. Solutions are color-mapped to x_2 values. (Color figure online)

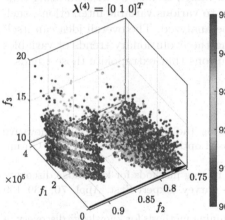

Fig. 10. Objective space for FMC. Solutions are color-mapped to x_8 values. (Color figure online)

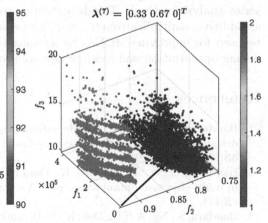

Fig. 11. Objective space for FMC. Solutions are color-mapped to x_9 values. (Color figure online)

throughput (f_2) is highly dependent on x_8. The same is true for x_6 and x_7 whose color-mapped values in the objective space look very similar to Fig. 10. Though f_2 is a function of all variables, trend mining reveals that x_6, x_7 and x_8 are more important than others for decision making. Note that these variables are of the same nature as x_4 and x_5, yet somehow x_4 and x_5 don't influence the objective space as much as x_6, x_7 and x_8.

According to the heatmap, variable x_9 has the next highest interestingness score. Selecting cell $(9, 7)$ gives Fig. 11. Here, we see that the two layered regions in the objective space separated roughly by $f_2 \approx 0.85$, are caused by variable

x_9. Again, even though x_3 is also a categorical variable like x_9, it is the latter that defines the macro-structure of the objective space.

5 Conclusions

In this paper, we have proposed trend mining, a procedure that can reveal interesting relationships between the decision and objective spaces to support the decision making process. Specifically, the technique analyzes how different variables change along different directions in the objective space and assigns an interestingness score to each variable-direction pair. Trend mining is easy to use as it involves no non-intuitive parameters and conveys information through visualization in the form of heatmaps, trend lines and scatter plots. Through a few problems, we demonstrated how trend mining can help find which variables influence the structure of the objective space and in what way. Trend mining can be extended in many ways. For example, in this paper we only looked at increasing or decreasing trends. The technique can be enhanced to recognize other interesting trends, such as cyclic, step, constant or mixed trends, by incorporating time series analysis methods. Trends pertaining to various variable interactions, such as additive, multiplicative etc., can also be analyzed. The overall idea can itself be used for improving MOEAs by learning the evolutionary trends of variables during optimization and creating new solutions that extrapolate those trends.

References

1. Bandaru, S., Deb, K.: Towards automating the discovery of certain innovative design principles through a clustering-based optimization technique. Eng. Optim. **43**(9), 911–941 (2011)
2. Bandaru, S., Ng, A.H.C., Deb, K.: Data mining methods for knowledge discovery in multi-objective optimization: part A - survey. Expert Syst. Appl. **70**, 139–159 (2017)
3. Bandaru, S., Ng, A.H.C., Deb, K.: Data mining methods for knowledge discovery in multi-objective optimization: part B - new developments and applications. Expert Syst. Appl. **70**, 119–138 (2017)
4. Chan, W.W.Y.: A survey on multivariate data visualization. Sci. Technol. **8**(6), 1–29 (2006)
5. Das, I., Dennis, J.E.: Normal-boundary intersection: a new method for generating the Pareto surface in nonlinear multicriteria optimization problems. SIAM J. Optim. **8**(3), 631–657 (1998)
6. Deb, K., Jain, H.: An evolutionary many-objective optimization algorithm using reference-point based non-dominated sorting approach, part I: solving problems with box constraints. IEEE Trans. Evol. Comput. **18**(4), 577–601 (2014)
7. Filipivc, B., Tusar, T.: Visualization in multiobjective optimization. In: Proceedings of the Genetic and Evolutionary Computation Conference Companion, pp. 858–879. ACM (2018)
8. Huband, S., Hingston, P., Barone, L., While, L.: A review of multiobjective test problems and a scalable test problem toolkit. IEEE Trans. Evol. Comput. **10**(5), 477–506 (2006)

9. Keim, D.: Information visualization and visual data mining. IEEE Trans. Vis. Comput. Graph. **8**(1), 1–8 (2002)
10. Lotov, A.V., Miettinen, K.: Visualizing the Pareto frontier. In: Branke, J., Deb, K., Miettinen, K., Słowiński, R. (eds.) Multiobjective Optimization. LNCS, vol. 5252, pp. 213–243. Springer, Heidelberg (2008). https://doi.org/10.1007/978-3-540-88908-3_9
11. Miettinen, K.: Nonlinear Multiobjective Optimization, vol. 12. Springer, Berlin (2012)
12. Trivedi, A., Srinivasan, D., Sanyal, K., Ghosh, A.: A survey of multiobjective evolutionary algorithms based on decomposition. IEEE Trans. Evol. Comput. **21**(3), 440–462 (2017)

IRA-EMO: Interactive Method Using Reservation and Aspiration Levels for Evolutionary Multiobjective Optimization

Rubén Saborido[1(✉)], Ana B. Ruiz[2], Mariano Luque[2], and Kaisa Miettinen[3]

[1] Department of Computer Science & Software Engineering, Concordia University, 1455 De Maisonneuve Blvd West, Montréal H3G 1M8, Canada
`ruben.saborido-infantes@polymtl.ca`
[2] Department of Applied Economics (Mathematics), Universidad de Málaga, C/Ejido 6, 29071 Málaga, Spain
`{abruiz,mluque}@uma.es`
[3] Faculty of Information Technology, University of Jyvaskyla, P.O. Box 35 (Agora), 40014 University of Jyvaskyla, Finland
`kaisa.miettinen@jyu.fi`

Abstract. We propose a new interactive evolutionary multiobjective optimization method, IRA-EMO. At each iteration, the decision maker (DM) expresses her/his preferences as an interesting interval for objective function values. The DM also specifies the number of representative Pareto optimal solutions in these intervals referred to as regions of interest one wants to study. Finally, a real-life engineering three-objective optimization problem is used to demonstrate how IRA-EMO works in practice for finding the most preferred solution.

Keywords: Evolutionary multi-objective optimization ·
Reference point · Region of interest · Interactive methods ·
Preferences

1 Introduction

Many real-world problems involve dealing with several conflicting criteria, which must be optimized simultaneously. These problems, called *multiobjective optimization problems*, are defined by objective functions which model the criteria, and by constraints and bounds for variables which define the feasible set. In order to solve multiobjective optimization problems and to decide which solution is the final one, a person, called decision maker (DM), is usually involved in the solution process in order to choose the solution which best suits her/his preferences (*the most preferred solution*).

Recently, *Evolutionary Multiobjective Optimization* (EMO) methods that include preferences and interactive EMO algorithms have received attention due

© Springer Nature Switzerland AG 2019
K. Deb et al. (Eds.): EMO 2019, LNCS 11411, pp. 618–630, 2019.
https://doi.org/10.1007/978-3-030-12598-1_49

to the reduction of the computational load and applicability [1]. Within interactive algorithms, the elicitation of preferences can be done in different ways [12]. Many methods use a so-called *reference point*, which is formed by desirable aspiration levels for the objective functions that the DM would like to reach. A key feature in interactive methods is that only one or few solutions are shown at each iteration in order not to overwhelm the DM with too much information.

We have earlier proposed an interactive version of the preference-based EMO algorithm WASF-GA [15], called Interactive WASF-GA [13], where the DM expresses preferences as aspiration levels (i.e. a reference point) and the number of solutions one wants to see. Then, the algorithm generates exactly the number of nondominated solutions the DM desires, reflecting the preferences in the reference point given.

In this paper, we propose a new interactive EMO algorithm called *Interactive Reservation and Aspiration points-based EMO (IRA-EMO)* method, which uses two kinds of reference points to generate nondominated solutions. In addition to aspiration levels, also reservation levels given by the DM are used. For a minimization problem, reservation levels are values above which the objective function values are not admissible. Thus, we consider preferences expressed as lower and upper bounds for the objective functions defining a region of interest in the Pareto optimal front and generate solutions within it. It is important to properly represent all the possible trade-offs among the objectives in the region of interest to let the DM have an idea of which nondominated solutions can be achieved based on preferences. To this aim, a variant of WASF-GA called *Modified WASF-GA* is proposed to be used within IRA-EMO to approximate the region of interest.

The rest of the paper is organized as follows. In Sect. 2, we introduce the main concepts and notations used, including a brief overview of interactive EMO algorithms. In Sect. 3, the IRA-EMO method and Modified WASF-GA are described. The usefulness of IRA-EMO in practice is demonstrated in Sect. 4, with a real-world problem. Finally, conclusions are drawn in Sect. 5.

2   Background

We consider *multiobjective optimization problems* of the form:

$$\text{minimize} \quad \{f_1(\mathbf{x}), \ldots, f_k(\mathbf{x})\}$$
$$\text{subject to} \quad \mathbf{x} \in S, \tag{1}$$

where $f_i : S \rightarrow \mathbf{R}$, for $i = 1, \ldots, k$ $(k \geq 2)$, are *objective functions* to be minimized simultaneously over the *feasible set* S in the decision space $\mathbf{R}^n$, which is formed by *solutions* or *decision vectors* $\mathbf{x} = (x_1, \ldots, x_n)^T$. In the objective space $\mathbf{R}^k$, the solutions are *objective vectors* $\mathbf{f}(\mathbf{x}) = (f_1(\mathbf{x}), \ldots, f_k(\mathbf{x}))^T$, for $\mathbf{x} \in S$, belonging to the *feasible objective region* $Z = \mathbf{f}(S)$.

Because of the degree of conflict among the objective functions, it is very unlikely to find a single solution where all of them can reach their individual optima. Therefore, we consider so-called *Pareto optimal solutions*, at which no objective function can be improved without deteriorating, at least, one of the others. A solution $\mathbf{x} \in S$ (and its objective vector $\mathbf{f}(\mathbf{x})$) is said to be *Pareto optimal* if there is no other $\mathbf{x}' \in S$ such that $f_i(\mathbf{x}') \leq f_i(\mathbf{x})$ for all $i = 1, \ldots, k$ and $f_j(\mathbf{x}') < f_j(\mathbf{x})$ for, at least, one index j. The set of all Pareto optimal solutions is called the *Pareto optimal set E*, and its image in the objective space is referred to as the *Pareto optimal front* $\mathbf{f}(E)$. The *nadir objective vector* $\mathbf{z}^{\mathrm{nad}} = (z_1^{\mathrm{nad}}, \ldots, z_k^{\mathrm{nad}})^T$ and the *ideal objective vector* $\mathbf{z}^\star = (z_1^\star, \ldots, z_k^\star)^T$ provide upper and lower bounds for the objective functions in E, respectively. Their components are $z_i^{\mathrm{nad}} = \max_{\mathbf{x} \in E} f_i(\mathbf{x})$ and $z_i^\star = \min_{\mathbf{x} \in E} f_i(\mathbf{x})$ $(i = 1, \ldots, k)$. While $\mathbf{z}^\star$ can be easily obtained, $\mathbf{z}^{\mathrm{nad}}$ can usually only be approximated [3].

Many preference-based EMO algorithms and interactive methods are based on the use of reference points [18]. A *reference point* is a vector $\mathbf{q} = (q_1, \ldots, q_k)^T$ consisting of desirable objective function values q_i for the DM (*aspiration levels*). We say that $\mathbf{q}$ is *achievable* if $\mathbf{q} \in Z + \mathbf{R}_+^k$ (where $\mathbf{R}_+^k = \{\mathbf{y} \in \mathbf{R}^k \mid y_i \geq 0$ for $i = 1, \ldots, k\}$), that is, if either $\mathbf{q} \in Z$ or if $\mathbf{q}$ is dominated by a Pareto optimal objective vector; otherwise, $\mathbf{q}$ is said to be *unachievable*. Using a reference point, an *achievement scalarizing function* (ASF) can be built and minimized over the feasible set to find the Pareto optimal solution that best satisfies the DM's expectations. For a reference point $\mathbf{q}$ and a vector of weights $\boldsymbol{\mu} = (\mu_1, \ldots, \mu_k)^T$, with $\mu_i > 0$ $(i = 1, \ldots, k)$, we consider the following ASF proposed in [18]:

$$s(\mathbf{q}, \mathbf{f}(\mathbf{x}), \boldsymbol{\mu}) = \max_{i=1,\ldots,k} \left\{ \mu_i (f_i(\mathbf{x}) - q_i) \right\} + \rho \sum_{i=1}^{k} \mu_i (f_i(\mathbf{x}) - q_i). \tag{2}$$

The parameter ρ has a real positive value which ensures that the solution which minimizes (2) over S is a Pareto optimal solution to the original problem (1). Actually, any Pareto optimal solution of (1) can be obtained by minimizing (2) over S and varying the reference point and/or the weight vector [12].

We use two types of reference points: in addition to those consisting of aspiration levels, we also use reference points formed by *reservation levels* acceptable for the DM, i.e. values above which the objective functions are not admissible.

As explained later, our proposal borrows some ideas from (a) the preference-based EMO algorithm WASF-GA [15], which approximates a *region of interest of the Pareto optimal front defined by a reference point* $\mathbf{q}$ as defined in [15], and from (b) the EMO algorithm Global WASF-GA [16], which approximates the whole Pareto optimal front by using both a utopian (a vector slightly better than the ideal objective vector) and a nadir objective vector as reference points.

In the literature, many preference-based and interactive EMO methods have been proposed. In R-NSGA-II [5], the DM gives one or several reference points and the crowding distance used in NSGA-II [4] is replaced by a preference distance. PBEA [17] considers a reference point to modify the binary quality indicator of IBEA [20]. In [8], an interactive version of MOEA/D [19] is suggested, where a DM selects one among a set of solutions shown at intermediate generations. The algorithm with a controllable accuracy proposed in [10] is also based on reference points. Regarding the use of both aspiration and reservation levels in the EMO field, to the best of our knowledge, few references can be found. In [7], a preference-based EMO method is proposed for the selection of direct load control actions in electrical distribution networks, in which preferences are elicited in a similar way. The nondominated solutions violating some reservation level are penalized and those closer to the aspiration levels according to the Euclidean distance are rewarded. Furthermore, [9] suggests an interactive EMO method based on RVEA [2], where the DM can specify, if desired, preferred ranges for the objectives (i.e. aspiration and reservation levels), which are used to adjust the set of reference vectors in RVEA. These two methods follow a solution process which is different from our proposal, as described next.

3   IRA-EMO for Decision Making

In this section, we describe the *Interactive Reservation and Aspiration points-based EMO* (IRA-EMO) method. At each iteration it of IRA-EMO, the DM indicates her/his preferences by specifying desirable bounds for the objective functions in the form of aspiration and reservation levels, denoted by $q_i^{a,it}$ and $q_i^{r,it}$, respectively, with $z_i^\star \le q_i^{a,it} < q_i^{r,it} \le z_i^{\mathrm{nad}}$, for $i = 1, \ldots, k$. Thus, an aspiration point $\mathbf{q}^{a,it} = (q_1^{a,it}, \ldots, q_k^{a,it})^T$ and a reservation point $\mathbf{q}^{r,it} = (q_1^{r,it}, \ldots, q_k^{r,it})^T$ can be formed. The DM also sets the number of solutions (s)he wants to analyze at each iteration, denoted by N_S^{it}. Then, N_S^{it} nondominated solutions are shown to the DM, whose objective values are between the levels.

For an aspiration and a reservation points, $\mathbf{q}^a$ and $\mathbf{q}^r$, respectively, with $z_i^\star \le q_i^a < q_i^r \le z_i^{\mathrm{nad}}$ for $i = 1, \ldots, k$, we denote by R^a and R^r the regions of interest they define, respectively. For this preference information, the DM is interested in Pareto optimal solutions which are in $R = R^a \cap R^r$. Figure 1 represents examples of different situations for a biobjective minimization problem, where the subset R is highlighted in bold. We have assumed that $\mathbf{q}^a$ is unachievable and $\mathbf{q}^r$ is achievable, which is the most logical situation when the DM gives both points. In case $\mathbf{q}^a$ and $\mathbf{q}^r$ are both unachievable, $R = R^a \cap R^r = R^r$, and if both are achievable, $R = R^a \cap R^r = R^a$.

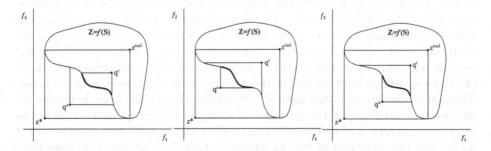

Fig. 1. Region of interest R when $\mathbf{q}^a$ is unachievable and $\mathbf{q}^r$ is achievable.

The ideal and the nadir objective vectors can be estimated if needed, and the aspiration and the reservation levels are supposed to satisfy $z_i^* \leq q_i^a < q_i^r \leq z_i^{\text{nad}}$ for $i = 1, \ldots, k$. It may occur that $\mathbf{z}^*$ does not dominate $\mathbf{q}^a$ and/or $\mathbf{z}^{\text{nad}}$ is not dominated by $\mathbf{q}^r$. If $\mathbf{q}^a$ dominates $\mathbf{z}^*$ and $\mathbf{q}^r$ dominates $\mathbf{z}^{\text{nad}}$, we have $R = R^r$. In case $\mathbf{z}^{\text{nad}}$ dominates $\mathbf{q}^r$ and $\mathbf{z}^*$ dominates $\mathbf{q}^a$, then $R = R^a$. If $\mathbf{q}^a$ dominates $\mathbf{z}^*$ and $\mathbf{z}^{\text{nad}}$ dominates $\mathbf{q}^r$, then R is the whole Pareto optimal front. When $\mathbf{q}^a$ and $\mathbf{z}^*$ do not dominate each other, and also $\mathbf{q}^r$ and $\mathbf{z}^{\text{nad}}$ do not dominate each other, R is constituted by a part of the Pareto optimal front.

We want to emphasize that IRA-EMO does not depend on the availability of estimations for $\mathbf{z}^*$ and $\mathbf{z}^{\text{nad}}$. If no estimations are known, in practice, the DM may need to take a few iterations of IRA-EMO at the beginning of the solution process just to gain an idea of the possible objective function ranges and to fine-tune her/his preferences. Also, $\mathbf{z}^*$ and $\mathbf{z}^{\text{nad}}$ may be used for normalizing the objective function values in (2) if they are in different scales. In case they are not available, any other normalization approach can be used.

To generate nondominated solutions in the subset R defined by the aspiration and the reservation points at each iteration it, we propose *Modified WASF-GA* as an internal part of IRA-EMO. It is based on the working procedure of Global WASF-GA for using two reference points, instead of only one as in WASF-GA.

3.1   Modified WASF-GA

Let us consider an aspiration and a reservation points, $\mathbf{q}^a$ and $\mathbf{q}^r$, and N_μ vectors of weights representing the weight vector space $(0, 1)^k$. Let us denote by N the population size, P^{final} the final set of nondominated solutions generated by Modified WASF-GA and h the generation counter. At each generation h, P^h is the population of individuals, Q^h is the offspring population, Z^h is the population of parents and offspring, and F_n^h is the n-th front. The number of elements in a set A is denoted by $\#(A)$.

At each generation h, Modified WASF-GA selects solutions which best match with $\mathbf{q}^a$ and $\mathbf{q}^r$. The population of parents and offspring Z^h is divided into several fronts according to the values they take on the ASF (2) for both $\mathbf{q}^a$ and $\mathbf{q}^r$ at the same time. The lower the values of (2) reached by a solution for one of these two reference points, the more this solution is highlighted. To be more precise, once Z^h is formed (of size $2N$), the division of the individuals into different fronts is performed as follows. The first front is formed by the solutions in Z^h with the lowest values of (2) for $\mathbf{q}^a$ taking into account a half of the N_μ weight vectors (the odd order ones), and by the solutions in Z^h with the lowest values of (2) for $\mathbf{q}^r$ using the other half of the N_μ weight vectors (the even order ones). These solutions are removed from Z^h. Similarly, the second front is formed by the solutions in Z^h with the next lowest values of (2) for $\mathbf{q}^a$ and a half of the N_μ weight vectors, and by the solutions in Z^h with the next lowest values of (2) for $\mathbf{q}^r$ and the other half of the N_μ weight vectors. This process continues until every individual in Z^h has been classified. The set of nondominated solutions generated by Modified WASF-GA, P_{final}, consists of the N_μ solutions in the first front of the last generation. So far, these solutions are the best ones with respect to the weight vectors and the aspiration and the reservation points used. Note that P_{final} approximates the regions of interest defined by $\mathbf{q}^a$ and $\mathbf{q}^r$, that is, $R^a \cup R^r$. Since the DM wants to see solutions in $R = R^a \cap R^r$ at each iteration of IRA-EMO, P_{final} is later filtered to select the solutions belonging to this set.

Modified WASF-GA minimizes, at each generation, the ASF (2) for both the aspiration and the reservation points. Therefore, in practice, Modified WASF-GA projects the aspiration and the reservation points onto the Pareto optimal front simultaneously, using the set of projection directions defined by the weight vectors. Thus, it is important that the N_μ weight vectors used define a well-spread set of projection directions in order to preserve diversity.

3.2   Algorithm of IRA-EMO

At each iteration it with the DM, we denote by $\boldsymbol{\mu}^{it,j}$ the weight vectors used in Modified WASF-GA ($j = 1, \ldots, N_\mu$), $\bar{P}^{it}$ the outcome of Modified WASF-GA, and P^{it} the subset of solutions of $\bar{P}^{it}$ whose objective function values are within the given aspiration and reservation levels. The main steps of IRA-EMO are:

Step 1. Initialization. Set $it = 1$. Show $\mathbf{z}^{nad}$ and $\mathbf{z}^\star$ to the DM (if available). Ask the DM how many solutions (s)he would like to see, denoted by N_S^{it}.

Step 2. Preference information I. If $it = 1$, ask the DM to specify aspiration and reservation levels for the objective functions, which define $\mathbf{q}^{a,it}$ and $\mathbf{q}^{r,it}$, respectively. If $it > 1$, $\mathbf{q}^{a,it}$ and $\mathbf{q}^{r,it}$ are set according to the preference information the DM wants to give as follows:

- The DM is asked if the current reservation point is to be updated. If so, the DM specifies a new reservation point, $\mathbf{q}^{r,it}$. If not, let $\mathbf{q}^{r,it} = \mathbf{q}^{r,it-1}$.
- The DM is asked if the current aspiration point is to be updated. If so, the DM specifies a new aspiration point, $\mathbf{q}^{a,it}$. If not, let $\mathbf{q}^{a,it} = \mathbf{q}^{a,it-1}$.

Step 3. Preference information II. Ask if the DM wants to change the number of solutions to be obtained, update N_S^{it} accordingly. Next, define $N_\mu = 2N_S^{it}$. If $it > 1$ and $N_S^{it} = N_S^{it-1}$, set $\boldsymbol{\mu}^{it,j} = \boldsymbol{\mu}^{it-1,j}$ for all $j = 1, \ldots, N_\mu$ and go to Step 5. Otherwise, continue.

Step 4. Generation of the weight vectors. Following the procedure described in [15], generate N_μ weight vectors $\boldsymbol{\mu}^{it,j}$, with $j = 1, \ldots, N_\mu$.

Step 5. Generation of solutions. Generate N_μ nondominated solutions by applying Modified WASF-GA using $\mathbf{q}^{a,it}$, $\mathbf{q}^{r,it}$, and the set of weight vectors $\boldsymbol{\mu}^{it,j}$, with $j = 1, \ldots, N_\mu$. Let $\bar{P}^{it}$ be the set formed by these N_μ solutions.

Step 6. Set $P^{it} = \{\mathbf{x} \in \bar{P}^{it} | q_i^{a,it} \leq f_i(\mathbf{x}) \leq q_i^{r,it} \text{ for all } i = 1, \ldots, k\}$. If $\#(P^{it}) > N_S^{it}$, show N_S^{it} representative solutions in P^{it} to the DM. If $\#(P^{it}) = N_S^{it}$, then show all the solutions in P^{it} to the DM. Otherwise, if $\#(P^{it}) < N_S^{it}$, complete P^{it} with individuals $\mathbf{x} \in \bar{P}^{it} \setminus P^{it}$ which satisfy $f_i(\mathbf{x}) \leq q_i^{r,it}$ for every $i = 1, \ldots, k$ and whose objective vectors are closer to $\mathbf{q}^{a,it}$ regarding the Euclidean distance, until $\#(P^{it}) = N_S^{it}$. Then, show all the solutions in P^{it} to the DM. Let $\{\bar{\mathbf{x}}_1^{it}, \ldots, \bar{\mathbf{x}}_{N_S^{it}}^{it}\}$ be the set of solutions shown to the DM.

Step 7. Solutions closer to the aspiration point (or to the reservation point). Ask the DM if (s)he desires to see solutions with objective function values closer to their aspiration levels (respectively, to their reservation levels). If no, go to Step 8. If yes, ask which of the N_S^{it} solutions (s)he would like to replace. Let $\{\bar{\mathbf{x}}_1^{it}, \ldots, \bar{\mathbf{x}}_t^{it}\}$ (with $t < N_S^{it}$) be the set of solutions to replace and $\{\bar{\mathbf{x}}_{t+1}^{it}, \ldots, \bar{\mathbf{x}}_{N_S^{it}}^{it}\}$ the set of solutions to maintain. Remove from $\bar{P}^{it}$ the solutions $\{\bar{\mathbf{x}}_1^{it}, \ldots, \bar{\mathbf{x}}_{N_S^{it}}^{it}\}$. Set $P^{it} = \{\mathbf{x} \in \bar{P}^{it} | q_i^{a,it} \leq f_i(\mathbf{x}) \text{ for all } i = 1, \ldots, k\}$ (respectively, $P^{it} = \{\mathbf{x} \in \bar{P}^{it} | f_i(\mathbf{x}) \leq q_i^{r,it} \text{ for all } i = 1, \ldots, k\}$) and find t solutions in P^{it} whose objective vectors are the closest ones to $\mathbf{q}^{a,it}$ (respectively, the furthest ones to $\mathbf{q}^{r,it}$) regarding the Euclidean distance. Let us denote these solutions by $\{\bar{\mathbf{x}}_1^{it}, \ldots, \bar{\mathbf{x}}_t^{it}\}$. Show $\{\bar{\mathbf{x}}_1^{it}, \ldots, \bar{\mathbf{x}}_t^{it}\} \cup \{\bar{\mathbf{x}}_{t+1}^{it}, \ldots, \bar{\mathbf{x}}_{N_S^{it}}^{it}\}$ to the DM.

Step 8. Optional ordering. Ask if the DM wants to order the solutions:

- According to some of the objective functions f_r, with $r \in \{1, \ldots, k\}$: in this case, show the solutions $\{\bar{\mathbf{x}}_1^{it}, \ldots, \bar{\mathbf{x}}_{N_S^{it}}^{it}\}$ in a descending order with respect to their values for f_r.
- According to their ASF values for the aspiration point $\mathbf{q}^{a,it}$: in this case, show the solutions $\{\bar{\mathbf{x}}_1^{it}, \ldots, \bar{\mathbf{x}}_{N_S^{it}}^{it}\}$ in an ascending order based on their values for $s(\mathbf{q}^{a,it}, \mathbf{f}(\mathbf{x}), \mathbf{w}^{it})$, where $\mathbf{w}^{it} = \left(\frac{1}{f_1^{max} - f_1^{min}}, \ldots, \frac{1}{f_k^{max} - f_k^{min}}\right)$, with $f_j^{min} = \min_{l=1,\ldots,N_S^{it}} f_j(\bar{\mathbf{x}}_l^{it})$ and $f_j^{max} = \max_{l=1,\ldots,N_S^{it}} f_j(\bar{\mathbf{x}}_l^{it})$.

Step 9. Termination rule. Ask the DM to select the most preferred solution from the set $\{\bar{\mathbf{x}}_1^{it}, \ldots, \bar{\mathbf{x}}_{N_S^{it}}^{it}\}$ and denote it by $\mathbf{x}^{it}$. If the DM is satisfied enough with this solution and (s)he wishes to *Stop*, the solution process concludes with $\mathbf{x}^{it}$ as the final solution and $\mathbf{f}(\mathbf{x}^{it})$ as the final objective vector. Otherwise, set $it = it + 1$ and go to Step 2.

In Step 1, estimations of $\mathbf{z}^\star$ and $\mathbf{z}^{nad}$ are shown to the DM to give her/him an idea of the objective function ranges for giving the reservation and aspiration levels. If they are not available, (s)he sets the levels based on her/his intuition.

In Steps 2 and 3, the DM expresses her/his initial preferences. Then, a set of solutions is generated and filtered in Steps 4–6. In Step 5, new $N_\mu = 2N_S^{it}$ solutions are generated using Modified WASF-GA and, in Step 6, N_S^{it} solutions are selected from its outcome and shown to the DM. Once the first solutions have been generated, the interaction with the DM starts from Step 7 onwards. Step 7 allows the DM to fine-tune the solutions shown by replacing some of them by others with objective function values closer to either aspiration or reservation levels. The idea is to let her/him freely explore different trade-offs among the objectives given the current preferences. With Step 8, the main aim is to support the DM in analyzing the solutions, but note that it can be skipped if desired.

In Step 6, if needed, we complete P^{it} with solutions not worsening the reservation levels and violating the aspiration ones as little as possible. It is very unlikely, but there may be no solutions that meet the reservation levels. In this case, the

DM must be informed, so that (s)he can decide whether to give new reservation levels, analyze just the available solutions satisfying $q_i^{a,it} \leq f_i(\mathbf{x}) \leq q_i^{r,it}$ $(i = 1, \ldots, k)$, or even just solutions with $q_i^{a,it} \leq f_i(\mathbf{x})$ $(i = 1, \ldots, k)$.

Internally, to accelerate the speed of the solution process, the final population P^{it} generated by Modified WASF-GA at an iteration it can be used as its initial population at the next iteration $it+1$. Also, if desired, the local Pareto optimality of the final solution can be assured by minimizing (2) using its objective function values as a reference point with some local optimization method.

4   Numerical Example

Next, we illustrate the performance of IRA-EMO with the three-objective optimization problem proposed in [14]. The aim is to identify the most convenient combination of improvements in the auxiliary services of a 1100 MW thermal power plant, to maximize the energy saving (denoted by f_1, in MWh), to minimize the economic investment required (denoted by f_2, in € million) and to maximize the Internal Rate of Return (IRR) of the investment (denoted by f_3, in %). The problem has 13 continuous and 20 binary decision variables and is modelled using a black-box simulator. We present objective values in their original form (and not using a minimization formulation) to make the interaction with the DM more understandable. The Pareto optimal front of this problem is discontinuous and formed by several disconnected subsets of solutions [14]. We have approximated $\mathbf{z}^* = (47526.37, 0.0, 100.0)$ and $\mathbf{z}^{\mathrm{nad}} = (0.0, 9.28, 0.0)$.

IRA-EMO and Modified WASF-GA have been implemented using jMetal [6], a Java-based framework for multiobjective optimization.[1] The parameter setting used in Modified WASF-GA is the same used in [14]. We use the simulated binary crossover (SBX) operator and a polynomial distribution mutation operator for continuous variables and the binary crossover and the binary mutation for integer variables. The crossover and mutation distribution indices used are 2 and 25, respectively. For all variables, the crossover and the mutation probabilities are 0.9 and $1/n$, respectively, where n is the number of binary or continuous variables. We have used a population size of 50 individuals and 100 generations.

Next, we describe the interactive solution process, i.e. how the DM used IRA-EMO to identify his most preferred solution. At the first iteration, the DM decided to generate five solutions ($N_S^1 = 5$) using $\mathbf{q}^{a,1} = (47526.37, 0.0, 60.0)$ and $\mathbf{q}^{r,1} = (0.0, 5.0, 0.0)$. Initially, he wished to study which type of trade-offs were possible by setting the aspiration and the reservation levels as their ideal and nadir objective values, except the aspiration level for f_3 (having a 60% IRR was profitable enough for him) and the reservation level for f_2 (the nadir value for f_2 was too much money for him and, although his budget limit was €2 million, initially he wanted to see solutions needing more expensive investments, such as €5 million, at most, to study the "price" to pay for a lower investment). In Fig. 2 (a), the solutions generated by IRA-EMO are plotted using a value path

---

[1] The source code is freely available at https://github.com/rsain/IRA-EMO.

and a table shows the objective vectors. This representation enables the DM to analyze the objective values reached within the aspiration and the reservation levels, and to see how wide the given ranges are in comparison to their maximum ranges (i.e., from the ideal to the nadir values), helping him to broaden or to shrink them if he wants to relax or to further limit the objective values.

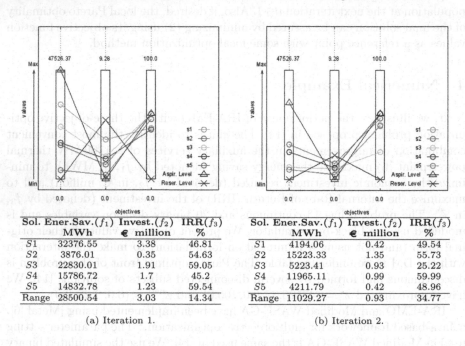

Sol.	Ener.Sav.(f_1) MWh	Invest.(f_2) € million	IRR(f_3) %
$S1$	32376.55	3.38	46.81
$S2$	3876.01	0.35	54.63
$S3$	22830.01	1.91	59.05
$S4$	15786.72	1.7	45.2
$S5$	14832.78	1.23	59.54
Range	28500.54	3.03	14.34

(a) Iteration 1.

Sol.	Ener.Sav.(f_1) MWh	Invest.(f_2) € million	IRR(f_3) %
$S1$	4194.06	0.42	49.54
$S2$	15223.33	1.35	55.73
$S3$	5223.41	0.93	25.22
$S4$	11965.11	0.99	59.99
$S5$	4211.79	0.42	48.96
Range	11029.27	0.93	34.77

(b) Iteration 2.

Fig. 2. Solution process of IRA-EMO.

All the solutions generated at the first iteration reached objective values within the specified ranges. Overall, limiting the investment to €5 million enabled very high energy savings to be obtained, but still far from the desired aspiration level. However, all the solutions achieved IRR values near to 60% (the aspiration level), reflecting their high profitability. Next, the DM wanted to know if it was possible to get the same profitability but limiting the investment to €2 million. At a new iteration, five solutions ($N_S^2 = 5$) were obtained with $q^{a,2} = (33000.0, 0.0, 60.0)$ and $q^{r,2} = (0.0, 2.0, 0.0)$, shown in Fig. 2 (b). Here he also relaxed the aspiration level for f_1 until a value near the highest energy saving obtained at the first iteration. The solutions found required an investment bellow €2 million, but the energy savings were not as high as at the first iteration. This highlights the conflict among the two objectives. Solution $S2$ needed the highest investment (still far from the budget limit), but reached the best energy saving value, and the second best IRR value. Observe that nearly all the solutions attained IRR values close to the aspiration level. Furthermore, solution $S4$ reached a 59.99% IRR, having the second best value for the energy saving. Based on this, the DM mainly liked solutions $S2$ and $S4$.

Although his budget was limited by €2 million, he wished to check what happened when relaxing the reservation level for f_2 a bit, with the condition of having a 20–70% IRR and energy savings up to 10000 MWh. A new iteration was carried out, with $\mathbf{q}^{a,3} = (40000.0, 0.0, 70.0)$, $\mathbf{q}^{r,3} = (10000.0, 3.0, 20.0)$, and $N_S^3 = 5$. Figure 3 depicts the five new solutions. Most of them improved the energy saving and the IRR values achieved in the second iteration, requiring to invest less than €2 million. Finally, the DM selected $S2$ as his most preferred solution. Although it needed the highest investment, it was bellow his budget limit and it reached the highest energy saving, with a satisfactory IRR.

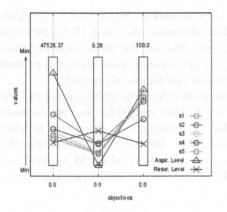

Sol.	Ener.Sav.(f_1) MWh	Invest.(f_2) € million	IRR(f_3) %
$S1$	11883.99	0.97	60.48
$S2$	22358.39	1.89	58.69
$S3$	14712.55	1.14	64.30
$S4$	16020.19	1.83	42.54
$S5$	13066.12	1.03	62.73
Range	10474.40	0.92	21.76

Fig. 3. Solution process of IRA-EMO (iteration 3).

To evaluate the performance of IRA-EMO, we compare it with R-NSGA-II [5], which can also use more than one reference point. In R-NSGA-II, a parameter ϵ controls the extent of the distribution of solutions. We set $\epsilon = 0.001$ (the same value used in [5] for two engineering design problems). To perform a fair comparison, we executed R-NSGA-II with the parameter setting used for IRA-EMO and performed three iterations with the same aspiration and reservation points. Thus, for $t = 1, 2, 3$, we generated nondominated solutions with $\mathbf{q}^{a,t}$ and $\mathbf{q}^{r,t}$. From the final population generated at each iteration, we first selected the solutions in the region of interest. Then, we applied the k-means clustering [11] to select the five most representative solutions. They are shown in Table 1.

Let us then analyze the dominance relations among the five solutions provided to the DM by each algorithm at each iteration. In other words, we compare objective vectors in the tables of Figs. 2 and 3 (solutions of IRA-EMO) with the ones in Table 1 (solutions of R-NSGA-II). We can see that, at any iteration, no solution of R-NSGA-II dominates any solution IRA-EMO. However, at the first iteration, solution $S1$ of IRA-EMO dominates solutions $S3$ and $S5$ of R-NSGA-II, and $S3$ of IRA-EMO dominates $S4$ of R-NSGA-II. Furthermore, at the second iteration, both $S1$ and $S5$ of IRA-EMO dominate $S3$ of R-NSGA-II, $S2$ of IRA-EMO dominates $S5$ of R-NSGA-II, and $S4$ of IRA-EMO domi-

Table 1. Solutions generated by R-NSGA-II.

Sol.	f_1 MWh	f_2 €million	f_3 %	f_1 MWh	f_2€million	f_3%	f_1 MWh	f_2 €million	f_3%
	Iteration 1			Iteration 2			Iteration 3		
$S1$	5601.56	1.07	22.85	5407.63	0.99	24.15	16330.36	1.99	39.62
$S2$	25101.18	2.11	58.87	15439.39	1.54	49.11	24979.27	2.04	60.71
$S3$	29555.99	3.41	42.08	4167.04	0.59	33.49	24670.84	2.01	60.74
$S4$	16123.46	1.93	40.47	4828.34	0.76	29.15	16136.67	1.89	41.37
$S5$	28021.10	3.40	39.70	15136.85	1.44	51.85	24844.48	2.03	60.76
Range	23954.44	2.34	36.02	11272.34	0.96	27.70	8842.60	0.15	21.14

nates $S1$ of R-NSGA-II. Finally, at the third iteration, $S2$ of IRA-EMO dominates both $S1$ and $S4$ of R-NSGA-II. This demonstrates that IRA-EMO was able to produce higher quality solutions (regarding the Pareto dominance) than R-NSGA-II. Thus, IRA-EMO generated sets of solutions which better represent the trade-offs existing among the objectives in the region of interest.

Comparing interactive EMO algorithms from a decision-making point of view is a research topic of its own and deserves further research. Because we have a real-world problem, we reported results for a single run, since this is what usually happens when interacting with real DMs. Actually, this constitutes the main difference between using EMO methods to solve real-world applications instead of testing with benchmark problems designed to study performance in general.

5   Conclusions

In this paper, we have proposed the IRA-EMO method for solving multiobjective optimization problems. At each iteration of IRA-EMO, very easy to understand preference information is asked from the DM: (a) aspiration and reservation levels for the objective functions and (b) the number of solutions (s)he wishes to analyze. According to this, the desired number of solutions is generated to represent the part of the Pareto optimal front bounded by the aspiration and the reservation levels given. Such solutions are internally generated at each iteration with the DM by means of Modified WASF-GA. The applicability of IRA-EMO has been described with a real three-objective optimization problem. The DM could analyze the solutions obtained with objective values within the aspiration and the reservation levels given at each iteration until finding a suitable final solution.

The region of interest could be approximated by methods such as R-NSGA-II using appropriate preference information. However, the interactive solution process of our method is different, including specific steps for decision making purposes. In addition, IRA-EMO assures that the region of interest bounded by the aspiration and the reservation points is approximated in just one run, generating the number of solutions the DM desires to see. These two features are not so easily controllable by other methods. We found that IRO-EMO was able to produce better solutions than R-NSGA-II for a real three-objective optimization problem. Actually, IRA-EMO generated solutions that better represent the trade-off existing among the objectives in the region of interest.

In the future, we plan to investigate how to configure IRA-EMO to explore, in the same iteration, solutions with objective function values within several regions of interest. We would also like to define and develop a way to compare the performance of interactive EMO methods from a decision-making perspective.

Acknowledgements. This research is funded by the Spanish Government (ECO2017-88883-R and ECO2017-90573-REDT), the Andalusian Regional Government (SEJ-532) and the Academy of Finland (project 287496). Ana B. Ruiz thanks the post-doctoral fellowship "Captación de Talento para la Investigación" at the Univ. of Málaga. The research is related to thematic research area DEMO (Univ. of Jyvaskyla).

References

1. Branke, J., Deb, K., Miettinen, K., Słowiński, R. (eds.): Multiobjective Optimization. LNCS, vol. 5252. Springer, Heidelberg (2008). https://doi.org/10.1007/978-3-540-88908-3
2. Cheng, R., Jin, Y., Olhofer, M., Sendhoff, B.: A reference vector guided evolutionary algorithm for many-objective optimization. IEEE Trans. Evol. Comput. **20**(5), 773–791 (2016)
3. Deb, K., Miettinen, K., Chaudhuri, S.: Towards an estimation of nadir objective vector using a hybrid of evolutionary and local search approaches. IEEE Trans. Evol. Comput. **14**(6), 821–841 (2010)
4. Deb, K., Pratap, A., Agarwal, S., Meyarivan, T.: A fast and elitist multiobjective genetic algorithm: NSGA-II. IEEE Trans. Evol. Comput. **6**(2), 182–197 (2002)
5. Deb, K., Sundar, J., Ubay, B., Chaudhuri, S.: Reference point based multi-objective optimization using evolutionary algorithm. Int. J. Comput. Intell. Res. **2**(6), 273–286 (2006)
6. Durillo, J.J., Nebro, A.J.: jMetal: a java framework for multi-objective optimization. Adv. Eng. Softw. **42**, 760–771 (2011)
7. Gomes, A., Antunes, C.H., Martins, A.G.: A multiple objective approach to direct load control using an interactive evolutionary algorithm. IEEE Trans. Power Syst. **22**(3), 1004–1011 (2007)
8. Gong, M., Liu, F., Zhang, W., Jiao, L., Zhang, Q.: Interactive MOEA/D for multi-objective decision making. In: 13th Annual Conference on Genetic and Evolutionary Computation, GECCO 2011, pp. 721–728 (2011)
9. Hakanen, J., Chugh, T., Sindhya, K., Jin, Y., Miettinen, K.: Connections of reference vectors and different types of preference information in interactive multiobjective evolutionary algorithms. In: IEEE Symposium Series on Computational Intelligence (SSCI), pp. 1–8 (2016)
10. Kaliszewski, I., Miroforidis, J., Podkopaev, D.: Interactive multiple criteria decision making based on preference driven evolutionary multiobjective optimization with controllable accuracy. Eur. J. Oper. Res. **216**(1), 188–199 (2012)
11. MacQueen, J.B.: Some methods for classification and analysis of multivariate observations. In: 5-th Berkeley Symposium on Mathematical Statistics and Probability, vol. 1, pp. 281–297. University of California Press, Berkeley (1967)

12. Miettinen, K.: Nonlinear Multiobjective Optimization. Kluwer, Boston (1999)
13. Ruiz, A.B., Luque, M., Miettinen, K., Saborido, R.: An interactive evolutionary multiobjective optimization method: interactive WASF-GA. In: Gaspar-Cunha, A., Henggeler Antunes, C., Coello, C. (eds.) EMO 2015. LNCS, vol. 9019, pp. 249–263. Springer, Cham (2015). https://doi.org/10.1007/978-3-319-15892-1_17
14. Ruiz, A.B., Luque, M., Ruiz, F., Saborido, R.: A combined interactive procedure using preference-based evolutionary multiobjective optimization. Application to the efficiency improvement of the auxiliary services of power plants. Expert Syst. Appl. **42**(21), 7466–7482 (2015)
15. Ruiz, A.B., Saborido, R., Luque, M.: A preference-based evolutionary algorithm for multiobjective optimization: the weighting achievement scalarizing function genetic algorithm. J. Global Optim. **62**(1), 101–129 (2015)
16. Saborido, R., Ruiz, A.B., Luque, M.: Global WASF-GA: an evolutionary algorithm in multiobjective optimization to approximate the whole Pareto optimal front. Evol. Comput. **25**(2), 309–349 (2017)
17. Thiele, L., Miettinen, K., Korhonen, P., Molina, J.: A preference-based evolutionary algorithm for multi-objective optimization. Evol. Comput. **17**(3), 411–436 (2009)
18. Wierzbicki, A.P.: The use of reference objectives in multiobjective optimization. In: Fandel, G., Gal, T. (eds.) Multiple Criteria Decision Making, Theory and Applications, vol. 177, pp. 468–486. Springer, Berlin (1980). https://doi.org/10.1007/978-3-642-48782-8_32
19. Zhang, Q., Li, H.: MOEA/D: a multiobjective evolutionary algorithm based on decomposition. IEEE Trans. Evol. Comput. **11**(6), 712–731 (2007)
20. Zitzler, E., Künzli, S.: Indicator-based selection in multiobjective search. In: Yao, X., et al. (eds.) PPSN 2004. LNCS, vol. 3242, pp. 832–842. Springer, Heidelberg (2004). https://doi.org/10.1007/978-3-540-30217-9_84

Progressive Preference Learning:
Proof-of-Principle Results in MOEA/D

Ke Li[(✉)] [iD]

Department of Computer Science, University of Exeter, Exeter EX4 4QF, UK
k.li@exeter.ac.uk

Abstract. Most existing studies on evolutionary multi-objective optimisation (EMO) focus on approximating the whole Pareto-optimal front. Nevertheless, rather than the whole front, which demands for too many points (especially when having many objectives), a decision maker (DM) might only be interested in a partial region, called the region of interest (ROI). Solutions outside this ROI can be noisy to the decision making procedure. Even worse, there is no guarantee that we can find DM preferred solutions when tackling problems with complicated properties or a large number of objectives. In this paper, we use the state-of-the-art MOEA/D as the baseline and develop its interactive version that is able to find solutions preferred by the DM in a progressive manner. Specifically, after every several generations, the DM is asked to score a limited number of candidates. Then, an approximated value function, which models the DM's preference information, is learned from the scoring results. Thereafter, the learned preference information is used to obtain a set of weight vectors biased towards the ROI. Note that these weight vectors are thus used in the baseline MOEA/D to search for DM preferred solutions. Proof-of-principle results on 3- to 10-objective test problems demonstrate the effectiveness of our proposed method.

Keywords: Interactive multi-objective optimisation ·
Preference learning · MOEA/D

1   Introduction

The multi-objective optimisation problem (MOP) considered in this paper is formulated as:

$$
\begin{aligned}
\text{minimize} \quad & \mathbf{F}(\mathbf{x}) = (f_1(\mathbf{x}), \cdots, f_m(\mathbf{x}))^T \\
\text{subject to} \quad & \mathbf{x} \in \Omega
\end{aligned}
\tag{1}
$$

where $\mathbf{x} = (x_1, \cdots, x_n)^T$ is a n-dimensional decision vector and $\mathbf{F}(\mathbf{x})$ is an m-dimensional objective vector. Ω is the feasible set in the decision space $\mathbb{R}^n$ and $\mathbf{F} : \Omega \to \mathbb{R}^m$ is the corresponding attainable set in the objective space

Supported by Royal Society under grant IEC/NSFC/170243.

© Springer Nature Switzerland AG 2019
K. Deb et al. (Eds.): EMO 2019, LNCS 11411, pp. 631–643, 2019.
https://doi.org/10.1007/978-3-030-12598-1_50

$\mathbb{R}^m$. Without considering the DM's preference information, given two solutions $\mathbf{x}^1, \mathbf{x}^2 \in \Omega$, $\mathbf{x}^1$ is said to dominate $\mathbf{x}^2$ if and only if $f_i(\mathbf{x}^1) \leq f_i(\mathbf{x}^2)$ for all $i \in \{1, \cdots, m\}$ and $\mathbf{F}(\mathbf{x}^1) \neq \mathbf{F}(\mathbf{x}^2)$. A solution $\mathbf{x} \in \Omega$ is said to be Pareto-optimal if and only if there is no solution $\mathbf{x}' \in \Omega$ that dominates it. The set of all Pareto-optimal solutions is called the Pareto-optimal set (PS) and their corresponding objective vectors form the Pareto-optimal front (PF). Accordingly, the ideal point is defined as $\mathbf{z}^* = (z_1^*, \cdots, z_m^*)^T$, where $z_i^* = \min\limits_{\mathbf{x} \in PS} f_i(\mathbf{x})$.

Evolutionary algorithms, which work with a population of solutions and can approximate a set of trade-off solutions simultaneously, have been widely accepted as a major tool for solving MOPs. Over the past two decades and beyond, many efforts have been devoted to developing EMO algorithms, e.g. NSGA-II [7], IBEA [23] and MOEA/D [22]. The ultimate goal of multi-objective optimisation is to help the DM find solutions that meet at most her/his preference. Supplying a DM with a large amount of trade-off points not only increases her/his workload, but also provides many irrelevant or even noisy information to the decision making process. Moreover, due to the curse of dimensionality, approximating the whole high-dimensional PF not only becomes computationally inefficient (or even infeasible), but also causes a severe cognitive obstacle for the DM to comprehend the high-dimensional data. To facilitate the decision making process, it is more practical to incorporate the DM's preference information into the search process. By doing so, it allows the computational efforts to be concentrated on the ROI and thus has a better approximation therein. Generally speaking, preference information can be incorporated *a priori*, *posteriori* or *interactively*. Note that the traditional EMO just goes along the posteriori way whose disadvantages have been described before. When the preference information is elicited a priori, it is used to guide the solutions towards the ROI. However, it is non-trivial to faithfully model the preference information before solving the MOP at hand. In practice, articulating the preference information in an interactive manner, which has been studied in the multi-criterion decision making (MCDM) field for over four decades, seems to be interesting. This enables DMs to progressively learn and understand the characteristics of the MOP at hand and adjust their preference information. As a consequence, the solutions are effectively driven towards the ROI.

In the past decade, the development for hybrid EMO-MCDM schemes, where the DM's preference information is integrated into EMO either a priori or interactively, have become increasingly popular. Generally speaking, their ideas can be briefly summarised as the following five categories.

1. The first one employs weight information, e.g. relative importance order [12], to model DM's preference information. However, it is difficult to control the guidance of the search towards the ROI and there is no obvious motivation to utilise weights in an interactive manner.
2. The second sort modifies the trade-off information by either classifying objectives into different levels and priorities or expressing DM's preference information via fuzzy linguistic terms according to different aspiration levels, e.g. [19]. This is method is interesting yet complicated, especially when the number of

objectives becomes large [20]. In addition, using such approach interactively increases the DM's burden.

3. The third category tries to bias the density of solutions towards the ROI by considering DM's preference information, e.g. [4]. However, density/diversity management itself in EMO is difficult, especially in a high-dimensional space.

4. The fourth class, as a recent trend, combines DM's preference information with performance indicators in algorithm design, e.g. [21]. Nevertheless, the computational cost of certain popular performance indicator, e.g. hypervolume [1] increases exponentially with the number of objectives.

5. The last one uses aspiration level vector, which represents the DM's desired values of each objective, to assist the search process, e.g. [10,13]. As reported in [3], aspiration level vector have been recognised as one of the most popular ways to elicit DM's preference information. Without a demanding effort from the DM, she/he is able to guide the search towards the ROI even when encountering a large number of objectives.

Take MOEA/D, a state-of-the-art EMO algorithm, as the baseline, this paper develops a simple yet effective progressive preference learning paradigm. It progressively learns an approximated value function (AVF) from the DM's behaviour in an interactive manner. The learned preference information is thus used to guide the population towards the ROI. Generally speaking, the progressive preference learning paradigm consists of the following three modules.

- *Optimisation module*: it uses the preference information elicited from the preference elicitation module to find the preferred solutions. In principle, any EMO algorithm can be used as the search engine while this paper takes MOEA/D for proof-of-principle purpose.
- *Consultation module*: it is the interface by which the DM interacts with the optimisation module. It supplies the DM with a few incumbent candidates to score. Thereafter, the scored candidates found so far are used to form the training data, based on which a machine learning algorithm is applied to find an AVF that models the DM's preference information.
- *Preference elicitation module*: it aims at translating the preference information learned from the consultation module in the form that can be used in MOEA/D. In particular, the learned preference information is used to obtain a set of weight vectors biased towards the ROI.

In the remaining paragraphs, the technical detail of the progressive preference learning for MOEA/D will be described step by step in Sect. 2. Proof-of-principle experiments, shown in Sects. 3 and 4, demonstrate the effectiveness of our proposed algorithm for finding DM preferred Pareto-optimal solutions on benchmark problems with 3 to 10 objectives. At the end, Sect. 5 concludes this paper and provides some future directions.

2   Proposed Method

Generally speaking, the method proposed in this paper is a generic framework for progressive preference learning. It consists of three interdependent modules, i.e. consultation, preference elicitation and optimisation. For proof-of-principle purpose, this paper uses the state-of-the-art MOEA/D as the search engine in the optimisation module. It uses the preference information provided by the preference elicitation module to approximate DM's preferred solutions. In addition, it periodically supplies the consultation module with a few incumbent candidates to score. Since no modification has been done upon MOEA/D, we do not intend to delineate its working mechanism here while interested readers are suggested to refer to [22] for details. The following paragraphs will focus on describing the consultation and preference elicitation modules.

2.1 Consultation Module

The consultation module is the interface where the DM interacts with, and expresses her/his preference information to the optimisation module. In principle, there are various ways to represent the DM's preference information. In this paper, we assume that the DM's preference information is represented as a value function. It assigns a solution a score that represents its desirability to the DM. The consultation module mainly aims to progressively learn an AVF that approximates the DM's 'golden' value function, which is unknown *a priori*, by asking the DM to score a few incumbent candidates. We argue that it is labor-intensive to consult the DM every generation. Furthermore, as discussed in [2], consulting the DM at the early stage of the evolution might be detrimental to the decision-making procedure, since the DM can hardly make a reasonable judgement on poorly converged solutions. In this paper, we fix the number of consultations. Before the first consultation session, the EMO algorithm runs as usual without considering any DM's preference information. Afterwards, the consultation session happens every $\tau > 1$ generations.

There are two major questions to address when we want to approximate the DM's preference information: (1) which solutions can be used for scoring? and (2) how to learn an appropriate AVF?

Scoring. To relieve the DM's cognitive load and her/his fatigue, we only ask the DM to score a limited number (say $1 \leq \mu \ll N$) of incumbent candidates chosen from the current population. Specifically, we use the AVF learned from the most recent consultation session to score the current population. The μ solutions having the best AVF values are used as the incumbent candidates, i.e. deemed as the ones that are satisfied by the DM most. However, if it is at the first consultation session, no AVF is available for scoring. In this case, we first initialise another μ 'seed' weight vectors, which can either be generated by the Das and Dennis' method [6] or chosen from the weight vectors initialised in the optimisation module. Afterwards, for each of these 'seed' weight vectors, we find the nearest neighbour from the weight vectors initialised in the optimisation module. Then, the solutions associated with these selected weight vectors are used as the initial incumbent candidates.

Learning. In principle, many off-the-shelf machine learning algorithms can be used to learn the AVF. In this paper, we treat it as a regression problem and use the Radius Basis Function network (RBFN) [5] to serve this purpose. In particular, RBFN, a single-layer feedforward neural network, is easy to train and its performance is relatively insensitive to the increase of the dimensionality.

Let $\mathcal{D} = \{(\mathbf{F}(\mathbf{x}^i), \psi(\mathbf{x}^i))\}_{i=1}^M$ denote the dataset for training the RBFN. The objective values of a solution $\mathbf{x}^i$ are the inputs and its corresponding value function $\psi(\mathbf{x}^i)$ scored by the DM is the output. In particular, we accumulate every μ solutions scored by the DM to form $\mathcal{D}$. An RBFN is a real-valued function $\Phi : \mathbb{R}^m \rightarrow \mathbb{R}$. Various RBFs can be used as the activation function of the RBFN, such as Gaussian, splines and multiquadrics. In this paper, we consider the following Gaussian function:

$$\varphi = \exp(-\frac{\|\mathbf{F}(\mathbf{x}) - \mathbf{c}\|}{\sigma^2}), \tag{2}$$

where $\sigma > 0$ is the width of the Gaussian function. Accordingly, the AVF can be calculated as:

$$\Phi(\mathbf{x}) = \omega^0 + \sum_{i=1}^{\mathrm{NR}} \omega^i \exp(-\frac{\|\mathbf{F}(\mathbf{x}) - \mathbf{c}^i\|}{\sigma^2}), \tag{3}$$

where NR is the number of RBFs, each of which is associated with a different centre $\mathbf{c}^i$, $i \in \{1, \cdots, \mathrm{NR}\}$. ω^i is the network coefficient, and ω^0 is a bias term, which can be set to the mean of the training data or 0 for simplicity. In our experiment, we use the RBFN program newrb provided by the Neural Network Toolbox from the MATLAB[1].

2.2   Preference Elicitation Module

The basic idea of MOEA/D is to decompose the original MOP into several sub-problems and it uses a population-based technique to solve these subproblems in a collaborative manner. In particular, this paper uses the Tchebycheff function [16–18] to form a subproblem as follows:

$$\begin{aligned} \text{minimize } g(\mathbf{x}|\mathbf{w}, \mathbf{z}^*) &= \max_{1 \leq i \leq m} |f_i(\mathbf{x}) - z_i^*|/w_i \\ \text{subject to } \quad \mathbf{x} &\in \Omega \end{aligned} \tag{4}$$

where $\mathbf{z}^*$ is the ideal point and $\mathbf{w}$ is the weight vector associated with this subproblem. Since the optimal solution of each subproblem is a Pareto-optimal solution of the original MOP, MOEA/D can in principle approximate the whole PF with a necessary diversity by using a set of evenly distributed weight vectors $W = \{\mathbf{w}^i\}_{i=1}^N$, where N is the population size. When considering the DM's preference information, the ROI becomes a partial region of the PF. A natural

---

[1] https://uk.mathworks.com/help/nnet/ug/radial-basis-neural-networks.html.

idea, which translates the DM's preference information into the form that can be used in MOEA/D, is to adjust the distribution of weight vectors. Specifically, the preference elicitation module uses the following four-step process to achieve this purpose.

Step 1: Use $\Phi(\mathbf{x})$ learned in the consultation module to score each member of the current population P.

Step 2: Rank the population according to the scores assigned in Step 1, and find the top μ solutions. weight vectors associated with these solutions are deemed as the promising ones, and store them in a temporary archive $W^U := \{\mathbf{w}^{Ui}\}_{i=1}^{\mu}$.

Step 3: For $i = 1$ to μ do

Step 3.1: Find the $\lceil \frac{N-\mu}{\mu} \rceil$ closest weight vectors to $\mathbf{w}^{Ui}$ according to their Euclidean distances.

Step 3.2: Move each of these weight vectors towards $\mathbf{w}^{Ui}$ according to

$$w_j = w_j + \eta \times (w_j^{Ui} - w_j), \tag{5}$$

where $j \in \{1, \cdots, m\}$.

Step 3.3: Temporarily remove these weight vectors from W and go to Step 3.

Step 4: Output the adjusted weight vectors as the new W.

In the following paragraphs, we would like to make some remarks on some important ingredients of the above process.

- In MOEA/D, each solution should be associated with a weight vector. Therefore, in Step 2, the rank of a solution also indicates the importance of its associated weight vector with respect to the DM's preference information. The weight vectors stored in W^U are indexed according to the ranks of their associated solutions. In other words, $\mathbf{w}^{U1}$ represents the most important weight vector, and so on.
- Step 3 implements the adjustment of the distribution of weight vectors according to their satisfaction to the DM's preference information. Specifically, each of those μ promising weight vectors is used as a pivot, towards which its closest $\lceil \frac{N-\mu}{\mu} \rceil$ neighbours are moved according to Eq. 5.
- η in Eq. 5 controls the convergence rate towards the promising weight vector. For proof-of-principle purpose, we set $\eta = 0.5$ in this paper.
- Step 3 is similar to a clustering process, while we give the weight vector, which has a higher rank, a higher priority to attract its companions.

To better understand this preference elicitation process, Fig. 1 gives an intuitive example in a two-objective case. In particular, three promising weight vectors are highlighted by red circles. $\mathbf{w}^{U1}$ has the highest priority to attract its companions, and so on. We can observe that the weight vectors are biased towards those promising ones after the preference elicitation process.

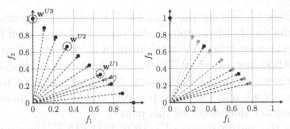

(a) Original distribution. (b) Adjusted distribution.

Fig. 1. Illustration of the preference elicitation process.

3   Experimental Settings

To validate the effectiveness of our proposed algorithm, dubbed as I-MOEA/D-PLVF, for approximating the DM preferred solutions, the widely used DTLZ [11] test problems are chosen to form the benchmark suite. Note that the DTLZ problems are scalable to any number of objectives. The parameter settings of our proposed progressive preference learning paradigm are summarised as follows:

- number of incumbent candidates presented to the DM for scoring: $\mu - 2m + 1$ at the first consultation session and $\mu = 10$ afterwards;
- number of generations between two consecutive consultation sessions: $\tau = 25$;
- number of weight vectors, population size settings and number function evaluations (FEs) are set as suggested in [15]. Due to the page limit, they can be found in the supplementary document[2] of this paper.
- the simulated binary crossover [8] is used as the crossover operator while its probability and distribution index are set as: $p_c = 1.0$ and $\eta_c = 30$;
- the polynomial mutation [9] is used as the mutation operator while its probability and distribution index are set as: $p_m = \frac{1}{n}$ and $\eta_m = 20$;

As discussed in [14], the empirical comparison of interactive EMO methods is tricky since a model of the DM's behavior is required yet unfortunately sophisticated to represent. In this paper, we use a pre-specified 'golden' value function, which is unknown to an interactive EMO algorithm, to play as an artificial DM. Specifically, the DM is assumed to minimise the following nonlinear function:

$$\psi(\mathbf{x}) = \max_{1 \leq i \leq m} |f_i(\mathbf{x}) - z_i^*|/w_i^*, \tag{6}$$

where $\mathbf{z}^*$ is set to be the origin in our experiments, and $\mathbf{w}^*$ is the utopia weights that represents the DM's emphasis on different objectives. We consider two types of $\mathbf{w}^*$: one targets the preferred solution on the middle region of the PF while the other targets the preferred solution on one side of the PF, i.e. biased towards a particular extreme. Since a m-objective problem has m extremes, there are

---

[2] https://coda-group.github.io/emo19-supp.pdf.

m different choices for setting the biased $\mathbf{w}^*$. In our experiments, we randomly choose one for the proof-of-principle study. Since the Tchebycheff function is used as the value function and the analytical forms of the test problems are known, we can use the method suggested in [15] to find the corresponding Pareto-optimal solution (also known as the DM's 'golden' point) with respect to the given $\mathbf{w}^*$. Detailed settings of $\mathbf{w}^*$ and the corresponding DM's 'golden' point can be found in the supplementary document of this paper.

To evaluate the performance of I-MOEA/D-PLVF for approximating the ROI, we consider using the approximation error of the obtained population P with respect to the DM's 'golden' point $\mathbf{z}^r$ as the performance metric. Specifically, it is calculated as:

$$\mathbb{E}(P) = \min_{\mathbf{x} \in P} \text{dist}(\mathbf{x}, \mathbf{z}^r) \tag{7}$$

where $\text{dist}(\mathbf{x}, \mathbf{z}^r)$ is the Euclidean distance between $\mathbf{z}^r$ and a solution $\mathbf{x} \in P$ in the objective space.

To demonstrate the importance of using the DM's preference information, we also compare I-MOEA/D-PLVF with its corresponding baseline algorithms without considering the DM's preference information. In our experiments, we run each algorithm independently 21 times with different random seeds. In the corresponding table, we show the results in terms of the median and the interquartile range (IQR) of the approximation errors obtained by different algorithms. To have a statistical sound comparison, we use the Wilcoxon signed-rank test with a 95% confidence level to validate the significance of the better results.

4   Empirical Results

From the results shown in Table 1, as we expected, I-MOEA/D-PLVF shows overwhelming superiority over the baseline MOEA/D for approximating the DM's

Table 1. Performance comparisons of the approximation errors (median and the corresponding IQR) obtained by I-MOEA/D-PLVF versus the baseline MOEA/D on DTLZ1 to DTLZ4 test problems.

		DTLZ1		DTLZ2	
m	ROI	I-MOEA/D-PLVF	MOEA/D	I-MOEA/D-PLVF	MOEA/D
3	c	4.213E-4(2.87E-3)	3.104E-2(3.18E-3)	1.026E-2(1.78E-2)	1.030E-1(6.35E-3)
	b	1.471E-3(2.87E-3)	3.103E-2(3.30E-3)	8.832E-3(1.09E-2)	9.103E-2(2.56E-2)
5	c	4.173E-3(1.73E-2)	5.262E-2(1.90E-2)	1.721E-2(2.86E-2)	2.417E-1(1.90E-2)
	b	1.082E-2(2.09E-2)	7.648E-2(1.65E-2)	5.082E-2(4.73E-2)	2.049E-1(1.45E-2)
8	c	2.130E-3(1.71E-2)	1.484E-2(2.21E-3)	1.625E-2(1.79E-1)	2.615E-1(1.52E-2)
	b	1.012E-2(1.03E-1)	5.534E-2(1.12E-2)	4.185E-2(1.10E-1)	1.250E-1(1.05E-2)
10	c	1.269E-1(2.71E-1)	1.789E-1(1.10E-3)	1.087E-1(1.62E-1)	7.386E-1(8.54E-2)
	b	1.543E-1(1.77E-1)	2.634E-1(5.05E-3)	1.183E-1(2.08E-1)	2.596E-1(2.88E-2)
		DTLZ3		DTLZ4	
m	ROI	I-MOEA/D-PLVF	MOEA/D	I-MOEA/D-PLVF	MOEA/D
3	c	7.214E-4(7.26E-3)	1.055E-1(1.59E-3)	1.303E-2(2.78E-2)	1.042E-1(1.89E-3)
	b	2.811E-3(1.09E-2)	8.678E-2(7.75E-3)	7.634E-3(8.76E-3)	9.469E-2(8.07E-3)
5	c	1.128E-2(8.77E-2)	2.442E-1(4.62E-2)	2.762E-2(5.74E-2)	2.569E-1(2.37E-2)
	b	1.792E-2(1.53E-1)	2.162E-1(2.35E-2)	3.717E-2(6.28E-2)	2.121E-1(6.66E-3)
8	c	6.821E-2(2.78E-1)	4.277E-1(9.56E-3)	6.538E-2(8.62E-2)	7.236E-1(1.07E-2)
	b	8.697E-2(1.63E-1)	1.574E-1(1.32E-2)	1.271E-1(1.86E-1)	2.164E-1(1.69E-2)
10	c	2.168E-1(5.71E-1)	7.365E-1(2.81E-2)	1.927E-1(2.63E-1)	8.676E-1(1.07E-1)
	b	1.629E-1(2.55E-1)	3.344E-1(6.99E-2)	1.018E-1(3.28E-1)	2.055E-1(4.21E-2)

The ROI column gives the type of the DM supplied utopia weights. c indicates the preference on the middle region of the PF while b indicates the preference on an extreme. All better results are with statistical significance according to Wilcoxon signed-rank test with a 95% confidence level, and are highlighted in bold face with a grey background.

'golden' solution. In particular, they obtain statistically significantly better metric values (i.e. smaller approximation error) on all test problems. In the following paragraphs, we discuss the results from the following aspects.

– Due to the page limit, we only plot some results on 3- and 10-objective scenarios in Figs. 2, 3, 4 and 5, while more comprehensive results can be found in the supplementary document. From these plots, we can observe that I-MOEA/D-PLVF is always able to find solutions that well approximate the unknown DM's 'golden' point with a decent accuracy as shown in Table 1. In contrast, since the baseline MOEA/D is designed to approximate the whole PF, it is not surprised to see that most of their solutions are away from the DM's 'golden' point. Although some of the solutions obtained by the baseline MOEA/D can by chance pass the ROI, i.e. the vicinity of the DM's 'golden'

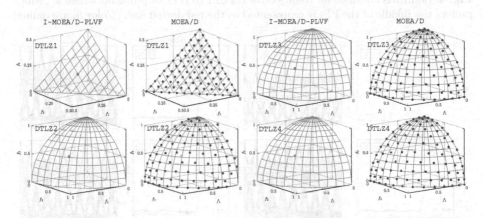

Fig. 2. Solutions obtained on 3-objective DTLZ1 to DTLZ4 problems where $\mathbf{z}^r$, which prefers the middle region of the PF, is represented as the red dotted line. (Color figure online)

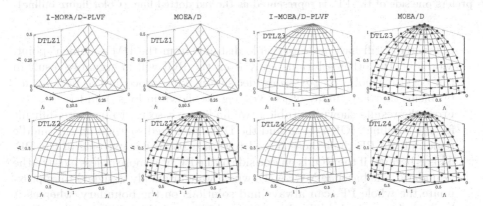

Fig. 3. Solutions obtained on 10-objective DTLZ1 to DTLZ4 problems where $\mathbf{z}^r$, which prefers one side of the PF, is represented as the red dotted line. (Color figure online)

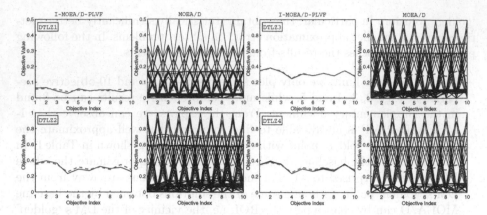

Fig. 4. Solutions obtained on 10-objective DTLZ1 to DTLZ4 problems where z^r, which prefers the middle of the PF, is represented as the red dotted line. (Color figure online)

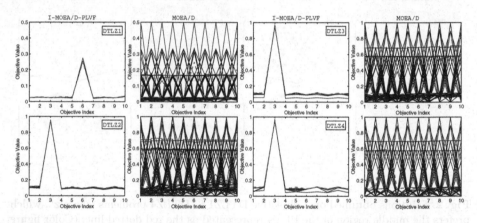

Fig. 5. Solutions obtained on 10-objective DTLZ1 to DTLZ4 problems where z^r, which prefers one side of the PF, is represented as the red dotted line. (Color figure online)

point, they still have an observable distance from the DM's 'golden' point. Moreover, the other solutions away from the ROI will unarguably result in the cognitive noise to *posteriori* decision-making procedure, especially for problems that have many objectives, e.g. as shown in Figs. 4 and 5.

- From the results shown in Table 1, we find that it seems to be more difficult for the baseline MOEA/D to find the DM's preferred solution on the middle region of the PF than those biased toward a particular extreme of the PF. This is because if the ROI is on one side of the PF, it is more or less close to the boundary. The baseline MOEA/D, which were originally designed to approximate the whole PF, can always find solutions on the boundary, whereas it becomes increasingly difficult to find solutions on the middle region of the PF with the increase of the number of objectives. Therefore, the approximation error to a DM's 'golden' point on one side of the PF seems to be better than those on the middle region of the PF. In contrast, since our proposed I-MOEA/D-PLVF can progressively learn the DM's preference information and adjust the search direction, it well approximates the ROI in any part of the PF.

5 Conclusions

This paper has proposed a simple yet effective paradigm for progressively learning the DM's preference information in an interactive manner. It consists of three modules, i.e. optimisation, consultation and preference elicitation. For proof-of-principle purpose, this paper uses the state-of-the-art MOEA/D as the baseline algorithm in the optimisation module. The consultation module aims to progressively learn an AVF that models the DM's preference information. In particular, during the consultation session, the DM is presented with a few incumbent candidates for scoring according her/his preference. Once the AVF is learned, the preference elicitation module translates it into the form that can be used in the optimisation module, i.e. a set of weight vectors that are biased towards the ROI. Proof-of-principle results on 3- to 10-objective test problems demonstrate the effectiveness of our proposed I-MOEA/D-PLVF for approximating the DM's preferred solution(s).

In principle, the progressive preference learning paradigm proposed in this paper is a generic framework which can be used to help any EMO algorithm to approximate DM preferred solution(s) in an interactive manner. For proof-of-principle purpose, we use MOEA/D as the search engine in the optimisation module. Therefore, the learned preference information is translated as a set of biased weight vectors in the preference elicitation module. One of the future directions is to adapt this to other formats according to the characteristics of the baseline algorithm. In addition, this paper assumes that the DM's preference information is represented as a monotonic value function. However, in practice, it is not uncommon that the DM judges some of the alternatives to be incomparable. How to discriminate the order information from incomparable comparisons? Moreover, instead of assigning a scalar score to a solution, it is interesting to study how to derive the preference information through holistic comparisons among incumbent candidates. Although this paper has restricted the value function to be the form as Eq. 6, other more value function formulations can also be considered. Furthermore, it is interesting to further investigate the robustness consideration in deriving the AVF. More studies are required to investigate the side effects brought by the inconsistencies in decision-making and the ways to mitigate that. Last but not the least, there are a couple of parameters associated with the proposed progressive preference learning paradigm, i.e. those listed in Sect. 3. It is important to investigate the effects of these parameters as a part of future work.

642   K. Li

References

1. Auger, A., Bader, J., Brockhoff, D., Zitzler, E.: Hypervolume-based multiobjective optimization: theoretical foundations and practical implications. Theor. Comput. Sci. **425**, 75–103 (2012)
2. Battiti, R., Passerini, A.: Brain-computer evolutionary multiobjective optimization: a genetic algorithm adapting to the decision maker. IEEE Trans. Evol. Comput. **14**(5), 671–687 (2010)
3. Bechikh, S., Kessentini, M., Said, L.B., Ghédira, K.: Chapter four - preference incorporation in evolutionary multiobjective optimization: a survey of the state-of-the-art. Adv. Comput. **98**, 141–207 (2015)
4. Branke, J., Deb, K.: Integrating user preferences into evolutionary multi-objective optimization. In: Jin, Y. (ed.) Knowledge Incorporation in Evolutionary Computation, vol. 167, pp. 461–477. Springer, Berlin (2005). https://doi.org/10.1007/978-3-540-44511-1_21
5. Buhmann, M.D.: Radial Basis Functions. Cambridge University Press, Cambridge (2003)
6. Das, I., Dennis, J.E.: Normal-boundary intersection: a new method for generating the pareto surface in nonlinear multicriteria optimization problems. SIAM J. Optim. **8**, 631–657 (1998)
7. Deb, K., Pratap, A., Agarwal, S., Meyarivan, T.: A fast and elitist multiobjective genetic algorithm: NSGA-II. IEEE Trans. Evol. Comput. **6**(2), 182–197 (2002)
8. Deb, K., Agrawal, R.B.: Simulated binary crossover for continuous search space. Complex Syst. **9**, 1–34 (1994)
9. Deb, K., Goyal, M.: A combined genetic adaptive search (GeneAS) for engineering design. Comput. Sci. Inform. **26**, 30–45 (1996)
10. Deb, K., Sundar, J., Bhaskara, U., Chaudhuri, S.: Reference point based multiobjective optimization using evolutionary algorithms. Int. J. Comput. Intell. Res. **2**(3), 273–286 (2006)
11. Deb, K., Thiele, L., Laumanns, M., Zitzler, E.: Scalable test problems for evolutionary multiobjective optimization. In: Abraham, A., Jain, L., Goldberg, R. (eds.) Evolutionary Multiobjective Optimization. Advanced Information and Knowledge Processing, pp. 105–145. Springer, London (2005). https://doi.org/10.1007/1-84628-137-7_6
12. Jin, Y., Okabe, T., Sendho, B.: Adapting weighted aggregation for multiobjective evolution strategies. In: Zitzler, E., Thiele, L., Deb, K., Coello Coello, C.A., Corne, D. (eds.) EMO 2001. LNCS, vol. 1993, pp. 96–110. Springer, Heidelberg (2001). https://doi.org/10.1007/3-540-44719-9_7
13. Li, K., Chen, R., Min, G., Yao, X.: Integration of preferences in decomposition multiobjective optimization. IEEE Trans. Cybern. **48**(12), 3359–3370 (2018)
14. Li, K., Deb, K., Yao, X.: R-metric: evaluating the performance of preference-based evolutionary multi-objective optimization using reference points. IEEE Trans. Evol. Comput. (2017). accepted for publication
15. Li, K., Deb, K., Zhang, Q., Kwong, S.: An evolutionary many-objective optimization algorithm based on dominance and decomposition. IEEE Trans. Evol. Comput. **19**(5), 694–716 (2015)
16. Li, K., Fialho, Á., Kwong, S., Zhang, Q.: Adaptive operator selection with bandits for a multiobjective evolutionary algorithm based on decomposition. IEEE Trans. Evol. Comput. **18**(1), 114–130 (2014)

17. Li, K., Kwong, S., Zhang, Q., Deb, K.: Interrelationship-based selection for decomposition multiobjective optimization. IEEE Trans. Cybern. **45**(10), 2076–2088 (2015)
18. Li, K., Zhang, Q., Kwong, S., Li, M., Wang, R.: Stable matching based selection in evolutionary multiobjective optimization. IEEE Trans. Evol. Comput. **18**(6), 909–923 (2014)
19. Parmee, I.C., Cvetkovic, D.: Preferences and their application in evolutionary multiobjective optimization. IEEE Trans. Evol. Comput. **6**(1), 42–57 (2002)
20. Said, L.B., Bechikh, S., Ghédira, K.: The r-dominance: a new dominance relation for interactive evolutionary multicriteria decision making. IEEE Trans. Evolut. Comput. **14**(5), 801–818 (2010)
21. Wagner, T., Trautmann, H., Brockhoff, D.: Preference articulation by means of the $R2$ indicator. In: Purshouse, R.C., Fleming, P.J., Fonseca, C.M., Greco, S., Shaw, J. (eds.) EMO 2013. LNCS, vol. 7811, pp. 81–95. Springer, Heidelberg (2013). https://doi.org/10.1007/978-3-642-37140-0_10
22. Zhang, Q., Li, H.: MOEA/D: a multiobjective evolutionary algorithm based on decomposition. IEEE Trans. Evol. Comput. **11**, 712–731 (2007)
23. Zitzler, E., Künzli, S.: Indicator-based selection in multiobjective search. In: Yao, X., et al. (eds.) PPSN 2004. LNCS, vol. 3242, pp. 832–842. Springer, Heidelberg (2004). https://doi.org/10.1007/978-3-540-30217-9_84

A Dichotomous Approach to Reduce Rank Reversal Occurrences in PROMETHEE II Rankings

Erica Berghman, Yves De Smet[✉], Jean Rosenfeld, and Dimitri Van Assche

Computer and Decision Engineering Department, SMG Research Unit,
Ecole Polytechnique de Bruxelles, Université Libre de Bruxelles, Brussels, Belgium
yves.de.smet@ulb.ac.be

Abstract. Multicriteria methods based on pairwise comparisons suffer from rank reversal occurrences when the set of alternatives is modified. In this paper, we consider an alternative PROMETHEE-based method to compute pre-orders. This should lead to face less rank reversal (RR) occurrences than in the traditional PROMETHEE II ranking. It is based on hierarchical clustering using a dichotomous process. The property on which is based the dichotomous separation is explained. The two methods are compared. At first, we show that the rankings obtained are similar. Then, a comparison between the frequency of rank reversal occurrences in both methods is conducted.

Keywords: Multicriteria decision aid · PROMETHEE II method · Rank reversal

1 Introduction

PROMETHEE II is a Multiple Criteria ranking method based on pairwise comparisons. It provides a complete ranking of different alternatives based on net flow scores. It has been proved that methods based on pairwise comparisons suffer from rank reversal problems [5,7,8]. For instance, the deletion of an alternative can alter the relative rank of two other alternatives. De Keyzer and Peeters [4] were the first to underline this issue in the context of the PROMETHEE I method. Rank reversal is at the origin of long debates about its legitimacy. Here we will not address this question. We will rather consider that it is intrinsic to pairwise comparison methods. On the contrary, we will rather investigate if it is possible to *manage* rank reversal. Therefore, we consider a new ranking method based on a property of the PROMETHEE's net flow scores [6] with the aim to reduce rank reversal occurrences.

The paper is organized as follows; the first section is dedicated to a reminder of PROMETHEE II. The new ranking procedure - a dichotomous method based on a hierarchical clustering procedure - is introduced in the next section. Then, a comparison between the two methods is conducted on several points. Firstly,

© Springer Nature Switzerland AG 2019
K. Deb et al. (Eds.): EMO 2019, LNCS 11411, pp. 644–654, 2019.
https://doi.org/10.1007/978-3-030-12598-1_51

we compare the rankings obtained, based on the Kendall's tau. Secondly, we investigate the indifference relation in the second method. Then we tackle the topic of rank reversal. We show that the threshold obtained by Mareschal et al. [5] for PROMETHEE II is not valid anymore for the dichotomous method.

2   Promethee II

This section is a short overview of PROMETHEE II. We refer the interested reader to [1,2] for a detailed description. Let $A = \{a_1, a_2, \ldots, a_n\}$ be a set of n alternatives and $F = \{f_1, f_2, \ldots, f_m\}$ be a set of m criteria. We assume that each alternative a_i is characterized by an evaluation for every criterion, denoted $f_k(a_i)$. For each couple of alternatives (a_i, a_j) and each criterion f_k, the difference $d_k(a_i, a_j) = f_k(a_i) - f_k(a_j)$ is computed. This difference is then transformed into an uni-criterion preference degree denoted $P_k(d_k(a_i, a_j))$ (also referred as $P_k(a_i, a_j)$ in the rest of the paper). The preference function used is normalized ($P_k \in [0,1]$) and is a non-decreasing function.

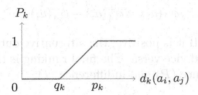

Fig. 1. Linear preference function with an indifference q_k and a preference p_k thresholds.

The most common preference function is a linear function with an indifference and a preference threshold, mathematically expressed in Eq. 1 and represented on Fig. 1. The indifference (q_k) and preference (p_k) thresholds for each criterion f_k are assumed to be given by the Decision Maker (DM).

$$P_k(a_i, a_j) = \begin{cases} 0 & d_k(a_i, a_j) < q_k \\ \frac{d_k(a_i, a_j) - q_k}{p_k - q_k} & q_k \leq d_k(a_i, a_j) < p_k \\ 1 & d_k(a_i, a_j) \geq p_k \end{cases} \qquad (1)$$

Now we are able to compute a pairwise comparison matrix π in which each element $\pi(a_i, a_j)$ is the weighted sum of the preference functions on each criterion:

$$\pi(a_i, a_j) = \sum_{k=1}^{m} w_k . P_k(a_i, a_j) \qquad (2)$$

The weights w_k associated to each criterion f_k are positive parameters given by the DM and are assumed to be normalized. The π matrix shows the global

preference of an element over another element of the set. Obviously, we have:
$\pi(a_i, a_j) \geq 0$ and $\pi(a_i, a_j) + \pi(a_j, a_i) \leq 1$.

Once the pairwise comparison matrix has been obtained, the next step is to compute outranking flow scores. The positive flow score ϕ^+ of an alternative a_i expresses its average preference over the other alternatives. Similarly, the negative flow score ϕ^- expresses how the other alternatives are being preferred on average.

$$\phi_A^+(a_i) = \frac{1}{n-1} \sum_{j=1}^{n} \pi(a_i, a_j) \tag{3}$$

$$\phi_A^-(a_i) = \frac{1}{n-1} \sum_{j=1}^{n} \pi(a_j, a_i) \tag{4}$$

The net flow score of an alternative is the balance between the positive and negative flow scores.

$$\phi_A(a_i) = \phi_A^+(a_i) - \phi_A^-(a_i) \tag{5}$$

It varies from -1 to 1. If it is positive, the alternative outranks more on average than it is outranked and vice versa. The final ranking is then obtained following the relations of preference (P) and indifference (I):

$$\phi_A(a_i) > \phi_A(a_j) \Rightarrow a_i P a_j$$

$$\phi_A(a_i) = \phi_A(a_j) \Rightarrow a_i I a_j$$

This gives us a complete pre-order of the set of alternatives. At this stage, one can already notice that indifference relations will be rather uncommon (due to the associated strict constraint).

3   Dichotomous Method

In this section, we present an alternative method to obtain a pre-order on the basis of a preference matrix. This is based on a hierarchical clustering procedure using a dichotomous process that has been proposed recently by De Smet [6]. The idea is to separate the n alternatives into two subgroups that are ordered. Then, one of the obtained group is divided once again. The procedure is repeated until the desired number of clusters is obtained. In this paper, we will consider the degenerate case of a number of clusters being equal to the number of alternatives. To do so, we have to find a way to separate a group of alternatives into two subsets. Let us try to separate A in two complementary subsets B and $\overline{B}$ while respecting two conditions:

1. The global preference of B over $\overline{B}$ is as high as possible;
2. The global preference of $\overline{B}$ over B is as low as possible.

To respect the first condition, we have to maximize the global preference of all the elements of B over all the elements of $\overline{B}$. As explained before, the global preference of an element a_i over another element a_j is expressed by $\pi_{ij} = \pi(a_i, a_j)$. Thus we have to find groups such that Eq. (6) is maximized.

$$\sum_{i \in B} \sum_{j \in \overline{B}} \pi_{ij} \tag{6}$$

Similarly, to meet the second condition, we have to minimize the global preference of all the elements of $\overline{B}$ over all the elements of B. Thus we have to find groups such that Eq. (7) is minimized.

$$\sum_{i \in B} \sum_{j \in \overline{B}} \pi_{ji} \tag{7}$$

To meet both conditions, we need to find groups such that Eq. (6) is maximized and Eq. (7) is minimized. A compromise solution can be found by detecting groups which maximize Eq. (8).

$$\sum_{i \in B} \sum_{j \in \overline{B}} (\pi_{ij} - \pi_{ji}) \tag{8}$$

We must therefore find B^* such that:

$$B^* = argmax \sum_{i \in B} \sum_{j \in \overline{B}} (\pi_{ij} - \pi_{ji}) \tag{9}$$

The proposition made by De Smet [6] states that B^* is determined by the set of alternatives which have positive net flow scores. Indeed, by symmetry, we have $\sum_{i,j \in B}(\pi_{ij} - \pi_{ji}) = 0$. As a consequence:

$$\sum_{i \in B} \sum_{j \in \overline{B}} (\pi_{ij} - \pi_{ji}) = \sum_{i \in B} \sum_{j \in A} (\pi_{ij} - \pi_{ji}) = (n-1) \sum_{i \in B} \phi_A(a_i).$$

By construction, each element of B is better ranked than elements of $\overline{B}$. The resulting subgroups are thus ordered. We process in the same way for each subset until either the cardinality of the subset equals one or the net flow scores of all the elements of a given subset equal zero. In the last case, we have, for a given subset $C \subseteq A, \forall a_i \in C : \phi_C(a_i) = 0$. Therefore elements belonging to C cannot be differentiated on the basis on the net flow scores. In this case, we face an indifference relation between the elements of the considered subset and all the alternatives have the same rank.

4   Comparison of the Two Methods

In order to compare the methods, we will use three data sets from 2012; the Human Development Index (HDI), the Environmental Performance Index (EPI) and the Academic Ranking of World Universities (Shanghai).

Firstly, we computed the rankings obtained with the two methods based on 50 randomly chosen alternatives of HDI. Tests have been repeated 10,000 times. For each criterion, the indifference and preference thresholds are respectively the first and the third quartiles of all the evaluations differences between the countries present in the HDI. The criteria are life expectancy at birth ($w_1 = 1/3$), mean years of schooling ($w_2 = 1/3$), expected years of schooling ($w_3 = 1/6$) and GNI per capita ($w_4 = 1/6$). We calculated Kendall's τ_B coefficient to investigate the correlation of the two rankings. We used the τ_B to be able to manage ties. Clearly, these correlations show a high degree of compatibility (see Fig. 2).

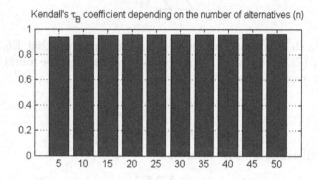

Kendall's τ_B coefficient depending on the number of alternatives (n)

Fig. 2. Kendall's τ_B coefficient (50 randomly chosen alternatives of the HDI).

The next step is to check that when two alternatives a_i and a_j are indifferent in the dichotomous method ($a_i I a_j$), the net flow scores in PROMETHEE II of these two alternatives are close. Therefore, we computed the difference $\Delta \phi_{ij} = \phi(a_i) - \phi(a_j)$ when two alternatives have the same ranking in the dichotomous method. Among all tests, the maximal value observed is $\Delta \phi = 0.23$. However, most of the values are less than 0.05 as shown in Fig. 3. Finally, Fig. 4 shows an histogram of the ranks differences between the two methods.

5   Rank Reversal

The notion of rank reversal (RR) covers different definitions. In this paper, we used the same as in Mareschal et al. [5]: *"Rank reversal occurs whenever the relative ranking of two alternatives in the global ranking is reversed when a third alternative is removed from A"*.

For all tests, we consider n alternatives and order them to get a first ranking C_0. We then remove one alternative and compute a new ranking. C_i denotes the

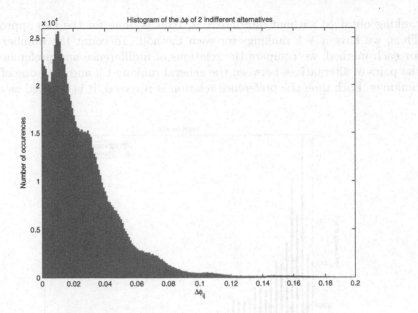

Fig. 3. Differences of net flow scores when two alternatives are indifferent according to the dichotomous method (50 randomly chosen alternatives of the HDI).

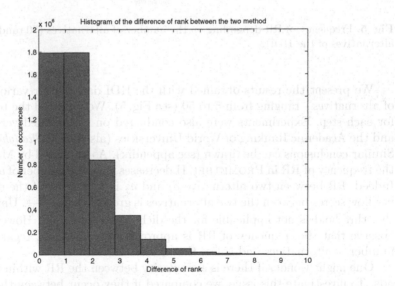

Fig. 4. Differences of ranks between the rankings of each alternative in the two methods (50 randomly chosen alternatives of the HDI).

ranking obtained without alternative a_i (this is done for the two approaches). Thus, we have $n + 1$ rankings for each method. To count the number of RR for each method, we compare the relations of indifference and preference of all the pairs of alternatives between the general ranking C_0 and each one of the C_i rankings. Each time the preference relation is reversed, it is counted as a RR.

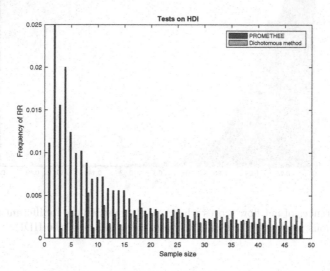

Fig. 5. Frequency of RR depending on the number of alternatives (50 randomly chosen alternatives of the HDI).

We present the results obtained with the HDI data set for various numbers of alternatives - ranging from 5 to 50 (see Fig. 5). We repeated the test 50 times for each step. Experiments were also conducted on a simplified version of EPI and the Academic Ranking of World Universities (also called *Shanghai* ranking). Similar conclusions can be drawn (see appendix). As predicted by Mareschal [5], the frequency of RR in PROMETHEE II decreases with the number of alternatives. Indeed, RR between two alternative a_i and a_j is impossible if the difference of net flow scores between the two alternatives is greater than $\frac{2}{(n-1)}$. Unfortunately, this threshold is not applicable for the dichotomous method. However, we can observe that the frequency of RR is approximately constant, regardless of the number of alternatives and is less that 0.5%.

One might wonder if there is a similarity between the RR within the 2 methods. To investigate this issue, we compared if they occur between the same pair of alternatives in Table 1, where Column 4 is the mean of the number of RR observed for each test. Table 1 may suggest that the RR are not similar between the methods.

Finally, we investigated how intra-preference parameters influence the number of rank reversal occurrences in both methods. Indeed, having small values for q and p leads to further discriminate the alternatives. Tests have been conducted

Table 1. Table of concordance and discordance pairs of RR between the 2 approaches.

Number of alternatives n	Concordant pairs (%)	Discordant pairs (%)	Number of RR (mean)
5	2.07	97.93	0.34
10	1.93	98.07	2.34
15	1.81	98.19	6.95
20	1.75	98.25	15.42
25	1.48	98.52	28.39
30	1.35	98.65	48.02
35	1.15	98.85	66.81
40	1.10	98.90	103.34
45	0.97	99.03	132.60
50	0.98	99.02	174.81

on the HDI data set for values q (respectively p) representing successively the 10% (90%), 30%(70%) and 45% (55%) decile of all the differences (see Figs. 6, 7 and 8). Clearly, we observe that these parameters have an influence on the number of rank reversal occurrences (especially in PROMETHEE II rankings). This observation was expected but deserves more attention.

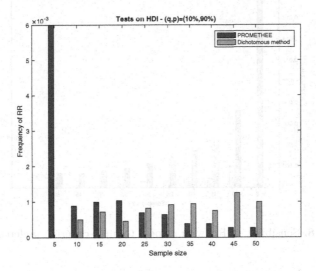

Fig. 6. Sensibility of RR with respect to intra-criterion preferences.

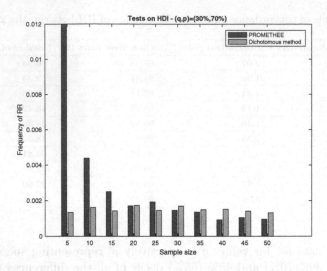

Fig. 7. Sensibility of RR with respect to intra-criterion preferences.

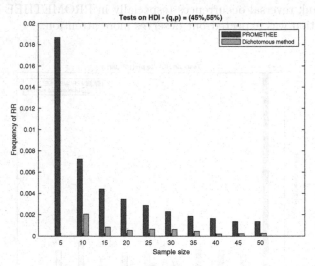

Fig. 8. Sensibility of RR with respect to intra-criterion preferences.

6   Conclusion

In this paper, we presented a new way to rank alternatives using a net flow score property. The new method, based on a dichotomous procedure, was compared to the traditional PROMETHEE II ranking. We showed that for small set of alternatives, rank reversal is less frequent than in PROMETHEE II. Then the frequency of rank reversal becomes negligible for both approaches.

Of course a number of open questions remains. At first it is interesting to study the performances of the new method. Indeed, it is more demanding than the traditional PROMETHEE methods. Regarding the experiments that were conducted, execution time was never an issue. In addition, let us mention the recent work of Calders and Van Assche [3] that allows to improve the computation performances of PROMETHEE-like rankings. The difference between the *type* of rank reversal in both approaches has to be further investigated. Finally, experiments seem to show that a bound exist regarding the frequency of RR on the new approach. If this observation is true, the determination of this analytical bound remains a challenge.

Appendix: Results Related to EPI and Shanghai

Results related to EPI and the Academic Ranking of World Universities are shown on Figs. 9 and 10.

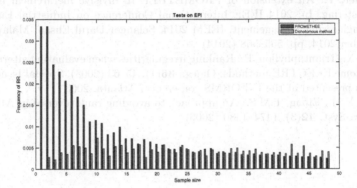

Fig. 9. Frequency of RR depending on the number of alternatives (50 randomly chosen alternatives of the EPI (simplified)).

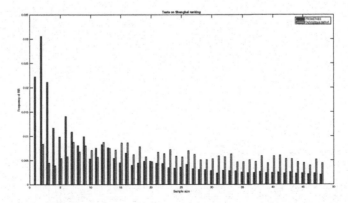

Fig. 10. Frequency of RR depending on the number of alternatives (50 randomly chosen alternatives of the Shanghai).

References

1. Brans, J.P., De Smet, Y.: PROMETHEE methods. In: Greco, S., Ehrgott, M., Figueira, J. (eds.) Multiple Criteria Decision Analysis: State of the Art Surveys. ISOR, vol. 233, pp. 187–219. Springer, New York (2016). https://doi.org/10.1007/978-1-4939-3094-4_6
2. Brans, J.-P., Vincke, P., Mareschal, B.: How to select and how to rank projects: the PROMETHEE method. Eur. J. Oper. Res. **24**, 228–238 (1986)
3. Calders, T., Van Assche, D.: PROMETHEE is not quadratic: an $o(qnlog(n))$ algorithm. Omega **76**, 63–69 (2018)
4. De Keyser, W., Peeters, P.: A note on the use of PROMETHEE multicriteria methods. Eur. J. Oper. Res. **89**(3), 457–461 (1996)
5. Mareschal, B., De Smet, Y., Nemery, P.: Rank reversal in the PROMETHEE II method: some new results. In: 2008 IEEE International Conference on Industrial Engineering and Engineering Management, pp. 959–963, December 2008
6. De Smet, Y.: An extension of PROMETHEE to divisive hierarchical multicriteria clustering. In: 2014 IEEE International Conference on Industrial Engineering and Engineering Management, IEEM 2014, Selangor Darul Ehsan, Malaysia, 9–12 December 2014, pp. 555–558 (2014)
7. Wang, X., Triantaphyllou, E.: Ranking irregularities when evaluating alternatives by using some ELECTRE methods. Omega **36**(1), 45–63 (2008). Special Issue Section: Papers presented at the INFORMS conference, Atlanta, 2003
8. Wang, Y.M., Elhag, T.M.S.: An approach to avoiding rank reversal in AHP. Decis. Support Syst. **42**(3), 1474–1480 (2006)

A Viability Study of Renewables and Energy Storage Systems Using Multicriteria Decision Making and an Evolutionary Approach

Carolina G. Marcelino[1,3]([✉]), Carlos E. Pedreira[1], Manuel Baumann[2], Marcel Weil[2], Paulo E. M. Almeida[3], and Elizabeth F. Wanner[3,4]

[1] Intelligent Systems Laboratory, PESC - COPPE/UFRJ, Rio de Janeiro, Brazil
carolimarc@cos.ufrj.br
[2] Institute for Technology Assessment and Systems Analysis,
KIT, Karlsruhe, Germany
manuel.baumann@kit.edu
[3] Intelligent Systems Laboratory, CEFET-MG, Belo Horizonte, Brazil
[4] School of Engineering and Applied Sciences, Aston University, Birmingham, UK

Abstract. Renewable energy technologies use natural sources, such as wind and solar, to produce electricity. Nowadays, there is a global sustainable electric power generation pressure to alleviate environmental impacts caused by the usage of fossil fuels. Energy market is focused on improving those technologies by meeting customer needs, but it proves to be challenging. Renewable power production integrated with a Hybrid Micro-Grid System (HMGS), a power distribution system composed of one or more distributed sources, may provide a reliable and cost-effective solution. This paper proposes a grid-connected HMGS model able of planning energy production and operating in parallel autonomously or connected on a public grid. The optimization of such HMGS is done using a swarm evolutionary approach and the results are obtained using different battery technologies. A life cycle assessment model and a multicriteria decision making approach are carried out to perform a viability study of the battery technologies. Wind and solar meteorological data from four regions in the Minas Gerais state, Brazil, were used as input for the model. Results show that lithium ion batteries are the most recommendable ones, ensuring not only the minimal cost and losses in the system but also minimizing the environmental impact.

Keywords: Optimization · Evolutionary swarm ·
Multicriteria decision making · Life Cycle Assessment ·
Renewable energy · Smart grids

1 Introduction

A secure energy supply is a basic service that a community needs to guarantee comfort, to improve life quality, economy, and crucial aspects of life. However,

© Springer Nature Switzerland AG 2019
K. Deb et al. (Eds.): EMO 2019, LNCS 11411, pp. 655–668, 2019.
https://doi.org/10.1007/978-3-030-12598-1_52

specially in a rural area, energy supply on a continuous and reliable way is not an easy task. The lack of energy services has hampered economic progress since power access can be seen as the engine towards development of any society. Not only connection to distant grids is expensive to be cost-effective for rural areas in large countries, but also to electrify those areas usually inhabited by poor people may not be a priority.

Renewable energy sources based on Hybrid Micro Grid Systems (HMGS) represent a cost-effective way for solving the energy supply issue in rural areas which are located near from the grids. The installation of HMGS – several parallel connected distributed resources with electronic controlled strategies – can be seen as a possibility to integrate such distributed electricity sources into the public grid or to enable safe standalone power systems. HMGS can offer an optimal and reliable service using, as example, clusters of small generators, loads and battery energy storage systems connected through a local electricity network, controlled by a power management system to optimize power flows [1].

A challenge for those systems is to provide a good balance between generation and load while maintaining frequency and voltage levels. Batteries systems (BESS) represent an important component for HMGS, balancing load and generation from the energy sources within seconds. There are distinct BESS technologies available, each one showing advantages and disadvantages. This work presents a viability study to compare different BESS architectures with focus on finding the best option to a HMGS installation from an environmental perspective. Typical goals related to HGMS energy planning can be posed as:

1. the minimization of the electricity cost;
2. the minimization of the power losses probability or breakdowns;
3. the maximization of the sustainable source usage in the HMGS;
4. the minimization of the environmental impacts.

When choosing the components to design renewable systems and the available storage technologies, those questions must be taken into account. Several attempts have been done to design and operate such small electrical systems in an efficient and sustainable way. In [2], a particle swarm optimization (PSO) algorithm was used to optimize the network topology of a HMGS and to maximize its total net present worth. This is achieved by optimizing the amount of BESS, photovoltaic (PV) units and wind generators over the considered project planning horizon. Similar approaches can be found in [1–6].

This work proposes an optimization model that contains a fitness function combining the goals 1, 2 and 3. The fourth goal is obtained from the optimization results in a separate environmental model. The contribution of this work is threefold: (i) to optimize the operation planning and (ii) to provide decision aid for the optimal choice of different BESS technologies in a HMGS through a Life Cycle Assessment (LCA) approach, and (iii) a Multicriteria Decision Making (MCDM) procedure. The wind and solar meteorological data of four regions in the Minas Gerais state, located at the southeast part of Brazil, are used as input for the model. An evolutionary swarm approach, the Canonical Differential Evolutionary Particle Swarm Optimization (C-DEEPSO) algorithm [7], is used to solve the optimization problem. Optimized goals serve as a baseline for an exploratory LCA and all results are used to determine the most suitable BESS for the smart grid in a multicriteria decision approach.

This paper is organized as follows. Section 2 presents the proposed HMGS model and the pre-processing of the meteorological data. Section 3 describes the decision making methods used in the approach. The LCA is described in Sect. 4. Section 5 discusses a case study, experiments and the results. Finally, Sect. 6 concludes the paper and gives an outlook for future work.

2   The HMGS Model

This work improves a power dispatch model presented in [3] aiming to provide electrical energy in a HMGS. The model has two main goals: the minimization of total production cost and the minimization of power losses in the grid. A simplified scheme of a hybrid system is given in Fig. 1 (left). The diagram shows typical equipments of a HMGS: photovoltaic panels, BESS, wind turbines and a control system. A HMGS can also be connected to a public grid. In this work, we carry out a viability study of a HMGS for four regions in Minas Gerais state, Brazil. This area is chosen due to the good seasonal conditions existing on the aforemention region. It is important to say that our approach can be applied to any other locality with similar weather characteristics.

Minas Gerais, showed in Fig. 1 (right), is one of out 26 Brazilian states, and it ranks as the fourth largest state by area and the second most populous one. This state has an area of 586.528 km^2 and the landscape is mainly marked by plateaus, hills and mountains. The rugged landscape gives the state a privileged amount of water resources. The hydroelectric potential estimated by Eletrobras (the Brazilian energy company), in the state of Minas Gerais, is 24.710 MW, making it the third largest in the country.

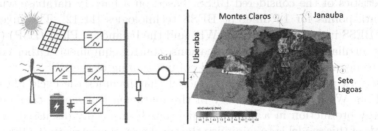

Fig. 1. Left: Simple diagram of HMGS – Right: Promising regions to implement HMGS in Minas Gerais [8].

The predominant vegetation and roughness of the terrain contribute to seasonal influence of aerodynamics in the displacement of winds. Energy planning in Brazil depends on the mapping of areas according to the legislation for use and occupation of the Brazilian soil. The deployment of wind farms is not only commonly affected by many environmental issues but also by the requirement of large areas for installation. The state of Minas Gerais has four wind potential areas and those areas can also be seen in Fig. 1(right). The cities of Janauba, Montes Claros, Sete Lagoas, and Uberaba are not only situated in areas of wind potential but also present high annual solar irradiation.

In this work, new boundaries, side conditions and techno-economic values are taken into account to improve the model presented in [3]. Additionally, the model is tested with four available BESS technologies, two Li-Ion batteries: Lithium Iron phosphate (LFP) and Lithium Nickel Cobalt Aluminum Oxide (NCA); one Valve Regulated Lead Acid battery (VRLA), and one high temperature battery: Sodium Nickel Chloride (NaNiCl) also referred as Zebra battery.

Yearly time series for a typical standard load profile, solar radiation [9], hourly wind speed and ambient temperatures of each city shown in Fig. 1, serve as inputs for the model. A test scenario, with a maximum number of 100 households, is used in the simulation. HMGS model includes the following economic parameters representing the general characteristics of the components:

- Inverter: efficiency = 92%, life time = 24 years and initial cost = 643 \$/kW;
- PV: regulator efficiency = 95%; life time = 24 years; initial cost = 3400 \$/kW, rated power = 7.3 kW, and PV regulator cost = \$1500;
- Wind: rated speed = 9.5 m/s, rated power = 5 kW, price = 6985 \$/kW, life time = 24 years, swept area = 128.6 m, wind regulator cost = \$1000, blades diameter = 6.4 m, efficiency = 95%, cut out and cut in [25, 2.5] m/s;
- Economic parameters: public grid energy cost = \$0.31, discount rate = 8%, real interest = 13%, inflation rate = 5%, O&M+running cost = 20%.

Techno-economic values and performance curves for wind turbines are taken from [10]. Power control system costs (inverter) are scale dependent and are taken from [11]. The cost for public grid energy is averaged from the energy company in Minas Gerais. Table 1 gives a brief overview of the main techno-economic characteristics of the considered BESS, based on a battery database with over 5.000 data points for 14 different BESS technologies [11,12]. The investment costs of BESS include the cells (\$/kWh) and the balance of Plant (BoP) (\$/kW) includes auxiliary devices, communications, control equipment. Other costs are related to installation, permitting and commissioning of the BESS.

This work proposes an improved model that includes the cycle life time of each BESS. We present a summary of the mathematical model that describes the energy production in a HMGS, which has been deeply discussed in [3,10]. The goals of this model is to minimize the total cost of electricity (COE) and the loss of power supply probability (LPSP), and to maximize the use of renewable sources in HMGS. The COE (\$/kWh) can be obtained by,

$$COE = \frac{Total_{costs}}{\sum_{h=1}^{h=8640} P_{load}(h)(kWh)} \times CRF, \tag{1}$$

in which $Total_{costs}$ represents the sum of the initial cost (IC) given by personnel costs, installation and connections, periodic costs PW_p given by maintenance of

Table 1. Input data for BESS based on median values [11].

Factor	Unit	VRLA	LFP	NCA	NaNiCl
Cost	$/kWh	230.36	308.87	212.5	220.3
Cycles	-		1400	5000	3000	3000
Efficiency	%	77	92	92	86
Life time	years	18	10	10	14
BoP	$/kW			374	374	
Other cost	$/kW			328	328	

photovoltaic panels, maintenance of wind generator, among others, and non-recurrent cost PW_{np} characterized as the cost of BESS replacement.

Power consumption over time is given by P_{load} and CRF represents the present value of all components equally distributed over the project life time. There are some new improvements in the model presented in [3]. The previous model considers a standalone system meaning that the majority of the energy resources comes from renewable sources and, in case of intermittent nature of wind and solar energy, the diesel generators are used to overcome it.

The proposed model considers a grid-connected systems allowing the wind turbines and photovoltaic cells to be used in synchronised connection with a public grid supply. Another novelty in the proposed model is the inclusion of a new factor called degradation cost related to the BESS. The degradation cost c_d [13] is introduced to provide a more realistic scenario to calculate COE. It considers battery degradation in terms of available cycle lifetime L_c at a certain depth of discharge (DoD) related to total battery cell costs c_{bat} as indicated,

$$c_d = \frac{c_{bat}}{L_c E_s DoD}.$$ (2)

Using the degradation cost, in our proposed model, $Total_{cost}$ is calculated,

$$Total_{cost} = IC + PW_p + PW_{np} + \sum_{h=1}^{8640} c_d,$$ (3)

and the total cost of electricity, COE, can be obtained in terms of $/kWh as,

$$COE = \frac{IC + PW_p + PW_{np} + \sum_{h=1}^{8640} c_d}{\sum_{h=1}^{8640} P_{load}} \times CRF,$$ (4)

Statistical techniques and chronological simulation approaches are used to calculate the LPSP. Another novelty in the proposed model is given by the inclusion of two factors in the calculation of LPSP: (i) the power generated by the public grid which, in some situations, may generate exceeding energy. The excess power may be used to charge the BESS. It provides a more environmentally friendly approach; and (ii) the state of charge of the battery. The state of charge of the BESS is a measure of the short term capacity of the battery and it changes over time since the battery capacity gradually reduces as it ages. This inclusion provides a more realistic model. For calculating the LPSP, time series data in a given period are based on the energy accumulative effect of BESS as expressed,

$$LPSP(\%) = \frac{\sum P_{load} - P_{pv} - P_{wind} + P_{soc_{min}} + P_{grid}}{\sum P_{load}}, \tag{5}$$

in which P_{load} is the hourly power consumption, P_{pv} and P_{wind} are the power generated by PV and by the wind generator respectively, $P_{soc_{min}}$ is the minimum state of charge of the battery, and P_{grid} is the power generated from public grid.

The amount of energy generated in the renewable source (RWF) system is used as a boundary to determine the amount of energy coming from a public grid as compared to the renewable generation. An ideal system based on renewable resources only would have a RWF of 100%. The higher renewable factor, the lower the environmental impact. The RWF is calculated as,

$$RWF(\%) = \left(1 - \frac{\sum P_{grid}}{\sum P_{pv} + \sum P_{wind}}\right) \times 100. \tag{6}$$

The resultant optimization problem is a bi-objective one aiming to minimize the total cost of electricity (Eq. (4)) and to minimize the loss of power supply probability (Eq. (5)). The boundary constraints (lower and upper bounds) of the decision variables are given by: nominal power of PV ([10, 150] in kW); autonomy grade for the BESS ([1, 3] in hours); number of wind turbines ([1, 10]), and nominal power of the public grid ([10, 200] in kW).

In HGMS, the COE and the $LPSP$ are equally important since the obtained system must guarantee reliable and uninterrupted energy supply at a competitive cost. Furthermore, the system must be as closest as possible to an ideal system based on renewable sources only. Therefore, the normalized fitness function adopted in this work is,

$$\min F = 0.5 \times COE + 0.5 \times LPSP + \rho \sum_{i=1}^{n} \max [0, RWF]^2, \tag{7}$$

in which ρ is a penalty factor associated to the RWF constraint.

The mono-objective optimization problem given by Eq. (7) is solved using the C-DEEPSO [7], a hybrid single-objective metaheusristic which incorporates distinct features of Evolutionary Programming, Particle Swarm Optimization, and Differential Evolution. Swarm evolutionary approaches have been widely applied for solving power systems optimization problems [3,7,14] to name a few. It is worthwhile to mention that the focus of this paper lies on the proposed multicriteria decision making based approach to conduct the viability study of the available BESS. Any other swarm algorithm, or other evolutionary approach, could have been applied to solve the problem.

The optimization problem (Eq. (7)) must be solved for each available BESS technology: Lithium Iron phosphate (LFP), Lithium Nickel Cobalt Aluminum

Oxide (NCA), Valve Regulated Lead Acid battery (VRLA), and Sodium Nickel Chloride (NaNiCl). The viability study using the Life Cycle Assessment is then carried out using the optimal values for each battery.

3 Decision Making Methods: AHP+TOPSIS

Multicriteria Decision Making (MCDM) methods can be roughly schematized by a construction phase (input data and the modeling phase including the interface with stakeholders) and an exploitation phase (the aggregation and calculation leading to recommendations). There are three sets of methods available: (i) elementary methods (e.g., weighted sum method); (ii) single synthesizing methods, (Analytic Hierarchy Process – AHP [15], Technique for Order Preference by Similarity to Ideal Solution – TOPSIS [16]); and (iii) hybrid approaches as AHP+TOPSIS [17,18]. Each of these methods has its strengths and weaknesses and should be selected with care for each assessment. In this work, AHP is used for the construction phase to elicit weights and TOPSIS is considered as a suitable approach for the exploitation phase (to rank considered alternatives).

The AHP is based on judgments on comparative elements. It measures the relative importance through pairwise comparison matrices. These can then be recombined to achieve a overall rating of alternatives. Inconsistency is a consequence of the attempt to derive a priority through the comparison of two objects at the same time. Inside AHP, aiming to avoid inconsistency, the geometric consistency index (φ) is applied [19]. TOPSIS lacks a procedure to determine the importance of considered criteria. The AHP represents such procedure, but is less efficient in dealing with tangible attributes and number of alternatives to be addressed [18]. Therefore, TOPSIS represents an efficient and easy way for criteria aggregation. It is based on the idea of [16] that a chosen alternative should have a minimum distance to the positive ideal solution A^*, and a maximum distance to the negative ideal solution A^-. Finally, it is necessary to compute the distance to ideal solution (ς) for ranking the alternatives. The terms φ and ς can be seen in Eq. (8),

$$\varphi = \frac{2}{(n-1)(n-2)} \sum_{i<j} \log^2 e_{ij} \quad , \quad \varsigma_i = \frac{D_i^-}{D_i^* + D_i^-} \quad i = 1, 2, ..., n, \quad (8)$$

in which, given a pairwise comparison matrix $M = (a_{ij})$ with $i, j = 1, ..., n$ and the vector of priorities, w, the $e_{ij} = a_{ij} w_j / w_i$ is considered as the error obtained when the ratio w_i / w_j is approximated by a_{ij} (see [15,19]). The φ values corresponds to a Consistence Ratio (CR)$\leq$0.1 are: $\varphi = 0.31$ for $n = 3$ and $\varphi = 0.35$ for $n = 4$ according to [19]. The best solution is presented by $\varsigma_j^* = 1$ if $(A_j = A^*)$ and the worst by $\varsigma_j^* = 0$ if $(A_j = A^-)$. Ranking is carried out by the descending order of ς_j, where the highest value represents the better performance [17]. TOPSIS inhibits the danger of rank reversal [20]. In our approach, the AHP results are used as weights for TOPSIS, merging the construction and exploitation phases as described in [17]. The process is summarized in Fig. 2.

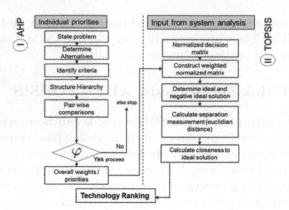

Fig. 2. Scheme of the adopted MCDM-model for evaluating sustainability of BESS.

4   Life Cycle Assessment

The production phase of the batteries is very energy consuming. Moreover, at the end of their useful life, the batteries must be disposed of and the residues must be dealt with. The life cycle assessment (LCA) allows to assess the environmental impacts and resources used throughout a product life cycle, from raw material acquisition, via production and use phases, to waste management [21]. There are several impact categories available including global warming potential (GWP, in kgCO2-eq), Human Toxicity - cancer effects (CTUh) and Particulate matter (kgPM 2.5-eq).

In this work, the optimization results for each battery technology (VRLA, LFP, NCA, and NaNiCl) provide the input for LCA in the form of renewable generation ratios, yearly operation hours and storage cycles. In this way, it is possible to analyze environmental impacts (EI) over the entire life cycle of the system equipment, such as BESS, PV, and wind turbines related to the four considered locations. Every LCA requires the definition of system boundaries and a functional unit [12]. In this case, EIs are based on the functional unit per converted kWh. All EIs related to a kWh are summed up and divided by the sum of generated power energy (P_n) in the system as indicated by Eq. (9),

$$EI_{kWh} = \frac{\sum_{n=1}^{n=8640} EI_n}{\sum_{n=1}^{n=8640} P_n}. \tag{9}$$

The global warming potential (GWP) is used as an example for the potential environmental impacts of considered system components. The end-of-life stage is excluded in the present study due to the lack of data for recycling processes and the corresponding environmental impacts or benefits for certain battery types and small wind turbines. The GWP of producing 1 kg of battery or 1 kWh of storage capacity, respectively, are given in Table 2.

Table 2. GWP CO_{2-eq} values for each BESS technology [11,12]

	VRLA	LFP	NCA	NaNiCl
Kg	2.7	15.0	18.6	15.2
kWh	51.6	158.0	115.7	115.7

Additionally, the system boundaries are used to analyze potential implications resulting in a single technology perspective scenario as follows. EI of the BESS itself only considers internal losses due to energy conversion losses in the storage unit and the inverter.

5   Experiments and Results

C-DEEPSO algorithm was used to optimize the HMGS model using an empirical parameter initialization based on previous works: population size (50 individuals), communication rate (0.9), mutation rate (0.5), number of generations (50). The hybrid evolutionary swarm method was executed 30 times for each city/BESS to obtain averaged results (shown in Table 4) that were used as input to LCA.

Figure 3 provides an exemplary insight of HMGS operation as a result of C-DEEPSO optimization using Sete Lagoas data. It can be observed how the BESS was charged and discharged, at least once a day, over the period of one week. The public grid was used when there was not enough available energy coming from PV, Wind and BESS.

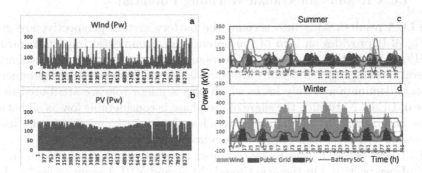

Fig. 3. Typical results of the HMGS optimization obtained by C-DEEPSO. (a) power production by PV - kW. (b) power production by wind turbines - kW. (c) week production in HMGS ongrid at summer. (d) week production in HMGS on grid at winter.

The HMGS economic viability can be shown in a COE analysis presented in Table 3. The power provided by the public grid costs \$0.31 kWh and the community of 100 houses needs a load of 200 kWh. In one year, the cost of electrical energy supply using only public grid is \$535.680. Analyzing only cost results from COE, using LFP and NCA technologies, it is possible to note that the aggregated costs of Uberaba, Montes Claros, Sete Lagoas and Janauba using LFP for annual period are: \$345.600, \$328.320, \$293.760 and \$241.920, respectively. In contrast, the cost values using NCA for those cities are: \$328.320, \$311.040, \$276.480 and \$241.920, respectively. However, our study contemplates other criteria, such as minimization of losses and environmental impacts as the renewable sources maximization described in Sects. 5.1 and 5.2.

Table 3. Optimization results for each city/BESS. Mean (m)/ Standard deviation (s).

Goal		UBERABA				MONTES CLAROS			
		VRLA	LFP	NCA	NaNiCl	VRLA	LFP	NCA	NaNiCl
LPSP (%)	m	0.02	0.04	0.04	0.03	0.02	0.04	0.04	0.03
	s	0.05	0.03	0.03	0.04	0.05	0.03	0.03	0.04
COE (\$/kWh)	m	0.29	0.20	0.19	0.21	0.26	0.19	0.18	0.21
	s	0.00	0.01	0.01	0.00	0.01	0.01	0.01	0.00
RWF (%)	m	0.69	0.77	0.77	0.74	0.70	0.79	0.79	0.76
	s	0.01	0.01	0.01	0.00	0.00	0.00	0.00	0.00
Goal		SETE LAGOAS				JANAUBA			
		VRLA	LFP	NCA	NaNiCl	VRLA	LFP	NCA	NaNiCl
LPSP (%)	m	0.02	0.03	0.03	0.03	0.02	0.03	0.03	0.03
	s	0.04	0.03	0.03	0.04	0.04	0.03	0.03	0.04
COE (\$/kWh)	m	0.23	0.17	0.16	0.17	0.20	0.14	0.14	0.15
	s	0.03	0.02	0.03	0.02	0.05	0.04	0.04	0.03
RWF (%)	m	0.75	0.83	0.83	0.80	0.79	0.86	0.86	0.84
	s	0.03	0.05	0.05	0.03	0.01	0.01	0.01	0.01

5.1   LCA Results for Global Warming Potential

The LCA results taking into account the battery system perspective are given in Fig. 4. It provides an idea of the potential environmental impacts regarding GWP related to the use of different storage systems in the distinct locations. In general, rankings for all regions are comparable on battery electric production impacts dominate and only losses which play a minor role are attributed to life cycle of BESS. Naturally, the share of use phase is considerable low as all battery types have relatively high efficiency grades of over 70% and due to the use of renewable energies which have a low environmental burden.

LFP and NaNiCl seem to be the most recommendable technologies in this type of perspective. On the other hand, NCA and VRLA seem to lead to high impacts due low life time and related high rate of battery exchange. Additionally, one can observe that there are differences when the four locations are compared to each other. Montes Claros and Uberaba show a change in the performance of LFP and NCA as a result of a slightly lower cell exchange demand for NCA due to a smaller number of operation cycles, leading to a lower GWP.

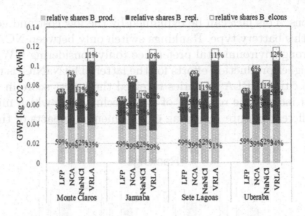

Fig. 4. GWP for single technology perspective on BESS in the 4 locations.

5.2 MCDM Results and Discussions

Six experts from academia with a focus on electrical engineering and mathematical modeling were involved to carry out the AHP. For this purpose, an automated VBA sheet with a consistency check was provided and was distributed among the experts. All experts provided consistent preferences. An overview of all preferences is given in Table 4. It has been mentioned that aggregating single preferences to a group preference deso not automatically indicate group consensus (one can,e.g., observe one conflict of goals between expert 1 and two regarding the importance of GWP.

Table 4. Overview of single and group preferences.

	Exp.1	Exp.2	Exp.3	Exp.4	Exp.5	Exp.6	$\bar{\varphi}$
COE	0.03	0.10	0.16	0.09	0.06	0.17	0.09
RS-factor	0.20	0.26	0.16	0.43	0.28	0.42	0.27
LPSP	0.08	0.45	0.39	0.03	0.10	0.05	0.12
GWP	0.56	0.14	0.03	0.43	0.51	0.27	0.23

Thus, an additional sensitivity analysis was conducted to depict the impact of different group weights. Three weight scenarios were used for this purpose as follows: (i) a strong cost perspective where COE was weighted the highest score (nine) and the other criteria were equally weighted (score of one), (ii) the same perspective for global warming potential and renewable shares, and (iii) all criteria were set to equal importance. The criterion of LPSP was not analyzed in the sensitivity analysis as all battery types provide very similar results here. The ranking of technologies was the same for the considered cities. Figure 5 shows the averaged results in which LFP can be considered as the best choice for the locations. This fact can be explained by the low cost and good technical properties of this battery type. Rankings switch only between NCA and LFP in the case of strong environmental preference (only considering GWP) which is a result of the higher production efforts for the latter. The NaNiCl is the third-best option for the cities. VRLA batteries received the lowest score in this analysis. Results do not include the recycling of these technologies which might lead to a completely different picture regarding environmental impacts of these systems.

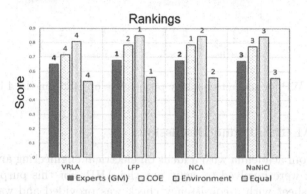

Fig. 5. TOPSIS results based on C-DEEPSO, AHP and LCA results.

6   Conclusion

The proposed Hybrid Micro Grid Systems model, that served as a starting point to make a viability analysis of regions to provide sustainable power energy, was improved in several ways including lifetime cycle to battery systems, being grid connected, using real seasonal time series and others. New techno-economic assumptions, four battery types (LFP - Lithium Iron Phosphate, NCA - Lithium Nickel Cobalt Aluminium, VRLA - Valve Regulated Lead Acid and NaNiCl - Sodium Nickel Chloride), new side conditions and calculations as well as new time series have been introduced to test the seasonal data of four distinct cities in Minas Gerais/Brazil. The optimization problem was solved using C-DEEPSO for all BESSs and cities, and the results were used to provide the viability study applying Life Cycle Assessment, AHP, and TOPIS were used as a multicriteria decision-making approach. The best choice after the LCA approach, focusing only on global warming potential, ranked LFP in first, NCA in second place, in most cases NaNiCl in third and VRLA on the fourth. The MCDM approach showed the viability HMGS implementation in Janauba is potentially more efficient with the use of LFP batteries. In an annual projection, the HMGS installation in this region using the LFP technology would save $290.000 in comparison with the usage of electricity via the public grid only. Further work should include the LCA using selected impact categories directly into a multiobjective optimization. Also, the AHP-TOPSIS approach should be directly integrated into C-DEEPSO to replace the weighted sum approach. Other MCDA methods as ELECTRE or MAVT should also be considered as alternative to the proposed compensatory approach. It would allow a deep understanding concerning which technology represents the most suitable one, not only using a techno-economic viewpoint but also environmental and systemic perspectives.

Acknowledgement. The authors would like to thank Brazilian research foundations CAPES, CNPq, FAPERJ, FAPEMIG and Helmholtz-Project Energy System 2050.

References

1. Levron, Y., Guerrero, J.M., Beck, Y.: Optimal power flow in microgrids with energy storage. IEEE Trans. Power Syst. **28**(3), 3226–3234 (2013)
2. Mohammadi, M., Hosseinian, S., Gharehpetian, G.: Optimization of hybrid solar energy sources/wind turbine systems integrated to utility grids as microgrid (MG) under pool/bilateral/hybrid market using PSO. Sol. Energy **86**, 112–125 (2012)
3. Borhanazad, H., Mekhilef, S., Ganapathy, V., et al.: Optimization of micro-grid system using MOPSO. Renewable Energy **71**, 295–306 (2014)
4. Barelli, L., Bidini, G., Bonucci, F.: A micro-grid operation analysis for cost-effective battery energy storage and RES plants integration. Energy **113**, 831–844 (2016)
5. Moghaddam, A., Seifi, A., Niknam, T., Pahlavani, M.: Multi-objective operation management of a renewable MG (micro-grid) with back-up micro-turbine/fuel cell/battery hybrid power source. Energy **36**, 6490–6507 (2011)
6. Marcelino, C., Baumann, M., Carvalho, L., et al.: A combined optimization and decision-making approach for battery-supported HMGS. JORS, Accepted 2018
7. Marcelino, C.G., Almeida, P.E.M., Wanner, E.F., et al.: Solving security constrained optimal power flow problems: a hybrid evolutionary approach. Appl. Intell. **48**, 3672–3690 (2018). https://doi.org/10.1007/s10489-018-1167-5
8. CEMIG: Atlas eólico de minas gerais. Technical report (2010)
9. SoDa: Time series of solar radiation data - for free (2016). http://www.soda-pro. com/
10. Baumann, M., Marcelino, C., Peters, J., et al.: Environmental impacts of different battery technologies in renewable hybrid micro-grids. In: ISGT Europe, pp. 1–6 (2017)
11. Baumann, M., Peters, J., Weil, M., et al.: CO2 footprint and life cycle costs of electrochemical energy storage for stationary grid applications. Energy Technol. **5**, 1071–1083 (2016)
12. Peters, J., Baumann, M., Zimmermann, B., Braun, J., Weil, M.: The environmental impact of Li-Ion batteries and the role of key parameters - a review. Renew. Sustain. Energy Rev. **67**, 491–506 (2017)
13. Kemptin, W., Tomic, J.: Vehicle-to-grid power fundamentals: calculating capacity and net revenue. J. Power Sources **144**, 268–279 (2005)
14. Baños, R., et al.: Optimization methods applied to renewable and sustainable energy: a review. Renew. Sustain. Energy Rev. **15**, 1753–1766 (2011)
15. Saaty, T.L.: The Analytic Hierarchy Process: Planning, Priority Setting, Resource Allocation. RWS Publications, Pittsburgh (1990)
16. Hwang, C.-L., Yoon, K.: Multiple Attribute Decision Making: Methods and Applications; A State-of-the-Art-Survey, p. 1981. Springer, Berlin (1981). https://doi. org/10.1007/978-3-642-48318-9

17. Zaidan, A.A., Zaidan, B.B., Al-Haiqi, A., Kiah, M.L.M., Hussain, M., Abdulnabi, M.: Evaluation and selection of open-source EMR software packages based on integrated AHP and TOPSIS. J. Biomed. Inform. **53**, 390–404 (2015)
18. Chakladar, N.D., Chakraborty, S.: A combined TOPSIS-AHP-method-based approach for non-traditional machining processes selection. J. Eng. Manuf. **12**, 1613–1623 (2008)
19. Aguaron, J., Moreno-Jimenez, J.M.: The geometric consistency index: approximated thresholds. Eur. J. Oper. Res. **147**, 137–145 (2003)
20. Garcia-Cascalesa-Cascales, M.S., Lamata, M.T.: On rank reversal and TOPSIS method. Math. Comput. Model. **56**(5–6), 123–132 (2012)
21. Finnveden, G., et al.: Recent developments in life cycle assessment. J. Environ. Manag. **91**, 1–21 (2009)

Applications

Analysing Optimisation Data
for Multicriteria Building Spatial Design

Koen van der Blom[1]([envelope]) [iD], Sjonnie Boonstra[2] [iD], Hèrm Hofmeyer[2], and
Michael Emmerich[1]

[1] Leiden Institute of Advanced Computer Science, Leiden University,
Niels Bohrweg 1, 2333 CA Leiden, The Netherlands
{k.van.der.blom,m.t.m.emmerich}@liacs.leidenuniv.nl
[2] Department of the Built Environment, Eindhoven University of Technology,
P.O. Box 513, 5600 MB Eindhoven, The Netherlands
{s.boonstra,h.hofmeyer}@tue.nl

Abstract. Domain experts can benefit from optimisation simply by getting better solutions, or by obtaining knowledge about possible trade-offs from a Pareto front. However, just providing a better solution based on objective function values is often not sufficient. It is desirable for domain experts to understand design principles that lead to a better solution concerning different objectives. Such insights will help the domain expert to gain confidence in a solution provided by the optimiser. In this paper, the aim is to learn heuristic rules on building spatial design by data-mining multi-objective optimisation results. From the optimisation data a domain expert can gain new insights that can help engineers in the future; this is termed *innovization*. Originally used for applications in mechanical engineering, innovization is here applied for the first time for optimisation of building spatial designs with respect to thermal and structural performance.

Keywords: Data analysis · Building spatial design ·
Multicriteria optimisation · Mixed integer optimisation ·
Evolutionary algorithms

1 Introduction

During early phases of building design, decisions are made that significantly influence the quality of the final design. As such, optimising during the early stages can have substantial benefits. One of the first design steps entails capturing the building spatial design (BSD). Since numerous disciplines are involved in building design, multiple objectives have to be considered in the optimisation process as well. Here the focus lies on the structural and thermal performance.

Up till now, a mixed-integer representation was defined for the BSD problem in [6,8]. Based on this representation, multi-objective evolutionary algorithms

© Springer Nature Switzerland AG 2019
K. Deb et al. (Eds.): EMO 2019, LNCS 11411, pp. 671–682, 2019.
https://doi.org/10.1007/978-3-030-12598-1_53

have been devised, along with specialised operators in [4,5]. Further, the application of the hypervolume indicator gradient [11], to improve local search, was studied in [3], which resulted in a considerable amount of optimisation data.

Despite all this progress, the transfer of an optimisation result to a design expert is not merely a matter of stating "this solution is better than the previous one". The design expert needs to be convinced that the optimisation result is based on sensible design rules. Therefore, the optimisation result needs to be made explainable. In addition to being able to provide an optimised design, and being able to explain why it works well, it may also be possible to learn new design rules from an optimised design. Once the designer has obtained a set of proven design rules, they may apply these to similar problems, without having to endure another lengthy optimisation process. Additionally, such rules can also be integrated in co-evolutionary design algorithms like those considered in [7].

The process of learning innovative design rules from optimisation data was introduced in [10], and termed *innovization*. This concept has since been applied to a variety of problems such as clutch brake design in [10], and truss design in [1]. Later, the learning process was interleaved with the optimisation process in [13], and further automated in [1,9]. Furthermore, in [2] it was studied how an optimiser learns new concepts over time. Here it is investigated if simple techniques used to verify optimisation results may also lead to innovative insights.

This work is a first step in applying innovization in BSD. The following contributions are made: Optimisation results are verified through data analysis of a subset of the 800,000 solutions found by multi-objective optimisation in [3]. Handling a dataset of this size also results in new challenges. With this in mind, here simple and computationally inexpensive analysis techniques are applied.

From here on, this paper first introduces the problem of finding heuristic rules for building spatial design in Sect. 2. Following that, in Sect. 3 features are defined to enable the discovery of such rules. The preparation of the considered dataset is then described in Sect. 4. Section 5 evaluates the results from analysis of the data, and the implications that follow. Finally, Sect. 6 briefly summarises the study, draws conclusions, and proposes possible directions for future work.

2 Problem

Building spatial design (BSD) constitutes the arrangement of the internal and external divisions of a building. These divisions together form a number of *spaces*. A space is similar to a room. However, it also encapsulates concepts such as open kitchen-living room combinations that are not structurally separated, or hallways. This work considers the multi-objective optimisation of a BSD for structural and thermal performance. Structural performance is measured by means of compliance. This measure aggregates the total strain energy in the structural elements that constitute the structural model related to the BSD. Whereas thermal performance is taken as the total heating and cooling energy required to maintain a comfortable temperature in a given BSD.

For an optimised BSD to be used, the solution must be trusted by the design expert. To inspire such confidence in the optimised design, the optimised results should be made explainable. This can be achieved by learning heuristic design rules from the optimisation data. Given such rules, it becomes clear why the design is effective. Ideally, not only known rules that experts trust and understand are obtained, but also new insights. By combining known and new design rules it is possible for experts, and automated (e.g. co-evolutionary [7]) design systems, to improve their design process. These improved design processes can then be applied to similar problems, without another lengthy optimisation procedure.

3   Features

The supercube representation introduced in [6,8] is a mixed-integer representation of the building spatial design (BSD) problem, consisting of binary and positive real numbers. Raw data in this format is difficult to interpret in terms of building properties, making it difficult to learn directly from this data. To ease this process, this section introduces elementary features that allow building engineers to characterise a BSD. Such features are necessarily domain specific. However, the same process may be applied in other domains.

Since the supercube representation is key to understanding the dataset and features it is briefly introduced in the following. Since it is used for BSD, the supercube representation considers a number of spaces that together form the BSD. Each space is defined as a cuboid (3D rectangle), such that the whole building consists of rectangular surfaces, like in Fig. 1. Additional constraints ensure that the floors of all spaces are connected with the soil via other spaces, that is, in the given representation no floating or overhanging spaces may exist.

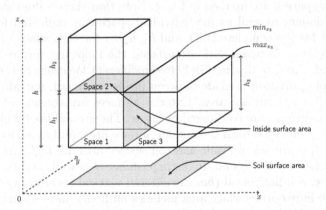

Fig. 1. Example building spatial design, annotated with a selection of features

All considered features are listed in Table 1 with their definitions and explanations. Except for the last three, all other features are computed both for the building and for individual spaces. Since the ordering of spaces is arbitrary,

Table 1. Features, definitions, and explanations

Feature	Definition	Explanation
vol	$w \times d \times h$	Volume of the space, or sum of spaces for the full BSD
short	$min(w, d)$	Shortest horizontal edge, indicator of span
long	$max(w, d)$	Longest horizontal edge, indicator of span
height	$max_z - min_z$	Height of the space or the full BSD
out	$sum(out\_area)$	Outside surface area, indicator of energy flow
in	$sum(in\_area)$	Inside surface area, indicator of energy flow
soil	$sum(soil\_area)$	Soil (ground floor) surface area, indicator of spread
horz	$sum(horz\_area)$	Horizontal surface area, indicator of total wall area
vert	$sum(vert\_area)$	Vertical surface area, indicator of floor and roof area
in_out	$in/(in + out)$	Ratio between inside- and outside surface area
out_vol	out/vol	Ratio between outside surface area and volume
long_short	$long/(long + short)$	Ratio between longest- and shortest horizontal edge
meanh	$sum(h \times roof\_area)/soil$	Mean height of the building
meanh_h	$meanh/height$	Ratio between the mean height and the height
height_soil	$height/soil$	Ratio between the height and the soil area

including values for each of them in the feature set would be of little use. Therefore, statistics are taken over all spaces in a building for each feature. In particular, the min, max, mean, median, range, standard deviation and Gini index (average deviation from the mean) are considered. Since the last three features in Table 1 do not make sense for individual spaces (e.g. mean height of a space is equal to its height), they are only computed for the building as a whole.

Values for w, d, h are found by taking $max_* - min_*$, where $*$ corresponds to x, y, z respectively. In other words, they are simply the distance between the minimal and maximal coordinates of a given dimension. For example, the min_x and max_x of space 3 are marked in Fig. 1. Note that these values are computed for the full design, as well as for individual spaces, as indicated for height in Fig. 1 with h for the complete BSD, and h_1, h_2, h_3 for each space.

To differentiate between various surfaces, the following surface area definitions are used. First, to distinguish between different locations of the surfaces, a non-overlapping division is made between inside (in_area), outside (out_area), and soil ($soil\_area$) surface area. Exterior surfaces are considered as outside, while interior surfaces are considered as inside. The ground floor which connects with the soil is excluded from the outside surface area, and taken as soil surface area. In Fig. 1, examples of inside and soil surface area are highlighted (the rest is outside surface area). Second, to distinguish between walls and floors/ceilings, a division between horizontal ($horz\_area$) and vertical ($vert\_area$) surface area is made. The horizontal surface area includes all floors and ceilings, so also the ground floor, while the vertical surface area consists of all walls, regardless of them being interior or exterior. Finally, the $roof\_area$ considered for $meanh$ is a part of the roof area in the BSD positioned at equal height.

Note that when considering a building as a whole, each surface is counted only once per considered distinction (e.g. horizontal/vertical). However, on the space level, surfaces are sometimes counted twice. That is, for two neighbouring spaces, both count their connecting surface as being part of, for instance, their horizontal surface area. As a result, the sum of the surface areas of all spaces is not (necessarily) equal to the total surface area of the building.

A few different features measure the same thing. For instance, the outside surface area of a building has an equal distribution (but not value) to the mean outside surface area of the spaces. Despite this, such features are kept to simplify data processing. In the analysis, only one representative should be used for these equivalent features, unless the differing values provide additional insights.

Additionally, some features may result in distributions similar to each other. This is particularly common for the range, standard deviation, and Gini index. However, even small differences may make one of them more valuable in distinguishing between solution classes than the other. Since, a priori, it is not known which is more useful in which situation, all of them are included.

Finally, it is noted that NAN values may appear in a few cases. Some spaces may be disconnected (meaning they do not share a wall with another space). As a result, it can occur in a building design that none of the spaces has a neighbour, from which it follows that their inside surface area is zero. In these cases, the Gini indices of the interior surface area, and of the ratio between inside and outside surface areas will be undefined and marked as NAN (the Gini index divides by the sum of the set of spaces, which is zero in this case). However, since these are very low quality solutions, they are not considered in the analysis in the rest of the paper. This will become clear in the next section.

4   Data Preparation

In order to learn heuristic rules for building spatial design, the dataset from the optimisation experiments in [3] is used. The dataset is a Pareto front and an archive from a building design optimisation that aimed for a BSD consisting of three spaces, with a total volume of $300\,\mathrm{m}^3$. Note that while these may seem like simple BSDs, they already require 9 continuous and 81 binary variables to encode with the supercube representation [6,8], leading to a large search space. The optimisation runs resulted in a dataset of around 800,000 solutions. Here the data is prepared for analysis in the following five steps. First, classes are defined to enable the discovery of different qualities in different groups of solutions. Second, the non-dominated (ND) set is identified. Third, the knee point solution is identified. Fourth, solutions are assigned labels to link them to a class, based on the previously identified ND set and knee point. Fifth, a procedure is described to equalise the number of solutions in each class for those analysis techniques that demand this. Note that all steps are defined such that they should at least be generalisable for two-dimensional convex Pareto fronts.

To be able to learn from the features defined in the previous section, the data is split into different classes. This is accomplished based on objective values, rather than features. Classification based on objective values allows for the verification of the optimisation procedure: Do design experts agree that the designs with good objective values are indeed good? In addition, it is often a combination of features that indicate a certain quality in the BSD, making feature based classification more complex. Further, by classifying on known good qualities of a BSD, finding innovative design rules would become very unlikely. Here, four categories of solutions are considered: the knee point area (KP), good in the compliance objective (F1), good in the heating/cooling energy objective (F2), and relatively bad solutions (BD). The aim is to data-mine for heuristic design rules that make it possible to differentiate between all of these distinct classes. For more objectives additional classes F* can be added as needed.

The classification considers two primary aspects: (1) It should clearly distinguish between the classes in the objective space, and (2) It should be computationally efficient to enable processing of the large dataset of ca. 800,000 points. The computational efficiency should also allow the proposed methods to generalise to larger BSDs than those considered here.

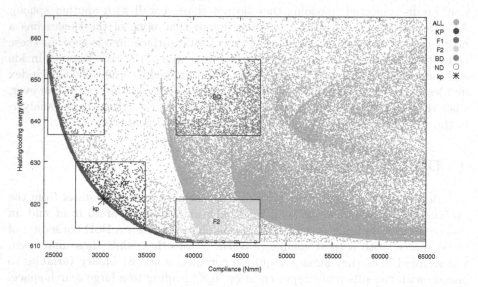

Fig. 2. Division of data into different classes: All points (ALL), knee point area (KP), objective one (F1), objective two (F2), bad solutions (BD); relative to the non-dominated set (ND), and the knee point (kp). A subset of the full dataset is shown

Since the considered classes are defined based on the non-dominated (ND) set and the knee point, these have to be identified first. For ND set computation the well-known log-linear time algorithm based on sorting is employed [12]. Based on the ND set, the knee point is derived as follows. First the objective values of the ND set are normalised to a $[0, 1]$ range, where outliers beyond 1.5 times the interquartile range are set to the appropriate boundary value. Next, the Euclidean distance to the origin $(0, 0)$ is computed for each normalised ND point. The point with the smallest distance is then taken as the knee point (indicated with 'kp' in Fig. 2), which is a reasonable approximation for the given dataset.

The data is then classified based on the knee point $p = (p_1, p_2)$, and the ND set. For this, the ND set is first reduced to the ND points that were not considered an outlier after normalisation, but the non-normalised values are used. In order to classify in a computationally efficient manner, each class is defined by a bounding box. These bounding boxes are found based on the range of the ND set in objective one r_1, and objective two r_2. For the knee point area class (KP) the lower bound of the box is set to $(0, 0)$, while the upper bound is set to $(p_1 + r_1 \times 0.2, p_2 + r_2 \times 0.2)$. For class F1 a lower bound of $(p_1 + r_1 \times 0.35, 0)$, and an upper bound of $(p_1 + r_1 \times 0.75, p_2)$ are taken. Similarly, F2 is found with the bounds $(0, p_2 + r_2 \times 0.35)$, and $(p_1, p_2 + r_2 \times 0.75)$. Lastly, BD uses the bounds $(p_1 + r_1 \times 0.35, p_2 + r_2 \times 0.35)$, and $(p_1 + r_1 \times 0.75, p_2 + r_2 \times 0.75)$. Following this, points are assigned a label based on the box they fall in. Any remaining unlabelled points are excluded from the analysis.

The result of the classification process is visualised in Fig. 2. Note that gaps are left between the different classes to improve the chances of being able to distinguish between them. If the classes would directly neighbour each other, points on the border are likely to have very similar features. This would impede learning what makes a solution perform well (or not) in one objective or the other. Future work could study how these points can be included in the analysis.

In Fig. 3 a randomly selected example of a BSD is shown for each class. Although the examples for KP and F1 look similar, the design for F1 is far more elongated. This result can be expected, as the short spans (here coupled with elongated spaces) allow F1 designs to reduce the strain energy, at the cost of a larger surface area, reducing thermal efficiency. The F2 design shows the reverse, with a much more compact design. Finally, the BD design is not as well arranged in the spatial sense, and shows relatively poor performance in both objectives.

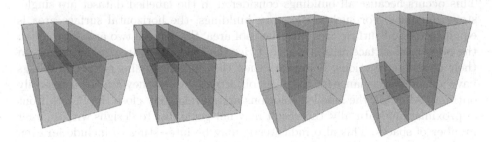

Fig. 3. Typical examples of the different classes, from left to right: KP, F1, F2, BD

After processing the dataset[1] 70,088 of the 806,430 solutions are labelled. With 5978 KP, 3400 F1, 48,482 F2, and 12,228 BD solutions respectively. Given the mixed-integer nature of the representation, multiple discrete subspaces can be seen in Fig. 2, indicated by the different curves. Since the dataset is not homogeneous, the resulting classes do not have an equal number of points. For

---

[1] The dataset is available under http://moda.liacs.nl/index.php?page=code.

some types of analysis, however, it is critical to have equally distributed classes. In such situations excess solutions are removed from the larger classes uniformly at random. In all other situations, all labelled data is used.

5   Results

Two techniques are used for data analysis, box plots and decision trees. Box plots give insight into the distribution of feature data for different solution classes. As such, it may be possible to identify features that allow for a clear distinction between two or more classes. Further, the decision tree can provide information about distinguishing features as well, since it generates clear rules based on such features. Moreover, it gives confidence measures for the classification of solutions to different classes. Finally, by using the learned decision tree on new data, it is possible to validate whether those rules can indeed be used reliably.

5.1   Box Plots

To generate box plots all labelled data is used, with each feature normalised to a $[0, 1]$ range, without removing outliers. In the plots, each class is then visualised by an individual box, such that any differences become clearly visible.

In Fig. 4 a subset of the features is shown that appears to allow for a significant amount of distinction between the different classes. Notice, for example, how the mean of the most extended horizontal edge (long.mean) enables differentiation between objective one (F1), and objective two (F2).

Surprisingly the soil surface area (soil.mean), and the horizontal surface area (not in the figure) showed exactly the same distribution in all of their features. This occurs because all buildings considered in the labelled dataset are single-story buildings. For such single-story buildings, the horizontal surface area is equal to the soil surface area plus the roof area. Since these two areas are equal, the horizontal surface area is exactly twice the soil surface area, which results in their equal distributions. It appears then that in general single-story buildings have a good performance for the given objectives, even if they are not necessarily optimal. After all, the labelled solutions are all relatively close the Pareto front approximation. Naturally, this result may not generalise to designs with a larger number of spaces. This also indicates it may be interesting to include an even worse class of solutions in future analysis to see how things differ with even worse solutions. Additionally, a feature indicating the number of stories a BSD has could be useful as well in this case. Even if just to identify this type of situation more easily.

5.2   Decision Trees

In order to use decision trees to their full potential, the data should be equally distributed among the classes. As such, this is carried out as described previously. Since the smallest class contains 3400 solutions, the other classes are reduced to

KP ⬜ F1 ▨ F2 ▨ BD ▨

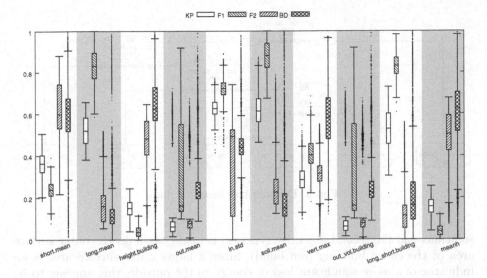

Fig. 4. Boxplot of a selection of distinguishing features

the same number of data points, resulting in a total of 13,600 solutions. This total is split into a training set of 10,200 solutions and a test set of 3400 solutions by sampling uniformly at random. Note that as a result, the representation of each class is not necessarily exactly equal in either of the training and test sets, but still sufficiently close. The training and test sets then consist of approximately 2550, respectively 850 solutions per class. Only labelled solutions are used, no normalisation is applied, and no outliers of individual features are removed. In the future it may be of interest to do the same study with unlabelled solutions to see if the generated rules generalise.

Given the prepared dataset, the decision tree in Fig. 5 was generated with the R package *rpart* [14]. From this figure, it can be found that the longest horizontal edge, the outer surface area, the ratio between the longest and shortest horizontal edge, and the ratio between the inner and outer surface area provide important information to distinguish between different classes of solutions.

These rules indicate properties of a building that contribute to qualities present in different solution classes. The first split shows that relatively long buildings (long.build) are likely to be efficient in objective one (compliance). This split intuitively makes sense, since buildings that are more stretched out are likely to have short spans. Note that this is under the assumption that not just the building is stretched out, but the spaces as well (e.g. F1 in Fig. 3).

In the other branch buildings are a bit more compact. Additionally, it can be seen that buildings where the minimal ratio of the spaces between the longest and shortest horizontal edge (long_short.min) is relatively high, are very likely to be solutions in the knee point area. This indicates that although the building as a whole is more compact, the individual spaces remain somewhat elongated to balance between the two objectives. The primary split between low quality

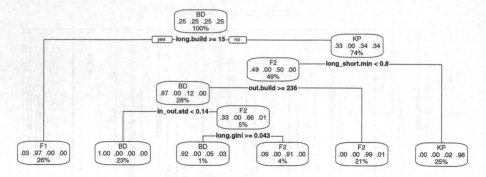

Fig. 5. Decision tree based on data

solutions and the second objective (energy) is made based on the outer surface area of the entire building (out.build). Since a larger outer surface area is an indicator of a more significant loss of energy to the outside, this appears to be a sensible rule. Further, these rules provide clear pointers on how to navigate towards the PF. It may be possible to incorporate this in specialised operators to speed up the optimisation process.

From the decision tree in Fig. 5 it appears classification of solutions is possible with high precision. To validate this, the tree was used to classify the 3400 solutions in the test set. Table 2 shows the resulting predictions. All assignments were made with a confidence of at least 90%, showing that it is possible to classify designs quite reliably. A particularly notable result is the classification of the majority of the solutions in the F2 and KP classes, which, for this dataset, is done with near perfect confidence. Not only does this provide confidence in the optimisation process, but these rules could even be useful during optimisation. By classifying new solutions based on these rules it may be possible to identify which solutions are more likely to perform well, such that expensive simulations might only be needed for those.

Table 2. Decision tree results on the test set. Columns relate to the predicted probability of belonging to a specific class, whereas rows refer to classes. Each cell then contains the number of solutions that belong to a solution class, with a particular probability.

Predictions	0.0000	0.0008	0.0046	0.0048	0.0051	0.0162	0.0276	0.0345	0.0483	0.0926	0.9074	0.9241	0.9655	0.9784	0.9949	0.9952
BD	1758	0	0	0	728	0	54	0	0	0	0	0	0	860	0	0
F1	1649	860	0	0	0	0	0	0	0	0	0	0	891	0	0	0
F2	891	0	0	748	0	860	0	0	54	0	119	0	0	0	728	0
KP	728	0	860	0	0	0	0	891	0	119	0	54	0	0	0	748

6   Conclusion

In summary, the results show that by using predefined features and decision trees, it is possible to apply innovization to large datasets from Pareto optimisation in building spatial design (BSD), and to obtain meaningful results from an engineering perspective. Furthermore, the obtained rules allow for high precision ($\geq 96\%$) classification of solutions.

Besides in generating insight, the design rules could also be useful in steering the multi-objective optimisation process. For future work, it would be interesting to investigate which moves in the optimisation process result in improvements. In other words, given an existing design, what changes to its features will, with high probability, result in an improved design. Furthermore, it may be possible to apply learned rules in co-evolutionary design processes [7]. Or, as mechanism to determine for which solutions to use expensive simulations during optimisation.

The current work analyses data for a specific type of building. To generalise the conclusions, the same methods should be evaluated on a larger variety of building types. Given the computationally efficient nature of the used approach, it is probable that larger BSDs can be handled. This must, however, still be verified. Additionally, currently only a subset of the optimisation data is labelled. As a result, it is unclear whether the learned rules generally allow the identification of, for instance, solutions that perform well in objective one. It may be the case that some areas of the objective space, that have not been considered here, have similar characteristics in some features. This should be studied in the future. A challenge here is how to do proper analysis with both sparse and dense areas in the objective space.

Based on first discussions with a design expert good heuristics are learned that accurately describe high quality BSDs. However, it remains difficult to foresee the consequences of changes in feature values with respect to the objective values. In order to improve this visual aids would be helpful. For instance, a slider controlling the weights of the structural and thermal objectives could be used to change the spatial design in real-time.

Acknowledgments. This work is part of the TTW-Open Technology Programme with project number 13596, which is (partly) financed by the Netherlands Organisation for Scientific Research (NWO).

References

1. Bandaru, S., Deb, K.: Automated innovization for simultaneous discovery of multiple rules in bi-objective problems. In: Takahashi, R.H.C., Deb, K., Wanner, E.F., Greco, S. (eds.) EMO 2011. LNCS, vol. 6576, pp. 1–15. Springer, Heidelberg (2011). https://doi.org/10.1007/978-3-642-19893-9_1
2. Bandaru, S., Deb, K.: Temporal innovization: evolution of design principles using multi-objective optimization. In: Gaspar-Cunha, A., Henggeler Antunes, C., Coello, C.C. (eds.) EMO 2015. LNCS, vol. 9018, pp. 79–93. Springer, Cham (2015). https://doi.org/10.1007/978-3-319-15934-8_6

3. van der Blom, K., Boonstra, S., Wang, H., Hofmeyer, H., Emmerich, M.T.M.: Evaluating memetic building spatial design optimisation using hypervolume indicator gradient ascent. In: Trujillo, L., Schütze, O., Maldonado, Y., Valle, P. (eds.) NEO 2017. SCI, vol. 785, pp. 62–86. Springer, Cham (2019). https://doi.org/10.1007/978-3-319-96104-0_3

4. van der Blom, K., Boonstra, S., Hofmeyer, H., Bäck, T., Emmerich, M.T.M.: Configuring advanced evolutionary algorithms for multicriteria building spatial design optimisation. In: 2017 IEEE Congress on Evolutionary Computation (CEC), pp. 1803–1810. IEEE (2017). https://doi.org/10.1109/CEC.2017.7969520

5. van der Blom, K., Boonstra, S., Hofmeyer, H., Emmerich, M.T.M.: Multicriteria building spatial design with mixed integer evolutionary algorithms. In: Handl, J., Hart, E., Lewis, P.R., López-Ibáñez, M., Ochoa, G., Paechter, B. (eds.) PPSN 2016. LNCS, vol. 9921, pp. 453–462. Springer, Cham (2016). https://doi.org/10.1007/978-3-319-45823-6_42

6. van der Blom, K., Boonstra, S., Hofmeyer, H., Emmerich, M.T.M.: A super-structure based optimisation approach for building spatial designs. In: Papadrakakis, M., Papadopoulos, V., Stefanou, G., Plevris, V. (eds.) VII European Congress on Computational Methods in Applied Sciences and Engineering - ECCOMAS VII, vol. 2, pp. 3409–3422. National Technical University of Athens, Athens (2016). https://doi.org/10.7712/100016.2044.10063

7. Boonstra, S., van der Blom, K., Hofmeyer, H., Emmerich, M.T.M.: Combined super-structured and super-structure free optimisation of building spatial designs. In: Koch, C., Tizani, W., Ninić, J. (eds.) Digital Proceedings of the 24th EG-ICE International Workshop on Intelligent Computing in Engineering, pp. 23–34. Curran Associates Inc., Red Hook (2017)

8. Boonstra, S., van der Blom, K., Hofmeyer, H., Emmerich, M.T.M., van Schijndel, J., de Wilde, P.: Toolbox for super-structured and super-structure free multidisciplinary building spatial design optimisation. Adv. Eng. Inform. **36**, 86–100 (2018). https://doi.org/10.1016/j.aei.2018.01.003

9. Deb, K., Bandaru, S., Greiner, D., Gaspar-Cunha, A., Tutum, C.C.: An integrated approach to automated innovization for discovering useful design principles: case studies from engineering. Appl. Soft Comput. **15**, 42–56 (2014). https://doi.org/10.1016/j.asoc.2013.10.011

10. Deb, K., Srinivasan, A.: Innovization: innovating design principles through optimization. In: Proceedings of the 8th Annual Conference on Genetic and Evolutionary Computation, GECCO 2006, pp. 1629–1636. ACM, New York (2006). https://doi.org/10.1145/1143997.1144266

11. Emmerich, M., Deutz, A.: Time complexity and zeros of the hypervolume indicator gradient field. In: Schuetze, O., et al. (eds.) EVOLVE - A Bridge between Probability, Set Oriented Numerics, and Evolutionary Computation III. Studies in Computational Intelligence, pp. 169–193. Springer, Heidelberg (2014). https://doi.org/10.1007/978-3-319-01460-9_8

12. Kung, H.T., Luccio, F., Preparata, F.P.: On finding the maxima of a set of vectors. J. ACM **22**(4), 469–476 (1975). https://doi.org/10.1145/321906.321910

13. Ng, A.H.C., Dudas, C., Boström, H., Deb, K.: Interleaving innovization with evolutionary multi-objective optimization in production system simulation for faster convergence. In: Nicosia, G., Pardalos, P. (eds.) LION 2013. LNCS, vol. 7997, pp. 1–18. Springer, Heidelberg (2013). https://doi.org/10.1007/978-3-642-44973-4_1

14. Therneau, T., Atkinson, B.: rpart: Recursive Partitioning and Regression Trees (2018). https://CRAN.R-project.org/package=rpart. r package version 4.1-13

Constrained Multi-objective Optimization Method for Practical Scientific Workflow Resource Selection

Courtney Powell[1(✉)], Katsunori Miura[2], and Masaharu Munetomo[1]

[1] Hokkaido University, Sapporo 060–0811, Japan
kotoni@ist.hokudai.ac.jp
[2] Otaru University of Commerce, Otaru 047–8501, Japan

Abstract. This paper presents and evaluates a constrained multi-objective optimization method for scientific workflow resource selection that uses equivalent transformation for constraint handling. Two different approaches are compared using a case study of optimal cloud resource configuration selection for a practical genomic analytics workflow. In the first approach, called the nondominated sorting equivalent transformation (NSET) method, feasible configurations are generated via equivalent transformation and the Pareto-optimal configurations are selected from among them via a process of nondominated sorting, reference points association, and niching/elitism. In the second approach, Pareto-optimal configurations are generated via the nondominated sorting genetic algorithms II/III (NSGA-II/III) and feasible configurations are generated via equivalent transformation. Then, the configurations that are common to both processes are considered to be both feasible and optimal. Preliminary results based on the Pareto-optimal configuration sets generated by NSGA-II/III indicate that NSET is feasible for constrained multi-objective optimization of practical scientific workflow resource selection problems.

Keywords: Constrained multi-objective optimization ·
Equivalent transformation · Genomic analytics ·
Nondominated sorting genetic algorithm · Pareto-optimal set ·
Scientific workflow

1 Introduction

Scientific workflows are increasingly being executed in the cloud as the computing requirements of their various constituent tools can be dynamically satisfied by combining components from the virtually limitless array of resources that are available to cloud users. Superficially, this is highly beneficial for users; however, the more choices there are, the more difficult it is for users to choose the resource configurations that are most appropriate for their applications. In addition, users typically have constraints, such as number of CPU/GPU cores, amount of memory and storage, minimum reliability, and maximum cost. These constraints may also be linked to multiple conflicting objectives—such as minimization of operating cost and application response time/makespan while maximizing system reliability and system availability—for which

© Springer Nature Switzerland AG 2019
K. Deb et al. (Eds.): EMO 2019, LNCS 11411, pp. 683–694, 2019.
https://doi.org/10.1007/978-3-030-12598-1_54

optimal compromise solutions are difficult to find. For example, choosing high-performance virtual machines (VMs) can reduce the response time or makespan of an application but usually increases its financial running cost.

Multiobjective evolutionary algorithms (MOEAs) are used extensively to solve complex problems involving conflicting objectives because they can simultaneously optimize the conflicting objectives and find and maintain multiple solutions in one single simulation run [1]. However, because MOEAs are inherently unconstrained population-based heuristics, they require retrofitting with one or more constraint compliance techniques to deal with real-world constrained multi-objective problems [2–4].

In cloud resource scenarios, various methods have been proposed for satisfying constraints [5], and dealing with multiple objectives [6–8]. However, whereas the proposed methods can easily deal with soft (quantifiable) constraints such as deadlines and limits on quality of service (QoS) parameters, they have difficulty dealing efficiently with hard (non-quantifiable) constraints such as allocation restrictions and precedence constraints [9]. Further, MOEAs still have difficulty dealing with problems in which optimal solutions lie close to or on the boundaries of the feasible region, and general constraints that result in non-contiguous solution spaces. In the case of constrained optimization, in which not all solutions in the search space are feasible, knowledge of the feasible region could help to reduce the computation time required to generate an optimal set of feasible solutions [10].

This paper presents a constrained multi-objective optimization method for scientific workflow resource selection that uses equivalent transformation for constraint handling; i.e., to find the feasible solutions in the search space. The proposed method, called the nondominated sorting equivalent transformation (NSET) method, is based on the nondominated sorting genetic algorithm III (NSGA-III) [11], but is not an evolutionary algorithm as it replaces the evolutionary operators in NSGA-III (mutation and crossover) with an equivalent transformation (ET) operator that facilitates exploration of the solution space via ET state transitions. Further, whereas MOEAs semi-randomly generate solutions via evolution, then apply constraint-compliance techniques retroactively to the generated solutions, NSET applies micro-constraints to partial solutions, resulting in only feasible solutions being generated in the optimizer. In addition, whereas MOEAs require a relatively large population to avoid inbreeding [12], in NSET, the only "restriction" on the lower limit of the number of final solutions is that it should be sufficient to generate well-distributed reference points.

In this paper, the feasibility of the NSET method is evaluated by comparing the feasible Pareto-optimal solutions it generates to those generated using NSGA-II/III for a case study of optimal cloud resource configuration selection for a genomic analytics workflow.

2 The Feasible Configurations Generation Process

In the proposed NSET constrained optimization method, the equivalent transformation (ET) computation model [13] is employed as a constraint satisfaction solver to generate the universe of feasible configurations based on requirements prepared in a predicate logic-defined specification (PLS) [14] document. In the ET computation model, computation is carried out by sets of prioritized rewriting rules that successively reduce a problem using meaning-preserving transformations with respect to given background knowledge [13]. In this paradigm, the problem is given as a definite clause set [15] that is then transformed to unit answer clauses using generated rewriting rules. For example, in the append problem-solving trace depicted in Fig. 1, the problem given in the form of the initial (topmost) clause set involves finding all pairs of lists, *x and *y, that can combine to give a resulting list containing the integers "1" and "2." In the rewriting rule, *a = [*d|*e] denotes a non-empty list *a with head element *d and tail list *e. Further, *b = *c signifies unification of *b and *c. Variable *c contains the original list and the resulting lists are first stored in *a and *b and subsequently output to *x and *y, respectively. This ET rule has two bodies, which cause multiple clauses to be generated when it is applied. The rule states that given three variables, *a, *b, and *c, if *a is an empty list, then *b = *c. Otherwise, the solution is a combination of the first element of the list and some other list, which is then found by branching. Using this rewriting rule, the three solutions shown comprising the solution set are obtained.

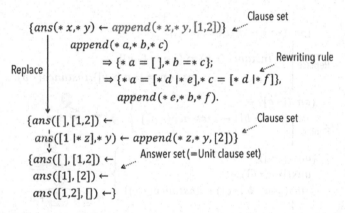

Fig. 1. Finding all pairs of lists that can concatenate to produce the list [1, 2] using the ET-based constraint satisfaction method.

In logic programming, a search tree is "a formalism for considering all possible computation paths" [16], i.e., generating all feasible solutions to a given problem. The ET computation model incorporates the logic programming paradigm. Thus, the trace in Fig. 1 can also be viewed as a search tree, in which the computation to find the solution set to the append problem based on the rewriting rule proceeds as follows. The problem clause is the root of the tree and each branch therefrom is a successive transformation of the original problem by the given rewriting rule. The leaves of the tree are success nodes, where unit (*ans*) clauses have been produced. These nodes correspond to solutions to the root of the tree (i.e., the problem clause). As in the logic programming paradigm, ET search trees contain multiple success nodes if the query has multiple solutions [16].

As demonstrated in [15, 17], each ET rule can also be supplied with "micro-constraints" to check a section of a solution. Thus, instead of waiting for the solution to be completely generated, then checking its feasibility, a check is performed on the partially formed solution. If the partially formed solution does not satisfy the micro-constraint, then the partial solution is discarded and the corresponding solution is not generated as it would be infeasible. In this manner, the ET operator rapidly prunes the search tree. Thus, the search space is rapidly traversed.

The simple procedure outlined above enables this ET-based process to generate all feasible solutions based on constraints simply by branching the state of the clause. For example, a constrained resource allocation problem in which the objective is to find all combinations of resources satisfying a given set of constraints can be represented by the topmost definite clause shown in Fig. 2, where the head of the clause (*ans* portion) represents the set of resources found and the body represents the constraints on those resources (e.g., deadline, maximum cost, minimum availability). By successive application of the rewriting rule to branch this clause, the universe of feasible resource configurations is generated. This is analogous to finding all combinations of lists for the given list *"[1, 2]"* in the append scenario presented in Fig. 1.

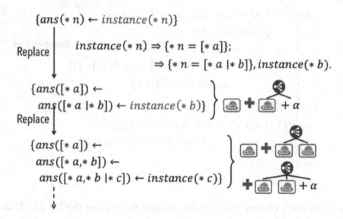

Fig. 2. Generation of feasible resource configurations by the ET-based constraint satisfaction method.

3 Two Alternative Feasible Pareto-Optimal Solutions Selection Techniques

In this paper, we consider two techniques for Pareto-optimal solutions set selection/generation and compare the overall results obtained. In the first technique, NSET, the feasible Pareto-optimal solutions are obtained via elite nondominated sorting and reference points association of the universe of feasible solutions generated by the ET constraint solver. In the second technique, NSGA-II/III [11, 18] are respectively used to generate Pareto-optimal solutions that are compared to the universe of feasible solutions to determine the solutions that are both feasible and optimal.

3.1 Technique 1: Nondominated Sorting Equivalent Transformation (NSET)

The process of finding the Pareto-optimal solutions to a multi-objective problem can be divided into two phases that both occur in each generation/iteration: (1) search/ exploration and (2) selection. In MOEAs, phase (1) is accomplished via the evolutionary operators crossover and mutation [1]. However, in NSET, these genetic operators are removed and exploration is instead carried out by equivalent transformation. Figure 3 presents a flowchart for NSET. For simple problems, the constraint satisfaction process is able to generate all the feasible solutions fairly quickly and then terminate. This, in essence, results in sequential operation, with the lowest decision box in the flowchart not being traversed. Conversely, problems that require significant amounts of computation and problems with large solution spaces result in concurrent operation by the constraint satisfaction process and the optimization process, with the feasible solutions being generated over an extended period and the entire flowchart being traversed.

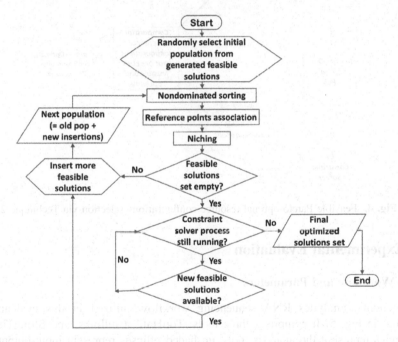

Fig. 3. Flowchart for feasible Pareto-optimal resource solutions selection via NSET.

3.2 Technique 2: Feasible Solutions Common to NSGA-II/III Results

Figure 4 illustrates the process employed in Technique 2. In this process, NSGA-II/III and the ET constraint solver employ common datasets to generate unconstrained Pareto-optimal solutions and feasible configurations, respectively. Subsequently, the solutions that are common to both sets of results are selected as being feasible Pareto-optimal solutions.

This technique can be considered a simpler form of the push and pull strategy presented by Fan et al. [19]. In this case, NSGA-II/III is the MOEA used in the "push" phase to explore the search space without constraints to quickly traverse both feasible and infeasible regions and approach the unconstrained Pareto surface. However, instead of using a constrained MOEA in the "pull" phase, the set of Pareto-optimal solutions found by NSGA-II/III in the "push" phase is simply compared with the set of feasible solutions generated by the ET constraint solver. The solutions common to both sets are then regarded as feasible Pareto-optimal solutions.

For problems for which numerous Pareto-optimal solutions exist in the search space, multiple runs with the same MOEA or using an ensemble approach [20, 21] can be carried out in the push phase and the Pareto-optimal solutions found combined to obtain the overall unconstrained Pareto-optimal solutions set.

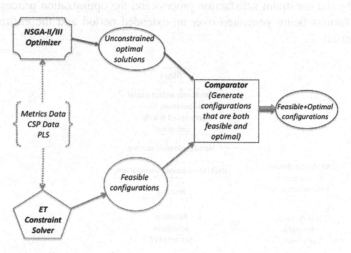

Fig. 4. Feasible Pareto-optimal resource configurations selection via Technique 2.

4   Experimental Evaluation

4.1   Workflow and Parameters

The genomic analytics RNA sequencing workflow utilized in this evaluation is depicted in Fig. 5. It comprises three tools: TopHat2, Cufflinks, and StringTie. The rectangles represent the analysis tools, unshaded ellipses represent input-output data, and the shaded ellipse represents input data from users. The optimizers were implemented in Jupyter notebook [22] using the Python-based Optima library [23].

Requirements and Constraints. The primary requirements and constraints were as follows (where USD signifies United States dollars): Running cost ≤ 10 USD/h; AWS Regions: Tokyo, Virginia; Policy: Single cloud service provider.

Parameters and Objectives. Table 1 lists the general parameters and objectives employed in the evaluation, whereas Table 2 presents the parameters used with the optimizers.

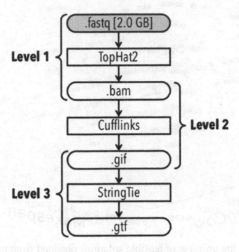

Fig. 5. Genomic analytics RNA sequencing workflow employed in the evaluation.

Table 1. General parameters and objectives.

Parameter	Value
Feasible resource configurations (solutions) from ET constraint solver	6116
User objectives	Availability *(maximize)*, Cost, Makespan *(minimize)*
AWS regions	Tokyo, Virginia
AWS instances families	C4, M4, R4

Table 2. Parameter values used in NSET, NSGA-II, and NSGA-III.

Parameter	NSET	NSGA-II	NSGA-III
No. of reference points	21	–	91
Population size	21	100	100
No. of generations	–	100	100
SBX probability	–	–	1

4.2   Evaluation Results

Figure 6 shows the distribution of the feasible solutions universe obtained from the ET constraint solver for the requirements and constraints presented in Sect. 4.1. It can be clearly seen that the feasible solutions are liberally distributed over the search space.

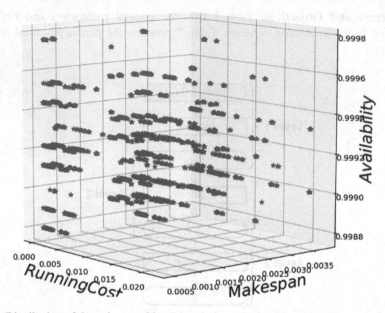

Fig. 6. Distribution of the universe of feasible solutions obtained from the ET constraint solver.

For the parameters and objectives presented in Tables 1 and 2, NSGA-II and III obtained the same set of 25 unconstrained Pareto-optimal (i.e., nondominated sorting rank = 1) solutions. These Pareto-optimal solutions are presented in Table 3 and depicted graphically in Fig. 7. The Pareto-optimal surface is clearly visible in the distribution.

Table 3. Unconstrained Pareto-optimal solutions obtained via NSGA-II/III.

Solution No.	Objective values [Cost, Makespan, Availability]	Solution No.	Objective values [Cost, Makespan, Availability]
1	[1.01599e−05, 0.00055, 0.99912]	14	[2.33127e−05, 0.00053, 0.99922]
2	[2.87957e−05, 0.00054, 0.99944]	15	[6.04085e−05, 0.00353, 0.99976]
3	[2.25656e−05, 0.00055, 0.99922]	16	[6.34191e−05, 0.00072, 0.99976]
4	[1.00850e−05, 0.00072, 0.99912]	17	[1.07787e−05, 0.00195, 0.99922]
5	[6.37040e−05, 0.00033, 0.99944]	18	[4.12014e−05, 0.00054, 0.99954]
6	[9.09370e−06, 0.00353, 0.99944]	19	[6.22217e−05, 0.00053, 0.99944]
7	[1.21044e−05, 0.00072, 0.99944]	20	[8.14040e−06, 0.00055, 0.99880]
8	[2.14995e−05, 0.00353, 0.99954]	21	[8.01104e−05, 0.00054, 0.99976]
9	[1.09070e−05, 0.00053, 0.99912]	22	[6.20934e−05, 0.00195, 0.99954]
10	[8.75930e−06, 0.00195, 0.99890]	23	[1.03699e−05, 0.00033, 0.99880]
11	[2.45101e−05, 0.00072, 0.99954]	24	[8.88750e−06, 0.00053, 0.99880]
12	[7.07430e−06, 0.00353, 0.99912]	25	[2.47951e−05, 0.00033, 0.99922]
13	[1.23893e−05, 0.00033, 0.99912]		

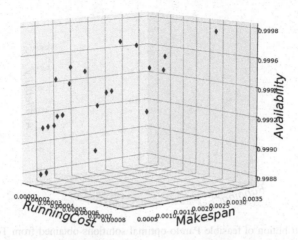

Fig. 7. Distribution of unconstrained Pareto-optimal solutions obtained from NSGA-II/III.

The distribution of feasible (i.e., constrained) Pareto-optimal solutions obtained using NSET and Technique 2 are illustrated in Figs. 8 and 9, respectively. The actual objective values for the feasible Pareto-optimal solutions obtained are compared in Table 4.

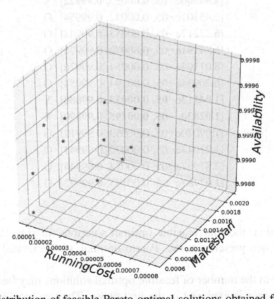

Fig. 8. Distribution of feasible Pareto-optimal solutions obtained from NSET.

Figures 8 and 9 and Table 4 show that all the feasible optimal solutions obtained from Technique 2 are present in the feasible optimal solutions set generated by NSET. However, whereas Technique 2 generated only nine feasible optimal solutions, NSET

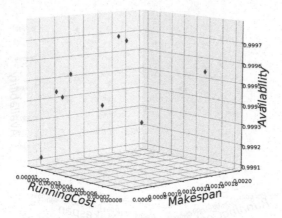

Fig. 9. Distribution of feasible Pareto-optimal solutions obtained from Technique 2.

Table 4. Comparison of solutions obtained by NSET and Technique 2 (O = present, X = absent).

NSET Solution No.	Objective values [Cost, Makespan, Availability]	Technique 2
1	**[6.62948e−05, 0.00054, 0.99954]**	X
2	**[4.84060e−05, 0.00053, 0.99922]**	X
3	[2.45101e−05, 0.00072, 0.99954]	O
4	[6.22217e−05, 0.00053, 0.99944]	O
5	[6.34191e−05, 0.00072, 0.99976]	O
6	[8.01104e−05, 0.00054, 0.99976]	O
7	[6.20934e−05, 0.00194, 0.99954]	O
8	**[1.04113e−05, 0.00053, 0.99880]**	X
9	**[1.02830e−05, 0.00195, 0.99890]**	X
10	[1.09070e−05, 0.00053, 0.99912]	O
11	[1.21044e−05, 0.00072, 0.99944]	O
12	[2.87957e−05, 0.00054, 0.99944]	O
13	[1.07787e−05, 0.00195, 0.99922]	O

generated 13. Further, the four feasible optimal solutions missing from the results of Technique 2 are not present in the Pareto-optimal set generated by NSGA-II/III (Table 3).

The difference in the number of feasible optimal solutions may be due to the way in which the search space is explored in the first phase. The two most important goals in Pareto-based multi-objective optimization is to find a set of solutions that is (1) as close as possible to the Pareto-optimal front and (2) as diverse as possible. This ensures a good set of optimal tradeoff solutions among objectives. To achieve these goals, MOEAs utilize an evolution phase in which three operators, crossover, mutation, and reproduction/selection, play a central role [1]. However, evolution is a stochastic process, in which the solutions found are highly dependent on the settings of the evolutionary operators [24].

In contrast, in NSET, exploration of the search space is achieved via transformation of the original problem by correct semantics-preserving rewriting rules [13]. This is, in essence, a correct state transition system, the like of which has previously produced promising results when compared with NSGA-II [25]. Thus, instead of evolving feasible solutions based on information from infeasible solutions (as done in CMOEAs), NSET applies ET to find all feasible solutions (states) based on information about the problem. This results in the complete set of feasible Pareto-optimal solutions being systematically generated. The disparity can also be considered from a "no free lunch" [26] viewpoint, where it has been established that for any algorithm, high performance on one class of problems is offset by its performance on another class of problems.

5 Conclusion

The results obtained in this study indicate that the proposed method (NSET) can actually generate solutions that are both feasible and optimal for a practical scientific workflow resource selection problem. This is verified by the fact that more than 69% of the solutions presented as being feasible Pareto-optimal solutions exist in the Pareto-optimal sets generated by NSGA-II/III. However, further investigation is required to determine whether the additional four solutions presented by NSET as feasible Pareto-optimal solutions are indeed optimal but were missed by both NSGA-II and NSGA-III.

Acknowledgment. This work is supported by CREST, Japan Science and Technology Agency (Grant No. JPMJCR1501).

References

1. Deb, K.: Multi-Objective Optimization using Evolutionary Algorithms. Wiley, Hoboken (2001)
2. Li, X., Zeng, S., Li, C., Ma, J.: Many-objective optimization with dynamic constraint handling for constrained optimization problems. Soft. Comput. **21**(24), 7435–7445 (2017)
3. Mezura-Montes, E., Coello, C.A.C.: Constraint-handling in nature-inspired numerical optimization: past, present and future. Swarm Evol. Comput. **1**(4), 173–194 (2011)
4. Jordehi, A.R., Hashemi, N., Dezfouli, H.N.: Analysis of the strategies in heuristic techniques for solving constrained optimisation problems. J. Am. Sci. **8**(10), 345–350 (2012)
5. Shaw, R., Howley, E., Barrett, E.: Predicting the available bandwidth on intra cloud network links for deadline constrained workflow scheduling in public clouds. In: Maximilien, M., Vallecillo, A., Wang, J., Oriol, M. (eds.) ICSOC 2017. LNCS, vol. 10601, pp. 221–228. Springer, Cham (2017). https://doi.org/10.1007/978-3-319-69035-3_15
6. Pietri, I., Chronis, Y., Ioannidis, Y.: Multi-objective optimization of scheduling dataflows on heterogeneous cloud resources. In: Proceedings of the IEEE Big Data (Big Data), pp. 361–368 (2017)

7. Wang, X., Yeo, C.S., Buyya, R., Su, J.: Optimizing the makespan and reliability for workflow applications with reputation and a look-ahead genetic algorithm. Future Gener. Comput. Syst. **27**(8), 1124–1134 (2011)
8. Frey, S., Fittkau, F., Hasselbring, W.: Search-based genetic optimization for deployment and reconfiguration of software in the cloud. In: Proceedings of the ICSE 2013, Piscataway, NJ, USA, pp. 512–521 (2013)
9. Yassa, S., Sublime, J., Chelouah, R., Kadima, H., Jo, G., Granado, B.: A genetic algorithm for multi-objective optimisation in workflow scheduling with hard constraints. Int. J. Metaheuristics **2**(4), 415–433 (2013)
10. Okabe, T., Jin, Y., Sendhoff, B.: A critical survey of performance indices for multi-objective optimisation. In: CEC 2003, vol. 2, pp. 878–885 (2003)
11. Deb, K., Jain, H.: An evolutionary many-objective optimization algorithm using reference-point-based nondominated sorting approach, part I: solving problems with box constraints. IEEE Trans. Evol. Comput. **18**(4), 577–601 (2014)
12. Tanabe, R., Ishibuchi, H., Oyama, A.: Benchmarking multi-and many-objective evolutionary algorithms under two optimization scenarios. IEEE Access **5**, 19597–19619 (2017)
13. Akama, K., Nantajeewarawat, E.: Formalization of the equivalent transformation computation model. JACIII **10**(3), 245–259 (2006)
14. Miura, K., Munetomo, M.: A predicate logic-defined specification method for systems deployed by intercloud brokerages. In: IC2EW, pp. 172–177 (2016)
15. Powell, C.: A formal methodology for concurrent componentwise development of rich internet applications. Ph.D. thesis. Hokkaido University (2011)
16. Sterling, L., Shapiro, E.Y.: The Art of Prolog: Advanced Programming Techniques. MIT Press, Cambridge (1994)
17. Powell, C., Akama, K., Nakamura, K.: Componentwise modelling and synthesis of dynamic interactive systems using the equivalent transformation framework. IJICIC **7**, 4067–4081 (2011)
18. Deb, K., Pratap, A., Agarwal, S., Meyarivan, T.: A fast and elitist multiobjective genetic algorithm: NSGA-II. IEEE Trans. Evol. Comput. **6**(2), 182–197 (2002)
19. Fan, Z., et al.: Push and pull search for solving constrained multi-objective optimization problems. Swarm Evol. Comput. **44**, 665–679 (2018)
20. Zhou, Y., Wang, J., Chen, J., Gao, S., Teng, L.: Ensemble of many-objective evolutionary algorithms for many-objective problems. Soft. Comput. **21**(9), 2407–2419 (2017)
21. Wang, W., et al.: An effective ensemble framework for multi-objective optimization. IEEE Trans. Evol. Comput. (2018)
22. Jupyter Notebook. http://jupyter.org/. Accessed 30 Nov 2018
23. Optima. https://github.com/bigfatnoob/optima/. Accessed 30 Nov 2018
24. Eiben, A.E., Smit, S.K.: Evolutionary algorithm parameters and methods to tune them. In: Hamadi, Y., Monfroy, E., Saubion, F. (eds.) Autonomous Search, pp. 15–36. Springer, Heidelberg (2011). https://doi.org/10.1007/978-3-642-21434-9_2
25. Zhou, J., Zhou, X., Yang, C., Gui, W.: A multi-objective state transition algorithm for continuous optimization. In: CCC, pp. 9859–9864, July 2017
26. Wolpert, D.H., Macready, W.G.: No free lunch theorems for optimization. IEEE Trans. Evol. Comput. **1**(1), 67–82 (1997)

Simulation Optimization of Water Usage and Crop Yield Using Precision Irrigation

Proteek Chandan Roy$^{(\boxtimes)}$ ⓘ, Andrey Guber, Mohammad Abouali,
A. Pouyan Nejadhashemi, Kalyanmoy Deb ⓘ, and Alvin J. M. Smucker

Michigan State University, East Lansing, MI 48824, USA
{royprote,akguber,mabouali,pouyan,kdeb,smucker}@msu.edu
http://www.coin-laboratory.com/

Abstract. Sustainable agriculture maximizes crop production with minimal use of resources, such as water and energy. Subsurface water retention technology (SWRT) uses an impermeable membrane in the soil to hold more water for plants. Optimal production of crops requires not only the optimal irrigation rate but also optimal shapes and placements of SWRT. Furthermore, some uncertain factors, e.g. incoming solar energy, plant transpiration rate temperature, climatic conditions, and genetics of crops are also important in crop production, thereby making the optimization process complicated. In this paper, we propose a computationally fast approach that utilizes HYDRUS-2D software for water flow simulation and DSSAT crop simulation software with an evolutionary multi-objective optimization (EMO) procedure in a coordinated manner to minimize water utilization and maximize crop production. Our method simulates SWRT in HYDRUS-2D software and calibrates and validates DSSAT model parameters according to the HYDRUS-2D simulation. Then it finds the best irrigation schedules to produce maximum crop production and water use efficiency by DSSAT. Our results show that HYDRUS-DSSAT calibration produces less than 5% validation error and the optimization procedure introduces 99% water use efficiency with the help of rainfall water and as much as 6 times increase of corn production compared to a non-optimized, random irrigation schedule without any SWRT membrane.

Keywords: Precision irrigation · HYDRUS-2D · DSSAT ·
Optimization · Crop · Water use efficiency ·
Subsurface water retention technology

1 Introduction

Sustainable technologies for crop production by using minimal water and energy have become very essential in today's world. Water is vital for irrigation of crops but water is also scarce and today's food and biomass producers are obligated to use judicial consumption of water for irrigation. In order to have a sustainable crop production system, we need to use minimal water yet holding most of them

© Springer Nature Switzerland AG 2019
K. Deb et al. (Eds.): EMO 2019, LNCS 11411, pp. 695–706, 2019.
https://doi.org/10.1007/978-3-030-12598-1_55

in the soil for plant growth. Sandy soil has much less holding capacity and large water permeability. Traditional approach using asphalt barrier has proved to be efficient and is widely accepted for sandy soil [4,15]. Being costly and labor intensive, new polyethylene membranes are also used. The proper membrane design and installation depth in specific soil and weather condition has been studied in [4] using two-dimensional modeling of water flow using HYDRUS-2D software [14] in partially sandy soils. They have investigated a profile distribution of water in an irrigated sand lysimeter with installed SWRT membranes at different depths. Based on their experiments, it is evident that HYDRUS-2D model with membrane geometry can produce the same water content (after calibration) as in practice with a considerable accuracy.

Water content simulation, nutrient flow modeling or crop yield prediction often need well-developed software [5,7,8]. In a previous study [15], an integrated model of water-flow and nutrient transport simulation using HYDRUS-2D was combined with an evolutionary multi-objective optimization (EMO) algorithm to obtain optimal membrane geometry and placement in soil profile along with prescriptive irrigation scheduling under two conflicting objectives.

Although HYDRUS-2D can predict the water and nutrient accumulation at the root zone of a plant, it cannot simulate the crop growth, which is a direct measurable outcome of the irrigation process. Besides a continuous supply of water at the root level either through an optimal irrigation pattern or through rainfall, the growth of crop and eventual crop yield depend on many other factors, such as incoming solar energy, plant transpiration rate, temperature, climatic conditions, type of crop, etc. Thus it is necessary to take help of another computational simulator that can explicitly provide an estimate of crop yield for given soil-water mix, nutrient content, and other parameters in a time-series manner. For this study, we use DSSAT (Decision Support System for Agrotechnology Transfer), a widely accepted tool for agronomists [6].

The main contribution of this paper is as follows. We propose an integrated methodology to find daily irrigation schedules that optimize crop production while using minimum water using an evolutionary multi-objective optimization methodology. Our results show that, we can achieve best corn production with 99% water use efficiency in sandy soil with the help of SWRT technology.

2   Two Simulation Software: HYDRUS-2D and DSSAT

Here we briefly introduce two software systems used in this study: HYDRUS and DSSAT. HYDRUS-2D/3D software was developed to simulate two and three-dimensional movement of water, heat, and multiple solutes by solving Richards Equation for saturated-unsaturated water flow, and the Fickian-based convection-dispersion equation for heat and solute transport [14]. HYDRUS-2D/3D uses van Genuchten (1980)'s model [16] (along with other recent models) to produce soil-water content, which is a unit-less volumetric ratio of amount of water, and other contents in the soil.

Figure 1 shows a 2D mesh design of SWRT membrane with water content values (after irrigation) on the right. We observe that much of the water is contained inside the membrane just after irrigation. Since HYDRUS-2D cannot model crop growth involving plant genetics, root growth, transpiration and other complicated processes we use a crop simulation software called DSSAT (Decision Support System for Agrotechnology Transfer) that simulates growth and development of plants over time in an one-dimensional arrangement with its own soil-water, carbon, and nutrient processes. The software is capable of simulating integrated crop models, namely, maize, wheat, barley, etc. and calculate yield at the end of crop growing period. A tipping bucket model [6] is used to produce soil-water content in a one dimensional space (only depth). The model used by DSSAT, namely, tipping bucket model, is computationally more efficient compare to van Genuchten's model used in HYDRUS-2D. The key parameters for DSSAT is the wilting point ($SLLL$) (lower limit), saturation ($SSAT$), and field capacity (drained upper limit $SDUL$). In the simplified equation of HYDRUS's model, soil water retention ($\theta(h)$, where h is pressure head) depends on residual and saturated water content (θ_r and θ_s) and saturated hydraulic conductivity (K_s).

Hydrus-2D mesh   Water contents after irrigation

Fig. 1. (a) A 2D mesh created in HYDRUS-2D software where two impermeable membrane (AR ratio 2:1) are placed in 30 and 45 cm depth. (b) Water content is shown on the membrane after irrigation.

3   Calibration of Simulation Models

Calibration of two models has one single purpose, namely, the adjusted parameters of DSSAT should produce the same water content as HYDRUS-2D simulating water flow under SWRT membranes. Water holding capacity is different for different types of soils and this can be controlled by changing the parameter values of DSSAT. Our assumption is that we can simulate the effect of SWRT membrane by only changing the soil parameters of DSSAT. Because of the installation of SWRT, the water content would be different at different depths. Therefore, we divide the soil domain (overall 120 cm depth) into ten asymmetric layers. Layers are $L = \{0$–$8, 8$–$13, 13$–$18, 18$–$23, 23$–$28, 28$–$33, 33$–$38, 38$–$43, 43$–$48, 48$–$120\}$ cm depth and label them as layer 1 to layer 10. In HYDRUS-2D, we create a 2D mesh with 30 cm width and 120 cm depth. Our calibration objective becomes minimizing mean squared error of between water contents (θ) of HYDRUS-2D and DSSAT over crop growing season d for different irrigation pattern $\mathcal{S}$.

$$\underset{\forall l \in L,}{\text{Min}} \quad F(l) = \sum_{s \in \mathcal{S}} \sum_{j=1}^{d} |\theta_{HYDRUS}^{j}(l,s) - \theta_{DSSAT}^{j}(l,s)| \tag{1}$$

subject to, fixed soil evaporation rate.

Here, $\theta_{DSSAT}^{j}(l,s)$ and $\theta_{HYDRUS}^{j}(l,s)$ are the average water contents of DSSAT and HYDRUS-2D simulators, respectively, for the corresponding layer l and irrigation pattern $s \in \mathcal{S}$ and we optimize for $d = 110$ simulation days. Note that, criteria other than absolute error, namely, MSE and Nash-Sutcliffe efficiency index (NSE [9]) can also be used in this regard. We have used exhaustive search for finding the best parameters for this calibration.

4   Optimization of Water Use Efficiency and Crop Yield

As mentioned before, our goal of this study to maximize crop yield using the minimum amount of water. For this study, we have collected five years (2011–2015) of weather data from nearby weather station in East Lansing (MSUHort), Michigan. We optimize average crop yield in these five years, some of which are hugely dry and others are wet. In this study, we have selected a particular type of corn DECALB XL71, a popular crop produced in north America. Inside DSSAT, we turn off the effect of nutrients and study the effect of water on crop growth alone. Irrigation schedule starts just after plantation and crop is assumed to be harvested at maturity. Before we discuss our optimization procedure, we highlight a previous study by the authors of [15], which considered the effect of water retention through SWRT membranes simulated by HYDRUS-2D software (Fig. 2).

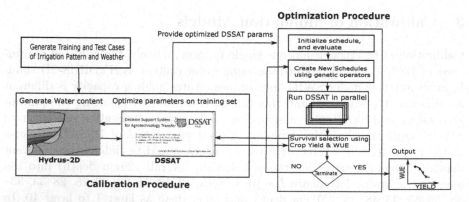

Fig. 2. Overall procedure for finding best crop production and water use efficiency is provided in a schematic diagram.

4.1   A Computationally Fast Approach

A previous study [15] considered two objectives related to water retention and utilization at the root system: (1) water use efficiency (WEF) and (2) root uptake efficiency (REF) by using HYDRUS-2D software for different aspect ratios of SWRT membranes. Each evaluation of a solution by HYDRUS-2D took about 1.5 to 2 min of computational time on a high-performance desktop computer requiring 1.5 days for on complete optimization run. To speed up the process, we procure a high performance server machine having an Intel Xeon E5-2697 v3 processor with 28 node and 56 available threads for this study. NSGA-II optimization procedure [2] with faster non-dominated sorting method [11,13] is used to handle two water retention related objectives – WUE and RUE, as discussed in the previous work. For our initial simulations here, we have used a population size of 32, maximum number of high-fidelity evaluations of 4,000 (a four-time increase from the prior study) with other parameters similar to previous study. Figure 3(a) shows the computational speed-up obtained by increasing the number of threads in our multi-core computer. Beyond 30 threads, the overhead of inter-thread communications increase and the marginal rate of time saving diminishes. Figure 3(b) and (c) shows the trade-off solutions obtained for four aspect ratios by executing the entire optimization run from the prior study [15] with (a) a single thread (b) 32 threads. For each aspect ratio set-up of the SWRT membrane, we perform an independent run and the final trade-off objective vectors (WUE and RUE) are shown in the same figure.

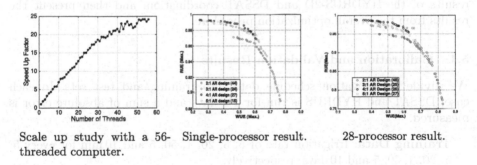

Scale up study with a 56-   Single-processor result.   28-processor result.
threaded computer.

Fig. 3. (a) Speed up using 28 cores, (b) A single processor result and (c) 28-processor result of root uptake efficiency and water use efficiency with different aspect ratios.

4.2   Optimization of Crop Production and Water Use

In this paper, we redefine water use efficiency (WUE) to be the ratio between plant transpiration and total input. Total input consists of irrigation and precipitation. Our second objective (crop production) is readily obtained from the DSSAT software after running a simulation on a particular irrigation schedule. The unit of this objective is kg per hectare. We take an average over 5 years of production to compute the objective function. Irrigation schedule contains starting date of irrigation, day and amount of irrigation in the crop growing season. We declare the daily irrigation amount as a variable and this makes the number of variables same as the number of days (110 used in this study). To constraint the irrigation pattern to be practical, we limit each irrigation amount to be less than or equal to 50 mm and the gap between two consecutive irrigation should not be more than 10 days. In our formulation, for each 10 day interval, we have a variable that tells us the amount of daily irrigation for those days. Thus, there will be a total of 11 variables for the entire 110-day simulation period. This formulation allows uncertainties of precipitation to be included in the study by averaging the rain-fall in every 10 consecutive days. Apart from this formulation, we have also performed experiments by defining 110 variables (one for each day) and two variables (irrigation rate and interval between two irrigation). We use a derivative-free global optimization method: NSGA-II [3], since it was used in previous studies [15]. Population size is kept 128 and the number of generations is fixed at 2,000. We have used similar parameters as before (Sect. 4.1) for SBX recombination and polynomial mutation operators. We ran our algorithm in server machine with 32 threads in parallel. This optimization is run on the soil parameters optimized for each design aspect ratio of SWRT.

5   Experimental Results

In this section, we summarize our results. First, we present calibration-validation results of the HYDRUS-2D and DSSAT coordination, and then present the results from the overall optimization procedure.

5.1   Calibration and Validation Results

We divide the irrigation schedule data into training and test sets. In each case, DSSAT and HYDRUS is run for 110 days and a sum of absolute error is measured.

– **Training Data:** Irrigation rate of 5, 5, 20, 1, 50, 5 and 10 mm for every 5, 5, 20, 1, 50, 5 and 10 days, respectively.
– **Test or Validation Data:** Irrigation rate of 2, 5, 7, 2, 10 and 1 mm for every 5, 30, 40, 10, 30 and 40 days, respectively.

In Fig. 4, we show the comparison of volumetric water content between HYDRUS and DSSAT softwares under the same irrigation and soil evaporation rate after the training is completed. The results show that the water content dynamics of DSSAT closely match to that obtained with HYDRUS. This suggests that we can use the obtained parameter settings of DSSAT with confidence. For six of our test cases, we get an average RMSE (Root mean squared error) of water content per layer of 10 layers to be {0.14, 0.14, 0.15, 0.14, 0.11, 0.07, 0.06, 0.11, 0.11, 0.12} and standard deviation of average RMSE is {0.05, 0.04, 0.07, 0.06, 0.03, 0.02, 0.01, 0.03, 0.03, 0.04}, respectively for total 110 days. The average RMSE is only about 5% per day of the volumetric water content values.

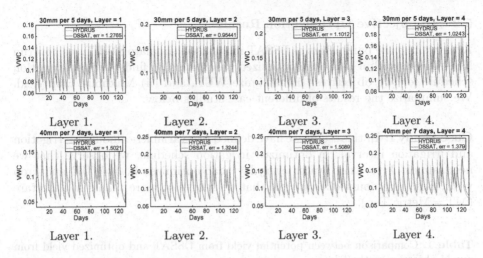

| Layer 1. | Layer 2. | Layer 3. | Layer 4. |

| Layer 1. | Layer 2. | Layer 3. | Layer 4. |

Fig. 4. Comparison of VWC (Volumetric water content) between DSSAT and HYDRUS-2D (with 2:1 AR design) for validation case.

In Fig. 5, we show cumulative water fluxes from HYDRUS and DSSAT. We show fluxes for infiltration (total input), run-off, drainage, and soil evaporation. In each test case, we observe that water fluxes of DSSAT effectively match with that obtained by HYDRUS-2D, wherever SWRT membrane is present.

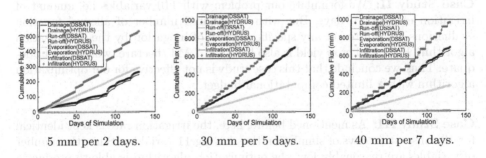

| 5 mm per 2 days. | 30 mm per 5 days. | 40 mm per 7 days. |

Fig. 5. Cumulative fluxes from DSSAT and HYDRUS for three different validation irrigation schedules.

The above validation procedure between HYDRUS-2D and DSSAT has also provided us with optimized parameter values for DSSAT parameters for each aspect ratio of the SWRT membrane. These optimized DSSAT parameters have matched volumetric and cumulative water flux values obtained by the HYDRUS-2D software with SWRT membrane embedded in the soil. Now, we are ready to use the DSSAT software with the obtained optimized parameter to simulate the crop growth (which HYDRUS-2D can not do!) and perform our overall bi-objective optimization study.

5.2 Bi-objective Optimization Results

After we find the parameters of DSSAT model that simulate SWRT, we now run NSGA-II algorithm using the optimized parameters to find the best irrigation schedule for two objectives: (i) maximize WUE, and (ii) Maximize crop yield. We summarize the results of different case studies below.

Case Study I: In this study, we have defined two variables- amount of irrigation and irrigation interval. In optimized irrigation schedule, we observe that we need almost 1 to 4 mm water everyday that optimizes both WUE and crop yield. But the amount of crop yield that we get is not more than 9 MT/hectare (MT = Metric Ton).

Table 1. Comparison between potential yield from DSSAT and optimized yield from our bi-objective methodology.

Year	2011	2012	2013	2014	2015
Potential yield (kg/ha)	11,324	13,059	11,929	10,837	10,300
Actual yield (kg/ha)	11,315	12,976	11,041	10,767	10,292

Case Study II: We formulate our problem with 110 variables i.e. amount of irrigation each of 110 days. But with the limited number of allowed function evaluations, we fail to get any specific pattern of irrigation by the optimization algorithm. Produced crop yield is also less than 7 MT/hectare, which is not adequate. Thus, we conclude that this case study is not effective for our optimization algorithm with a limited computational budget.

Case Study III: As mentioned before, here, the irrigation rate is kept identical for 10 consecutive days of simulation thus having 11 variables. Since the number of variables are reasonably low, the optimization algorithm is able to produce a well-distributed set of Pareto-optimal solutions trading-off WUE and crop yield objectives. We investigate the nature of trade-off solutions below. Potential and actual yield (MT/hectare) is given in Table 1.

Optimization Without SWRT: To compare, we create a similar project in HYDRUS-2D with same domain size (30 cm × 120 cm) and same coarse sand soil but without any impermeable SWRT membrane. We then optimize soil parameters of DSSAT so that it matches water contents of HYDRUS-2D simulation to imitate properties of that soil. After that, we optimize WUE and crop production using NSGA-II with the optimized parameters. In Fig. 6(a) and (e), we present the obtained non-dominated front without SWRT. It can be observed that at most 2MT/ha corn is possible to be grown on average, while gaining as

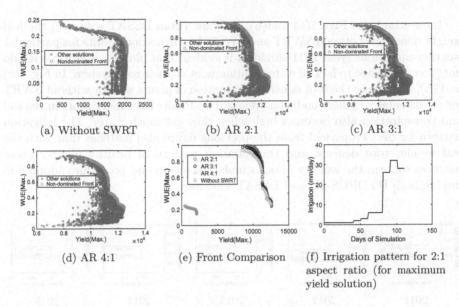

(a) Without SWRT   (b) AR 2:1   (c) AR 3:1

(d) AR 4:1   (e) Front Comparison   (f) Irrigation pattern for 2:1
aspect ratio (for maximum
yield solution)

Fig. 6. Obtained non-dominated solutions are shown after completion of whole optimization process. (a) is performed without SWRT and (b), (c) and (d) are for AR design 2:1, 3:1 and 4:1, respectively. Comparison among non-dominated fronts are shown in (e). Daily irrigation pattern for a solution (2:1 AR design) that optimizes the crop yield is shown in (f).

much as 25% in WUE. Since water conductivity is very high in sandy soil, it is expected that we obtain much less corn production in dry years.

Optimization with SWRT: In Fig. 6(b)–(e), the non-dominated fronts are shown. From the figures it is clear that, even the dominated solutions of this case are much better than the non-dominated solutions obtained without SWRT. This is because SWRT retains more water, thereby helping plant growth at their crucial stages. We repeat this experiment with 2:1, 3:1 and 4:1 AR design. There is a slight variation of non-dominated fronts obtained from these three different SWRT configurations. It is observed that 2:1 AR design is able to retain much water in the soil compared to other designs.

We also observe that one of the obtained non-dominated solutions produces 99% water use efficiency on average over the years 2011–2015. The crop yield is also as high as 12,630 kg/ha. Irrespective of AR in the membrane design, a small amount of daily water is needed initially. Amount of water varies significantly when crop becomes more mature. We can divide the entire non-dominated front into three parts. The first part produces 99% WUE in which the amount of daily irrigation is limited by 7 to 8 mm. In the second part (middle), we obtain better crop yield by increasing the irrigation rate to 15 to 25 mm over the vegetation period. In the last part (after around 40–50 days), we need to increase the water supply to around 40 to 50 mm per day. When the crop reaches maturity, the harvest time begins and we need less amount of water, as depicted in Fig. 6(f).

Interestingly, in Figs. 7(a) and(b), we show (from DSSAT software) the leaf weight difference between SWRT and SWRT-less irrigation results for particular weather conditions (year 2011–2015). It is evident that plants without SWRT do not grow much due to lack of water containment at their root system. In Fig. 8(a) and (b), we also see the root density information of plants with or without SWRT for a particular weather condition (year 2011). Plants have more roots in the soil and transpiration also becomes high when they get much water. The irrigation pattern for best crop yield from the previous figure also matches that with the leaf weight, root density, and transpiration pattern of healthy plants. These patterns confirm the validity of our simulation set-up and procedure adopted in linking both HYDRUS-2D and DSSAT softwares.

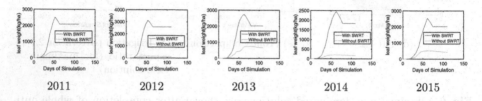

| 2011 | 2012 | 2013 | 2014 | 2015 |

Fig. 7. Comparison of leaf weight of simulated plants with or without SWRT for year 2011 and 2012.

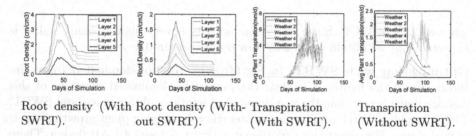

Root density (With SWRT).   Root density (Without SWRT).   Transpiration (With SWRT).   Transpiration (Without SWRT).

Fig. 8. Comparison of simulated growth of root density and transpiration with or without SWRT for a single weather condition (year 2011) for the 2:1 Aspect ratio membrane.

Analysis of Bi-objective Optimization Results. Table 1, discussed before, shows the potential and actual yields produced by DSSAT software. The potential yield is the maximum possible yield under standardized parameter settings (with SWRT 2:1 AR design) which is taken care by DSSAT internally. On the other hand, the actual yield is obtained by finding the best irrigation schedules using our proposed optimization procedure (same SWRT). The results show that, we can achieve maximum possible yield just by providing perfect irrigation schedule. We observe that in two years (2011 and 2012), the crop yield produced by our optimization procedure is 99% to that of the potential yield. In other three years, we have achieved a better yield than that of the potential yields provided by DSSAT itself. We present the total time savings by our proposed methodology in Table 2.

Table 2. Time Savings in HYDRUS-2D and DSSAT due to parallel evaluation of solutions.

Method/Time	Initial study (HYDRUS)	Calibration process (HYDRUS-DSSAT)	EMO-HYDRUS-DSSAT
1 Processor (Estimated)	6,000 min (100 hr)	416.67 min (6.94 hr)	177.78 hr
32 Threads	125 min (2.08 hr)	13 min	5.56 hr

6   Conclusions and Future Work

In this paper, we have proposed a computational approach to find an optimum irrigation schedule that not only minimizes water usage but also maximizes crop production. In order to simulate soil water movement under embedded SWRT membranes, we have used HYDRUS-2D simulation software in a two-dimensional setting and DSSAT software for predicting the crop yield. Between the two softwares, HYDRUS-2D simulation is exceedingly more time consuming, hence we have developed a computationally fast parallel procedure to interlink the two softwares so that both conflicting criteria can be optimized in a reasonable computational time.

We have proposed a new calibration-validation methodology to run our optimization method using computationally quick DSSAT software yet taking the advantage of HYDRUS-2D software's SWRT analysis capability. The optimization is carried out over different weather conditions to predict the amount of crop production. Our method not only reduces computational effort by multi-threaded implementation of solution evaluations, but also greatly cuts down the human effort of optimizing parameters by introducing an EMO algorithm in the precision irrigation field. The results presented here clearly indicates the promise of our proposed approach and it should facilitate further research in predicting yield of different crops at different soils with a suitable effect of a SWRT.

The study also spurs a number of future research directions. In addition to crop yield and water usage, fertilizer usage can also be considered for minimizing ground water contamination due to deep perlocation of fertilizers and pesticides by simulating nutrient flow and accumulation under SWRT using the two software in a similar linking procedure. To make the overall procedure more pragmatic, a three-dimensional water and nutrient flows may be considered. Given the expensive nature of HYDRUS-2D/3D based optimization, use of metamodel based optimization [1,10,12] can also be considered in future.

Acknowledgement. This material is based in part upon work supported by the National Science Foundation under Cooperative Agreement No. DBI-0939454. Any opinions, findings, and conclusions or recommendations expressed in this material are those of the author(s) and do not necessarily reflect the views of the National Science Foundation.

References

1. Deb, K., Hussein, R., Roy, P.C., Toscano, G.: A taxonomy for metamodeling frameworks for evolutionary multi-objective optimization. IEEE Trans. Evol. Comput. (in press)
2. Deb, K., Pratap, A., Agarwal, S., Meyarivan, T.: A fast and elitist multiobjective genetic algorithm: NSGA-II. IEEE Trans. Evol. Comput. **6**(2), 182–197 (2002)
3. Deb, K.: Multi-objective Optimization using Evolutionary Algorithms. Wiley, Hoboken (2001)
4. Guber, A.K., Smucker, A.J.M., Berhanu, S., Miller, J.M.L.: Subsurface water retention technology improves root zone water storage for corn production on coarse-textured soils. Vadose Zone J. **14**(7) (2015)
5. Han, M., Zhao, C., Simunek, J., Feng, G.: Evaluating the impact of groundwater on cotton growth and root zone water balance using Hydrus-1D coupled with a crop growth model. Agric. Water Manag. **160**, 64–75 (2015)
6. Hoogenboom, G., et al.: Decision support system for agrotechnology transfer (DSSAT) version 4.6 (2015). http://dssat.net
7. Iqbal, S., Guber, A.K., Khan, H.Z.: Estimating nitrogen leaching losses after compost application in furrow irrigated soils of pakistan using HYDRUS-2D software. Agric. Water Manag. **168**, 85–95 (2016)
8. Liu, H., et al.: Simulating water content, crop yield and nitrate-N loss under free and controlled tile drainage with subsurface irrigation using the DSSAT model. Agric. Water Manag. **98**(6), 1105–1111 (2011)
9. Nash, J., Sutcliffe, J.: River flow forecasting through conceptual models part I a discussion of principles. J. Hydrol. **10**(3), 282–290 (1970)
10. Roy, P., Deb, K.: High dimensional model representation for solving expensive multi-objective optimization problems. In: IEEE CEC, pp. 2490–2497 (2016)
11. Roy, P.C., Deb, K., Islam, M.M.: An efficient nondominated sorting algorithm for large number of fronts. IEEE Trans. Cybern. 1–11 (2018)
12. Roy, P., Hussein, R., Deb, K.: Metamodeling for multimodal selection functions in evolutionary multi-objective optimization. In: GECCO 2017. ACM Press (2017)
13. Roy, P.C., Islam, M.M., Deb, K.: Best order sort: a new algorithm to non-dominated sorting for evolutionary multi-objective optimization. In: GECCO 2016 Companion, pp. 1113–1120. ACM (2016)
14. Simunek, J., Genuchten, V., Sejna, M.: Hydrus technical manual version 2. University of California Riverside, Riverside, Technical report (2012)
15. Tutum, C.C., Guber, A.K., Deb, K., Smucker, A., Nejadhashemi, A.P., Kiraz, B.: An integrated approach involving EMO and HYDRUS-2D software for SWRT-based precision irrigation. In: 2015 IEEE Congress on Evolutionary Computation (CEC), pp. 885–892, May 2015
16. Van Genuchten, M.T.: A closed-form equation for predicting the hydraulic conductivity of unsaturated soils. Soil Sci. Soc. Am. J. **44**(5), 892–898 (1980)

Optimum Wind Farm Layouts:
A Many-Objective Perspective
and Case Study

Kalyan Shankar Bhattacharjee, Hemant Kumar Singh, and Tapabrata Ray[✉]

School of Engineering and Information Technology,
The University of New South Wales, Canberra, Australia
k.bhattacharjee@student.adfa.edu.au,
{h.singh,t.ray}@adfa.edu.au

Abstract. A steady increase in the prices of non-renewable energy sources, their environmental impact, and the ever-increasing energy demands have made it imperative to explore alternative, renewable energy options. Wind energy is one of the prominent alternatives, and for onshore installations, optimal placement of wind turbines is necessary to harness maximum power. This optimal placement problem, referred to as *wind-farm layout optimization*, has received significant research attention with regards to output power maximization. However, in practice, apart from maximization of power, a number of other key factors need to be considered, such as cabling cost, maintenance cost and noise levels. Furthermore, the wind farm itself may have irregular boundaries and within the area there may be several protected areas due to existing archaeological deposits, water bodies, bird feeding areas, etc. In this paper, we present a framework to support practical layout optimization of wind farms. In the proposed approach, a variable discretization scheme is employed to deal with irregular land boundaries and a many-objective formulation is used to identify the set of trade-off solutions. The utility of the approach is highlighted using a case study resembling the Capital wind farm located in New South Wales, Australia. We hope that this study will motivate use of such tools to solve practical wind farm layout optimization problems.

1 Introduction

Wind power is one of the prominent sources of large scale renewable energy. In 2015, Australia's wind farms produced 33.7 per cent of the country's clean energy and supplied 4.9 per cent of Australia's overall electricity [3]. A typical onshore wind farm contains several turbines installed over a large land area. Each turbine has an individual capacity of producing a certain amount of energy. However, if installed too close to each other, the turbulence and wake effects cause a reduction in wind speed and consequently in the power generated at the downstream turbines. Consequently, a wind farm tends to span expansive land area, which in turn increases its interference with natural habitats [15,23]. Additionally, noise

© Springer Nature Switzerland AG 2019
K. Deb et al. (Eds.): EMO 2019, LNCS 11411, pp. 707–718, 2019.
https://doi.org/10.1007/978-3-030-12598-1_56

generated by the turbines also need appropriate consideration. The noise levels at nearby residential areas should be below prescribed levels to limit potential health hazards [8,28,30]. Wind farm layout design is thus a challenging optimization problem with a number of practical considerations.

Optimum design of wind farm layouts has been attempted in the past with various levels of simplification in the model. For example, in [18] the wind farm was assumed to be of a square shape and represented using a 10×10 grid (100 cells in total), where turbines could only be located at the center of the cells. This discrete location model relied on the size of the cells to inherently enforce proximity constraints (of a minimum distance between any two turbines). Continuous location models have also been used, for example in [13], where turbines could be located anywhere within the area. In reality, a wind farm often spans across areas belonging to multiple land owners and there may be a limit on the number of turbines installed within each block depending on agreements with the respective owners. The model in [32,33] considered straight-line boundaries between these blocks. More realistic models that consider regulatory land use [27] or variation in elevation also appear in recent literature. Irregular boundaries have rarely been considered, although this would be the most likely scenario in a practical wind farm design.

In terms of estimation of power, simplifications range from considering unidirectional wind at constant speed through to models that consider wind speed/ direction variation with turbine interactions [22]. Turbine interaction models also vary in complexity ranging from Jensen wake model [12] through to three dimensional wake models [25]. There are also a variety of optimization formulations reported in the literature which range from energy maximization as the sole objective [10,18], energy maximization subject to proximity and noise level constraints [2,9,13] or even bi-objective formulations that consider noise levels and energy maximization simultaneously [14]. Further extensions that consider energy maximization, cable length minimization and enclosed land area minimization have also been reported in [31]. Since the underlying optimization problem is NP-hard, a range of stochastic algorithms (NSGA-II [14], CMA-ES [11], SPEA [13,25], IBEA [17]) have been used for solving it.

Given that the adoption of wind farms clearly depends on a number of factors apart from the power maximization, it is useful to seek and present a rich trade-off set of solutions to the stakeholders. In view of the existing research on this problem discussed above and the associated limitations, the key highlights of this study are listed below.

- Firstly, we offer an optimization framework to deal with wind farm layout optimization involving realistic objectives such as (a) maximization of wind power, (b) minimization of cable length connecting the turbines, (c) minimization of enclosed land area of the layout to reduce maintenance and inspection costs and (d) minimization of noise level. While some of these objectives have been considered individually in the works before, in its essence the problem is a multi-objective optimization problem; also referred to colloquially as a *many*-objective optimization problem (MaOP) when the number of objectives are more than three. MaOP has been a highly active research area in the past decade [16]; but most studies predominantly use mathematical benchmark functions that may not capture some of the real-life modeling challenges. The formulation of the windfarm layout optimization problem as a MaOP, with its unique set of challenges, not only offers opportunity to obtain realistic trade-off solutions, but the case-study can also serve as an application benchmark from an algorithm development perspective.

- Secondly, the proposed approach can deal with realistic constraints such as (a) multiple infeasible regions where turbines cannot be placed due to environmental regulations (archaeological deposits, bird feeding areas, natural flora, etc.) and/or landowners' restrictions and (b) proximity constraints on the turbine placements. Notably, (a) is handled using a novel solution representation through triangulation of the given irregular land area, while (b) is handled through assistance of infeasibility driven constraint handling approach.
- Thirdly, the proposed approach incorporates a mechanism capable of generating a feasible cable layout, i.e., rerouting of cables to avoid all infeasible areas.
- Lastly, the above contributions are demonstrated through a case-study conducted on a problem resembling the Capital wind farm located near Lake George, New South Wales, Australia.

The remainder of the paper is organized as follows. The details of the wind farm layout problem are discussed in Sect. 2. The algorithm is briefly outlined in Sect. 3, followed by the results obtained on the case study in Sect. 4. Concluding remarks and future directions are given in Sect. 5.

2   Wind Farm Layout Optimization Problem

2.1   Generic Problem Formulation

A typical onshore wind farm contains several turbines located over a significant stretch of land. It is well known that the total generated power of a wind farm is significantly less than the summation of the rated power of the individual turbines [26]. This is due to the *wake effect*, where flow past a turbine affects other turbines located downstream from it [13]. The magnitude of wake and in turn the efficiency of energy production depends primarily on the layout of the wind turbines. While maximization of energy production is a key consideration, there are several other factors which affect the design of the layout such as minimization of noise, minimization of cable length, minimization of enclosed area of the layout, etc. Furthermore, there might be several prohibited zones inside the layout where no turbines can be placed due to environmental regulations or landowners' requirements. There are also a number of practical constraints e.g. (a) distance between any two turbines should be more than a prescribed distance (generally 8 times the turbine rotor radius) to minimize wake losses and hazardous loads on the turbines (b) cables between two turbines cannot pass through restricted areas, etc. Thus the layout problem is best represented as a constrained many-objective optimization problem presented in Eq. 1.

$$\text{Minimize: } \mathbf{F}(\mathbf{X}) = f_i(\mathbf{X}), i = 1, 2, \ldots M$$
$$\text{Subject to}$$
$$c_j(\mathbf{X}) \leq 0, j = 1, 2, \ldots p$$
$$h_j(\mathbf{X}) = 0, j = 1, 2, \ldots q \tag{1}$$
$$\mathbf{X}^{(L)} \leq \mathbf{X} \leq \mathbf{X}^{(U)}$$

Here, $f_1(\mathbf{X}), f_2(\mathbf{X}), f_3(\mathbf{X}), \ldots f_M(\mathbf{X})$ are the M objective functions to be optimized; considered here in a minimization sense. The number of inequality and equality constraints are denoted by p and q, respectively. The upper and lower bounds of the variables are denoted as $\mathbf{X}^{(U)}$ and $\mathbf{X}^{(L)}$, respectively. For every solution, the sum of constraint violations is denoted by CV, where $CV = 0$ indicates a feasible solution. For every pair of turbine locations, a violation value is computed if the distance between them is less than 8 times the turbine rotor radius. Sum of these violations correspond to CV. As for the constraints on cable routing, it is managed through a repair (re-routing), as will be discussed shortly. The constraints of being in feasible irregular boundaries is handled implicitly through the solution representation itself.

2.2   Solution Representation

Contrary to some of the regular geometries considered in the literature, e.g. [18], wind farms could typically have geometries with very irregular boundaries. Further, the areas prohibited for turbine installation may be irregular too. In this study, we propose a simple representation that can place the turbines within these irregular boundaries and avoid the prohibited zones. In order to do this, we discretize the boundaries of the allowable areas using the technique proposed in [20] and construct a triangular mesh within the feasible zones. Figure 1(a) illustrates an example of a discretized feasible area with irregular geometry and several infeasible areas with irregular geometries within it such as bird feeding area, private property and a water body. Once the N_t triangles have been obtained through this step, the location of an ith turbine can be represented using the set of variables (A^i, w^i), where A^i is one of the triangles, and w^i is a set of weights such that $\sum_{j=1}^{3} w_j{}^i = 1$. The weights corresponding to a set of uniformly distributed N_w points on a given triangle generated using normal boundary intersection (NBI) [4]. All these points on any given triangle are considered candidate locations for the turbine. The representation is given in Eq. 2.

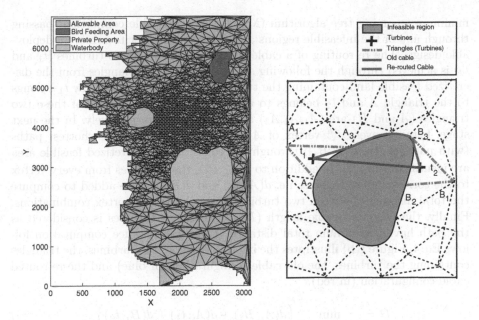

Fig. 1. Discretization of area and cable re-routing (Color figure online)

$$\mathbf{X} = \{A^i, w^i\}$$
where   (2)
$$i \in [1, N],$$
$$A^i \in \{A^1, A^2, \dots, A^{N_t}\}, \quad w^i \in \{w^1, w^2, \dots, w^{N_w}\}$$

Let's say, the vertices of A^i are $(A_1{}^i, A_2{}^i, A_3{}^i)$, the x-coordinate of the vertices of A^i are denoted by $\{x_{A_j{}^i} \mid j = 1, \dots, 3\}$ and the y-coordinates are denoted by $\{y_{A_j{}^i} \mid j = 1, \dots, 3\}$. The Cartesian coordinates of the ith turbine location can thus be simply computed as:

$$x_i = \sum_{j=1}^{3} w_j{}^i x_{A_j{}^i}; \quad y_i = \sum_{j=1}^{3} w_j{}^i y_{A_j{}^i} \qquad (3)$$

2.3 Objectives and Computation Models

In this study, the performance of a candidate layout is assessed using four objectives: (a) energy production(maximize), (b) total cable length (minimize), (c) enclosed area of the layout (minimize) and (d) the maximum noise level (minimize). The energy production model is based on [13] which considers wind speed and directional variations. Turbines are connected via cables and the total length of the cable configuration contributes to the levelised annual cost of a farm. Given the location of the turbines, the cable routes are derived using minimum spanning tree algorithm (MST). However, cable connections passing through any of the infeasible regions are re-routed in order to achieve a deployable design. The re-routing of a cable connection between two turbines (t_1 and t_2) is achieved through the following stages. At first, the triangles from the discretized feasible land containing the turbines are identified. Let's say t_1 belongs to the triangle A and t_2 belongs to the triangle B. The vertices of these two triangles (A and B) are (A_1,A_2,A_3) and (B_1,B_2,B_3), respectively. In the next stage, between every i^{th} vertex of A and j^{th} vertex of B, the shortest paths (with path length $\tilde{d}(A_i, B_j)$) through the edges of the discretized feasible area are identified using [7]. In addition to $\tilde{d}(A_i, B_j)$, the distances from every vertex to the corresponding turbines i.e. $d(A_i, t_1)$ and $d(B_j, t_2)$ are added to compute the total distances between two turbines through various vertex combinations. Finally, the re-routed cable length (D) between two turbines is considered as the path having minimum total distance. The total distance computation follows Eq. 4. Figure 1(b) illustrates the infeasible region, the turbines, the triangles containing the turbines, the old cable configuration (in blue) and the re-routed cable configuration (in red).

$$D = \min_{1 \leq i \leq 3, 1 \leq j \leq 3} \left(\tilde{d}(A_i, B_j) + d(A_i, t_1) + d(B_j, t_2) \right) \qquad (4)$$

The maintenance cost of a wind farm is proportional to the enclosed area of the wind farm layout. The enclosed area is computed based on the convex hull bounded by the turbine locations. Generation of noise by the wind farm is one of the most important environmental concerns. In general, the sound level is measured at the receptors at the nearby residences [29]. In this study, the ISO-9613-2 standard has been followed to compute the noise generation at the receptor locations and the maximum noise level generated among all receptors is considered as an objective.

2.4   Case Study Description

The application is based on Capital wind farm located in New South Wales, Australia. The wind farm has three different regions: Groses hill, Ellenden and Hammonds hill. It has a total of 67 turbines out of which 17 are placed in Groses hill, 21 are within Ellenden and 29 turbines are located in Hammonds hill region. There are several infeasible regions, such as woodland vegetation, secondary grassland, wattle woodland, yellowbox woodland, she-oak region and nearby residences as shown in Fig. 2. Due to environmental regulations, turbines cannot be placed in these infeasible regions. The turbines are of same make and model, i.e., Suzlon S88 and the parameters related to the turbines are listed in Table 1. The wind scenario used in this study is constructed from the wind rose provided for the Capital wind farm project in [19].

This problem is solved as a constrained many-objective optimization problem. The need to place 67 turbines translates to 134 variables and results in $67(67 - 1)$, i.e., 4422 proximity constraints. The discretization of the allowable land generates a total of $N_t = 10476$ triangles among which 2167 triangles belong

Table 1. Turbine related parameters

Turbine parameters	Value	Turbine parameters	Value
Make and model	Suzlon S88	Rated power (P_{rated})	2100 kW
Rotor radius (R)	44 m	Hub height	80 m
Cut-in wind speed (v_{cut-in})	4 m/s	Rated wind speed (v_{rated})	14 m/s
Cut-out wind speed ($v_{cut-out}$)	25 m/s	Slope of the power curve (λ)	262.5
Intercept of the power curve (η)	−1050	Thrust coefficient (C_T)	0.9

to Groses hill, 4639 triangles to Ellenden and 3670 to Hammonds hill. The number of combinations of weights (N_w) for each triangle is set to be 8001. Take note that the discretization does lead to triangles with different sizes and the choice of (N_w) is just to ensure appropriate discretization for the largest triangle. Among the objectives, the noise level at the residences shown in Fig. 2 are computed using the parameters listed in Table 2.

Table 2. Turbine related parameters

Noise parameters	Value
Noise generation (L_w)	105.9 dBA
Residence height	1.5 m
Directivity correction (D_s)	0
Average temperature	10°C
Average humidity	80%
Ground factor ($G = 0$: hard, $G = 1$: porous)	0
Nominal midband frequency (f)	{63, 125, 250, 500, 1000, 2000, 4000, 8000} Hz
Atmospheric attenuation coefficient (α)	{0, 0, 1, 2, 4, 9, 29, 104}
A-weighted factors (A_f)	{−26.2, −16.1, −8.6, −3.2, 0, 1.2, 1, −1.1}

3   Algorithm

The optimization algorithm is based on a ($\mu + \lambda$) evolutionary model, where μ parents are recombined to generate λ offspring and the best μ solutions are selected as parents for the next generation. The pseudo-code of the proposed method is presented in Algorithm 1 and uses *decomposition* of objective space, a strategy commonly used in the contemporary algorithms for solving MaOPs.

The algorithm has a framework similar to reference vector based evolutionary algorithm (RVEA) [1], but there are two key modifications. The first relates to parent selection scheme. While random parents are selected from a neighborhood in RVEA for recombination, we opt to use ranking that prefers marginally infeasible solutions over feasible solutions. Such a ranking scheme was introduced in [21,24] and demonstrated to perform better than strictly feasibility first schemes for constrained optimization problems. Secondly, in order to utilize the advantages of both differential evolution crossover [5] and simulated

---

Algorithm 1. Proposed algorithm used for windfarm layout optimization

---

Input: Gen_{max} (Maximum number of generations), W (Number of Reference points/population size), Crossover and mutation parameters
 1: $i=1$. {Generation counter}
 2: **Generate** W reference points using Systematic Sampling.
 3: **Construct** W reference directions; Straight lines joining origin and W reference points.
 4: θ_{th}: Compute the minimum angle between a reference direction and all others.
 5: **Initialize** the population using LHS sampling P^i; $|P^i| = W$.
 6: **Assign** individuals of P^i to the reference directions.
 7: **while** ($i \leq Gen_{max}$) **do**
 8:     **Create** C offspring from P^i via recombination.
 9:     **Assign** P^{i+1} individuals from $P^i + C$ to W reference directions
10:     $i=i+1$.
11: **end while**

---

binary crossover [6], at each generation both types of crossover are employed for alternate base parents. Similarly during the evolution, reverse order of crossover types are used in alternate generations. Thus, if at generation 1, first reference direction uses differential evolution crossover and second reference direction uses simulated binary crossover, at generation 2 the first reference direction will use simulated binary crossover and differential evolution crossover will be used for the second. The intent is to improve convergence by adopting the high quality solutions generated using the two types of crossovers, while also reducing bias towards either of them. Due to space constraints, we omit the detailed description of the algorithm, but the interested readers are referred to [1].

4   Results

A single optimization run has been performed to solve this problem due to the computationally expensive nature of the underlying simulations. A population size of 220 solutions was evolved over 600 generations to obtain the final layout of the turbines. The run-time was approximately 75 h on a 2.30 GHz, 32 cores with 128 GB of memory. A total of 27 feasible solutions were obtained at the end of evaluation budget, which highlights that the problem is highly constrained. Out of the feasible solutions, 9 solutions were nondominated and there were 3 unique extreme solutions. The extreme solutions in the context of minimum cable length and minimum enclosed area were the same. The obtained values of the maximum energy production, the minimum cable length, the minimum enclosed area and the minimum noise level were 49.16 MW, 53.38 km, 71.07 km^2 and 55.54 dBA, respectively.

The complete set of trade-off solutions are presented in Fig. 3. Since there are only 9 solutions under consideration, the stakeholders can collectively work to select the most preferred option. The layouts corresponding to each extreme solutions including the noise level on each residence is plotted in Fig. 3. In the current state since such modeling/optimization tools are either not readily accessible or well understood by communities at large, there is very limited understanding of the benefits and the impacts of wind farms. The considerations of multiple criteria and resulting visualization can help in an informed decision-making about

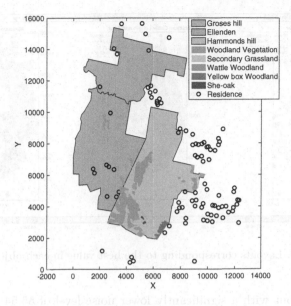

Fig. 2. Wind farm land

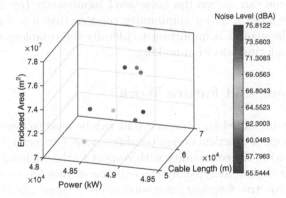

Fig. 3. Nondominated solutions

trade-offs between various designs. For example, it can be observed from Fig. 4 that the location of the turbines corresponding to minimum noise level are away from the residential areas. Among the feasible solutions, there were variations of 1.56%, 14.66%, 8.55% and 26.73% in terms of power generation, cable length, land area and noise level, respectively (calculated as $\frac{\max(|f|)-\min(|f|)}{\max(|f|)}$). This observation raises an interesting and practically relevant design consideration - if one opted to solve the above problem as a single objective power maximization problem, the best solution would correspond to a total power of 49.16 MW with a noise level of 75.81 dBA. On the other hand, using a multi-objective approach one can identify alternatives and opt for a layout with marginally lower power

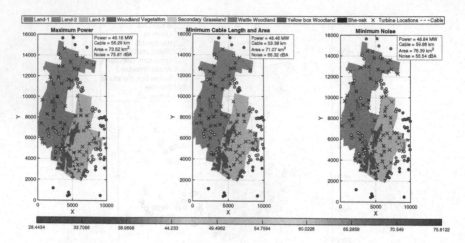

Fig. 4. Layouts corresponding to the best value in each objective

of 48.84 MW but with a significantly lower noise level of 55.54 dBA. That is, by maintaining almost the same level of power output (i.e. only 1.56% lower than the best), one can reduce the noise level significantly (by 26.73%). Since the noise level is often a major community concern that has a major bearing on adoption of the plan, it is important to identify the complete set of trade-off solutions for an informed decision-making.

5   Conclusions and Future Work

In this paper, we presented an approach that can be used to develop wind farm layouts with a range of practical design considerations. Currently, there is limited understanding within the community with respect to the trade-offs involved in a wind farm layout design, as typically the existing studies have solved the problem as a single-objective formulation involving power maximization. In absence of the consideration of multiple objectives and constraints relevant to the environment, the obtained designs may not reflect realistic layouts, which in turn affects the uptake and exploitation of wind energy. In this paper, we presented an approach that can be used to analyze or design potential wind farm layouts with appropriate level of details such as irregular land boundaries, multiple land owners, consideration of protected areas, noise levels at nearby residential dwellings, etc. It also offers an opportunity to view alternative layouts while considering maintenance costs, cable layouts, noise levels and power generation simultaneously. The utility of the approach is highlighted using a case study resembling the Capital wind farm located in New South Wales, Australia. We hope that this study will motivate use of such formulations and tools to identify optimal wind farm layouts.

While in our current analysis we did not impose an upper limit on maximum noise level constraints, it could be a straightforward extension in the future to the existing proximity constraints. Furthermore, variation in elevation of the wind farm was ignored in the model which can be easily incorporated in power estimation models. Apart from its utility as a tool, the underlying problem is also interesting as an application problem for research in evolutionary many-objective optimization as it represents a highly constrained optimization problem with modest number of variables.

References

1. Cheng, R., Jin, Y., Olhofer, M., Sendhoff, B.: A reference vector guided evolutionary algorithm for many-objective optimization. IEEE Trans. Evol. Comput. **20**(5), 773–791 (2016)
2. Chowdhury, S., Zhang, J., Messac, A., Castillo, L.: Characterizing the influence of land configuration on the optimal wind farm performance. In: Proceedings of the ASME International Design Engineering Technical Conferences & Computers and Information in Engineering Conference (2011)
3. Clean Energy Council. https://www.cleanenergycouncil.org.au/policy-advocacy/renewable-energy-target.html
4. Das, I., Dennis, J.E.: Normal-boundary intersection: a new method for generating the pareto surface in nonlinear multicriteria optimization problems. SIAM J. Optim. **8**(3), 631–657 (1998)
5. Das, S., Suganthan, P.N.: Differential evolution: a survey of the state-of-the-art. IEEE Trans. Evol. Comput. **15**(1), 4 31 (2011)
6. Deb, K., Pratap, A., Agarwal, S., Meyarivan, T.: A fast and elitist multiobjective genetic algorithm: NSGA-II. IEEE Trans. Evol. Comput. **6**(2), 182 197 (2002)
7. Dijkstra, E.W.: A note on two problems in connexion with graphs. Numer. Math. **1**, 269–271 (1959)
8. Environmental Protection Heritage Council: National Wind Farm Development Guidelines-Draft (2010)
9. Fagerfjäll, P.: Optimizing wind farm layouts: more bang for the buck using mixed integer linear programming. Master's thesis, Chalmers University of Technology, Gothenberg University, Gothenberg, Sweden (2010)
10. Grady, S.A., Hussaini, M.Y., Abdullah, M.M.: Placement of wind turbines using genetic algorithms. Renewable Energy **30**(2), 259–270 (2005)
11. Hansen, N., Müller, S.D., Koumoutsakos, P.: Reducing the time complexity of the derandomized evolution strategy with covariance matrix adaptation (CMA-ES). Evol. Comput. **11**(1), 1–18 (2006)
12. Jensen, N.O.: A note on wind turbine interaction. Technical report M-2411, Risoe National Laboratory (1983)
13. Kusiak, A., Song, Z.: Design of wind farm layout for maximum wind energy capture. Renewable Energy **35**(3), 685–694 (2010)
14. Kwong, W.Y., Zhang, P.Y., Romero, D., Moran, J., Morgenroth, M., Amon, C.: Multi-objective wind farm layout optimization considering energy generation and noise propagation with NSGA-II. J. Mech. Des. **136**(9), 091010:1–091010:11 (2014)
15. Leung, D.Y.C., Yang, Y.: Wind energy development and its environmental impact: a review. Renew. Sustain. Energy Rev. **16**(1), 1031–1039 (2012)
16. Li, B., Li, J., Tang, K., Yao, X.: Many-objective evolutionary algorithms: a survey. ACM Comput. Surv. (CSUR) **48**(1), 13 (2015)

17. Li, W., Özcan, E., John, R.: Multi-objective evolutionary algorithms and hyper-heuristics for wind farm layout optimisation. Renewable Energy **105**, 473–482 (2017)
18. Mosetti, G., Poloni, C., Diviacco, B.: Optimization of wind turbine positioning in large windfarms by means of a genetic algorithm. J. Wind Eng. Ind. Aerodyn. **51**(1), 105–116 (1994)
19. NSW Planning & Environment: (2006). http://majorprojects.planning.nsw.gov. au/index.pl?action=view_job&job_id=670
20. Persson, P.O., Strang, G.: A simple mesh generator in MATLAB. SIAM Rev. **46**(2), 329–345 (2004)
21. Ray, T., Singh, H.K., Isaacs, A., Smith, W.F.: Infeasibility driven evolutionary algorithm for constrained optimization. In: Mezura-Montes, E. (ed.) Constraint-Handling in Evolutionary Optimization. Studies in Computational Intelligence, vol. 198, pp. 145–165. Springer, Heidelberg (2009). https://doi.org/10.1007/978-3-642-00619-7_7
22. Saavedra-Moreno, B., Salcedo Sanz, S., Paniagua Tineo, A., Prieto, L., Portilla Figueras, A.: Seeding evolutionary algorithms with heuristics for optimal wind turbines positioning in wind farms. Renewable Energy **36**(11), 2838–2844 (2011)
23. Saidur, R., Rahim, N., Islam, M.R., Solangi, K.H.: Environmental impact of wind energy. Renew. Sustain. Energy Rev. **15**(5), 2423–2430 (2011)
24. Singh, H.K., Ray, T., Sarker, R.: Optimum oil production planning using infeasibility driven evolutionary algorithm. Evol. Comput. **21**(1), 65–82 (2013)
25. Song, Z., Zhang, Z., Chen, X.: The decision model of 3-dimensional wind farm layout design. Renewable Energy **85**, 248–258 (2016)
26. Sorensen, P., Nielsen, T.: Recalibrating wind turbine wake model parameters-validating the wake model performance for large offshore wind farms. In: Proceedings of the European Wind Energy Conference and Exhibition (2006)
27. Sorkhabi, S.Y.D., et al.: The impact of land use constraints in multi-objective energy-noise wind farm layout optimization. Renewable Energy **85**, 359–370 (2016)
28. South Australia Environment Protection Authority, Adelaide: wind farms environmental noise guidelines (2009)
29. International Organization for Standardization: ISO 9613-2 (acoustics)-attenuation of sound during propagation outdoors-Part 2: general method of calculation (1996)
30. Standards Australia, Sydney: AS4959 (acoustics)-measurement, prediction and assessment of noise from wind turbine generators (2010)
31. Tran, R., Wu, J., Denison, C., Ackling, T., Wagner, M., Neumann, F.: Fast and effective multi-objective optimisation of wind turbine placement. In: Proceedings of the International Conference on Genetic and Evolutionary Computation, pp. 1381–1388 (2013)
32. Wang, L., Tan, A.C.C., Cholette, M.E., Gu, Y.: Optimization of wind farm layout with complex land divisions. Renewable Energy **105**, 30–40 (2017)
33. Wang, L., Tan, A.C.C., Gu, Y., Yuan, J.: A new constraint handling method for wind farm layout optimization with lands owned by different owners. Renewable Energy **83**, 151–161 (2015)

Comparison of Multi-objective Optimization Methods Applied to Electrical Machine Design

William R. Jensen$^{(\boxtimes)}$, Thang Q. Pham, and Shanelle N. Foster

Michigan State University, East Lansing, MI 48824, USA
{jensenwi,phamtha1,hogansha}@egr.msu.edu

Abstract. Electric machine design optimization is a growing topic of interest. Using a genetic algorithm for optimization, an efficient search of the design possibilities can be performed. Finite element analysis software includes optimization tools for machine designs. The exact operation of these genetic algorithms is unknown to the user, and the operating parameters of the built-in genetic algorithm are not completely configurable. Finite element analysis software in most cases allow for machine designers to link user-defined optimization algorithms. Given this option, the designer has the ability to select and tune an optimization algorithm to achieve diverse solutions that converge close to the true Pareto-optimal front. In this work, the benefit of user-defined optimization algorithms is demonstrated through optimizing design of a linear permanent magnet synchronous machine and evaluating the obtained Pareto-optimal fronts.

Keywords: Linear permanent magnet machine · Optimization · Finite Element Analysis · Genetic Algorithm · NSGA-II

1 Introduction

The use of electric machines is expanding across industries. In 2011, it was estimated that electric motor driven systems account for at least 43% of the global electricity consumption [1]. Several manufacturers offer a wide range of "off-of-the-shelf" electric machine solutions; however, many electric machines selected by end-users for their intended application are oversized, although motors are designed to operate most efficiently at a certain torque and speed. Some applications require customized electric machines to meet strict operating requirements. Performance, cost and reliability are only a few of the considerations that electric motor designers have to meet through optimization. Solving such complex multi-variable, multi-objective and robust optimization problems requires careful selection of the model and search algorithm, as well as strategic formulation of the optimization problem.

Modeling the performance of an electric machine involves a trade-off between accuracy and computation time. Electric machines can be modeled analytically

© Springer Nature Switzerland AG 2019
K. Deb et al. (Eds.): EMO 2019, LNCS 11411, pp. 719–730, 2019.
https://doi.org/10.1007/978-3-030-12598-1_57

or numerically. Finite elements models of electric machines provide accuracy to optimization techniques; however, its evaluation may negatively impact computation time. Some machine topologies require 3D finite elements analysis (FEA) for accuracy, which lengthens computation time. Hybrid modeling techniques, as well as simplified finite element (FE) models that exploit symmetry of the magnetic and electric circuit, are considered computationally efficient [2, 3]. Surrogate models, in lieu of finite element models, were used in [4] to reduce computation time; however, development of accurate meta-models require significant training data and knowledge. Analytical machine models have been used with optimization algorithms to reduce total time to reach Pareto-optimal solutions [5, 6]. While the machine performance is calculated quickly, accurate calculation of the losses and torque/force harmonics are difficult to include in analytical models. Several recently published techniques use very detailed analytical models for optimization [7–10]. This procedure allows evaluation of many cases without significantly increasing total simulation time. Only the final optimal design is evaluated with FEA to provide confidence; however, it is not certain that the accuracy translates to all designs. Any discrepancies may lead to the omission of non-dominated designs from the Pareto-optimal front.

Machines with complex geometries can have a large number of constraints. Many constraints can lead to discontinuities in the Pareto-optimal front. The selected algorithm must properly penalize constraint violations to avoid infeasible solutions but also reach the Pareto-optimal front. Deterministic and gradient-based optimization algorithms have been used for electric machine design; however, due to the nonlinearities in electric machines, these algorithms have difficulty finding the global optimum [2]. Unlike gradient-based algorithms, evolutionary multi-objective optimization algorithms are able to randomly search a large decision space and arrive at Pareto-optimal solutions quicker. The selection, crossover and mutation involved in evolutionary multi-objective optimization algorithms significantly affects the diversity of the solutions. Therefore, an evolutionary multi-objective optimization algorithm is best suited for optimizing an electric machine.

Over the last decade, optimization tools have been included with many FEA software packages. This allows machine designers to optimize the geometry completely within a single software package. Similarly, some FEA software packages link to other software packages that offer optimization tools. Built-in optimization algorithms are quite convenient but lack documentation detailing the algorithm as well as user control of the algorithmic operating parameters. Although a user-defined optimization algorithm may be more challenging to setup, it is beneficial for electric machine designers to have an understanding of the optimization algorithms. As stated best by the "No Free Lunch" theorem [2], there is no single optimization algorithm that is most efficient at solving every problem. A user-defined algorithm can provide more flexibility to select an efficient optimization algorithm that provides diversity in its solutions, better handle constraints to avoid infeasible solution and define stopping criteria to avoid unnecessary FEA simulations.

In this paper, it is demonstrated that a built-in optimization algorithm can lead to infeasible solutions, low diversity of solutions and solutions that are dominated by a user-defined algorithm. Here, a genetic algorithm (GA) included with a FEA software is compared to a user-defined GA for multi-objective optimization of a linear permanent magnet synchronous machine (LPMSM).

2   Problem Definition

A linear PMSM is similar to a rotating PMSM in its design and operation. PMSMs are advantageous because of their high torque density, and in the case of a linear PMSM, a high force density.

Typically, maximizing the produced force is an objective for electric machine design. One approach to increase average force is to shape the geometry of the stator and rotor [11]. However, modifying certain geometric parameters that increase average force lead to a significant increase in the harmonic content. Higher harmonics create a ripple which increases the noise and vibrations produced and decreases the operating lifetime of the machine.

Obtaining an optimal trade-off between average force and force ripple is a common challenge in machine design. Even small geometric changes produce significant changes in electromagnetic performance, which makes geometric optimization difficult.

The machine selected as the target for optimization is the LPMSM model presented in [12]. This machine was designed to achieve high acceleration by light-weighting the moving mass without significant reduction of the force produced. Although it's not an optimal design, it has been experimentally characterized and compared to its FEA model.

Force ripple reduction can be accomplished by shaping the stator teeth [13]. Instead of open slots, a trapezoidal wedge is added in the slot opening of the stator teeth. The distance from the center of the tooth to the end of the wedge is one geometric variable (d_1 in Fig. 1b) and the other variable is half of the thickness of the tooth (d_2 in Fig. 1b). The positive and negative allowable change from initial values of the variables is provided in Table 1.

Table 1. Allowable percent change from initial values for geometric variables

Variable	Allowable decrease	Allowable increase
d_1	−30%	+90%
d_2	−30%	+30%

In the initial design, a thermal analysis of the motor, including a water-cooled aluminum flange, showed that the current level of 12 A produced a steady-state temperature lower than the permanent magnets and winding insulation temperature limits. A size 18 AWG wire, which has a maximum current limit of

16 A, was used in the original design. Higher current will lead to higher force; however, higher current may also lead to temperatures that are too high for the selected cooling method. Therefore, the peak current is selected to vary between 12 A and 16 A.

The multi-objective optimization problem is represented by:

$$maximize \quad F \tag{1}$$
$$minimize \quad F_{pp} \tag{2}$$

Subject to

$$d_1 \geq d_2 \tag{3}$$
$$\frac{N\hat{I}}{A_{slot}} = J \leq 9.5 \tag{4}$$

Here F is the average force, F_{pp} is the peak-to-peak force ripple, J is the current density, N is the number of turns, A_{slot} is the stator slot area and $\hat{I}$ is the peak current.

3   LPMSM Model

Evaluating the small geometric changes for optimization requires high-resolution FEA. Tubular LPMSM typically requires 3D FEA simulation; however, a 2D axisymmetric model of the LPMSM was shown to accurately calculate the average force and force ripple of interest for this optimization problem. Using a 2D model reduces the required computation time for the optimization process. A full and cross-sectional view of the LPMSM are shown in Fig. 2. Geometric parameters and performance metrics are provided in Tables 2 and 3 respectively.

Table 2. Geometric parameters of initial design

Parameter	Symbol	Value (mm)	Parameter	Symbol	Value (mm)
Stator length	L_{stator}	354	Stator height	h_{stator}	18.7
Stator outer diameter	R_{out}	35	Insulation length	L_{INS}	0.5
Aluminum height	h_{Al}	6.3	Slot pitch	τ_s	18.88
PM height	h_{PM}	8	Fillet radius	R_f	0
Length of axial PM	L_{AX}	10.625	Back iron height	h_{BI}	4
Pole pitch	τ_p	21.25	Half wedge height	d_1	4.72
Air gap length	g	2	Half tooth width	d_2	4.72

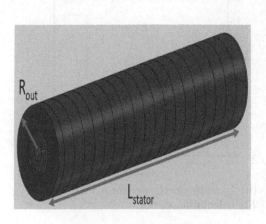

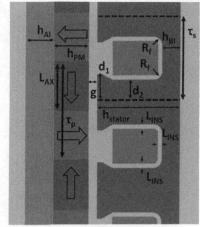

(a) Full view of LPMSM

(b) 2D cross-sectional view of LPMSM

Fig. 1. LPMSM geometry with labeled geometric parameters

Table 3. Operating characteristics of initial machine design

Performance measure	Value	Performance measure	Value
Average force	266 N	Rated speed	26.2 m/s
Force ripple	47.88 N	Rated voltage	480 V$_{ll}$
Number of turns per coil	80	Rated power	5 kW

Figure 2a shows the mesh used in the axisymmetric FEA model of the LPMSM. The mesh in the air gap region is important because that is where the energy transfer takes place. Five layers of elements with a 0.4 mm length are used in the air gap mesh to obtain an accurate calculation of the force ripple. Further away from the air gap region, the mesh size becomes less critical for force and force ripple calculation. To avoid forces due to the end effects that occur in short secondary tubular LPMSM [14], the initial position is two pole-pitches (τ_p) away from the end of the stator and the total travel distance is limited to one electrical cycle. The force waveform from one design solution is shown in Fig. 2b. The maximum and minimum of the force waveform are used to calculate the force ripple F_{pp}.

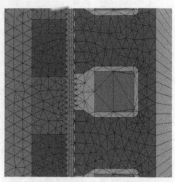

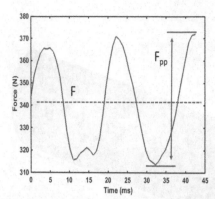

(a) FEA mesh used in simulation

(b) Example force waveform with F and
F_{pp} calculation shown

Fig. 2. Mesh in FEA setup and example force waveform used to evaluate objectives

4   Optimization Algorithm

A GA built into the FEA software and a user-defined GA are used to optimize the
LPMSM. The parameters, constraints and objectives used for both algorithms
are listed in Table 4.

Table 4. Variable limits and GA parameters used in optimization

Variable	Value
d_1 range	3.3 mm to 8.5 mm
d_2 range	3.3 mm to 6.1 mm
$\hat{I}$ range	12 to 16
Number of generations	10
Population size	200

One of the optimization algorithms used is built into the FEA software package. Details regarding the optimization processes used in the built-in GA are
masked from the user. There are few parameters that the user can modify to
improve the results of the built-in algorithm, such as population size, number of
generations and the weights on each objective. Unfortunately, documentation for
the built-in algorithms is limited and does not detail how the algorithm handles
diversity, elite preservation, and constraints. Knowing such details is a primary
advantage of a user-defined algorithm.

Most FEA software packages allows for the user to custom optimization algorithms. In this work, NSGA-II is selected as the user-defined algorithm. NSGA-II
is a well-documented GA in literature [15] and has shown to perform well for two-
objective optimization problems. It is well-known that with NSGA-II, the elite

solutions are kept and diversity in the objective space is maintained throughout the generations.

4.1 Procedure

Genetic algorithms are stochastic processes, so multiple executions of the same algorithm for the same problem produces different solutions. Although more runs improve the statistical significance of the results of each GA, due to the use of a high-fidelity model, both algorithms are executed four times. For each GA, the Pareto-optimal front is created from the non-dominated solutions of the last generation of all four runs. In three of the four optimization runs, the population of the initial generation is randomly selected. In the remaining run the population of the initial generation in each algorithm are identical.

The Pareto-optimal front is generated in the process shown in Fig. 3. For electric machine design, strict adherence to the constraints is required. Solutions that violate the constraints are not considered to be on the Pareto-optimal front and are removed.

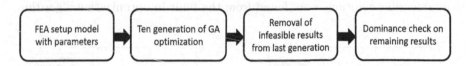

Fig. 3. Flow chart showing how Pareto front solutions were obtained

4.2 Evaluation Metrics

The true Pareto-optimal front is not known for this problem. To compare built-in GA to the user-defined GA, four quantifiable metrics are used; normalized hypervolume, Set Convergence Metric, Spacing, and total number of Pareto-optimal points. The normalized hypervolume and Set Convergence Metric quantify the convergence to the Pareto-optimal front. Spacing is used to quantify the diversity of solutions. The number of Pareto-optimal solutions quantifies the number of design options available for machine designers.

To calculate hypervolume, a reference point and ideal point are used to normalize the area of the Pareto front. The reference point is selected as a combination of the worst objective function values and the ideal point is selected as the combination of best objective function values. The objective value of each solution on the Pareto front is normalized and the hypervolume is calculated as the total area covered by the solutions [15]. A larger hypervolume is a good indicator of convergence to the true Pareto-optimal front. The Set Convergence Metric [15] compares the non-dominated solutions on two Pareto-fronts. A percentage of the non-dominated solutions that are dominated by at least one solution from the other Pareto-front is calculated. A lower percentage indicates that fewer solutions found by that algorithm were dominated by a solution found from the other algorithm.

Diversity of solutions on the Pareto-optimal front is compared by calculating the Spacing [15], as given in Eq. (5). Q is the total number of non-dominated solutions, d_i represents the minimum distance between point "i" and all of the other points, and $\bar{d}$ represents the average distance between all points. A smaller number for S is desired, which means there is small variation in the space between the non-dominated solutions.

$$S = \sqrt{\frac{1}{|Q|} \sum_{i=1}^{|Q|} (d_i - \bar{d})^2} \tag{5}$$

5   Optimization Results

The complete solution space from the built-in GA is provided in Fig. 4a, and that of the user-defined GA is shown in Fig. 4b. As indicated by the different markers, more infeasible solutions were found from the built-in optimization algorithm.

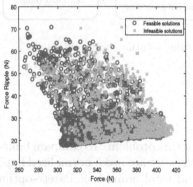

(a) Entire solution space of every generation using the built-in GA

(b) Entire solution space of every generation using the user-defined GA

Fig. 4. All 200 members of 10 generations in the four combined runs of each GA where the infeasible points that violate constraints are indicated

The final Pareto-optimal front from each algorithm is created from the non-dominated, feasible solutions from the solution spaces and is provided in Fig. 5.

The tenth generation of each GA were used to create the Pareto-front. Some of the solutions in the tenth generation of the results from the built-in GA violated the constraint on current density, as shown in Eq. (4). These results

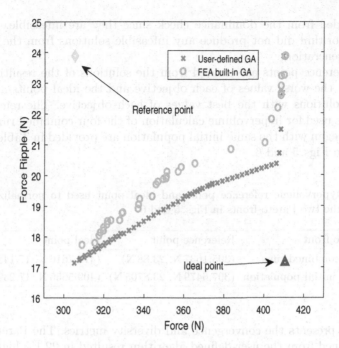

Fig. 5. Pareto-optimal front generated from four total runs of each optimization algorithm

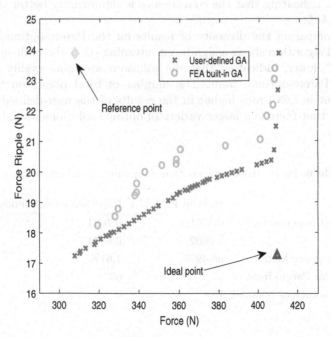

Fig. 6. Pareto-optimal front generated from each algorithm starting with same initial population

were excluded from the dominance check since they are infeasible. The user-defined algorithm did not produce any infeasible solutions from the second to the tenth generation.

The reference points are created from the solutions of the resulting Pareto fronts with the worst values of each objective and the ideal points are created from the solutions with the best values of each objective. The reference and ideal points used for hypervolume calculation of the four combined runs and the runs that began with the same initial population are provided in Table 5 and are displayed in Figs. 5 and 6.

Table 5. Hypervolume reference point and ideal point used to normalize the area covered by the two Pareto-fronts in Figs. 5 and 6

Pareto front	Reference point	Ideal point
Four combined runs	(305.1637 N, 23.88 N)	(410.7616 N, 17.1483 N)
Same initial population	(307.6479 N, 23.8705 N)	(409.5655 N, 17.2399 N)

Table 6 presents the convergence and diversity metrics. The Pareto-optimal front obtained from the user-defined algorithm resulted in 22.1% higher hypervolume and only 1.61% of its solutions are dominated by any solution from the built-in GA, indicating that the convergence is significantly better than that of the built-in GA.

When comparing the diversity of results on the Pareto-optimal front, the user-defined algorithm shows a significant advantage over the built-in GA. Spacing is 71.3% lower, indicating that the solutions are more evenly distributed across the Pareto-optimal front. The number of total points on the Pareto-optimal front is 1.68 times higher in the results of the user-defined algorithm. This means that there is a larger variety of optimal solutions available for decision making.

Table 6. Pareto front analysis from four runs combined for each GA

Result	Built-in FEA GA	User-defined GA linked to FEA
Normalized hypervolume	0.5751	0.7024
Spacing	3.0602	0.8774
Set Convergence Metric	86.49%	1.61%
Solutions on Pareto front	37	62
Infeasible solutions removed	78	0

The Pareto fronts from each algorithm with the same initial population is presented in Fig. 6 and the analysis of the Pareto front is provided in Table 7.

Convergence is significantly better in the Pareto front found with the user-defined GA where the hypervolume is 28.9% higher. Also, the Set Convergence Metric shows all of the Pareto-optimal points of the built-in GA are completely dominated by the Pareto front of the user-defined GA. Diversity is also significantly better where the spacing is 82% lower in the Pareto front from the user-defined GA and there are 2.88 times more optimal points for decision making. Using the same population members in the initial generation did not return a different result in the convergence or diversity of points on the Pareto-optimal front between the user-defined and built-in GAs.

The final row in Tables 6 and 7 show the number of infeasible results that were removed from the final generation from each GA. With the built-in GA the constraint handling method is not known, and as a result, there were a significant number of solutions in the final generation that violated constraints.

Table 7. Pareto front analysis where each GA is set with the same population members in the initial generation

Result	Built-in FEA GA	User-defined GA linked to FEA
Normalized hypervolume	0.5512	0.7105
Spacing	4.2935	0.7701
Set Convergence Metric	100%	0%
Solutions on Pareto front	17	49
Infeasible solutions removed	9	0

6   Conclusion

Results obtained from a built-in optimization tool in FEA software and a user-defined optimization algorithm from a electric machine design problem show several quantitative and qualitative advantages of a user-defined algorithm. It is shown that:

1. The user-defined algorithm results in more Pareto-optimal solutions
2. The user-defined algorithm has more solutions that dominate those resulting from the built-in GA optimization
3. Solutions from the user-defined algorithm are more diverse
4. Infeasible solutions were avoided using the user-defined algorithm

Considering the advantages evaluated in this work, it is important for electric machine designers to be knowledgeable of optimization algorithms rather than relying on undocumented, built-in algorithms. As different machine designs have different constraints and requirements to meet, more customized optimization tools are needed for efficient design of all possible machine topologies.

References

1. Waide, P., Brunner, C.U.: Energy-efficiency policy opportunities for electric motor-driven systems. OECD Publishing, IEA Energy Papers 2011/7 (2011). https://EconPapers.repec.org/RePEc:oec:ieaaaa:2011/7-en
2. Bramerdorfer, G., Zavoianu, A., Silber, S., Lughofer, E., Amrhein, W.: Possibilities for speeding up the FE-based optimization of electrical machines - a case study. IEEE Trans. Ind. Appl. **52**(6), 4668–4677 (2016)
3. Sizov, G.Y., Ionel, D.M., Demerdash, N.A.O.: A review of efficient FE modeling techniques with applications to PM AC machines. In: 2011 IEEE Power and Energy Society General Meeting, pp. 1–6, July 2011
4. Lim, D.K., Woo, D.K., Yeo, H.K., Jung, S.Y., Ro, J.S., Jung, H.K.: A novel surrogate-assisted multi-objective optimization algorithm for an electromagnetic machine design. IEEE Trans. Magn. **51**(3), 1–4 (2015)
5. Sindhya, K., Manninen, A., Miettinen, K., Pippuri, J.: Design of a permanent magnet synchronous generator using interactive multiobjective optimization. IEEE Trans. Ind. Electron. **64**(12), 9776–9783 (2017)
6. Akiki, P., et al.: Multi-physics modeling and optimization of a multi-V-shape IPM with concentrated winding. In: 2017 IEEE International Electric Machines and Drives Conference (IEMDC), pp. 1–7, May 2017
7. Jing, L., Qu, R., Kong, W., Li, D., Huang, H.: Genetic-algorithm-based analytical method of SMPM motors. In: 2017 IEEE International Electric Machines and Drives Conference (IEMDC), pp. 1–6, May 2017
8. Islam, M.Z., Choi, S.: Design optimization of rare-earth free PM-assisted synchronous reluctance motor to improve demagnetization prevention capability. In: 2017 IEEE International Electric Machines and Drives Conference (IEMDC), pp. 1–6, May 2017
9. Takbash, A., Pillay, P.: Design optimization of a new spoke type variable-flux motor using AlNiCo permanent-magnet. In: 2017 IEEE International Electric Machines and Drives Conference (IEMDC), pp. 1–6, May 2017
10. Baek, J., Rahimian, M.M., Toliyat, H.A.: Optimal design and comparison of stator winding configurations in permanent magnet assisted synchronous reluctance generator. In: 2009 IEEE International Electric Machines and Drives Conference, pp. 732–737, May 2009
11. Lopez, C.A., Strangas, E.G.: Optimization of PMSM performance with torque ripple reduction and loss considerations. In: 2018 XIII International Conference on Electrical Machines (ICEM), September 2018, pp. 899–905 (2018)
12. Jensen, W.R., Pham, T.Q., Foster, S.N.: Linear permanent magnet synchronous machine for high acceleration applications. In: 2017 IEEE International Electric Machines and Drives Conference (IEMDC), May 2017, pp. 1–8 (2017)
13. Jahns, T.M., Soong, W.L.: Pulsating torque minimization techniques for permanent magnet AC motor drives-a review. IEEE Trans. Ind. Electron. **43**(2), 321–330 (1996)
14. Pham, T.Q., Jensen, W.R., Foster, S.N.: Stator incipient fault identification in short secondary linear permanent magnet synchronous machines. In: 2017 IEEE International Electric Machines and Drives Conference (IEMDC), May 2017, pp. 1–7 (2017)
15. Deb, K.: Multi-Objective Optimization Using Evolutionary Algorithms. Wiley, Hoboken (2003)

Designing Solar Chimney Power Plant Using Meta-modeling, Multi-objective Optimization, and Innovization

Fateme Azimlu(✉) [ID], Shahryar Rahnamayan, Masoud Makrehchi [ID], and Pedram Karimipour-Fard [ID]

Nature Inspired Computational Intelligence (NICI) Lab,
University of Ontario Institute of Technology (UOIT),
2000 Simcoe Street, North Oshawa, ON L1H 7K4, Canada
{fateme.azimlushanjani,shahryar.rahnamayan,masoud.makrehchi,
pedram.karimipourfard}@uoit.ca

Abstract. Generally speaking, the couples of simulator-optimizer are very common combination in optimizing the systems in science and engineering fields; but when the simulation process is an expensive one, then each fitness evaluation of the optimizer can take several hours even days, which makes the mentioned process impractical to run in given time budget. In this paper, we replace the solar chimney power plant (SCPP) simulator with a bi-objective meta-model which is created by Genetic Programming (GP) and we compared the created model with neural network and regression models to be sure it is accurate enough. Then, we have utilized a genetic algorithm based multi-objective algorithm to solve the created bi-objective optimization problem resulted from the meta-modeling phase. After finding optimal Pareto-front (PF), we use a GP-based innovization technique to discover engineering design knowledge and rules for the designed optimal power plant. The created models have been validated with results of the 4 corresponding simulator, a promising error rate (<4%) has been obtained. This work can be evaluated as a successful energy application of GP-based meta-modeling, evolutionary multi-objective optimization, and GP-based innovization.

Keywords: Solar chimney power plant · Optimization ·
Genetic algorithm · Genetic programming · Meta-modeling ·
Innovization

1 Introduction

Simulation techniques have an important role in engineering and science. They facilitate design and prediction of complex systems by reducing their development cost and time. Furthermore, they have liberated research works from a number of limitations, such as, time and space restrictions by offering performance assessment environments which otherwise could be difficult or impossible

© Springer Nature Switzerland AG 2019
K. Deb et al. (Eds.): EMO 2019, LNCS 11411, pp. 731–742, 2019.
https://doi.org/10.1007/978-3-030-12598-1_58

to achieve. For example, the functionality of instruments at different conditions can be checked using a simulator software, such as, Ansysfluent[1] or Comsol Multiphysics[2]. These are examples of numerical simulators which engineers and scientists widely use to simulate devices and processes in all fields of engineering, and scientific research but are time consuming. Therefore, to reduce running time, we can use parallel processing techniques and super computers. For these computationally expensive cases meta-modeling techniques can be utilized to reduce the time and cost. In fact, meta-models are built based on small numbers of running simulator and then replacing the simulator for creating further results for required experiments. For example, if an optimizer and a simulator coupled to optimize the decision parameters, then meta-model can continue to work with the optimizer to accelerate the processioning time. In this case study, we have optimized designing of a solar chimney power plant [1]. However, our proposed framework is not limited to the current application. The world currently relies heavily on fossil fuels, which are non-renewable and finite, as the main energy supply source. In addition, they are not environmentally friendly. In contrast, many types of renewable energy resources such as wind and solar energy are constantly replenished. The most common methods for collecting solar energy are the reflective collectors and photo-voltaic solar planes which are both expensive and require a costly maintenance. Solar chimneys can be used in generating electricity with the added benefit that they do not waste land because the land under the green house component can also be used for agricultural proposes. The first solar chimney was designed by Schlaich in 1970s, and its practical test on a primary model revealed the potential of this instrument in generating electricity and clean energy [2,3]. After testing the prototype model, other pilot-scale solar chimney power plants were construct by Krisst [4], Kulunk [5], Zhou et al. [8], Pasumarthi and Sherif [6,7]. Moreover, Bernardes et al. [9] through a new analytical method studied performance of solar chimneys. Additionally, they added thermal energy storage capability to the system to improve its performance and increase its time of use during the day and night. Zhou et al. [10] designed a new models to find the optimal chimney height and achieve the maximum output power using an analytical method. Analyses were also performed to examine the effect of different time intervals of atmospheric temperatures and the optimum size of the Chimney parameters [10]. Moreover, Sangi et al. [12] designed a simulation model to calculate the effect of the collector and curvature radius on the temperature, pressure and air movement and velocity in the chimney and collector sections. Guo [11] carried out another numerical simulation for the output power and other feature variations and different solar radiation values. Patel et al. [13] proposed novel numerical simulation model to study the effect of collector height, collector diameter, and chimney throat diameter on air velocity, temperature and output power in a SCPP. The performance of a solar chimney power plant also can be optimized

---

[1] http://www.ansys.com/Products/Simulation+Technology/Fluid+Dynamics/
Fluid+Dynamics+Products/ANSYS+Fluent.

[2] https://www.comsol.com/comsol-multiphysics.

using electric corona wind [14]. Sciuto et al. [15] designed an optimized solar chimney power plant by utilizing finite elements based on analysis and cascade neural networks. Majority of previous attempts for optimizing SCPP parameters employed numerical methods or prevalent instrument simulator software. In this study, we utilize genetic programming (GP) for meta-modeling proposing, and then multi-objective genetic algorithm to optimize our designed power plant; furthermore, finally GP has been used for innovization proposed which is a successful attempt to discover the engineering design rules after the optimization. Section two reviews the structure of a solar chimney power plant and defines its corresponding design variables. Section three includes procedures to develop equations which determine the output power and efficiency dependence on the SCPP structural variables; in fact, GP is using results of set of simulation results to create meta-model to replace the simulator. In addition, we discuss the optimization of the SCPP structural variables. In section four, analyzes the results and applying innovization technique [16] using GP to discover engineering design rules, and also utilizing them for finding more Pareto-front solutions and finally section five concludes the current research work.

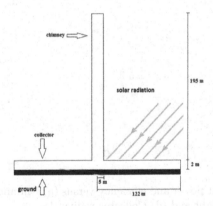

Fig. 1. A typical solar chimney power plant [18].

2 Structure of Solar Chimney Power Plant

A SCPP consists of three main parts: a collector, a turbine, and a chimney. A collector is a cylinder constructed of glass or plastic that allows solar radiation to pass through it and retains its heat. The chimney is a high cylindrical Structure placed in the centre of a collector with the turbine located in its entrance, Fig. 1. The greenhouse effect in the collector heats the incoming air from the bottom. The hot air ascends due to its lower density and by passing through the collector propels the turbine inside the chimney [1]. The output power and efficiency depend on the size of the chimney and the collector. Karimipour et al. [17]

calculated the correlation between solar chimney parameters, the output power and efficiency. They simulated many SCPPs [18] by defining collector height, chimney height, collector radius, and chimney radius as input variables of their simulation. In addition, they investigated the correlation among these variables and the output power and the efficiency, Fig. 2. The base for the Karimipour simulation is the solar chimney power plant in Manzanares, Spain, which the collector Height is 5 m, chimney height is 195 m, collector Radius is 122 m, and chimney radius is 2 m. We have used their simulation results to make our meta-model using the GP algorithm.

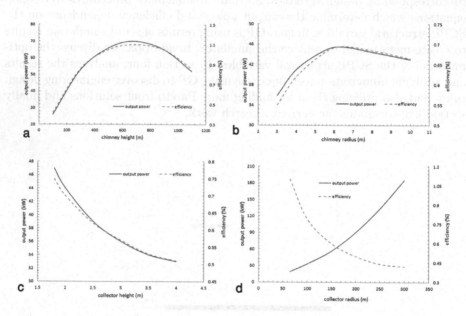

Fig. 2. Simulator output power and efficiency versus (a) Chimney height, (b) Chimney radius, (c) Collector height and (d) Collector radius [18].

3   Proposed Framework

In conventional simulation models, (Branch A, Fig. 5) simulator creates hundreds of sample points which are used for discovering the optimum points. In our proposed model, (Branch B, Fig. 5), first we use simulator to create a few sample points then we use them to create our meta-model using GP, which results two objective functions; Power and efficiency. Then we Utilize multi objective genetic algorithm, MOGA, to find optimum Pareto-front set for SCCP's decision variables. At the end, again we apply GP on the Pareto-front set to discover engineering design knowledge using innovization. The details of our proposed method are provided in the next subsections.

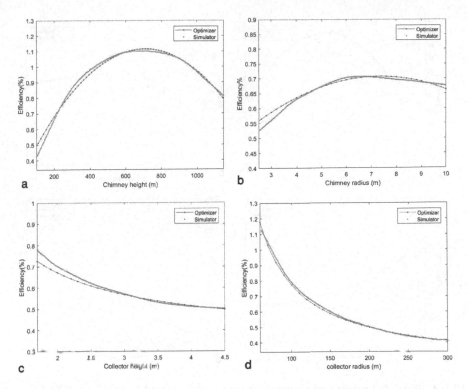

Fig. 3. Simulated efficiency and optimizer output versus (a) Chimney height, (b) Chimney radius, (c) Collector height and (d) Collector radius.

3.1   Meta-modeling Using Genetic Programming

Genetic programming, GP, is a useful and effective technique for creating symbolic equations or computer programs or models using the process similar to natural selection [19]. GP is a method of programming which employs the notion of biological evolution and natural selection to solve complicated problems. Different pieces of programs which include mathematical functions contest each other and only the most proper programs can survive and compete or concatenate with other piece of programs in the next step. As a result, trough many iterations it continuously approaches to the best solution which has the best fitness to our data. Therefore, GP can be used in creating a model or data-driven formula. In this study, we extracted 1640 data points, from simulator with four variables: collector height (h), chimney height (H), collector radius (r), and chimney radius (R) and used them as the inputs of Eureqa[3] which we used for generating the equations. The utilized GP created the following equations for power (P) and efficiency (E) with R-squared Goodness of Fit bigger than 99%.

$$P = a/h + b \times H \times R + c \times r^2 - d - f \times H^2 - g \times R^2 \qquad (1)$$

---

[3] https://www.nutonian.com/.

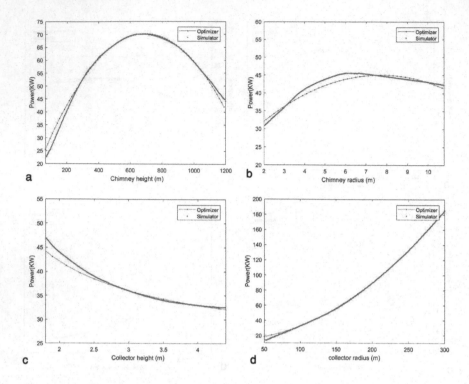

Fig. 4. Simulated power and optimizer output versus (a) Chimney height, (b) Chimney radius, (c) Collector height and (d) Collector radius.

where $a = 36.68, b = 0.031, c = 0.002, d = 21.02, f = 0.0001, g = 0.39$.

$$E = a + b/h + c/r^2 + (d \times H \times R - f \times H^2)/r - g \times R^2 \tag{2}$$

where $a = 0.086, b = 0.62, c = 442.10, d = 0.057, f = 0.0002, g = 0.006$.

Table 1, shows the accuracy of these equations. Figures 3 and 4 show how good the generated power and efficiency fit the simulated ones. In facts, If the optimizer is connected directly to the simulator, the results would be more accurate. Simulation is very expensive as it is very time consuming. By generating these equations, actually, meta-modeling is performed to replace the expensive simulation process to work with multi-objective optimization algorithm. Moreover, we used neural network, linear and polynomial regressions to estimate efficiency and power based on the extracted data points from the simulator. Surprisingly, we discovered that none of them could reach the accuracy of using GP as shown in Table 1.

Table 1. Comparison of accuracy for different modeling for estimating power and efficiency.

	Power	Efficiency
GP Goodness of Fit	**0.9983**	**0.9913**
ANN Goodness of Fit	0.9387	0.9032
Linear regression Goodness of Fit	0.7207	0.7064
Polynomial regression Goodness of Fit	0.8135	0.7826

3.2   Multi-objective Optimization: Maximization of Output Power and Efficiency

Now, we have our meta-model for our two objectives, independent form the simulator, we can use them as our objectives for our multi objective optimization. Optimization problems are ubiquitous in science and engineering. The wide usage of optimization methods and it's potential of generalization has opened a wide area in computer science. An optimization problem has a function that should be maximized or minimized, considering given constraints. There are various types of algorithms available which can be used for optimization. In these cases, Evolutionary algorithms such as Differential Evolution (DE) or Genetic Algorithm (GA) methods [20], can be useful to tackle our multi objective optimization problem. DE and GA are population based evolutionary directed search methods. Similar to other evolutionary algorithms, First they generate an initial population vector, then, by crossover and mutation a new generation is created and compete with the old ones. Only the best members with a better fitness value can survive and progress to the next generation. In this method, by creating new generations with better member's solution variables approach the best possible solution and after stopping criteria, the best solution is reported.

3.3   Multi Objective Optimization

In SCPP there are two important functions; power and efficiency. Both are dependent on collector height (h), chimney height (h), collector radius (r) and chimney radius (R) Fig. 2. In most plots they have the same pattern but Figure 2. d shows that by increasing the collector radius the power increases, In contrast, the efficiency decreases. It concludes that considering the collector radius as a variable, the power and the efficiency objectives are in conflict. Therefore, we should optimize both functions simultaneously. In this case we used MATLAB Optimtool [21] which provides multi objective genetic algorithm (MOGA) where our objectives are two equations created in previous step by Genetic programming, the result of optimization (Pareto-front, PF) is presented in Fig. 6. A decision maker can choose the best point that meets the specific conditions.

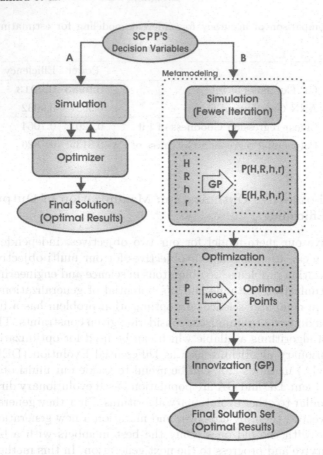

Fig. 5. Comparison between the steps of A: conventional simulation method, and B: the proposed method. Where P is power and E is efficiency. h, H, r and R are collector height, chimney height, collector radius, and chimney radius respectively.

4   Experimental Results and Analysis

Using Multi objective Genetic algorithm, we got a Pareto-front with 18 non-dominated solutions, listed in Table 2, Fig. 6. The final results show that the collector height (h) has the value of approximately 1 m, and the chimney height and radius for the optimal case, should be around 846 m and 9.9 m collector. But we see variations in the collector radius (r) that should be selected based on other important constrains such as the costs or available land. In addition, r can be higher than 380 m, which is comparable with industrial greenhouses. Solar chimneys are best suited to the environments with plenty of sun and low price lands. Farmlands are good examples of areas where solar chimneys can be implemented.

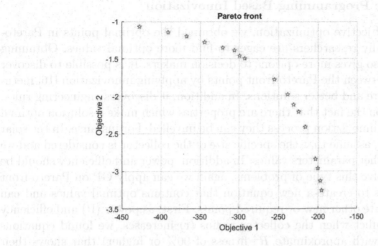

Fig. 6. The Pareto-front obtained from GA optimization for two objective functions P (Objective 1) and E (Objective 2).

Table 2. Pareto-front solutions. H, R, h and r are chimney height, chimney radius, collector height, and collector radius, respectively.

Index	Power (kW)	Efficiency	H (m)	R (m)	h (m)	r (m)
1	439.669	0.9878	847.6680	9.9999	1.00000	382.2812
2	412.619	1.0345	847.0569	9.9998	1.00043	363.3272
3	397.758	1.0624	846.490	9.9897	1.00110	352.6529
4	356.229	1.1591	845.3852	9.9944	1.00350	320.4110
5	341.417	1.20350	845.5928	9.9979	1.0029	307.961
6	324.695	1.25912	845.4907	9.9982	1.00334	293.4009
7	300.394	1.3586	845.0810	9.9998	1.00181	270.7871
8	256.400	1.62486	846.2752	9.9920	1.00297	224.2411
9	239.554	1.78018	846.5982	9.9996	1.0081	203.6074
10	218.981	2.06906	847.1745	9.9996	1.00158	174.6673
11	189.792	2.8852	844.8878	9.9979	1.00149	123.9666
12	185.293	3.12000	844.7817	9.9964	1.00435	114.3901
13	181.517	3.37254	844.7613	9.9925	1.0052	105.6496
14	176.263	3.89191	844.6083	9.9984	1.0055	91.4651
15	173.603	4.31226	845.2680	9.9996	1.0007	82.6093
16	170.276	4.9689	844.5295	9.9998	1.00103	71.6368
17	168.573	5.47448	844.5259	9.9998	1.0005	65.0409
18	168.573	5.4744	844.5259	9.9998	1.0006	65.0403

4.1   Genetic Programming Based Innovization

Using Multi objective optimization, we obtained the optimal points in Pareto-front but usually researchers are eager to find more optimal values. Obtaining more points also gives more options to decision makers. It is possible to discover correlations between the Pareto-front points by applying innovization [16] methods to find more and better solutions. In addition, it discovers engineering rules. This is based on the fact that there are properties which make a solution optimal and trough an innoviztion process they can be unveiled. For instance, In the solar chimney study, assume that the specific size of the collector is considered and we need to find other parameters values, In addition, power and efficiency should be optimal. To solve this type of problems, again we can apply GP on Pareto-front points features to create a new equation that contains optimal values and can be used to create other new or required points. First, as power (P) and efficiency (E) are in conflict when the collector radius (r) increases, we found equations for P and E (with approximate R^2 fitness of 90% or higher) that shows their correlation together and correlation with r.

$$P = a + b \times E^2 + c \times \cos(d + f \times E) - g \times E - i \times E^3 \tag{3}$$

Where $a = 1285, b = 273, c = 379.46, d = 4.14, f = 0.71, g = 1155.4, i = 17.57$.

$$E(P) = a + b \times c^{(d \times P - f)} \exp(g \times P - i) - j \times P \tag{4}$$

Where $a = 2.6, b = 0.0004, c = 0.03, d = 0.01, f = 2.8, g = 0.06, i = 16.3, j = 0.003$.

$$P(r) = a + b \times r^2 - c \times r - d \times r^3 - f \times \cos(g + i \times r^2 - j \times r) \tag{5}$$

Where $a = 173.65, b = 0.002, c = 0.12, d = 6.2e - 7, f = 2.81, g = 2.35, i = 0.0005, j = 0.2115$.

$$E(r) = a \times r + b/r + c \times \tan(d \times r) + f \times \tan(g \times r) - i \tag{6}$$

Where $a = 0.0002, b = 376.01, c = 0.0004, d = 0.21, f = 0.0002, g = 0.23, i = 0.06$.

More over, we discovered correlation between P, E and all features:

$$E(H, R, h, r) = a \times R + b \times H + c \times \sin(d \times H) - f - g \times h - i \times r^2 - \\ j \times \arctan 2(\arctan 2(\sin(k \times H), h), h) \tag{7}$$

Which arctan 2 is 2-argument arctan and $a = 15.79, b = 0.67, c = 7.18, d = 23.345, f = 771.87, g = 5.6, i = 2.76e - 5, j = 13.35, k = 23.34$.

$$P(H, R, h, r) = a \times h + b \times r + c \times \tan(d \times R) - f - h^g \times \tan(i \times h) - j \times \sin(k \times r - l \times R) \tag{8}$$

where $a = 24.57, b = 1.21, c = 5.23, d = 11151, f = 14.71, g = 8.38, i = 1062, j = 118, k = 1.28, l = 11407$.

In addition, if we consider P and E as variables we can discover the relation between all features and the other dependent objective (P or E).

$$P(E, H, R, h, r) = a \times E + b \times R + c \times r + d \times f^{(g \times E - i)} - j - k \times h \\ - l \times E \times r - m \times E \times R - n \times (\tan(\sin(q + s \times E))) \quad (9)$$

where $a = 388, b = 116, c = 0.09, d = 3.68, f = 0.09, g = 0.9, i = 2.72, j = 935, k = 41.83, l = 0.01, m = 38.74, n = 19.36, q = 4.88, s = 0.46$.

As this equations are generated based on optimal values, they are applicable to find other optimal values. The derived rules show the relation between different features and can be used in construction of the real model. Finally, to be sure that our model works properly and the GP output equations and MOGA output points are acceptable, we checked the final results (created optimal points) with a simulator. We set the optimal points values (H, R, h and r) which are listed in Table 1, as the input of Ansysfluent simulator and compared the output (Power and Efficiency) for each point with our generated model output in Table 2. Then we calculated the absolute relative error of this two outputs. Fortunately, the average error was less than 4% which reveals the high accuracy in our proposed method. Considering Fig. 2, that the simulation approximately matches the generated model, the low relative error was expected.

5   Conclusion Remarks

In this paper, we followed three main phases to optimal design of solar chimney power plant, namely, (a) replacing the simulator with GP-based meta-modeling approach, (b) utilizing equations of bi-objective (i.e., power and efficiency) from previous phase in our GA-based optimization approach to create optimal Pareto-front, and finally, (c) using GP-based innovization to discover engineering design knowledge after optimization process. The obtained results are well-validated by the simulator results. The proposed framework can be investigated on other applications too. In addition, if we have any constraint for our model it is applicable in multi objective optimization algorithms such as NSGA-II. It seems, GP is a proper candidate for both meta-modeling and innovization, after data-driven discovery of the multi objectives, then we save time of expensive simulations for running our optimization algorithm. As future work, we will try to design other renewable energy systems with the proposed framework in this paper, then we will have strong evidences to support evolution in action.

References

1. Zhou, X., Wang, F., Ochieng, R.M.: A review of solar chimney power technology. Renew. Sustain. Energy Rev. **14**, 2315–2338 (2010)
2. Haaf, W., Friedrich, K., Mayr, G., Schlaich, J.: Solar chimneys, part i: principle and construction of the pilot plant in Manzanares. Int. J. Sustain. Energy **2**(1), 3–20 (1983)

3. Haaf, W.: Solar chimneys. Part II: preliminary test results from the Manzanares pilot plant. Int. J. Sustain. Energy **2**(2), 141–161 (1984)
4. Krisst, R.J.K.: Energy transfer system. Alternat Sour. Energy **63**, 8–11 (1983)
5. Kulunk, H.: A prototype solar convection chimney operated under Izmit conditions. In: Veiroglu, T.N. (ed.) Proceedings of the Seventh MICAS, Turkey, vol. 162 (1985)
6. Pasumarthi, N., Sherif, S.A.: Experimental and theoretical performance of a demonstration solar chimney model part I: mathematical model development. Int. J. Energy Res. **22**, 277–288 (1998a)
7. Pasumarthi, N., Sherif, S.A.: Experimental and theoretical performance of a demonstration solar chimney model part II: experimental and theoretical results and economic analysis. Int. J. Energy Res. **22**, 443–461 (1998b)
8. Zhou, X.P., Yang, J.K., Xiao, B., Hou, G.X.: Experimental study of temperature field in a solar chimney power setup. Appl. Therm. Eng. **27**, 2044–2050 (2007)
9. Bernardes, M.A.S., Dos, S., Voß, A., Weinrebe, G.: Thermal and technical analyses of solar chimneys. Solar Energy **75**, 511–524 (2003)
10. Zhou, X., Yang, J., Xiao, B., Hou, G., Xing, F.: Analysis of chimney height for solar chimney power plant. Appl. Thermal Eng. **29**, 178–185 (2009)
11. Guo, P.H., Li, G.Y., Wang, Y.: Numerical simulation of solar chimney power plant with radiation. Renew. Energy **62**, 24–30 (2004)
12. Sangi, R., Amidpour, M., Hosseinizadeh, B.: Modeling and numerical simulation of solar chimney power plants. Solar Energy **85**, 829–838 (2011)
13. Patel, S.K., Prasad, D., Ahmed, M.R.: Computational studies on the effect of geometric parameters on the performance of a solar chimney power plant. Energy Convers. Manag. **77**, 424–431 (2014)
14. Nasirivatan, S., Kasaeian, A., Ghalamchi, M., Ghalamchi, M.: Performance optimization of solar chimney power plant using electric/corona wind. J. Electrostatic **78**, 22–30 (2015)
15. Sciuto, G.L., Cammarata, G., Capizzi, G., Coco, S., Petrone, G.: Design optimization of solar chimney power plant by finite elements based numerical model and cascade neural networks. In: International Symposium on Power Electronics, Electrical Drives, Automation and Motion (SPEEDAM) (2016)
16. Deb, K., Srinivasan, A.: Innovization: discovery of innovative design principles through multi objective evolutionary optimization. In: Knowles, J., Corne, D., Deb, K., Chair, D.R. (eds.) Multi Objective Problem Solving from Nature. NCS, pp. 243–262. Springer, Heidelberg (2008). https://doi.org/10.1007/978-3-540-72964-8_12
17. KarimiPour-Fard, P., Beheshti, H., Baniasadi, E.: Energy and exergy analyses of a solar chimney power plant with thermal energy storage. Int. J. Exergy **20**(2) (2016)
18. Karimipour-Fard, P., Beheshti, H.: Performance enhancement and environmental impact analysis of a solar chimney power plant: twenty-four-hour simulation in climate condition of Isfahan, Iran. IJE Trans. B: Appl. **30**(8), 1260–1269 (2017)
19. Koza, J.R.: Evolution of subsumption using genetic programming. In: Conference Preceding (NM) (1991)
20. Kumar, M., Husian, M., Upreti, N., Gupta, D.: Genetic algorithm: review and application. Int. J. Inf. Technol. Knowl. Manag. **2**(2), 451–454 (2010)
21. Chipperfield, A.J., Fleming, P.J., Pohlheim, H., Fonseca, C.M.: The MATLAB genetic algorithm toolbox. In: ICSE, Tenth International Conference on Systems Engineering, pp. 200–207 (1994)

Neuroevolutionary Multiobjective Methodology for the Optimization of the Injection Blow Molding Process

Renê Pinto(✉), Hugo Silva, Fernando Duarte, Joao Nunes, and Antonio Gaspar-Cunha

IPC - Institute for Polymer and Composites,
University of Minho, Guimarães, Portugal
b8057@dep.uminho.pt

Abstract. Injection blow molding process is widely used in the industry to produce plastic parts. One of the main challenges in optimizing this process is to find the best manufacturing thickness profiles which provides the desirable mechanical properties to the final part with minimal material usage. This paper proposes a methodology based on a neuroevolutionary approach to optimize this process. This approach focuses on finding the optimal thickness distribution for a given blow molded product as a function of its geometry. Neural networks are used to represent thickness distributions and an evolutionary multiobjective optimization algorithm is applied to evolve neural networks in order to find the best solutions, i.e., to obtain the best trade-off between material usage and mechanical properties. Each solution is evaluated through finite element analysis simulation considering the design of an industrial bottle. The results showed that the proposed technique was able to find good solutions where the material was distributed along the most critical regions to maintain adequate mechanical properties. This approach is general and can also be applied to different geometries.

Keywords: Blow molding · MOEA · Neuroevolutionary

1 Introduction

One of the most important processes to manufacture plastic parts in industry is the injection blow molding process, which is widely used in the production of several kinds of container products, such as bottles, jars and containers to hold different types of liquids (laundry detergents, oil, water, among others). In general, this process comprises the injection of molten material (to form a preform, also called parison) into a mold which is inflated with gas (usually air). The pressure imposed by the gas pushes the melted material towards the mold, leading the plastic material to acquire the shape of the mold. After cooling, the plastic is pulled out, producing the final part.

The total costs of blow molded products are heavily influenced by the amount of material used in manufacturing and therefore can be reduced by minimizing material usage. However, several mechanical properties are also dependent on this feature.

© Springer Nature Switzerland AG 2019
K. Deb et al. (Eds.): EMO 2019, LNCS 11411, pp. 743–754, 2019.
https://doi.org/10.1007/978-3-030-12598-1_59

Thus, this requires a trade-off between production costs and quality criteria, once the reduction of material can affect important properties of the final product [1].

A common approach to optimize blow molding process is reducing the material empirically, but good results will rely on expert experience. In this context, numerical models can help to reduce the number of empirical trials or even eliminate real productions needs by using simulations during the optimization process. Several numerical approaches, such as Finite Element Methods (FEMs), neural networks, gradient-based and stochastic search techniques have been used in blow molding design [1–4]. The major challenge to optimize this process is to find the best geometry and thickness profile of injected preform in order to obtain the final part with all desirable mechanical and weight properties satisfied.

Artificial Neural Networks (ANNs) has been used in several studies to describe blow molding process with high accuracy. In [2] the authors use a neural network to predict wall thickness distribution of a container from the parison (preform) thickness distribution. In [5] the preform diameter and thickness swells were predicted by an ANN from operation parameters. In [3] the authors use ANNs to model a parison extrusion process based on experimental data.

Besides ANNs, genetic algorithms and other kind of optimization techniques have been used as well. In [6] the authors use a genetic algorithm to find the optimal thickness distribution for a preform in order to produce a blow molded bottle with desired wall thickness distribution. In [7] ANNs and particle swarm optimization are used to modeling nonlinear relationships between power lamp settings and output temperature in infrared ovens used to heat PET (Polyethylene terephthalate) preforms during injection blow molding process.

This study proposes a new methodology for injection blow molding optimization which merges several methods into a neuroevolutionary approach. Wall thicknesses distributions are modeled through ANNs, the injection blow molding process are simulated using Finite Element Models (FEM) and evolutionary multiobjective optimization algorithms are applied to find optimal solutions, i.e., thickness distributions which gives the best trade-off between the total amount of used material and suitable mechanical properties. Although this approach can be applied to all stages of injection blow molding process, this study focuses on the final stage, aiming at finding the thickness profile of the final part which satisfies required mechanical properties. The methodology is applied to an industrial bottle model.

2 Injection Blow Molding Optimization

2.1 Process Overview

In general, the injection blow molding process can be summarized into five phases which are illustrated in Fig. 1. These phases are (P1) Injection, (P2) Stretching, (P3) Blowing, (P4) Mold opening and (P5) Blow molded part, respectively.

The process starts in P1, where the polymer should be melted at right conditions considering injection molding parameters. This phase is performed at the injection machine which has a heated barrel with a rotating screw that helps to mix molten

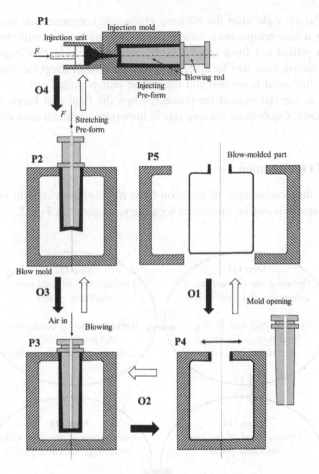

Fig. 1. Injection blow molding process overview.

material, distribute heat and drive material forward. The molten material is injected into a heated cavity to form the preform and then is clamped around a blowing rod.

The next phase (P2) comprises stretching the preform. This phase might be unnecessary for certain products or even be executed simultaneously with phase P3. Stretching the preform allow the maximization of the amount of material at the bottom of final part. Temperature should be controlled to avoid deformation or damage to the material during stretching. The geometry of blowing rod should be optimized to facilitate material flow.

Phase P3 comprises the injection of air (at a certain pressure and velocity) inside the preform resulting from the previous phase, pushing the material towards the mold and leading it to match its internal shape. The preform thickness profile and the mold geometry will determine the thickness profile and hence mechanical properties of the final part. Thus, optimization process should find thickness profiles which lead to less material utilization at the same time that required mechanical properties are accomplished.

Phase P4 starts right after the blowing phase and comprises the wait for material cool down at a safe temperature, where the plastic are rigid enough to not break or deform when pulled out from mold. The thickness profile is also important for this phase since cooling time will be different across parts with irregular geometries. After cooling time, the mold is opened and the plastic part is pulled out.

Phase P5 is the last step of the process, when the final part keeps cooling and is ready for storage. Controlling cooling rate is important to obtain uniform properties in the final part.

2.2   Global Optimization

In this study, the optimization of injection blow molding process will be divided into four major steps that can be optimized separately, as show in Fig. 2.

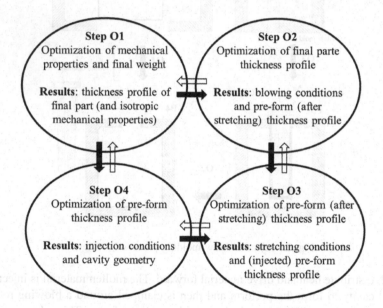

Fig. 2. Global optimization steps for injection blow molding.

The optimization process can be started by taking into account a costumer specification for a blow molded product, i.e., which properties should be accomplished by the final part. Then, the proposed optimization methods should be applied to each phase of the blow molding process in order to find the best settings that will produce the desirable final part at the end of the process. It should be clear in each optimization which objectives to be accomplished and variables to be optimized.

This study proposes four major steps to compose the global optimization process of injection blow molding (Fig. 2). In this methodology, the bests results of a given step will be the objectives of the next (starting by the end of manufacturing process). Each step can be summarized as follows:

- (O1) Optimize the mechanical properties and weight of the final part. This step aims at find the optimal thickness profile of the final part which gives the best trade-off between mechanical properties and the total weight. Decision variables are the wall thickness profile of the final part, which is composed by the thickness values for each point of the mesh that represents the final part.
- (O2) Optimize the final part thickness profile. This step aims at find the best preform geometry which gives the optimal final part thickness profiles that was obtained in the previous step. Decision variable are the blowing conditions and the preform thickness profile (after stretching, when applicable).
- (O3) Optimize the preform thickness profile after stretching. This step should be done when stretching is applicable. Decision variables are stretching conditions and the preform thickness profile. The optimization process is analogous as the previous step, but this step aims at find the best solutions which produces the optimal preform thickness profile obtained in step (O2).
- (O4) Optimize the preform thickness profile (before stretching). Decision variables are injection conditions and cavity geometry. This step aims at find the best solutions which produces the optimal preform thickness profile (before stretching) obtained in previous step.

It is important to point out that the optimization algorithms and procedures used in each step are exactly the same except by decision variables and results considered in each of them. Since the results of a given step is used by the next one, the optimization should follow the chain during its execution, but at any time it is possible to go back to previous steps to reformulate the results. In this case, further steps should be executed again to update the results. The optimization workflow is indicated by white arrows in Fig. 2.

2.3 Proposed Methodology

Injection blow molding simulations are done through finite element methods hence all parts are modeled in 3D meshes where each mesh point is supposed to have a certain thickness value. One of the main issues concerning the optimization is how to handle the different sizes and geometries of each mesh. For instance, a simple bottle mesh can be composed by thousands of points (to have good accuracy). Furthermore, considering each point as a decision variable will lead to a huge search space for optimization algorithms.

The proposed methodology follows the described global optimization to optimize injection blow molding process. To reduce the search space and handle different kinds of models (and meshes), thickness profiles are treated as a function of the container's geometry and neural networks are used to compute the wall thickness at any location of the mesh based on the respective coordinates. It is important to point out that no supervised training method are used, the parameters for the networks are determined by the evolutionary multiobjective optimization process. As a result of optimizations, there will be a neural network which gives the optimal thickness profile of the corresponding optimization phase, that might be the profile of final part or the parison thickness profile, for example.

The proposed neuroevolutionary approach is illustrated in Fig. 3. It begins by reading input parameters and generating the initial population randomly. Each population is composed by a set of individuals, each of them representing a neural network that models a wall thickness distribution. The information of ANNs (weights and biases) is encoded in a chromosome of real numbers. Thus, the size of each chromosome will be directly related with the topology of the network. Figure 4 illustrates the thickness calculation process. The coordinates of each point of a given mesh are fed into an ANN that will output the wall thickness value for each point, respectively. The network is composed of three layers where the number of neurons in the hidden layer can be fixed or vary during optimization. Due to computational resources and time constraints, in this study two fixed topologies were considered: 3-20-1 (20 neurons at the hidden layer) and 3-5-1 (5 neurons at the hidden layer). These topologies were previously determined by empirical experiments.

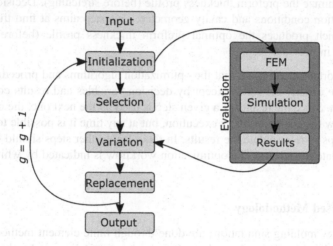

Fig. 3. Neuroevolutionary optimization workflow

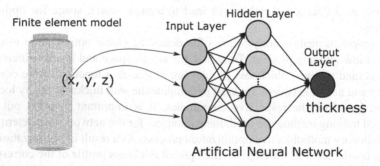

Fig. 4. Thickness calculation using neural network

When initialization is done, the algorithm performs a predefined numbers of generations of a steady-state variant of evolutionary process based on the SMS-EMOA multiobjective optimization algorithm [8]. A single offspring is produced in each generation. Selection is done by a uniform distribution (each member of the population has the same chance to be selected). Variation is performed by SBX-Crossover operator, which is adequate to work with real number representations and replacement strategy is based on Pareto front and *hypervolume* measure.

After being generated, each individual is evaluated by a procedure that comprises assembly the neural network from chromosome information and fed into the network the coordinates of each point of the finite element model. As a result of this step, the thickness of each point of the mesh will be provided, creating the thickness profile that will be considered in the simulation process.

At the end of optimization process there will be a set of optimal solutions, i.e., wall thickness profiles modeled by neural networks, each one giving different trade-off between the considered objectives.

3 Experiments and Results

3.1 Experimental Setup

The proposed methodology was applied to optimize injection blow molding of an industrial plastic bottle model. Figure 5a shows the geometry of the model, which is 45 mm in diameter and 182 mm height, composed by a plastic material with mass density of 1.15 g/cm^3 and Poisson's ration of 0.4. The ratio between the applied blowing pressure and Young's modulus is 0.0027.

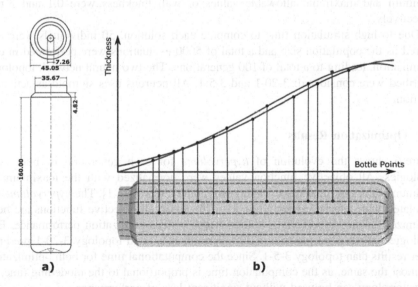

Fig. 5. Bottle model with wall thickness distribution illustrated (dimensions in millimeters)

The wall thickness distribution is composed by thickness values of each point of the mesh considering vertical lines from bottleneck to the bottom on the model. Figure 5b illustrates a thickness distribution plot for two vertical lines, but the points of all vertical lines should be plotted on the same graph, being possible to visualize how the thickness changes along the bottle. Note that for uniform distributions, lines will be overlapped.

Numerical simulations were carried out by finite element analysis software ANSYS Workbench version 18.1 to simulate an internal pressure applied to the final bottle. The objective of the optimization in this phase is to find optimal thickness distributions which provide the best relationships between the total mass and maximum strain supported by the bottle.

Since the ANNs are not aware of the geometry of the final product, non-uniform thickness distributions can be found by the optimization algorithm. However, for the model considered in this study, uniform distributions are desirable. Thus, an objective which takes into account the uniformness of thickness distributions was considered. Three objectives were chosen for the optimization: (i) the total mass of final product (f_1), (ii) the maximum strain suffered (f_2) and (iii) the maximum difference between each vertical line in the thickness distribution (f_3). The difference between two vertical lines is calculated using the root-mean-square error index (RMSE), given by Eq. (1).

$$RMSE = \sqrt{\frac{1}{n}\sum_{i=1}^{n}(\widehat{y}_i - y_i)^2} \qquad (1)$$

In Eq. (1) $\widehat{y}_i$ and y_i represents the thickness points of two different vertical lines and n are the total number of points in each distribution. All vertical lines are compared to each other and f_3 is the maximum calculated RMSE, that is to be minimized. The minimum and maximum allowable values of wall thickness were 0.1 and 3 mm, respectively.

Due to high simulation time to compute each solution, 50 individuals were considered as the population size and a total of 5000 evaluations were performed in each optimization, leading to a total of 100 generations. The two neural network topologies described were considered: 3-20-1 and 3-5-1. All neurons uses sigmoid as activation function.

3.2   Optimization Results

Figure 6 shows the evolution of *hypervolume* for each generation of both ANN topologies. All objective function values were normalized with the maximum and minimum values of the dataset, staying within the interval [0, 1]. The *hypervolume* was calculated with reference point (1.0, 1.0, 1.0). Once all objective functions are being minimized, higher *hypervolume* values means better optimization performance. Both topologies converge at generation 30 (approximately) and topology 3-20-1 presented better results than topology 3-5-1. Since the computational time for both optimizations is almost the same, as the computation time is proportional to the modelling time, the better topology can be used without significant loss of performance.

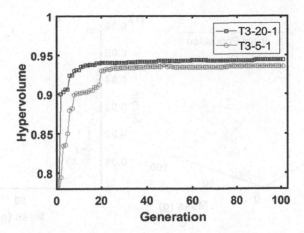

Fig. 6. Evolution of *hypervolume* for different ANN topologies

Figure 7 shows the Pareto front (for 100th generation) of both topologies. Topology 3-20-1 provides more different optimal solutions than topology 3-5-1, especially for the objectives f_1 (mass) and f_2 (maximum strain), where the Pareto front is much more distributed. Thus, the final results were selected from this front.

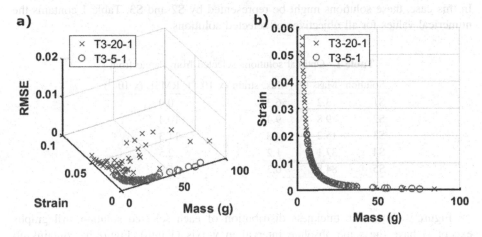

Fig. 7. Pareto front of each topology. In (a) all objectives are plotted whereas in (b) only objectives f_1 (mass) and f_2 (maximum strain) are shown

Figure 8a shows the evolution between the initial and last populations in the optimization process for topology 3-20-1. All objective functions were clearly minimized forming the Pareto front, which is shown in Fig. 8b (for f_1 and f_2).

The five selected solutions in Fig. 8b were selected along the Pareto front to obtain different trade-offs between the total mass of used material and the maximum strain supported by the bottle for the imposed pressure. For example, solution S2 gives a

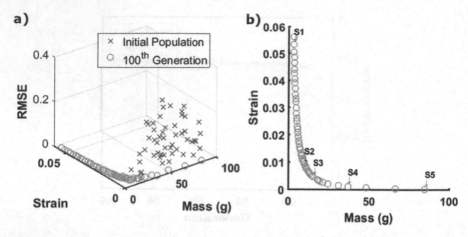

Fig. 8. (a) Initial and last population for topology 3-20-1. (b) Pareto front of last population where only objectives f_1 (mass) and f_2 (maximum strain) are shown. Five selected solutions are highlighted

certain thickness distribution for the final bottle which leads to a maximum strain of 9.4×10^{-3} with 9.8 g in weight. Considering the conflicting nature of the objectives, best relationships will be provided by solutions usually located at the knee of the curve. In this case, these solutions might be represented by S2 and S3. Table 1 contains the numerical values for all objectives of selected solutions.

Table 1. Optimal solutions selected from Pareto front

Solution	Mass (g)	Max. strain (x 10^{-3})	RMSE (x 10^{-3})
S1	3.2	56.3	0.9
S2	9.8	9.4	10.4
S3	15.2	4.8	11.3
S4	37.1	1.2	2.5
S5	84.8	0.3	5.9

Figure 9 shows the thickness distribution of each selected solution. All graphs except f) have the same absolute interval in y-axis (1 mm). Figure 9f contains all distributions on the same graph. Although each solution has a different value for RMSE, when considering absolute 1 mm interval (which is a high precision for an industrial blow molding process) there is no significant differences in distributions concerning the uniformity. However, each solution presents different thickness average levels, which can be seen in Fig. 9f. For instance, solution S5 has a mean thickness of 2.8 mm, giving lowest strain (0.3×10^{-3}) but with higher weight (84.8 g).

Looking at Table 1 values, solution S3 (15.2 g in total weight with maximum strain 4.8×10^{-3}) can be considered the general optimal solution, i.e., which gives the best trade-off between material utilization (mass) and the minimum strain suffered by the

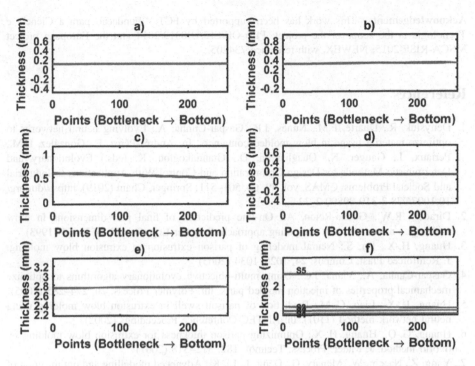

Fig. 9. Thickness distributions of selected solutions (a) S1, (b) S2, (c) S3, (d) S4 and (e) S5, respectively. (f) shows all distributions on the same graph

bottle. Thus, the thickness profile provided by S3 can be used in further optimizations of the global optimization process.

4   Conclusions

Optimization of injection blow molding is a great asset in industry since it can decrease production cost and improve manufacturing process. This paper proposes a new methodology based on a neuroevolutionary approach to optimize the injection blow molding process. Neural networks are used to model wall thickness distributions and evolutionary multiobjective optimization algorithms are applied to find optimal solutions, giving the best trade-offs between material utilization and mechanical properties. The methodology has been successfully applied to an industrial bottle model to find the best relationship between total mass and maximum strain when pressure is applied. As the result, a set of optimal thickness profiles has been found, providing less strain under pressure with less material utilization.

Optimization experiments provided by this study were applied to one phase of injection blow molding process. As a future work, the proposed methodology will be applied to other phases as well.

Acknowledgements. This work has been supported by FCT - Fundação para a Ciência e Tecnologia in the scope of the project: PEst-OE/EEI/UI0319/2014 and the European project MSCA-RISE-2015, NEWEX, with reference 734205.

References

1. Denysiuk, R., Duarte, F.M., Nunes, J.P., Gaspar-Cunha, A.: Evolving neural networks to optimize material usage in blow molded containers. In: Andrés-Pérez, E., González, L.M., Periaux, J., Gauger, N., Quagliarella, D., Giannakoglou, K. (eds.) Evolutionary and Deterministic Methods for Design Optimization and Control With Applications to Industrial and Societal Problems. CMAS, vol. 49, pp. 501–511. Springer, Cham (2019). https://doi.org/10.1007/978-3-319-89890-2_32
2. Diraddo, R.W., Garcia-Rejon, A.: On-line prediction of final part dimensions in blow molding: a neural network computing approach. Polymer Eng. Sci. **33**, 653–664 (1993)
3. Huang, H.-X., Lu, S.: Neural modeling of parison extrusion in extrusion blow molding. J. Reinforced Plast. Compos. **24**, 1025–1034 (2005)
4. Gaspar-Cunha, A., Viana, J.C.: Using multi-objective evolutionary algorithms to optimize mechanical properties of injection molded parts. Int. Polymer Process. **20**, 274–285 (2005)
5. Huang, H.-X., Liao, C.-M.: Prediction of parison swell in extrusion blow molding using neural network method (1101). In: ANTEC-Conference Proceedings (2002)
6. Huang, G.-Q., Huang, H.-X.: Optimizing parison thickness for extrusion blow molding by hybrid method. J. Mater. Process. Technol. **182**, 512–518 (2007)
7. Yang, Z., Naeem, W., Menary, G., Deng, J., Li, K.: Advanced modelling and optimization of infared oven in injection stretch blow-moulding for energy saving. IFAC Proc. Vol. **47**, 766–771 (2014)
8. Beumea, N., Naujoksa, B., Emmerich, M.: SMS-EMOA: multiobjective selection based on dominated hypervolume. Eur. J. Oper. Res. **181**(3), 1653–1669 (2007)

Author Index

Printed in the United States
By Bookmasters

Printed in the United States
By Bookmasters